A TREATISE

ON

ASTRONOMY,

SPHERICAL AND PHYSICAL;

WITH

ASTRONOMICAL PROBLEMS,

AND

SOLAR, LUNAR, AND OTHER ASTRONOMICAL TABLES.

FOR THE USE OF

COLLEGES AND SCIENTIFIC SCHOOLS.

BY

WILLIAM A. NORTON, M.A.,

PROFESSOR OF CIVIL ENGINEERING IN YALE COLLEGE.

FOURTH EDITION.

REVISED, REMODELLED, AND ENLARGED.

NEW YORK:

JOHN WILEY & SON,

15 ASTOR PLACE.

1874.

PREFACE TO THE REVISED EDITION.

In the preparation of the present edition the work has been entirely remodelled. The chapters which treat of Astronomical Instruments, Comets, the Fixed Stars, and the Tides, and the portion of the chapter on the Sun, that treats of the Sun's Spots and Physical Constitution, and the Zodiacal Light, have been wholly, or mostly, rewritten. Several changes of plan and arrangement have been made with the view of facilitating study and class instruction. The more difficult investigations of astronomical formulæ, occurring in the text of the former editions, have been transferred to the Appendix. On the other hand, the text has been enlarged by giving a more extended description of astronomical facts and appearances, and a more complete discussion of physical phenomena, including a detail of the important results of recent investigations concerning the physical constitution of the different classes of heavenly bodies, and a succinct exposition of the physical theories that have been generally received, or explain the phenomena most satisfactorily. Such theoretical discussions are kept distinct from the universally recognized truths of the science. The results of the author's own investigations on the physical constitution and phenomena of Comets, and on the physical constitution of the Sun, and the origin of the Sun's Spots, are briefly given in the same connection. New theoretical views are offered, in a note in the Appendix, on the possible development of sidereal systems from primordial nebulous masses; under the operation of recognized material forces, originated and sustained by the Creator, which unceasingly execute His will. Some prominence is given to the author's theory of the variable intensity of the repulsive force of the Sun, acting on different portions of cometic matter, as the operative cause of the lateral dispersion of the nebulous matter that makes up the train of a comet. This is believed to have been substantiated by a detailed discussion and comparison with observations; and as recent astronomical treatises, published in this country and in Europe, have advocated it without making mention of its previous publication and mathematical discussion, it is but just and proper that it should be distinctly set forth in the present work.

The Astronomical Problems in Part III. remain substantially the same as in the last edition. The table of Latitudes and Longitudes of Places, the tables of the Planetary Elements, and the table of the Mean Places of Fixed Stars, have been replaced by others that are more accurate and more extended. The tables of the Sun's and Moon's Epochs have been extended to 1884. Many new illustrative figures, and several plates of telescopic appearances, have been added.

One of the most important of the special improvements introduced consists in the adoption of the new and more accurate determination of the Sun's parallax, and mean distance from the earth. This is now generally adopted, as one evidence of which may be mentioned its introduction into the computations of the English Nautical Almanac for 1870. It brings with it a more accurate determination of the distances of all the planets from the Sun, and of the satellites from their primaries, and the dimensions and densities of the planets. The present is the first American treatise in which this important advance in exact astronomical science has been incorporated.

Another improvement is the insertion of a brief description of the methods used in the United States Coast Survey in determining from astronomical observations the latitude and longitude of a place. These may be characterized as the American methods, as they were devised and perfected by American astronomers and engineers; and are superior to all others that have yet been tried.

Without further specification of alterations and supposed improvements, it is hoped that the work will be found, in all its features, a true exposition, within the limits necessarily prescribed, of the present condition of the sublime science of Astronomy; from both the theoretical and practical point of view.

A large number of astronomical treatises and scientific periodicals have been consulted. Professor Chauvenet's admirable work on Spherical and Practical Astronomy, should be particularly mentioned as having been especially consulted in preparing the chapter on Instruments. In the mention of new discoveries and theoretical views, as well as of the signal advances which modern Astronomy has made, the name of the discoverer, or author, is generally given. The history of Astronomy cannot properly be wholly omitted from a text-book on the science, although it may be simpler to present the science as a body of admitted truths, without making mention of their discovery. The author takes occasion here to acknowledge his obligations to Professor C. S. Lyman, of Yale College, for important advice, and valuable assistance frequently rendered.

TABLE OF CONTENTS.

PART I.

SPHERICAL ASTRONOMY.

CHAPTER I.

CHAPTER II.

CHAPTER III.

CHAPTER IV.

CHAPTER V.

CHAPTER VI.

CHAPTER VII.

CHAPTER VIII.

CHAPTER IX.

CHAPTER X.

CHAPTER XI.

CHAPTER XII.

CHAPTER XIII.

CHAPTER XIV.

CHAPTER XV.

CHAPTER XVI.

PART II.

PHYSICAL ASTRONOMY.

PART III.

ASTRONOMICAL PROBLEMS.

APPENDIX.

ASTRONOMY.

PART I.

SPHERICAL ASTRONOMY.

CHAPTER I.

Definitions and Fundamental Conceptions; General Phenomena of the Heavens.

1. The sun, moon, and stars—the luminous bodies disseminated through *the heavens*, or indefinite space surrounding the earth—are called *Heavenly Bodies*. The heavenly bodies, considered collectively, are often termed *the Heavens*. The science which treats of the heavenly bodies is called *Astronomy*. It is divided into *Theoretical* and *Practical Astronomy*. Theoretical Astronomy is divided into *Spherical* and *Physical Astronomy*.

2. *Spherical Astronomy* treats of the positions, motions, and distances of the heavenly bodies; and of their appearance, magnitude, form, and structure. It comprises the theory of the methods of observation and calculation by which the positions, motions, etc., of the heavenly bodies have been determined; and the whole body of exact knowledge thus acquired, which is often termed *Descriptive Astronomy*.

Physical Astronomy investigates the general physical cause of the motions and constitution of the bodies of the material universe, and deduces from this general cause, called the force of *universal gravitation*, all the details of the celestial mechanism.

Practical Astronomy treats of astronomical instruments, and astronomical observation; practical determinations, as of the latitude or longitude of a place, from instrumental observation; and the solution of astronomical problems with the aid of tables.

3. Form of the Earth. We learn from the following circumstances that the earth is a body of a globular form, insulated in space.

(1.) When a vessel is receding from the land, an observer, from a point on the coast, first loses sight of the hull, then of the lower parts of the sails, and lastly of the topsails. It will be readily perceived, on glancing at Fig. 1, that no part of the earth could become interposed between the hull, and then the lower portions of the sails of a distant vessel, and the eye of the observer, if the sea were really what it appears to be, an indefinitely extended plane; also that if the earth be round, a receding ship should disappear in the manner it is actually observed to do, as the hull, mainsail, and topsails pass in succession below the line of sight tangent to the surface of the sea. If the observer take a more elevated position the ship should begin to sink out of sight at a greater distance, because the line of sight will touch the sea at a more distant point.

FIG. 1.

(2.) At sea the *visible horizon*, or the line bounding the visible portion of the earth's surface, is everywhere a circle, of a greater or less extent according to the altitude of the point of observation, and is on all sides equally depressed. To illustrate this proof, let BOA (Fig. 2) represent a portion of the earth's sur-

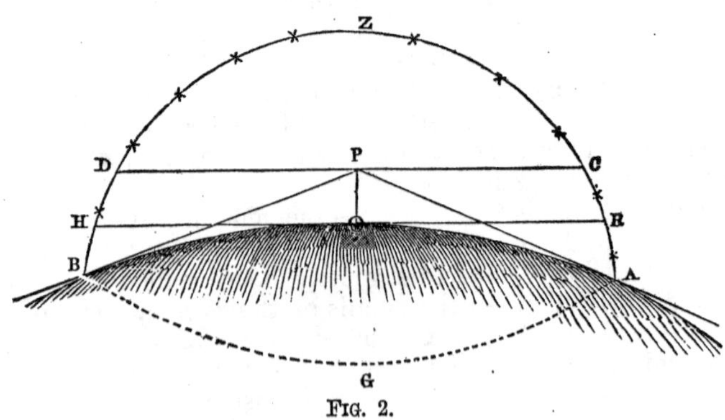

FIG. 2.

face supposed to be spherical, P the position of the eye of the observer, and DPC a horizontal line. If we conceive lines, such as PA and PB, to be drawn through the point of observation P, tangent to the earth in every direction, it is plain that these lines will all touch the earth at the same distance from the observer, and therefore that the line AGB, conceived to be traced through all the points of contact, A, B, etc., which would be the visible horizon, is a circle. It is also manifest that the angles of depression CPA, DPB, etc., of the horizon in different directions, will be equal; and that a greater portion of the earth's surface will be seen, and thus that the horizon will increase in extent, in proportion as the altitude of the point of observation, P, increases.

(3.) Navigators, as it is well known, have sailed entirely around the earth.

These facts prove the surface of the sea to be convex, and the surface of the land conforms very nearly to that of the sea; for the elevations of the highest mountains bear an exceedingly small proportion to the dimensions of the whole earth.

4. Visible and Invisible Portions of the Heavens. If an indefinite number of lines, PA, PB, etc., be conceived to be drawn through the point of observation P, (Fig. 2,) touching the earth on all sides, a conical surface will be formed, having its vertex at P, and extending indefinitely into space. All heavenly bodies, which at any time are situated below this surface, have the earth interposed between them and the eye of the observer, and therefore cannot be seen. All bodies that are above this surface, which send sufficient light to the eye, are visible. That portion of the heavens which is above this surface, presents the appearance of a solid vault or canopy, resting upon the earth at the visible horizon, (see Fig. 2;) and to this vault the sun, moon, and stars seem to be attached. It is hardly necessary to remark that this is an optical illusion. It will be seen in the sequel that the heavenly bodies are distributed through space at various distances from the earth, and that the distances of all of them are very great in comparison with the dimensions of the earth.

It will suffice, in the conception of phenomena, to suppose the eye of the observer to be near the earth's surface, and that the imaginary conical surface above mentioned, which separates the visible from the invisible portion of the heavens, is a horizontal plane, confounded for a certain distance with the visible part of the earth. This is called the plane of the horizon, and sometimes the horizon simply.

5. *Up* and *down*, at any place on the earth's surface, are from and towards the surface; and thus at different places have every variety of absolute direction in space.

6. The Sky. The earth is surrounded with a transparent gaseous medium, called the *earth's atmosphere*, estimated to be some fifty miles in height; through which all the heavenly bodies are seen. The atmosphere is not perfectly transparent, but shines throughout with light received from the heavenly bodies, and reflected from its particles; and thus forms a luminous canopy over our heads by day and by night. This is called the *Sky*. It appears blue because this is the color of the atmosphere; that is, because the particles of the atmosphere reflect the blue rays more abundantly than any other color. By day the portion of the atmosphere which lies above the horizon is highly illuminated by the sun, and shines with so strong a light as to efface the stars.

7. Diurnal Motion of the Heavens. The most conspicuous of the celestial phenomena, is a continual motion common to all the heavenly bodies, by which they are carried around the earth in regular succession. The daily circulation of the sun and moon about the earth is a fact recognised by all persons. If the heavens be attentively watched on any clear evening, it will soon be seen that the stars have a motion precisely similar to that of the sun and moon. To describe the phenomenon in detail, as witnessed at night:—if, on a clear night, we observe the heavens, we shall find that the stars, while they retain the same situations with respect to each other, undergo a continual change of position with respect to the earth. Some will be seen to ascend from a quarter called the *East*, being replaced by others that come into view, or *rise;* others, to descend towards the opposite quarter, the *West*, and to go out of view, or *set:* and if our observations be continued throughout the night, with the east on our left, and the west on our right, the stars which rise in the east will be seen to move in parallel circles, entirely across the visible heavens, and finally to set in the west. Each star will ascend in the heavens during the first half of its course, and descend during the remaining half. The greatest heights of the several stars will be different, but they will all be attained towards that part of the heavens which lies directly in front, called the *South*. If we now turn our backs to the south, and direct our attention to the opposite quarter, the *North*, new phenomena will present themselves. Some stars will appear, as before, ascending, reaching their greatest heights, and descending; but other stars will be seen, further to the north, that never set, and which appear to revolve in circles, from east to west, about a certain star that seems to remain stationary. This seemingly stationary star is called the *Pole Star;* and the stars which revolve about it, and never set, are called *Circumpolar Stars*. It should be remarked, however, that the pole star, when accurately observed by means of instruments, is found not to be strictly stationary, but to describe a small circle about a point at a little distance from it as a fixed centre. This point is called the *North Pole*. It is, in reality, about the north pole, as thus defined, and not the pole star, that the apparent revolutions of the stars at the north are performed. At the corresponding hours of the following night the aspect of the heavens will be the same, from which it appears that the stars return to the same position once in about 24 hours. It would seem, then, that the stars all appear to move from east to west, exactly as if attached to the concave surface of a hollow sphere, which rotates in this direction about an axis passing through the station of the observer and the north pole of the heavens, in a space of time nearly equal to 24 hours. For the sake of simplicity this conception is generally adopted. This motion, common to all the heavenly bodies, is called their *Diurnal Motion.* It is ascertained, by certain accurate methods of observation and computation, that the diurnal motion of the stars is strictly *uniform* and *circular.*

8. Rotating Sphere of the Heavens. It is important to observe, that the conception of a single sphere to which the stars are supposed to be attached, will not represent their diurnal motion, *as seen from every part of the earth's surface*, unless the sphere be supposed to be of such vast dimensions that the earth is comparatively but a mere point at its centre.

A circle cut out of the heavens conceived to be a rotating sphere, by a plane passing through the axis of rotation, has a *north* and *south* direction.

9. Fixed Stars and Planets. The greater number of the stars constantly preserve the same relative positions, and are therefore called *Fixed Stars.* But there are also many stars which are perpetually changing their places in the heavens. These are called *Planets*, or wandering stars. Each planet has received a distinctive name. For convenience of designation they are divided into the two classes of *Planets*, and *Planetoids* or Minor Planets. The former class comprises the planets Mercury, Venus, Mars, Jupiter, Saturn, Uranus, and Neptune. The first five of these are visible to the naked eye; but Uranus and Neptune, and the planetoids, cannot be seen without the aid of a telescope; and have all been discovered since the year 1780. Table II. (*a*), p. 5, &c. contains a list of the planetoids at present known, with the date and place of discovery of each, and the name of the discoverer. The number of planetoids hitherto discovered is *ninety-one.* Every year adds one or more to the list.

10. Distinctive Peculiarities of Different Planets. The planets are distinguishable from each other, either by a difference of aspect, or by a difference of apparent motion with respect to the sun. Venus and Jupiter are the two most brilliant planets. They are quite similar in appearance, but their apparent motions with respect to the Sun are very different. Thus Venus never recedes beyond 40° or 50° from the Sun, while Jupiter is seen at every variety of angular distance from him. Mars is known by the ruddy color of his light. Saturn has a pale, dull aspect.

11. Apparent Motions of the Planets. The apparent motion of each of the planets, is generally directed towards the east; but they are occasionally seen moving towards the west. As their easterly prevails over their westerly motion, they all, in process of time, accomplish a revolution around the earth. The periods of revolution are different for each planet.

12. Apparent Motions of the Sun and Moon. The sun and moon, are also continually changing their places among the fixed stars. From repeated observations of its position among the stars, it is found that the moon has a progressive circular motion in the heavens from *west* to *east*, and completes a revolution around the earth in about 27 days.

The motion of the sun, is also constantly progressive, and directed from *west* to *east.* This will appear on observing for a number of successive evenings, the stars which first become visible in that part of the heavens where the sun sets. It will be found that the stars, which in the first instance were observed to set just after the sun, soon cease to be visible, and are replaced by others that were seen immediately to the east of them; and that these in their turn, give place to others situated still further to the east. The sun must then be continually approaching the stars that lie on the eastern side of him. To make this more evident, let us suppose that the small circle *aon* (Fig. 3) repre-

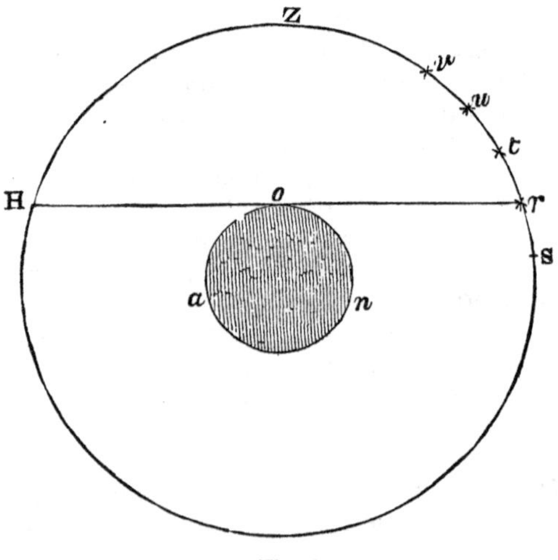

Fig. 3.

sents a section of the earth perpendicular to the axis of rotation of the imaginary sphere of the heavens, (8,*) conceived to pass through the earth's centre; the large circle H Z S a section of

* Numbers thus inclosed in a parenthesis refer to a previous article.

the heavens perpendicular to the same line, and passing through the sun; and the right line H *o r* the plane of the horizon at the station *o*. The direction of the diurnal motion is from H towards Z and S. Suppose that an hour or so after sunset the sun is at S, and that the star *r* is seen in the western horizon; also that the stars *t*, *u*, *v*, &c., are so distributed that the distances *rt*, *tu*, *uv*, &c., are each equal to S *r*. Then, at the end of two or three weeks, an hour after sunset the star *t* will be in the horizon; at the end of another interval of two or three weeks the star *u* will be in the same situation at the same hour; at the end of another interval, the star *v*, &c. It is plain, then, that the sun must at the ends of these successive intervals be in the successive positions in the heavens, *r*, *t*, *u*, &c.; otherwise, when it is brought by its diurnal motion to the point S, below the horizon, the stars *t*, *u*, *v*, &c., could not be successively in the plane of the horizon at *r*. Whence it appears that the sun has a motion in the heavens in the direction S *r t u v*, opposite to that of the diurnal motion; that is, towards the east.

Another proof of the progressive motion of the sun among the stars from west to east, is found in the fact that the same stars rise and set earlier each successive night, and week, and month during the year. At the end of six months the same stars rise and set 12 hours earlier; which shows that the sun accomplishes half a revolution in this interval. At the end of a year, or of 365 days, the stars rise and set again at the same hours, from which it appears that the sun completes an entire revolution in the heavens in this period of time.

It is to be observed that the sun does not advance *directly* towards the east. It has also some motion from south to north, and north to south. It is a matter of common observation that the sun is moving towards the north from winter to summer, and towards the south from summer to winter.

When the place of the sun in the heavens is accurately found from day to day by certain methods of observation, hereafter to be explained, it appears that his path is an exact circle, inclined about 23° to a circle running due east and west (8).

13. The Zodiac. The motions of the sun, moon, and planets, are for the most part confined to a certain zone, of about 18° in breadth, extending around the heavens obliquely from west to east, which has received the name of the *Zodiac*.

14. Comets. There is yet another class of bodies, called *Comets*, or *hairy Stars*, that have a motion among the fixed stars. They appear only occasionally in the heavens, and continue visible only for a few weeks or months. They shine with a diffusive nebulous light, and are commonly accompanied by a fainter divergent stream of similar light, called a *tail*.

The motions of the comets are not restricted to the zodiac.

These bodies are seen in all parts of the heavens, and moving in every variety of direction.

15. Satellites. By inspecting the planets with telescopes, it has been discovered that some of them are constantly attended by a greater or less number of small stars, whose positions are continually varying. These attendant stars are called *Satellites.* The planets which have satellites are Jupiter, Saturn, Uranus, and Neptune. The satellites are sometimes called *Secondary Planets;* the planets upon which they attend being denominated *Primary Planets.*

16. The Solar System. The sun and moon, the planets, (including the earth,) together with their satellites, and the comets, compose the *Solar System.*

From the consideration of the apparent motions and other phenomena of the solar system, several theories have been formed in relation to the arrangement and actual motions in space of the bodies that compose it. The theory, or *system*, now universally received, is, in its most prominent features, that which was taught by Copernicus in the sixteenth century, and which is known by the name of the *Copernican System.* It is as follows:

The sun occupies a fixed centre, about which the planets (including the earth) revolve from west to east, in planes that are but slightly inclined to each other, and in the following order: Mercury, Venus, the Earth, Mars, the Planetoids, Jupiter, Saturn, Uranus, and Neptune. The earth rotates from west to east, about an axis inclined to the plane of its orbit about $66\frac{1}{2}°$, and which remains continually parallel to itself as the earth revolves around the sun. The moon revolves from west to east around the earth as a centre; and in like manner the satellites circulate from west to east around their primaries. Without the solar system, and at immense distances from it are the fixed stars.

A motion in space from *west* to *east*, is a motion from *right* to *left*, as observed from a station within the orbit described, and on the north side of its plane. To obtain a clear conception of the motions of the planets, the reader should place himself in imagination at or near the centre of the system, and on the north side of the plane of the earth's orbit within which the planets may all, for the present, be conceived to revolve.

17. Symbols. The principal planets, and the sun and moon, are often designated by the following conventional characters or symbols.

The Sun, ...	☉	Jupiter, ...	♃
Mercury, ...	☿	Saturn, ...	♄
Venus, ...	♀	Uranus, ...	♅
The Earth, . . .	⊕	Neptune, ...	♆
Mars, ...	♂	The Moon, . . .	☽

18. Inferior, and Superior Planets. The two planets,

Mercury and Venus, whose orbits lie within the earth's orbit, are called *Inferior Planets.* The others are called *Superior Planets.* The terms inferior and superior as here used, have merely the signification of lower and higher in place, or in position with respect to the sun, as compared with the earth.

19. Vast Distance of the Fixed Stars. The angular distance between any two fixed stars, is found to be the same from whatever point of the earth's surface it is measured. It follows, therefore, that the diameter of the earth is insensible, when compared with the distance of the fixed stars; and that with respect to the region of space which separates us from those bodies, the whole earth is a mere point. Moreover, the angular distance between any two fixed stars, is the same at whatever period of the year it is measured. Hence, if the earth revolves around the sun, its entire orbit must be insensible in comparison with the distance of the stars.

20. Explanation of the Diurnal Motion of the Heavens. On the hypothesis of the earth's rotation, the diurnal motion of the heavens is a mere illusion occasioned by the rotation of the earth. To explain this, suppose the axis of the earth to be prolonged till it intersects the heavens considered as concentric with the earth. Conceive a great circle to be traced through the two points of intersection, and the point directly overhead, and let the position of the stars be referred to this circle. It will be readily perceived that the relative motion of this circle and the stars will be the same, whether the circle rotates with the earth from west to east, or, the earth being stationary, the whole heavens rotate about the same axis and at the same rate in the opposite direction. Now, as the motion of the earth is perfectly equable, we are insensible of it, and therefore attribute the changes in the situations of the stars with respect to the earth to an actual motion of these bodies. It follows, then, that we must conceive the heavens to rotate as above mentioned, since, as we have seen, such a motion would give rise to the same changes of situation as the supposed rotation of the earth. It was stated (7) that the sphere of the heavens appears to rotate about a line passing through the north pole and the station of the observer; but, as the radius of the earth is insensible in comparison with the distance of the stars, an axis passing through the centre of the earth will apparently pass through the station of the observer, wherever this may be upon the earth's surface.

21. Explanation of the Sun's apparent Motion. We in like manner infer that the observed motion of the sun in the heavens is only an apparent motion, occasioned by the orbital motion of the earth. Let E, E′ (Fig. 4) represent two positions of the earth in its orbit EE′E″ about the sun S. When the earth is at E, the observer will refer the sun to that part of the heavens marked *s*; but when the earth is arrived at E′, he will refer it to the part marked *s′*; and being in the mean time insensible of his own motion, the sun will appear to him to have described in the heavens the arc *s s′*, just the same as if it had

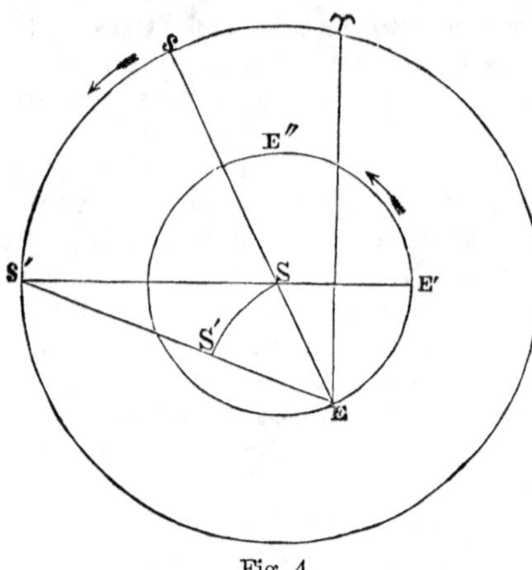

Fig. 4.

actually passed over the arc SS′ in space, and the earth had, during that time, remained quiescent at E. The motion of the sun from *s* towards *s′* will be from west to east, since the motion of the earth from E towards E′ is in this direction. Moreover, as the axis of the earth is inclined to the plane of its orbit under an angle of 66½° (16), the plane of the sun's apparent path, which is the same as that of the earth's orbit, will be inclined 23½° to a circle perpendicular to the earth's axis, or to a circle directed due east and west.

CHAPTER II.

Celestial and Terrestrial Spheres.

22. Celestial Sphere. In determining from observation the apparent positions and motions of the heavenly bodies, and, in general, in all investigations that have relation to their apparent positions and motions, astronomers conceive all these bodies, whatever may be their actual distance from the earth, to be referred to a spherical surface of an indefinitely great radius, having the station of the observer, or what comes to the very same thing, the centre of the earth, for its centre. This imaginary spherical surface is called the *Sphere of the Heavens*, or the *Celestial Sphere.* It is important to observe, that by reason of the great dimensions of this sphere, if two lines be drawn through any two points of the earth, and parallel to each other, they will, when indefinitely prolonged, meet it sensibly in the same point; and that, if two parallel planes be passed through any two points of the earth, they will intersect it sensibly in the same great circle. This amounts to saying that *the earth, as compared to this sphere, is to be considered as a mere point at its centre.*

Not only is the size of the earth to be neglected in comparison with the celestial sphere, but also the size of the earth's orbit. Thus the supposed annual motion of the earth around the sun, does not change the point in which a line conceived to pass from any station upon the earth in any fixed direction into space, pierces the sphere of the heavens; nor the circle in which a plane cuts the same sphere.

The fixed stars are so remote from the earth that observers, wherever situated upon the earth, and in the different positions of the earth in its orbit, refer them to the same points of the celestial sphere, (19.) The other heavenly bodies are referred by observers at different stations to points somewhat different.

Definitions. For the purposes of observation and computation, certain imaginary points, lines, and circles, appertaining to the celestial sphere, are employed, which we shall now proceed to define and explain.

(1.) The *Vertical Line*, at any place on the earth's surface, is the line of descent of a falling body, or the position assumed by a plumb-line when the plummet is freely suspended and at rest.

Every plane that passes through the vertical line is called a

Vertical Plane. Every plane that is perpendicular to the vertical line, is called a *Horizontal Plane.*

(2.) The *Sensible Horizon* of a place on the earth's surface, is the circle in which a horizontal plane drawn through the place, cuts the celestial sphere. As its plane is tangent to the earth, it separates the visible from the invisible portion of the heavens, (4.)

(3.) The *Rational Horizon* is a circle parallel to the former, the plane of which passes through the centre of the earth. The zone of the heavens comprehended between the sensible and rational horizon is imperceptible, or the two circles appear as one and the same at the distance of the earth.

(4.) The *Zenith* of a place is the point in which the vertical line prolonged upwards pierces the celestial sphere. The point in which the vertical line, when produced downwards, intersects the celestial sphere, is called the *Nadir.*

The zenith and nadir are the geometrical poles of the horizon.

(5.) The *Axis of the Heavens* is an imaginary right line passing through the north pole (7) and the centre of the earth. It is the line about which the apparent rotation of the heavens is performed. It is, also, on the hypothesis of the earth's rotation, the axis of rotation of the earth prolonged on to the heavens.

(6.) The *South Pole* of the heavens is the point in which the axis of the heavens meets the southern part of the celestial sphere.

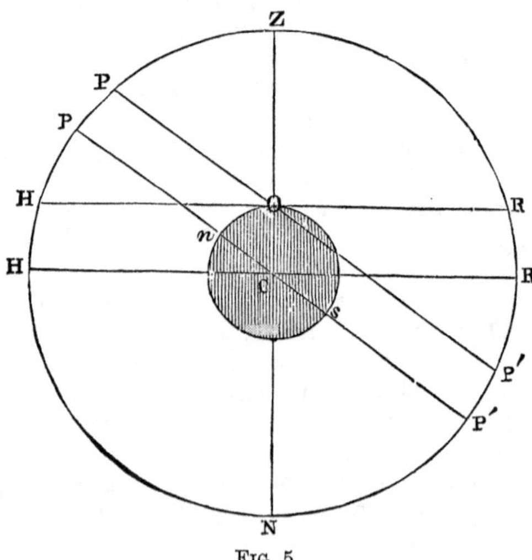

FIG. 5.

To illustrate the preceding definitions, let the inner circle *n*O*s* (Fig. 5) represent the earth, and the outer circle HZRN the sphere of the heavens; also let O be a point on the earth's surface, and OZ the vertical line at the station O.—Then HOR will be the plane of the sensible horizon, HCR the plane of the rational horizon, Z the zenith, and N the nadir; and if P be the north pole of the heavens, OP, or CP its parallel, will be the axis of the heavens, and P′ the south pole.

The lines CP and OP intersect the heavens in the same point, P; and the planes HOR, and HCR, in the same circle, passing through the points H and R.

Unless we are seeking for the exact apparent place in the heavens of some other heavenly body than a fixed star, we may conceive the observer to be stationed at the earth's centre, in which case OP will become the same as CP, and HOR the same as HCR; as represented in Fig. 6. In this diagram, the circle of the horizon being supposed to be viewed from a point above its plane, is represented by the ellipse HARa, Z and N are its geometrical poles. In the construction of Fig. 5, the eye is supposed to be in the plane of the horizon, and HARa is projected into its diameter HCR.

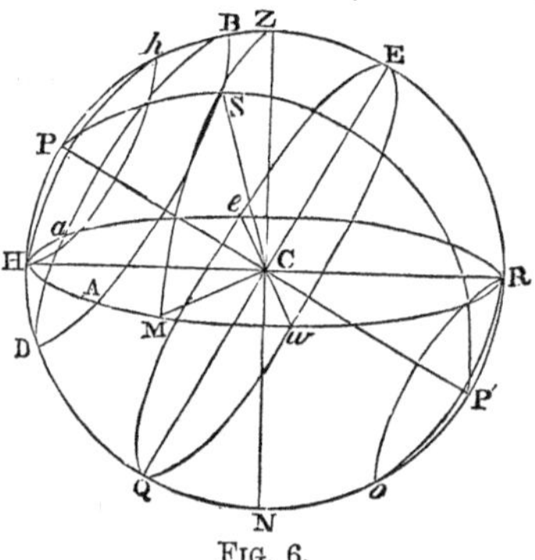

FIG. 6.

Every different place on the surface of the earth has a different zenith, and except in the case of diametrically opposite places, a

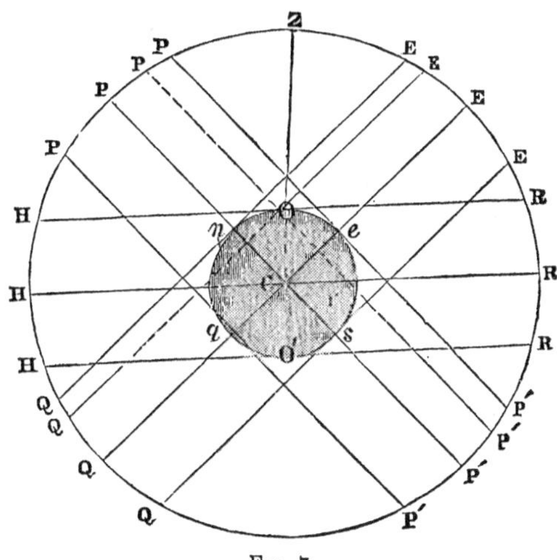

FIG. 7.

different horizon. To illustrate this, let *nesq* (Fig. 7) represent the earth, and HZRP′ the sphere of the heavens; then considering the four stations, *e*, O, *n*, and *q*, the zenith and horizon of the first will be respectively E and P*e*P′; of the second Z and HOR; of the third P and Q*n*E; of the fourth Q and P′*q*P. The diametrically opposite places O and O′ have the same rational horizon, viz. HCR. The same is true of the places *n* and *s*, and *e* and *q*. Their rational horizons are respectively QCE and PCP′.

(7.) *Vertical Circles* are great circles passing through the zenith and nadir. They cut the horizon at right angles, and their planes are vertical. Thus ZSM (Fig. 6) represents a vertical circle passing through the star S, called the *Vertical Circle of the Star.*

(8.) The *Meridian* of a place is that vertical circle which contains the north and south poles of the heavens. The plane of the meridian is called the *Meridian Plane.*

Thus, PZRP′ is the meridian of the station C. The half HZR, above the horizon, is termed the *Superior Meridian*, and the other half RNH, below the horizon, is termed the *Inferior Meridian.* The two points, as H and R, in which the meridian cuts the horizon, are called the *North and South Points* of the horizon; and the line of intersection, as HCR, of the meridian plane with the plane of the horizon, is called the *Meridian Line*, or the *North and South Line.*

(9.) The *Prime Vertical* is the vertical circle which crosses the meridian at right angles. It cuts the horizon in two points, as *e*, *w*, called the *East* and *West Points of the Horizon.*

(10.) The *Altitude* of any heavenly body is the arc of a vertical circle, intercepted between the centre of the body and the horizon, or the angle at the centre of the sphere, measured by this arc. Thus, SM or MCS is the altitude of the star S.

(11.) The *Zenith Distance* of a heavenly body is the arc of a vertical circle, intercepted between its centre and the zenith; or the distance of the centre of the body from the zenith as measured by the arc of a great circle. Thus, ZS, or ZCS, is the zenith distance of the star S.

It is obvious that the zenith distance and altitude of a body are *complements* of each other, and therefore when either one is known that the other may be found.

(12.) The *Azimuth* of a heavenly body is the arc of the horizon, intercepted between the meridian and the vertical circle passing through the centre of the body; or the angle comprehended between the meridian plane and the vertical plane containing the centre of the body. It is reckoned either from the north or from the south point, and each way from the meridian. HM or HCM represents the azimuth of the star S.

The *Azimuth and Altitude, or azimuth and zenith distance of a heavenly body, ascertain its position with respect to the horizon and meridian, and therefore its place in the visible hemisphere.* Thus, the azimuth HM determines the position of the vertical circle ZSM of the star S with respect to the meridian ZPH, and the altitude MS, or the zenith distance ZS, the position of the star in this circle.

(13.) The *Amplitude* of a heavenly body at its rising, is the arc of the horizon intercepted between the point where the body rises and the east point. Its amplitude at setting, is the arc of the horizon intercepted between the point where the body sets and the west point. It is reckoned towards the north, or towards the south, according as the point of rising or setting is north or south of the east or west point. Thus, if *a*BSA represents the circle described by the star S in its diurnal motion, *ea* will be its amplitude at rising, and *w*A its amplitude at setting.

(14.) The *Celestial Equator*, or the *Equinoctial*, is a great circle of the celestial sphere, the plane of which is perpendicular to the axis of the heavens. The north and south poles of the heavens are therefore its geometrical poles. The celestial equator is represented in Fig. 6 by E*w*Q*e*. This circle is also frequently called *the Equator*, simply.

(15.) *Parallels of Declination* are small circles parallel to the celestial equator. *a*BSA represents the parallel of declination of the star S.

The parallels of declination passing through the stars, are the circles described by the stars in their apparent diurnal motion. These, by way of abbreviation, we shall call *Diurnal Circles*.

(16.) *Celestial Meridians*, *Hour Circles*, and *Declination Circles*, are different names given to all great circles which pass through the poles of the heavens, cutting the equator at right angles. PSP′ is a celestial meridian. The angles comprehended between the hour circles and the meridian, reckoning from the meridian towards the west, are called *Hour Angles*, or *Horary Angles*.

(17.) The *Ecliptic* is that great circle of the heavens which the sun appears to describe in the course of the year.

(18.) The *Obliquity of the Ecliptic* is the angle under which the ecliptic is inclined to the equator. Its amount is $23\frac{1}{2}°$.

(19.) The *Equinoctial Points* are the two points in which the ecliptic intersects the equator. That one of these points which the sun passes in the spring is called the *Vernal Equinox*, and the other, which is passed in the autumn, is called the *Autumnal Equinox*. These terms are also applied to the *epochs* when the sun is at the one or the other of these points. These epochs are, for the vernal equinox the 21st of March, and for the autumnal equinox the 23d of September, or thereabouts.

(20.) The *Solstitial Points* are the two points of the ecliptic 90° distant from the vernal and autumnal equinox. The one that lies to the north of the equator is called the *Summer Solstice*, and the other the *Winter Solstice*. The epochs of the sun's arrival at these points are also designated by the same terms. The summer solstice happens about the 21st of June, and the winter solstice about the 22d of December.

(21.) The *Equinoctial Colure* is the celestial meridian passing through the equinoctial points; and the *Solstitial Colure* is the celestial meridian passing through the solstitial points.

(22.) The *Polar Circles* are parallels of declination at a distance from the poles equal to the obliquity of the ecliptic. The one about the north pole is called the *Arctic Circle;* the other, about the south pole, is called the *Antarctic Circle.*

The polar circles contain the geometrical poles of the ecliptic.

(23.) The *Tropics* are parallels of declination at a distance from the equator equal to the obliquity of the ecliptic. That which is on the north side of the equator is called the *Tropic of Cancer*, and the other the *Tropic of Capricorn.*

The tropics touch the ecliptic at the solstitial points.

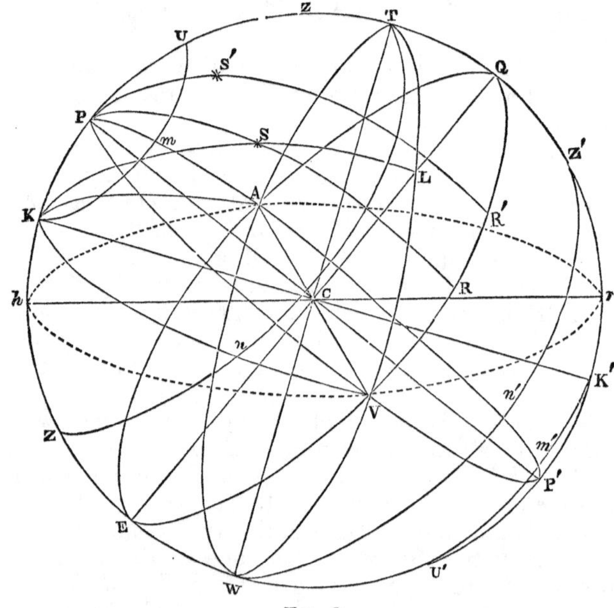

FIG. 8.

Let C (Fig. 8) represent the centre of the earth and sphere, PCP′ the axis of the heavens, EVQA the equator, WVTA the ecliptic, and K, K′, its poles. Then will V be the *vernal* and A the *autumnal equinox;* W the *winter*, and T the *summer solstice;* PVP′A the *equinoctial colure;* PKWK′T the *solstitial colure;* the angle TCQ, or its measure the arc TQ, the *obliquity of the ecliptic;* K*m*U, K′*m*′U′, the *polar circles;* and T*n*Z, W*n*′Z′, the *tropics.*

It is important to observe that, agreeably to what has been stated (p. 11), the directions of the equator and ecliptic, of the equinoctial points, and of the other points and circles just defined and illustrated, are the same at any station upon the surface of the earth as at its centre. Thus, the equator lies always in the plane passing through the place of observation, wherever this may be, and parallel to the plane which, passing through the earth's centre, cuts the heavens in this circle. In like manner the ecliptic lies, everywhere, in a plane parallel to that which is conceived to pass through the centre of the earth and cut the heavens in this circle, and so for the other circles.

(24.) The *Zodiac* (13) extends about 9° on each side of the ecliptic.

(25.) The ecliptic and zodiac are divided into twelve equal parts, called *Signs*. Each sign contains 30°. The division commences at the vernal equinox. Setting out from this point, and following around from west to east, the *Signs of the Zodiac*, with the respective characters by which they are designated, are as follows: Aries ♈, Taurus ♉, Gemini ♊, Cancer ♋, Leo ♌, Virgo ♍, Libra ♎, Scorpio ♏, Sagittarius ♐, Capricornus ♑, Aquarius ♒, Pisces ♓. The first six are called *northern signs*, being north of the equinoctial. The others are called *southern signs*.

The vernal equinox corresponds to the first point of Aries, and the autumnal equinox to the first point of Libra. The summer solstice corresponds to the first point of Cancer, and the winter solstice to the first point of Capricornus.

The mode of reckoning arcs on the ecliptic is by signs, degrees, minutes, &c.

A motion in the heavens in the order of the signs, or from west to east, is called a *direct* motion, and a motion contrary to the order of the signs, or from east to west, is called a *retrograde* motion.

(26.) The *Right Ascension* of a heavenly body is the arc of the equator intercepted between the vernal equinox and the declination circle which passes through the centre of the body, as reckoned from the vernal equinox towards the east. It measures the inclination of the declination circle of the body to the equinoctial colure. Thus, PSR being the declination circle of the star S, and V the place of the vernal equinox, VR is the right ascension of the star. It is the measure of the angle VPS. If PS′R′ be the declination circle of another star S′, SPS′, or RR′, will be their difference of right ascension.

(27.) The *Declination* of a heavenly body is the arc of a circle of declination, intercepted between the centre of the body and the equator. It therefore expresses the distance of the body from the equator. Thus, RS is the declination of the star S.

Declination is *North* or *South*, according as the body is north or south of the equator.

In the use of formulæ, a south declination is regarded as negative.

The right ascension and declination of a heavenly body are two co-ordinates, which, taken together, fix its position in the sphere of the heavens: for they make known its situation with respect to two circles, the equinoctial colure, and the equator. Thus, VR and RS ascertain the position of the star S with respect to the circles PVP′A and VQAE.

(28.) The *Polar Distance* of a heavenly body is the arc of a declination circle, intercepted between the centre of the body and the elevated pole. The polar distance is the complement of the declination, and, therefore, when either is known the other may be found.

(29.) *Circles of Latitude* are great circles of the celestial sphere, which pass through the poles of the ecliptic, and therefore cut this circle at right angles. Thus, KSL represents a part of the circle of latitude of the star S.

(30.) The *Longitude* of a heavenly body is the arc of the ecliptic, intercepted between the vernal equinox and the circle of latitude which passes through the centre of the body, as reckoned from the vernal equinox towards the east, or in the order of the signs. It measures the inclination of the circle of latitude of the body to the circle of latitude passing through the vernal equinox. Thus, VL is the longitude of the star S. It is the measure of the angle VKS.

(31.) The *Latitude* of a heavenly body is the arc of a circle of latitude, intercepted between the centre of the body and the ecliptic. It therefore expresses the distance of the body from the ecliptic. Thus, LS is the latitude of the star S.

Latitude is *North* or *South*, according as the body is north or south of the ecliptic.

In the use of formulæ, a south latitude is affected with the minus sign.

The longitude and latitude of a heavenly body are another set of co-ordinates, which serve to fix its position in the heavens. They ascertain its situation with respect to the circle of latitude passing through the vernal equinox, and the ecliptic. Thus, VL and LS fix the position of the star S, making known its situation with respect to the circles KVK′A and VTAW.

(32.) The *Angle of Position* of a star is the angle included at the star between the circles of latitude and declination passing through it. PSK is the angle of position of the star S.

(33.) The *Astronomical Latitude*, or *the Latitude* of a place, is the arc of the meridian intercepted between the zenith and the celestial equator. It is *North* or *South*, according as the zenith is north or south of the equator. ZE (Fig. 7) represents the latitude of the station O; QOE or QCE being the equator.

23. Terrestrial Sphere. The earth's surface, considered as spherical (which accurate admeasurement, upon principles that will be explained in the sequel, shows it to be, very nearly), is called the *Terrestrial Sphere.* The following geometrical constructions appertain to the terrestrial sphere, as it is employed for the purposes of astronomy. It will be observed that they correspond to those of the celestial sphere above described, and are used for similar objects.

Definitions. (1.) The *North* and *South Poles of the Earth* are the two points in which the axis of the heavens intersects the terrestrial sphere. They are also the extremities of the earth's axis of rotation.

(2.) The *Terrestrial Equator* is the great circle in which a plane passing through the centre of the earth, and perpendicular to the axis of the heavens and earth, cuts the terrestrial sphere. The *terrestrial* and the *celestial equator*, then, lie in the same plane. The poles of the earth are the geometrical poles of the terrestrial equator. The two hemispheres into which the terrestrial equator divides the earth are called, respectively, *the Northern Hemisphere* and *the Southern Hemisphere.*

(3.) *Terrestrial Meridians* are great circles of the terrestrial sphere, passing through the north and south poles of the earth, and cutting the equator at right angles. Every plane that passes through the axis of the heavens cuts the celestial sphere in a *celestial meridian*, and the terrestrial sphere in a *terrestrial meridian.*

Let PP′ (Fig. 9) represent the axis of the heavens, O the centre of the earth, and p and p' its poles. Then, elq will represent the *terrestrial equator* (ELQ representing the celestial equator); and pep' and psp' *terrestrial meridians* (PEP′ and PSP′ representing celestial meridians).

(4.) The *Reduced Latitude* of a place on the earth's surface is the arc of a terrestrial meridian, intercepted between the place and the equator, or the angle at the centre of the earth measured by this arc. Thus, oe, or the angle oOe, is the reduced latitude of the place o. Latitude is *North* or *South*, according as the place is north or south of the equator. The reduced latitude differs somewhat from the astronomical latitude, by reason of the slight deviation of the earth from a spherical form. Their difference is called the *Reduction of Latitude.*

(5.) *Parallels of Latitude* are small circles of the terrestrial sphere parallel to the equator. Every point of a parallel of latitude has the same latitude.

The parallels of latitude which correspond in situation with the polar circles and tropics in the heavens, have received the same appellations as these circles. (See defs. 22, 23, p. 16.)

(6.) The *Longitude* of a place on the earth's surface is the inclination of its meridian to that of some particular station, fixed upon as a circle to reckon from, and called the *First Meridian.* It is measured by the arc of the equator, intercepted between the first meridian and the meridian passing through the place, and is called *East* or *West,* according as the latter meridian is to the east

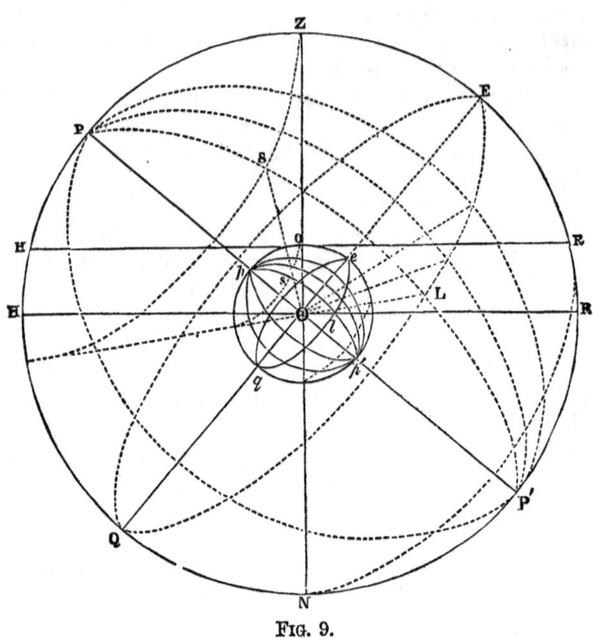

Fig. 9.

or to the west of the first meridian. Thus, if pqp' be supposed to represent the first meridian, the angle spq, or the arc ql, will be the longitude of the place s.

Different nations have, for the most part, adopted different first meridians. The English use the meridian which passes through the Royal Observatory at Greenwich, near London; and the French, the meridian of the Imperial Observatory at Paris. In the United States the longitude is, for astronomical purposes, reckoned from the meridian of Washington, or of Greenwich.

The longitude and latitude of a place designate its situation on the earth's surface. They are precisely analogous to the right ascension and declination of a star in the heavens.

24. Altitude of the Pole. The diagram (see Fig. 6) that was made use of in Art. 22, in illustrating the description of the circles of the celestial sphere, represents the aspect of this sphere at a place at which the north pole of the heavens is somewhere between the zenith and horizon. Such is the position of the north pole at all places situated between the equator and the north pole of the earth. For, let O (Fig. 10) represent a place on the earth's surface, HOR the horizon, OZ the vertical, HZR the meridian, and ZE the latitude. QOE will then represent the equinoctial, and P, 90° distant from E and on the superior meridian, the elevated pole. Now we have

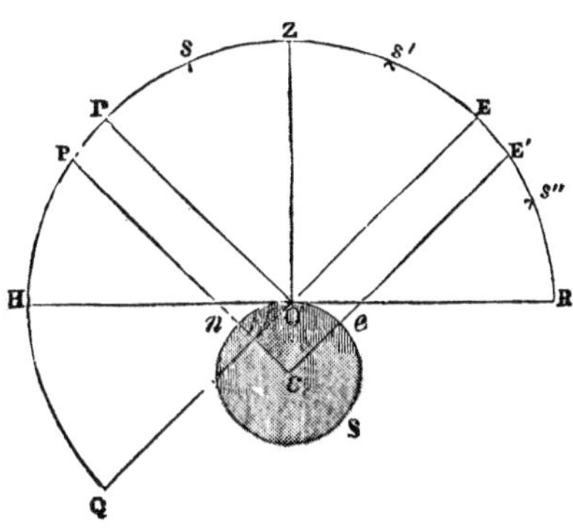

FIG. 10.

$$HP = ZH - ZP = 90° - ZP; \quad ZE = PE - ZP = 90° - ZP.$$
$$\text{Whence, } HP = ZE.$$

Thus, *the altitude of the pole is everywhere equal to the latitude of the place.* It follows, therefore, that in proceeding from the equator to the north pole, the altitude of the north pole of the heavens will gradually increase from 0° to 90°.

By inspecting Fig. 7, it will be seen that this increase of the altitude of the pole in going north, is owing to the fact that in following the curved surface of the earth the horizon, which is continually tangent to the earth, is constantly more and more depressed towards the north, while the absolute direction of the pole remains unaltered.

If the spectator is in the southern hemisphere, the elevated pole, as it is always on the opposite side of the zenith from the equator, will be the south pole. At corresponding situations of the spectator, it will obviously have the same altitude as the north pole in the northern hemisphere.

25. Oblique Sphere. Let us now inquire minutely into the principal circumstances of the diurnal motion of the stars, as it is seen by a spectator situated somewhere between the equator and the north pole. And in the first place, it is a simple corollary from the proposition just established, that the parallel of declination to the north, whose *polar distance is equal to the latitude of the place*, will lie entirely above the horizon, and just touch it at the north point. This circle is called the *circle of perpetual apparition;* the line aH (Fig. 11) represents its projection on the meridian plane. The stars comprehended between it and the north pole will *never set.* As the depression of the south pole is equal to the altitude of the north pole, the parallel of declination oR, at a distance from the south pole equal to the latitude of the place, will lie entirely below the horizon, and just touch it at the south point. The parallel thus situated is called the *circle of perpetual occultation.* The stars comprehended between it and the south pole will *never rise.*

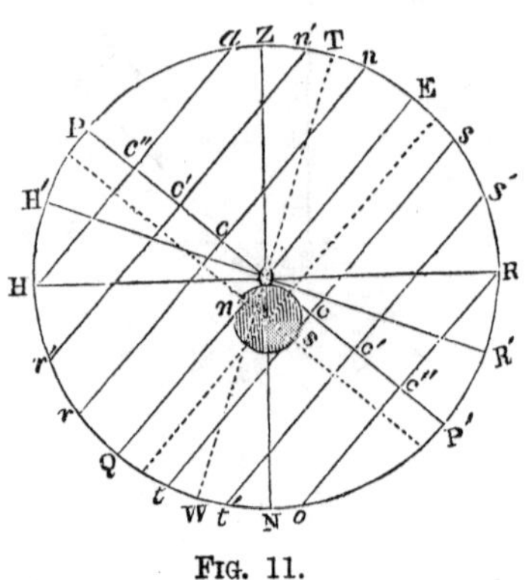

Fig. 11.

The celestial equator (which passes through the east and west points) will intersect the meridian at a point E, whose zenith distance ZE is equal to the latitude of the place (def. 33, Art. 22), and consequently, whose *altitude* RE *is equal to the co-latitude of the place.* Therefore, in the situation of the observer above supposed, the equator QOE, passing to the south of the zenith, will, together with the diurnal circles *nr*, *st*, etc., which are all parallel to it, be *obliquely* inclined to the horizon, making with it an angle equal to the co-latitude of the place. As the centres *c*, *c'*, etc., of the diurnal circles lie on the axis of the heavens, which is inclined to the horizon, all diurnal circles situated between the two circles of perpetual apparition and occultation, *a*H and *o*R, with the exception of the equator, will be divided unequally by the horizon. The greater parts of the circles *nr*, *n'r'*, etc., to the north of the equator, will be above the horizon; and the greater parts of the circles *st*, *s't'*, etc., to the south of the equator, will be below the horizon. Therefore, while the stars situated in the equator will remain an equal length of time above and below the horizon, those to the north of the equator will remain a longer time above the horizon than below it; and those to the south of the equator, on the contrary, a longer time below the horizon than above it. It is also obvious, from the manner in which the horizon cuts the different diurnal circles, that the disparity between the intervals of time that a star remains above and below the horizon will be the greater the more distant it is from the equator. Again, the stars will all *culminate*, or attain to their greatest altitude, in the meridian: for, since the meridian crosses the diurnal circles at right angles, they will have the least zenith distance when in this circle. Moreover, as the meridian bisects the portions of the diurnal circles which lie above the horizon, the stars will all employ the same length of time in passing from the eastern horizon to the meridian, as in passing from the meridian to the western horizon. The circumpolar stars will pass the meridian *twice* in 24 hours; once *above*, and once *below* the pole. These meridian passages are called, respectively,

Upper and *Lower Culminations*, or *Inferior* and *Superior Transits.*

It will be observed, that in travelling towards the north the circles of perpetual apparition and occultation, together with those portions of the heavens about the poles which are constantly visible and invisible, are continually on the increase.

It is evident, from what is stated in Art. 24, that the circumstances of the diurnal motion will be the same at any place in the southern hemisphere, as at the place which has the same latitude in the northern.

The celestial sphere in the position relative to the horizon which we have now been considering, which obtains at all places situated between the equator and either pole, is called an *Oblique Sphere,* because all bodies rise and set obliquely to the horizon.

26. Right Sphere. When the spectator is situated on the equator, both the celestial poles will be in his horizon (24), and therefore the celestial equator and the diurnal circles in general will be perpendicular to the horizon. This situation of the sphere is called a *Right Sphere*, for the reason that all bodies rise and set at right angles with the horizon. It is represented in Fig. 12. As the diurnal circles are bisected by the horizon, the stars will all remain the same length of time above as below the horizon.

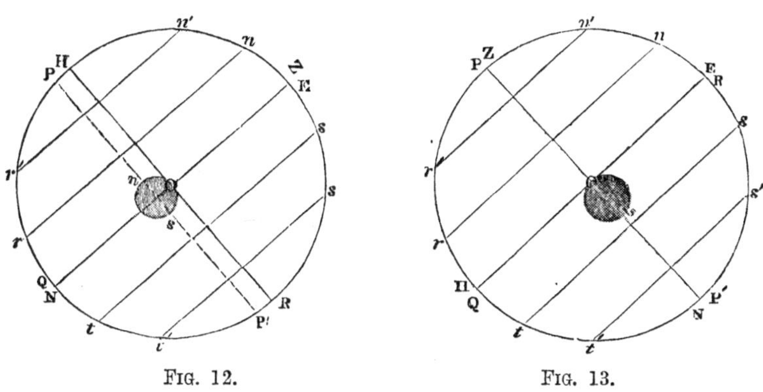

FIG. 12. FIG. 13.

27. Parallel Sphere. If the observer be at either of the poles, the elevated pole of the heavens will be in his zenith (24), and consequently the celestial equator will be in his horizon. The stars will move in circles parallel to the horizon, and the whole hemisphere, on the side of the elevated pole, will be continually visible, while the other hemisphere will be continually invisible. This is called a *Parallel Sphere.* It is represented in Fig. 13.

CHAPTER III.

Astronomical Instruments.—Astronomical Observation.

28. Astronomical instruments are used to measure arcs of the celestial sphere, or their corresponding angles at a station on the earth. They consist, essentially, of a refracting telescope turning upon an axis, and a graduated limb, or two graduated limbs at right angles to each other, to indicate the angle passed over by the telescope. When designed to measure angles in the meridian plane, the axis of rotation is horizontal, and a single vertical limb is used.

29. The Reticle. At the common focus of the object-glass and eye-glass of the telescope, is a piece of apparatus called a *reticle*, the design of which is to furnish a definite line of sight. In its simplest form it consists of a flat circular ring, attached to which are two very fine wires, or spider lines, crossing each other at right angles in its centre (Fig. 14). The line passing through the point of intersection of the cross wires and the centre of the object-glass, indefinitely prolonged, is the line of sight, or *Line of Collimation* of the telescope.

The reticle can be moved up or down, or to the right or left, by adjusting screws; and the line of collimation thus made perpendicular to the axis of rotation of the telescope. These screws are shown at *aa* and *bb*, Fig. 14. They pass through narrow slits in the tube of the telescope, so that they can be turned from without, and each pair of screws, *aa* or *bb*, gives a motion to the wire-plate. The line passing through the centre of the eye-glass and the centre of the object-glass, or the *optical axis* of the telescope, is perpendicular, or nearly so, to the axis of rotation. When it is in that precise position and the line of collimation accurately adjusted, the two lines will coincide. But it is not important in the use of instruments that this coincidence should be perfect. It is sufficient if the line of collimation is perpendicular to the axis of rotation.

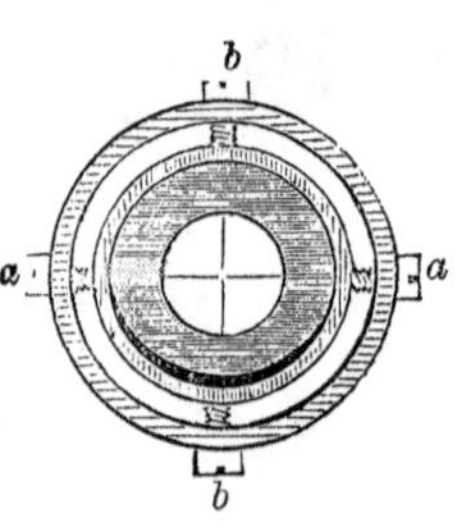

Fig. 14.

Reticle Tube. The reticle is placed in a tube, which slides in the lower end of the principal tube of the telescope. The eye-piece is inserted in the outer end of this tube; and can be pressed in or drawn out until the wires of the reticle are distinctly seen. In making an observation, the reticle tube with the eye-piece is moved out or in if necessary, by means of a milled head screw that works a pinion in a rack connected with the tube, until the image of the star, formed by the object-glass, falls upon the wires, when both the wires and the star will be distinctly seen. The reticle tube can also be turned around until the wires have the right direction in the field of view. A star is known to be on the line of collimation when it is bisected by each of the two cross wires.

30. Improved Form of Reticle. The form of reticle just described is now attached only to portable instruments. That which is adapted to the larger instruments of an observatory, differs from this in the number of the wires, the form of the wire-plate, and the mode of attaching the plate to its tube and of adjusting the wires. It has several parallel and equidistant wires, crossed at right angles by a single wire, or more commonly by two very close parallel wires (Fig. 15). In meridian instruments, and those for measuring altitudes, the single wire, or the equivalent pair of close parallel wires, is made horizontal. The middle wire of the others is brought into the meridian plane; these are called *transit wires.* The star is made to pass through the field between the two horizontal wires. The point of the middle transit wire that lies midway between the two horizontal ones, corresponds to the point of intersection of the two cross wires in Fig 14.

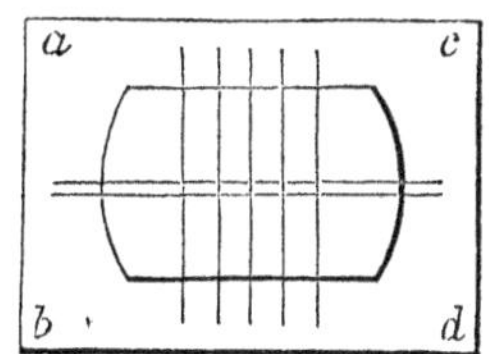

FIG. 15.

The wire-plate lies within a frame fastened across the outer end of the reticle tube (see Fig. 19, p. 31), and is adjusted by screws that act upon pieces projecting from its outer rim. The eye-piece is screwed into a plate that slides within the same transverse frame, and is moved by means of a screw. By turning this screw the eye-piece may be brought into such positions that the star observed is kept in the middle of the field of view.

31. Movable Micrometer Wire. In the focus of the eye-glass there is often fastened to a transverse sliding plate, and movable with it, a wire at right angles to the direction in which the plate is moved by a screw. The screw has the form of the micrometer-screw with graduated head, soon to be described. This wire is called the *movable micrometer-wire*, and the whole apparatus, being especially designed for the measurement of small angular distances, is called a *Micrometer.* The same name is sometimes, though improperly, given to the reticle alone, when the movable wire is not employed.

32. Reading off the Angle. The telescope, and the graduated limb which is perpendicular to the axis of rotation, are, in most instruments, firmly attached to each other, and turn together about this axis. The limb glides past a fixed index. The angle read off is that which is pointed out by the index. The limbs of even the largest instruments, are not divided into smaller parts than 2′; but by means of certain subsidiary contrivances, the angle may, with some instruments, be read off to within a fraction of a second. The principal contrivances in use for increasing the accuracy of the reading off of angles, are the *Vernier*, and the *Reading Microscope.*

33. The Vernier is simply the index-plate so graduated that a certain number of its divisions occupy the same space as a number one less on the limb. A division, or space on the vernier, will therefore be less, by a certain amount, say 1′, than a division on the limb. The index will, therefore, have moved 1′, or 2′, or 3′, etc., beyond the last line of division on the limb, passed before it became stationary, according as the first, second, third, etc., line of division of the vernier beyond the index coincides with a line of division on the limb.

In Fig. 16, MN represents a portion of the limb of an instrument, divided into degrees and 10′ spaces; V the Vernier, ten

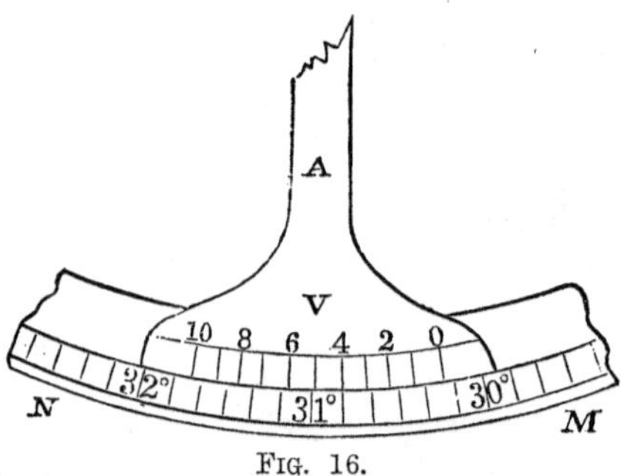

Fig. 16.

equal divisions of which have the same extent as nine of the 10′ spaces on the limb; and A the index-arm, which is here supposed to revolve with the telescope. The index-point, or zero of the vernier, is seen to be just beyond the point 30° 10′ on the limb; and on looking along the vernier, we perceive that the fourth line of division from the zero coincides with one of the lines of division of the limb. The zero of the vernier is, therefore, 4′ beyond the point 30° 10′ on the limb; and the whole reading is 30° 14′. It is here implied that one of the divisions of the vernier is less by 1′ than the 10′ space on the limb. Tc show this, let x = the number of minutes in a division of the vernier, then by what is stated above,

$$10\,x = 9 \times 10'; \text{ whence } x = 9', \text{ and } 10' - x = 1'.$$

By increasing the number of divisions of the vernier that corresponds to a number one less on the limb, the angle may be read off more accurately. For example, if sixty divisions of the vernier were made equal to fifty-nine of the limb, a division of the vernier would be 10″ less in value than a division of the limb; and the reading would be within 10″ of the exact position of the zero of the vernier on the limb.

But when the highest degree of accuracy is sought for, as in the large, fixed instruments of an observatory, the angle is read off by means of the Reading Microscope, instead of the vernier.

34. The Reading Microscope is a compound microscope, firmly fixed opposite to the limb, and furnished with cross wires in its focus, which are movable by a fine-threaded *Micrometer Screw.* This is a screw to the head of which is attached a graduated cylindrical head, that moves past a fixed index, to measure, by means of the turns and parts of a turn of the screw, the exact distance through which it is moved in the direction of its axis. In Fig. 17, AC is the microscope, and MN a portion of the limb seen edgewise. At D, on the optical axis, is the conjugate focus of the object-glass C; when the microscope is set at the proper distance from the limb, it is coincident with the focus of the eye-glass A. An image of a portion of the limb below C is formed at this point, and is seen distinctly through the eye-glass. ST is a box containing the sliding frame to which the cross-wires are attached; G, the milled head of the screw; EF, the graduated cylindrical head, called the *graduated head* of the screw; and *i* the fixed index.

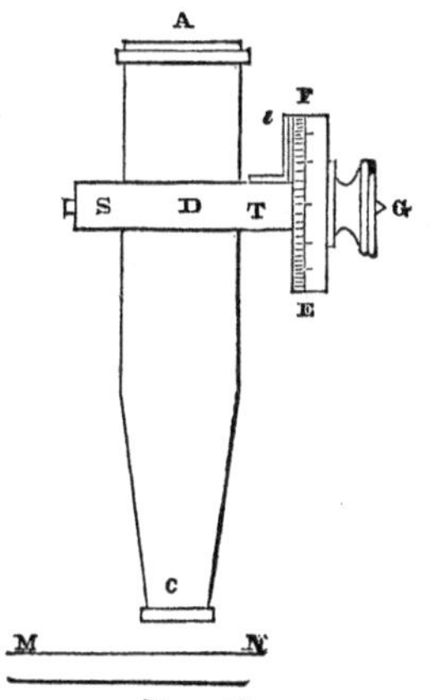

FIG. 17.

The cross-wires, with the connected apparatus for giving motion to them, and measuring the distance through which they are moved, is called a *Micrometer.*

Fig. 18 shows, upon an enlarged scale, the whole of the micrometer, as it would appear if viewed from A in Fig. 17. *aa* is the sliding frame to which the cross-wires are attached; *c* is the end of the screw working into this frame; and *bb*, spiral springs between the end of the frame and the end of the box, to prevent dead motion of the screw, and give more steadiness and regularity to the movements of the frame under the action of the screw.

The divisions of the limb are shown as short, heavy, equidistant lines. The cross-wires are the fine lines intersecting under an acute angle. A wire-pointer, not shown in the figure, in a position such that its prolongation would bisect this acute angle,

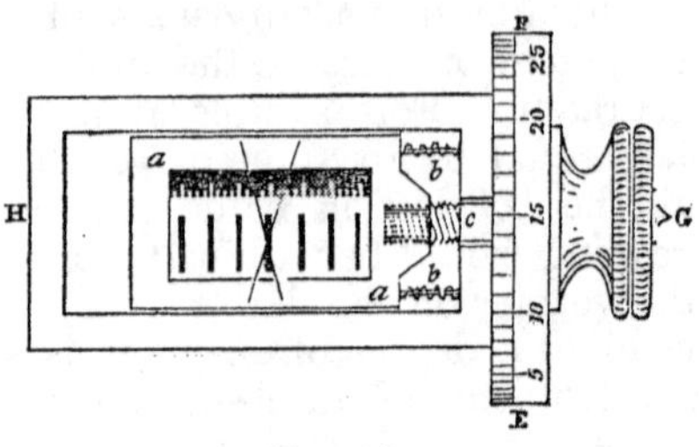

FIG. 18.

is generally used. On one side of the field is shown a notched scale of teeth, called a *comb-scale;* the distance from the middle of one notch to the middle of the next being the same as that between the threads of the screw. The wire-pointer is moved over this scale along with the cross-wires.

This scale is attached to the micrometer-box, and does not move with the cross-wires. The number of teeth passed by the intersection of the wires, therefore, shows the number of turns made by the screw; and the fractional part of a turn is indicated by the number of divisions of the graduated head that move past the index (*i*, Fig. 17), from the zero. If one revolution of the screw answers to a space of 1′ on the limb of the instrument, the number of teeth passed by the intersection of the wires will be the number of minutes of arc through which it is moved; and if the head of the screw is divided into sixty equal parts, the line of division opposite the fixed index will give the number of seconds to be added to the minutes, to determine the additional space moved over.

In reading off the angle the observer looks through the microscope at the limb. The point of intersection of the cross-wires of the microscope, when brought against the central notch of the scale, is a fixed point of reference, like the zero of a fixed vernier-plate. When the angle is to be read, this point will not, in general, fall upon one of the lines of division of the limb. By turning the micrometer-screw, the intersection of the wires is moved over the space which separates it from the line of division beyond which it falls; the number of teeth passed on the notched scale, will then be the number of minutes, and the number of the division of the screw-head opposite the index, will be the number of seconds, to be added to the angle taken from the limb.

To increase the accuracy of the reading, and determination of an angle, several microscopes are used, set opposite the limb at equally distant points. The fraction of a division in the reading is thus measured at different points of the circle, and the mean of the different measures is taken. Four reading microscopes, sometimes six, or even a greater number, are thus used. The whole degrees and minutes are read at only one of the microscopes.

35. Accuracy of Instruments. It is obvious that, other things being the same, instruments are accurate in proportion to the power of the telescope and the size of the limb. The large instruments now in use in astronomical observatories, are relied upon as furnishing angles to within a fraction of 1″.

36. Time is an essential element in astronomical observations. Three different kinds of time are employed by astronomers; *Sidereal*, *Apparent* or *True Solar*, and *Mean Solar Time.*

Sidereal Time is time as measured by the diurnal motion of the stars; or, as it is now considered, of the vernal equinox. A *Sidereal Day* is the interval between two successive meridian transits of a star; or, as now defined, the interval between two successive transits of the vernal equinox. It commences at the instant when the vernal equinox is on the superior meridian, and is divided into 24 *Sidereal Hours.*

Apparent, or *True Solar Time*, is deduced from observations upon the sun. An *Apparent Solar Day* is the interval between two successive meridian passages of the sun's centre, commencing when the sun is on the superior meridian. It appears from observation that it is a little longer than a sidereal day, and that its length is variable during the year. It is divided into 24 *Apparent Solar Hours.*

Mean Solar Time is measured by the diurnal motion of an imaginary sun, called the *Mean Sun*, conceived to move uniformly from west to east in the equator, with the real sun's mean motion in the ecliptic, and to have at all times a right ascension equal to the sun's mean longitude. A *Mean Solar Day* commences when the mean sun is on the superior meridian, and is divided into 24 *Mean Solar Hours.*

Since the mean sun moves uniformly and directly towards the east, the length of the mean solar day must be invariable.

The *Astronomical Day* commences at noon, and is divided into 24 hours; but the *Calendar Day* begins at midnight, and is divided into two portions of 12 hours each.

37. Astronomical Observations are, for the most part, made in the plane of the meridian. But some of minor importance are made out of this plane. The chief instruments employed for meridian observations, are the *Meridian Circle*, and the *Transit Instrument*, used in connection with the *Astronomical Clock.* These are the capital instruments of an observatory, inasmuch as they serve, as will soon be explained, for the determination of the places of the heavenly bodies, which are the fundamental data of astronomical science. The principal instruments used for making observations out of the meridian plane, are the *Altitude* and *Azimuth Instrument*, the *Equatorial*, and the *Sextant.*

THE TRANSIT INSTRUMENT.

38. The Transit Instrument, or Transit, is an instrument employed, in connection with a clock, for observing the passage of celestial objects across the meridian; either for the purpose of determining their right ascension, or obtaining the correct time. It is constructed of various dimensions, from a focal length of 20 inches, to one of 10 feet. The larger and more perfect instruments are permanently fixed in the meridian plane, and rest upon stone piers. The smaller ones are mounted on portable stands. Fig. 19 represents a fixed transit instrument in its most approved form. It is a sketch of the meridian transit instrument of the Washington Observatory, made by Ertel & Sons, Munich. The telescope has a focal length of 85 inches, with a clear aperture of 5.3 inches. TT is the telescope, firmly fixed to an inflexible axis, AA, at right angles to its length. The axis consists of two hollow cones, AA, proceeding from the opposite sides of a hollow cube, M; the whole being cast in one piece. The tube of the telescope is composed of two tubes, which are fastened by screws to the other two faces of the cube, M. The axis terminates in two steel pivots, V, accurately turned to the cylindrical shape, and of equal size. These pivots rest on two angular bearings, in form like the upper part of a Y, and called Y's. The Y's are notches cut in two blocks of metal, set in metallic boxes; the latter being imbedded in the tops of the stone piers PP. Sufficient play is given to the blocks in their boxes to allow one of the pivots to be raised or lowered, and the other to be moved to the right or left by means of adjusting screws, that give a motion to the blocks. To relieve the pivots of a portion of the weight of the telescope, a brass pillar, S, is firmly set upon the top of each pier, and furnishes a fulcrum to a lever, R, from one end of which depends a strong brass hook that supports the friction rollers X, under the end of the axis. A counterpoise, W, is adapted to the other end of the lever, which serves to sustain the greater part of the weight of the telescope, and leaves only a sufficient pressure at the pivots to secure a perfect contact with the Y's. This not only saves the pivots from wear, but gives the greatest possible freedom of motion to the telescope—the lightest touch of the finger being sufficient to rotate the instrument upon the friction rollers on which the axis chiefly rests.

Illumination of the Reticle-Wires. The pivots are perforated, to admit the light of a lamp placed on the top of either pier.

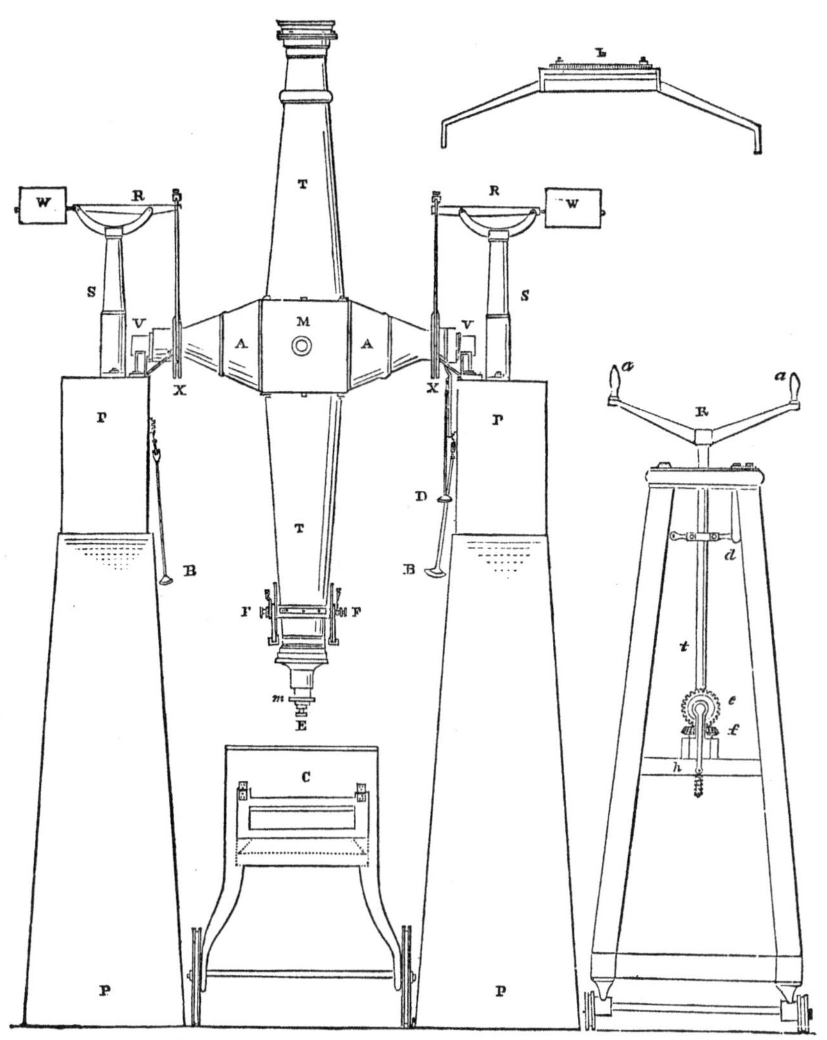

FIG. 19.

The light is received upon a plane metallic speculum, set within the hollow cube, M, at an angle of 45° to the axis of the telescope, and is reflected to the eye-glass; thus illuminating the field of view, and exhibiting the wires of the reticle, at *m*, as dark lines on a comparatively bright ground. The reflector has an elliptical opening at its centre, to permit the light that enters the telescope from a star, to pass on to the eye-glass. In observing small stars the wires are illuminated from the side of the eye-glass, by two small lamps (omitted in the drawing) suspended upon the telescope, near the eye-piece, which throw their light obliquely upon the wires, through openings in the eye-tube, without illuminating the field. The wires are thus made to appear as bright lines on a dark ground. The reticle has seven transit wires, placed at equal intervals, and two horizontal ones, between which the star is made to pass (30).

Finding Circles. On each side of the eye-end of the telescope, is fastened a small vertical graduated circle, F, about the centre of which turns freely an index-arm which carries a spirit-level and a vernier. This piece of apparatus is called a *Finding Circle*, or a *Finder*. An outline sketch of the finding circle, in one of its forms, is shown in Fig. 20; *a* is the index-arm, and *l*

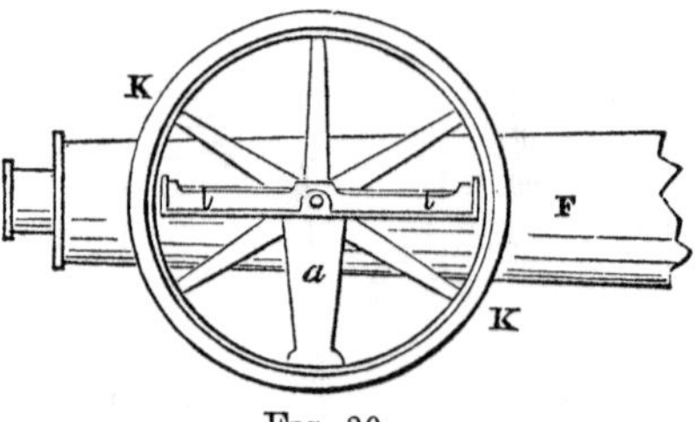

FIG. 20.

the level fastened at right angles to this, at the centre of the divided circle. Both turn freely about this centre. At the lower end of *a* is a vernier, and also a clamp and tangent-screw (not shown in the figure). The finding circles attached to the present instrument have a vernier at each end of a horizontal arm that carries the level; and the vertical arm serves only for clamping, and the tangent-screw motion.

By means of the finder, the telescope can be set to any given altitude or zenith distance, preparatory to an observation of the meridian passage of a star. This is done by setting the vernier of the finder to the given angle, and then depressing the eye-end of the telescope until the spirit-level is horizontal. In accomplishing this, the handles, BB and D, are used. The handle D acts upon a clamp that fastens the rotation axis. When the telescope has been depressed nearly to the required position, it is clamped by this handle, and the handles BB, which are connected with tangent-screws, serve to give the telescope a slow motion in altitude. By the same means, when the star to be observed enters the field of view of the telescope, it can be made to pass through the middle of the field.

A Reversing Apparatus, or Car, with which the instrument may be lifted from the Y's, and the rotation axis reversed, is shown at R*t*. It is mounted on grooved wheels that run upon two rails laid in the observatory floor, between the piers PP. The telescope having been placed in a horizontal position, the car is brought directly beneath the axis. By turning the crank *h*, acting upon two bevelled wheels, *e* and *f*, the latter of which has an internal screw engaging in an external screw upon the lower end of the vertical shaft *t*, two forked arms, *aa*, are lifted and brought into contact with the axis at AA; then, continuing the motion, the telescope is lifted sufficiently for the axis to clear the Y's and the friction rollers at XX. The car is then rolled out from between the piers, bearing the telescope with it; the instrument is turned half around upon the vertical shaft, the car rolled back to its former position, and the axis lowered into the Y's. The exact semi-revolution is determined by the stop, *d*.

An observing couch, C, runs on the rails between the piers. It is so arranged that the observer, reclining upon it, may give his head any required elevation; and thus promotes facility and accuracy of observation, by giving greater steadiness to the head, and relieving the observer of the fatigue of a constrained position when the telescope is directed upon stars at high altitudes.

L is a *striding level*, which is used in levelling the rotation axis.

39. Adjustments of the Transit. To secure accurate observations with the transit, three adjustments of the instrument are necessary:

1. The axis of rotation is to be brought into a horizontal position.

2. The line of collimation is to be made perpendicular to the axis of rotation.

3. The line of collimation is to be brought accurately into the meridian plane.

When these adjustments have been effected, the line of sight will lie in the plane of the meridian in every position given to the telescope.

40. First Adjustment. The first adjustment is effected by means of the striding level, L, which is applied to the pivots, VV; the feet of the level having the form of an inverted V for this purpose. By alternately working the screws that raise or depress one of the pivots, and the adjusting screws of the spirit-level, until the level is horizontal, whichever leg rests upon the eastern end of the axis, the axis may be made truly horizontal. Instead of attempting to secure in this way a perfect adjustment of the axis, it is found more convenient to determine the inclination of the axis to the horizon, by means of the scale marked off upon the tube of the spirit level, and calculate the error that is entailed from this cause, upon the observation.

41. Second Adjustment. The second, or *collimation adjustment*, is now generally made by means of special contrivances for the purpose, but it may also be accomplished in the following manner. Bring the telescope into a horizontal position, and direct it upon a well-defined point of a distant terrestrial object. Then, by means of the reversing apparatus raise the telescope from the Y's, and replace it with the ends of the axis reversed. Bring the telescope again into a horizontal position, and note whether it is directed upon the same point as before. If not, bring it half-way back to this point by the adjusting screws of the reticle, and the remaining distance by the screws that give a lateral motion to one end of the rotation axis. By one or more repetitions of this process, the desired adjustment may be effected.

The better plan, and the one ordinarily adopted by astronomical observers, is, after the *error of collimation* has been reduced to a small amount, to determine its value, and allow for it. This can readily be done when the reticle is provided with a movable micrometer-wire (31). It is only necessary to measure, with the micrometer, the distance of the point observed from the middle wire of the reticle in both positions of the telescope, convert each of the measured distances, expressed in revolutions of the screw-head, into their equivalent angular measures, and take the half difference of the two results. This will be the error of collimation.

The opportunity of reversing the instrument also enables the observer to determine the *correction for inequality of the pivots;* that is, the inclination of the mathematical axis of rotation to the horizon that may result from any such inequality. This correction is equal to one quarter of the difference between the inclinations of the line on which the feet of the level virtually rest, as determined by the level, in both positions of the telescope.

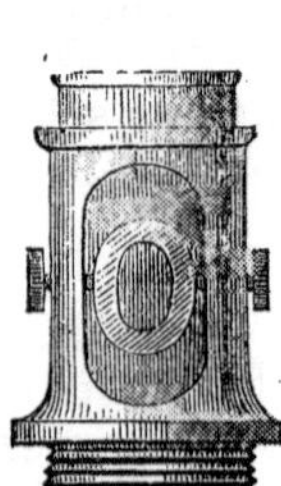
Fig. 21.

Collimating Eye-Piece. The most convenient method of determining the error of collimation is by making a certain observation with what is called the collimating eye-piece, substituted for the ordinary eye-piece of the telescope (Fig. 21). This differs from the common eye-piece in having an opening in one side of the tube, and a metallic reflector, of the form of an elliptical ring, set obliquely within the tube, to reflect the light of a lamp upon the wires of the micrometer. The observation to be made with it consists simply in looking vertically downward through the telescope at the image of the micrometer-wires, reflected from a basin of mercury placed on an immovable stone slab under the telescope. If the axis has been truly levelled, the error of collimation will be half the distance between the middle wire, as seen directly, and its reflected image. This distance can be measured by means of the movable wire of the micrometer. By working the adjusting screws of the reticle, and the vertical adjusting screws of the axis of rotation, the interval between the wire and its image may be made to disappear entirely; when the axis will be truly level, and the line of collimation in perfect adjustment.

42. Third Adjustment. The piers must first be established in such positions that the telescope, when the pivot ends of the axis have been placed in the Y's, and the axis levelled, will lie nearly in the meridian plane. This may be accomplished by bringing the telescope, after repeated trials, into such a position that it will be directed upon the pole-star when it is on the meridian. By referring to a map of the stars, it may be seen that the pole-star will be nearly on the meridian when a straight line from it to a point midway between the fifth and sixth stars, designated as ε and ζ, in the constellation of the Great Bear, is in a vertical position. The pole-star is also known to be on the meridian when it attains to its greatest, or least altitude. When the instrument has thus been approximately established, it may be more accurately adjusted to the meridian, with the aid of the screws that give a horizontal motion to one end of the axis. For this purpose observations may be made upon the pole-star at its upper and lower meridian transits, and the telescope moved in azimuth, until the interval between the upper and lower transit is made equal to that between the lower and upper transit.

The more convenient method is to ascertain from existing tables the time of the meridian passage of some known star, and bring the middle wire of the telescope upon the star at the instant of the transit. In order to effect this, the error of the timepiece must be known. If it indicates sidereal time, its error may be approximately determined with the instrument that is being established, by selecting a star that passes the meridian near the zenith, and noting the time of its transit across the middle wire of the telescope. This time should differ very little from the instant of the true meridian passage, as determined from astronomical tables; the difference will then be the error of the timepiece, nearly. The subsequent observations for adjustment to the meridian plane should be made upon stars remote from the zenith (the pole-star in preference). This process may be many times repeated, until the line of collimation of the transit telescope is brought, with all attainable accuracy, into the meridian plane. Or, the error of the adjustment may be calculated from the results of the observations upon the star near the zenith and the pole-star, and allowed for in subsequent observations. This method of adjustment is called *the method of high and low stars.* The final result obtained by it may be tested by the method of circumpolar stars already alluded to; which has the advantage of being independent of the error of the clock.

If the timepiece used in setting up the transit keeps mean solar time, its error may be determined by measuring an altitude of the sun with the transit or sextant, as will hereafter be explained.

43. The Time of the Meridian Passage of a Star is ascertained as follows: the telescope is first set by means of the finding circle, to the meridian altitude, or zenith distance of the star to be observed, and the instants of its crossing each of the parallel wires of the reticle noted. The sum of these observed times, divided by the number of the wires, will be the time of the star's crossing the middle wire; provided the wires are equidistant. The distances between the wires, in time, are called the *wire-intervals.* They can be determined, and their equality tested, by noting the intervals of time employed by a star situated on the celestial equator, in passing over them successively; these equatorial intervals, divided by the cosine of the declination of any star, will be the wire-intervals for that star. By means of these intervals the time of the star's passing either wire can be reduced to the middle wire. The mean of such reduced times obtained for all the wires, will be the time of the meridian transit of the star. The utility of having several wires, instead of one only, will be readily understood, from the consideration that a mean result of several observations is deserving of more confidence than a single one; since the chances are that an error which may have been made at one observation will be compensated by an opposite error at another.

If the body observed has a disc of perceptible magnitude, as in the cases of the *sun, moon,* and *planets,* the time of the passage of both the western and eastern limb across each of the parallel wires is noted, and reduced to the middle wire; the mean of all the results is then taken, which will be the instant of the meridian transit of the centre. We may, at the present day, obtain the time of the meridian passage of the centre of the sun, moon, or any planet, from an observation upon the western limb only, by adding "the sidereal time of the semi-diameter passing the meridian," taken from the Nautical Almanac, to the observed time. Or, the observation may be made upon the eastern limb, and the same quantity subtracted.

44. *Electro-Chronograph.* The accuracy of transit observations has recently been greatly increased, by the introduction of the electro-chronograph. This valuable contrivance consists of an electro-magnetic recording apparatus, put into communication with the pendulum of an astronomical clock, in such a manner that the circuit is broken at a certain point of each oscillation;

and, as a consequence, the seconds beat by the pendulum are designated by a series of equally distant breaks in a continuous line, upon a roll of paper to which an equable motion is given by machinery.

The observer holds in his hand a break-circuit key, by means of which he interrupts the circuit at the instant that the star is bisected by one of the wires in the field of the telescope, and thus makes a break in one of the short lines that answer to the successive seconds; as shown between 44s. and 45s., in Fig. 22.

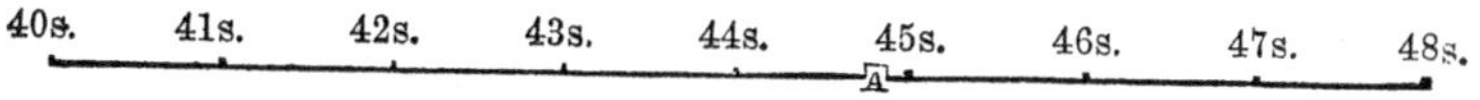

40s. 41s. 42s. 43s. 44s. 45s. 46s. 47s. 48s.

A

Fig. 22.

In this way, the instant of the transit across a single wire can be noted to within a much smaller fraction of a second than by the common method. Besides, the number of bisections in a single culmination of a star, by increasing the number of wires, may be augmented fivefold.

This method of observation was adopted at the Washington Observatory, in 1849, and soon after at the Observatory of Harvard College. It has since been introduced at the Greenwich and other principal observatories.

45. To determine the Right Ascension of a Star. When a star is on the meridian, its declination circle (def. 16, p. 15) coincides with the meridian; moreover, the arc of the equator which lies between the declination circles of two stars, measures their difference of right ascension. Thus, RR′ (Fig. 8) is the difference of right ascension of the stars S and S′; their absolute right ascensions being VR and VR′. In the interval between the transits of the two stars, the arc RR′, which is equal to their difference of right ascension, passes across the meridian at the rate of 15° to a sidereal hour. If, therefore, the times of their meridian transits be determined with the transit instrument and sidereal clock, the difference between these times, converted into degrees by allowing 15° to the hour, will be the difference of right ascension of the two stars. In this way, the difference between the right ascension of any standard star, S, fixed upon as a point of reference, and other stars, may be successively determined. This having been done, the absolute right ascensions of these stars will become known as soon as the position of the vernal equinox with respect to the standard star has been found. For, it is plain that RR′ being known, if VR be also determined, VR′ may be found by adding VR and RR′. The manner of determining the position of the vernal equinox, or the value of VR, will be explained in the chapter on the Apparent Motion of the Sun. Right ascensions are commonly expressed in time.

ASTRONOMICAL CLOCK.

46. The Astronomical Clock is provided with a pendulum so constructed that its length is unaffected by changes of temperature. The mercurial compensation pendulum, in which the ordinary brass bob is replaced by a glass jar containing a certain quantity of mercury, is generally employed. The clock is secured to a stone pier resting upon a firm foundation, which is disconnected from the floor of the observatory. It keeps sidereal time.

47. To Regulate a Sidereal Clock. When a clock is used for determining differences of right ascension (45), it is adjusted to sidereal time if it goes equably and marks out twenty-four hours in a sidereal day; it being altogether immaterial at what time it indicates 0h. 0m. 0s. To ascertain its *daily rate*, note by the clock the times of two successive meridian transits of the same star: the difference between the interval of the transits and twenty-four hours will be the *daily gain*, or *loss* (as the case may be), of the clock with respect to a perfectly accurate sidereal clock. If the gain or loss, when found in this manner, proves to be the same each day, then the mean rate of going is the same each day.

Error.—The sidereal clock now in use in astronomical observatories, is made to indicate 0h. 0m. 0s. when the vernal equinox is on the superior meridian; and it is necessary to know not only its *rate* but also its *error*. This may be found from day to day by noting the time of the transit of some known star, whose place has been accurately determined, and comparing this with its right ascension expressed in time. If the two are equal the clock is right; otherwise their difference will be its error. For greater accuracy in the determination of the error and rate, the successive transits of several standard stars should be noted. To facilitate these and other determinations, the apparent places of a large number of stars are given in nautical almanacs, and other similar works.

Clock Stars. The stars most favorably situated for determining the clock correction are those which pass the meridian near the zenith; or, next to these, the stars which cross the meridian between the zenith and equator. Stars considerably to the north of the zenith pass too slowly through the field of the telescope; and if the transit instrument has not been accurately adjusted to the meridian, the error in the time of the transit will be greater in proportion as the star observed is further from the zenith.

48. A Mean Solar Clock is usually regulated by observations upon the sun. The methods by which its error and rate are determined will be explained in the chapter on the Measurement of Time.

MERIDIAN CIRCLE.

49. The Meridian Circle is an instrument used to measure the zenith distance, or altitude of a heavenly body, at the instant of its arrival on the meridian. It is, in its general construction, a combination of the transit instrument and a graduated vertical circle; and is hence sometimes called the *Transit Circle.* In the larger observatories, it is mounted on two piers, like the transit. The graduated circle is firmly attached at right angles to the horizontal axis of rotation, and turns with it. The angle is read from the circle by a reading miscroscope, attached to the adjacent pier; or in some instances, to a frame which rests upon the axis itself. For greater accuracy four or six reading microscopes are used, at equally distant points of the limb. The degrees, minutes, and seconds, are read from one of the microscopes, and the seconds only from the others. If the seconds read from either microscope be added to the degrees and minutes obtained from the first, the result will be the reading of that microscope reduced to the first. By taking the mean of all the results, for the different microscopes, the errors from imperfect graduation, inaccurate centring, and unequal expansion of the limb, may be materially lessened.

50. Fig. 23 represents a meridian circle manufactured by Repsold, a celebrated German instrument-maker, and mounted in 1852 in the observatory of the United States Naval Academy. It has two graduated circles, CC and C′C′, of the same size, but only one of these, CC, is graduated finely; this is read by four microscopes, two of which are seen at RR. The microscopes are attached to the four corners of a square frame which is centred upon the rotation axis; but does not turn with it, being held in a fixed position by screws connected with the piers. Each horizontal side of the frame carries a spirit level, by which any change of inclination of the frame with respect to the horizon may be detected.

The second circle, constructed of the same size as the first, for the sake of symmetry, is graduated more coarsely, and is used only as a finder.

The counterpoises WW act at XX, to support the greater part of the weight of the instrument upon friction rollers, as in the case of the transit instrument. The inclination of the rotation axis is measured with a *hanging level*, LL.

A horizontal arm, FG, seen to the right of the telescope in the figure, extends out from the pier, and receives a vertical arm which is connected with a collar upon the rotation axis. By turning a screw, the head of which is at G, the telescope is clamped in the collar; and then a screw (not seen in the drawing), connected with the arm FG, and acting horizontally upon the

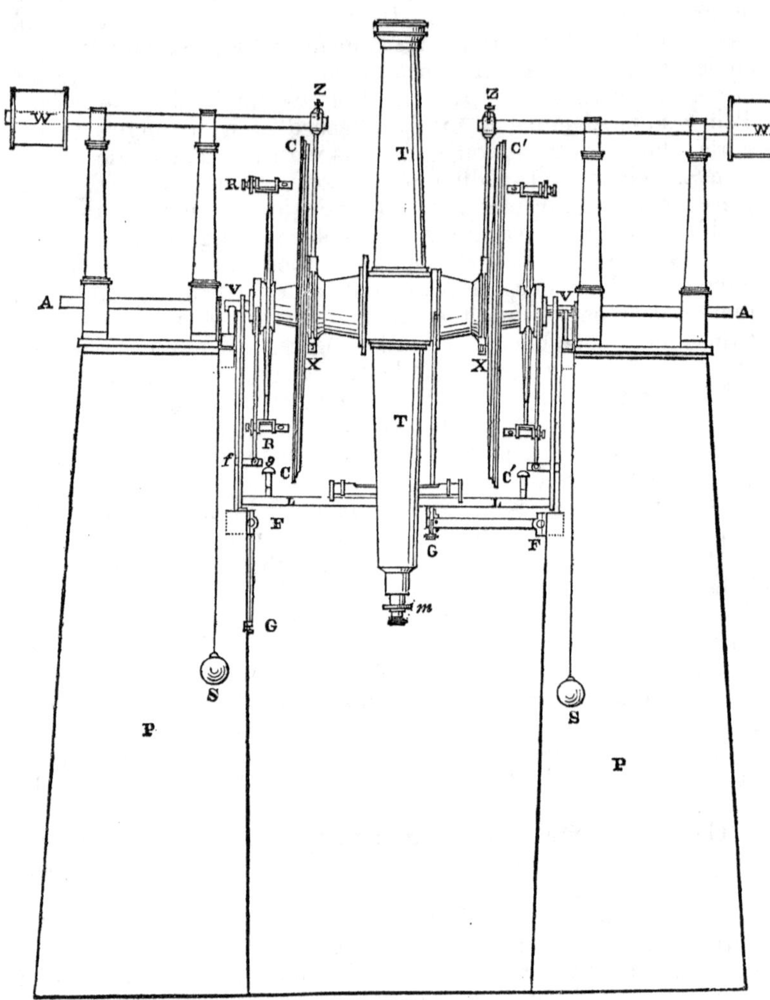

FIG. 23.

vertical arm, gives a fine motion to the telescope. FG turns upon a joint at F; and to the left of the telescope is shown in the position it takes when detached from the vertical arm, preparatory to a reversal of the instrument.

Another arm, *fg*, similar in its form and arrangement to FG, receives a vertical arm attached to the microscope frame. Screws connected with *fg*, and acting horizontally at *g* upon the vertical arm, serve to adjust the frame.

The field is illuminated by light thrown into the interior of the telescope through tubes at AA, and reflected towards the reticle by a mirror in the central cube. The quantity of light is regulated by revolving discs with eccentric apertures, at the extremities of the tubes nearest the Y's. These discs are revolved by means of a cord to which hangs a small weight, S.

The micrometer at *m* contains seven fixed transit threads, and three equally distant horizontal threads movable by a micrometer-screw. The more common form consists of a reticle with several stationary transit wires, or threads, and one stationary horizontal wire; in connection with one or more movable horizontal wires (31). The movable micrometer-wires serve for the measurement of small differences of declination.

51. Mural Circle. This is another form of the meridian circle that has been much used in large observatories. The graduated limb of the mural circle is secured to one end of a horizontal axis, which is let into a massive pier or wall of stone. Its axis, therefore, is not symmetrically supported, and it cannot be reversed. On these accounts it is inferior, for nice determinations, to the form of meridian circle just described.

Mural circles have been constructed as large as eight feet in diameter.

52. Adjustments. The same adjustments have to be effected with the meridian circle as with the transit; and the same methods may be adopted. But it is also necessary to determine with great accuracy what is called *the horizontal point of the limb.* This is the place of the index, or zero of the reading microscope, answering to a horizontal position of the line of collimation of the telescope.

53. To determine the Horizontal Point of the Limb. Direct the telescope upon any known star, at the time of its passing the meridian, and read off the angle on the limb. On the next night, when the star comes to the meridian, direct the telescope upon the image of the same star reflected from a basin of mercury, and note the angle as before. By a fundamental law of reflection the angle of depression of this image will be equal to the angle of elevation of the star. Accordingly the arc on the limb, which passes before the reading microscope, in moving the telescope from the star to its image, will be double the altitude of the star, and its point of bisection the horizontal point.

This method will not give an exact result unless a correction is applied for the difference in the values of the atmospheric refraction at the times of observation (81). The necessity of making this correction may be avoided, and a more reliable result obtained, whenever the instrument is provided with a micrometer having a movable horizontal wire. By a rapid manipulation an observation may then be made upon the star at the time it is crossing the first transit wire, and another observation taken upon its image, as it is crossing the last transit wire. The instrument is first set to the altitude of the star, as nearly known, and the correction to this altitude measured by bringing the movable horizontal wire upon the star at the instant it is crossing the first transit wire. In observing the image of the star, it is brought near the fixed horizontal wire, the limb clamped, and the observation completed by the tangent screw of the limb. The observer may then read, at his leisure, the microscopes for the last measured angle, and the micrometer correction to the first angle. To each of the angles measured, a small correction must be applied to reduce it to the meridian.

54. To measure the Altitude of a Heavenly Body. (1). *Of a fixed star.* Direct the telescope of the meridian circle upon the star, bring it on the horizontal wire of the reticle, and clamp the limb; then by means of the tangent screw that gives a small motion to the limb and telescope, bisect the star with the horizontal wire at the instant of its crossing the middle transit wire. Then read off the angle from the different microscopes, as already explained (49), and take the mean of the several results. This must be corrected for the deviation of the horizontal point from the zero of the limb, and all the detected errors that result from imperfect adjustments, or defects of construction.

If an observation be made upon the star at the time of its crossing any other than the central wire, it can readily be reduced to the meridian.

(2). *Of the sun, moon, or any planet.* Measure the altitudes of the upper and lower limbs, and take their half sum for the altitude of the centre, or measure the altitude of the upper or lower limb, and add or subtract the apparent semi-diameter of the body, taken from the Nautical Almanac. The observations are facilitated by using the movable micrometer wire in establishing the contact with the limb; then, by turning the micrometer screw, measuring the interval between the position of the movable and that of the parallel stationary wire, and adding this measured interval to the mean of the microscope readings.

55. To determine the Declination of a Heavenly Body. The meridian altitude, or zenith distance of a heavenly body, having been measured at a place the latitude of which is known, its declination may easily be found. For let s (Fig. 10, p. 21) represent the point of meridian passage of a star which crosses to the north of the zenith (Z), Es will be its declination (def. 27, p. 17), Zs its meridian zenith distance, and ZE the latitude of the place of observation (O), (def. 33, p. 18); and we obviously have

$$Es = ZE + Zs \ldots. (a)$$

If the star cross the meridian at some point s' between the zenith (Z) and the equator (E), we shall have $Es' = ZE - Zs'$, (b); and if its point of transit be some point s'' to the south of the equator (E), we shall have $Es''=Zs''-ZE$, and $-Es''=ZE-Zs''$, (*c*). The three formulæ (*a*), (*b*), and (*c*), may all be comprehended in one, viz.:

Declination = latitude + meridian zenith distance, . . . (1)

If we adopt the following conventional rules: (1) north latitude is +, south latitude —; (2) the zenith distance is *north*, or *south*, according as the star passes to the north or south of the zenith; and it has the same sign as the latitude when it has the same name, the contrary sign when it is of a contrary name; (3) north declination is +, south declination —.

The latitude which is here supposed to be known, may be found by measuring the meridian altitudes of a circumpolar star with the meridian circle, and taking their half sum. For, as the pole lies midway between the points at which the transits take place, its altitude will be the arithmetical mean, or the half sum of the altitudes of these points; and the altitude of the pole is equal to the latitude of the place (24).

It will be seen in the next Chapter, that certain corrections must be applied to all measured altitudes.

56. To determine the Longitude and Latitude of a Body. When the right ascension and declination of a heavenly body have been obtained from observation, with a transit instrument and circle (45, 55), its longitude and latitude may be computed. For, let S (Fig. 8) represent the place of the body, VRQE the equator, VLTW the ecliptic, and P, K, the north poles of the equator and ecliptic. In the spherical triangle PKS we shall know PS the complement of SR the declination, and the angle KPS = ER = EV + VR = 90° + right ascension; and if we suppose the obliquity of the ecliptic to be known, we shall know PK. We may therefore compute KS, and the angle PKS. But KS is the complement of SL, which is the latitude of the body S; and PKS = 180°—WKS = 180°—(WV + VL) = 180° —(90° + longitude) = 90° — longitude.

The obliquity of the ecliptic, which we have here supposed to be known, is, in practice, easily found; for it is equal to TQ, the sun's greatest declination.

ALTITUDE AND AZIMUTH INSTRUMENT.

57. This instrument consists essentially of a telescope mounted upon either a fixed or portable stand, and provided with both a vertical and a horizontal graduated limb. The telescope turns with the vertical limb about a horizontal axis, and the whole turns about the vertical axis of the horizontal limb. The instrument is so adjusted, that when the line of sight of the telescope is in the meridian plane, the zero of the reading microscope of the horizontal limb will answer to the zero of the limb, or nearly so. If they do not correspond, the distance between them will be the index error. This having been determined, if the telescope be directed upon a star out of the meridian, the reading of the horizontal limb, corrected for the index error, will be the azimuth of the star at the instant of the observation. The vertical circle serves to measure the altitude. The altitude and azimuth instrument is sometimes called the *Altazimuth;* also the *Astronomical Theodolite.*

58. The Meridian Line (def. 8, p. 14) at a place may easily be determined with the altitude and azimuth instrument, by a method called the *Method of Equal Altitudes.* Let O (Fig. 24) represent the place of observation, NPZ the meridian, and S, S′ two positions of the same star, at which the altitude is the same. Now, the spherical triangles ZPS and ZPS′ have the side ZP common, ZS=ZS′, and (allowing the stars to move in *circles*) PS=PS′. Hence they are equal, and consequently the angle PZS=PZS′; that is, *equal altitudes* of a star correspond to *equal azimuths.* Therefore, by bisecting the arc of the horizontal limb, comprehended between two positions of the vertical limb for which the observed altitude of a star is the same, we shall obtain the meridian line.

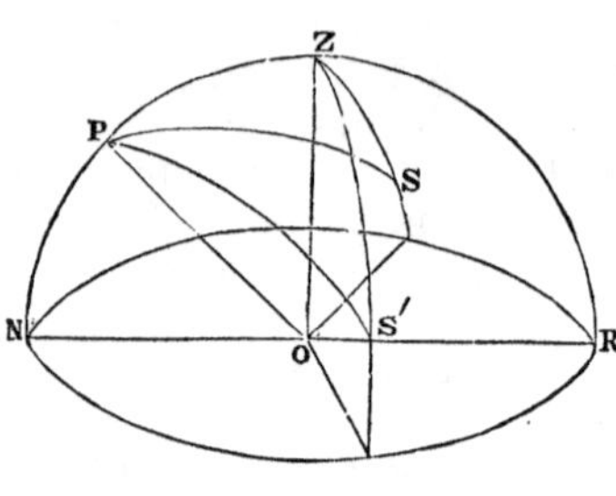

FIG. 24.

The meridian line may be approximately determined, by this method, with the common theodolite; the observations being made upon the sun. The result will be more accurate if they be made towards the summer or winter solstice, when the sun will have but a slight motion towards the north or south in the interval of the observations. It is, however, easy to determine and allow for the effect of the sun's change of place in the heavens.

When the time is accurately known, the north and south line may be found very easily by directing the telescope of any instrument that has a motion in azimuth, upon a star in the vicinity of the pole, at the instant of its arrival on the meridian.

59. Zenith Telescope. This may be regarded as a modified form of the portable altitude and azimuth instrument. It is of great value for the convenient and accurate determination of the latitude of a place; and has been used for this purpose with great success in the United States Coast Survey. Its chief peculiarities consist in the substitution of a finding circle with a delicate spirit level, similar to the finding circle of the transit instrument (38), for the ordinary vertical limb of the altitude and azimuth instrument, and the adaptation to the telescope of a micrometer with a movable horizontal wire.

If such a micrometer be adapted to a transit instrument, that instrument may be successfully used as a substitute for the zenith telescope, for the accurate determination of the latitude of a station.*

EQUATORIAL.

60. The equatorial consists of a telescope mounted with two axes of motion, at right angles to each other, one of which is parallel to the axis of the earth, and of the celestial sphere. The angular movement about this axis is measured by a graduated circular limb at right angles to the axis, and therefore parallel to the plane of the equator; from which the instrument takes its name. This limb is called the *hour circle.* There is also a graduated circle, called the *declination* circle, adapted to the other axis; which lies, in every one of its positions, in the plane of a celestial meridian. The telescope turns in the plane of a celestial meridian about this axis; and can at the same time be made to rotate, in connection with it, about the other, or *polar axis.* It can thus be readily set upon any star, whose hour angle and declination are known; and when once directed towards it, can be made to follow the star in its diurnal motion, by simply producing a continuous movement about the polar axis. This motion is generally communicated by clock-work, without the use of the hand.

Plate I. represents the large equatorial telescope mounted under the dome of the observatory of Harvard College. It is connected with a bed-plate which is fastened by screw-bolts to the top of a granite block, in a position parallel to the axis of the heavens. This block is ten feet in height, aud rests upon a granite pier forty-two feet high. The clock-work is on the further side of the stone support, and does not appear in the figure. The instrument is so nicely counterpoised that it can be moved with the greatest ease by the pressure of the hand upon the end of one of the balance rods.

61. Uses of the Equatorial. A telescope thus *equatorially mounted*, and provided with a movable micrometer-wire, is especially adapted to the measurement of the apparent diameter of a heavenly body, the angular distance between stars in close proximity, and in general to all observations that require the telescope to be directed upon a body for a considerable interval of time. Accordingly the large telescope of every prominent observatory is mounted in this manner.

* This has been satisfactorily shown by Professor C. S. Lyman, of Yale College (see American Journal of Science, Vol. XXX., p. 52).

The zenith telescope is essentially the invention of Capt. Andrew Talcott, of the United States Corps of Engineers, who also devised a method of determining the latitude by this instrument which surpasses all others, both in simplicity and accuracy. This is now known as Talcott's method (Chauvenet's Spherical aud Practical Astronomy).

The equatorial can also be advantageously used for determining the unknown place of a fixed star, or planet, in the heavens, by measuring the angular distance and direction of the star from some known star seen with it in the field of the telescope; or by noting the interval of the transits, and measuring directly the difference of declination of the two stars. For this purpose the telescope is furnished with a certain form of micrometer, called the *Position Filar Micrometer;* with which the measurements in question can be made with great accuracy.

Differences of right ascension and declination can also be measured with the equatorial, by means of the hour and declination circles, but with much less accuracy than with the transit instrument and meridian circle.

62. *Position Filar Micrometer.* This piece of apparatus serves at the same time to measure small angular distances, and the angle included between the line connecting two stars in close proximity and the celestial meridian. This angle is called the *angle of position* of one of the stars with respect to the other. It is estimated from the S. round by the W. to 360°. The *Filar Micrometer*, designed for the measurement of small angles, is shown in Fig. 25. It is the same in principle as the micrometer employed in the reading microscope (34).

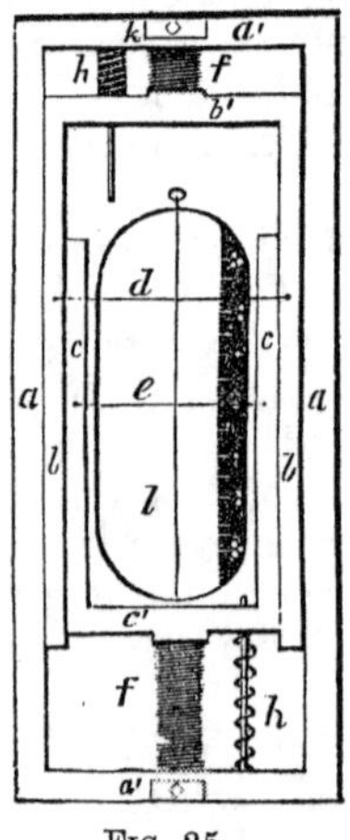

FIG. 25.

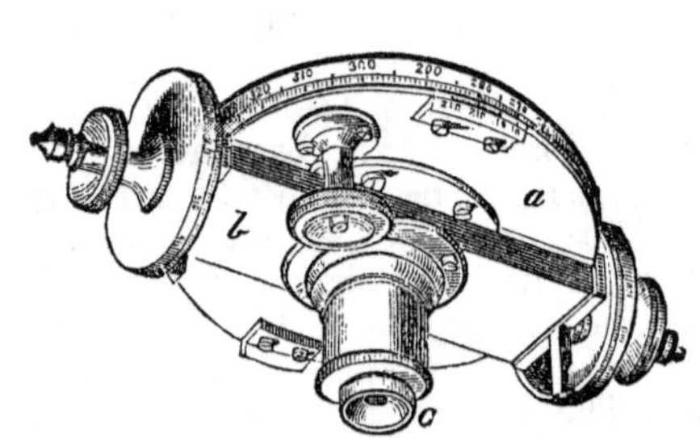

FIG. 26.

It consists of two forks of brass, *bb'b*, *cc'c*, sliding within a rectangular brass box, *aa'a*, and one within the other. Each of these forks carries a very fine wire, or spider line, stretched perpendicularly across from one prong to the other; they are movable, and the parallel wires which they carry, by micrometer screws passing through the ends of the box, and attached to the forks. A third and stationary wire, *l*, perpendicular to the other two, is attached to a diaphragm disconnected from the forks. The heads of the screws are not shown in the figure, but they may be seen in Fig. 26, in which *b* is the micrometer-box. The eyepiece is screwed into the micrometer-box, as shown in Fig. 26. The graduated screw-heads are connected with nuts which turn, without advancing, upon the screws that are fastened to the forks. Accordingly by turning the nuts, the forks may be moved either forwards or backwards.

A stationary *comb-scale* on one side of the box, indicates the number of revolutions of either screw, answering to any distance that the wires may be separated from each other; and the fractional part of a revolution is shown by the graduated head of the screw. The value of one revolution of the micrometer-screw may be found by bringing the two parallel wires into a position perpendicular to the celestial equator, separating them by a certain number of revolutions, and then noting the time taken by an equatorial star to traverse the interval between them. The interval of time thus obtained, converted into the equivalent angular space by allowing 15″ to 1^s, will be the number of seconds of arc answering to the assumed number of revolutions of the screw.

To adapt the filar micrometer to the measurement of *angles of position*, the micrometer-box, with its attached eye-piece, is so mounted as to admit of a rotation around the centre of a graduated circle (Fig. 26). The circle is fastened at the end of the reticle-tube, and in a plane perpendicular to the optical axis of the telescope. The revolving motion is produced by a milled-head screw *s*, which works on an interior toothed wheel; and the angle is read off upon the stationary graduated circle, by aid of the vernier movable with the plate *a*.

SEXTANT.

63. The instruments which have now been described are observatory instruments, the chief design of whose construction is to furnish the places of the heavenly bodies with all attainable exactness. That of which we are now to treat is much less exact, though still of great utility in effecting certain important astronomical determinations; as of the latitude or longitude of a place, and the time of day. It is chiefly used by navigators, and astronomical observers on land, who are precluded by their situation, or other circumstances, from using the more accurate instruments of an observatory. It is much more conveniently portable than any of these, and has not to be set up and adjusted at every new place of observation. Besides, as it is held in the hand, it can be used at sea, where by reason of the agitations of the vessel, no instrument supported in the ordinary way is of any service.

64. Construction:—Principle of Construction. The sextant may be defined, in general terms, to be an instrument which serves for the direct admeasurement of the angular distance between any two visible points. The particular quantities that may be measured with it, are; 1st, the altitude of a heavenly body; 2d, the angular distance between any two visible objects in the heavens or on the earth. Its essential parts are a graduated limb BC (Fig. 27), comprising about 60 degrees of the entire circle, which is attached to a triangular frame BAC; two mirrors, of which one (A) called the *Index Glass*, is movable in connection with an index, G, about A, the centre of the limb, and the other (D) called the *Horizon Glass*, is permanently fixed parallel to the radius AC drawn to the zero point of the limb, and is only half silvered (the upper half being transparent); and a small immovable telescope at E, directed towards the horizon-glass. *The principle of the construction* and use of the sextant may be understood from what follows: A ray of light SA from a celestial object S, which impinges against the index-glass, is reflected off at an equal angle, and striking the horizon-glass (D) is again reflected to E, where the eye likewise receives through the transparent

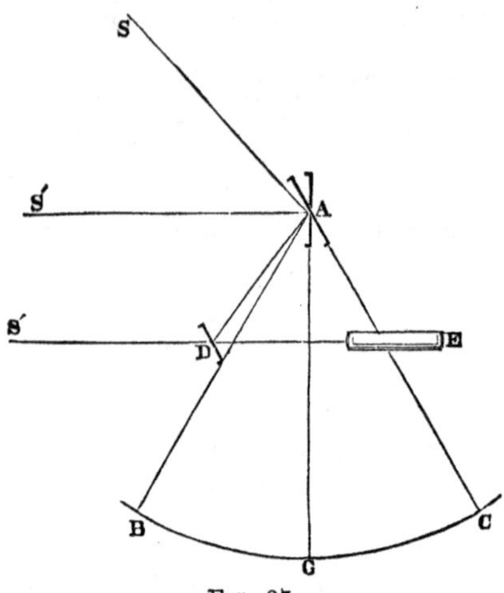

FIG. 27.

part of that glass a direct ray from another point or object S′. Now, if AS′ be drawn, directed to the object S′, SAS′, the angular distance between the two objects S and S′, is equal to double the angle CAG measured upon the limb of the instrument (AC being parallel to the horizon-glass). For, when the index-glass is parallel to the horizon-glass, and the angle on the limb is zero, AD, the course of the first reflected ray, will make equal angles with the two glasses, and therefore the angle SAD will become the angle S′AD, (=ADE;) and the observer, looking through the telescope, will see the same object S′ both by direct and reflected light. Now, if the index-glass be moved from this position through any angle, CAG, the angle made by the reflected ray which follows the direction AD, with this glass, will be diminished by an amount equal to this angle; for, we have DAG=DAC—CAG. Therefore the angle made with the index-glass by the new incident ray SA, which after reflection now pursues the same course ADE, and reaches the eye at E, as it is always equal to that made by the reflected ray, will be diminished by this amount. Consequently, the incident ray in question will on the whole, that is, by the diminution of its inclination to the mirror by the angle CAG, and by the motion of the mirror through the same angle, be displaced towards the right, or upwards, an angle S′AS equal to 2GAC. Thus, the angular distance SAS′ of two objects S, S′, seen in contact, the one (S′) directly, and the other (S) by reflection from the two mirrors, is equal to twice the angle CAG that the index-glass is moved from the position (AC) of parallelism to the horizon-glass.

Hence the limb is divided into 120 equal parts, which are *called* degrees; and to obtain the angular distance between two points, it is only necessary to sight directly at one of them, and then move the index until the reflected image of the other is brought into contact with it; the angle read off on the limb will be the angle sought.

To obtain the angular distance between two bodies which have a sensible diameter, bring the *nearest* limbs into contact, and to the angle read off on the limb *add* the sum of the apparent semi-diameters of the two bodies, or bring the *farthest* limbs into contact, and *subtract* this sum.

65. The Detail of the Construction of the Sextant is shown in Fig. 28. The limb, and the triangular frame to which it is

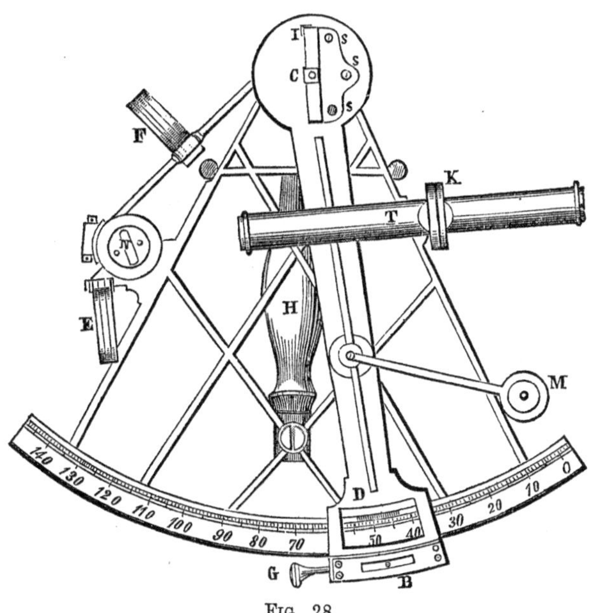

FIG. 28.

attached, are of hammered brass, and strengthened by cross-plates. The graduation is upon silver inlaid in the brass. Each degree is divided into six equal parts, of 10′. N is the horizon-glass, fastened to the frame in the position before stated; I the index-glass, in a brass frame, attached to the index-bar CD, by the screws s s s, and movable with it about the centre C of the graduated arc. These two mirrors are of plate-glass silvered. The upper half of the horizon-glass is left unsilvered, that the direct rays from the object towards which the small telescope, T, is directed may not be intercepted. The telescope is supported in a ring, K, attached to a stem underneath, which can be raised or lowered by a screw. By this means the relative brightness of the direct and reflected images can be regulated. M is a microscope, movable about a centre on the index-bar, used in reading the angle from the vernier at D. The vernier is so divided as to give the angle to within 10″. At B, under the index-bar, is a screw for clamping it to the limb; and G is a tangent screw for giving the bar, with the index-glass, a small motion, in securing the accurate contact or coincidence of the images. H is a wooden handle at the back of the sextant, by which it is held when an observation is taken. At E and F are colored glasses of different shades, to diminish the intensity of the light when the sun is observed. Those at F are interposed between the index-glass and the horizon-glass when the sun is seen by reflection from the index-glass. The others are used when the telescope is directed upon the sun.

66. Adjustments. The adjustments of the sextant consist in setting the index-glass and the horizon-glass, and bringing the line of sight of the telescope parallel to the plane of the graduated arc, and in determining the *index error.*

The index-glass may be adjusted by setting the index near the middle of the arc, placing the eye nearly in the plane of the sextant, and near the index-glass, and observing whether the arc seen directly and its reflected image form one continuous arc. If the reflected image does not appear to form a true continuation of the arc, the index-glass is not perpendicular to the plane of the sextant. It may be corrected by loosening the screws *s s s*, and inserting a piece of paper under the plate through which they pass.

The horizon-glass is adjusted by sighting through the telescope at a star, and moving the index until the direct and reflected image of the star pass each other. If, in passing, the two images can be made to coincide, the horizon-glass is perpendicular to the plane of the instrument. If any correction is necessary, it can be made by turning a small screw at the top or bottom of the horizon-glass.

To test the position of the line of sight of the telescope, select two objects, as two stars, 100° to 120° apart, and bring the reflected image of the one in contact with the direct image of the other, on the wire within the telescope that is nearest the plane of the sextant: if then, on moving the instrument, the contact remains when the images are thrown upon the other parallel wire of the telescope (although a separation occurs in the interval between them), no adjustment is required. It can be made, when necessary, by means of two small screws in the ring which supports the telescope.

To find the index error. Bring the direct and reflected images of the same point of a distant terrestrial object, or of the same star, into coincidence, and read off the arc. This reading will be the index error, and may be either positive or negative.

67. Taking an Angle. When observing with the sextant, it is held in the right hand by the handle, and the telescope directed upon one of the two objects whose angular distance is to be measured, generally the fainter one. It is then turned about the line of sight until the other object lies in its plane; and the index moved with the left hand until the reflected image of this object is brought, at the centre of the field of the telescope, into apparent contact with the object seen directly;—

the contact being finally effected by the use of the tangent screw. The angle is then read from the limb and vernier, with the microscope.

When the sextant is employed to take *the altitude of a heavenly body*, a horizontal reflector, called an *Artificial Horizon*, is placed in front of the observer. The angle between the body and its reflected image is then measured as if this image were a real object; the half of which will be the altitude of the body.

A small quantity of mercury, poured into a shallow vessel of tinned iron or copper, forms a very good artificial horizon.

In obtaining the altitude of a body *at sea*, its altitude above the visible horizon is measured, by bringing the lower limb into contact with the horizon. To this angle is added the apparent semi-diameter of the body, and from the result is subtracted the depression of the visible horizon below the horizontal line, called the *Dip of the Horizon.*

68. Hadley's Quadrant. Hadley's Quadrant differs from the sextant in having a graduated limb of 45°, instead of 60°, in real extent, and a sight-vane instead of a small telescope. It is not capable, then, of measuring any angle greater than about 90°, while the sextant will measure an angle as great as 120°; or even 140° (for the graduation generally extends to 140°). The quadrant is also inferior to the sextant in respect to materials and workmanship, and its measurements are less accurate.

69. Reflecting Circle. The Reflecting Circle is but an enlarged sextant. Its limb is a full circle, and the index-arm is prolonged in the other direction, and carries a vernier on each end. The angle is read from each vernier, and the mean of the two readings taken, to eliminate the error of eccentricity.

70. Prismatic Sextant. This is an improved form of sextant, recently introduced. It takes its name from the fact that a reflecting prism is used in place of the ordinary horizon-glass. This prism also occupies a different position with respect to the index-glass. The graduated limb extends 120°. The prismatic sextant can be used to measure an angular distance of 180°, and an altitude of 90°. It is also superior to the ordinary sextant in certain other peculiarities of construction.

Prismatic Reflecting Circles are also constructed which possess similar advantages over the ordinary reflecting circle.

ERRORS OF INSTRUMENTAL ADMEASUREMENT.

71. Whatever precautions may be taken, the results of instrumental admeasurement will never be wholly free from errors. Errors that arise from inaccuracy in the workmanship or adjustment of the instrument, may be detected and allowed for. But errors of *observation* are, obviously, undiscoverable. Since, however, the chances are, that an error committed at one observation, will be compensated by an opposite error at another, it is to be expected that a more accurate result will be obtained if a great number of observations, under varied circumstances, be made, instead of one, and the *mean* of the whole taken for the element sought. And accordingly, it is the uniform practice of astronomical observers to *multiply* observations as much as is practicable.

72. Instrumental Errors may be divided into three classes; viz. *errors of construction*, *errors of adjustment*, and *incidental errors*. Errors of construction, in the best instruments, result chiefly from *imperfect graduation*, an *eccentricity* of the limb, an *inequality* or an *ellipticity of the pivots*, and an *imperfect rigidity* of the telescope or axis. The effect of eccentricity and of the ellipticity of the pivot, may be eliminated by taking the mean of the readings of two microscopes, at opposite points of the limb. The error of graduation may be greatly reduced, by reading the angle from several equidistant points of the limb, and taking the mean of all the readings. When the construction of the instrument is such that the principle of *repetition* may be adopted—that is, the angle read off from all parts of the limb—the error of graduation may, theoretically speaking, be removed entirely.

It is not the practice of astronomical observers to strive to bring instruments into the nicest possible adjustment, but instead, after a good adjustment has been effected, to deduce, by a systematic series of observations, the several errors that remain, and derive from these the corrections to be applied to the quantity to be determined.

Incidental errors may arise from diverse effects produced by changes of temperature, especially an unequal expansion of different parts of the limb, and a derangement of the microscopes; from flexure produced by weight; and also from vibrations produced by passing vehicles, and other derangements from extraneous mechanical causes. All such errors may be mostly neutralized by making numerous measurements, under a great variety of circumstances.

THE TELESCOPE.

73. An observatory is not completely furnished unless it is supplied with a large telescope for examining the various classes of objects in the heavens; and one or more smaller ones for exploring the heavens and searching for particular objects invisible to the naked eye, as faint comets, and making observations upon occasional celestial phenomena, as eclipses of the sun and moon, occultations of the stars, etc. Telescopes are divided into the two classes of *Reflecting* and *Refracting Telescopes*. In the former class, the image of the object is formed by a concave speculum, and in the latter by a converging achromatic lens. This image is viewed and magnified by an eye-glass; or rather by an achromatic eye-piece consisting of two glasses. In the simplest form of the reflecting telescope, the Herschelian, the image formed by the concave speculum is thrown a little to one side, and near the open mouth of the tube, where the observer views it through the eye-glass, with his back turned towards the object.

74. *Magnifying power—illuminating power—space-penetrating power.* The magnifying power of a telescope is to be carefully distinguished from its illuminating, and space-penetrating power. A telescope magnifies by increasing the angle under which the object is viewed; it increases the light received from objects, and reveals to the sight remote stars, nebulæ, etc., by intercepting and converging to a point a much larger beam of rays. The magnifying power is measured by the ratio of the focal length of the object-glass, or speculum, to that of the eye-piece. The *illuminating power*, by which it reveals stars invisible to the naked eye, if we leave out of view the amount of light lost by reflection and absorption, is measured by the proportion which the area of the object-glass, or speculum, bears to that of the pupil of the eye. Since the quantity of light received from any luminous point, viewed at different distances by the naked eye, decreases in the same proportion that the square of the distance increases, and the quantity of light from the same point, conveyed to the eye by a telescope, is augmented in the ratio of the square of the diameter of its aperture to the square of the diameter of the pupil of the eye, it follows that the diminution of the light from an increase of distance, will be just supplied if the aperture of the telescope exceed in its diameter that of the pupil of the eye in the same ratio that the distance is augmented. The power of a telescope to penetrate into space, and discern stars, therefore, exceeds that of the naked eye in the same ratio that the diameter of its aperture exceeds that of the pupil of the eye (0.2 in.). In the larger reflecting telescopes, the space-penetrating power, calculated by this rule, requires to be diminished about one-fifth, in consequence of the loss of light incident to the use of the telescope.

Telescopes are provided with several *eye-glasses*, of various powers. The power to be used varies with the object to be viewed, and the purity and degree of tranquillity of the atmosphere. Of two telescopes of the same focal length, that which has the largest aperture will form the brightest image in the focus, and therefore, other things being equal, admit of the use of the most powerful eye-piece. In this way, it happens that the available magnifying power indirectly depends materially upon the size of the aperture. In all telescopes, there is a certain fixed ratio between the aperture and focal length, or at least limit to this ratio. In reflecting telescopes, it is one linear inch of aperture for every foot of focal length, and in refracting telescopes one inch of aperture for from one to two feet of focal length. Reflectors and refractors of the same focal length, have about the same actual magnifying and illuminating power.

The highest theoretical magnifying power that has yet been obtained is about 7,000. But the highest actually available power, in observing any celestial object, does not exceed 2,500. The higher powers can be used only upon double stars, and clusters of stars. With the best telescopes, a magnifying power of four or five hundred is the highest that can be applied to the moon and planets; owing to the great diminution of brightness that results from the enlargement of the image.

75. *Defining power.* Telescopes of equal size may differ materially in their defining power: that is, in their capability to show the planets, and other celestial objects which have a sensible disc, with a sharp outline, and all their peculiarities of appearance with distinctness, and to separate close double stars and clusters of stars. The excellence of telescopes in this respect, depends upon the precision of form and perfection of polish of the lenses, their freedom from chromatic and spherical aberrations, and other niceties of construction.

76. The *field of view* of telescopes diminishes in proportion as the magnifying power increases. It is stated that with a magnifying power of between 100 and 200 it is a circle not as large as the full moon; and with a power of 600 or 1,000 is nearly filled by one of the planets, while a star will pass across it in from two to three seconds.

The diminution of the field of view, and the trepidations of the image occasioned by the varying density of the atmosphere, and the unavoidable tremors of the instrument, must ever affix a practical limit to the magnifying power of telescopes. This limit, it is probable, is already nearly attained; for the highest powers of the best telescopes can now be used only in the most favorable states of the weather. The illuminating and space-penetrating power of telescopes may, however, yet be materially increased, and a greater distinctness and definiteness in the outline of objects obtained.

77. *Large Telescopes.* The largest reflecting telescope that has yet been constructed and directed to the heavens, is the great *Rosse Telescope*, devised and constructed by Lord Rosse, of Ireland. It has a focal length of 53 feet, and an aperture of 6 feet. Its illuminating power is about 78,000; and its space-penetrating power, for single stars, about 280 times the distance of the most remote star visible to the naked eye. The most powerful refractor yet constructed, is the great *Clark Telescope*, made by Clark & Sons, Cambridgeport, Mass., and recently set up in the Chicago Observatory. It has a clear aperture of 18½ inches, and a focal length of 23 feet. It has, by the adaptation of different eye-pieces, different magnifying powers, varying from 70 to about 2,000. The great telescope of the Observatory of Harvard College has an aperture of 15 inches, and a focal length of 22½ feet. Its highest magnifying power is 2,000. The refractor of the observatory at Pulkova, in Russia, is but slightly inferior to this in its dimensions and capabilities Refracting telescopes of large dimensions and great excellence, are mounted equatorially in all the prominent observatories in the United States and in Europe.

CHAPTER IV.

CORRECTIONS OF MEASURED ANGLES.

78. Angles measured at the earth's surface with astronomical instruments answer to the *Apparent Place* of a heavenly body, and are termed *Apparent* elements. In astronomical language the *True Place* of a heavenly body is its real place in the heavens, as it would be seen from the centre of the earth. Angles which relate to the true place are denominated *True* elements. The co-ordinates of the apparent place of a body are termed its *apparent co-ordinates*, and those of its true place its *true co-ordinates*.

79. Corrections. The apparent co-ordinates are reduced to the true, by the application of certain corrections, called *Refraction*, *Parallax*, and *Aberration*. Refraction and aberration are corrections for errors committed in the estimation of a star's place, while parallax serves to transfer the co-ordinates from the earth's surface to its centre. The object of the reduction of observations from the surface to the centre of the earth, is to render observations made at different places on the earth's surface directly comparable with each other. Observers occupying different stations upon the earth refer the same body, unless it be a fixed star, to different points of the celestial sphere. Their observations cannot, therefore, be compared together, unless they be reduced to the same point, and the centre of the earth is the most convenient point of reference that can be chosen.

REFRACTION.

80. Atmospheric Refraction. We learn from the principles of Pneumatics, as well as by experiments with the barometer, that the atmosphere gradually decreases in density from the earth's surface upwards. We learn also from the same sources, that it may be conceived to be made up of an infinite number of strata of decreasing density, concentric with the earth's surface. From the known pressure and density of the atmosphere at the surface of the earth, it is computed, that by the laws of the equilibrium of fluids, if its density were throughout the same as immediately in contact with the earth, its altitude would be about 5 miles. Certain facts, hereafter to be mentioned, show that its actual altitude is not far from 50 miles. Now, it is an established principle of Optics, that light in passing from a vacuum into a transparent medium, or from a rarer into a denser medium, is bent or *refracted* towards the perpendicular to the surface at the point of incidence. It follows, therefore, that the light which comes from a star, in passing into the earth's atmosphere, or in passing from one stratum of atmosphere into another, is refracted towards the radius drawn from the centre of the earth to the point of incidence.

Path of a ray of light. Let M*mn*N, N*no*O, O*oq*Q, (Fig. 29)

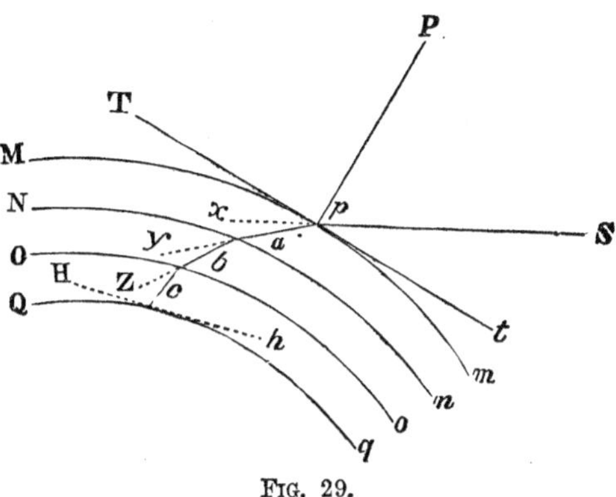

Fig. 29.

represent successive strata of the atmosphere. Any ray, S*p*, will then, instead of pursuing a straight course, S*px*, follow the broken line *pabc*; being bent downwards at the points *p*, *a*, *b*, *c*, &c., where it enters the different strata. But, since the number of strata is infinite, and the density increases by infinitely small degrees, the deflections *apx*, *bay*, &c., as well as the lengths of the lines *pa*, *ab*, &c., are infinitely small; and therefore *pabc*, the path of the ray, is a broken line of an infinite number of parts or a curved line concave towards the earth's surface, as it is represented in Fig. 30. Moreover, it lies in the vertical plane containing the original direction of the ray; for this plane is perpendicular to all the strata of the atmosphere, and therefore the ray will continue in it in passing from one to the other.

81. Astronomical Refraction. The line OS′ (Fig. 30) drawn tangent to *pa*O, the curvilinear path of the light, at its lowest point, will represent the direction in which the light enters the eye, and therefore the apparent line of direction of the star. If, then, OS be the true direction of the star, the angle SOS′ will be the displacement of the star produced by *Atmospheric Refraction.* This angle is called the *Astronomical Refraction*, or simply the *Refraction* of the star.

Since *pa*O is concave towards the earth, OS′ will lie above

OS; consequently, *refraction makes the apparent altitude of a star greater than its true altitude, and the apparent zenith distance of a star less than its true zenith distance.* (We here speak of the true altitude and true zenith distance, as estimated from the station

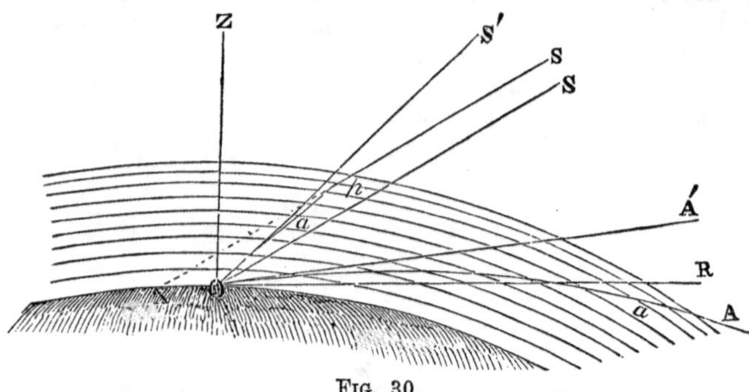

FIG. 30.

of the observer upon the earth's surface.) Thus, to obtain the *true altitude* from the apparent, we must *subtract* the refraction; and to obtain the *true zenith distance* from the apparent, we must *add* the refraction. As refraction takes effect wholly in a vertical plane (80), it *does not alter the azimuth of a star.*

The amount of the refraction varies with the apparent zenith distance. In the zenith it is zero, since the light passes perpendicularly through all the strata of the atmosphere: and it is the greater, the greater is the zenith distance; for, the greater the zenith distance of a star, the more obliquely does the light which comes from it to the eye penetrate the earth's atmosphere, and enter its different strata, and therefore, according to a well-known principle of optics, the greater is the refraction.

82. To find the Amount of the Refraction for a given Zenith Distance or Altitude. Let us first show a method of resolving this problem by the general theory of refraction. According to this theory, the amount of the refraction, except so far as the convexity of the strata of the atmosphere may have an effect, depends wholly upon the absolute density of the air immediately in contact with the earth, and not at all upon the law of variation of the density of the different strata; that is, the actual refraction is the same that would take place if the light passed from a vacuum immediately into a stratum of air of the density which obtains at the earth's surface. Let us suppose, then, that the whole atmosphere is brought to the same density as that portion of it which is in contact with the earth, and let bah (Fig. 31) represent its surface; also let O represent the station of the observer upon the earth's surface, and Sa a ray incident upon the atmosphere at a. Denote the angle of refraction OaC by p, and the refraction Oax by r. The angle of incidence

$$Z'aS = Z'aS' + S'aS = OaC + Oax = p + r.$$

Now if we represent the index of refraction of the atmosphere by m, we have, by a law of refraction,

$$\sin Z'aS = m \sin OaC, \text{ or } \sin (p + r) = m \sin p;$$

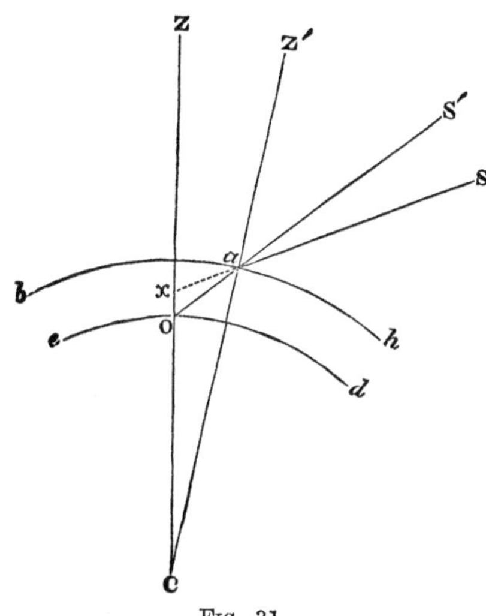

FIG. 31.

developing, (App. For. 15,)

$$\sin p \cos r + \cos p \sin r = m \sin p;$$

or, dividing by $\sin p$,

$$\cos r + \cot p \sin r = m.$$

But, as r is small, we may take $\cos r = 1$, and $\sin r = r = r''$ $\sin 1''$, (App. 47.)

Whence, $1 + \cot p . r'' \sin 1'' = m$, or $r'' = \frac{m - 1}{\sin 1''} \times \frac{1}{\cot p} = A \text{ tang} p;$ putting $A = \frac{m - 1}{\sin 1''}$. Let $ZCa = C$; and $ZOa = Z$. $OaC = ZOa - ZCa$, or $p = Z - C$. Substituting, we have $r'' = A$ tang $(Z - C)$; or, omitting the double accent, and considering r as expressed in seconds,

$$r = A \text{ tang } (Z - C) \dots\dots (2)$$

When the zenith distance is not great, C is quite small compared with Z. If we neglect it, we have

$$r = A \text{ tang } Z \dots\dots (3);$$

which is the expression for the refraction, answering to the supposition that the surface of the earth is a plane, and that the light is transmitted through a stratum of uniformly dense air, parallel to its surface. We perceive, therefore, that the *refraction, except in the vicinity of the horizon, varies nearly as the tangent of the apparent zenith distance.*

It has been ascertained by experiment that m, the index of refraction (the barometer being = 29.6 inches, and the thermometer = 50°), = 1.0002803. Substituting in equation (3), after having restored the value of A, and reducing, there results

$$r = 57''.8 \text{ tang } Z \ldots\ldots (4).$$

83. Formulæ of Refraction. With the aid of this formula, or of others purely theoretical, astronomers have sought to determine the precise amount of the refraction at various zenith distances from observation, and by collating the results of their observations to obtain empirical formulæ that are more exact.

One of the simplest methods of accomplishing this is the following: When the latitude or co-latitude of a place, and the polar distance of a star which passes the meridian near the zenith, have been determined, the refraction may be found for all altitudes from observation simply. For, let P (Fig. 32) be the elevated pole, Z the zenith, PZE the meridian, HOR the horizon, S the true place of a star, and S′ its apparent place. Suppose the apparent zenith distance ZS′ to have been measured. Now, in the triangle ZPS, ZP the co-latitude and PS the polar distance are known by hypothesis, and the angle P is the sidereal time which has elapsed since the star's last meridian transit, (or, if the star be to the east of the meridian, the difference between this interval and 24 sidereal hours,) converted into degrees by allowing 15° to the hour. Therefore we may compute the true zenith distance ZS, and subtracting from it the apparent zenith distance ZS′, we shall have the refraction. For the solution of this problem, the polar distance may be found by taking the complement of the declination computed from an observed meridian zenith distance (55); and, since the upper and lower transits of a circumpolar star take place at equal distances from the pole, the co-latitude may be found by taking the half sum of the greatest and least zenith distances of the pole-star. But it is obvious that neither of these quantities can be accurately determined, unless the measured zenith distances be corrected for refraction. When, however, the zenith distances in question differ considerably from 90°, the corresponding refractions may be at first ascertained with considerable accuracy by means of equation (4). When more correct formulæ have been obtained by this or any other process, the latitude and polar distance, and therefore the refraction answering to the measured zenith distance, will become more accurately known.

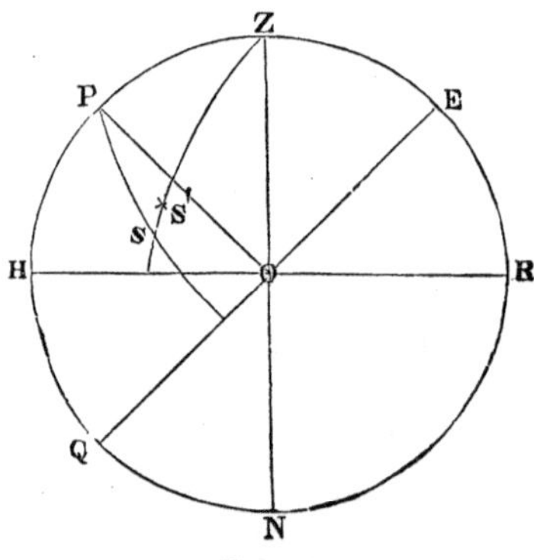

FIG. 32.

The various formulæ of refraction having been tested by numerous observations, it is found that they are all, though in different degrees, liable to material errors when the zenith distance exceeds 80°, or thereabouts. At greater zenith distances than this the refraction is *irregular*, or is frequently different in amount when the circumstances on which it is supposed to depend are the same.

84. Mean Refractions.—Corrections for the varying density of the Air. The refractive power of the air varies with its density, and hence the refractions must vary with the height of the barometer and thermometer. The refractions which have place when the barometer stands at 30 inches and the thermometer at 50°, are called *mean refractions*. The refractions corresponding to any other height of the barometer or thermometer, are obtained by seeking the requisite *corrections* to be applied to the mean refractions in consequence of the difference between the actual density of the air and its assumed mean density.

Tables of Refraction. To save astronomical observers the trouble of calculating the refraction whenever it is needed, the mean refractions corresponding to various zenith distances, or altitudes, are computed from the formula, as also the corrections for various heights of the barometer and thermometer, and inserted in tables. (See Tables VIII. and IX.)

On inspecting Table VIII., it will be seen that the refraction amounts to about 34′ when a body is in the apparent horizon, and to about 58″ when it has an altitude of 45°.

85. Other Effects of Atmospheric Refraction. Atmospheric refraction makes the apparent distance of any two heavenly bodies less than the true; for it elevates them in vertical circles which continually approach each other from the horizon till they meet in the zenith.

Refraction also gives to the discs of the sun and moon an eliptical form when near the horizon. As it increases with an increase of zenith distance, the lower limb of the sun or moon is more refracted than the upper, and thus the vertical diameter is shortened, while the horizontal diameter remains the same, or very nearly so. This effect is greatest near the horizon, for the reason that the increase of the refraction is there the most rapid; and it is most observable at sea, as the sun and moon, at their rising or setting, can there be seen in closer proximity to the horizon than at most stations on land. The difference between the vertical and horizontal diameters may amount to $\frac{1}{8}$ part of the whole diameter.

When a star appears to be in the horizon, it is actually 34′ below it (84): refraction, then, retards the setting and accelerates the rising of the heavenly bodies.

Having this effect upon the rising and setting of the sun, it must increase the length of the day.

The apparent diameter of the sun is about 32′; as this is less than the refraction in the horizon, it follows, that when the sun appears to touch the horizon it is actually entirely below it. The same is true of the moon, as its apparent diameter is nearly the same with that of the sun.

PARALLAX.

The correction for atmospheric refraction having been applied, the zenith distance of a body is reduced from the surface of the earth to its centre, by means of a correction called *Parallax.*

86. Definitions. Parallax is, in its most general sense, the angle made by the lines of direction, or the arc of the celestial sphere comprised between the places of an object, as viewed from two different stations. It may also be defined to be the angle subtended at an object by a line joining two different places of observation. Let S (Fig. 33) represent a celestial object, and A B two places from which it is viewed. At A it will be referred to the point *s* of the celestial sphere, and at B to the point *s′*; the angle BSA, or the arc *ss′*, is the parallax. The arc *ss′* is taken as the measure of the angle BSA, on the principle that the celestial sphere is a sphere of an indefinitely great radius, so that the point S is not sensibly removed from its centre.

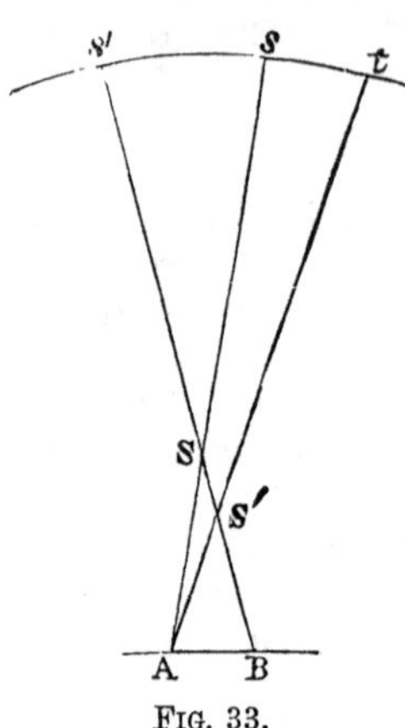

FIG. 33.

The term parallax is, however, generally used in Astronomy in a limited sense only, namely, to denote the angle included between the lines of direction of a heavenly body, as seen from a point on the earth's surface and from its centre; or the angle subtended at a heavenly body by a radius of the earth. If C (Fig. 34) is the centre of the earth, O a point on its surface, and S a heavenly body, OSC is the parallax of the body. When there is occasion to distinguish this angle from other angles of parallax, it is termed the *Geocentric Parallax.*

The parallax of a heavenly body above the horizon is called *Parallax in Altitude.*

The parallax of a body at the time its apparent altitude is zero, or when it is in the plane of the horizon, is called the *Horizontal Parallax* of the body. Thus, if the body S (Fig. 34) be supposed to cross the plane of the horizon at S′, OS′C will be its horizontal parallax. OSC is a parallax in altitude of this body.

It is to be observed, that the definition just given of the horizontal parallax, answers to the supposition that the earth is of a spherical form. In point of fact, the earth (as will be shown in the sequel) is a spheroid, and accordingly the vertical and the radius at any point of its surface are inclined to each other; as

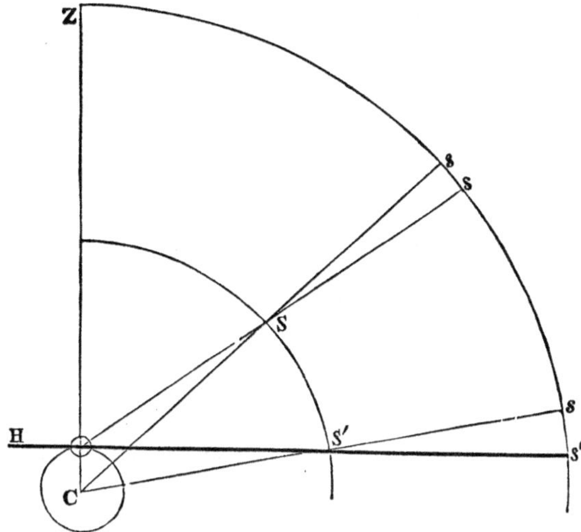

FIG. 34.

represented in Fig. 35, where OC is the radius, and OC′ the vertical. The points *Z* and *z*, in which the vertical and radius

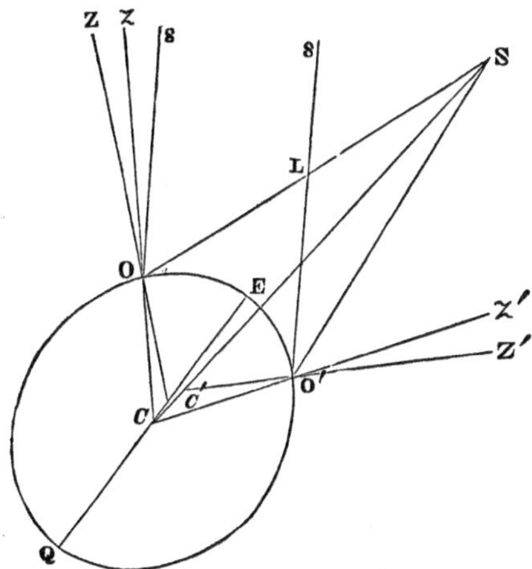

FIG. 35.

pierce the celestial sphere, are called, respectively, the *Apparent Zenith* and the *True*, or *Central Zenith*. In perfect strictness, the horizontal parallax is the parallax at the time *z*OS, the apparent distance from the true zenith, is 90°. But no material error will be committed in supposing the earth to be spherical, except when the question relates to the parallax of the *moon.*

87. True Zenith Distance. Let the apparent zenith distance ZOS = Z, (Fig. 34,) the true zenith distance ZCS = z, and the parallax OSC = p. Since the angle ZOS is the exterior angle of the triangle OSC, we have

ZOS = ZCS + OSC, and hence also ZCS = ZOS — OSC;

or,

$$Z = z + p, \text{ and } z = Z - p \quad \ldots\ldots (5).$$

Thus, to obtain the *true* zenith distance from the apparent, we have to *subtract* the parallax; and to obtain the *apparent* zenith distance from the true, to *add* the parallax.

Parallax, then, takes effect wholly in a vertical plane, like the refraction, but in the inverse manner; depressing the star, while the refraction elevates it. Thus, the refraction is added to Z, but the parallax is subtracted from it.

88. To find an Expression for the Parallax in Altitude, in terms of the apparent zenith distance. In the triangle SOC (Fig. 34) the angle OSC = parallax in altitude = p, OC = radius of the earth = R, CS = distance of the body S = D, and COS = 180° — ZOS = 180° — apparent zenith distance = 180° — Z; and we have by Trigonometry the proportion

$$\sin OSC : \sin COS :: CO : CS;$$

whence,

$$\sin p : \sin (180° - Z) :: R : D;$$

and

$$D \sin p = R \sin Z;$$

or,

$$\sin p = \frac{R}{D} \sin Z \quad \ldots\ldots (6).$$

This equation shows that the parallax p depends for any given zenith distance Z upon the distance of the body, and is less in proportion as this distance is greater: also, that for any given distance of the body it increases with an increase in the zenith distance. When Z = 90°, p has its maximum value, and then = horizontal parallax = H; and equa. (6) gives

$$\sin H = \frac{R}{D} \quad \ldots\ldots (7);$$

substituting, we have

$$\sin p = \sin H \sin Z \quad \ldots\ldots (8).$$

This equation may be somewhat simplified. The distances of the heavenly bodies are so great, that p and H are always very small angles; even for the moon, which is much the nearest, the value of H does not at any time exceed 62′. We may, therefore, without material error, replace sin p and sin H by p and H. This being done, there results,

$$p = \text{H} \sin \text{Z} \ldots . (9).$$

Wherefore, *the parallax in altitude equals the product of the horizontal parallax by the sine of the apparent zenith distance.*

If we take notice of the deviation of the earth's form from that of a sphere, Z, in equation (8), will represent the apparent distance from the true zenith, (86,) and H the horizontal parallax as it is defined in Art. 86.

In order to be able to compute the parallax in altitude by means of formula (9) it is necessary to know H, the horizontal parallax.

89. To find an Expression for the Horizontal Parallax, in terms of measurable quantities. Let O, O′ (Fig. 35) represent two stations upon the same terrestrial meridian OEO′, and remote from each other, Z, Z′ their apparent zeniths, and z, z' their true zeniths, QCE the equator, and S the body (supposed to be in the meridian) the parallax of which is to be found. Let the angle OSO′ = A, zOS = Z, z'O′S = Z′; also let CO = R, CO′ = R′, CS = D, the parallax in altitude OSC = p, and the parallax in altitude O′SC = p'. Now, by equation (6), replacing the sine of the parallax by the parallax itself, (88,)

$$p = \frac{\text{R}}{\text{D}} \sin \text{Z}, \text{ and } p' = \frac{\text{R}'}{\text{D}} \sin \text{Z}';$$

whence

$$p + p' = \frac{\text{R}}{\text{D}} \sin \text{Z} + \frac{\text{R}'}{\text{D}} \sin \text{Z}' = \frac{\text{R} \sin \text{Z} + \text{R}' \sin \text{Z}'}{\text{D}}; \, . \, (10);$$

and, (equ. 7,)

$$\text{H} = \frac{\text{R}}{\text{D}}, \text{ or } \text{D} = \frac{\text{R}}{\text{H}}.$$

Substituting this value of D, and deducing the value of H, we have

$$\text{H} = \frac{\text{R}\,(p + p')}{\text{R} \sin \text{Z} + \text{R}' \sin \text{Z}'} = \frac{\text{R} \times \text{A}}{\text{R} \sin \text{Z} + \text{R}' \sin \text{Z}'} \; . \; . \; . \; (11).$$

It remains now to find an expression for A in terms of measurable quantities. Let Os and O′s (Fig. 35) be the directions, at O and O′, of a fixed star which crosses the meridian nearly at the same time with the body. Owing to the immense distance of the star, these lines will be sensibly parallel to each other (19). Let the angle SOs, the difference between the meridian zenith distances of the body and star, as observed at O, be represented by d, and let the same difference SO′s for the station O′, be represented by d'. Now,

$$\text{OSO}' = \text{OLO}' - \text{SO}'s = \text{SO}s - \text{SO}'s, \text{ or } \text{A} = d - d'.$$

If the body be seen on different sides of the star by the two observers, we shall have

$$A = d + d'.$$

Substituting in equation (11), there results,

$$H = \frac{R\,(d \pm d')}{R \sin Z + R' \sin Z'} \quad . \ . \ . \ (12).$$

If we regard the earth as a sphere, $R = R'$, and dividing by R, we have

$$H = \frac{d \pm d'}{\sin Z + \sin Z'} \quad . \ . \ . \ . \ (13).$$

90. To Determine the Horizontal Parallax of a body, from Observation; by means of this formula. Let each of the two observers measure the meridian zenith distance of the body, and also of a star which crosses the meridian nearly at the same time with the body, and correct the measured distances for refraction. The difference of the two will be, respectively, the value of d and d'; and the corrected zenith distances of the body will be the values of Z and Z′. If formula (12) be used, the measured zenith distances of the body must still be corrected for the reduction of latitude, (Art. 23, def. 4.)

It is not necessary that the two stations should be on precisely the same meridian; for if the meridian zenith distance of the body be observed from day to day, its daily variation will become known; then, knowing also the difference of longitude of the two places, the following simple proportion will give the change of zenith distance during the interval of time employed by the body in moving from the meridian of the most easterly to that of the most westerly station, viz.: as interval (T) of two successive transits : diff. of long., expressed in time, (t) :: variation of zenith dist. in interval T : its variation in interval t. This result, applied to the zenith distance observed at one of the stations, will reduce it to what it would have been if the observation had been made in the same latitude on the meridian of the other station.

The horizontal parallax of the moon has been determined by this process with sufficient accuracy. The parallaxes of the sun and planets, which are very small, have been determined by much more accurate methods. The importance of having recourse to methods of the greatest possible accuracy, in the case of the sun and planets, will appear in the sequel.

91. Horizontal Parallax in Different Latitudes. In consequence of the spheroidal form of the earth, the horizontal parallax of a body is somewhat different in different latitudes. Let H and H′ denote the horizontal parallaxes of the same body, at the distance D, and R and R′ the radii of the earth at two different latitudes; then, by equ. 7,

$$\sin H = \frac{R}{D}, \text{ and } \sin H' = \frac{R'}{D};$$

$$\text{whence, } \sin H : \sin H' :: \frac{R}{D} : \frac{R'}{D} :: R : R'.$$

Also, as H and H′ are small, we have very nearly,

$$H : H' :: R : R'.$$

Thus the horizontal parallax is greatest at the equator, and decreases nearly in the same ratio with the radius of the earth from the equator to the poles. The horizontal parallax of the moon is about 11″ greater at the equator than at the poles. In the case of the sun, or of any planet, the difference is in every instance less than $\frac{1}{8}''$.

92. Equatorial Parallax. The horizontal parallax of a body, for a station on the equator, is called its equatorial horizontal parallax, or simply its equatorial parallax.

The equatorial parallax of the moon varies from 52′ 50″ to 61′ 32″, according to the distance of the moon from the earth. At the mean distance its value is 57′ 3″.

The equatorial horizontal parallax of the sun, at the earth's mean distance, is 8″.95. The sun's horizontal parallax varies with the earth's distance less than $\frac{1}{8}''$.

The horizontal parallaxes of the planets, at their varying distances from the earth, are comprised between the limits 34″ and 0.″3. The greater limit is the parallax of Venus when nearest the earth, and the smaller limit is the parallax of Neptune when farthest from the earth.

The fixed stars have no geocentric parallax.

Tables of Parallax. In the present condition of astronomical science, the horizontal parallax of the sun, moon, or any planet, may be calculated for any particular time from the results of astronomical observations, or may more readily be obtained by the aid of tables that have been computed for the purpose of facilitating its determination. It may also be obtained by simple inspection, from the *Nautical Almanac.* The American, or English Nautical Almanac, is a collection of data to be used in nautical and astronomical calculations, published annually, two or three years in advance of the year for which it is calculated.

93. Parallax in Right Ascension and in Declination. Since the parallax of a body displaces it in its vertical circle, which is generally oblique to the equator, it will alter its right ascension and declination. The consequent corrections to be applied to the right ascension and declination are called, respectively, parallax in right ascension, and parallax in declination.

For a similar reason the parallax of a body, generally alters both its longitude and latitude; and the requisite corrections are termed *parallax in longitude*, and *parallax in latitude.*

5

Formulæ for calculating the parallax in right ascension, and in declination, as well as in longitude and latitude, are investigated in the Appendix.

ABERRATION.

94. The celebrated English astronomer, Dr. Bradley, commenced in the year 1725 a series of accurate observations, with the view of ascertaining whether the apparent places of the fixed stars were subject to any direct alteration in consequence of the continual change occurring in the earth's position in space. The observations showed that there had been in reality, during the period of observation, small changes in the apparent places of each of the stars observed, which, when greatest, amounted to about 40″; but they were not such as should have resulted from the orbital motion of the earth. These phenomena Dr. Bradley undertook to examine and reduce to a general law. After repeated trials, he at last succeeded in discovering their true explanation. His theory is, that they are different effects of one general cause, a progressive motion of light in conjunction with the orbital motion of the earth.

95. Aberration of Light. Let us conceive the observer to be stationed at the earth's centre; and let ACB (Fig. 36) be a

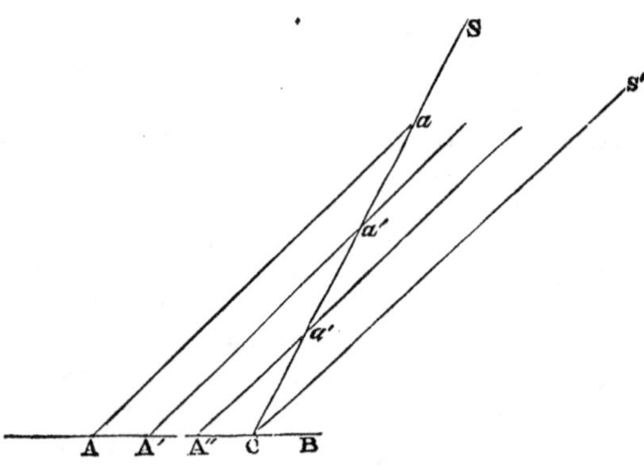

Fig. 36.

portion of the earth's orbit, so small that it may be considered a right line, CS the true direction of a fixed star as seen from the point C, AC the distance through which the earth moves in some small portion of time, and aC the distance traversed by a wave of light, in the same time. Then, a ray of light, which, coming from the star in the direction SC, is at a at the same time that the earth is at A, will arrive at C at the same time that the earth does. Suppose that Aa is the position of the axis or central line of a telescope, when the earth is at A, and that, continuing parallel to itself, it takes up, by virtue of the earth's motion, the successive positions A′a′, A″a″......CS′. A ray of light which follows the line SC in space will descend along this axis: for aa' is to AA′ and aa'' is to AA″, as aC is to AC, that is, as the velocity of light is to the velocity of the earth; consequently, when the earth is at A′ the ray of light is on the axis at a', and when the earth is at A″ the ray is on the axis at a'', and so on for all the other positions of the axis, until the earth arrives at C. The apparent direction of the star S, as far, at least, as it depends upon the cause under consideration, will therefore be CS′.

The angle SCS′, which expresses the change in the apparent place of a star S, produced by the motion of light combined with the motion of the spectator, is called the *Aberration* of the star; and the phenomenon of the change of the apparent course of the light coming from a star, thus produced, is called *Aberration of Light*, or simply *Aberration.*

The phenomenon of the aberration of light may be familiarly illustrated by taking falling drops of rain instead of supposed particles of light, and a vessel in motion at sea instead of the earth moving through space; and considering what direction must be given to a small tube by a person standing upon the deck of the vessel, so as to permit the drops falling perpendicularly to pass through the tube. It is plain that if the tube had a precisely vertical position, its forward motion would bring the back part of the tube against the drop; and that the only way to prevent this is to incline the upper end of the tube forward, or draw the lower end backward, whereby the back part of it would be made to pass through a greater distance before it comes up to the line of descent of the drop. The quantity that it is made to deviate in direction from this line, must depend upon the relative velocities of the falling drop and moving tube. To the observer, unconscious of his own motion, the drop will appear to fall in the oblique direction of the tube.

96. Angle of Aberration. If through the point a (Fig. 37)

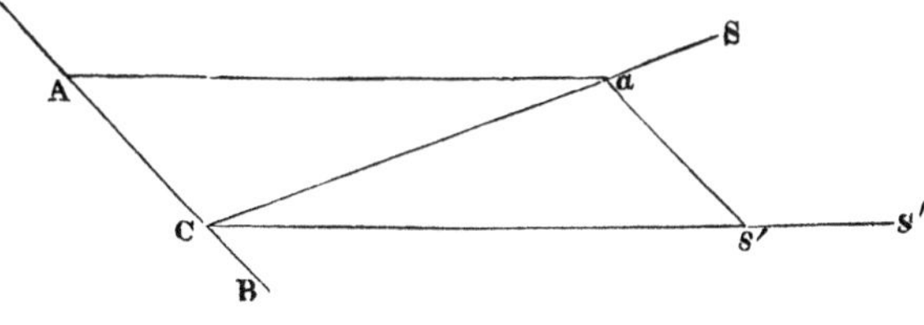

FIG. 37.

a line, as', be drawn parallel to AC, and terminating in CS′, the figure Aas'C will be a parallelogram, and therefore as' will be equal to AC. Hence it appears, that if on CS, the line of direction of a star S, a line Ca be laid off, representing the velocity of light, and through a a line, as', be drawn, having the same direction as the earth's motion and equal to its velocity, the line joining s' and C will be the apparent line of direction of the star, the point S′ its apparent place in the heavens, and the angle aCs' its aberration. We conclude, therefore, that by virtue of aberration a star is seen in advance of its true place, in the plane passing through the line of direction of the star and the line of the earth's motion.

The amount of the aberration of a star is always very small (never greater than about 20″), because of the very great disproportion between the velocity of light and the velocity of the earth. It is very much exaggerated in Figs. 36 and 37.

The aberration is the same when a star is viewed with the naked eye as when it is seen through a telescope. For, let aC, the velocity of the light, be decomposed into two velocities, of which one, AC, is equal and parallel to the velocity of the earth, the other will be represented by $s'C$. Now, since the velocity AC is equal and parallel to the velocity of the earth, it will produce no change in the relative position of a supposed particle of light and the eye, and therefore the relative motion of the light and the eye will be the same that it would be if the earth were stationary and the light had only the velocity $s'C$; accordingly, the light entering the eye just as it would do if it actually came in the direction $s'C$, and the eye were at rest, Cs' will be the apparent direction of the star from which it proceeds.

If we regard the observer as situated upon the earth's surface, instead of being at its centre, the aberration resulting from the earth's motion of revolution will be still the same, for all points of the earth advance at the same rate and in the same direction with the centre. The motion of rotation will produce an aberration proper to itself, but it is so small that there is no occasion to take it into account.

97. To find a General Expression for the Aberration. We have by Trigonometry (Fig. 37),

$$\sin AaC : \sin CAa :: CA : Ca :: \text{vel. of earth} : \text{vel. of light};$$

whence,

$$\sin AaC = \sin CAa \frac{CA}{Ca}; \text{ or, since } AaC = SCS',$$

$$\sin \text{aberr.} = \sin CAa \frac{\text{vel. of earth}}{\text{vel. of light}} \ldots (14).$$

When CAa is 90°, the aberration has its maximum value, and this has been found by observation to be 20″.445; whence,

$$\sin 20''.445 = \frac{\text{vel. of earth}}{\text{vel. of light}} \ldots (15):$$

substituting, and taking sin BCa for sin CAa, to which it is very nearly equal, we have

$$\sin \text{aberr.} = \sin BCa \sin 20''.445 \ldots . (16).$$

We may conclude from this equation, that the aberration increases with the angle BCa made by the direction of the star with the direction of the earth's motion; that it is equal to zero when this angle is zero, and has its maximum value of 20''.445 when this angle is 90°.

98. Annual Curve of Aberration. Let us now inquire into the entire effect of aberration in the course of a year. Let S (Fig. 38) be the sun; E the earth; E*fg* its orbit; ZTV that orbit

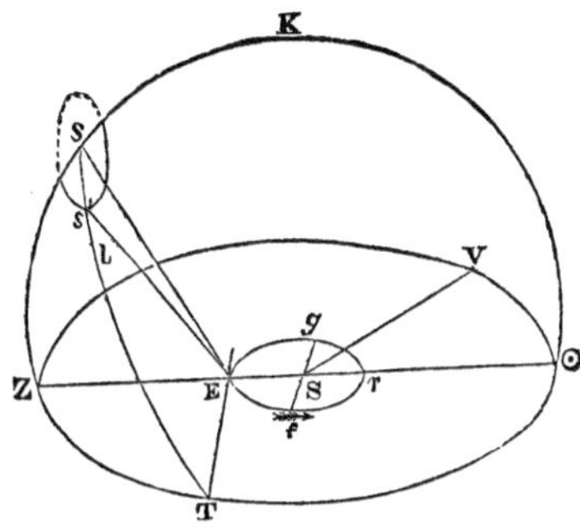

FIG. 38.

extended to the fixed stars, or the ecliptic (p. 15, def. 17); ET a tangent to the earth's orbit at E; ⊙ the place of S among the fixed stars or in the ecliptic, as seen from the earth; *s* a fixed star; *sl*T the arc of a great circle passing through *s* and T. Then, by what has preceded (96), the earth moving in the direction E*fg*, the apparent place of the star may be represented by *s'* and the aberration by *s*E*s'*. Thus, the effect of aberration at any one time is to displace the star by a small amount, directly towards the point T of the ecliptic, which is 90° behind the sun. As the earth moves, the position of the point T will vary; and in the course of a year, while the earth describes its entire orbit in the direction E*fg*, this point will move in the same direction entirely around the ecliptic. In this period of time, therefore, *ss'*, the small arc of aberration, will revolve entirely around *s*, the true position of the star; from which we conclude, that in consequence of aberration a star appears to describe a closed curve in the heavens around its true place.

As the inclination of the direction of the star to the direction of the earth's motion will vary during a revolution of the earth, the aberration will also vary during this period (97), and hence the curve in question will not be a circle. It appears upon investigation that it is an ellipse, having the true place of the star for its centre, and of which the semi-major axis is constant and equal to 20''.445, and the semi-minor axis variable and expressed by 20''.445 sin λ (λ denoting the latitude of the star). Each star, then, describes an ellipse which is the more eccentric in proportion as the star is nearer to the ecliptic; for, the expression for the minor axis shows that the smaller the latitude the less will be this axis. For a star situated in the ecliptic the minor axis will be zero, and the ellipse will be reduced to a right line. For a star in the pole of the ecliptic the minor axis will be equal to the major, and the ellipse therefore becomes a circle.

In following the motion of the star in its ellipse, it is to be observed that the orbit of the earth is a mere point at the centre of the celestial sphere, and the angle *s*ET as the earth moves forward, decreases from 90° at E to its minimum value at *f*, and then increases to 90° at *r;* and that similar changes occur while the earth is describing the other half of its orbit. When the earth is at E, the star is at one extremity of the major axis of its ellipse, and when the earth is arrived at *r*, the star is at the opposite extremity of the major axis. The points *f* and *g*, where the angle *s*ET has its minimum value, answer to the extremities of the minor axis.

99. *Aberration of the sun.—Displacement of the moon and planets.* Since the motion of the earth is at all times in a direction perpendicular, or nearly so, to the line followed by the light which comes from the sun to the earth, the aberration of the sun, which takes place only in longitude, is continually equal to about 20''.44. Thus the sun's apparent place is always about 20''.44 behind its true place.

The apparent displacement of a planet, resulting from the progressive motion of light, differs from that of a fixed star in a similar position. As a planet changes its place during the interval of time that a ray of light is passing from it to the earth, it would, if the earth were stationary, appear to be as far behind its true place as it has moved during this interval. This angular displacement, dependent upon the motion of the planet, combined with the aberration proper due to the earth's motion, constitutes the actual angular displacement of the planet from the cause under consideration.

The apparent change of place caused by the motion of the moon around the earth is very small.

100. Aberration in Right Ascension and in Declination. Since aberration causes the apparent place of a star, that has been corrected for refraction, to differ slightly from its true place, the true and apparent co-ordinates will differ somewhat from each other. The effects of the aberration of light upon the right ascension and declination of a star are called, respectively, the *aberration in right ascension* and *the aberration in declination.* These are to be determined and applied as corrections to the apparent right ascension and declination; the result will be the true co-ordinates, which will define the actual place in the heavens of the body observed.

Formulæ for computing the aberrations of a star in right ascension and declination, are investigated in the Appendix.

101. Proof of the Progressive Motion of Light. If the apparent places of a star, found at various times, be corrected for aberration, the same result for the true place of the star is obtained. Again, the deductions of Art. 98 agree in every particular with the observed phenomena of the apparent displacement of the stars, first discovered by Dr. Bradley. These facts show that the aberration of light is the true cause of these phenomena, and consequently establish at the same time the fact of the progressive motion of light, and that of the orbital motion of the earth.

Although Bradley derived from the phenomena of aberration decisive proof of the progressive motion of light, it was first discovered by Roemer, a Danish astronomer, in 1675, from a comparison of observations upon the eclipses of Jupiter's satellites.

Velocity of Light. We have by equation (15),

$$\text{vel. of earth} : \text{vel. of light} :: \sin 20''.445 : 1 :: 1 : 10{,}088.8\,;$$

and taking the velocity of the earth in its orbit at 65,460 miles per hour, or 18.1833 miles per second, we obtain for the velocity of light 183,448 miles per second. The orbital velocity of the earth here used is that which answers to the recent more accurate determination of the earth's distance from the sun (viz. 91,328,100 miles). The result obtained for the velocity of light is nearly 8,000 miles per second less than the former determination, in which the mean distance of the earth from the sun was taken a little over 95,000,000 miles.

Light traverses the distance from the sun to the earth in 8m. 18s.

CHAPTER V.

FIGURE AND DIMENSIONS OF THE EARTH.—LATITUDE AND LONGITUDE OF A PLACE.

102. ALTHOUGH it is in general sufficient for astronomical purposes to regard the earth as a sphere, still it is necessary in some cases of astronomical observation and computation, when accurate results are desired, to take notice of its deviation from the spherical form. No account need, however, be taken of the irregularities of its surface, occasioned by mountains and valleys, as they are exceedingly minute when compared with the whole extent of the earth. It is to be understood, then, that by the figure of the earth is meant the general form of its surface, supposing it to be smooth, or that the surface of the land corresponds with that of the sea.

103. Method of determining the Form of a Terrestrial Meridian. The figure of the earth is ascertained from an examination of the form of the terrestrial meridians.

A *Degree* of a terrestrial meridian is an arc of it corresponding to an inclination of 1° of the vertical lines at the extremities of the arc. It is also called a *Degree of Latitude.* Thus, if QNE (Fig. 39) represent a terrestrial meridian, *ab* will be a degree of it if it be of such length that the angle *a*C*b* between the vertical lines Z′*a*C, Z*b*C, is 1°.

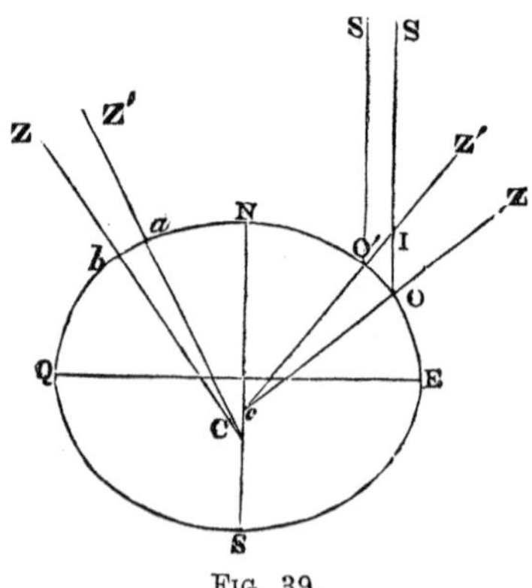

FIG. 39.

The length of a degree at any place will serve as a measure of the curvature of the meridian at that place; for it is obvious, from considerations already presented (3), that the earth, if not strictly spherical, must be nearly so, and therefore that a degree ab (Fig. 39) may, with but little if any error, be considered as an arc of 1° of a circle which has its centre at C, the point of intersection of the verticals Ca, Cb, at the extremities of the arc. The curvature will then decrease in the same proportion as the radius of this circle increases, and therefore in the same proportion as the length of a degree increases. Wherefore, the form of a meridian may be determined by measuring the length of a degree at various latitudes.

104. To determine the Length of a Degree of a Terrestrial Meridian. To accomplish this, we have,

(1.) *To run a meridian line;* an operation which is performed in the following manner. An altitude and azimuth instrument (or some other instrument adapted to meridian observations) is first placed at the point of departure, and accurately adjusted to the meridian. A new station is then established by sighting forward with the telescope. To this station the instrument is removed, and is there adjusted to the meridian by sighting back to the first station. A third station is then established by sighting forward with the telescope as before, to which the instrument is removed. By thus continually establishing new stations, and carrying the instrument forward, the meridian line may be marked out for any required distance. The meridian adjustments may be corrected from time to time by astronomical observations (42, 58).

(2.) *To find the length of the arc passed over.* When the ground is level, the length of the arc may be directly measured. In case the nature of the ground is such as not to allow of a direct measurement, it may be determined with great precision by means of a base line and a chain of triangles, the angles of which are measured.

(3.) *To find the inclination of the verticals at the extreme stations.* This angle may be obtained by measuring the meridian zenith distances of the same fixed star at the two stations, correcting them for refraction, and taking their difference. For, let O, O′ (Fig. 39) be the two stations in question, Z, Z′ their zeniths, and OS, O′S, the directions of a fixed star, and we shall have

$$OcO' = ZOI - OIc = ZOS - Z'IS = ZOS - Z'O'S;$$

that is, the angle comprised between the verticals equal to the difference of the meridian zenith distances of the same star.

(4.) *The length of an arc of the meridian, either somewhat greater or less than a degree, having been found by the foregoing operations, thence to compute the length of a degree.* Let N denote the number of degrees and parts of a degree in the measured arc, A its length, and x the length of a degree. Then, allowing that the earth for an extent of several degrees does not differ sensibly from a sphere, we may state the proportion

$$N : A :: 1^\circ : x; \text{ whence } x = \frac{1^\circ \times A}{N} \ldots . (17).$$

105. Results of the Measurements of Degrees. Degrees have been measured with the greatest possible care, at various latitudes and on various meridians. Upon a comparison of the measured degrees, it appears that *the length of a degree increases as we proceed from the equator towards either pole.* It follows, therefore (103), that the curvature of a meridian is greatest at the equator, and diminishes as the latitude increases; and consequently, that *the earth is flattened at the poles.*

The fact of the decrease of the curvature of a terrestrial meridian from the equator to the poles, leads to the supposition that it is an ellipse, having its major axis in the plane of the equator and its minor axis coincident with the axis of the earth. Analytical investigations, founded on the lengths of a degree in different latitudes, and on different meridians, have established that a meridian is, in fact, very nearly an ellipse, and that the earth has very nearly the form of an *oblate spheroid.* The same investigations have also made known the dimensions of the earth. The amount of the oblateness at the poles is measured by the ratio of the difference of the equatorial and polar diameters to the equatorial diameter, which is technically termed the *Oblateness* of the earth.

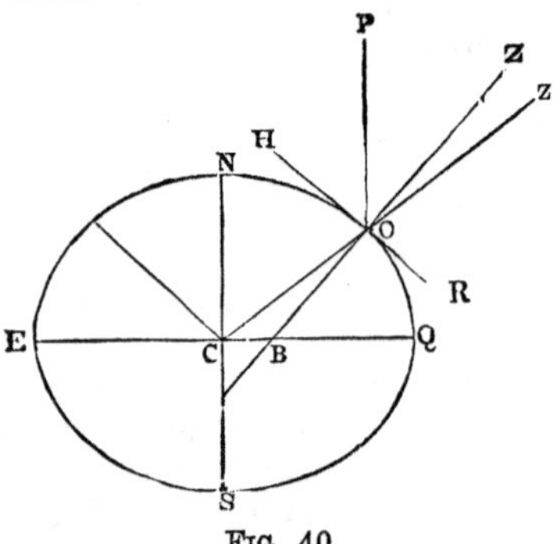

FIG. 40.

The form of the earth has also been determined by other methods, which cannot here be explained. All the results of measurements, taken together, indicate an oblateness of $\frac{1}{299}$.

The following are the dimensions of the earth in miles:

Radius at the equator................	3,962.80 miles.
Radius at the pole..................	3,949.55 "
Difference of equatorial and polar radii.	13.25 "
Radius at 45° latitude...............	3,956.20 "
Mean length of a degree of meridian...	69.048 "
The fourth part of a meridian.........	6,214.33 "

106. Inclination of Radius to Vertical Line. Owing to the elliptical form of a terrestrial meridian, the radius and vertical line at a place do not coincide. Let ENQS (Fig. 40) represent a terrestrial meridian. For any point O situated on this meridian, CO will be the radius, and the normal line ZOB the vertical. The position of the vertical line will always be such that the apparent zenith Z will lie between the true zenith z and the elevated pole P. The inclination of the radius to the vertical line, or the angle COB, called the reduction of latitude, is greatest at the latitude 45°, and is there equal to about $11\frac{1}{2}'$.

DETERMINATION OF THE LATITUDE AND LONGITUDE OF A PLACE.

107. The latitude and longitude of a place ascertain its situation upon the earth's surface, and are essential elements in many astronomical investigations.

108. To find the Latitude of a Place.

(1.) *By the zenith distances or altitudes of a circumpolar star, at its upper and lower transits.* The principle of this method has already been stated (55), and represented to be a particular case of a well-known principle of arithmetical proportions; the following is a detailed proof of it. Let Z (Fig. 41) represent the zenith, HOR the horizon, P the pole, and S, S′ the points at which the upper and lower transits of a circumpolar star take place; HP will be equal to the latitude (24), and ZP will be equal to the co-latitude. Now, we have

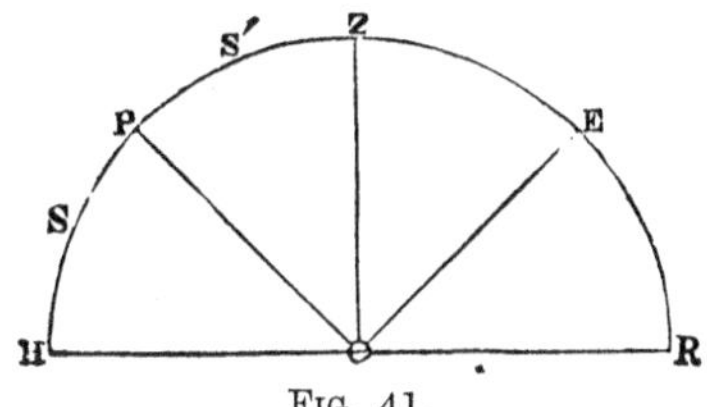

FIG. 41.

$$HP = HS + PS, \text{ and } HP = HS' - PS' = HS' - PS;$$

$$\text{whence, } 2HP = HS + HS', \text{ or, } HP = \frac{HS+HS'}{2} \ldots . (18).$$

In like manner we obtain,

$$ZP = \frac{ZS+ZS'}{2} \ldots . (19).$$

Wherefore, let the altitudes of a circumpolar star at its upper and lower transits be measured and corrected for refraction, and their half sum will be the latitude; or, let the zenith distances be measured, and corrected for refraction, and their half sum subtracted from 90° will be the latitude. Stars should be selected that have a considerable altitude at their inferior transit, for, the greater is the altitude the less is the uncertainty as to the amount of the refraction. On this principle the pole-star is to be preferred to all others.

(2.) *By a single meridian altitude or zenith distance.* Let s, s', s'' (Fig. 10, p. 21) be the points of meridian passage of three different stars, the first to the north of the zenith, the second between the zenith and equator, and the third to the south of the equator ZE = the latitude, and we have for the three stars,

$$ZE = sE - Zs,\ ZE = s'E + Zs',\ ZE = Zs'' - s''E.$$

Thus, if the zenith distance be called north or south, according as the zenith is north or south of the star when on the meridian, in case the zenith distance and declination are of the same name their sum will be equal to the latitude; but if they are of different names their difference will be the latitude, of the same name with the greater.

This method supposes the declination of the body observed to be known. The declination of a star or of the sun at any time is, *in practice*, obtained for the solution of this and other problems, by the aid of tables, or is taken by inspection from the American Nautical Almanac, or other similar work. If the time of the meridian transit be known, the altitude may be measured by a sextant (67). The observed altitude must be corrected for refraction, and also for parallax if the body observed be the sun, or moon, or either one of the planets.

This method of finding the latitude is the one most generally employed at sea, the sun being the object observed. As the time of noon is not known with accuracy, several altitudes about the time of noon are taken, and the meridian altitude is *deduced* from these.

(3.) *By the difference of the meridian zenith distances of two stars that cross the meridian near the zenith, on opposite sides.* This is *Talcott's Method* alluded to in connection with the subject of the zenith telescope (59). It is to be preferred to all other methods of determining the latitude, when the observer is provided with a zenith telescope.

Let z be the true zenith distance of the star that passes to the south of the zenith, and δ its declination; z' and δ' the true zenith distance and declination of the other star; and l the latitude of the station: we then have

$$l = \delta + z, \text{ and } l = \delta' - z',$$

and therefore,

$$l = \tfrac{1}{2}(\delta + \delta') + \tfrac{1}{2}(z - z') \ldots . (a).$$

Also, let Z denote the apparent zenith distance of the star that passes to the south of the zenith, r its refraction, and Z′, r' the corresponding quantities for the other star; then,

$$z = Z + r, \text{ and } z' = Z' + r';$$

and substituting in equation (a) we obtain,

$$l = \tfrac{1}{2}(\delta + \delta') + \tfrac{1}{2}(Z - Z') + \tfrac{1}{2}(r - r') \ldots . (b).$$

As we may suppose the declinations of the two stars to be known, it is then only necessary to determine the values of $Z - Z'$, and $r - r'$. Now, if two stars be selected whose zenith distances are nearly equal, their difference, $Z - Z'$, can be directly measured by the micrometer of the zenith telescope, and thus a result obtained for the latitude free from the instrumental errors that attend all methods in which the absolute zenith distances are measured. Also, if the selected stars pass the meridian near the zenith, their refractions will be small, and the amount of their difference, $r - r'$, very minute, and liable to no appreciable uncertainty. If m and m' denote the micrometer readings in observing the two stars, converted into their equivalent angular values, equation (b) becomes,

$$l = \tfrac{1}{2}(\delta + \delta') + \tfrac{1}{2}(m - m') + \tfrac{1}{2}(r - r') \ldots . (c).$$

It is here tacitly supposed that the micrometer reading increases with an increase of zenith distance. If the reverse be true, the second term should be affected with the negative sign.

The only instrumental correction that is to be applied to the result given by this formula, is for any error that may occur in the position of the vertical axis of the zenith telescope, when either star is observed. This is determined by means of a horizontal level, attached to the instrument in a position perpendicular to the horizontal axis of rotation of the telescope; and therefore turning with the instrument around the vertical axis.

The method of making the observations is briefly as follows: the instrument having been previously adjusted to the meridian, the observer, by means of the finding circle (p. 32), sets the telescope to the mean of the zenith distances of the selected pair of stars, and when the preceding star has entered the field follows it with the movable micrometer wire, and bisects it as it reaches the meridian. He then reads the micrometer, and also the level; and turns the instrument around its vertical axis, 180° in azimuth. When the second star enters the field of the telescope, it is bisected, like the first, with the micrometer-wire as it reaches the meridian. The micrometer and level are then read as before. The micrometer readings multiplied by the angular value of one revolution of the micrometer-screw, are the values of m and m' in equation (c).

Both the north and south ends of the bubble of the level are read in each observation, and the south end reading subtracted from the north end reading. Half the difference multiplied by the value of one division of the level in seconds of arc, will be the inclination of the level to a horizontal line, in each observation. The half algebraic sum of these inclinations for the two observations, will be the correction to be applied, according to its sign, to the result obtained by equation (c), for the deviation of the vertical axis from the truly vertical position.

It is found that the probable error, from all causes, of a single determination, by a practised observer, does not exceed 1″; and that by continuing the observations upon a series of pairs of suitably selected stars, for a number of nights, the latitude of a station can be determined with a probable error of only 0″.1, which answers to a distance on the meridian of only ten feet.

Reduced Latitude. The astronomical latitude being known, the reduced latitude (p. 19, def. 4) may be obtained by subtracting from it the reduction of latitude. For if OC (Fig. 40) represents the radius, and OB the vertical, at any place O, and ECQ represents the terrestrial equator, OBQ will be the astronomical latitude, OCQ the reduced latitude, and COB the reduction of latitude; and we have,

$$OBQ = OCQ + COB, \text{ and } OCQ = OBQ - COB \ldots\ldots (20).$$

(For the practical method of resolving this problem, see Problem XV.)

109. Longitude of a Place: — General Principle. There are various methods of finding the longitude of a place, nearly all of which rest upon the following principle:

The difference at any instant between the local times (whether sidereal or solar), at any place and on the first meridian, is the longitude of the place expressed in time; and consequently, also, the difference between the local times at any two places is their difference of longitude in time.

The truth of this principle is easily established. In the first place, we remark that the longitude of a place contains the same number of degrees and parts of a degree as the arc of the celestial equator comprised between the meridian of Greenwich and the meridian of the place. Now, it is 0h. 0m. 0s. of mean solar time, or mean noon, at any place, when the mean sun (36) is on the meridian of that particular place. Therefore, as the mean sun, moving in the equator, recedes from the meridian towards the west at the rate of 15° per mean solar hour, when it is mean noon at a place to the *west* of Greenwich, it will be as many hours and parts of an hour *past* mean noon at Greenwich, as is expressed by the quotient of the division of the arc of the celestial equator, or its equal the longitude, by 15. If the place be to the *east* instead of to the west of Greenwich, when it is mean noon there, it will be as much *before* mean noon at Greenwich as is expressed by the longitude of the place converted into time (as above). In either situation of the place, then, the principle just stated will be true.

It is plain that the equality between the difference of the times and of the longitudes will subsist equally if sidereal instead of solar time be used.

110. To find the Longitude of a Place.

(1.) *Let two observers, stationed one at Greenwich and the other at the given place, note the times of the occurrence of some phenomenon which is seen at the same instant at both places;* the difference of the observed times will be the longitude in time. The same observations made at any two places will make known their *difference* of longitude. If the stations are not distant from each other, a signal, as the flashing of gunpowder, or the firing of a rocket, may be observed. When they are remote from each other, celestial phenomena must be taken. Eclipses of the satellites of Jupiter and of the moon, are phenomena adapted to the purpose in question. But as in these eclipses the diminution of the light of the body is not sudden, but gradual, the longitude cannot be obtained with very great accuracy from observations made upon them.

(2.) *Transport a chronometer which has been carefully adjusted to the local time at Greenwich, to the place whose longitude is sought, and compare the time given by the chronometer with the local time of the place.* In the same way, by transporting a chronometer from any one place to another, their difference of longitude may be obtained. The error and rate of the chronometer must be determined at the outset, and as often afterwards as circumstances will admit, that the error at the moment of the observation may be known as accurately as possible. To insure greater certainty and precision in the knowledge of the time, a number of chronometers are often taken, instead of one only.

This method is much used at sea; the local time being obtained from an observation upon the sun or some other heavenly body, in a manner to be hereafter explained.

(3.) *Let the Greenwich time of the occurrence of some celestial phenomenon be computed, and note the time of its occurrence at the given place.*

Eclipses of the sun and moon, and of Jupiter's satellites, occultations of the stars by the moon, and the angular distance of the moon from some one of the heavenly bodies, are the phenomena employed. The Greenwich times of the beginning and end of the eclipses of Jupiter's satellites, are published for the solution of the problem of the longitude in the English Nautical Almanac. When the longitude is estimated from Washington, the Washington times of the occurrence of the same phenomena may be taken from the American Nautical Almanac.

Eclipses of the sun, and occultations of the stars, furnish the most exact determinations of the longitude, but they cannot be used for this purpose unless the longitude is already approximately known.

The *method of lunar distances* is chiefly used at sea, and is given in detail in treatises on navigation and nautical astronomy.

(4.) *Another and more accurate method* of determining the difference of longitude of two places, has recently been introduced and perfected by American astronomers. It consists in the use of the electric telegraph for the transmission of signals from one station to the other, and the introduction of the electro-chronograph into the circuit, to measure off and record, at each station, the beats of a sidereal clock. The clock may be at either station, or at some other astronomical station in the circuit. Its beats are electrically transmitted, and recorded upon a moving roll of paper, adapted to the registers at each station, in a series of equally distant dots, or in a succession of equally distant breaks in a continuous line (see Fig. 22, p. 37). The signals adopted are the passages of a star across the wires of a transit instrument.

The observer at the most easterly station strikes his break-circuit key as the star passes each of the wires in succession. As the result, the instants of these successive transits are shown upon the roll of paper at each station, by breaks in the line of seconds, falling between those which indicate the seconds. When the star reaches the meridian of the other station, a similar set of observations are made by the other observer; and the instants of the successive transits are recorded as before, upon the roll of paper at each station. It then only remains for each observer to remove the roll upon which the instants of the passage of the star across the wires of the transit instrument at each station are noted, and carefully measure the distance between each break in the time-line, obtained by the one set of observations, from the corresponding break obtained by the other set; then convert this into the equivalent interval of time, and take the mean of all the intervals. This will be his determination of the difference of longitude of the two stations, in time. The mean of the results thus obtained by the two observers, is then to be taken as more reliable than either of the single determinations.

For greater accuracy a number of selected stars should be observed. The observations should also be many times repeated; the clocks at the two stations being alternately thrown into the circuit. The result obtained is free from the errors that may exist in the tabular places of the stars observed, and from the clock error; since neither of these errors will affect the intervals of time employed by the stars in passing from the meridian of the one station to that of the other. But each observer should carefully determine and allow for the errors of adjustment of his transit instrument.

The longitudes of the principal observatories in the United States, and of several important stations of the United States Coast Survey, have been very accurately determined by this method.

CHAPTER VI.

Apparent Motion of the Sun in the Heavens.

111. The sun's declination and the difference between the right ascension of the sun and that of some fixed star, found from day to day (45 and 55) throughout a revolution, are the elements from which the circumstances of the sun's apparent motion are derived.

The curve on the sphere of the heavens, passing through all the successive positions thus determined from day to day, is the *Ecliptic.* If we suppose it to be a circle, as it appears to be, its position will result from the position of the equinoctial points and its obliquity to the equator.

112. To find the Obliquity of the Ecliptic. Let EQA (Fig. 42) represent the equator; ECA the ecliptic; and OC, OQ,

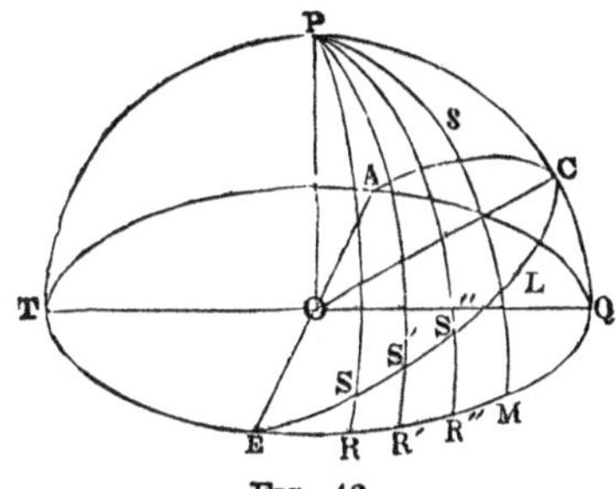

FIG. 42.

lines drawn through O, the centre of the earth, and perpendicular to the line of the equinoxes, AOE: then the angle COQ will be the obliquity of the ecliptic. This angle has for its measure the arc CQ, and therefore *the obliquity of the ecliptic is equal to the greatest declination of the sun.* It can but rarely happen that the time of the greatest declination will coincide with the instant of noon at the place where the observations are made, but it must fall within at least twelve hours of the noon for which the observed declination is the greatest. In this interval the change of declination cannot exceed 4″, and therefore the greatest observed declination cannot differ more than 4″ from the obliquity. A formula has been investigated, which gives in terms of determinable quantities the difference between any of the greater declinations and the maximum declination. By *reducing*, by means of this formula, a number of the greater declinations to the maximum declination, and taking the mean of the individual results, a very accurate value of the obliquity may be found.

The obliquity of the ecliptic changes slightly from year to year. It is also subject to a slight diminution from century to century. Its mean value at the present date (Jan., 1867) is 23° 27′ 24″.

113. To find the Position of the Vernal or Autumnal Equinox.

(1.) On inspecting the observed declinations of the sun, it is seen that about the 21st of March the declination changes in the interval of two successive noons from south to north. The vernal equinox occurs at some moment of this interval. Let RS, R′S′ (Fig. 43) represent the declinations at the noons between

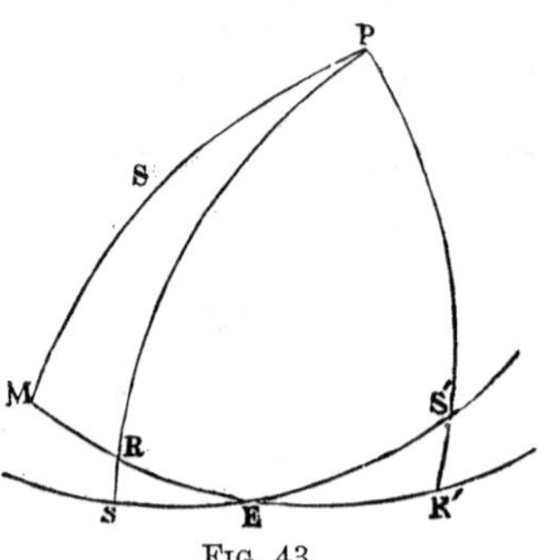

FIG. 43.

which the equinox occurs: as one is north and the other south, their sum (S) will be the daily change of declination at the time of the equinox. Denote the time from noon to noon by T. Now, to find the interval (x) between the noon preceding the equinox and the instant of the equinox, state the proportion

$$S : RS :: T : x = \frac{T \times RS}{S};$$

on the principle that the declination changes, for a day or more, proportionally to the time. Next, take the daily change in right ascension (RR′) on the day of the equinox and compute the value of RE, by the proportion

$$T : x, \text{ or } \frac{T \times RS}{S} :: RR' : RE;$$

add RE to MR, the observed difference of right ascension (111) on the day preceding the equinox, and the sum ME will be the distance of the equinox from the meridian of the star observed in connection with the sun; if the star be to the west of the sun, as in the figure.

The position of the autumnal equinox may be found by a similar process, the only difference in the circumstances being that the declination changes from north to south instead of from south to north.

If the value of x which results from the first proportion be added to the time of noon on the day preceding the equinox, the result will be the *time* of the equinox.

(2.) In the triangle RES (Fig. 42) we have the angle RES $= \omega$ the obliquity of the ecliptic, and RS $=$ D the declination of the sun, both of which we may suppose to be known, and we have by Napier's first rule (Appendix),

$$\sin \text{ER} = \tan (\text{co. RES}) \tan \text{RS} = \cot \omega \tan \text{D} \ldots . (21),$$

whence we can find ER. And by taking the sum or difference of ER and MR, according as the star observed is on the opposite side of the sun from the equinox or the same side, we obtain ME as before. If this calculation be effected for a number of positions, S, S′, S″, etc., of the sun on different days, and a mean of all the individual results be taken, a more exact value of ME will be obtained.

ME being accurately known, the precise time of the equinox may readily be deduced from the observed daily variation of right ascension on the day of the equinox.

The calculations just mentioned rest upon the hypothesis that the ecliptic is a great circle. The close agreement which is found to subsist between the values of ME deduced from observations upon the sun in different positions, S, S′, S″, etc., establishes the truth of this hypothesis. It is also confirmed by the fact that the right ascensions of the vernal and autumnal equinox differ by 180°, since we may infer from this that the line of the equinoxes passes through the centre of the earth.

114. Longitude of the Sun. The longitude of the sun may be expressed in terms of the obliquity of the ecliptic and the right ascension or declination. In the triangle ERS (Fig. 42), ES ($=$ L) represents the longitude of the sun supposed to be at S, ER ($=$ R) its right ascension, and RS ($=$ D) its declination. Now, by Napier's first rule,

$$\cos \text{RES} = \tan \text{ER} \cot \text{ES}, \text{ or } \cot \text{ES} = \frac{\cos \text{RES}}{\tan \text{ER}} = \cos\text{RES} \cot \text{ER};$$

thus,

$$\cot \text{L} = \cos \omega \cot \text{R}, \text{ or } \tan \text{L} = \frac{\tan \text{R}}{\cos \omega} \ldots . (22).$$

Also (Napier's second rule, Appendix),

$\sin RS = \cos(\text{co. RES}) \cos(\text{co. ES})$; whence, $\sin ES = \frac{\sin RS}{\sin RES}$;

or,

$$\sin L = \frac{\sin D}{\sin \omega} \ldots . (23).$$

With these formulæ the longitude of the sun may be computed from either its right ascension or declination. (See Prob. XII., Part III.)

Formulæ (22) and (23) may be written thus,

$$\tan R = \tan L \cos \omega; \ \sin D = \sin L \sin \omega \ldots . (24).$$

These formulæ will make known the right ascension and declination of the sun, when its longitude is given. (See Prob. XI.) It will be seen in the sequel that in the present condition of astronomical science, the longitude of the sun at any assumed time may be computed from the ascertained laws and rate of the sun's motion.

115. Tropical Year. The interval between two successive returns of the sun to the same equinox, or to the same longitude, is called a *Tropical Year.*

The interval between two successive returns of the sun to the same position with respect to the fixed stars, is called a *Sidereal Year.*

It appears from observation that the length of the tropical year is subject to slight periodical variations. The period from which it deviates periodically and equally on both sides, is called the *Mean Tropical Year.* As the changes in the length of the true tropical year are very minute, the length of the mean tropical year is obviously very nearly equal to the mean length of the true tropical year, in an interval during which this passes one or more times through all its different values. In point of fact, it may be found with a very close approximation to the truth by comparing two equinoxes observed at an interval of 60 or 100 years.

According to the most accurate determinations, the length of the mean tropical year, expressed in mean solar time, is 365d. 5h. 48m. 46.1s.

116. Sun's Daily Motion in Longitude. In a mean tropical year the sun's mean motion in longitude is 360°; hence, to find his *mean daily motion in longitude* we have only to state the proportion

$$365\text{d. } 5\text{h. } 48\text{m. } 46\text{s.} : 1\text{d.} :: 360 : x = 59' 8''.33.$$

If from the right ascension or declination of the sun, found on two successive days, the corresponding longitudes be deduced (equs. 22, 23), and their difference taken, the result will be the sun's daily motion in longitude at the date of the observations.

The sun's daily motion in longitude is not the same throughout the year, but, on the contrary, is continually varying. It gradually increases during one-half of a revolution, and gradually decreases during the other half, and at the end of the year has recovered its original value. Thus, the greatest and least daily motions occur at opposite points of the ecliptic. They are, respectively, 61′ 10″ and 57′ 12″.

The exact law of the sun's unequable motion, can only be obtained by taking into account the variation of his distance from the earth; for the two are essentially connected by the physical law of gravitation, which determines the nature of the earth's motion of revolution around the sun.

That the distance of the sun from the earth is in fact subject to a variation, may be inferred from the observed fact that his apparent diameter varies. On measuring with the micrometer the apparent diameter of the sun from day to day throughout the year, it is found to be the greatest when the daily angular motion, or in longitude, is the greatest, and the least when the daily motion is the least; and to vary gradually between these two limits. Accordingly the sun is nearest to us when its daily angular motion is the most rapid, and farthest from us when its daily motion is the slowest. The greatest apparent diameter of the sun is 32′ 36″; and the least apparent diameter 31′ 32″.

CHAPTER VII.

PRECESSION OF THE EQUINOXES.—NUTATION.

117. Proof of an Annual Precession of the Equinoxes. The determination of the position of the vernal equinox, consists in deducing from the results of certain observations the difference between the right ascension of the equinox and that of one of the fixed stars (113). This difference is represented by ME, in Fig. 42, and by VR in Fig. 8. We have seen (45) that when this has become known, the absolute right ascensions of all the stars may be determined. We have seen also (56), that when the right ascension and declination of a star are known, its longitude and latitude may be computed. Now, if the position of the vernal equinox be determined at two epochs separated by a number of years, it is found that the value of ME has materially increased, if the star *s*, observed with the sun, is to the east of the equinox; and decreased if the star lies to the west of the equinox. From this fact we may conclude that the equinox has a retrograde motion, or towards the west, from year to year.

Again, if the longitudes and latitudes of the same fixed stars, obtained as above, at different periods, be compared, it is found that their latitudes continue very nearly the same, but that their longitudes all increase at the same mean rate of about 50″ per year. Thus, EL (Fig. 42) represents the longitude of the star *s*, and *s*L its latitude, and it is found that *s*L remains the same, but that EL increases at the mean rate of 50″ per year. It follows, therefore, that the vernal equinox must have an annual motion of about 50″ along the ecliptic, in a direction contrary to the order of the signs, or from east to west. As it has been ascertained that the autumnal is always at the distance of 180° from the vernal equinox, it must have the same motion. This retrograde motion of the equinoctial points is called the *Precession of the Equinoxes.*

118. Ecliptic Stationary. As the latitude of a star is its angular distance from the ecliptic, it follows from the circumstance of the latitudes of all the stars continuing very nearly the same, that the ecliptic remains fixed, or very nearly so, with respect to the situations of the fixed stars.

The ecliptic being stationary, it is plain that the precession of the equinoxes must result from a continual slow motion of the equator in one direction. It appears from observation that the obliquity of the ecliptic, or the inclination of the equator to the ecliptic, remains in the course of this motion very nearly the same.

119. Progressive Motion of the Pole of the Heavens. Since the equator is in motion, its pole must change its place in the heavens. Let VLA (Fig. 44) represent the ecliptic; K its

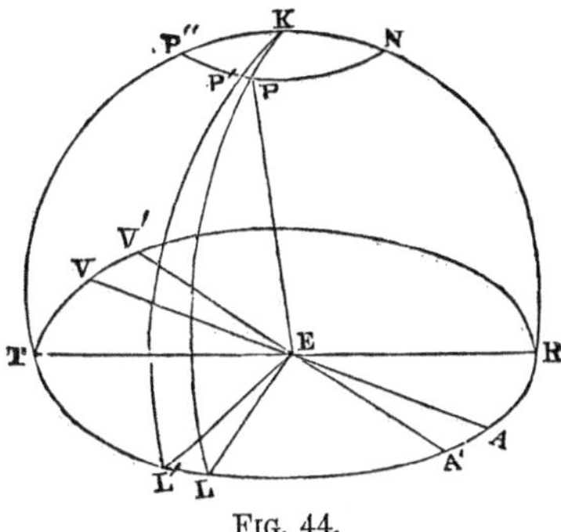

FIG. 44.

stationary pole; P the position of the north pole of the equator, or of the heavens, at any given time, and VEA the corresponding position of the line of the equinoxes: KPL represents the circle of latitude passing through P, or the solstitial colure. Now, the point V being at the same time in the ecliptic and equator, it is 90° distant from the two points K and P, the poles of these circles; therefore, it is the pole of the circle KPL passing through these points, and hence VL = 90°. It follows from this, that when the vernal equinox has retrograded to any point V′, the pole of the equator, originally at P, will be found in the circle of latitude KP′L′ for which V′L′ equals 90°: it will also be at the distance KP′ from the pole of the ecliptic, equal to KP. Whence it appears that the pole of the equator has a retrograde motion in a small circle about the pole of the ecliptic, and at a distance from it equal to the obliquity of the ecliptic. As the motion of the equator which produces the precession of the equinoxes is uniform, the motion of the pole must be uniform also; and as the pole will accomplish a revolution in the same time with the equinox, its rate of motion must be the same as that of the equinox, that is, 50″ of its circle in a year. The period of revolution of the equinox and the pole of the equator, is about 24,500 years.

It is an interesting consequence of this motion of the pole of the equator and heavens, that *the pole-star, so called, will not always be nearer to the pole than any other star.* The pole is at the present time approaching it, and it will continue to approach it until the present distance of 1½° becomes reduced to less than ½°, which will happen about the year 2100: after which it will begin to recede from it, and continue to recede, until about the year 3200 another star will come to have the rank of a pole-star. The motion of the pole still continuing, it will, in the lapse of centuries, pass in the vicinity of several pretty distinct stars in succession, and in about 12,000 years will be within a few degrees of the star Vega, in the constellation of the Lyre, the brightest star in the northern hemisphere.

The present pole-star has held that rank since the time of the celebrated astronomer Hipparchus, who flourished about 120 B. C. In very ancient times, a pretty bright star in the constellation of the Dragon (α Draconis) was the pole-star.

The motion of the equator which produces the precession of the equinoxes, must also produce *changes in the right ascensions and declinations of the stars.* These changes will be different according to the situations of the stars with respect to the equator and equinoctial points.

120. Effect of Precession on the Length of the Year. The precession of the equinoxes makes the tropical year shorter than the sidereal year. For, since the precession is a retrograde movement of each equinox of $50''.24$ per year, when the sun has returned to the same equinox, it will not have accomplished a sidereal revolution into $50''.24$. The excess of the sidereal over the tropical year results from the proportion

$$59'\ 8''.33 :: 50''.24 :: 1\text{d} : x = 20\text{m}.\ 23.3\text{s}.$$

Thus the length of the mean sidereal year, expressed in mean solar time, is 365d. 6h. 9m. 9.4s.

121. Secular Diminution of the Obliquity of the Ecliptic. The ecliptic, although very nearly stationary, as stated in Art. 118, is not strictly so. By comparing the values of the obliquity of the ecliptic, found at distant periods, it is ascertained that it is subject to a gradual diminution of $46''$ from century to century. It appears from observation that there are minute secular changes in the latitudes of the stars, which establish that the progressive diminution of the obliquity of the ecliptic arises from a slow displacement of the plane of the ecliptic, or of the earth's orbit, in space.

It remains for us now to take notice of a minute inequality in the motion of the equator and its pole, which we have thus far overlooked.

NUTATION.

122. Discovery of Nutation. Dr. Bradley, in observing the polar distance of a certain star (γ Draconis) with the view of verifying his theory of aberration, discovered that the observed polar distance did not agree with the polar distance as computed from the results of previous observations, by allowing for the change due to the precession in the interval; the proper corrections for refraction and aberration having been applied in both cases. On continuing his observations he found that the polar distance alternately increased and diminished, and that it returned to the same value in about 19 years. These phenomena led him to suppose that the pole, instead of moving uniformly in a circle around the pole of the ecliptic, oscillated from the one side to the other of a point conceived to move in this manner.

123. Ellipse of Nutation. If the pole has such a motion it is plain that, allowing the fact of the earth's rotation, it must result from a vibratory motion of the earth's axis. To this supposed vibration of the axis of the earth, and consequently of that of the heavens, Dr. Bradley gave the name of *Nutation*. Upon a detailed examination of all his observations, it appeared that the oscillation of the pole did not take place in a right line, but in a minute ellipse. The motion may accordingly be regarded as a motion of revolution in an ellipse around its centre. This central point, about which the pole revolves, is the mean position of the pole, and is called the *Mean Pole*. The direction of the motion of revolution is retrograde, or from east to west, and the period is about 19 years.

In Fig. 45, $pgf'g'$ represents the *ellipse of nutation*, and P the

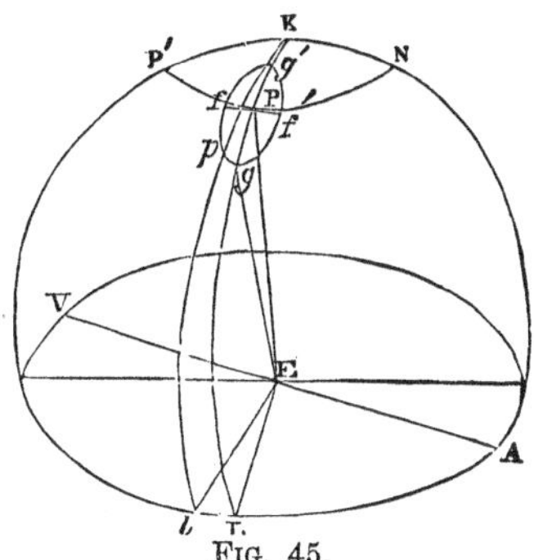

FIG. 45.

mean pole; the direction of the motion of revolution being from p towards f. The major axis gg' lies in the solstitial colure KPL, and is equal to 19″; and the minor axis ff' is equal to 14″. While the true pole revolves in its ellipse about the mean pole P, the mean pole has a uniform retrograde movement in a circle NPP′, around the pole of the ecliptic K. Accordingly the pole has two cotemporaneous motions; one in a minute ellipse, and about its centre, and another in a circle of 23½° radius, about the pole of the ecliptic. Its actual motion must therefore be in a slightly waving curve, passing alternately from one side to the other of this

FIG. 46.

circle, as shown in Fig. 46; in which, however, the deviations from the circle are greatly exaggerated. The ellipse of nutation is also greatly exaggerated in Fig. 45.

124. Effects of Nutation. As the equator must move with the axis of the earth or heavens, *nutation must change the position of the equinox and the obliquity of the ecliptic.* It is plain that its effect upon the position of the equinox will be to make it oscillate periodically, and by equal degrees, from one side to the other of the position which corresponds to the mean pole; and that its effect upon the obliquity of the ecliptic will be to make it alternately greater and less than the obliquity corresponding to the mean pole. The position of the equinox which corresponds to the mean pole is called the *Mean Equinox;* and the obliquity corresponding to the mean pole is called the *Mean Obliquity. Mean Equator* has a like signification. The real equinox and the real equator are called, respectively, the *True Equinox* and the *True Equator.* The actual obliquity of the ecliptic is termed the *Apparent* Obliquity.

In like manner, the right ascension, declination, etc., of a star, referred to the mean equator and mean equinox, are designated the *mean* right ascension, *mean* declination, etc.; to distinguish them from the corresponding elements referred to the true equator and true equinox. The distance of the true from the mean equinox in longitude, is called the *equation of the equinoxes in longitude.*

CHAPTER VIII.

MEASUREMENT OF TIME.

DIFFERENT KINDS OF TIME.

125. IN Astronomy, as we have already stated, three kinds of time are used—*Sidereal*, *True or Apparent Solar*, and *Mean Solar Time;* sidereal time being measured by the diurnal motion of the vernal equinox, true or apparent solar time by that of the sun, and mean solar time by that of an imaginary sun called the *Mean* sun, conceived to move uniformly in the equator with the real sun's mean motion in right ascension or longitude.

126. True Solar Day. The sidereal day and the mean solar day are each of uniform duration, but *the length of the true solar day is variable*, as we will now proceed to show.

The sun's daily motion in right ascension, expressed in time, is equal to the excess of the solar over the sidereal day. Now this arc, and therefore the true solar day, varies from two causes, viz.:

(1.) *The inequality of the Sun's daily motion in longitude.*

(2.) *The obliquity of the ecliptic to the equator.*

If the ecliptic were coincident with the equator, the daily arc of right ascension would be equal to the daily arc of longitude, and therefore would vary between the limits 57′ 12″ and 61′ 10″, which would answer, respectively, to the apogee and perigee. But, owing to the obliquity of the ecliptic, the inclination of the daily arc of longitude to the equator is subject to a variation; and this, it is plain (see Fig. 42), will be attended with a variation in the daily arc of right ascension. The tendency of this cause is obviously to make the daily arc of right ascension least at the equinoxes, where the obliquity of the arc of longitude is greatest, and greatest at the solstices, where the obliquity is least.

127. Mean Solar Time. As the length of the apparent solar day is variable, it cannot conveniently be employed for the expression of intervals of time; moreover, a clock, to keep apparent solar time, requires to be frequently adjusted. These inconveniences attending the use of apparent solar time, led astronomers to devise a new method of measuring time, to which they gave the name of mean solar time. By conceiving an imaginary sun to move uniformly in the equator with the real sun's mean motion, a day was obtained of which the length is invariable, and equal to the mean length of all the apparent solar days in a tropical year. The point and time of departure of this fictitious sun, were also so chosen that its distance from the mean equinox would always be equal to the sun's mean longitude; the time deduced from its position with respect to the meridian, was thus made to correspond very nearly with apparent solar time.

To find the excess of the mean solar day over the sidereal day, we have the proportion

$$360° : 24 \text{ sid. hours} :: 59' 8''.33 : x = 3\text{m. } 56.555\text{s.}$$

A mean solar day, comprising 24 mean solar hours, is therefore 24h. 3m. 56.555s. of sidereal time. Hence, a clock regulated to sidereal time will gain 3m. 56.555s. in a mean solar day.

To find the expression for the sidereal day in mean solar time, we must use the proportion

$$24\text{h. } 3\text{m. } 56.555\text{s.} : 24\text{h.} :: 24\text{h.} : x = 23\text{h. } 56\text{m. } 4.092\text{s.}$$

The difference between this and 24 hours is 3m. 55.908s.; and therefore, a mean solar clock will lose with respect to a sidereal clock, or with respect to the fixed stars, 3m. 55.908s. in a sidereal day, and proportionally in other intervals. This is called the *daily acceleration* of the fixed stars.

To express any given period of sidereal time in mean solar time, we must subtract for each hour $\frac{3\text{m. } 55.91\text{s.}}{24} = 9.83\text{s.}$, and for minutes and seconds in the same proportion. And, on the other hand, to express any given period of mean solar time in sidereal time, we must add for each hour $\frac{3\text{m. } 56.55\text{s.}}{24} = 9.86\text{s.}$, and for minutes and seconds in the same proportion.

It is the practice of astronomers to adjust the sidereal clock to the motions of the *true* instead of the mean equinox. The inequality of the diurnal motion of this point is too small to occasion any practical inconvenience. Sidereal time, as determined by the position of the true equinox, will not deviate from the same as indicated by the position of the mean equinox, more than 2.3s. in 19 years.

CONVERSION OF ONE SPECIES OF TIME INTO ANOTHER.

128. The difference between the apparent and mean time is called the *Equation of Time.* The equation of time, when known, serves for the conversion of mean time into apparent, and the reverse.

129. To find the Equation of Time. The hour angle of the sun (p. 15, def. 16) varies at the rate of 360° in a solar day, or 15° per solar hour. If, therefore, its value at any moment be divided by 15, the quotient will be the apparent time at that moment. In like manner, the hour angle of the mean sun, divided by 15, gives the mean time. Now, let the circle

VSD (Fig. 47) represent the equator, V the vernal equinox, M the point of the equator which is on the meridian, and VS the right ascension of the sun ; and we shall have,

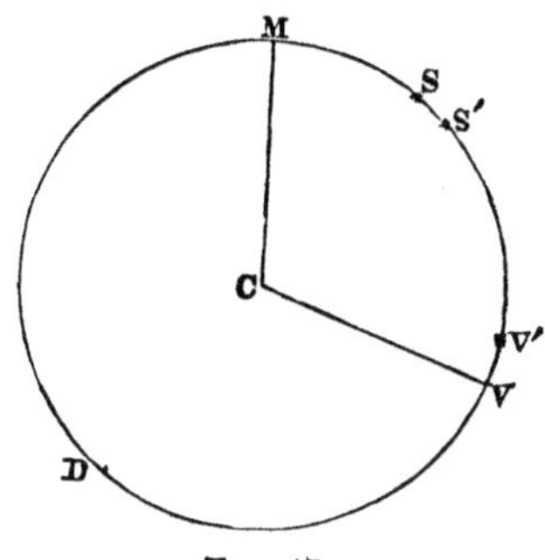

FIG. 47.

$$\text{appar. time} = \frac{\text{MS}}{15} = \frac{\text{VM} - \text{VS}}{15}.$$

Again, if we suppose S′ to be the position of the mean sun (VS′ being equal to the mean longitude of the sun), we shall have

$$\text{mean time} = \frac{\text{MS}'}{15} = \frac{\text{VM} - \text{VS}'}{15}:$$

$$\text{thus, equa. of time} = \text{mean time} - \text{ap. time} = \frac{\text{VS} - \text{VS}'}{15} \quad ..(25);$$

or, *the equation of time is equal to the difference between the sun's true right ascension and mean longitude, converted into time.*

This rule will require some modification if very great accuracy is desired; for, in seeking an expression for the mean time, the circle VSD ought properly to be considered as the mean equator, answering to the mean pole (124), and the mean longitude of the sun is really estimated from the mean equinox V′, and ought therefore to be corrected by the arc VV′, or the equation of the equinoxes in right ascension.

The value of the equation of time, determined from formula (25), is to be applied with its sign to the apparent time to obtain the mean, and with the opposite sign to the mean time to obtain the apparent.

A formula has been investigated, and reduced to a table, which makes known the equation of time by means of the sun's mean longitude. (See Table XII.; also Art. 158.) The value of the equation of time at noon, on any day of the year, is also to be found in the tables of calculations for the sun, published in the Nautical Almanac. If its value for any other time than noon be desired, it may be obtained by simple proportion.

The equation of time is zero, or mean and true time are the same four times in the year, viz. about the 15th of April, the 15th of June, the 1st of September, and the 24th of December.

Its greatest additive value (to apparent time) is about 14½ minutes, and occurs about the 11th of February; and its greatest subtractive value is about 16¼ minutes, and occurs about the 3d of November.

130. To convert Sidereal Time into Mean Time, and vice versa.—Making use of Fig. 47, already employed, the arc VM, called the *Right Ascension of Mid-Heaven*, expressed in time, is the sidereal time; VS′ is the right ascension of the mean sun, estimated from the true equinox, or the mean longitude of the sun corrected for the equation of the equinoxes in right ascension (124); and MS′ expressed in time, is the mean time. Let the arcs VM, MS′, and VS′, converted into time, be denoted respectively by S, M, and L. Now,

$$VM = MS' + VS';$$

or, $$S = M + L \,..(26);\ \text{and}\ M = S - L \,..(27).$$

If M + L in equation (26) exceeds 24 hours, 24 hours must be subtracted; and if L exceeds S in equation (27), 24 hours must be added to S, to render the subtraction possible.

This problem may in practice be solved most easily by means of an ephemeris of the sun (220), which gives the value of S, or the sidereal time, at the instant of mean noon of each day, together with a table of the acceleration of sidereal on mean solar time, and the corresponding table of the retardation of mean on sidereal time.

The conversion of apparent into sidereal time, or sidereal into apparent time, may be effected by first obtaining the mean time, and then converting this into sidereal or apparent time, as the case may be.

DETERMINATION OF THE TIME AND REGULATION OF CLOCKS BY ASTRONOMICAL OBSERVATIONS.

131. The regulation of a clock consists in finding its *error* and its *rate*.

132. Mean Solar Clock. The error of a mean solar clock is most conveniently determined from observations with a transit instrument of the time, as given by the clock, of the meridian passage of the sun's centre. The time noted will be the *clock-time* at apparent noon, and the exact mean time at apparent noon may be obtained by applying to the apparent time (24h., or 0h. 0m. 0s.) the equation of time with its proper sign, which may for this purpose be taken from the Nautical Almanac by simple inspection. A comparison of the clock time with the exact mean time, will give the error of the clock.

The daily rate of a mean solar clock may be ascertained by finding as above the error at two successive apparent noons. If the two errors are the same and lie the same way, the clock goes accurately to mean solar time; if they are different, their difference or sum, according as they lie the same or opposite ways, will be the daily gain or loss, as the case may be.

133. Sidereal Clock. The methods of determining the error and rate of a sidereal clock have already been explained (47). In practice, the apparent right ascension of the clock star to be observed, is taken from the table of the apparent places of stars, in the Nautical Almanac, as already intimated. The method of calculating such apparent places is given in Prob. XXI.

134. Time by Observations out of the Meridian. In default of a transit instrument, the time may be obtained and time-keepers regulated by observations made out of the meridian. There are two methods by which this may be accomplished, called, respectively, the method of *Single Altitudes*, and the method of *Double Altitudes*, or of *Equal Altitudes*. These we will now explain.

(1.) *To determine the time from a measured altitude of the sun, or of a star, its declination, and also the latitude of the place being given.*

Let us first suppose that the altitude of the sun is taken; correct the measured altitude for refraction and parallax, and also, if the sextant is the instrument used, for the semi-diameter of the sun. Then, if Z (Fig. 48) represents the zenith, P the elevated pole, and S the sun; in the triangle ZPS we shall know ZP = co-latitude, PS = co-declination, and ZS = co-altitude, from which we may compute the angle ZPS (= P), which is the angular distance of the sun from the meridian or, if expressed in time, the time of the observation from apparent noon; by the following equations (App., Resolution of oblique-angled spherical triangles, Case 1),

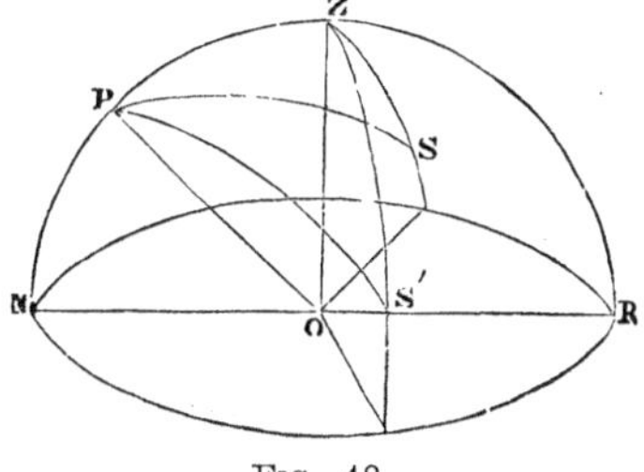

FIG. 48.

$$2k = \text{ZP} + \text{PS} + \text{ZS} = \text{co-lat.} + \text{co-dec.} + \text{co-alt.} \dots (28);$$

$$\sin^2 \tfrac{1}{2}\text{P} = \frac{\sin(k - \text{ZP}) \sin(k - \text{PS})}{\sin \text{ZP} \sin \text{PS}} \dots (29),$$

or, $$\sin^2 \tfrac{1}{2}\text{P} = \frac{\sin(k - \text{co-lat.}) \sin(k - \text{co-dec.})}{\sin(\text{co-lat.}) \sin(\text{co-dec.})} \dots (30).$$

The value of P being derived from these equations and converted into time (see Prob. III.), the result will be the apparent time at the instant of the observation, if it was made in the afternoon; if not, what remains after subtracting it from 24 hours will be the apparent time. The apparent time being found, the mean time may be deduced from it by applying the equation of time.

A more accurate result will be obtained if several altitudes be measured, the time of each measurement noted, and the mean of all the altitudes taken and regarded as corresponding to the mean of the times. The correspondence will be sufficiently exact if the measurements be all made within the space of 10 or 12 minutes, and *when the sun is near the prime vertical.* If an even number of altitudes be taken, and alternately of the upper and lower limb, the mean of the whole will give the altitude of the sun's centre, without it being necessary to know his apparent semi-diameter. In practice, the declination of the sun may be taken for the solution of this problem from an ephemeris of the sun. For this purpose, the time of the observation and the longitude of the place must be approximately known.

Example. On March 20, 1867, the following double altitudes of the sun were taken with a prismatic sextant, at New Haven; upper limb, 64° 12′ 0″, 64° 21′ 35″, 64° 33′ 0″,—lower limb, 63° 38′ 50″, 63° 51′ 0″, 63° 58′ 5″; the corresponding times of observation, noted by a watch, were 9h. 6m. 49s. A.M., 9h. 7m. 20.5s., 9h. 7m. 56s., 9h. 8m. 29.5s., 9h. 9m. 7.7s., 9h. 9m. 31s.; the barometer stood at 30.47in., and the thermometer at 34°. What was the mean time answering to the mean of the times of observation?

Mean of times of observation.........9h. 8m. 12.3s. A.M.
Long. of station of observer, west of Greenwich,..................4 51 42

Corresponding Greenwich time.......1 59 54.3. P.M.

Sun's dec. at that time, Am. Naut. Alm... 0° 11′ 37″ S
Sun's co-dec., or N. P. dist.............90 11 37

Mean of measured double altitudes......64° 5′ 45″
Index error.............................. —1 3

2) 64 4 42

Appar. alt. of sun's centre.............32° 2′ 21″
Refraction (Tables VIII., and IX.)..... —1 37.3

True alt. of sun's centre...............32 0 43.7

Lat. of station..41° 18′ 37″

Co-lat.........48 41 23......ar. co. sin. 0.124276
Co-dec.........90 11 37......ar. co. sin. 0.000003
Co-alt.........57 59 16.3

2) 196 52 16.3

k.............98 26 8
k — co-dec..... 8 14 31............sin. 9.156408
k — co-lat.....49 44 45............sin. 9.882630

2) 19.163317

½P = 22° 26′ 7″.8.......9.581658
P = 44 52 15 .6
.4

179m. 29s. 2‴

2h. 59m. 29.03s.
12

	h.	m.	s.	
	9h.	0m.	30.97s.	A. M.
Equa. of time.		+ 7	41.37	
M. time sought	9	8	12.34	A. M.
Time by watch	9	8	12.3	
Error of watch			— 0.04s.	

The error of the watch, as estimated from transit observations, was less than 1s.

On the same date, the following measurements were made:

	Double Altitudes of Sun.			Times of Observation.			
L. L.	64°	11′	45″	9h.	10m.	15s.	A. M.
	64	18	35	9	10	36	
	64	25	20	9	10	58	
U. L.	65	41	40	9	11	37	
	65	49	50	9	12	4.5	
	65	59	50	9	12	36	

Barometer, thermometer, and index error, same as above. The error of the watch, as determined from these data, was + 0.08s.

In case the altitude of a star is taken, the value of P derived from formula (30), when converted into time, will express the distance in time of the star from the meridian; and being added to the right ascension of the star, if the observation be made to the westward of the meridian, or subtracted from the right ascension (increased by 24h., if necessary) if the observation be made to the eastward, will give the *sidereal* time of the observation.

(2.) *To determine the time of noon from equal altitudes of the sun, the times of the observations being given.*

If the sun's declination did not change while he is above the horizon, he would have equal altitudes at equal times before and after apparent noon. Hence, if to the time of the first observation one-half the interval of time between the two observations should be added, the result would be the time of noon, as shown by the clock or watch employed to note the times of the observations. The deviation from 12 o'clock would be the error of the clock with respect to apparent time. The difference between this error and the equation of time would be the error of the clock with respect to mean time.

But, as in point of fact the sun's declination is continually changing, equal altitudes will not have place precisely at equal times before and after noon, and it is therefore necessary, in order to obtain an exact result, to apply a correction to the time thus obtained. This correction is called the *Equation of Equal Altitudes.* Tables have been constructed by the aid of which the equation is easily obtained. This is at the same time a very simple and quite accurate method of finding the time, and the error of a clock.

If equal altitudes of a star should be observed, it is evident that half the interval of time elapsed would give the time when the star passed the meridian, without any correction. From this the error of the clock (if keeping sidereal time) may be found, as explained in Art. 133.

THE CALENDAR.

135. Natural Periods of Time. The apparent motions of the sun, which bring about the regular succession of day and night and the vicissitude of the seasons, and the motion of the moon to and from the sun in the heavens, attended with conspicuous and regularly recurring changes in her disc, furnish three natural periods for the measurement of the lapse of time: viz., 1, the period of the apparent revolution of the sun with respect to the meridian, comprising the two natural periods of day and night, which is called *the solar day;* 2, the period of the apparent revolution of the sun with respect to the equator, comprehending the four seasons, which is called *the tropical year;* 3, the period of time in which the moon passes through all its phases and returns to the same position relative to the sun, called *the lunar month.* The day is arbitrarily divided into twenty-four equal parts, called hours; the hours into sixty equal parts, called minutes; and the minutes into sixty equal parts, called seconds. The tropical year contains 365d. 5h. 48m. 46s. The lunar month consists of about 29½ days. The week, consisting of seven days, has its origin in Divine appointment alone. A Calendar is a scheme for taking note of the lapse of time, and fixing the dates of occurrences, by means of the four periods just specified, viz., the day, the week, the month, and the year, or periods taken as nearly equal to these as circumstances will admit. Different nations have, in general, had calendars more or less different: and the proper adjustment or regulation of the calendar by astronomical observation has in all ages, and with all nations, been an object of the highest importance. We propose, in what follows, to explain only the Julian and Gregorian Calendars.

136. The Julian Calendar divides the year into 12 months, containing in all 365 days. Now, it is desirable that the calendar should always denote the same parts of the same season by the same days of the same months: that, for instance, the summer and winter solstices, if once happening on the 21st of June and 21st of December, should ever after be reckoned to happen on the same days; that the date of the sun's entering the equinox, the natural commencement of spring, should, if once, be always on the 20th of March. For thus the labors of agriculture, which really depend on the situation of the sun in the heavens, would be simply and truly regulated by the calendar.

This would happen if the civil year of 365 days were equal to the astronomical; but the latter is greater; therefore, if the calendar should invariably distribute the year into 365 days, it would fall into this kind of confusion, that in process of time, and successively, the vernal equinox would happen on every day of the civil year. Let us examine this more nearly.

Suppose the excess of the astronomical year above the civil to be exactly 6 hours, and on the noon of March 20th of a certain year, the sun to be in the equinoctial point; then, after the lapse of a civil year of 365 days, the sun would be on the meridian, but not in the equinoctial point; it would be to the west of that point, and would have to move 6 hours in order to reach it, and to complete the astronomical or tropical year. At the completions of a second and a third civil year, the sun would be still more and more remote from the equinoctial point, and would be obliged to move for 12, and 18 hours, respectively, before he could rejoin it and complete the astronomical year.

At the completion of a fourth civil year the sun would be more distant than on the two preceding ones from the equinoctial point. In order to rejoin it, and to complete the astronomical year, he must move for 24 hours; that is, for *one whole day*. In other words, the astronomical year would not be completed till the beginning of the next astronomical day; till, in civil reckoning, the *noon of March* 21*st*.

At the end of four more common civil years, the sun would be in the equinox on the noon of March 22d. At the end of 8 and 64 years, on March 23d and April 6th, respectively; at the end of 736 years, the sun would be in the vernal equinox on September 20th; and in a period of 1460 years, the sun would have been in every sign of the zodiac on the same day of the calendar, and in the same sign on every day.

If the excess of the astronomical above the civil year were really what we have supposed it to be, 6 hours, this confusion of the calendar might be very easily avoided. It would be necessary merely to make every fourth civil year to consist of 366 days; and for that purpose to interpose, or to *intercalate*, a day in a month previous to March. By this *intercalation*, what would have been March 21st is called March 20th, and accordingly the sun would be still in the equinox on the same day of the month.

This mode of correcting the calendar was adopted by Julius Cæsar. The fourth year into which the intercalary day is introduced was called *Bissextile;* it is now frequently called *Leap* year. The correction is called the *Julian* correction, and the length of a mean Julian year is 365d. 6h.

By the Julian Calendar, every year that is divisible by 4 *is a leap year, and the rest common years.*

137. Reformation of the Calendar.—Gregorian Calendar. The astronomical year being equal to 365d. 5h. 48m. 46.1s, it is less than the mean Julian by 11m. 13.9s., or 0.007800d. The Julian correction, therefore, itself needs correction. The calendar regulated by it would, in process of time, become erroneous, and would require *reformation.*

The intercalation of the Julian correction being too great, its effect would be to *antedate* the happening of the equinox. Thus (to return to the old illustration) the sun, at the completion of the fourth civil year, now the Bissextile, would have passed the equinoctial point by a time equal to four times 0.007800d.; at the end of the next Bissextile, by eight times 0.007800d.; at the end of 130 years, by about one day. In other words, the sun would have been in the equinoctial point 24 *hours previously*, or on the *noon of March* 19*th.*

In the lapse of ages this error would continue and be increased. Its accumulation in 1300 years would amount to 10 days, and then the vernal equinox would be reckoned to happen on March 10th.

The error into which the calendar had fallen, and would continue to fall, was noticed by Pope Gregory XIII., in 1582. At his time the length of the year was known to greater precision than at the time of Julius Cæsar. It was supposed equal to 365d. 5h. 49m. 16.23s. Gregory, desirous that the vernal equinox should be reckoned on or near March 21st (on which day it happened in the year 325, when the Council of Nice was held), ordered that the day succeeding the 4th of October, 1582, instead of being called the 5th, should be called the 15th: thus suppressing 10 days, which, in the interval between the years 325 and 1582, represented nearly the accumulation of error arising from the *excessive intercalation of the Julian correction.*

This act *reformed* the calendar. In order to correct it in future ages, it was prescribed that, at certain convenient periods, the intercalary day of the Julian correction should be omitted. Thus the centurial years 1700, 1800, 1900, are, according to the Julian Calendar, Bissextiles, but on these it was ordered that the intercalary day *should not be inserted;* inserted again in 2000, but not inserted in 2100, 2200, 2300; and so on for succeeding centuries. *By the Gregorian Calendar,* then, *every centurial year that is divisible by* 400 *is a Bissextile or Leap year, and the others common years.* For other than centurial years, the rule is the same as with the Julian Calendar.

This is a most simple method of regulating the calendar. It corrects the insufficiency of the Julian correction, by omitting in the space of 400 years 3 intercalary days. It is easy to estimate the degree of its inaccuracy; for the real error is 0.007800d. in one year, and 400 × 0.007800d., or 3.1200d. in 400 years. Consequently 0.1200d., or 2h. 52m. 48s. in 400 years, or 1 day in 3333 years, is the measure of the degree of inaccuracy of the Gregorian correction.

The Gregorian Calendar was adopted immediately on its promulgation, in all Catholic countries, but in those where the Protestant religion prevailed it did not obtain a place till some time after. In England, "the change of style," as it was called, took place after the 2d of September, 1752, eleven nominal days being then struck out; so that the last day of *Old Style* being the 2d, the first of *New Style* (the next day) was called the 14th, instead of the 3d. The same legislative enactment which established the Gregorian Calendar in England, changed the time of the beginning of the year from the 25th of March to the 1st of January. Thus the year 1752, which by the old reckoning would have commenced with the 25th of March, was made to begin with the 1st of January; so that the number of the year is, for dates falling between the 1st of January and the 25th of March, one greater by the new than by the old style. In consequence of the intercalary day omitted in the year 1800, there is now, for all dates, 12 days difference between the old and new style.

Russia is at present the only Christian country in which the Gregorian Calendar is not used.

The calendar months consist, each of them, of 30 or 31 days, except the second month, February, which, in a common year, contains 28 days, and in a Bissextile, 29 days; the intercalary day being added to the last of this month.

To find the number of days comprised in any number of civil years, multiply 365 by the number of years, and add to the product as many days as there are Bissextile years in the period.

CHAPTER IX.

MOTIONS OF THE SUN, MOON, AND PLANETS, IN THEIR ORBITS.

KEPLER'S LAWS.

138. The celebrated astronomer, Kepler, by examining the observations upon the planets that had been made by the renowned Danish observer, Tycho Brahé, discovered, early in the seventeenth century, that the motions of these bodies were in conformity with the following laws:

(1.) *The areas described by the radius-vector of a planet* (or a line from the sun to the planet) *are proportional to the times.*

(2.) *The orbit of a planet is an ellipse, of which the sun occupies one of the foci.*

(3.) *The squares of the periods of revolution of the planets are proportional to the cubes of their mean distances from the sun, or of the semi-major axes of their orbits.*

These laws are known by the denomination of *Kepler's Laws.* They were announced by Kepler as the fundamental laws of the planetary motions, after a partial examination only of these motions. They have since been completely verified, and shown to hold good for all the planets, including the earth. We shall adopt the first two laws for the present, as *hypotheses*, and show in the sequel that they are verified by the results deducible from them. These laws being established, the third is obtained by simply comparing the known major axes and periods of revolution.

139. Motion of the Sun in its Apparent Orbit. The apparent motion of the sun in space must be subject to Kepler's first two laws; for the apparent orbit of the sun is of the same form and dimensions as the actual orbit of the earth, and the law and rate of the sun's motion in its apparent orbit are the same as the law and rate of the earth's motion. To establish these two principles, let EE′A (Fig. 49) represent the elliptic orbit of the earth, and S the position of the sun in space. If the earth move from E to E′, as it seems to remain stationary at E, it is plain that the sun will appear to move from S to S′, on the line ES′ drawn parallel to E′S the actual direction of the sun from the earth; and at a distance ES′ equal to E′S the actual distance of the sun from the earth. Thus, for every position of the earth in its orbit, the corresponding apparent position of the sun is obtained by drawing a line parallel to the radius-vector of the earth, and equal to it. It follows, therefore, that the area SES′ apparently described by the radius-vector of

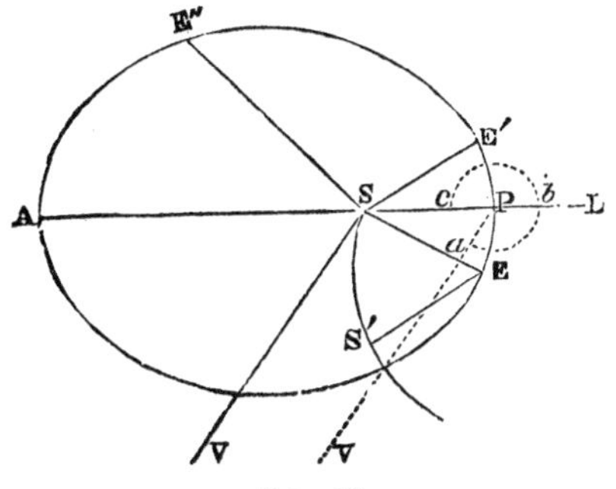

FIG. 49.

the sun (or a line drawn from the sun to the earth) in any interval of time, is equal to the area ESE′ actually described by the radius-vector of the earth in the same time; and consequently that the arc SS′ apparently described by the sun in space, is equal to the arc EE′ actually described in the same time by the earth. Whence we conclude, that the apparent motion of the sun in space, and the actual motion of the earth, are the same in every particular.

140. It has been ascertained that *the motion of the moon* in its revolution around the earth, is subject to the same laws as the motion of a planet in its revolution around the sun. We shall assume this to be a fact, and show that the hypothesis is verified by the results to which it leads.

141. Perihelion.—Aphelion. That point of the orbit of a planet, which is nearest to the sun, is called the *Perihelion*, and that point which is most distant from the sun, the *Aphelion*. The corresponding points of the moon's orbit, or of the sun's apparent orbit, are called, respectively, the *Perigee* and the *Apogee*.

These points are also called *Apsides;* the former being termed the *Lower Apsis*, and the latter the *Higher Apsis*. The line joining them is denominated the *Line of Apsides.*

The orbits of the sun, moon, and planets, being regarded as ellipses, the perigee and apogee, or the perihelion and aphelion, are the extremities of the major axis of the orbit.

142. Law of the Angular Motion of a Planet. The law of the *angular* motion of a planet about the sun may be deduced from Kepler's first law. Let $PpAp''$ (Fig. 50) represent the orbit of a planet, considered as an ellipse, and p, p' two positions of the planet at two instants separated by a short interval of time; and let n be the middle point of the arc pp'. With the radius Sn describe the small circular arc lnl', and with the radius Sa, equal to unity, describe the arc ab. It is plain that the

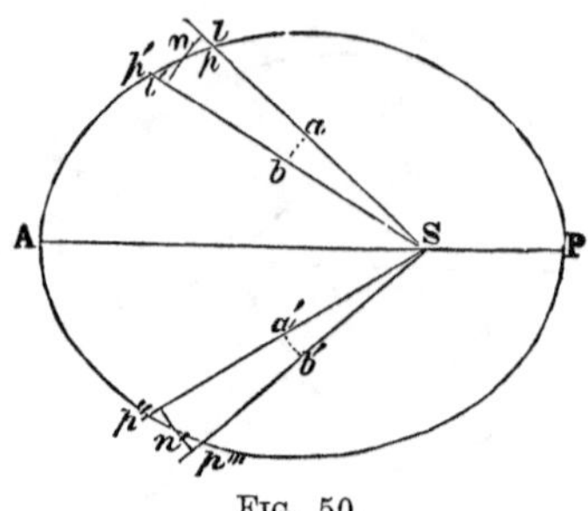

FIG. 50.

two positions p, p' may be taken so near to each other, that the area Spp' will be sensibly equal to the circular sector Sll'. If we suppose this to be the case, as the measure of the sector is $\frac{1}{2}lnl' \times$ S$n = \frac{1}{2}ab \times \overline{Sn}^2$ (substituting for lnl' its value, $ab \times$ Sn), we shall have

$$\text{area } Spp' = \tfrac{1}{2}ab \times \overline{Sn}^2.$$

When the planet is at any other part of its orbit, as n', if S$p''p'''$ be an area described in the same interval of time as before, we shall have

$$\text{area } Sp''p''' = \tfrac{1}{2}a'b' \times \overline{Sn'}^2.$$

But these areas are equal according to Kepler's first law: hence,

$$\tfrac{1}{2}ab \times \overline{Sn}^2 = \tfrac{1}{2}a'b' \times \overline{Sn'}^2 \ldots . (31);$$

and $$ab : a'b' :: \overline{Sn'}^2 : \overline{Sn}^2;$$

that is, *the angular motion of a planet about the sun for a short interval of time, is inversely proportional to the square of the radius-vector.*

It results from this that the angular motion is greatest at the perihelion, and least at the aphelion, and the same at corresponding points on either side of the major axis: also, that it decreases progressively from the perihelion to the aphelion, and increases progressively from the aphelion to the perihelion.

143. Mean Place.—True Place. Now to compare the true with the mean angular motion, suppose a body to revolve in a circle around the sun, with the mean angular motion of a planet, and to set out at the same instant with it from the perihelion. Let PMAM′ (Fig. 51) represent the elliptic orbit of the planet, and PBaB′ the circle described by the body. The position B of this fictitious body at any time, will be the *mean place* of the planet as seen from the sun. The two bodies will accomplish a semi-revolution in the same period of time, and therefore be, respectively, at A and a at the same instant; for it is obvious that the fictitious body will accomplish a semi-revolution in half the period of a whole revolution, and by Kepler's law of areas, the planet will describe a semi-ellipse in half the time of a revolution. At the outset, the motion of the planet is the most rapid (142), but it continually decreases until the planet reaches the aphelion, while the motion of the body remains constantly equal to the mean motion. The planet will therefore take the lead, and its angular distance pSB from the body will increase until its motion becomes reduced to an equality with the mean motion; after which it will decrease until the planet has reached the aphelion A, where it will be zero. In the motion from the aphelion to the perihelion, the angular velocity of the planet will at first be less than that of the body (142), but it will continually increase, while that of the body will remain unaltered: thus, the body will now get in advance of the planet, and their angular distance p'SB$'$ will increase, as before, until the motion of the planet again attains to an equality with the mean motion, after which it will decrease as before, until it again becomes zero at the perihelion.

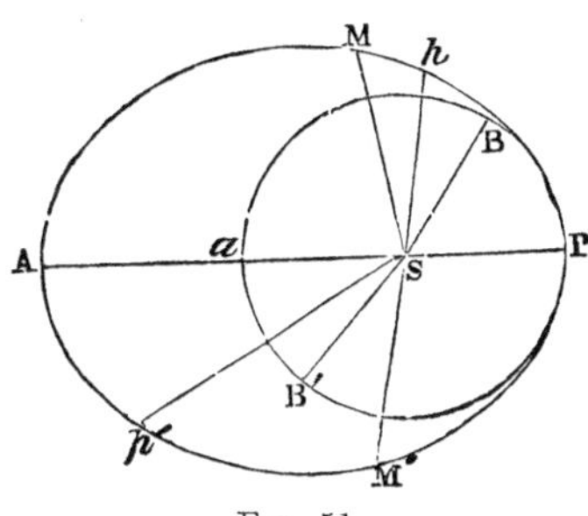

FIG. 51.

It appears, then, that *from the perihelion to the aphelion the true place is in advance of the mean place;* and that *from the aphelion to the perihelion*, on the contrary, *the mean place is in advance of the true place.*

The angular distance of the true place of a planet from its mean place, as it would be observed from the sun, is called the *Equation of the Centre.* Thus, pSB is the equation of the centre corresponding to the particular position p of the planet. It is evident, from the foregoing remarks, that the equation of the centre is zero at the perihelion and aphelion, and greatest at the two points, as M and M$'$, where the planet has its mean motion. The greatest value of the equation of the centre is called the *Greatest Equation of the Centre.*

As the laws of the motion of the moon (140), and of the apparent motion of the sun (139), are the same as those of a planet, the principles established in the two preceding articles are as applicable to these bodies in their revolution around the earth, as to a planet in its revolution around the sun.

DEFINITIONS OF TERMS.

144. (1.) The *Geocentric Place* of a body is its place as seen from the earth.

(2.) The *Heliocentric Place* of a body is its place as it would be seen from the sun.

(3.) *Geocentric Longitude* and *Latitude* appertain to the geocentric place, and *Heliocentric Longitude* and *Latitude* to the heliocentric place.

(4.) Two heavenly bodies are said to be *in Conjunction* when their longitudes are the same, and to be *in Opposition* when their longitudes differ by 180°. When any one heavenly body is in conjunction with the sun, it is, for the sake of brevity, said to be *in Conjunction;* and when it is in opposition to the sun, to be *in Opposition.*

The planets Mercury and Venus, allowing that their distances from the sun are each less than the earth's distance (18), can never be in opposition. But they may be in conjunction, either by being between the sun and earth, or by being on the opposite side of the sun. In the former situation they are said to be in *Inferior Conjunction*, and in the latter in *Superior Conjunction.*

(5.) A *Synodic Revolution* of a body is the interval between two consecutive conjunctions or oppositions.

For the planets Mercury and Venus a synodic revolution is the interval between two consecutive inferior or superior conjunctions.

(6.) The *Periodic Time* of a planet is the period of time in which it accomplishes a revolution around the sun.

(7.) The *Nodes* of a planet's orbit, or of the moon's orbit, are the points in which the orbit cuts the plane of the ecliptic. The node at which the planet passes from the south to the north side of the ecliptic is called the *Ascending* Node, and is designated by the character ☊. The other is called the *Descending* Node, and is marked ☋.

(8.) The *Eccentricity* of an elliptic orbit is the ratio which the distance between the centre of the orbit and either focus bears to the semi-major axis.

145. To illustrate these Definitions, let EE′E″ (Fig. 52) represent the orbit of the earth; C′DC the orbit of Venus, or Mercury, which we will suppose, for the sake of simplicity, to lie in the plane of the ecliptic or of the earth's orbit; LNP a part of the orbit of Mars, or of any other planet more distant from the sun S than the earth is; and ANB a part of the projection of this orbit on the plane of the ecliptic. N or ☊ will represent the ascending node of the orbit; and the descending node will be diametrically opposite to this in the direction S*n*′.

Also let SV be the direction of the vernal equinox, as seen from the sun, and EV, E′V the parallel directions of the same point, as seen from the earth in the two positions E and E′; and P being supposed to be one position of Mars in his orbit, let *p* be

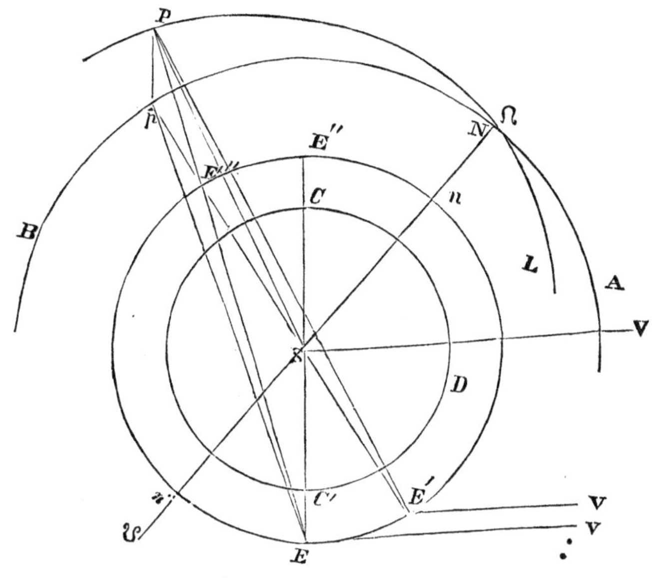

FIG. 52.

the projection of that position on the plane of the ecliptic. The *heliocentric longitude* and *latitude* of Mars, in the position P, are respectively VS*p* and PS*p*; and if the earth be at E, his *geocentric longitude* and *latitude* are respectively VE*p* and PE*p*. If we suppose that when Mars is at P the earth is at E′, he will be in *conjunction*; and if we suppose the earth to be at E‴ he will be in *opposition*. Again, if we suppose the earth to be at E, and Venus at C, she will be in *superior conjunction*; but if we suppose that Venus is at C′ at the time that the earth is at E, she will be in *inferior conjunction*. The term *inferior* is used here in the sense of lower in place, or *nearer the earth*; and *superior* in the sense of higher in place, or *farther from* the earth. Since the earth and planets are continually in motion, it is manifest that the positions of conjunction and opposition will recur at different parts of the orbit, and in process of time in every variety of position. The time employed by a planet in passing around from one position of conjunction or opposition to another, called the *synodic revolution*, is, for the same reason, longer than the *periodic time*, or time of passing around from one point of the orbit to the same again.

ELEMENTS OF THE ORBIT OF A PLANET.

146. To have a complete knowledge of the motions of the planets, so as to be able to calculate the place of any one of them at any assumed time, it is necessary to know for each planet, in addition to the laws of its motion discovered by Kepler, the position and dimensions of its orbit, its mean motion, and its place at a specified epoch. These necessary particulars of information are subdivided into seven distinct elements, called the *Elements of the Orbit of a Planet*, which are as follows:

(1.) The longitude of the ascending node.

(2.) The inclination of the plane of the orbit to the plane of the ecliptic, called the inclination of the orbit.

(3.) The mean distance of the planet from the sun, or the semi-major axis of its orbit.

(4.) The eccentricity of the orbit.

(5.) The heliocentric longitude of the perihelion.

(6.) The epoch of the perihelion passage of the planet, or instead, the mean longitude of the planet at a given epoch.

(7.) The periodic time of the planet.

The first two ascertain the *position of the plane* of the planet's orbit; the third and fourth, the *dimensions* of the orbit; the fifth, the *position of the orbit in its plane;* the sixth, the *place of the planet at a given epoch;* and the seventh, its *mean rate of motion.*

The elements of the earth's orbit, or of *the sun's apparent orbit*, are but *five* in number; the first two of the above-mentioned elements being wanting, as the plane of the orbit is coincident with the plane of the ecliptic.

The elements of the moon's orbit are the same with those of a planet's orbit, it being understood that the perigee of the moon's orbit answers to the perihelion of a planet's orbit, and that the *geocentric* longitude of the perigee and the *geocentric* longitude of the node of the moon's orbit answer, respectively, to the heliocentric longitude of the perihelion and the heliocentric longitude of the node of a planet's orbit.

147. The Linear Unit adopted, in terms of which the semi-major axes and radius-vectors of the planetary orbits are expressed, is the mean distance of the sun from the earth, or the semi-major axis of the earth's orbit. When thus expressed, these lines are readily obtained in known measures whenever the mean distance of the sun becomes known. The lines of the moon's orbit are found in terms of the moon's mean distance from the earth, as unity.

DETERMINATION OF THE ELEMENTS OF THE SUN'S APPARENT ORBIT, OR OF THE EARTH'S REAL ORBIT.

MEAN MOTION.

148. The sun's mean daily motion in longitude results from the length of the mean tropical year obtained from observation (115).

SEMI-MAJOR AXIS.

149. As we have just stated, the semi-major axis of the sun's apparent orbit, is the linear unit in terms of which the dimensions of the planetary orbits are expressed. Its absolute length is computed from the mean horizontal parallax of the sun.

The Horizontal Parallax of a body being given, to find its Distance from the Earth. We have (equation 7, Art. 88)

$$D = \frac{R}{\sin H};$$

where H represents the horizontal parallax of the body, D its distance from the centre of the earth, and R the radius of the earth. The parallax of all the heavenly bodies, with the exception of the moon, is so small, that it may, without material error, be taken in this equation in place of its sine. Thus,

$$D = \frac{R}{\sin H} = R \times \frac{1}{H}; \ldots . (32).$$

Again, since 6.2831853 is the length of the circumference of a circle of which the radius is 1, and 1296000 is the number of seconds in the circumference, we have 6.2831853 : 1 :: 1296000″ : $x = 206264.''806$ = the length of the radius (1) expressed in seconds. Hence, if the value of H be expressed in seconds,

$$D = R\,\frac{206264.''806}{H} \ldots . (33).$$

150. Determination of the Sun's Mean Horizontal Parallax. In the determination of the sun's parallax, by the process of Art. 90, an error of 2″ or 3″, equal to about one-fourth of the whole parallax, may be committed, so that the distance of the sun, as deduced by equation (33) from his parallax found in that manner, may be in error by an amount equal to one-fourth or more of the true distance. There are more accurate methods of obtaining the sun's parallax. By one method, which will be noticed in another connection, the equatorial parallax of the sun (92) was deduced from certain observations made upon Venus, when seen to pass between the sun and earth, in 1761 and 1769, and the value 8″.58 obtained. This is the value of the sun's equatorial horizontal parallax which has been universally adopted until within a very few years. Quite recently, several different determinations have been made of this important element, by independent astronomical methods. The different values obtained fall between 8″.93 and 8″.97, the mean of which is 8″.95. One of these has been the deduction of the solar parallax, by the process of Art. 90, from the parallax of Mars determined by direct observations at the opposition of this planet, in 1862, when its distance from the earth attained its minimum value. This deduction was easily effected, since, as will appear in the next Chapter, the theory of the orbital motions of the planets would give the distance of Mars from the earth at the epoch of the observations, in terms of the mean distance of the sun from the earth as the linear unit (147). The mean of two results obtained from the observations made by two sets of observers, at localities remote from each other, is 8″.95. This value, which is the mean of all the results, has been definitively fixed upon in the most approved Solar Tables (Leverrier's); and has since been adopted in the English Nautical Almanac for 1870. It may be relied upon as exact to within a small fraction of a second.

151. Calculation of Sun's Mean Distance. We have, then, for *the sun's mean distance from the earth*, or the semi-major axis of its orbit,

$$D = R\frac{206264''.806}{H} = 23046.347\,R = 91{,}328{,}064 \text{ miles};$$

taking for R the equatorial radius of the earth, 3962.80 miles.

ECCENTRICITY.

152. First Method. *By the greatest and least daily motions in longitude.* We have already explained (116) the mode of deriving from observation the sun's motion in longitude from day to day. Now, let $v =$ the greatest daily motion in longitude; $v' =$ the least daily motion in longitude; $r =$ the least or perigean distance of the sun; and r' the greatest or apogean distance; and we shall have, by the principle of Art. 142,

$$r : r' :: \sqrt{v'} : \sqrt{v};$$

whence, $r' + r : r' - r :: \sqrt{v} + \sqrt{v'} : \sqrt{v} - \sqrt{v'}$,

or, $\dfrac{r' + r}{2} : r' - r :: \dfrac{\sqrt{v} + \sqrt{v'}}{2} : \sqrt{v} - \sqrt{v'}$:

but,

$\dfrac{r' + r}{2} =$ semi-major axis $= 1$; and $r' - r = 2(\text{eccentricity}) = 2e$;

thus, $$1 : 2e :: \frac{\sqrt{v} + \sqrt{v'}}{2} : \sqrt{v} - \sqrt{v'},$$

and $$e = \frac{\sqrt{v} - \sqrt{v'}}{\sqrt{v} + \sqrt{v'}} \ldots (34).$$

The greatest and least daily motions are, respectively (at a mean), 61'.167 and 57'.200. Substituting, we have

$$e = 0.016761.$$

The eccentricity may also be obtained from the *greatest and least apparent diameters*, by a process similar to the foregoing, on the principle that the distances of the sun at different times are inversely proportional to its corresponding apparent diameters (116).

153. Second Method. *By the greatest equation of the centre.*

(1.) *To find the greatest equation of the centre.* Let L = the true longitude, and M = the mean longitude, at the time the true and mean motions are equal between the perigee and apogee (143); L' = the true longitude and M' = the mean longitude, when the motions are equal between the apogee and perigee; and E = the greatest equation of the centre. Then (143)

$$L = M + E, \text{ and } L' = M' - E;$$

whence, $$L' - L = M' - M - 2E,$$

and $$E = \frac{(M' - M) - (L' - L)}{2} \ldots (35).$$

About the time of the greatest equation the sun's true motion, and consequently the equation of the centre, continues very nearly the same for two or three days; we may therefore, with but slight error, take the noon, when the sun is on either side of the line of apsides, that separates the two days on which the motions in longitude are most nearly equal to 59' 8'', as the epoch of the greatest equation.

The longitude L or L' at either epoch thus ascertained, results from the observed right ascension and declination. M' — M = the mean motion in longitude in the interval of the epochs, and is found by multiplying the number of mean solar days and fractions of a day comprised in the interval by 59' 8''.330, the mean daily motion in longitude.

For example: from observations upon the sun, made by Dr. Maskelyne, in the year 1775, it is ascertained in the manner just explained, that the sun was near its greatest equation at noon, or at 0h. 3m. 35s. mean solar time, on the 2d April, and at noon on the 31st, or at 23h. 49m. 35s. mean solar time, on the 30th of September. The observed longitudes were, at the first period 12° 33' 39''.06, and at the second 188° 5' 44''.45. The interval of time between the two epochs is 182d. — 14m.

Mean motion in 182d. — 14m.	179°	22'	41''.56
Difference of two longitudes	175	32	5 .39
Difference ....2)	3	50	36 .17
Greatest equation of centre	1	55	18 .08

More accurate results are obtained by reducing observations made during several days before and after the epoch of the greatest equation, and taking the mean of the different values of the greatest equation thus obtained. According to M. Delambre, the greatest equation was in 1775, 1° 55' 31''.66.

(2.) The eccentricity of an orbit may be derived from the greatest equation of the centre by means of the following formula:

$$e = \frac{K}{2} - \frac{11\ K^3}{3.2^8} - \frac{587\ K^5}{3.5.2^{16}} - \&c. \ldots (36),$$

in which K stands for the expression $\frac{E}{57^\circ.2957795}$ (E being the greatest equation of the centre). In the case of the sun's orbit, K being a small fraction, all its powers beyond the first may be omitted. Thus, retaining only the first term of the series, and taking $E = 1^\circ\ 55'\ 31''.66$ the greatest equation in 1775, we have

$$e = \frac{K}{2} = \frac{1^\circ\ 55'\ 31''.66}{2 \times 57^\circ.2957795} = .016803.$$

154. Equation of Centre depends on Eccentricity. It appears from the law of the angular velocity of a revolving body, investigated in Art. 142, that the amount of the proportional variation of this velocity, which obtains in the course of a revolution, depends altogether upon the amount of the proportional variation of distance, or, in other words, upon the eccentricity of the orbit (def. 8, p. 106). It follows, therefore, that the amount of the greatest deviation of the true place from the mean place, that is, of the greatest equation of the centre (143), must depend upon the value of the eccentricity. If the eccentricity be great, the greatest equation of the centre will have a large value; and if the eccentricity be equal to zero, that is, if the orbit be a circle, the equation of the centre will also be equal to zero, or the true and mean place will continually coincide.

If either of the two quantities, the greatest equation and the eccentricity, be known, the other will then become determinate; and formulæ have been investigated which make known either one when the other is given. Equation 36 is the formula for the eccentricity.

From observations made at distant periods it is discovered that the equation of the centre, and consequently the eccentricity, is subject to a continual slow diminution. The amount of the diminution of the greatest equation, in a century, is 17''.6.

LONGITUDE AND EPOCH OF THE PERIGEE.

155. Methods of Determination. As the sun's angular velocity is the greatest at the perigee, the longitude of the sun at the time its angular velocity is greatest will be the longitude of the perigee. The time of the greatest angular velocity may be easily obtained, within a few hours, by means of the daily motions in longitude derived from observation (116).

The more accurate method of determining the longitude and epoch of the perigee, rests upon the principle that the apogee and perigee are the only two points of the orbit whose longitudes differ by 180°, in passing from one to the other of which the sun employs half a year. This principle may be inferred from Kepler's law of areas, for it is a well known property of the ellipse, that the major axis is the only line drawn through the focus that divides the ellipse into equal parts, and, by the law in question, equal areas correspond to equal times.

156. Progressive Motion of the Perigee. By a comparison of the results of observations made at distant epochs, it is discovered that the longitude of the perigee is continually increasing at a mean rate of 61″.7 per year. As the equinox retrogrades 50″.2 in a year, the perigee must then have a direct angular motion of 11″.5 per year.

It will be seen that as a consequence the interval between the times of the sun's passage through the apogee and perigee, is not, strictly speaking, half a sidereal year, but exceeds this period by the interval of time employed by the sun in moving through an arc of 5″.7, the sidereal motion of the apogee and perigee in half a year.

According to the most exact determinations, *the mean longitude of the perigee* of the sun's orbit at the beginning of the year 1800, was 279° 29′ 56″. It is now $280\frac{2}{3}$°.

157. The Heliocentric Longitude of the Perihelion of the Earth's Orbit, is equal to the geocentric longitude of the perigee of the sun's apparent orbit minus 180°. For, let AEP (Fig. 49, p. 103) be the earth's orbit, and PV the direction of the vernal equinox. When the earth is in its perihelion, P, the sun is in its perigee, S, and we have the heliocentric longitude of the perihelion, VSP = VPL = angle *abc* — 180° = geocentric longitude of the sun's perigee — 180°. It is plain that the same relation subsists between the heliocentric longitude of the earth and the geocentric longitude of the sun in every other position of the earth in its orbit.

158. The Mean Longitude of the Sun, at any assumed epoch, may be obtained by means of the mean motion in longitude (116), the epoch and mean longitude of the perigee of the sun's orbit having once been found.

DETERMINATION OF THE ELEMENTS OF THE MOON'S ORBIT.

LONGITUDE OF THE NODE.

159. In order to obtain the longitude of the moon's ascending node, we have only to find the longitude of the moon at the time its latitude is zero, and the moon is passing from the south to the north side of the ecliptic. This may be deduced from the longitudes and latitudes of the moon, derived from observed right ascensions and declinations (56); by methods precisely analogous to those by which the right ascension of the sun, at the time its declination is zero, and it is passing from the south to the north side of the equator, or the position of the vernal equinox, is ascertained (113).

INCLINATION OF THE ORBIT.

160. Among the latitudes computed from the moon's observed right ascensions and declinations, the greatest measures the inclination of the orbit. It is found to be about 5°; sometimes a little greater, and at other times a little less.

MEAN MOTION.

161. Tropical Revolution. With the longitudes of the moon, found from day to day, it is easy to obtain the interval from the time at which the moon has any given longitude till it returns to the same longitude again. This interval is called a *Tropical Revolution* of the moon. It is found to be subject to considerable periodical variations, and thus one observed tropical revolution may differ materially from the mean period. In order to obtain the mean tropical revolution, we must compare two longitudes found at distant epochs. Their difference augmented by the product of 360° by the number of revolutions performed in the interval of the epochs, will be the mean motion in longitude in the interval; from which the mean motion in 100 years, or 36,525 days, called the *Secular* motion, may be obtained by simple proportion. The secular motion being once known, it is easy to deduce from it the period in which the motion is 360°, which is the mean tropical revolution.

It should be observed, however, that to find the precise mean secular motion in longitude, it is necessary to compare the mean longitudes instead of the true. Now, the true longitude of the moon at any time having been found, the mean longitude at the same time is derived from it by correcting for the equation of the centre and certain other periodical inequalities of longitude hereafter to be noticed. But this cannot be done, even approximately, until the theory of the moon's motions is known with more or less accuracy.

The longitude of the moon, at certain epochs, may be very conveniently deduced from observations upon lunar eclipses. For, the time of the middle of the eclipse is very near the time of opposition, when the longitude of the moon differs 180° from that of the sun, and the longitude of the sun results from the known theory of its motion. The recorded observations of the ancients upon the times of the occurrence of eclipses, are the only observations that can now be made use of for the direct determination of the longitude of the moon at an ancient epoch.

162. Mean Daily Motion in Longitude. The mean tropical revolution of the moon is found to be

27.321582d. or 27d. 7h. 43m. 4.7s. (5s. nearly).

Hence, 27.321582d. : 1d. : : 360° : 13°.17639 = 13° 10′ 35″.0 = moon's mean daily motion in longitude.

163. Sidereal Revolution. Since the equinox has a retrograde motion, the sidereal revolution of the moon must exceed the tropical revolution, as the sidereal year exceeds the tropical year. The excess will be equal to the time employed by the moon in describing the arc of precession answering to a revolution of the moon. Thus,

365.25d. : 50″.2 : : 27.3d. : 3″.752 = the arc of precession,

and 13°.176 : 1d. : : 3″.752 : 6.8s. = *excess*.

Wherefore, the mean sidereal revolution of the moon is 27d. 7h. 43m. 11.5s.

164. Secular Acceleration of Moon's Motion. It has been found, by determining the moon's mean rate of motion for periods of various lengths, that it is subject to a continual slow acceleration. This acceleration will not, however, be indefinitely progressive; Laplace investigated its physical cause, and showed, from the principles of Physical Astronomy, that it is really a periodical inequality in the moon's mean motion, which requires an immense length of time to go through its different values.

The mean motion given in Art. 162 answers to the commencement of the present century.

LONGITUDE OF THE PERIGEE, ECCENTRICITY, AND SEMI-MAJOR AXIS.

165. The methods of determining these elements of the moon's orbit are similar to those by which the corresponding elements of the sun's orbit are found.

166. Orbit Longitudes. The only essential difference in the methods adopted, is that in place of the longitudes of the sun, which are laid off in the plane of the ecliptic, in the case of the moon corresponding angles are laid off in the plane of its orbit. These angles are reckoned from a line drawn making an angle with the line of nodes equal to the longitude of the ascending node, and are called *Orbit Longitudes*. The orbit longitude is equal to the moon's angular distance from the ascending node plus the longitude of the ascending node. Thus, let VNC (Fig. 53) represent the plane of the ecliptic, and V′NM a portion of the moon's orbit; N being the ascending node; also let EV be the direction of the vernal equinox, and let EV′ be drawn in the plane of the moon's orbit, making an angle V′EN with the line of the nodes equal to VEN, the longitude of the ascending node N. The orbit longitudes lie in the plane of the moon's orbit, and are estimated from this line, while the ecliptic longitudes lie in the plane of the ecliptic, and are estimated from the line EV. Thus, V′EM, or its measure V′NM, is the orbit longitude of the moon in the position M; and VEm is the ecliptic longitude; that is, the longitude as it has been hitherto considered. V′NM = V′N + NM = VN + NM; that is, orbit long. = long. of ☊ + ☽'s distance from ☊.

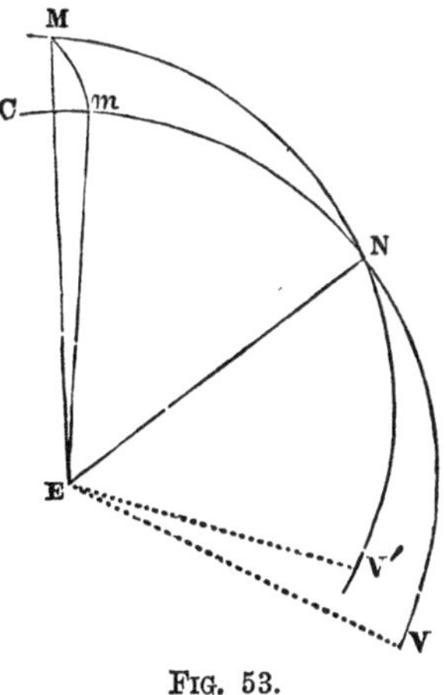

FIG. 53.

The orbit longitudes are calculated from the ecliptic longitudes; these being derived from observed right ascensions and declinations.

167. *The ecliptic longitude of the moon at any time being given, to find the orbit longitude.* As we may suppose the longitude of the node to be given (159), the equation of the preceding article will make known the orbit longitude so soon as MN, the moon's distance from the node, becomes known: now, by Napier's first rule we have

$$\cos \mathrm{MN}m = \cot \mathrm{NM} \tan \mathrm{N}m;$$

or,

$$\cot \mathrm{NM} = \cos \mathrm{MN}m \cot \mathrm{N}m.$$

$\mathrm{N}m$ = ecliptic long. — long. of node; and $\mathrm{MN}m$ = inclination of orbit.

168. The Horizontal Parallax of the Moon, like almost every other astronomical element, is subject to periodical changes of value. It varies not only during one revolution, but also from one revolution to another. The fixed and mean parallax, about which the true parallax may be conceived to oscillate, answers to the mean distance, that is, the distance about which the true distance varies periodically, and is called the *Constant of the Parallax.* It is, for the equatorial radius of the earth, 57′ 2″.7, from which we find by equation (32) the mean distance of the moon from the earth to be 238,824 miles.

169. The Eccentricity of the moon's orbit is more than three times as great as that of the sun's apparent orbit. Its greatest equation exceeds 6° (154).

MEAN LONGITUDE AT AN ASSIGNED EPOCH.

170. We have already explained (161) the principle of the determination of the mean longitude of the moon from an observed true longitude. Now, when the mean longitude at any one epoch whatever becomes known, the mean longitude at any assigned epoch is easily deduced from it by means of the mean motion in longitude.

DETERMINATION OF THE ELEMENTS OF A PLANET'S ORBIT.

171. Heliocentric Longitude and Radius-Vector of the Earth. The methods of determining the elements of the planetary orbits, suppose the possibility of finding the heliocentric longitude and radius-vector of the earth for any given time. Now, the elements of the earth's orbit having been found by the processes heretofore detailed, the longitude may be computed by means of Kepler's first law; and the radius-vector from the polar equation of the orbit, as given in treatises on Analytical Geometry. The manner of effecting such computation will be considered hereafter; at present the possibility of effecting it will be taken for granted.

HELIOCENTRIC LONGITUDE OF THE ASCENDING NODE.

172. First Method. When the planet is in either of its nodes, its latitude is zero. It follows, therefore, that the longitude of the planet at the time its latitude is zero, is the geocentric longitude of the node at the time the planet is passing through it. Now, if the right ascension and declination of the planet be observed from day to day, about the time it is passing from one side of the ecliptic to the other, and converted into longitude and latitude, the time at which the latitude is zero, and the longitude at that time, may be obtained by a proportion. When the planet is again in the same node, the geocentric longitude of the node may again be found in the same manner as before. On account of the different position of the earth in its orbit, this longitude will differ from the former.

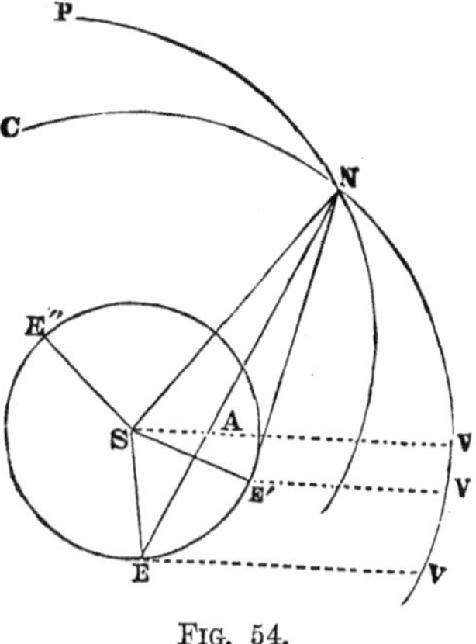

FIG. 54.

Now, *if two geocentric longitudes of the same node be found, its heliocentric longitude may be computed.* Let S (Fig. 54) be the sun, N the node, and E one of the positions of the earth for which the geocentric longitude of the node (VEN) is known. Denote this angle by G, the sun's longitude VES by S, and the radius-vector SE by r. Also, let E′ be the other position of the earth, and denote the corresponding quantities for this position, VE′N, VE′S, and SE′, respectively, by G′, S′, and r'. Let the radius-vector of the planet when in its node, or SN = V; and the heliocentric longitude of the node, or VSN = X. The triangle SNE gives

$$\sin \text{SNE} : \sin \text{SEN} :: \text{SE} : \text{SN};$$

but $$\text{SEN} = \text{VES} - \text{VEN} = \text{S} - \text{G},$$

and $$\text{SNE} = \text{VAN} - \text{VSN} = \text{VEN} - \text{VSN} = \text{G} - \text{X};$$

hence, $$\sin(\text{G} - \text{X}) : \sin(\text{S} - \text{G}) :: r : \text{V},$$

or, $$r \sin(\text{S} - \text{G}) = \text{V} \sin(\text{G} - \text{X}) \ldots. (37).$$

In like manner, $$r' \sin(\text{S}' - \text{G}') = \text{V} \sin(\text{G}' - \text{X}).$$

Dividing, $$\frac{r \sin(\text{S} - \text{G})}{r' \sin(\text{S}' - \text{G}')} = \frac{\sin(\text{G} - \text{X})}{\sin(\text{G}' - \text{X})},$$

or, $$\frac{r \sin(\text{S} - \text{G})}{r' \sin(\text{S}' - \text{G}')} = \frac{\sin \text{G} \cos \text{X} - \sin \text{X} \cos \text{G}}{\sin \text{G}' \cos \text{X} - \sin \text{X} \cos \text{G}'} = \frac{\sin \text{G} - \cos \text{G} \tan \text{X}}{\sin \text{G}' - \cos \text{G}' \tan \text{X}};$$

whence,

$$\tan \text{X} = \frac{r \sin(\text{S} - \text{G}) \sin \text{G}' - r' \sin(\text{S}' - \text{G}') \sin \text{G}}{r \sin(\text{S} - \text{G}) \cos \text{G}' - r' \sin(\text{S}' - \text{G}') \cos \text{G}} \ldots. (38).$$

Equation (37) gives $$\text{V} = \frac{r \sin(\text{S} - \text{G})}{\sin(\text{G} - \text{X})} \ldots. (39).$$

173. Second Method. The longitude of the node may also be found approximately from observations made upon the planet at the time of conjunction or opposition. It will happen in process of time that some of the conjunctions and oppositions will occur when the planet is near one of its nodes; the observed longitude of the sun at this conjunction or opposition, will either be approximately the heliocentric longitude of the node in question, or will differ 180° from it. This will be seen on inspecting

Fig. 55. If at a certain time the earth should be at E, crossing the line of nodes, and the planet in conjunction, it will be in the node N, and VES, the longitude of the sun, will be equal to VSN, the heliocentric longitude of the node. If the earth should be

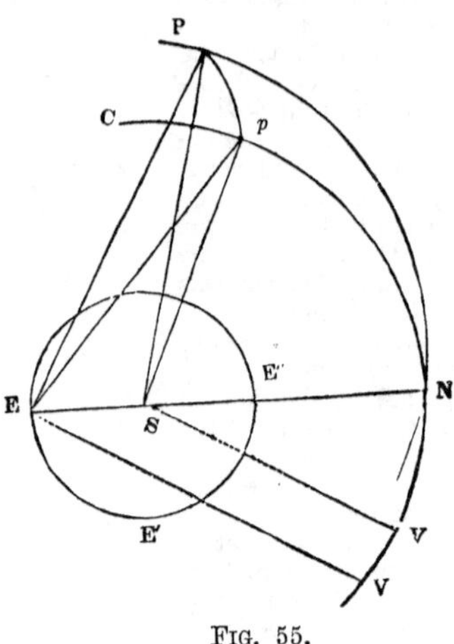

FIG. 55.

at E″ and the planet in opposition, the longitude of the sun would be VE″S = VE″N + 180° = VSN + 180° = hel. long. of node + 180°.

If the daily variations of the latitude of the planet should be observed about the time of the supposed conjunction or opposition near the node, the time when the latitude becomes zero, or the planet is in its node, could approximately be calculated by simple proportion; and then so soon as the rate of the angular motion about the sun becomes known (176) the longitude of the node could be more accurately determined.

INCLINATION OF THE ORBIT.

174. The longitude of the node having been found by the preceding, or some other method, compute the day on which the sun's longitude will be the same or nearly the same: the earth will then be on the line of the nodes. Observe on that day the planet's right ascension and declination, and deduce the geocentric longitude and latitude. Let ENp (Fig. 55) be the plane of the ecliptic, V the vernal equinox, S the sun, N the node, E the earth on the line of the nodes, and P the planet as referred to the celestial sphere, from the earth. Let λ denote the geocentric latitude PEp; E the arc Np = Vp — VN = geo. long. of planet — long. of node; and I the inclination PNp. The right-angled triangle PNp gives

$$\sin Np = \tan Pp \cot PNp = \tan \lambda \cot I;$$

hence,
$$\cot I = \frac{\sin E}{\tan \lambda}, \text{ and } \tan I = \frac{\tan \lambda}{\sin E};$$

$$\text{or, } \tan \text{ inclination} = \frac{\tan \text{ lat.}}{\sin (\text{long.} - \text{long. of node})} \ldots (40).$$

It will be understood, that to obtain an exact result, we must compute the precise time of day at which the longitude of the sun is the same as that of the node, and then, by means of their observed daily variations, correct the longitude and latitude of the planet for the variations in the interval between the time thus ascertained and the time of the observation above mentioned.

PERIODIC TIME.

175. The interval from the time the planet is in one of its nodes till its return to the same, gives the periodic time or sidereal revolution.

Another and more accurate method is to observe the length of a synodic revolution and compute the periodic time from this. If we compare the time of a conjunction which has been observed in modern times, with that of a conjunction observed by the earlier astronomers, and divide the interval between them by the number of synodic revolutions contained in it, we shall have the mean synodic revolution with great exactness; from which the mean periodic time may be deduced, as will be shown hereafter.

176. Mean Daily Motion. The periodic time being known, the mean daily motion around the sun may be found by dividing 360° by the periodic time expressed in days and parts of a day.

TO FIND THE HELIOCENTRIC LONGITUDE AND LATITUDE, AND THE RADIUS-VECTOR, FOR A GIVEN TIME.

177. General Problem. The earth being in constant motion in its orbit, and being thus at different times very differently situated with regard to the other planets, as well in respect to distance as direction, it is necessary for the purpose of comparing the observations made upon these bodies with each other, to refer them all to one common point of observation. As the sun is the fixed centre about which the revolutions of the planets are performed, it is the point best suited to this purpose, and accordingly it is to the sun that the observations are in reality referred. The reduction of observations from the earth to the sun, as it is actually performed, consists in the deduction of the heliocentric longitude and latitude from the geocentric longitude and latitude; these being calculated from the observed right ascension and declination.

The requisite formulæ for effecting this reduction are investigated in the Appendix.

178. Special Cases. The heliocentric longitude, or radius-vector of a planet, may be more readily obtained if the observations be made upon it when it is in certain favorable positions.

Case I. *When the planet is in conjunction or opposition*, its heliocentric longitude will then either be equal to the geocentric longitude, or differ 180° from it.

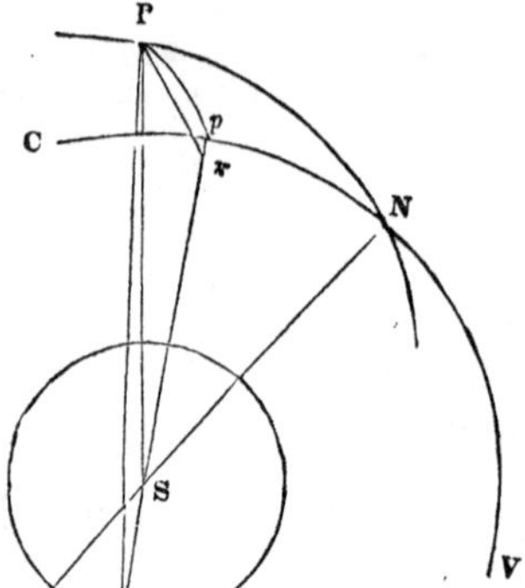

FIG. 56.

When the heliocentric longitude is thus found, *the latitude* for the same time may be obtained by solving the triangle PNp (Fig. 56). For, by Napier's first rule,

$$\sin Np = \cot PNp \tan Pp,$$

$$\text{or } \tan Pp = \sin Np \tan PNp;$$

where Pp is the latitude sought, PNp the known inclination of the orbit, and Np = VNp — VN = long. of planet — long. of node, both of which may be considered as known.

The radius-vector may be computed for the same time from the triangle ESP; for the side SE, the radius-vector of the earth, is known, as well as the angle SEP, the geocentric latitude of the planet, and the angle ESP = 180° — PSp = 180° — heliocentric latitude.

Case II. *When an inferior planet is at its maximum elongation from the sun.* The radius-vector of either of the inferior planets at the time of maximum elongation, or greatest angular distance from the sun, may be approximately deduced from the amount of the greatest elongation determined from observation. The elongation which obtains at any time, may be found by ascertaining from instrumental observations the places of the planet and sun in the heavens, and connecting these by an arc of a great circle, and with the pole by other arcs. In the triangle PSp (Fig. 57) thus formed, there will be known the two polar dis-

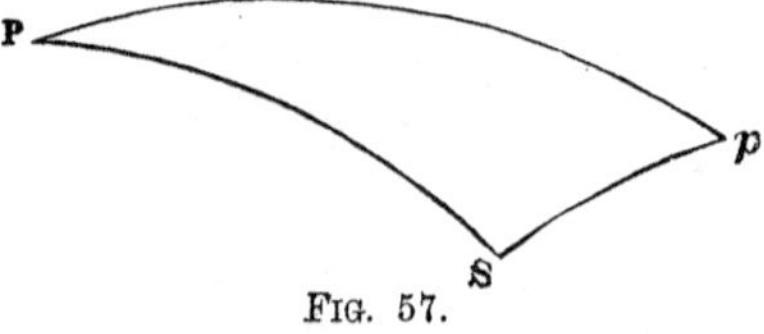

FIG. 57.

tances PS and Pp, which are the complements of the observed declinations, and the angle SPp the difference of their observed right ascensions, from which the angular distance Sp between the two bodies may be calculated. The maximum elongation being then supposed to be known, let NPP′ (Fig. 58) represent the orbit of the inferior planet. The line EP drawn from the earth to the planet will, at the time of maximum elongation, be perpendicular to SP, the radius-vector of the planet; and thus we shall have in the right-angled triangle EPS, the line ES, and the angle SEP, from which the radius-vector SP may be computed.

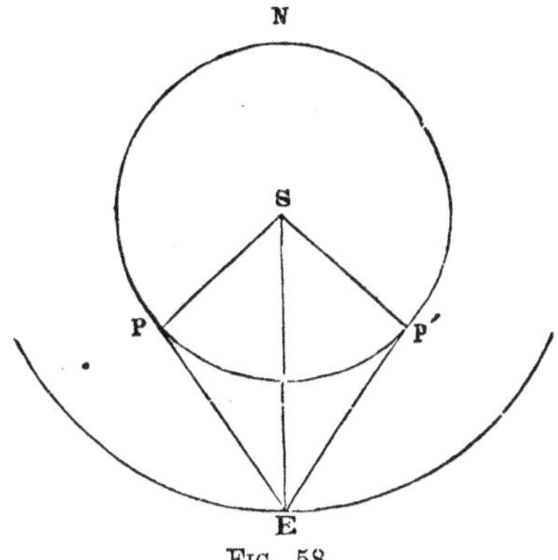

FIG. 58.

As the earth and planet are in motion, the greatest elongation will occur at different points of the planet's orbit, and therefore we may find by the foregoing process different radius-vectors.

179. The Orbit Longitude of a Planet may be derived from the ecliptic longitude in the same manner that the orbit longitude of the moon is calculated from its ecliptic longitude (166). The orbit longitude and radius-vector, when found for a given time, ascertain the position of the planet in the plane of its orbit at that time.

LONGITUDE OF THE PERIHELION, ECCENTRICITY, AND SEMI-MAJOR AXIS.

180. These elements may be calculated from the heliocentric orbit longitude and radius-vector, found for three different times.

Let SP, SP′, SP″ (Fig. 59), be the three given radius-vectors; V′SP, V′SP′, V′SP″, the three given longitudes; and AB the line of apsides of the planet's orbit. Let the angles PSP′, PSP″, which are known, be represented by m and n, and the angle BSP, which is unknown, by x; and let the radius-vectors SP, SP′, SP″, be denoted by v, v', v'', the semi-major axis AC by a, and the eccentricity by e. Then the three unknown quantities which are to be determined are a, e, and the angle x; and the general polar equation of the ellipse furnishes for their determination the three equations,

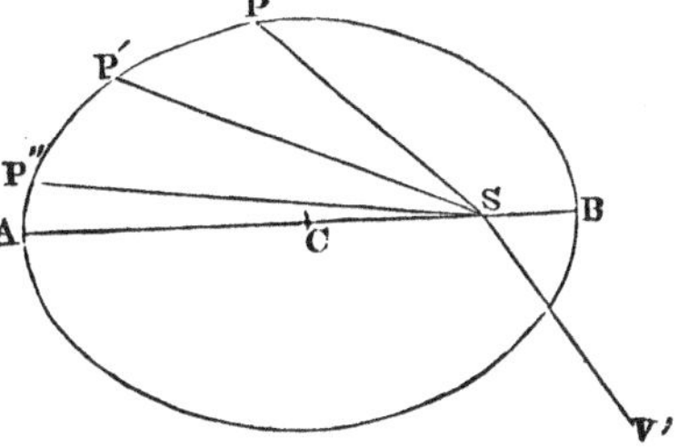

FIG. 59.

$$v = \frac{a(1 - e^2)}{1 + e \cos x}, \; v' = \frac{a(1 - e^2)}{1 + e \cos (x + m)}, \; v'' = \frac{a(1 - e^2)}{1 + e \cos (x + n)}.$$

The process of solution is given in the Appendix. When x has been found, by subtracting it from V′SP we obtain V′SB, the longitude of the perihelion.

181. Other Methods of Determining the Semi-Major Axis. The semi-major axis, or mean distance from the sun, may also be had by taking the mean of a great number of values of the radius-vector found for every variety of position of the planet in its orbit (178).

Now that Kepler's third law has been established by investigations in Physical Astronomy, it furnishes the most accurate method of finding the mean distance of a planet from the sun. Thus let P denote the periodic time of a planet, and a its mean distance from the sun; then the length of the sidereal year being 365.256359 days (120),

$$(365.256359\text{d.})^2 : P^2 :: 1^3 : a^3;$$

whence,
$$a = \left(\frac{P}{365.256359\text{d.}}\right)^{2/3} \ldots .(41).$$

182. Longitude of Perihelion, and Eccentricity, by Approximate Methods. If a great number of values of the radius-vector, in a great variety of positions of the planet in its orbit, be found by the method explained in Art. 178, the longitude of the planet at the time when its calculated radius-vector is the least, will be approximately the longitude of the perihelion; or, if it chances that among the calculated radius-vectors there are two equal to each other, the position of the line of apsides may be found by bisecting the angle included between these.

The ratio of the difference between the greatest and least calculated radii to the mean of the whole, will be the approximate value of the eccentricity.

EPOCH OF THE PERIHELION PASSAGE.

183. From several observations upon the planet about the time it has the same longitude as the perihelion, the correct time of its being at the perihelion may be easily determined by proportion.

The Mean Longitude at an assigned epoch is obtained on the same principles as the mean longitude of the sun or moon (158, 170).

REMARKS.

184. The foregoing methods of determining the elements of a planet's orbit suppose observations to be made at two or more successive returns of the planet to its node; but it is not necessary to wait for the passage of a planet through its node. Soon after the planet Uranus was discovered by Sir William Herschel, Laplace contrived methods by which the elements of its elliptic orbit were determined from four observations within little more than a year from its first discovery by Herschel. After the discovery of Ceres, Gauss invented another general method of calculating the orbit of a planet from three observations, and applied it to the determination of the orbit of Ceres, and subsequently to the determination of the orbits of Pallas, Juno, and Vesta. This method can be more readily employed in practice than that of Laplace, or than any of the solutions which other mathematicians have given of the same problem, and is now generally used in computing the orbit of a newly discovered planet.

TRUE AND MEAN ELEMENTS.

185. *True elements and their variations.* The elements of the planetary orbits, obtained by the foregoing processes, are the true elements at the periods when the observations are made. Upon determining them at different periods, it appears that they are subject to minute variations. A comparison of the values found at various distant epochs shows that they are slowly changing from century to century, and that the changes experienced during equal long periods of time are very nearly the same. The amount of the variation of an element in a period of 100 years is called its *Secular Variation.* Upon reducing the elements, found at different times, to the same epoch, by allowing for the proportional parts of the secular variations, the different results for each element are found to differ slightly from each other, which shows that the elements are also subject to slight periodical variations. These variations being very minute, the true elements can never differ much from the mean, or those from which they deviate periodically and equally on both sides.

186. *Mean Elements and their Secular Variations.* The mean elements at an assigned epoch may be had by finding the true elements at various times, and reducing them to the given epoch, by making allowance for the proportional parts of the secular variations, and then taking for each element the mean of all the particular values obtained for it.

A comparison of the mean values of the same element, found at distant epochs, makes known the variation of its mean value in the interval between them, from which the *secular variation* may be deduced by simple proportion.

187. *Variations of Elements of Moon's Orbit.* The elements of the moon's orbit are also subject to continual variations. These are, for the most part. periodic, and are far greater than the variations of the corresponding elements of a planet's orbit. It will be seen, then, that in determining the mean elements, a much greater number of observations will be required than in the case of a planetary orbit. The mean node and perigee have a rapid and nearly uniform motion. Their motions, in connection with the mean motion of revolution of the moon, are subject to minute secular variations. The mean eccentricity, and inclination of the orbit, are constant.

188. *Verifications.* The mean elements which have been derived as above, directly from observation, have subsequently been verified and corrected by comparing the computed with the observed places of the planet; and for this purpose many thousands of observations have been made.

189. Tables II. and III. contain the elements of the orbits of the principal planets, and of the moon's orbit, together with their secular variations for the beginning of the year 1850. Table II. (*a*) contains the mean distances, sidereal revolutions, and eccentricities of the orbits of the planetoids.

If an element be desired for any time different from the epoch of the table, we have only to allow for the proportional part of the secular variation, in the interval between the given time and the epoch of the table.

190. Secular Variations. It will be seen on inspecting Table II., that the mean distances of the planets from the sun, or the semi-major axes of their orbits, are the only elements that are invariable. The rest are subject to minute secular variations. The nodes have all retrograde motions. The perihelia, on the contrary, have direct motions, with the single exception of the perihelion of the orbit of Venus, which has a retrograde motion. The eccentricities of some of the orbits are increasing; of others, diminishing. That of the earth's orbit is diminishing.

The node of the moon's orbit has a retrograde motion, and the perihelion a direct motion. The former accomplishes a tropical revolution in about 18 years and 224 days, and the latter in about 8 years and 309 days. The mean motion of the node and the mean motion of the perigee are both subject to a slow secular diminution.

191. Eccentricities and Inclinations. It will be seen also, that the orbits of the principal planets are ellipses of small eccentricity, or which differ but slightly from circles; and that they are inclined under small angles to the plane of the ecliptic. The eccentricity is in almost every instance so small that, if a representation of the orbit were accurately delineated, it would not differ perceptibly from a circle. The most eccentric orbits are those of Mercury and Mars; and the least eccentric those of Venus, Neptúne, and the earth. The eccentricity of Mercury's orbit is 12 times that of the earth's, of Mars 6 times, of Venus less than $\frac{1}{2}$. The eccentricities of the orbits of Jupiter, Saturn, and Uranus, are each about three times that of the earth's orbit.

The orbit of Mercury is more inclined to the ecliptic than the orbit of any other of the eight principal planets; and the orbit of Uranus is less inclined than that of any other planet. The inclination of the latter is $\frac{3}{4}$°, of the former 7°.

The orbits of the planetoids are in general more eccentric, and more inclined to the plane of the ecliptic than those of the other planets. The inclination of the orbit of Pallas is nearly 35°.

192. The Mean Distances of the Planets from the Sun, expressed in miles, are in round numbers as follows: Mercury 35 millions, Venus 66 millions, the earth 91 millions, Mars 139 millions, Juno 244 millions, Jupiter 475 millions, Saturn 871 millions, Uranus 1752 millions, Neptune 2,743 millions. The range of distance is from 1 to 77½. The distance of Neptune is 30 times the earth's distance.

193. The Approximate Periods of Revolution of the planets are: of Mercury 3 months, Venus 7½ months, Mars 1⅞ years, Juno 4⅗ years, Jupiter a little less than 12 years, Saturn 29½ years, Uranus 84 years, Neptune 164½ years. The periods and mean distances are more exactly given in Table II. (For the planetoids, see Table II. (*a*)).

194. Bode's Law. A remarkable empirical law, called *Bode's Law of the Distances*, was announced in 1772 by Professor Bode, of Berlin, as connecting the distances of the planets from the sun. It is as follows: If we take the numbers 0, 3, 6, 12, 24, 48, 96, 192, and add 4 to each one of them, so as to obtain 4, 7, 10, 16, 28, 52, 100, 196, this series of numbers will express the order of distance of the planets from the sun. This law embodies the following curious relation between the distances of the orbits from one another, viz.: setting out from Venus, the distance between two contiguous orbits increases nearly in a duplicate ratio as we recede from the sun; that is, the distance from the orbit of the earth to the orbit of Mars, is twice the distance from the orbit of Venus to the orbit of the earth, and one-half the distance from the orbit of Mars to the orbits of the planetoids.

Previous to the discovery of the planetoids, to complete the above law a planet was wanting between Mars and Jupiter. It was, on this account, surmised by Bode that another planet might exist between these two. Instead of one such planet, however, no less than ninety-one have since been discovered, revolving at pretty nearly the distance from the sun that Bode had derived from his law for the distance of the supposed planet; some at a little greater, and others at a little less distance.

Bode's law, though it holds good for the planets in general, fails in the case of the planet Neptune; the error for this planet being more than one-fourth the whole distance. The error is one-twentieth of the distance for Mars, and also for the planetoids. For Mercury, Venus, and Saturn it is about one-thirtieth. For Uranus and Jupiter it is a still smaller fraction.

195. Dimensions of the Solar System. A better idea of the dimensions of the solar system than is conveyed by the statement of distances above given, may be gained by reducing its scale sufficiently to bring it within the range of familiar distances. Thus, if we suppose the earth to be represented by a ball only one inch in diameter, the distance of Mercury from the sun will be represented, on the same scale, by 370 feet, the distance of Venus by 700 feet, that of the earth by 960 feet, that of Mars by 1,500 feet, that of Juno by half a mile, that of Jupiter by 1 mile, that of Saturn by 1¾ miles, that of Uranus by 3½ miles, and that of Neptune by 5½ miles. On the same scale, the distance of the moon from the earth would be only 2½ feet.

CHAPTER X.

Determination of the Place of a Planet, or of the Sun or Moon, for a Given Time, by the Elliptic Theory.—Verification of Kepler's Laws.

PLACE OF A PLANET IN ITS ORBIT.

196. True and Mean Anomaly. The angle contained between the line of apsides of a planet's orbit and the radius-vector, as reckoned from the perihelion towards the east, is called the *True Anomaly*. Thus, let BPAP′ (Fig. 60) represent the

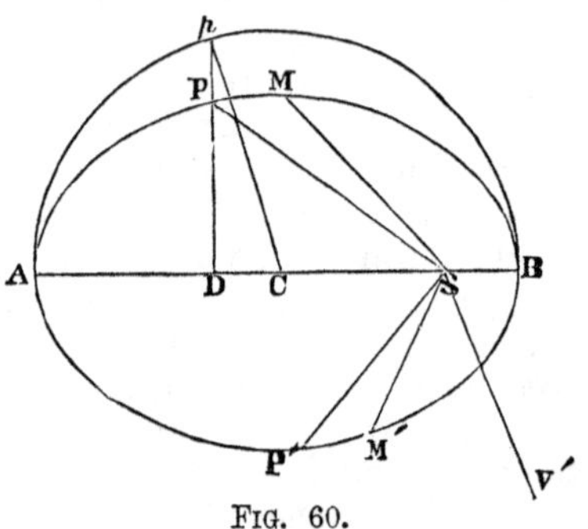

Fig. 60.

orbit, B the perihelion, and P the position of the planet; then BSP is its true anomaly. The angle contained between the line of apsides and the mean place of the planet, also reckoned from the perihelion towards the east, is called the *Mean Anomaly*. Thus, let M be the mean place of a planet at the time P is its true place, and BSM will be its mean anomaly. The difference between the true anomaly BSP and the mean anomaly BSM, is the angular distance MSP between the true and mean place of the planet, or the equation of the centre (143).

Describe a circle B*p*A on the line of apsides as a diameter; through P draw *p*PD perpendicular to the line of apsides, and join *p* and C; the angle BC*p*, which the line thus determined makes with the line of apsides, is called the *Eccentric Anomaly*.

The corresponding angles appertaining to the sun's apparent orbit, and to the moon's orbit, have received the same appellations.

197. Anomalistic Revolution. The interval between two

consecutive returns of a body to either apsis of its orbit, is called the *Anomalistic Revolution.* The anomalistic revolution of the earth, or of the sun in its apparent orbit, is termed, also, the *Anomalistic Year.*

The periodic time, or the mean motion of a body, and the motion of the apsis of its orbit, being known, the anomalistic revolution may be easily computed. Let $m =$ the sidereal motion of the apsis answering to the periodic time, and $M =$ the mean daily motion of the planet; then,

$M : 1d. :: m : x =$ diff. of anomalistic rev. and periodic time.

When the epoch of any one passage of a planet through its perihelion, or of the sun or moon through its perigee, has been found, we may, by means of the anomalistic revolution, deduce from it the epoch of every other passage.

The length of the anomalistic year exceeds that of the sidereal year by 4m. 39s.

198. Calculation of Mean Anomaly. From the anomalistic revolution, and the epoch of the last passage through the perihelion or perigee (as the case may be), we may derive the mean anomaly for any given time. Let $T =$ the anomalistic revolution, $t =$ the time that has elapsed since the last passage through the perihelion or perigee, and $A =$ the mean anomaly; then,

$$T : 360° :: t : A = 360° \frac{t}{T} \dots (42).$$

199. The Place of a Body in its Elliptic Orbit is ascertained by finding its true anomaly. The problem which has for its object the determination of the true anomaly from the mean, was first resolved by Kepler, and is called *Kepler's Problem.* Another and more convenient method of obtaining the true anomaly, is to compute the equation of the centre from the mean anomaly, and add it to the mean anomaly, or subtract it from it, according to the position of the body in its orbit (143). (See Appendix, Solution of Kepler's Problem.)

HELIOCENTRIC PLACE OF A PLANET.

200. The Place of a Planet in the Plane of its Orbit is designated by its orbit longitude (166) and radius-vector. To find the orbit longitude we have the equation $V'SP = V'SB + BSP$ (see Fig. 60); or,

long. = long. of perihelion + true anomaly.

The orbit longitude may also be deduced from the mean longitude, by adding or subtracting the equation of the centre; for,

$$V'SP = V'SM + MSP,$$

or, true long. = mean long. + equa. of centre:
also, $V'SP' = V'SM' - M'SP'$,

or, true long. = mean long. — equa. of centre.

The radius-vector results from the polar equation of the elliptic orbit, viz.:

$$V = \frac{a\,(1 - e^2)}{1 + e \cos x} \dots (43).$$

in which x denotes the true anomaly, e the eccentricity, and a the semi-major axis.

201. To find the Heliocentric Longitude and Latitude, which ascertain the position of the planet with respect to the ecliptic, the triangle NPp (Fig. 56, p. 120) gives

$$\sin Pp = \sin NP \sin PNp;$$

or, sin lat. = sin (orbit long.—long. of node) × sin (inclin.)..(44); and

$$\cos PNp = \tan Np \cot NP, \text{ or } \tan Np = \tan NP \cos PNp,$$

or,

tan (long. — long. of node) = tan (orbit long. — long. of node) × cos (inclination)....(45).

GEOCENTRIC PLACE OF A PLANET.

202. The theoretical determination of the place of a planet, as it would be seen from the centre of the earth, consists in deducing its geocentric longitude and latitude, and its distance from the earth, from its heliocentric longitude and latitude and radius-vector; the latter having been calculated by the methods just explained. (For the detail of the solution of this problem see Appendix.)

PLACES OF THE SUN AND MOON.

203. The place of the sun, as seen from the earth, may be easily deduced from the heliocentric place of the earth; for the longitude of the sun is equal to the heliocentric longitude of the earth plus 180° (157), and the radius-vector of the earth's orbit is the same as the distance of the sun from the earth. But it is more convenient to regard the sun as describing an orbit around the earth, and compute its true anomaly (199); and thence the longitude and radius-vector, by the equation

long. = true anomaly + long. of perigee,

and the polar equation of the orbit.

204. The Orbit Longitude and the Radius-vector of the Moon, are found by the same process as the longitude and radius-vector of the snn. The orbit longitude being known the ecliptic longitude and the latitude may be determined by a process precisely similar to that by which the heliocentric longitude and latitude of a planet are found (201).

VERIFICATION OF KEPLER'S LAWS.

205. If Kepler's first two laws be true, then the geocentric places of the planets, computed by the process that we have described (202), which is founded upon them, ought to agree with the true geocentric places as obtained for the same times by direct observation; or, the heliocentric places computed from the observed geocentric places (177), ought to agree with the same as computed by the elliptic theory (200, 201). Now, a great number of comparisons have been made between the observed and computed places, and in every instance a close agreement between the two has been found to subsist. We infer, therefore, that the motions of the planets must be very nearly in conformity with these laws.

The truth of the third law has been established by a direct comparison of the mean distances of the different planets with their periodic times.

Kepler's laws have been verified for the sun and moon, in a similar manner.

206. The Relative Distances of the Sun or Moon, at different times, result for this purpose, from measurements of the apparent diameter, upon the principle that any two distances are inversely proportional to the corresponding apparent diameters. Let $\Delta =$ semi-diameter corresponding to the mean distance, and $\delta =$ semi-diameter corresponding to any distance D: then

$$\delta : \Delta :: 1 : D; \text{ whence, } D = \frac{\Delta}{\delta} \ldots. (46);$$

an equation which, when Δ has been found, will make known the distance corresponding to any observed semi-diameter δ, in terms of the mean distance as a unit.

Now, to find Δ, denote the greatest and least semi-diameters, respectively, by δ', δ'', and the corresponding distances by D' and D'', and we have

$$D' = \frac{\Delta}{\delta'}, \; D'' = \frac{\Delta}{\delta''};$$

and thence,

$$\tfrac{1}{2}(D' + D'') = \tfrac{1}{2}\left(\frac{\Delta}{\delta'} + \frac{\Delta}{\delta''}\right), \text{ or, } 1 = \tfrac{1}{2}\left(\frac{\Delta}{\delta'} + \frac{\Delta}{\delta''}\right);$$

whence,

$$\Delta = \frac{2\,\delta'\,\delta''}{\delta' + \delta''} \ldots. (47).$$

CHAPTER XI.

INEQUALITIES OF THE MOTIONS OF THE PLANETS AND OF THE MOON; TABLES FOR FINDING THE PLACES OF THESE BODIES.

207. Gravitation. It is a general law of nature, discovered by Sir Isaac Newton, that bodies tend or *gravitate* towards each other, with a force directly proportional to their mass, and inversely proportional to the square of their distance. The force which causes one body to gravitate towards another, is supposed to arise from a mutual attraction existing between the particles of the two bodies, and is hence called the *Attraction of Gravitation.* This force of attraction, common to all the bodies of the Solar System, is the general physical cause of their motions. The sun's attraction retains the planets in their orbits, and the planets, by their mutual attractions, slightly alter each other's motions. The reasoning by which *Newton's Theory of Universal Gravitation* is established, appertains to Physical Astronomy, and will be presented in Part II.

208. Perturbations;—Inequalities. If a planet were acted on by no other force than the attraction of the sun, it is proved that its orbit would be accurately an ellipse, and the areas described by its radius-vector, in equal times, would be precisely equal. But it is in reality attracted by the other planets, as well as the sun, and therefore its actual motions cannot be in strict conformity with the laws of Kepler. In fact, if we descend to great accuracy, the agreement between the observed and computed places, noticed in Art. 205, is found not to be exact. The deviations from the elliptic motion, which are produced by the attractions of the planets, are called *Perturbations*, or, in Spherical Astronomy, *Inequalities.* Although, as we have just seen, the fact of the existence of inequalities in the motions of the planets is discoverable from observation, their laws cannot be determined without the aid of theory.

209. Disturbing Force. In treating of the perturbations in the motions of one planet, resulting from the attractions of another, the attracting planet is called the *Disturbing Body*, and the force which produces the perturbations the *Disturbing Force.* To find the disturbing force, let P (Fig. 61) be the planet, S the sun, and M the disturbing body; and let PD represent the attraction of M for the planet. Decompose PD into two forces,

PE and PF, one of which, PE, is equal and parallel to SG, the attraction of M for the sun; the other, PF, will be known in position and intensity. The two forces, PE and SG, being equal and parallel, they cannot alter the relative motion of the sun

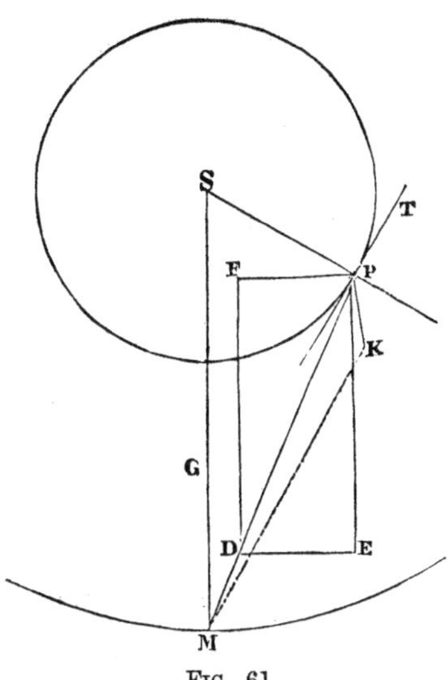

FIG. 61.

and planet, and accordingly may be left out of account: there remains, therefore, the component PF, which will be wholly effective in disturbing this motion. This, then, is the disturbing force.

It happens in the case of each planet, that the distances of some of the other planets are so great that their disturbing forces are insensible. The attractions of these bodies for the sun and planet, when they are exterior to the planet, are sensibly equal and parallel. Owing to the great distance of the planets from each other, and the smallness of their mass compared with that of the sun, the disturbing force is in every instance very minute in comparison with the sun's attraction.

210. Components of Disturbing Force;—their Effects. It is plain that the disturbing force will, in general, be obliquely inclined to the perpendicular to the plane of the orbit, PK, the tangent to the orbit, PT, and the radius-vector, PS; and may, therefore, be decomposed into forces acting along these lines. The component along the perpendicular will alter the latitude, and the two others both the longitude and radius-vector; that along the tangent by changing the velocity of the planet, and that along the radius-vector by changing the gravity towards the sun. It appears, therefore, that the disturbing force produces at the same time perturbations or *inequalities of longitude*, *of latitude*, and *of radius-vector*.

211. Determination of Inequalities. Let us now consider how these inequalities may be determined. In the first place, the inequalities produced by each disturbing body may be separately investigated upon mechanical principles, as if the other bodies did not exist; for the reason that the effect of each disturbing body is sensibly the same that it would be if the other bodies did not act. That this is very nearly, if not quite true, may be at once inferred from the minuteness of the whole disturbance produced by the joint action of all the disturbing forces of the system. The problem which has for its object the determination of the inequalities in the motions of one body, in its revolution around a second, produced by the attraction of a third, is called the *Problem of the Three Bodies.* If, in the case of any one planet, this problem be solved for each of the other bodies of the system which occasion sensible perturbations, all the inequalities to which the motion of the planet is subject will become known.

The general solution of the problem of the three bodies, that is, for any mass and distance of the disturbing body, or any intensity of the disturbing force, cannot be effected in the existing state of the mathematical sciences. But the problem has been solved for the case that presents itself in nature, in which the disturbing force is very minute in comparison with the central attraction. The results obtained by the analysis are certain analytical expressions for the perturbations in longitude, latitude, and radius-vector, involving variables and constants.

212. Equations of Specific Inequalities of Longitude. The general expression for the whole perturbation in longitude, due to the action of any one disturbing body, is of the form

$$C \sin A + C' \sin A' + C'' \sin A'', \text{ etc.},$$

in which C, C′, C″, etc., are constants, and A, A′, A″, etc., angles depending upon the positions of the disturbing and disturbed planets, with respect to each other and the sun, and also, in some cases, with respect to the nodes and perihelia of their orbits. Each of the terms, C sin A, C′ sin A′, etc., is technically called an *Equation*, and is considered as representing a specific inequality. The variable angle whose sine enters into the term is called the *Argument* of the inequality, and the constant is called the *Coefficient* of the inequality. As the greatest value of the sine of the argument is unity, the coefficient is equal to the greatest value of the inequality.

213. Calculation of Inequalities. The value of each argument may be derived for any assumed time, from the elliptic theory of the planetary motions; and the coefficients of all the inequalities may be calculated by making repeated determinations of the difference between the observed and computed longitude of the disturbed planet. By putting the entire expression, C sin A + C′ sin A′, etc., equal to each one of the differences of longitude so determined, we may form as many equations as there are unknown quantities, C, C′, etc., from which their values may be deduced.

The coefficient of any inequality being known, the value of the inequality, at any particular time, will become known if that of the argument be found. This value will be the correction for that inequality, to be applied to the elliptic place of the planet computed for the assumed time.

214. Inequalities of Latitude and Radius-vector. The theory of these inequalities, and of their computations, is similar to that of the inequalities of longitude just explained.

215. Inequalities are Periodic. We have seen that the arguments of the inequalities are angles depending on the configurations of the disturbing and disturbed planets with respect to each other and the sun, or with respect to the nodes or perihelia of their orbits. Whenever these configurations become the same, as they will periodically, the arguments, and therefore the inequalities themselves, will have the same value. It follows, therefore, that the inequalities in question are *periodic.*

The interval of time in which an inequality passes through all its gradations of positive and negative value, is called the *Period* of the inequality. It is manifestly equal to the interval of time employed by the argument in increasing from zero to 360°; for, in this interval sin A or cos A takes all its values, both positive and negative, and at the expiration of it recovers the same value again.

216. Inequalities of Elliptic Elements. It has been stated that the elements of the elliptic orbits of the planets are, for the most part, subject to a slow variation from century to century. Investigations in Physical Astronomy have established that the variations of the elements are due to the action of the disturbing forces of the planets, and that they are not progressive (except in the cases of the longitude of the node and the longitude of the perihelion), but are really periodic inequalities whose periods comprise many centuries. From the great lengths of their periods these inequalities are termed *Secular Inequalities,* in order to distinguish them from the inequalities of the elliptic motion, denominated *Periodic Inequalities,* the periods of which are comparatively short.

Physical Astronomy furnishes expressions called *Secular Equations,* which give the value of an element at any assumed time.

217. The Inequalities of the Moon's Motion arise from the disturbing action of the sun. The attractions of each of the planets for the moon and earth are sensibly equal and parallel.

The lunar inequalities are investigated upon the same principle as the planetary, and are represented by *equations* of the same general form, that is, consisting of a constant coefficient and the sine or cosine of a variable argument. They far exceed in number and magnitude those of any single planet.

There are three lunar inequalities of longitude which are prominent above the rest, and were early discovered by observation.

The most considerable is called the *Evection*, and was discovered by Ptolemy in the first century of the Christian era. It has for its argument double the angular distance of the moon from the sun minus the mean anomaly of the moon, and amounts when greatest to 1° 20′ 30″.

The second is called the *Variation*, and was discovered in the sixteenth century by Tycho Brahé. Its argument is double the angular distance of the moon from the sun, and its maximum value is 35′ 42″.

The third is denominated the *Annual Equation*, from the circumstance of its period being an anomalistic year. Its argument is the mean anomaly of the sun.

The discovery of the other lunar inequalities (with the exception of one inequality of latitude), is due to Physical Astronomy.

218. Calculation of Exact Heliocentric Place of a Planet. To present now at one view the entire process of calculating the co-ordinates of the exact heliocentric place of a planet, or of the geocentric place of the moon, at any assumed time,—

(1). Seek the elements of the elliptic orbit from a table of elements, such as Table II. or III., allowing for the proportional part of the secular variation; or (more exactly) obtain them from their secular equations (216).

(2). Compute the longitude, latitude, and radius-vector, by the elliptic theory (200, 201).

(3). Compute the values of the inequalities in longitude, latitude, and radius-vector, by means of their equations (212, 213, 214), and apply them individually, with their proper signs, as corrections to the elliptic values of the longitude, latitude, and radius-vector.

When the exact heliocentric place of a planet has been found, its geocentric place may be determined by the process referred to in Art. 202.

Geocentric Place of the Sun. The elements of the sun's apparent orbit are the same as those of the earth's actual orbit, except that the geocentric longitude of the perigee of the one exceeds the heliocentric longitude of the perihelion of the other by 180°. From these elements the longitude and radius-vector are obtained as in Art. 203. The values of the inequalities resulting from the earth's motion are then to be applied to these as corrections.

TABLES OF THE SUN, MOON, AND PLANETS.

219. The calculation of the co-ordinates of the place of the sun, moon, or any planet, for any assumed time, may be greatly facilitated by the use of tables. The principle and mode of construction of tables adapted to this purpose are explained in Part III. We will only remark here that the tables save the necessity of calculating the equations of the inequalities (218); since they make known their values corresponding to the values of the arguments at the time supposed. These values of the arguments are also readily obtained from tables especially designed for this purpose.

Tables of the sun, moon, and of each of the principal planets, have been calculated by different astronomers, and are now in general use.

220. Ephemeris. With the aid of these tables an ephemeris of each body is computed, and published for each year in advance, in the American and English Nautical Almanacs. An *Ephemeris* of a heavenly body is a collection of tables exhibiting the longitude, latitude, right ascension, declination, parallax, semi-diameter, etc., of the body, at stated periods of time, as at noon of each day throughout the year.

CHAPTER XII.

MOTIONS OF THE COMETS.

221. Apparent Motions. When first seen, a comet is ordinarily at some distance from the sun in the heavens, and moving towards it. After this, it continues to approach the sun, for a certain time, and then recedes to a greater or less distance, and finally disappears. In many instances comets have come so near the sun, as to be for a time lost in its beams. It has sometimes happened that a comet has not made its appearance in the firmament until after the time of its nearest apparent approach to the sun, and when it is receding from him in the heavens. This was the case with the great comet of 1843. It was first seen, in this country, in open day, on the 28th of February, in the immediate vicinity of the sun; and after this moved away from it, and, gradually diminishing in brightness, in about a month became invisible.

Comets resemble the planets in their changes of apparent place among the fixed stars, but they differ from them in never having been observed to perform an entire circuit of the heavens. Their apparent motions are also more irregular than those of the planets, and they are confined to no particular region of the heavens, but traverse indifferently every part.

222. Orbits of Comets. Sir Isaac Newton, from observations that had been made upon the remarkable comet of 1680, ascertained that this comet described a parabolic orbit, having the sun at its focus, or an elliptic orbit of so great an eccentricity as to be undistinguishable from a parabola, and that its radius-vector described equal areas in equal times. Since then, the orbits of 240 comets have been computed, and found to be, the majority of them, of a parabolic form, or sensibly so.

It was demonstrated by Newton, on the theory of gravitation, that a body projected into space may describe about the sun as a focus either one of the conic sections, and that the form of the orbit will depend upon the projectile velocity alone. With one particular velocity the orbit will be a parabola; with any less velocity it will be an ellipse or circle; and with any greater velocity it will be an hyperbola. Now, as there is but one velocity from which a parabolic orbit will result, and as any comet, which may have originally moved in a hyperbola, must have passed its perihelion, and receded beyond the limits of the solar system, it may be inferred, with great probability, that the orbits of the comets whose observed courses are not distinguishable from parabolic arcs, are in fact ellipses of great eccentricity. This is the theory of the cometary motions proposed by Newton.

The orbits of some of the comets are known from observation to be very eccentric ellipses.

223. Elements of Parabolic Orbit. The elements of the parabolic path conceived to be traced by a comet during the period in which it remains visible, are: the longitude of the ascending node, the inclination of the orbit, the longitude of the perihelion, and the epoch of the perihelion passage. Assuming that the radius-vector describes areas proportional to the times, these elements may be computed from three observed geocentric places. But the problem is one of considerable difficulty.

224. Entire Elliptic Orbits.—Periods of Revolution. Astronomers do not in general seek to deduce, from the observations made during one appearance of a comet, its entire elliptic orbit. It is impossible, from such observations, to compute the major-axis of its orbit and its period with any accuracy, inasmuch as in the interval during which they are made, the comet describes but a small portion of its entire orbit. As examples of the uncertainty of such determinations, four periods have been found by Bessel for the comet of 1807, of which the least is 1,483 years and the greatest 1,952 years; and for the great comet of 1811 the two periods, 2,301 years and 3,056 years, have been computed. The uncertainty becomes much less when the period of revolution is short.

The only mode of obtaining the period of a comet's revolution with certainty is by directly comparing the times of its successive perihelion passages. A comet cannot be recognized at a second appearance by its aspect; for this is liable to great alterations. But it may be identified by means of the elements of its parabolic orbit (223), as it is extremely improbable that the elements of the orbits of two different comets will agree throughout. This method of identifying a comet may sometimes fail of application, inasmuch as the orbit of a comet may experience great alterations from the attractions of the planets.

225. Comets of Known Period. Owing to the great lengths of the periods of revolution of most of the comets, and the comparatively short intervals of time during which their motions have been carefully observed, there are but eight comets whose periods and entire orbits have been determined with certainty. These have all reappeared, and in some instances repeatedly, and verified the determinations of their paths through space, and the predictions of their return to their perihelia. A comet usually receives the name of the astronomer who first determines its orbit and period of revolution. The comets just alluded to are designated as *Halley's*, *Encke's*, *Biela's*, *Faye's*, *De Vico's*, *Brorson's*, *D'Arrest's*, and *Winnecke's*. The last seven are known as *Comets of Short Period;* their periodic times being comprised within the limits of 3.3 years and 7½ years. Their mean distances from the sun are less than that of Jupiter, and they revolve within the orbit of Saturn. Halley's comet, in its recess from the sun, passes beyond the limits of the solar system, and its period approximates to that of Uranus. Fig. 62 shows the relative dimensions and positions of the orbits of Halley's, Encke's, and Biela's comets.

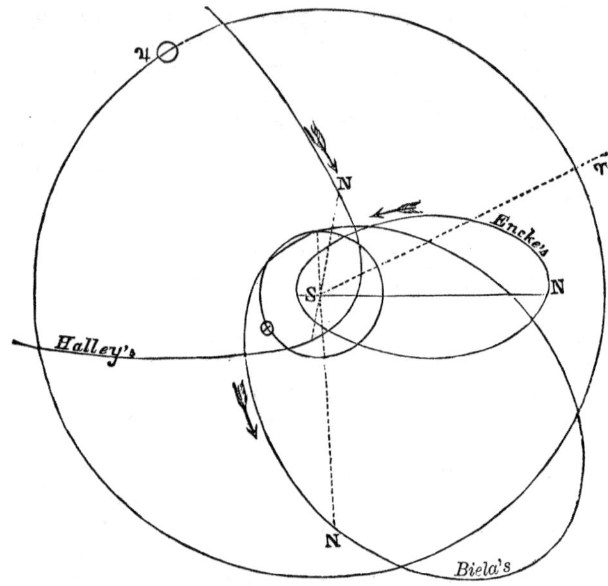

FIG. 62.

226. Comets whose Periods have been Approximately Calculated. There are a number of cometary bodies whose periods of revolution and elliptic orbits have been approximately deduced, by calculation, from observations made at the periods of their first discovery, but which have not since been seen. Five of these belong to the class of comets of comparatively short period, and small mean distance from the sun; their computed periods being from five to seven years. Two have periods of 10 years and 16 years, respectively. Five form, with Halley's comet, a distinct class; their periodic times are all about 75 years, and their mean distances from the sun nearly equal to that of Uranus.

There are also more than twenty comets whose entire elliptic orbits are believed to have been ascertained with a certain degree of approximation to the truth. Their mean distances exceed the limits of the solar system, and their periods are much longer than that of the most distant planet. The same is known to be true of the mean distances and periods of all the remaining comets that have been carefully observed.

227. All the Comets of Comparatively Short Period (viz., from 3.3 years to 16 years) revolve around the sun in the same direction as the planets, and like the planetoids, in planes inclined less than 35° to the plane of the ecliptic. But their orbits are much more eccentric than the orbits of the minor planets. They form a group of bodies whose orbits bear a striking resemblance to each other, and occupy a position, in respect to their orbital motions, intermediate between the planetoids and the comets of long period (75 years and more). They are comparatively faint objects, and have generally been visible only with the aid of a telescope. All the other comets, whose mean distance from the sun does not exceed that of the most distant planet, with the exception of Halley's, also have a direct motion. Some of these, on their return to their perihelia, have become visible to the naked eye; Halley's comet conspicuously so.

228. Comets of Long Period. Of 220 observed comets, whose mean distances from the sun exceed that of Neptune, about an equal number have a direct and a retrograde motion. The perihelia of more than two-thirds of the orbits fall within the orbit of the earth. The aphelia lie far beyond the orbit of Neptune. There is little reason to doubt that many comets recede tens of thousands of millions of miles before they begin to return to the sun again; and that the periods of most of them include a number of centuries, and of many of them even tens of centuries. The planes of their orbits are inclined under every variety of angle to the plane of the ecliptic.

229. Comets of Small Perihelion Distance. Some comets come into close proximity to the sun. The great comet of 1680, according to the computation of Newton, came 166 times nearer the sun than the earth is. The no less remarkable comet of 1843 approached still nearer; when at its perihelion, it was less than 70,000 miles from the sun's surface. Its orbital velocity at that time was 350 miles per second; and it accomplished a semi-revolution around the sun (from n to n', Fig. 63) in the astonishingly short interval of 2 hours.

230. Number of Comets. The number of recorded appearances of comets is about 800, but the actual number of cometary bodies connected with the solar system is undoubtedly far greater than this. This list of recorded appearances comprises, for the great number of years which precede the date of the invention of the telescope, only those comets which were very conspicuous to the naked eye; giving, for example, only three in the thirteenth, and three in the fourteenth century; and, since the heavens have been attentively examined with telescopes, from two to three comets, on an average, have made their appearance every year, of which the great majority were telescopic. The periods of these, as well as of the others, are in general of such vast length that probably not more than half the whole number of comets have returned twice to their perihelia during the last two thousand years. From these considerations it appears, that, had the heavens been attentively surveyed with the telescope during the last two thousand years, as many as 2,500 different cometary bodies would have been seen. But, as there are various causes which may tend to prevent a comet from being seen when present in our firmament,—as continued proximity to the sun in the heavens, too great distance from the sun and earth, want of intrinsic lustre, etc.,—it is highly probable that there are, in fact, many thousands of these bodies.

HALLEY'S COMET.

231. Halley's comet is so called from Sir Edmund Halley, Second Astronomer Royal of England, who ascertained its period, and correctly predicted its return. From a comparison of the elements of the orbits described by the comets of 1531, 1607, and 1682, he concluded that the same comet had made its appearance in these several years, and predicted that it would again return to its perihelion towards the end of 1758 or the beginning of 1759. Previous to its appearance, Clairaut, a distinguished French astronomer, undertook the arduous task of calculating its perturbations from the disturbing actions of the planets during this and the preceding revolution. He found, that, from this cause, it would be retarded about 618 days,—100 days from the effect of Saturn, and 518 days from the action of Jupiter,—and predicted that it would reach its perihelion within a month, one way or the other, of the middle of April, 1759. It actually passed its perihelion on the 12th of March, 1759. Assuming the earth's mean distance from the sun to be unity, the perihelion distance of this comet is 0.6, and aphelion distance 35.4. Accordingly it approaches the sun to within about one-half the distance of the earth, and recedes from him to nearly twice the distance of Uranus. (See Fig. 62.) Its period is about 76 years, but is liable to a variation of a year or more from the effect of the attractions of the planets. The inclination of its orbit is 18°, and its motion is retrograde. The last perihelion passage took place on the 16th of November, 1835, within a few days of the predicted time. The next will occur in the year 1911. It is to be expected that the perturbations will now be determined with such increased accuracy that the error in the prediction of its next perihelion passage will be less than one day.

Probable repeated appearances of this comet have been traced as far back as the year 11 B. C. It seems to have been particularly conspicuous in the years 1066 and 1456.

ENCKE'S COMET.

232. This comet is remarkable for its short period of revolution, which is only 3.3 years. It moves in an orbit inclined only 13° to the plane of the ecliptic, and whose perihelion is at the distance from the sun of the planet Mercury, and aphelion at a distance somewhat less than that of Jupiter (see Fig. 62). Its period and elliptic orbit were determined on the occasion of its fourth recorded appearance, by Professor Encke, of Berlin. Since then it has returned a number of times to its perihelion, and in every instance very nearly as predicted. At some of its returns it has become visible to the naked eye. Its last return took place in 1865; the next will be in September, 1868.

233. Disturbing Effects of a Resisting Medium. The motions of this comet present the anomalous fact in the solar system of a period continually diminishing, and an orbit slowly contracting, from the operation of some other cause than the disturbing actions of the other bodies of the system. Professor Encke found that after allowance had been made for all the perturbations produced by the planets, the actual time of each perihelion passage anticipated the time calculated from the duration of the previous revolution about 2⅔ hours; and that the comet now arrives at its perihelion about 2⅔ days sooner than it would if the period had remained unaltered since the comet was first seen in 1786. This continual acceleration of the time of the perihelion passage, discovered by Encke, could not be attributed to the disturbing attraction of some unknown body, because this attraction would produce other effects, which have not been noticed. He conceived that it could arise from no other cause than the action of a resisting medium, or *ether* in space. The immediate effect of such a medium subsisting in the regions of space traversed by the comet, would be to diminish the velocity in the orbit, which it would at first seem should delay the time of the perihelion passage; but the velocity being diminished, the centrifugal force is weakened, and consequently the comet is drawn nearer to the sun, and moves in an orbit lying within the orbit due to the sun's attraction alone; its mean distance is therefore diminished, and its period shortened. A similar phenomenon to this is presented in the oscillations of a pendulum freely suspended. It is well known that the arc of vibration of the pendulum shortens, and consequently its rapidity of oscillation increases, under the influence of the resistance of the air.

BIELA'S COMET.

234. In February, 1826, M. Biela, of Josephstadt, in Bohemia, detected a telescopic comet in the constellation Aries; and subsequently made repeated observations upon its varying position in the heavens. From the results of his observations, he calculated the elements of its supposed parabolic orbit, and found on inspecting a catalogue of comets that the computed elements bore a striking resemblance to those of the comets of 1772 and 1805. He also ascertained that the entire observed path of the comet could not be accurately represented by a parabolic orbit, and proceeded to compute from his observations the elements of an elliptic orbit. He found the period of revolution to be 6.7 years, and that it accorded with the supposition that the same comet had been previously seen in 1772 and 1805. The period, as since more accurately determined, is 6.6 years. Its orbit is inclined $12\frac{1}{2}°$ to the plane of the ecliptic; and the perihelion lies just within the orbit of the earth, while the aphelion falls beyond the orbit of Jupiter (Fig. 62). By a remarkable coincidence, the orbit of this comet very nearly intersects the orbit of the earth. At the return of the comet in 1832, Dr. Olbers found that in going through its descending node it would pass within 20,000 miles of the earth's orbit, on the inside, and that a portion of the orbit would fall within the filmy mass of the comet. The earth was more than 60,000,000 miles distant from the comet at the time of the nodal passage, and did not reach the point of nearest approach of the two orbits until one month after the comet had passed by it. In 1805 the same comet passed within 6,000,000 miles of the earth.

According to calculation, the last return of Biela's comet to its perihelion took place in February, 1866; but the comet escaped detection. The next return will be in September, 1872.

FAYE'S COMET.

235. This comet was discovered and its orbit determined by M. Faye, of the Paris Observatory. Its period of revolution is $7\frac{1}{2}$ years. The eccentricity of its orbit (0.556) is less than that of any other known cometary body, although nearly twice as great as that of the most eccentric planetary orbit.

The return of this comet to its perihelion appears to be accelerated, like that of Encke's comet, and in a much greater degree, by the operation of a resisting medium in space. As the perihelion distance of this comet is much greater than that of Encke's, it seems probable that the resistance encountered by these comets is due to a collision with meteoric bodies, or some other form of cosmical matter.

The remaining comets of short period need not be specially noticed.

LEXELL'S COMET OF 1770.

236. It has already been intimated that the motions of the comets are liable to great derangements, from the operation of the attractive forces of the planets. This results from the elongated form of the cometary orbits, in consequence of which the comets, while pursuing their course within the limits of the planetary system, may come into proximity to the planets, and be strongly attracted by them. Halley's comet has already furnished an illustration of this general fact. Lexell's comet offers a still more striking example of the disturbances to which the cometary motions are exposed. From observations made upon this comet in the year 1770, Lexell made out that its period was 5½ years; still, though a very bright comet, it has not since been seen. Burckhardt, an eminent French calculator, undertook to investigate the cause of this phenomenon, and found that on its return to the perihelion in 1776, the comet was so situated with regard to the earth and sun as to be continually hid by the sun's rays; and that in 1779, before its next return, it passed so near the planet Jupiter, that his attraction was very many times greater than the attraction of the sun. The consequence was that its orbit was greatly enlarged, so that it no longer comes near enough to the earth to be visible.

Another fact to be accounted for was, that the comet had not been seen previous to the year 1770. In seeking for its explanation it was discovered, by tracing back the orbit of the comet, that in 1767 it must have passed near Jupiter, and that the action of his attractive force must have altered its orbit from one of large dimensions to the comparatively small orbit, with short period, of the comet as seen in 1770. While describing, previous to 1767, an orbit with a large perihelion distance, it could not have come near enough to the earth and sun to be visible.

This comet is also remarkable as having made a nearer approach to the earth than any other on record. On July 1, 1770, its distance from the earth was less than 1,500,000 miles.

THE GREAT COMET OF 1843.

237. This comet has already been alluded to as remarkable for having made a nearer approach to the sun than any other comet. Its parabolic path is represented in Fig. 63. The positions of the

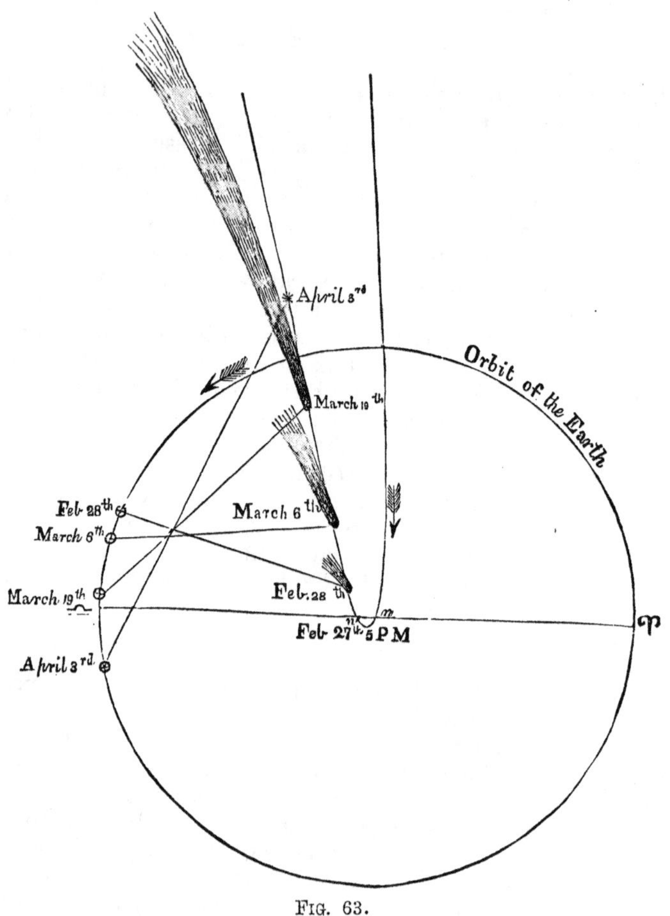

FIG. 63.

comet at several different dates, with the corresponding positions of the earth, are also indicated; n is the ascending and n' the descending node. The perihelion is within 500,000 miles of the sun's centre, and nearly midway between n and n'. The inclination of the orbit is 36°. The comet passed its perihelion on February 27, at about 5 P.M. (Philadelphia time). On the

28th it was observed in full daylight in various parts of New England, in Mexico, at several places in Italy, and off the Cape of Good Hope. It was then about 3° distant from the sun, and of a dazzling brightness. Its great lustre at that time doubtless resulted in part from its tail being foreshortened by the obliquity under which it was seen. After the 28th it showed itself with great distinctness early in the evening, over the western horizon; and though growing fainter from night to night, as it receded from the sun, continued visible to the naked eye until about the 3d of April.

This comet is believed to move in an elliptic orbit answering to a period of 175 years.

DONATI'S COMET.

238. This is the great comet that made its appearance in 1858. It was first seen by Donati at Florence, on the 2d of June, 1858. It was then but a faint nebulosity, discernible only with a telescope. Although becoming more distinct in the field of the telescope from week to week, it did not become visible to the naked eye until near the 1st of September. It attained to its greatest size and splendor after the perihelion passage on September 30, after which it decreased in brightness as it receded from the sun and earth, moved off rapidly towards the south, and finally disappeared from view in March, 1859, in

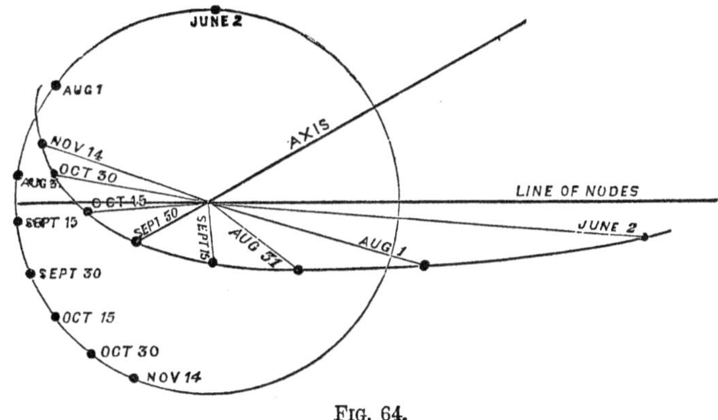

FIG. 64.

the southern heavens. Fig. 64 represents a portion of the orbit of the comet, as projected on the plane of the earth's orbit, and several corresponding positions of the comet and earth. The plane of the orbit is inclined to that of the earth's orbit under an angle of 63°, the portion of the orbit containing the perihelion lying on the north side of the plane of the earth's orbit. When first seen, on June 2, the comet was about 240,000,000 miles from the earth. At the perihelion (September 30) the distance was less than 70,000,000 miles. It was at its least distance from the earth (nearly 52,000,000 miles) on October 10, but attained its greatest brilliancy five days earlier.

The period of revolution of Donati's comet has not been determined; but it is estimated to exceed 1,600 years.

CONSPICUOUS COMETS OF THE PRESENT CENTURY.

239. These are, in addition to Donati's comet, and the great comet of 1843, the great comet of 1811, the bright comets of 1819, 1825, and 1835 (Halley's comet), and the great comet of 1861. The comet of 1811 affords an instance of a large and bright comet, with a perihelion distance exceeding the earth's distance from the sun.

CHAPTER XIII.

Motions of the Satellites.

240. As before stated, the planets which have satellites are Jupiter, Saturn, Uranus, and Neptune. The number of Jupiter's satellites is four, of Saturn's eight, of Uranus' eight, of Neptune's one.

241. The Satellites of Jupiter are perceptible with a telescope of very low power. It is found, by repeated observations, that they are continually changing their positions with respect to one another and the planet; being sometimes all to the right of the planet, and sometimes all to the left of it, but more frequently some on each side. They are distinguished from each other by the distance to which they recede from the planet; that which recedes to the least distance being called the *First Satellite*, that which recedes to the next greater distance the *Second*, and so on.

The satellites of Jupiter were discovered by Galileo, in the year 1610.

The Satellites of Saturn, Uranus, and Neptune cannot be seen, except through excellent telescopes. They experience changes of apparent position, similar to those of Jupiter's satellites.

242. The Satellites Revolve around the Planet. The apparent motion of Jupiter's satellites alternately from one side to the other of the planet, leads to the supposition that they actually revolve around the planet. This inference is confirmed by other phenomena. While a satellite is passing from the eastern to the western side of the planet, a small dark spot is frequently seen crossing the disc of the planet in the same direction; and again, while the satellite is passing from the western to the eastern side, it often disappears, and, after remaining for a time invisible, reappears at another place. These phenomena are easily explained, if we suppose that the planet and its satellites are opake bodies illuminated by the sun, and that the satellites revolve around the planet from west to east. On this hypothesis, the dark spot seen traversing the disc of the planet is the shadow cast upon it by the satellite on passing between the planet and the sun; and the disappearance of the satellite is an *eclipse*, occasioned by its entering the shadow of the planet.

As the transit of the shadow occurs during the passage of the satellite from the eastern to the western side of the planet, and the eclipse of the satellite during its passage from the western to the eastern side, the direction of the motion must be from west to east.

Analogous conclusions may be drawn from similar phenomena exhibited by the satellites of Saturn. The satellites of Uranus also revolve around their primary; but the direction of their motion, as referred to the ecliptic, is from east to west. The satellite of Neptune revolves around the planet from west to east.

243. Eclipses.—Transits of Shadows. Let us now examine into the principal circumstances of the eclipses of Jupiter's satellites, and of the transits of their shadows across the disc of the primary. Let EE′E″ (Fig. 65) represent the orbit of the

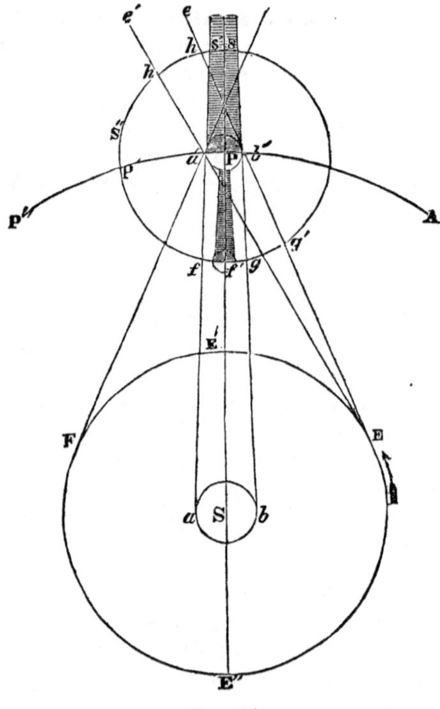

FIG. 65.

earth, PP′P″ the orbit of Jupiter, and $ss's''$ that of one of its satellites, supposed to lie in the plane of Jupiter's orbit. Suppose that E is the position of the earth, and P that of the planet, and conceive two lines, aa', bb', to be drawn tangent to the sun and planet: then, while the satellite is moving from s to s' it will be eclipsed; and, while it is moving from f to f', its shadow will fall upon the planet. Again, if Ee, Ee' represent two lines drawn from the earth tangent to the planet on either side, the satellite will, while moving from g to g', traverse the disc of the planet, and, while moving from h to h', be behind the planet, and thus concealed from view. It will be seen on an inspection of the figure, that, during the motion of the earth from E$''$, the position of heliocentric opposition, to E$'$ that of conjunction, the disappearances or *immersions* of the satellite will take place on the western side of the planet; and that the *emersions*, if visible at all, can be so only when the earth is so far from opposition and conjunction that the line Es', drawn from the earth to the point of emersion, will lie to the west of Ee. It will also be seen, that, during the passage of the earth from E$'$ to E$''$ the emersions will take place on the eastern side of the planet, and that the immersions cannot be visible, unless the line Fs, drawn from the earth to the point of immersion, passes to the east of the planet. It appears from observation that the immersion and emersion are never both visible at the same period, except in the case of the third and fourth satellites.

If the orbits of the satellites lay in the plane of Jupiter's orbit an eclipse of each satellite would occur every revolution, but, in point of fact, they are somewhat inclined to this plane, from which cause the fourth satellite sometimes escapes an eclipse.

244. Periods.—Mean Motions.—Mean Distances. The periods and other particulars of the motions of the satellites, result from observations upon their eclipses. The middle point of time between the instants when the satellite enters and emerges from the shadow of the primary, is the time when the satellite is in the direction, or nearly so, of a line joining the centres of the sun and primary. If the latter continued stationary, then the interval between this and the succeeding central eclipse would be the periodic time of the satellite. But, the primary planet moving in its orbit, the interval between two successive eclipses is a synodic revolution. The synodic revolution, however, being observed, and the period of the primary being known, the periodic time of the satellite may be computed.

The mean motions of the satellites differ but little from their true motions; and hence the forms of their orbits must be nearly circular. The orbit, however, of the third satellite of Jupiter has a small eccentricity; that of the fourth, a larger.

The distances of the satellites from their primary, are determined from micrometrical measurements of their apparent distances at the times of their greatest elongations.

A comparison of the mean distances of Jupiter's satellites with their periodic times proves that Kepler's third law with respect to the planets applies also to these bodies; or, that the squares of their sidereal revolutions are as the cubes of their mean distances from the primary.

The same law also has place with the satellites of Saturn and Uranus.

245. The Computation of the Place of a Satellite for a given time, is effected upon similar principles with that of the place of a planet. The mutual attractions of Jupiter's satellites occasion sensible perturbations of their motions, of which account must be taken when it is desired to determine their places with accuracy.

246. Relations of Mean Motion and Position. Laplace has shown from the theory of gravitation, that, by reason of the mutual attractions of the first three of Jupiter's satellites, their mean motions and mean longitudes are permanently connected by the following remarkable relations.

(1.) *The mean motion of the first satellite, plus twice that of the third, is equal to three times that of the second.*

(2.) *The mean longitude of the first satellite, plus twice that of the third, minus three times that of the second, is equal to* 180°.

It follows, from this last relation, that the longitudes of the three satellites can never be the same at the same time, and consequently that they can never be all eclipsed at once.

CHAPTER XIV.

THE SUN, AND THE PHENOMENA ATTENDING ITS APPARENT MOTIONS.

INEQUALITY OF DAYS.*

247. Sun's Motion relative to the Equator. We will first give a detailed description of the sun's apparent motion with respect to the equator, the phenomenon upon which the inequality of days (as well as the change of seasons, soon to be treated of) immediately depends.

Let VEAQ (Fig. 66) represent the equator; VTAW (inclined to VEAQ, under the angle TOE, measured by the arc TE, equal to $23\frac{1}{2}°$), the ecliptic; T*n*X and W*n*′X′, the two tropics; POP′, the axis of the heavens; and PEP′Q the meridian, and HVRA the horizon, in one of their various positions with respect to the other circles. About the 21st of March the sun is in the vernal equinox V, crossing the equator in the oblique direction VS, towards the north and east. At this time its diurnal circle is identical with the equator; and it crosses the meridian at the point E, south of the zenith a distance ZE equal to the latitude of the place. Advancing towards the east and north, it takes up the successive positions S, S′, S″, etc., and from day to day crosses the meridian at r, r', etc., farther and farther to the north. Its diurnal circles will be, respectively, the northern parallels of declination passing through S, S′, S″, etc., and continually more and more distant from the equator. The distance of the sun, and of its diurnal circle from the equator, continues to increase until about the 21st of June, when he reaches the summer solstice T. At this point he moves for a short time parallel to the equator; his declination changes but slightly for several days, and he crosses the meridian from day

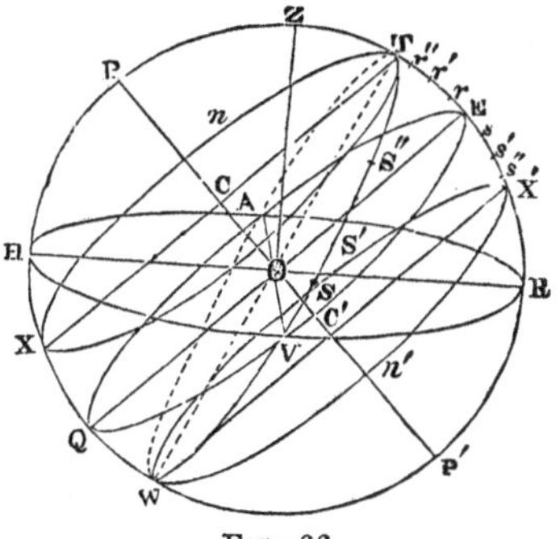

FIG. 66.

* The day here considered is the interval between sunrise and sunset.

to day at nearly the same place. It is on this account,—viz., because the sun seems to stand still for a time with respect to the equator, when at the point 90° distant from the equinox,—that this point has received the name of solstice.* The diurnal circle described by the sun is now identical with the tropic of Cancer, TnX; which circle is so called because it passes through T the beginning of the sign Cancer, and when the sun reaches it he is at his northern goal, and *turns about* and goes towards the south.† The sun is, also, when at the summer solstice, at its point of nearest approach to the zenith of every place whose latitude ZE exceeds the obliquity of the ecliptic TE, equal to 23½°. The distance ZT = ZE — ET = latitude — obliquity of ecliptic. During the three months following the 21st of June, the sun moves over the arc TA, crossing the meridian from day to day at the successive points r'', r', etc., farther and farther to the south, and arrives at the autumnal equinox A about the 23d of September, when its diurnal circle again becomes identical with the equator. It crosses the equator obliquely towards the east and south, and during the next six months has the same motion on the south of the equator, that it has had during the previous six months on the north of the equator. It employs three months in passing over the arc AW, during which period it crosses the meridian each day at a point farther to the south than on the preceding day. At the winter solstice, which occurs about the 22d of December, it is again moving parallel to the equator, and its diurnal circle is the same circle as the tropic of Capricorn. In three months more it passes over the arc WV, crossing the meridian at the points s'', s', etc.; so that on the 21st of March it is again at the vernal equinox.

248. Explanation of Inequality of Days. The phenomenon of the inequality of days obtains at all places on the earth situated north or south of the equator. At all such places, the observer is in an oblique sphere; that is, the celestial equator and the parallels of declination are oblique to the horizon. This position of the sphere is represented in Fig. 11, p. 22, where HOR is the horizon, QOE the equator, and *ncr*, *sct*, etc., parallels of declination; WOT is the ecliptic. It is also represented in Fig. 66, from which Fig. 11 differs chiefly in this, that the horizon, equator, ecliptic, and parallels of declination, which are represented as ellipses in Fig. 66, are in Fig. 11 projected into right lines upon the plane of the meridian. Since the centres of the parallels of declination are situated upon the axis of the heavens, which is inclined to the horizon, it is plain that these parallels, as it is represented in the Figs., and as we have before seen (25), will be divided into unequal parts, and that the disparity between the parts will be greater in proportion as the parallel is more distant from the equator; also, that to the north

* From *Sol*, the sun, and *sto*, to stand. † From τρεπω, to turn.

of the equator the greater parts will lie above the horizon, and to the south of the equator below the horizon. Now, the length of the day is measured by the portion of the parallel to the equator, described by the sun, which lies above the horizon; and it is evident, from what has just been stated, that (as it is shown by the Fig.) this increases continually from the winter solstice W to the summer solstice T, and diminishes continually from the summer solstice T to the winter solstice W; whence it appears that the day will increase in length from the winter to the summer solstice, and diminish in length from the summer to the winter solstice.

249. Length of Day. As the equator is bisected by the horizon at the equinoxes, the day and night must be each twelve hours long. But, when the sun is north of the equator, the greater part of its diurnal circle lies above the horizon, in northern latitudes; and therefore, from the vernal to the autumnal equinox, the day is, in the northern hemisphere, more than 12 hours in length. On the other hand, when the sun is south of the equator, the greater part of its circle lies below the horizon, and hence from the autumnal to the vernal equinox the day is less than 12 hours in length.

In the latter interval, the nights will obviously, at corresponding periods, be of the same length as the days in the former.

250. Effects of Increase of Latitude. The variation in the length of the day, in the course of the year, will increase with the latitude of the place; for the greater is the latitude the more oblique are the circles described by the sun to the horizon, and the greater is the disparity between the parts into which they are divided by the horizon. This will be obvious, on referring to Fig. 11, p. 22, where HOR, H′OR′, represent the positions of the horizons of two different places with respect to these circles; H′OR′ being the horizon for which the latitude, or the altitude of the pole, is the least.

For the same reason, the days will be the longer as we proceed from the equator northward, during the period that the sun is north of the equinoctial, and the shorter, during the period that he is south of this circle.

251. Longest Day. *At the equator*, the horizon bisects all the diurnal circles (26); and, consequently, the day and night are there each 12 hours in length throughout the year.

At the arctic circle the day will be 24 hours long at the time of the summer solstice; for the polar distance of the sun will then be $66\frac{1}{2}°$, which is the same as the latitude of the arctic circle; whence it follows, that the diurnal circle of the sun, at this epoch, will correspond to the circle of perpetual apparition for the parallel in question.

On the other hand, when the sun is at the winter solstice, the night will be 24 hours long on the arctic circle.

To the north of the arctic circle, the sun will remain continually above the horizon during the period, before and after the summer solstice, that his north polar distance is less than the latitude of the place, and continually below the horizon during the period, about the winter solstice, that his south polar distance is less than the latitude of the place.

At the north pole, as the horizon is coincident with the equator (27), the sun will be above the horizon while passing from the vernal to the autumnal equinox, and below it while passing from the autumnal to the vernal equinox. Accordingly, at this locality there will be but one day and one night in the course of a year, and each will be of six months' duration.

252. In the Southern Hemisphere, the circumstances of the duration of light and darkness are obviously the same as in the northern, for corresponding latitudes and corresponding declinations of the sun.

253. Problem I. *The latitude of the place and the declination of the sun being given, to find the times of the sun's rising and setting and the length of the day.*

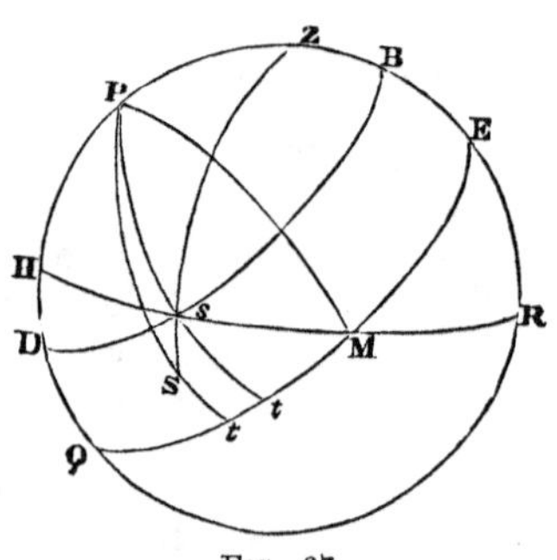

FIG. 67.

Let HPR (Fig. 67) be the meridian, HMR the horizon, and B*s*D the diurnal circle described by the sun. The hour angle EPt, or its measure Et, which, converted into time, expresses the interval between the rising or setting of the sun and his passage over the meridian, is called the *Semi-diurnal Arc.* Now,

$$Et = EM + Mt = 90° + Mt,$$

which gives

$$\cos Et = -\sin Mt;$$

and we have, by Napier's first rule,

$$\sin Mt = \cot tMs \tan ts = \tan PMH \tan EB = \tan PH \tan EB:$$

whence, $\cos Et = -\tan PH \tan EB,$

or, cos (semi-diurnal arc) = — tan lat. × tan dec. . . . (48).

The semi-diurnal arc (in time) expresses the apparent time of the sun's setting, and, subtracted from 12 hours, gives the apparent time of its rising. The double of it will be the length of the day.

In resolving this problem it will, in practice, generally answer to make use of the declination of the sun at noon of the given day, which may be taken from an ephemeris.

Exam. 1. Let it be required to find the apparent times of the sun's rising and setting, and the length of the day at New York, at the summer solstice.

Log. tan lat. (40° 42′ 40″)	9.93474 —
Log. tan dec. (23° 27′ 24″)	9.63740
Log. cos (semi-diurnal arc)	9.57214 —
Semi-diurnal arc	111° 55′ 26″
Time of sun's setting	7h. 27m. 42s.
Time of sun's rising	4 32 18
Length of day	14 55 24

Exam. 2. What are the lengths of the longest and shortest days at Boston; the latitude of that place being 42° 21′ 15″ N? Ans. 15h. 6m. 25s., and 8h. 53m. 35s.

Exam. 3. At what hours (apparent time) did the sun rise and set on May 1, 1866, at Charleston; the latitude of Charleston being 32° 47′, and the declination of the sun being 15° 9′ 30″ N? Ans. Time of rising, 5h. 19m. 48s.; time of setting, 6h. 40m. 12s.

254. Problem II. *To find the time of the sun's apparent rising or setting, the latitude of the place and the declination of the sun being given.*

At the time of apparent rising or setting, the sun, as seen from the centre of the earth, will be below the horizon a distance sS (Fig. 67) equal to the refraction minus the parallax. The mean difference of these quantities is 34′ 45″ (according to Bessel). Let it be denoted by R. Now, to find the hour angle ZPS (= P), the triangle ZPS gives (see Appendix),

$$k = \frac{\text{ZP} + \text{PS} + \text{ZS}}{2} = \frac{\text{co-lat.} + \text{co-dec.} + (90° + \text{R})}{2} \ldots (49)$$

and
$$\sin^2 \tfrac{1}{2}\text{P} = \frac{\sin(k - \text{ZP}) \sin(k - \text{PS})}{\sin \text{ZP} \sin \text{PS}},$$

or,
$$\sin^2 \tfrac{1}{2}\text{P} = \frac{\sin(k - \text{co-lat.}) \sin(k - \text{co-dec.})}{\sin(\text{co-lat.}) \sin(\text{co-dec.})} \ldots (50).$$

The value of P, in time, will be the interval between apparent noon and the time of the apparent rising or setting of the centre of the sun's disc; from which the apparent times of the apparent rising and setting are readily obtained. To obtain the mean times, these results must be corrected for the equation of time.

If the time of the rising or setting of the upper limb of the sun, instead of its centre, be required, we must take for R 34′ 45″ + sun's semi-diameter, or 50′ 47″.

Unless very accurate results are desired, it will be sufficient to take the declinations of the sun at 6 o'clock in the morning and evening. A more accurate calculation may be made by first computing the times of true rising and setting from equation (48), and making use of the declinations answering to these times.

TWILIGHT.

255. Explanation. When the sun has descended below the horizon, its rays still continue to fall upon a certain portion of the body of air that lies above it, and are thence radiantly reflected down to the earth, so as to occasion a certain degree of light; which gradually diminishes as the sun descends farther below the horizon, and the portion of air posited above the horizon, that is directly illuminated, becomes less. The same effect, though in a reverse order, takes place in the morning, previous to the sun's rising. The light thus produced is called the *Crepusculum* or *Twilight.* The explanation of twilight will be better understood on examining Fig. 68, where AON represents a portion of the earth's surface, H*k*R the surface of the atmosphere above it, and *km*S a line drawn touching the earth and passing through the sun. The unshaded portion, *kc*R, of the body of air which lies above the plane of the horizon, HOR, is still illuminated by the sun, and shines down, by reflection, upon the station of the observer at O. As the sun descends, this will decrease, until finally, when the sun is in the direction RNS′, it will illuminate directly none of that part of the atmosphere which lies above the horizon, and twilight will be theoretically at an end.

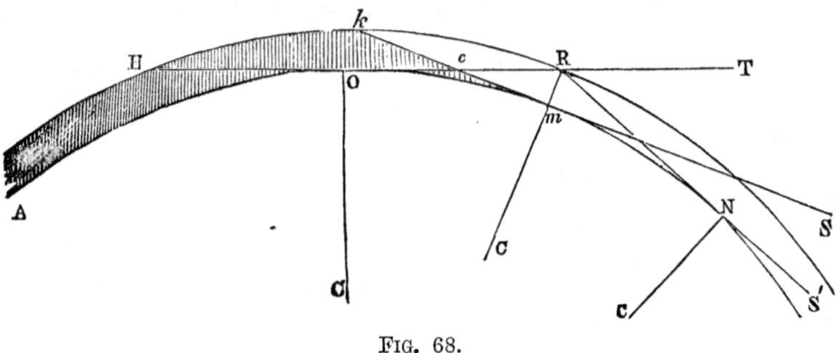

FIG. 68.

It is assumed that, when the sun has reached this position, in which no portion of air that lies above the horizon is directly illuminated, faint stars will become visible over the western horizon; and thus that the end of evening twilight is definitely marked by the appearance of such stars. In like manner, morning twilight is astronomically defined as beginning when faint stars situated in the vicinity of the eastern horizon begin to disappear. It has been ascertained from numerous observations that, at the beginning of the morning and end of the evening twilight, as thus defined, the sun is about 18° below the horizon.

256. Approixmate Determination of Height of Atmosphere. As we have just seen, at the end of evening twilight, the angle TRS′ (Fig. 68) is equal to 18°; H*k*R being the limit of that portion of the atmosphere which is capable of reflecting a sensible amount of light to the eye, in the direction RO. Now, if the vertical lines at O, *m*, and N, be produced to the centre of the earth, C, we shall have the angle OCN equal to TRS′, or 18°, and therefore OCR equal to 9°. If, then, we denote the radius of the earth C*m* by R, we shall have,

$$\text{height of atmos.} = mR = CR - Cm = R\sec 9^\circ - R = R(\sec 9^\circ - 1).$$

Making the calculation, we obtain for the height of the atmosphere, 49.3 miles. It is plain that the actual height of the atmosphere must be greater than this, since a stratum of air of considerable thickness may lie above *k*R, and yet not have sufficient density to send a sensible amount of reflected light to the eye at O, through the body of air lying on the line RO.

257. Problem. *The latitude of the place and the sun's declination being given, to find the time of the beginning or end of twilight.*

The zenith distance of the sun, at the beginning of morning or end of evening twilight, is 90° + 18°; we may therefore solve this problem by means of equations (49) and (50), taking R = 18°.

If the time of the commencement of morning twilight be subtracted from the time of sunrise, the remainder will be the *duration of twilight.*

258. Variable Duration of Twilight. The duration of twilight varies with the latitude of the place, and with the time of the year. In the northern hemisphere, the summer are longer than the winter twilights, and the longest twilights take place at the summer solstice; while the shortest occur when the sun has a small southern declination, different for each latitude. The summer twilights increase in length from the equator northward. In the southern hemisphere, the phenomena are similar for corresponding declinations of the sun.

These facts are consequences of the different situations with respect to the horizon of the centres of the diurnal circles described by the sun in the course of the year, and of the different sizes of these circles. To make this evident, let us conceive a circle to be traced in the heavens parallel to the horizon, and at the distance of 18° below it; this is called the *Crepusculum Circle.* The duration of twilight will depend upon the number of degrees in the arc of the diurnal circle of the sun, comprised between the horizon and the crepusculum circle, which, for the sake of brevity, we will call the arc of twilight: and this will vary from the two causes just mentioned. For, let *hkr* (Fig. 69) represent the equator, and *h′k′r′* a diurnal circle described by the sun when north of the equator; and let *hr*, *st*, and *h′r′*, *s′t′*, be the intersections of the equator and diurnal circle, respectively, with the planes of the horizon and the crepusculum circle. When the sun is in the equator, the arc of twilight is *hs*, and when he is on the parallel of declination *h′k′r′* it is *h′s′*. Draw the chords *hs*, *h′s′*, *mn*, and the radii, *cs*, *cs′*, *cr′*, *cn*, *cp*. The angle *r′h′s′* is the half of *r′cs′*, and the angle *pmn* is the half of *pcn*; but *r′cs′* is less than *pcn*, and therefore *r′h′s′* is less than *pmn*. Again, *chs* is the half of *rcs*, and therefore greater than *pmn*, the half of the less angle *pcn*. Whence it appears that the chord *h's'* is more oblique to the horizon, and therefore greater than the chord *mn*, and this more oblique and greater than the chord *hs*. It follows, therefore, that the arc *h's'* is greater, and contains a greater number of degrees than the arc *mn*, and that this arc is greater than *hs*. Thus, as the sun recedes from the equator towards the north, the arc of twilight, and therefore the duration of twilight, increases from two causes, viz.: 1st. The increase in the distance of the line of intersection of the horizon with the diurnal circle from the centre of the circle; and, 2d. The diminution in the size of the circle. The change will manifestly be greater in proportion as the latitude is greater.

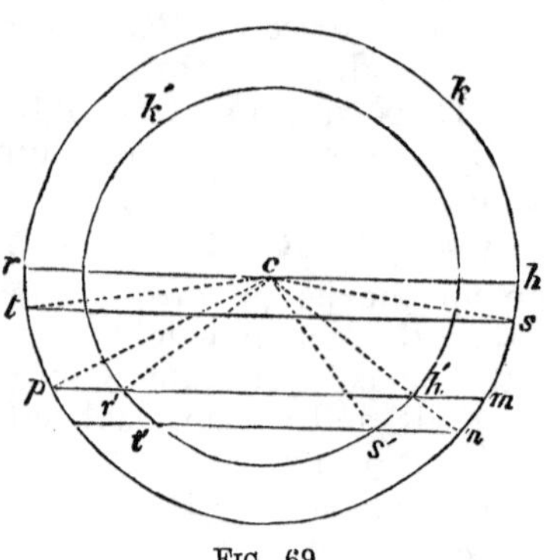

FIG. 69.

When the sun is south of the equator, twilight will, for the same declination, be shorter than when he is north of the equator, because, although the diurnal circle will be of the same size, and its intersection with the horizon at the same distance from its centre, on the opposite side, the intersection with the crepusculum circle will now fall between the intersection with the horizon and the centre, and therefore, by what has just been demonstrated, the arc of twilight will be shorter.

The shortest twilight occurs when the sun is somewhat to the south of the equator, because the arc of twilight, for a time, decreases by reason of the diminution of its obliquity to the horizon more than it increases in consequence of the decrease in the size of the diurnal circle. That the obliquity of the arc of twilight, or rather of the chord of the arc, to the horizon diminishes, for a time, when the sun gets to the south of the equator, will appear from this, viz., that the chord is perpendicular to the horizon when the centre of the diurnal circle is midway between the horizon and the crepusculum circle; which will happen when the sun is a certain distance south of the equator, varying with the inclination of the axis of the heavens to the plane of the horizon, and therefore with the latitude of the place.

The difference in the length of the summer and winter twilights, resulting from the causes above specified, is augmented by the inequality in the height of the atmosphere. Twilight also increases in length with the obliquity of the sphere.

259. Twilight in Low and Middle Latitudes. At the equator, the shortest astronomical twilight occurs at the equinoxes, and is 1h. 12m. in duration. At latitude 41°, it occurs when the sun is about 6° south of the equator, and continues about 1½ hours. At the polar circle it happens when the sun is 8½° south of the equator, and continues over 3 hours. The longest twilight at the equator is 1h. 19m.; and at latitude 41°, is 2h. 3m. in duration.

260. Twilight in High Latitudes. At the latitude 49°, the sun at the time of the summer solstice is only 18° below the horizon at midnight; for the altitude of the pole, on the parallel of 49°, differs only 18° from the polar distance of the sun, at this epoch. This may be illustrated by Fig. 66, p. 151, taking X as the point of passage of the sun across the inferior meridian, and supposing PH to be equal to 49°. At the summer solstice, PX = 67°; and thus the distance of the sun below the horizon at midnight $= HX = PX - PH = 67° - 49° = 18°$. At this latitude, therefore, evening and morning twilight will each continue half the night, at the summer solstice, and therefore nearly 4 hours.

At higher latitudes than 49°, twilight (evening and morning) will continue all night for a certain period of time before and after the summer solstice, during which the polar distance of the sun is less than the latitude augmented by 18°. At the polar circle, this will be the case for 2½ months before and 2½ months after the summer solstice.

To the north of the arctic circle, as far as 84° of latitude, during the long night that prevails before and after the winter solstice, there should be more or less of a twilight over the southern horizon, about the hour of noon of every day of 24 hours.

At either pole twilight commences about a month and a half before the sun appears above the horizon, and lasts about a month and a half after he has disappeared. For, since the horizon at the pole is identical with the celestial equator, the twilight which precedes the long day of six months will begin when the sun in approaching the equator, upon the other side, attains to a declination of 18°; and this will be about 50 days before he reaches the equator, and rises at the pole. The evening twilight will continue, in like manner, until the sun has descended 18° below the equator.

It should be observed, with reference to the above results, that the assumed limiting angle of depression of the sun below the horizon, 18°, having been determined from observations in the middle latitudes, is probably too great for high latitudes; and also that the astronomical twilight above considered, is much longer than what is ordinarily regarded as the period of twilight.

THE SEASONS.

261. General Explanation of Change of Seasons. The amount of heat received, at any place on the earth, directly from the sun, in the course of 24 hours, depends upon two operative causes; the length of time that the sun remains above the horizon, and the obliquity of its rays at noon. By reason of the obliquity of the ecliptic, both of these general circumstances vary materially in the course of the year; whence arises a variation of temperature, or a change of seasons. Since the obliquity of the ecliptic is a consequence of the inclination of the earth's axis to the perpendicular to the plane of the orbit, the inclined position of the axis is the primary cause of the change of seasons.

262. Climatic Zones. The tropics and polar circles divide the earth into five parts, called *Zones*, throughout each of which the yearly change of temperature is occasioned by a similar change in the circumstances of the sun's thermal action. The part contained between the two tropics is called the *Torrid Zone;* the two parts between the tropics and polar circles are called the *Temperate Zones;* and the other two parts, within the polar circles, are called *Frigid Zones.*

At all places in the north temperate zone, the sun will always pass the meridian to the south of the zenith; for the latitudes of such places exceed 23½°, the greatest declination of the sun (see Fig. 66). The meridian zenith distance will be greatest at the winter solstice, when the sun has its greatest southern declination; and it will vary continually between the values which obtain at the solstices. The day will be longest at the summer solstice, and shortest at the winter solstice, and will vary in length progressively from the one date to the other.

We infer, therefore, that throughout the zone in question the greatest amount of heat will be received from the sun at the summer solstice, and the least at the winter solstice; and that the amount received will gradually increase, or decrease, from one of these epochs to the other. The solstices are not, however, the epochs of maximum and minimum temperature, but are found from observation to precede these by about a month. The reason of this circumstance is, that the earth continues for a month, or thereabouts, after the summer solstice to receive during the day more heat than it loses during the night, and for about the same length of time after the winter solstice continues to lose during the night more heat than it receives during the day.

Within the torrid zone, the length of the day varies after the same manner as in the temperate zone, though in a less degree; but the motion of the sun with respect to the zenith is different. At all places in the torrid zone the sun passes the meridian during a certain portion of the year to the south of the zenith, and during the remaining portion to the north of it; for all places so situated have their zeniths between the tropics in the heavens, and the sun moves from one tropic to the other, and back again to its original position, in a tropical year. Throughout the torrid zone, therefore, the sun will be in the zenith twice in the course of the year, and will be at its maximum distance from it on the one side and the other at the solstices.

An inhabitant of the equator, or its vicinity, will have summer at the two periods when the sun is in the zenith, and winter (or a period of minimum temperature) both at the summer and winter solstice. Near the tropic, there will be but little variation in the daily amount of heat received, during the period that the sun is north of the zenith.

At the frigid zone, a new cause of a change of temperature exists; the sun remains continually above the horizon for a greater or less number of days about the summer solstice, and continually below it for the same number of days about the winter solstice.

263. General Effects of Increase of Latitude. The amount of the yearly variation of temperature increases with the latitude of the place; for the greater is the latitude the greater will be the variation in the length of the day. Also, the mean yearly temperature is lower as we recede from the equator and approach the poles; for since the sun is, in the course of the year, the same length of time above the horizon, at all places, the mean yearly temperature must depend altogether upon the mean obliquity of the sun's rays at noon, and this increases with the latitude.

264. Special Causes of Change of Climate. It is important to observe, that although, in the main, climate varies with the latitude, after the manner explained in the foregoing articles, it is still dependent, more or less, upon local circumstances, such as the vicinity of lakes, seas, or mountains, prevailing winds of some particular direction, etc. As the result of the operation of such special causes two places may be situated on the same parallel of latitude, and yet have climates quite different. Such differences of thermal condition are very marked on the opposite coasts of the Atlantic Ocean, in the middle and high latitudes.

265. Seasons Astronomically Defined:—Comparative Lengths. In the north temperate zone, *Spring*, *Summer*, *Autumn*, and *Winter*, the four seasons into which the year is divided, are considered as respectively commencing at the times of the *Vernal Equinox*, *Summer Solstice*, *Autumnal Equinox*, and *Winter Solstice.*

Let V (Fig. 70) represent the vernal, and A the autumnal

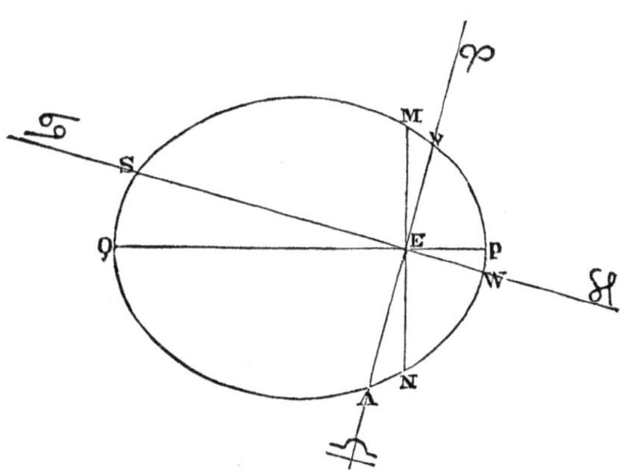

FIG. 70.

equinox; S the summer, and W the winter solstice. The perigee of the sun's apparent orbit is at present 10° 40′ to the east of the winter solstice. Let P denote its position. The lengths of the seasons are, agreeably to Kepler's law of areas, respectively proportional to the areas VES, SEA, AEW, and WEV. Thus, the winter is the shortest season, and the summer the longest; and spring is longer than autumn. Spring and summer, taken together, are about eight days longer than autumn and winter united.

266. Secular Variation of Length of Seasons. Since the perigee of the sun's orbit has a progressive motion, the relative lengths of the seasons above defined must be subject to a continual variation. At the beginning of the year 1800, the longitude of the sun's perigee was 279° 29′ 56″. If from this we take 180°, the longitude of the autumnal equinox, the remainder, 99° 29′ 56″, is the distance of the perigee from the autumnal equinox at that epoch. The motion of the perigee in longitude is, at the present date, at the rate of 61″.70 per year. Dividing 99° 29′ 56″ by 61″.70, the quotient is 5,805 years. Hence it appears that if the annual motion of the perigee had been constantly equal to 61″.7, about 5,800 years anterior to 1800 the perigee would have coincided with the autumnal equinox. But the motion of the perigee has in fact been very different in different centuries; and it appears from the calculations of Leverrier that 10,000 years before the beginning of the present century, the perigee was still 78° to the east of the autumnal equinox, and that the two points were in approximate coincidence 20,000 years earlier.

267. Secular Variations of Temperature. The eccentricity of the earth's orbit is so small (0.017) that the present annual change in the sun's distance from the earth has but little effect in producing a variation of temperature upon the earth's surface. The annual change of its heating power from this cause amounts to no more than one-fifteenth. So far as this cause operates, it makes the winters warmer and the summers colder in the northern hemisphere. But the eccentricity has not always had its present small value; it has been for ages slowly diminishing from a certain maximum value. Recent calculations made by Leverrier and other eminent computers, have made known its value at intervals of 10,000 years, or 50,000 years, for a period of 1,000,000 years previous to the beginning of the present century. From these results, it appears that it has increased and decreased during alternate periods comprising, in general, about 50,000 years; and that its recurring maximum value has fluctuated generally between the limits .05 and .075, while its minimum value has been about .01. The highest maximum value occurred 850,000 years since. The most recent maximum occurred 20,000 years ago, and was only .019. At the epoch of the highest maximum, the earth reached its perihelion during the summer (civil reckoning) in the northern hemisphere, and its aphelion during the winter. At that epoch the heating power of the sun was, by reason of the eccentricity of the earth's orbit, about one-fourth greater at the beginning of summer than at the beginning of winter; and the midwinter temperature, owing to the greater distance of the sun, was much lower than at present. In an article in the Philosophical Magazine for February, 1867, by James Croll, it is computed that the midwinter temperature of Scotland was not less than 45° F. lower than at present; and that, at the same epoch, the midsummer temperature was correspondingly higher. It is also maintained that, by a diversion of the gulf-stream, the midwinter temperature may have been reduced many degrees lower; and that this incidental effect of the great eccentricity of the earth's orbit, at that remote period, may have been the determining cause of the *glacial epoch* of the earth's geological history.

FORM AND DIMENSIONS OF THE SUN.

268. The sun presents the appearance of a luminous circular disc; but it does not follow from this that its surface must be really flat, for such is the appearance of all globular bodies when viewed at a great distance. It is ascertained from observations with the telescope that the sun has a rotatory motion; this being the fact, its surface must in reality be of a spherical form; for otherwise it would not, in presenting all its sides, always appear under the form of a circle.

No sensible difference between the equatorial and polar diameters of the sun can be detected by the nicest micrometrical measurements.

269. Dimensions of Sun. The sun's real diameter is calculated from his apparent diameter and horizontal parallax. Let ACB (Fig. 71) represent the sun, or other heavenly body, and E the place of the earth; and let δ = AEB, the sun's apparent diameter; d = 2AS, its real diameter; D = ES, its distance from the earth; and R = the radius of the earth. We have, from the triangle AES,

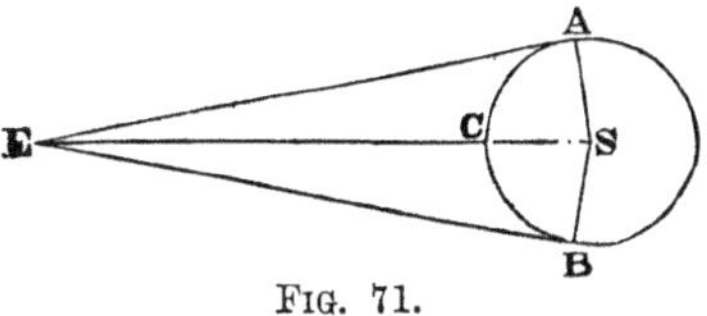

FIG. 71.

$$AS = ES \sin \tfrac{1}{2}AEB, \text{ or } 2AS = 2ES \sin \tfrac{1}{2}AEB;$$

and thus $$d = 2D \sin \tfrac{1}{2}\delta:$$

but (equa. 7), $$D = \frac{R}{\sin H},$$

whence, $$d = 2R\frac{\sin \frac{1}{2}\delta}{\sin H} = 2R\frac{\frac{1}{2}\delta}{H} = 2R\frac{\delta}{2H} \text{ (nearly)} \ldots . (51).$$

The apparent diameter of the sun at the mean distance is 32′ 0″, and the corresponding equatorial horizontal parallax is 8″.95. Accordingly we have, for the real diameter of the sun (by equa. 51, whether the sines be taken or the arcs)

$$d = 2R \times 107.263 = 7925^{m}.60 \times 107.263 = 850{,}123 \text{ miles.}$$

The mean diameter of the earth is 7,912.40 miles; the diameter of the sun exceeds this in the ratio of 107.442 to 1. The volume of the sun then exceeds that of the earth in the proportion of $(107.442)^3$ to 1^3, or 1,240,285 to 1. The surface of the sun bears to that of the earth the ratio of $(107.442)^2$ to 1^2, or 11,544 to 1.

If models were constructed to show the comparative dimensions of the sun and earth, and the earth were represented by a ball one inch in diameter, the sun would be represented by a globe nine feet in diameter. Perhaps a juster conception of the enormous bulk of the sun may be obtained from the consideration, that, if the centre of the sun were coincident with the centre of the earth, its mass would extend nearly 200,000 miles beyond the orbit of the moon.

270. General Principle. From equation (51) we may derive the proportion

$$d : 2R :: \delta : 2H.$$

Thus, *the real diameter of a heavenly body is to the diameter of the earth, as the apparent diameter of the body is to double its horizontal parallax.*

SUN'S SPOTS, AND ROTATION ON ITS AXIS.—PHYSICAL CONSTITUTION OF THE SUN.

271. When the sun is viewed with a good telescope, provided with colored glasses to protect the eye, black spots, or *maculæ*, of an irregular form, surrounded by a dark border of a nearly uniform shade, called a *penumbra*, are often seen on its disc (Fig. 72). Sometimes several spots are included within the same

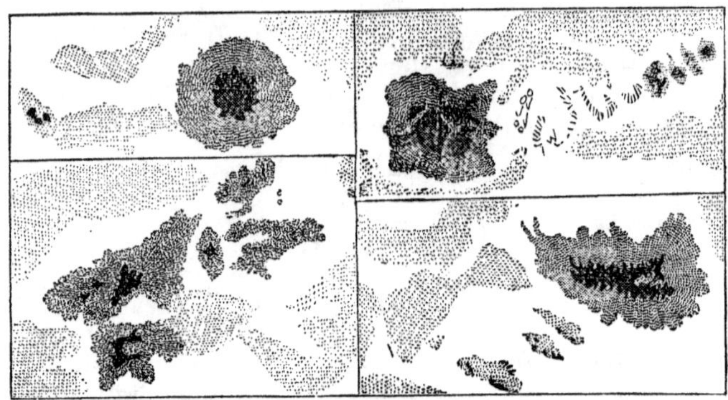

FIG. 72.

penumbra. On the other hand, a large penumbra has occasionally been seen without any central black spot. The spots usually appear in clusters, composed of various numbers, from two to sixty or seventy. It is even said, that as many as 200 have been counted, in one instance, in a single group.

In most of the individual spots, the central spot, or *umbra*, is not perfectly black; but a *black nucleus* is observed in the majority of the large and symmetrical spots, to occupy some part of the umbra, generally the centre. This distinction of shade in the umbra has been overlooked by most observers.

The penumbra has almost always a perceptibly darker shade at its outer edge than in any other part; and its light generally increases somewhat to its inner edge.

272. Magnitude of the Spots. The absolute magnitude of the solar spots is often very great. Spots are not unfrequently seen that subtend an angle of 1′, or 60″. Now the apparent diameter of the earth, as viewed at the distance of the sun, is equal to double the sun's horizontal parallax, or 18″; the breadth of such spots must therefore exceed three times the diameter of the earth, or 24,000 miles. Spots have been observed whose linear diameter was more than 45,000 miles, and which were therefore, in area, eight times as large as the entire surface of the earth. Some spots have attained to even a greater size than this, and become visible to the naked eye. A spot was seen in June, 1843, that continued visible to the naked eye for a whole week, and which, according to the measurements of M. Schwabe, of Dessau, had a breadth of 74,000 miles. A group of spots, with the penumbra surrounding it, will frequently cover a still larger portion of the sun's disc. One noticed in April, 1845, had a linear extent of 147,000 miles.

273. Variability of the Spots. The form and size of the spots are subject to rapid and almost incessant variations. When watched from day to day, or even from hour to hour, they are seen to enlarge or contract, and at the same time to change their form. They sometimes vanish in an incredibly short space of time, while others make their appearance as suddenly. Some spots disappear almost immediately after they become visible; others remain for weeks or even months. When a spot disappears, it usually contracts into a point, and vanishes before the penumbra, which gradually closes in upon it. When a new spot is developed, it is not till it has attained some measurable size that a penumbra begins to be perceived distinct from the umbra. The black nucleus within the umbra makes its appearance still later. The spot usually grows very rapidly, and often attains its full size in less than a day. During the period of increase, and while it remains without material change of size, its edges are sharply defined, and the penumbra exhibits a general uniformity of shade. Several observers have, however, distinctly noticed a radiated appearance in the penumbral fringes, as if they were traversed by bright veins diverging from the central spot. In the act of decreasing, the edges of the spot are less strongly defined, being apparently seen through a thin, luminous veil, which gradually extends over the spot. The process sometimes eventuates in the sudden appearance of a luminous line traversing the dark interval, which is then rapidly followed by the filling up and disappearance of the spot.

The velocity of the inward movement of the penumbral edges is found to have exceeded, in some of the larger spots, 44 miles per hour.

274. Periodicity of the Spots. It has been ascertained, by systematic observations upon the spots, that their number varies considerably in different years. It will sometimes happen that, on every clear day during a particular year, the sun's disc always contains one or more of them, while, in another year, for weeks or even months together, no spots of any kind can be perceived. After twenty-five years of continued observations, M. Schwabe discovered that there was a regular alternate increase and decrease in the varying numbers and sizes of the spots observed during successive years; the period from one maximum or one minimum to another being about ten years. More recently, Prof. Wolf, of Zurich, by a careful discussion of the observations of the solar spots made during the last one hundred years, has determined that the period of the spots has varied, during this interval, from 8 to 16 years, and that its mean value has been 11.11 years. 1860 was the last year of maximum. Agreeably to the mean period, the year 1866 should have been a year of minimum spots.

275. Faculæ. Curved or branching streaks more luminous than the general body of the sun, are frequently perceived upon parts of his disc, especially in the region of large spots, or of extensive groups of spots, or in localities where dark spots subsequently make their appearance. These are called *Faculæ*. They are chiefly to be seen near the margin of the disc. Adjacent bright spaces are also an invariable accompaniment of the spots. These, in the majority of instances, are most conspicuous behind the spots, or in a direction opposite to that of the sun's rotation.

It has recently been established by an observation made by Dawes, a distinguished English astronomer, that *the faculæ are ridges or masses of luminous matter, elevated above the general level of the sun's surface.* In 1859, he observed a bright streak at the very edge of the disc, which projected irregularly beyond the circular contour of the edge, like a low range of hills. For such elevations to have been distinctly perceptible, their actual height could not have been less than 500 miles, and was probably two or three times as great as this.

276. General Telescopic Appearance of Sun's Disc. The part of the sun's disc not occupied by spots is far from being uniformly bright. Inequalities of brightness prevail in all parts of the disc, which give it a coarsely mottled appearance. When more attentively scrutinized, its ground is seen to be finely mottled with minute dark dots or *pores*, which often appear to be in a state of change. It is also observed that the general luminous surface of the sun presents the appearance of bright *granules* scattered irregularly over it, and that, on the darker spaces between the granulated portions, the minute dark pores are especially prevalent. It is not yet decided whether these bright granules are to be regarded as distinct masses of greater brightness, or as merely different conditions of the luminous cloudy surface, diversified by elevated ridges or waves. This general granulation of the surface is entirely wanting on the faculæ, and on the luminous border of the penumbra of each of the dark spots. But lines of distinct, elongated, and comparatively bright masses are often seen projected on the penumbra, directed toward the centre of the spot, and even extending irregularly into the umbra, or central black spot.

277. Motions of the Spots:—Rotation of the Sun. When the positions of the spots on the disc are observed from day to day, it is perceived that they all have a common motion in a direction from east to west. Some of the spots close up and vanish before they reach the western limb; others disappear at the western limb, and are never afterwards seen; a few, after becoming visible at the eastern limb, have been seen to pass entirely across the disc, disappear from view at the western limb, and reappear again at the eastern limb. The time employed by a spot in traversing the sun's disc is about 14 days. About the same time is occupied in passing from the western to the eastern limb, while it is invisible. The motions of the spots are accounted for, in all their circumstances, by supposing that the sun has a motion of rotation from west to east, around an axis nearly perpendicular to the plane of the ecliptic; and that the spots are portions of the solid body of the sun. The truth of this explanation of the apparent motions of the sun's spots, is confirmed by the changes which are observed to take place in the magnitude and form of the more permanent spots during their passage across the disc. When they first come into view at the eastern limb, they appear as a narrow dark streak. As they advance towards the middle of the disc, they gradually open out and increase in magnitude; and after they have passed the middle of the disc, contract by the same degrees until they are again seen as a mere dark line upon the western limb.

A spot returns to the same position on the disc in about 27½ days. This is not, however, the precise period of the sun's rotation; for during this interval the sun has apparently moved forward nearly a sign in the ecliptic; the spot will therefore have accomplished that much more than a complete revolution, when it is again seen by an observer on the earth in the same position on the disc.

278. Period of Rotation. The apparent position of a spot with respect to the sun's centre may be accurately determined, from day to day, by observing, when the sun is crossing the meridian, the right ascension and declination both of the spot and centre. From three or more observations of this kind the period of the sun's rotation and the position of his equator may be ascertained.

The period of the sun's rotation, as determined from observations upon the spots, is found to increase with the latitude of the spot; from which it is to be inferred that the spots are not stationary, and have different rates of motion along the surface, in a direction parallel to the equator. If, as seems most probable, the general direction of motion is opposite to that of the rotation, then the spots nearest the equator have the slowest motion; and the period of rotation deduced from observations upon these spots approximates most nearly to the actual period of rotation of the body of the sun. The period in question is 25 days. Spots observed in the latitude 34° give a period of nearly 27 days. The inclination of the sun's equator to the ecliptic is about 7¼°; and the heliocentric longitude of the ascending node of the equator is about 74°.

279. Regions of the Spots. The solar spots are mostly confined to two zones parallel to the equator, and extending from 5° to 35° of latitude. Beyond 35° they are rarely seen, and in the polar regions never. The actual equator is also seldom, if ever, visited by spots. They are most abundant toward the middle of the spot-belts, and prevail more in the northern than in the southern hemisphere. It is observed that the spots have a tendency to form groups lying in lines or belts parallel to the equator. This is apparently the result of a tendency of new spots to break out behind the old ones.

280. Nature of the Spots. *The dark spots on the sun are depressions below the luminous surface.* This important fact was first established by Dr. Wilson, of Glasgow. He noticed that as a large spot, which was seen in November, 1769, came near the western limb, the penumbra on the side toward the centre of the disc contracted and disappeared, and that afterwards the luminous matter on that side seemed to encroach upon the central black spot, while in other parts the penumbra underwent but little change. On the reappearance of the spot at the eastern limb, he found that the penumbra was again wanting on the side toward the centre of the disc; and that when this part made its appearance, after the spot had advanced a short distance upon the disc, it was much narrower than the opposite part. These various appearances of the spot in question are represented in Fig. 73.

FIG. 73.

They show conclusively that both the black central spot and the penumbra were below the luminous surface of the sun. Dr. Wilson estimated the depth of the spot to be nearly 4,000 miles. It has since been observed that similar changes of appearance are experienced by the spots in general, in their passage across the disc.

281. Theories of Physical Constitution of Sun; and of Formation of Spots.

Wilson's Theory. Dr. Wilson drew from the various appearances of the spot observed by him in 1769 the natural conclusion that the solar spots were the dark body of the sun, seen through excavations made in the luminous matter at the surface. The luminous matter he conceived to have the consistence of a fog or cloud, rather than of a liquid; and suggested that openings might be made in it by the working of some sort of elastic vapor generated within the dark globe. The penumbra surrounding each black spot he conjectured to be the sloping sides of the opening in the stratum of luminous clouds.

Herschel's Theory. Sir William Herschel, after an assiduous study of the aspects and phenomena of the sun's spots, adopted substantially Dr. Wilson's views, but conceived it to be necessary in order to explain the uniform shade of the penumbra, to suppose the existence of an opake, non-luminous, cloudy stratum, posited between the luminous medium and the dark solid globe. On this hypothesis, the spots are accounted for by supposing that openings occasionally take place in both the luminous and non-luminous envelopes, through which the dark body of the sun is seen. The penumbra is the portion of the obscure envelope situated immediately around the opening made in it, and shining by reflected light only. Herschel supposed the openings to be made by the exertion of some sort of explosive energy from beneath; and that the same upheaving agency, when not of sufficient intensity to rend the luminous envelope, forced it up into masses or waves of hundreds of miles in height. The ridges of these waves he conceived to be the faculæ, which are distinctly seen only when near the margin of the disc, because the waves there appear in profile, and when near the middle of the disc are seen in front, or foreshortened. Sir John Herschel, who has also been an attentive observer of the sun's spots, has advanced the opinion that the agency by which the spots are formed, is exerted from above downwards, instead of from below upwards.

Recent observations made by Dawes indicate the existence of a second non-luminous envelope, posited below that which is seen in the penumbra of a spot; that this is seen in the umbra, and that the black nucleus often observed near the centre of the umbra is an opening made in this lower stratum.

282. Photosphere of the Sun. It is the received opinion among astronomers of the present day, that the sun, as maintained by Wilson and Herschel, consists of a comparatively dark globe, either solid or liquid, surrounded by one or more luminous envelopes, in a vaporous or nebulous condition, and some thousands of miles in total height. This exterior region pervaded by the medium which is the great source of the sun's light, is called the *photosphere* of the sun. That this luminous medium is really in the aëriform state, or in the condition of cloudy masses floating in a gaseous medium, may be inferred from its great mobility. The velocity of expansion and contraction of the spots, which often exceeds 40 miles per hour, is incompatible with the supposition of a liquid condition.

It does not necessarily follow from the fact that the solid or liquid globe appears perfectly dark, that it has no degree of luminosity; for it has been observed that intensely ignited solids appear only as black spots on the disc of the sun, when held between the sun and the eye. Professor Henry, and more recently Professor Secchi, of Rome, has established by experiment that the dark spots emit less heat than the luminous surface.

But it is to be observed that the result of the experiments does not give any decisive indication as to the comparative temperatures of the photosphere and dark body of the sun, since a considerable fraction of the heat radiated from the latter is no doubt intercepted by the sun's atmosphere, below the level of the luminous surface.

Depth of photosphere. Secchi has succeeded, from measurements made upon several spots at the time of the disappearance of the penumbra on the side toward the centre of the disc, in effecting an approximate determination of the depth of the photosphere, on the supposition that the penumbra is made up of the sloping sides of an opening in a single luminous envelope (281). He estimates the depth to be about one-third of the radius of the earth, or 1,300 miles. M. Faye has undertaken to determine the depth of the photosphere by a different method, and makes it about 4,000 miles.

283. Luminous Appearances exterior to the Photosphere. Whenever the sun becomes totally eclipsed by the moon interposed between it and the eye of the observer, the region exterior to the photosphere is seen to be pervaded, for a considerable distance, by luminous matter, which offers a variety of remarkable appearances. The principal of these are the *corona*, *luminous streamers*, or jets of light, and *rose-colored protuberances.*

284. The Corona is a ring or halo of white light encircling the sun, which becomes visible when the body of the sun is concealed from view. It is brightest next the dark limb of the moon, where it has a rosy tint, and gradually decreases in lustre until it becomes undistinguishable from the general light of the sky (See Plate II.). It has presented to observers more or less of a radiated appearance, as if made up of luminous radiations, or traversed by them. This appearance has been less distinct near the moon where the corona is brightest than at the more distant and fainter portions. In the total eclipse of July 18, 1860, the interruptions of continuity became distinctly perceptible at a distance from the photosphere of the sun equal to the sun's semi-diameter, or more than 400,000 miles. Beyond this, the corona was distinctly radiated. Its extreme breadth, both in that eclipse and the eclipse of September 7, 1858, exceeded the diameter of the sun, or 850,000 miles. The extreme outline of the corona is perceptibly elliptical in its form; the major axis lying in the plane of the sun's equator.

285. The Luminous Streamers, whether forming part of the corona, or distinct from it, have been seen to extend, at particular points, far beyond the general outline of the corona. In the eclipse of 1860, some of them were traced to a distance from the sun's photosphere equal to twice the diameter of the sun, or 1,700,000 miles (Plate II.). They present, in general, the appearance of radiations of luminous matter, in directions perpendicular to the sun's surface.

Observations made with the polariscope have established that the light of the corona and streamers is in part reflected light.

286. The Rose-colored Protuberances are regarded by observers as the most remarkable and beautiful phenomena witnessed in total solar eclipses. They consist of apparent cloudy masses, more or less tinged with red light, and of various forms and sizes, noticed just without the dark limb of the moon (Plate II.). In some instances, they have been seen entirely detached from the moon's limb. They are seen at various points of the limb, and in every variety of position with respect to the equator of the sun. The latter circumstance shows that they have no connection with the sun's spots, since these do not occur in high latitudes.

It has been repeatedly observed that, in the progress of a total eclipse of the sun, the protuberances which become visible at the eastern limb of the moon continually decrease in their apparent dimensions, as if the moon were screening more and more of them from view; while those seen at the western limb continually increase in their dimensions, as if they were more and more uncovered by the moon in its advance. These facts indicate that the protuberances in question are *luminous masses connected with the sun, and elevated above the photosphere.* Careful measurements have fully established this conclusion.

287. Height and Extent of the Protuberances. Some of the protuberances observed in the eclipse of 1860, had an apparent height of nearly 70,000 miles. The breadth of single protuberances is but a few minutes of arc, but they sometimes extend in a continuous chain for many degrees. Near the end of the eclipse of 1860, one chain of low elevations was observed by Secchi, just without that part of the moon's contour at which the sun was about to make its appearance, which extended 60°. About the middle of the eclipse, no less than ten distinct protuberances were counted, which were about regularly distributed around the disc. In view of these facts, it seems highly probable, as intimated by Secchi, that the rose-colored protuberances observed in that eclipse, were but the higher portions of cloudy masses that formed at a lower level, one continuous reddish envelope surrounding the sun. This envelope must have extended upwards from the sun's photosphere to the height of thousands of miles, and risen at some points into cloudy peaks of tens of thousands of miles in height.

288. Nature of the Corona. The corona is supposed by Sir John Herschel, and other astronomers, to be a gaseous solar atmosphere, extending above the sun's photosphere; but its vast extent (284) seems to be fatal to this explanation. Upon no reasonable supposition that can be made with regard to the effect of the sun's heat in expanding such an atmosphere, can it be supposed to have sufficient density to reflect a sensible amount of light, beyond a few thousand miles from the body of the sun. The natural indications of the phenomena are that the corona consists of luminous matter streaming off from the sun into space; and at the same time that the appearance of distinct radiations arises from an inequality of emanation from different parts of the sun's surface. The elliptical form of the corona (284) indicates that the emission of luminous matter is most abundant from the equatorial regions.

It has been suggested that the radiated appearance of the corona may result from a partial interception of the sun's light by clouds floating in the sun's atmosphere; but the course of the rays which pass unobstructedly through the spaces between the clouds, could not be recognized unless they encounter matter of sufficient density to reflect a sensible quantity of light to the eye, and such matter cannot extend to a distance of more than a million of miles from the sun (285), unless there are material emanations proceeding from the sun. We can only avoid this conclusion by assuming that there is a dense mass of meteoric bodies, or of cosmical matter, revolving around the sun within this distance.

It is maintained by some astronomers that the corona and streamers are phenomena of diffraction, resulting from the passage of the sun's light near the borders of the moon. But upon this explanation there ought to be certain attendant phenomena of variegated colors which are not in reality seen.

289. Temperature of the Sun's Surface. To an observer at the luminous surface of the sun, its disc would appear to cover an entire hemisphere in the heavens, and therefore the heat that falls upon a small area at the surface of the sun should exceed that received upon the same area at the distance of the earth, in the proportion that a hemisphere of the heavens exceeds the area occupied by the sun's disc as seen from the earth, or nearly in the ratio of 100,000 to 1. A heat many times less intense than this suffices to dissipate the most refractory metals in vapor.

That the calorific rays emitted from the sun have a far higher intensity than those which proceed from the hottest furnaces, or result from the most vivid ignition obtained by chemical or gal vanic processes, may be inferred from the fact that they penetrate glass with far greater facility.

Inequalities of temperature. Secchi has made a series of observations upon the comparative amounts of heat received from different parts of the sun's surface. He finds that the polar emit less heat than the equatorial regions; and that the two hemispheres separated by the equator have not exactly the same temperature. It also appears from his observations that the breaking out of a spot at any point of the sun's disc, occasions a fall of temperature there and at surrounding points; and that the faculæ do not sensibly augment the temperature of the points where they make their appearance.

Also, the calorific rays proceeding from the centre of the sun's disc have a higher intensity than those proceeding from the borders. The same is true of the luminous rays. From this fact it is inferred that the sun is surrounded by an atmosphere extending far above its photosphere, or by some form of matter, in a condition to intercept a large amount of light and heat.

290. Intensity of Sun's Light. The most intense artificial lights are the *Drummond Light*, produced by the flame of the oxyhydrogen lamp directed against a surface of chalk, and the *Electric Light*, generated by the passage of a galvanic current between two charcoal points. Fizeau and Foucault found, by ingenious and carefully conducted experiments, that the light of the sun's disc exceeded in intensity the Drummond light, in the ratio of 146 to 1; and that it exceeded the electric light from forty large plates of a Bunsen's battery in the ratio 2½ to 1. It appears, therefore, that the electric light is the only artificial light that approximates in intensity to the light of the sun.

291. Origin of the Sun's Heat. There would seem to be but two possible physical causes in operation that might be adequate to the development and maintenance of the high temperature of the sun. These are—

(1.) The contraction of the body of the sun from an original vaporous state to its present size and density.

(2.) The fall of meteoric masses into the photosphere of the sun.

Upon the hypothesis, hereafter considered, that comets and meteors were originally discharged from the surface of the sun during the successive stages of vaporous diffusion through which the sun's mass is supposed to have passed, these two causes are, physically speaking, essentially the same; since the heat ultimately developed must be the same, whether the subsidence of the matter is by gradual contraction, or by gravitation. Yet the fall of the revolving meteors one after another into the photosphere of the sun might now determine a much higher temperature than would have resulted from contraction alone. The heat due to their fall has been, as it were, stored up in these meteoric bodies, to be suddenly developed, instead of being gradually dissipated during the ages in which the sun has been going through its process of formation.

292. Results of Recent Investigations, concerning the sun's spots and physical constitution. The careful scrutiny and assiduous study, by several astronomers, of all the phenomena observable on the surface of the sun for a number of years past, has led to the following important discovery:

The sun's spots are for the most part developed by, or in some way connected with, the operation of a physical agency exerted by the planets upon the photosphere. This remarkable fact has been conclusively established by the observations of Schwabe, Carrington, Secchi, and others; and especially by the detailed discussion to which all the reliable observations upon the spots, made during the last 100 years, have been subjected by Professor Wolf, of Zurich. The planets which exercise the greatest influence are Jupiter and Venus. The planetary agency is directly recognized in the origination of spots on the parts of the sun's surface brought by the rotation into favorable positions, and in the subsequent changes experienced by the spots while subject to the direct action of the planet. It is also shown by the dependence of the epochs of the maximum and minimum of spots upon the positions of the planets, especially of Jupiter and Venus.

Effects of so marked a character, exerted by the planets upon the photosphere of the sun, cannot reasonably be attributed to their natural attractive action, and must apparently result from a repulsive or impulsive action exerted upon the photospheric matter.

Rotation of Spots. Some spots have been observed to have a motion of rotation around their centres; but according to Dawes, who has been a diligent observer of solar phenomena for many years, this is a phenomenon of exceedingly rare occurrence in the case of well-developed spots.

293. Theory of the Origin of the Sun's Spots. The following is a brief outline of a theory of the development of the sun's spots, based upon the principle of planetary action above stated.

(1.) The matter of the sun's photosphere, and for a certain distance beyond the luminous surface of the photosphere, is, either wholly or in part, in a magnetized state, and arranged in columns, or lines of magnetic polarization, like the auroral matter in the upper atmosphere of the earth.

(2.) A repulsive or impulsive action exerted by the planets upon the molecules of these columns, tends to disturb their electric and magnetic equilibrium, and induce electric discharges along certain of the upper columns, by which they are widely dispersed, and the mechanical equilibrium of the portions below disturbed. In this way, a vast column of expanding and ascending matter is originated in the photosphere, which in the process becomes more or less dissipated, and may reveal the body of the sun to view.

(3.) The matter dispersed, from a loss of magnetic intensity, or by the electric discharges, and certain portions of the vaporous matter of the column, as they rise above the photosphere, are brought into that subtle or nebulous condition observed in the matter of comets, in which it becomes subject to an effective repulsion from the sun, and so is expelled indefinitely into space. Other portions may become condensed above the photosphere, and subside into it.

(4.) The planets may be conceived to operate in two ways, to initiate the process of dispersion of the tops of the photospheric columns, and so develop spots on the sun; viz., by originating in the upper photosphere electric currents radiating in all directions from the region exposed to most direct action, or by developing electro-magnetic currents running in a direction opposite to that of the rotation. Such radial electric currents would be attended with an exaltation of the statical electric condition of the region exposed to planetary action; and such magnetic currents would tend to demagnetize, or magnetize, the upper photospheric columns, according as the upper or lower currents prevail.

It appears, from the results of observation, that the planets operate unequally in different parts of the ecliptic, and in different relative positions; and their effects are apparently modified, in certain positions, by the electro-magnetic currents developed in the sun's photosphere by the motion of the solar system through space.

(5.) The spots are more likely to occur in low than in high latitudes, because the induced magnetism of the photospheric columns has a lower intensity in proportion as the magnetic latitude is less; and spots do not make their appearance on the equator, nor in its immediate vicinity, because the columns of magnetic matter are there parallel to the surface of the sun.

(6.) In the supposed electric discharges along the magnetic columns, with the attendant accumulation of luminous matter in certain localities above the ordinary surface of the photosphere, we have at the same time an adequate explanation of the *faculæ*, and of the *rose-colored protuberances* at a still higher level.

When, in special localities, the discharge has attained to a sufficient intensity, or continued for a sufficient length of time, openings are made through the whole depth of the photosphere, and spots are seen in the region where the faculæ were before observed.

It may be added, in this connection, that the supposition of the distribution of the photospheric matter in separate columnar masses, accords with the granulated appearance presented by the sun's disc (276).

(7.) The photospheric matter dispersed by reason of the varying action of the planets, sufficiently to become subject to a repulsive action from the sun as it flows away into space, forms the *corona*, with its accompanying radiations and streamers, visible in total eclipses.

(8.) A portion of the attenuated matter thus expelled to an indefinite distance from the sun, is received into the upper atmosphere of the earth, and, by developing electric currents there, becomes one of the operating causes of the disturbances of the magnetic needle on the earth's surface; which are observed to increase and decrease, *pari passu* with the sun's spots. The impulses attendant upon the electric discharges occurring in the sun's photosphere, and propagated indefinitely into space, constitute another cause of magnetic disturbance upon the earth.

The escaping solar matter received into the earth's upper atmosphere, supplies the matter of terrestrial auroras, which also have the same periods as the sun's spots. (For a more complete exposition of the author's theoretical views, see Am. Journal of Science, Vol XLI., Nos. 121 and 122. See also Note in Appendix.)

ZODIACAL LIGHT.

294. At certain periods of the year a luminous appearance is observed in connection with the sun, extending upwards from the western horizon after evening twilight, and from the eastern horizon before daybreak, which is called the Zodiacal Light, from the circumstance of its being mostly comprehended within the zodiac. Its color is white, with a decided tinge of yellow at the lower altitudes. When most conspicuous, it has a striking brilliancy near the horizon, and fades upwards by imperceptible degrees. Its apparent form is nearly triangular, the base resting on the horizon, from which it tapers upwards to an indistinct vertex. The axis, or central line, lies nearly in the ecliptic. Its length varies with the season of the year, and the state of the atmosphere. As estimated from the sun, it is sometimes more than 100°, but ordinarily not more than 40° or 50°. Its breadth near the horizon varies from 8° to 30° or 40°. It is nowhere abruptly terminated, but gradually merges into the general light of the sky (Fig. 74).

Fig. 74.

295. It varies in Distinctness. The Zodiacal Light is seen most distinctly, in our northern latitudes, in February and March after sunset, and in October and November before sunrise. During the month of March it may be seen directed towards the star Aldebaran. In December, though fainter, it may often be seen both in the morning and evening. Also, towards the summer solstice, it is discernible, in a very pure state of the atmosphere, both in the morning and evening. The reason of the variations in the distinctness of the zodiacal light, from one season to another, is found in the change of its inclination to the horizon at the time of sunset or sunrise, together with the variation that occurs in the duration of twilight. As its length lies nearly in the plane of the ecliptic, its inclination to the horizon will be different, like that of this plane, according to the different positions of the sun in the ecliptic. At sunset, the zodiacal light will be most inclined to the horizon, and therefore extend farthest up in the heavens, towards the vernal equinox, when the ecliptic is at sunset most nearly perpendicular to the horizon; and at sunrise it will be most inclined to the horizon towards the autumnal equinox, when the inclination of the ecliptic to the horizon at sunrise is the greatest. The zodiacal light is much brighter and more frequently observed between the tropics than in these latitudes; because the ecliptic, in general, makes there a larger angle with the horizon, and twilight is of shorter duration.

According to Arago it appears, from the entire series of observations at Paris and Geneva, that the Zodiacal Light varies considerably from one year to another, and that the observed variations cannot result entirely from changes in the transparency of the atmosphere. Extraordinary changes of brightness and form, in the course of a single evening, have also been noticed by several observers, which were regarded as decided indications of a change in the intrinsic lustre or density of the substance of the Zodiacal Light. But all such abrupt changes may possibly be purely of atmospheric origin.

296. Recent Observations:—Important Results. The most valuable series of observations extant on the zodiacal light are those which were made in the years 1853–4–5, at various latitudes, from 41° 49′ N to 53° 28′ S, by Chaplain Jones, of the U. S. Navy; and by the same observer, in the years 1856–7, from the elevated station of Quito, very near the equator.

The discussion of these observations has furnished the following important results:

1. When the observer was in a position on either side of the plane of the ecliptic, the main body of the Zodiacal Light was on the same side of the ecliptic in the heavens; and when he was in the plane of the ecliptic, this light was equally divided by its circle in the heavens.

2. When the observer was carried by the earth's rotation rapidly towards or from the plane of the ecliptic, the change of the apex of the light, and of the direction of its boundary lines, was equally great, and corresponded to the change of place.

3. As the ecliptic changed its position with respect to the horizon, the entire shape of the Zodiacal Light became changed.

4. The entire luminous appearance consisted of a *stronger* light at the central part, and a much broader *diffuse* light extending beyond this on either side, and to a greater height. The stronger passed by degrees into the diffuse light, and the latter also gradually faded away. Yet there was a discernible line of greater suddenness of transition, that could be taken for the boundary of the former.

5. The light was visible, with more or less distinctness, on every favorable night during the entire period of the observations. The position of the observer, generally at sea, or in the lower latitudes when on land, was more favorable than that which most observers have had.

6. On favorable nights, when the ecliptic was nearly perpendicular to the horizon, at the observer's station, the Zodiacal Light was visible at midnight, over both the western and eastern horizons. This singular phenomenon was observed at sea, at certain stations within the tropics. At Quito, the light was seen every favorable night, and at all hours, to extend as a broad *Luminous Arch*, entirely from one horizon to the other. At midnight it had a pale and nearly uniform white lustre, from

one horizon to the other. The breadth was then nearly uniform, and about 30°.

297. Explanation. No generally received explanation of this singular phenomenon has yet been given. It was at one time supposed to be the atmosphere of the sun, but Laplace has shown that this explanation is at variance with the theory of gravitation. He found that at the distance of about sixteen millions of miles from the centre of the sun, the centrifugal force due to the sun's rotation balanced the gravity, and therefore that the solar atmosphere could not extend beyond this; but this distance is less than half the distance of Mercury from the sun, whereas the substance of the Zodiacal Light extends beyond the orbit of Venus, and even beyond the earth's orbit. The most plausible theory of the Zodiacal Light that has been advanced is that propounded by Laplace, that it consists of a broad ring, or lenticular mass of nebulous matter, encircling the sun in the plane of his equator. He supposed it to be maintained in a permanent condition by the revolution of its particles around the sun.

But, as intimated in former editions of this work, another conception of the mechanical condition of such a mass of nebulous matter may be formed, that accords equally well with the phenomena of the zodiacal light. We may regard the whole mass as made up of the streams of particles which we have recognised as continually in the act of flowing away from the sun (288 and 293), under the operation of a force of solar repulsion; or, in other words, that it is the indefinite continuation of the corona observed in total eclipses of the sun, with its attendant streamers. Upon this view it should appear elongated, like the faint outer boundary of the corona (284), in the plane of the sun's equator; and this elongation may be attributed to a more copious discharge of photospheric matter from the equatorial than from the polar regions of the sun. Or we may conceive, in accordance with the theoretical views of the probable condition of the sun's photosphere that have been presented (293), that the discharges may take place from the tops of the photospheric columns, in the direction of their prolongation. All such discharged particles would thus receive a projectile velocity oblique to the sun's surface, and toward the plane of the equator, and, being subsequently repelled by the sun, would move away into space in hyperbolic orbits, convex toward the sun. As a necessary consequence, there would be an augmentation of the quantity of escaping matter in the plane of the sun's equator, and an elongation of its visible portion in this plane. The light may vary in brightness from one year to another, with the varying activity of discharge from the sun's surface.

The appearance and phenomena of the Zodiacal Light indicate that the principal portion of the light received experiences *specular* reflection from the particles of its substance. In the report of the observations referred to in the last article, this idea is presented and advocated. Upon this view, the amount of light reflected to the eye from different directions will increase with the angle included between the directions of the incident and reflected rays, and with the density of the substance.

The lateral shiftings of position of the light, as the distance of the zenith from the ecliptic varies, may be satisfactorily explained by means of the following general considerations:

1. By reason of the small dimensions of the earth, as compared with the vast extent of the entire collection of matter flowing away from the sun to an indefinite distance, the density of this nebulous matter must be sensibly the same for considerable distances from the earth, in all directions. But beyond a certain distance, which we will call D, the density must begin sensibly to decrease as the line of sight makes a greater angle with the plane of the ecliptic (which is nearly coincident with the plane of the sun's equator, the supposed plane of greatest density of the solar emanations). This decrease of density will be more rapid as the portion

of nebulous matter considered is more remote from the earth. It is plain, then, that the light reflected to the earth, from all portions of this matter situated at distances from the earth greater than D, will decrease in intensity from the ecliptic, in both directions. The brightness of the light, at its different points, will also augment as the angular distance from the sun diminishes. The entire result from the light thus reflected should then be a luminous appearance similar to the observed Zodiacal Light, with its axis or line of greatest brightness lying in the ecliptic.

If we now take account of the absorptive action of the atmosphere upon the transmitted light, which increases in proportion as the direction of the ray makes a less angle with the plane of the horizon we perceive that the axis will be thrown to that side of the ecliptic on which the zenith lies, whenever the inclination of the ecliptic to the horizon is less than 90°.

The light received from all portions of the nebulous matter which are at a less distance than D from the earth, will produce a different result. Its intensity will be the same for all points at the same angular distance from the sun; that is, if we disregard atmospheric absorption. If this be taken into account, it will be seen that the actual appearance will be a diffuse luminosity, decreasing in both directions from the vertical circle passing through the sun below the horizon. From the station of the observer, within the shadow of the earth, the nearer portions of the shining nebulous matter will lie in the direction of this vertical circle. This circumstance will tend to augment the brightness along this circle, and in its vicinity, as compared with points at a distance from it.

The Zodiacal Light observed is the result of the combination of this light, which has its axis in the vertical circle through the sun, in its position below the horizon, with the stronger light reflected from the more remote regions, whose axis lies in the ecliptic.

2. The axes of these two different luminosities coincide whenever the ecliptic is perpendicular to the horizon; but when these circles are inclined to each other, a greater portion of the compound light falls on the side of the ecliptic on which the zenith and the vertical circle of the sun lie, than on the opposite side. When the inclination of the circles increases, the disparity between the portions of the light that lie on opposite sides of the ecliptic becomes greater, and the axis and the whole luminous appearance are displaced in the direction of the vertical circle. This displacement should certainly take place up to a certain limit of increase in the angle, which should be greatest at the lower altitudes, at which the observed displacement is greatest. It is also to be noticed that there is a secondary cause in operation, tending to augment this lateral displacement; which consists in the fact, that as the ecliptic becomes more oblique to the horizon, the sun when at the same distance as before, along the ecliptic from the horizon, will be nearer the horizon in the direction of the vertical circle, and therefore at equal heights above the horizon, the luminosity which has its axis in the vertical circle, will be increased in brightness, and so have a greater displacing effect on the boundary of the ecliptic light. It will also readily be perceived that up to a certain amount of deviation of the ecliptic from the vertical circle, the unequal atmospheric absorption of the light will operate to increase the displacement.

The *Diffuse Light* noticed in the last article, probably has its origin in the portions of the nebulous solar emanation that lie immediately beyond the distance D, the density of which will decrease from the ecliptic more slowly than that of the more remote portions.

The *Luminous Arch* seen at midnight in tropical regions (296) must be attributed, from the present point of view, to the radiant reflection, and feeble specular reflection at angles of incidence less than 45°, of the sun's rays, from the portion of the solar matter that extends indefinitely beyond the earth's orbit. The variations in the amount of light reflected from regions at different angular distances from the sun, in the density of the reflecting substance, and in the effects of atmospheric absorption, tend to equalize the light received from different directions lying in the plane of the ecliptic. The extent of the conical shadow cast by the earth is presumably small in comparison with that of the nebulous substance from which the light is received.

It will be perceived that atmospheric absorption plays the prominent part in the phenomenon of the lateral displacement of the Zodiacal Light, operating both directly and indirectly.

CHAPTER XV.

THE MOON AND ITS PHENOMENA.

PHASES OF THE MOON.

298. THE most conspicuous of the phenomena exhibited by the moon, is the periodical change that is observed to take place in the form and size of its disc. The different appearances which the disc presents are called the *Phases* of the moon.

The phenomenon in question is a simple consequence of the revolution of the moon around the earth. Let E (Fig. 75) represent the position of the earth, ABC the orbit of the

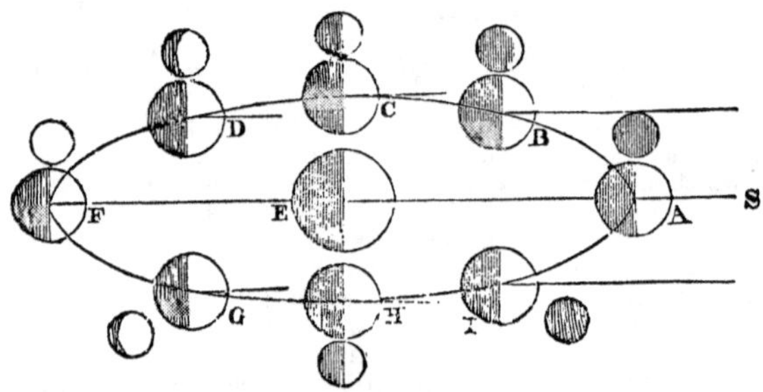

FIG. 75.

moon, which we will suppose for the present to lie in the plane of the ecliptic, and ES the direction of the sun. As the distance of the sun from the earth is about 400 times the distance of the moon, lines drawn from the sun to the different parts of the moon's orbit, may be considered, without material error, as parallel to each other. If we regard the moon as an opake non-luminous body, of a spherical form, that hemisphere which is turned towards the sun will be continually illuminated, and the other will be in the dark. Now, by virtue of the moon's motion, the enlightened hemisphere is presented to the earth under every variety of aspect in the course of a synodic revolution of the moon. Thus, when the moon is in conjunction, as at A, this hemisphere is turned entirely away from the earth, and it is invisible. Soon after conjunction, a portion of it on the right begins to be seen, and as this is comprised between the right half of the circle which limits the vision, and the right half of the circle which separates the enlightened and dark hemispheres of the moon, called the *Circle of Illumination*, it will obviously present the appearance of a crescent, with the horns turned from the sun, as represented at B. As the moon advances, more and more of the enlightened half becomes visible, and thus the crescent enlarges, and the eastern limb becomes less concave. At the point C, 90° distant from the sun, one-half of it is seen, and the disc is a semi-circle, the eastern limb being a right line. Beyond this point, more than half becomes visible; the nearer half of the circle of illumination falls to the left of the moon's centre, as seen from the earth, and thus becomes convex outward. This phase of the moon is represented at D. When the moon appears under this shape, it is said to be *Gibbous*. In advancing towards opposition, the disc will enlarge, and the eastern limb become continually more convex; and finally at opposition, where the whole illuminated face is seen from the earth, it will become a full circle. From opposition to conjunction, the nearer half of the circle of illumination will form the right or western limb, and this limb will pass in the inverse order through the same variety of forms as the eastern limb in the interval between conjunction and opposition. The different phases are delineated in the figure.

The moon's orbit is, in fact, somewhat inclined to the plane of the ecliptic, instead of lying in it, as we have supposed; but, it is plain that its inclination cannot change the order, nor the period of the phases, and that it can have no other effect than to alter somewhat the size of the disc, at particular angular distances from the sun. In consequence of the smallness of the inclination, this alteration is too slight to be noticed.

299. Definitions. When the moon is in conjunction, it is said to be *New Moon;* and when in opposition, *Full Moon.* At the time between new and full moon, when the difference of the longitudes of the moon and sun is 90°, it is said to be the *First Quarter.* And at the corresponding time between full and new moon, it is said to be the *Last Quarter.* In both these positions the moon appears as a semi-circle, and is said to be *dichotomized.* The two positions of conjunction and opposition are called *Syzigies;* and those of the first and last quarter, *Quadratures.* The four points midway between the syzigies and quadratures are called *Octants.*

300. Lunar Month. The interval from new moon to new moon again, is called a *Lunar Month*, and sometimes a *Lunation.*

The mean daily motion of the sun in longitude is 59′ 8″.33, and that of the moon 13° 10′ 35″.03; wherefore the moon separates from the sun at the mean rate of 12° 11′ 26″.70 per day;

and hence, to find the mean length of a lunar month, we have the proportion

$$12° \ 11' \ 26''.70 : 1\text{d.} :: 360° : x = 29\text{d. } 12\text{h. } 44\text{m. } 2.7\text{s.}$$

301. To Determine the Time of Mean New, or Full Moon, in any Given Month. Let the mean longitude of the sun, and also the mean longitude of the moon, at the beginning of the year, be found, and let the former be subtracted from the latter (adding 360° if necessary); the remainder, which call R, will be the mean distance of the moon to the east of the sun, at the beginning of the year. As the moon separates from the sun at the mean rate of 12° 11′ 26″.70 per day, $\frac{R}{12° \ 11' \ 26''.70}$ will express the number of days and fractions of a day, which at this epoch have elapsed since the last new moon. This interval is called the *Astronomical Epact.* If we subtract it from 29d. 12h. 44m. 2 7s. we shall have the time of mean new moon in January. This being known, the time of mean new moon in any other month of the year results very readily from the known length of a lunar month.

The time of mean new moon in any month being known, the time of mean full moon in the same month is obtained by the addition or subtraction, as the case may be, of half a lunar month.

This problem is in practice most easily resolved with the aid of tables. (See Problem XXVII.)

The time of true new moon differs from the time of mean new moon, for the same reasons that the true longitudes of the sun and moon differ from the mean. The same is true of the time of true full moon. For the mode of computing the time of true new or full moon from that of mean new or full moon, see Problem XXVII.

302. The Earth goes through the same Phases, as viewed from the moon, in the course of a lunar month that the moon does to an inhabitant of the earth. But, at any given time, the phase of the earth is just the opposite to the phase of the moon. About the time of new moon, the earth, then near its full, reflects so much light to the moon as to render the obscure part visible. (See Fig. 75.)

MOON'S RISING, SETTING, AND PASSAGE OVER THE MERIDIAN.

303. To find the Time of the Meridian Passage of the Moon on a Given Day.

Let S and M denote, respectively, the right ascension of the sun and the right ascension of the moon, at noon on the given day, and m, s, the hourly variations of the right ascension of the sun and moon; also let $t =$ the required time of the meridian passage. At the time t the right ascensions will be,

for the moon .... $M + tm$,
for the sun .... $S + ts$;

and, as the moon is on the meridian, the difference of these arcs will be equal to the hour angle t; whence,

$$t = M - S + t(m - s);$$

or, if all the quantities be expressed in seconds,

$$t = M - S + t\frac{m - s}{3600} \dots (52).$$

Thus, we find for the time of the meridian passage,

$$t = \frac{3600\,(M - S)}{3600 - (m - s)} \dots (53).$$

The quantities M, S, m, s, are, in practice, to be taken from ephemerides of the sun and moon.

Example. What was the time of the passage of the moon's centre over the meridian of New York on the 1st of August, 1837?

When it is noon at New York, it is 4h. 56m. 4s. at Greenwich. Now, by the English Nautical Almanac,

Aug. 1st, at 4h. ☽'s R. Ascen.	8h.	58m.	36.7s.
" at 5h. " "	9	0	38.3

1h. : 56m. 4s. : : 2m. 1.6s. : 1m. 53.6s.

Aug. 1st, at 4h. ☽'s R. Ascen.	8h.	58m.	36.7s.
Variation of R. Ascen. in 56m. 4s		1	53.6
☽'s R. Ascen. at M. Noon at N. York	9	0	30.3

Aug. 1st. ☉'s hourly Variation of R. Ascen. .... 9.704s.

1h. : 4h. 56m. 4s. : : 9.704s. : 47.8s.

Aug. 1st, M. Noon at Greenw., ☉'s R. Asc. ...	8h.	45m.	31.5s.
Variation of R. Ascen. in 4h. 56m. 4s			47.8
☉'s R. Ascen. at M. Noon at N. York	8	46	19.3

Aug. 1st, M. Noon at Greenw., ☽'s R. Asc. ...	8h.	50m.	27.7s.
Aug. 2d, " " "	9	38	18.7
	24)	47	51.0
Aug. 1st, ☽'s mean hourly Varia. of R. Asc. ...		1	59.6 (m)
" ☉'s " " " ...			9.7 (s)
	$m - s =$	1	49.9 = 109.9s.

By Nautical Almanac, equation of time = 5m. 59s.

1h. : 5m. 59s. : : 1m. 59.6s. : 11.9s.

☽'s R. Ascen. at M. Noon at N. York	9h.	0m.	30.3s.
Correction for equation of time			− 11.9
☽'s R. Ascen. at apparent Noon at N. York ...	9	0	18.4 (M)
☉'s " " " " ...	8	46	18.3 (S)
	M − S =	14	0.1 = 840.1s.

3600..................	log. 3.55630
M — S = 840.1	log. 2.92433
3600 — (m — s) = 3490.1............ar. co.	log. 6.45716

Apparent time of meridian passage, 14m. 26.5s. = 866.5s.. log. 2.93779
Equa. of time at merid. passage, 5 58

Mean time of meridian passage, 0h. 20m. 24s.

The Nautical Almanac gives the time of the moon's passage over the meridian of Greenwich for every day of the year. From this, the time of the passage across the meridian of any other place may easily be determined, as follows: subtract the time of the meridian passage at Greenwich on the given day, from that on the following day, and say, as 24h. : the difference : : the longitude of the place : a fourth term. The fourth term, added to the time of the meridian passage at Greenwich on the given day, will give the time of the meridian passage on the same day at the given place.

304. Moon's Rising and Setting. Since the moon has a motion with respect to the sun, the time of its rising and setting must vary from day to day. When first seen after conjunction, it will set soon after the sun. After this it will set (at a mean) about 50m. later every succeeding night. At the first quarter, it will set about midnight; and at full moon, will set about sunrise, and rise about sunset. During this interval it will rise in the daytime, and all along from sunrise to sunset. From full to new moon, it will rise at night and set during the day; and the time of the rising and setting will be about 50m. later on every succeeding night and day; thus, at the last quarter, it will rise about midnight, and set about midday.

305. Daily Retardation of Moon's Rising. The daily retardation of the time of the moon's rising is, as just stated, at a mean, about 50 minutes; but it varies in the course of a revolution from less than half an hour to one hour, in these latitudes. The retardation of the moon's rising at the time of full moon, varies from one full moon to another, in the course of the year, between the same limits. The reason of these variations is found in the fact, that the arc of the ecliptic (12° 11′) through which the moon moves away from the sun in a day, is variously inclined to the horizon, according to its situation in the ecliptic, and therefore employs different intervals of time in rising above the horizon. This fact may be very distinctly shown by means of a celestial globe. It will be seen that the arc in question will be most oblique to the horizon, and rise in the shortest time, in the signs Pisces and Aries. Accordingly, the full moons which occur in these signs will rise with the smallest retardation from day to day. These full moons occur when the sun is in the opposite signs, Virgo and Libra, that is, in September and October. They are called, the first the *Harvest Moon*, and the second the *Hunter's Moon.* The time of the moon's rising at these full moons will, for two or three days, be only about half an hour later than on the preceding day.

306. *To find the time of the moon's rising or setting on any given day.* Compute the moon's semi-diurnal arc from equation (50), or (48), according as it is the time of the apparent rising or setting, or the time of the true rising or setting, that is desired. Correct it for the moon's change of right ascension in the interval between the moon's passage over the meridian and setting, by the following proportion, 24h : 24h + $m - s$ :: semi-diurnal arc : corrected semi-diurnal arc; and add it to the time of the moon's meridian passage, found as explained in Art. 303. The result will be the time of the moon's setting: and if this be subtracted from 24 hours, the remainder will be the time of the moon's rising.

In consequence of the change of the moon's declination in the interval between its rising and setting, it would be more accurate to compute the semi-diurnal arc separately for the moon's rising. In computing the semi-diurnal arc by equation (48), the declination 6 hours before or after the meridian passage may be used at first; and afterwards, if a more accurate result be desired, the calculation may be repeated with the declination found for the computed approximate time. In equation (49), R = refraction — parallax = 34′ 54″ — 57′ 3″ (at a mean) = — 22′ 9″.

ROTATION AND LIBRATIONS OF THE MOON.

307. The moon presents continually nearly the same face towards the earth; for the same spots are always seen in nearly the same position upon the disc. It follows, therefore, that it rotates on its axis in the same direction, and with the same angular velocity, or nearly so, that it revolves in its orbit, and thus completes one rotation in the same period of time in which it accomplishes a revolution in its orbit.

308. Librations of the Moon. The spots on the moon's disc, although they constantly preserve very nearly the same situations, are not, however, strictly stationary. When carefully observed, they are seen alternately to approach and recede from the edge. Those that are very near the edge successively disappear and again become visible. This vibratory motion of the moon's spots is called *Libration*.

There are three librations of the moon, that is, a vibratory motion of its spots from three distinct causes.

(1.) The moon's motion of rotation being uniform, small portions on its east and west sides alternately come into sight and disappear, in consequence of its *unequal motion in its orbit*. The periodical oscillations of the spots in an easterly and westerly direction from this cause, are called the *Libration in Longitude*.

(2.) The lunar spots have also a small alternate motion from north to south. This is called the *Libration in Latitude*, and is accounted for by supposing that the moon's axis is not exactly perpendicular to the plane of its orbit, and that it remains continually parallel to itself. On this supposition, we ought sometimes to see beyond the north pole of the moon, and sometimes beyond the south pole.

(3.) Parallax is the cause of a third libration of the moon. The spectator upon the earth's surface being removed from its centre, the point towards which the moon continually presents the same hemisphere, he will see portions of the moon a little different according to its different positions above the horizon. The diurnal motion of the spots resulting from the parallax, is called the *Diurnal* or *Parallactic Libration.*

309. Equator of the Moon. The exact position of the moon's equator, like that of the sun's, is derived from accurate observations of the situations of the spots upon the disc. From calculations founded upon such observations, it has been ascertained that the plane of the moon's equator is constantly inclined to the plane of the ecliptic under an angle of 1° 32′, and intersects it in a line which is always parallel to the line of the nodes. It follows from the last mentioned circumstance, that if a plane be supposed to pass through the centre of the moon, parallel to the ecliptic, it will intersect the plane of the moon's equator and that of its orbit in the same line in which these planes intersect each other. The plane in question will lie between the plane of the equator and that of the orbit. It will make with the first an angle of 1° 32′, and with the second an angle of 5° 9′.

DIMENSIONS AND PHYSICAL CONSTITUTION OF THE MOON.

310. Diameter;—Surface;—Volume. The phases of the moon indicate that it is an opake spherical body. Its diameter is found by means of equ. (51), viz.:

$$d = 2R\frac{\delta}{2H},$$

in which d will denote the diameter sought, R the radius of the earth, δ the apparent diameter of the moon at a given distance, and H its horizontal parallax at the same distance.

The equatorial horizontal parallax of the moon, at the mean distance, is 57′ 2″.7, and its corresponding apparent diameter is 31′ 7″.0: thus we have

$$d = 2R\frac{31' \; 7''.0}{114' \; 5''.4} = 7925.6\text{m.} \times \frac{1867''}{6845''4} = 2161.6 \text{ miles.}$$

The ratio of the diameter of the moon to the mean diameter of the earth (7912.4m.) is 0.27319. This is very nearly equal to $\frac{3}{11}$, or a little more than $\frac{1}{4}$. The surface of the moon is therefore to the surface of the earth nearly as 3^2 to 11^2, or as 1 to $13\frac{1}{2}$; and the volume of the moon is to the volume of the earth nearly as 3^3 to 11^3, or as 1 to 49.

311. Telescopic Appearances:—Inferences. When the moon is viewed with a telescope, the edge of the disc which borders upon the dark portion of the face, is seen to be very irregular and serrated (see Fig. 76). It is hence inferred that the surface of the moon is diversified by mountains and valleys. The truth of this inference is confirmed by the fact that bright insulated spots are frequently seen on the dark part of the face, near the edge of the disc, which gradually enlarge until they become united to the disc. These bright spots are doubtless the tops of mountains illuminated by the sun, while the sur-

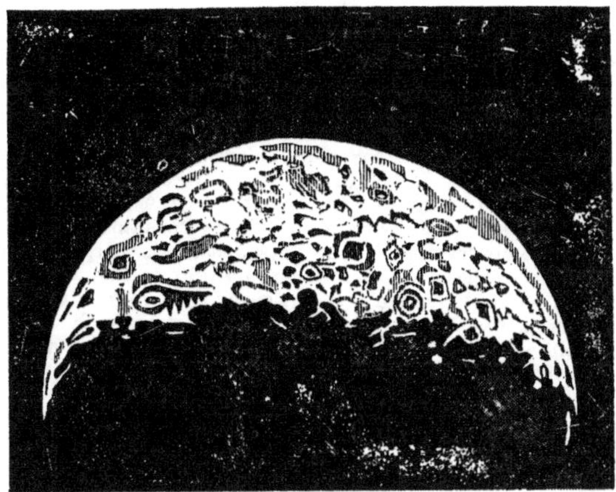

FIG. 76.

rounding regions that are less elevated are involved in darkness. The disc is also diversified with spots of different shapes and different degrees of brightness. The brighter parts are supposed to be elevated land, and the dark to be plains, and valleys, or cavities.

312. Lunar Mountains. The number of the lunar mountains is very great. Many of them, by their form and grouping, furnish decided indications of a volcanic origin.

From measurements made with the micrometer of the lengths of their shadows, or of the distance of their summits when first illuminated, from the adjacent boundary of the disc, the heights of a number of the lunar mountains have been computed. According to Herschel, the altitude of the highest is only about $1\frac{3}{4}$ English miles. But Schroeter, of Lilienthal, a distinguished Selenographist, makes the elevation of some of the lunar mountains to exceed 5 miles; and the more recent measurements of MM. Baer and Mädler, of Berlin, lead to similar results.

313. There are no Seas, nor other bodies of water, upon the surface of the moon. Certain dark and apparently level parts of the moon were for some time supposed to be extended sheets of water, and, under this idea, were named by Hevelius *Mare Imbrium*, *Mare Crisium*, etc.: but it appears that when the boundary of light and darkness falls upon these supposed seas, it is still more or less indented at some points and salient at others, instead of being, as it should be, one continuous regular curve;

besides, when these dark spots are viewed with good telescopes, they are found to contain a number of cavities, whose shadows are distinctly perceived falling within them. The spots in question are therefore to be regarded as extensive plains diversified by moderate elevations and depressions. The entire absence of water also from the farther hemisphere of the moon may be inferred from the fact that the moon's face is never obscured by clouds or mists.

314. Lunar Atmosphere. It has long been a question among Astronomers, whether the moon has an atmosphere. It is asserted, that, if it has any, it must be exceedingly rare, or very limited in its extent, since it does not sensibly diminish or refract the light of a star seen in contact with the moon's limb; for when a star experiences an occultation by reason of the interposition of the moon between it and the eye of the observer, it does not disappear or undergo any diminution of lustre until the body of the moon reaches it; and the duration of the occultation is as it is computed, without making any allowance for the refraction of a lunar atmosphere. But it is maintained, on the other hand, that these facts, if allowed, are not opposed to the supposition of the existence of an atmosphere of a few miles only in height; and that certain phenomena which have been observed afford indubitable evidence of the presence of a certain limited body of air upon the moon's surface. Thus, the celebrated Schroeter, in the course of some delicate observations made upon the crescent moon, perceived a faint grayish light extending from the horns of the crescent a certain distance into the dark part of the moon's face. This he conceived to be the moon's twilight, and hence inferred the existence of a lunar atmosphere. From the measurements which he made of the extent of this light he calculated the height of that portion of the atmosphere which was capable of affecting the light of a star to be about one mile. Again, in total eclipses of the sun, occasioned by the interposition of the moon, the dark body of the moon has been seen terminated by a luminous ring, which was at first most distinct at the part where the sun was last seen, and afterwards at the part where the first ray darted from the sun. This is supposed to have been a lunar twilight. A similar phenomenon was observed in the annular eclipse of 1836, just before the completion of the ring, at the point where the junction took place.

DESCRIPTION OF THE MOON'S SURFACE.

315. General Topographical Features. The surface of the moon, like that of the earth, presents the two general varieties of level and mountainous districts; but it differs from the earth's surface in having no seas or other bodies of water upon it, and in being more rugged and mountainous. The comparatively level regions occupy somewhat more than one-third of the nearer half of the moon's surface. These are, in general, the darker parts of the disc. The lunar plains vary in extent from 40 or 50 miles to 700 miles in diameter.

316. The Mountainous Formations of the other parts of the surface offer three marked varieties, viz.:

(1.) *Insulated Mountains*, which rise from plains nearly level, and which may be supposed to present an appearance somewhat similar to Mount Etna or the Peak of Teneriffe. The shadows of these mountains, in certain phases of the moon, are as distinctly perceived as the shadow of an upright staff when placed opposite to the sun.* The perpendicular altitudes of some of them, as determined from the lengths of their shadows, are between four and five miles. Insulated mountains frequently occur in the centres of circular plains. They are then called *Central Mountains.*

(2.) *Ranges of Mountains*, extending in length two or three hundred miles. These ranges bear a distinct resemblance to our Alps, Apennines, and Andes; but they are much less in extent, and do not form a very prominent feature of the lunar

* Dick's Celestial Scenery, p. 256.

surface. Some of them appear very rugged and precipitous, and the highest ranges are, in some places, above four miles in perpendicular altitude. In some instances they run nearly in a straight line from northeast to southwest, as in the range called the *Apennines;* in other cases they assume the form of a semicircle or a crescent.*

(3.) *Circular Formations.* The general prevalence of this remarkable class of mountainous formations is the great characteristic feature of the topography of the moon's surface. It is subdivided by late selenographists into three orders, viz.: *Walled Plains,* whose diameter varies from one hundred and twenty to forty or fifty miles, *Ring Mountains,* the diameter of which descends to ten miles; and *Craters,* which are still smaller. The term crater is sometimes extended to all the varieties of circular formations. They are also sometimes called *Caverns,* because their enclosed plains or bottoms are sunk considerably below the general level of the moon's surface.

The different orders of the circular formations differ essentially from each other only in size. The principal features of their constitution are, for the most part, the same, and they present similar varieties. Sometimes terraces are seen going round the whole ring. At other times ranges of concentric mountains encircle the inner foot of the wall, leaving intermediate valleys. Again, we have a few ridges of low mountains stretching through the circle contained by the wall, but oftener isolated conical peaks start up, and very frequently small craters having on an inferior scale every attribute of the large one.† The smaller craters, however, offer some characteristic peculiarities. Most of them are without a flat bottom, and have the appearance of a hollow inverted cone with the sides tapering towards the centre. Some have no perceptible outer edge, their margin being on a level with the surrounding regions: these are called *Pits.*

The bounding ridge of the lunar craters or caverns is much more precipitous within than without; and the internal depth of the crater is always much lower than the general surface of the moon. The depth varies from one-third of a mile to three miles and a half.

These curious circular formations occur at almost every part of the surface, but are most abundant in the southwestern regions. It is the strong reflection of their mountainous ridges which gives to that part of the moon's surface its superior lustre. The smaller craters occupy nearly two-fifths of the moon's visible surface.

* Dick's Celestial Scenery, p. 257.

† Nichol's Phenomena of the Solar System, p. 167.

CHAPTER XVI.

ECLIPSES OF THE SUN AND MOON.—OCCULTATIONS OF THE FIXED STARS.

317. AN eclipse of a heavenly body is a deprivation of its light, occasioned by the interposition of some opake body between it and the eye, or between it and the sun. Eclipses are divided, with respect to the objects eclipsed, into *eclipses of the sun*, *of the moon*, and *of the satellites;* and, with respect to circumstances, into *total*, *partial*, *annular*, and *central*. A *total* eclipse is one in which the whole disc of the luminary is darkened; a *partial* one is when only a part of the disc is darkened. In an *annular* eclipse the whole is darkened, except a ring or annulus, which appears round the dark part like an illuminated border; the definition of a central eclipse will be given in another place.

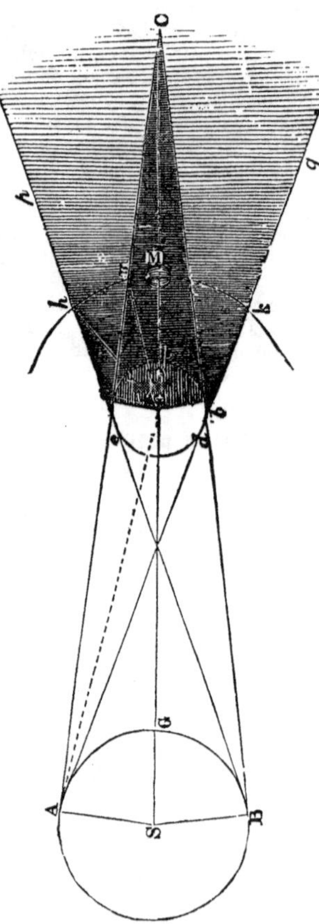

FIG. 77.

ECLIPSES OF THE MOON.

318. An eclipse of the moon is occasioned by an interposition of the body of the earth directly between the sun and moon, and thus intercepting the light of the sun; or the moon is eclipsed when it passes through part of the shadow of the earth, as projected from the sun. Hence it is obvious that lunar eclipses can happen only at the time of full moon, for it is then only that the earth can be between the moon and the sun.

319. Earth's Shadow. Since the sun is much larger than

the earth, the shadow of the earth must have the form of a cone, the length of which will depend on the relative magnitudes of the two bodies and their distance from each other. Let the circles AGB, *agb* (Fig. 77), be sections of the sun and earth by a plane passing through their centres S and E; A*a*, B*b*, tangents to these circles on the same side, and A*d*, B*c*, tangents on different sides. The triangular space *a*C*b* will be a section of the earth's shadow or *Umbra*, as it is sometimes called. The line EC is called the *Axis of the Shadow.* If we suppose the line *cp* to revolve about EC, and form the surface of the frustum of a cone, of which *pcdq* is a section, the space included within that surface and exterior to the umbra, is called the *Penumbra.* It is plain that points situated within the umbra will receive no light from the sun; and that points situated within the penumbra will receive light from a portion of the sun's disc, and from a greater portion the more distant they are from the umbra.

320. To find the Length of the Earth's Shadow. Let L = the length of the shadow; R = the radius of the earth; δ = the sun's apparent semi-diameter, and p = sun's parallax. The right-angled triangle E*a*C (Fig. 77) gives

$$\mathrm{EC} = \frac{\mathrm{E}a}{\sin \mathrm{EC}a}.$$

$\mathrm{E}a = \mathrm{R}$; and $\mathrm{EC}a = \mathrm{SEA} - \mathrm{EAC} = \delta - p$; whence,

$$\mathrm{L} = \frac{\mathrm{R}}{\sin(\delta - p)} \ldots (54).$$

As the angle $(\delta - p)$ is only about 16′, it will differ but little from its sine, and therefore,

$$\mathrm{L} = \mathrm{R}\frac{1}{\delta - p} \text{ (nearly)};$$

or, if δ and p be expressed in seconds,

$$\mathrm{L} = \mathrm{R}\frac{206264''.8}{\delta - p} \text{ (nearly)} \ldots (55).$$

The shadow will obviously be the shortest when the sun is nearest to the earth. We then have $\delta = 16'\ 18''$, and $p = 9''$, which gives L = 213 R. The greatest distance of the moon is 65R. It appears, then, that *the earth's shadow always extends to more than three times the distance of the moon.*

321. Circumstances under which an Eclipse occurs. Let *k*M*h* be a circular arc, described about E the centre of the earth, and with a radius equal to the distance between the centres of the earth and moon at the time of opposition. The angle ME*m*, the apparent semi-diameter of a section of the earth's shadow, made at the distance of the moon's centre, is called the *Semi-diameter of the Earth's Shadow.* And the angle ME*h*, the apparent semi-diameter of a section of the penumbra, at the same distance, is called the *Semi diameter of the Penumbra.*

Were the plane of the moon's orbit coincident with the plane of the ecliptic, there would be a lunar eclipse at every full moon; but, as it is inclined to it, an eclipse can happen only when the full moon takes place either in one of the nodes of the moon's orbit, or so near it that the moon's latitude does not exceed the sum of the apparent semi-diameters of the moon and of the earth's shadow. This will be better understood on referring to Fig. 78, in which N′C represents a portion of the ecliptic, and N′M a portion of the moon's orbit, N′ the descending node, E the earth, ES, ES′, ES″, three different directions of the sun, s, s', s'', sections of the earth's shadow in the three several positions corresponding to these directions of the sun, and m, m', m'', the moon in opposition. It will be seen that the moon will not pass into the earth's shadow unless at the time of opposition it is nearer to the node than the point m', where the latitude $m's'$ is equal to the sum of the semi-diameters of the moon and shadow.

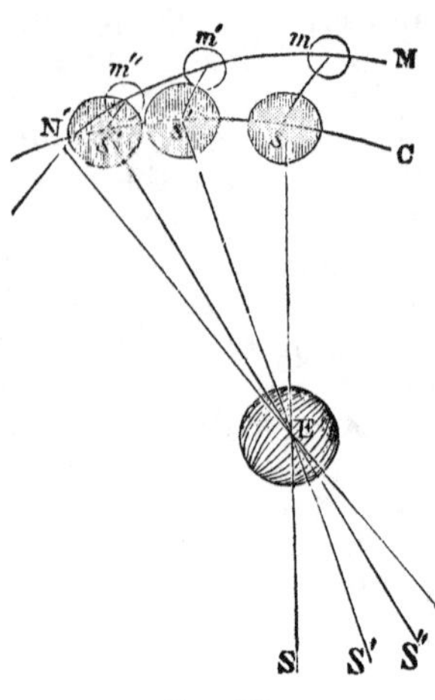

FIG. 78.

322. Calculation of Semi-diameter of Shadow. To determine the distance from the node, beyond which there can be no eclipse, we must ascertain the semi-diameter of the earth's shadow. Let this be denoted by A, and let P = the moon's parallax.

$$\text{ME}m = \text{E}ma - \text{EC}m \text{ (Fig. 77);}$$

but $\text{E}ma = \text{P}$ and $\text{EC}m = \delta - p$ (320); therefore,

$$\text{ME}m = \text{A} = \text{P} + p - \delta \ldots (56).$$

The semi-diameter of the shadow is the least when the moon is at its greatest and the sun is at its least distance, or when P has its minimum and δ its maximum value. In these positions of the moon and sun, $\text{P} = 52' 40''$, $\delta = 16' 18''$, and $p = 9''$. Substituting, we obtain for the least semi-diameter of the earth's shadow 36′ 31″, and for its least diameter 1° 13′ 2″. The greatest apparent diameter of the moon is 33′ 32″. Whence it appears that *the diameter of the earth's shadow is always more than twice the diameter of the moon.*

The means of the greatest and least values of P and δ are, respectively, 57′ 11″ and 16′ 2″; which gives for the mean semi-diameter of the earth's shadow, 41′ 18″.

323. Lunar Ecliptic Limits. If to $\text{P} + p - \delta$, the semi-diameter of the earth's shadow, we add d, the semi-diameter of the moon, the sum $P + p + d - \delta$ will give the greatest latitude of the moon in opposition, at which an eclipse can happen. It is easy for a given value of $P + p + d - \delta$, and a given inclination of the moon's orbit, to determine within what distance from the node the moon must be in order that an eclipse may take place. By taking the least and greatest inclinations of the orbit, the greatest and least values of $P + p + d - \delta$, and also taking into view the inequalities in the motions of the sun and moon, it has been found, that when at the time of mean full moon the difference of the mean longitudes of the moon and node exceeds 13° 21′, there cannot be an eclipse; but when this difference is less than 7° 47′ there must be one. Between 7° 47′ and 13° 21′ the happening of the eclipse is doubtful. These numbers are called the *Lunar Ecliptic Limits.*

To determine at what full moons in the course of any one year there will be an eclipse, find the time of each mean full moon (301); and for each of the times obtained find the mean longitude of the sun, and also of the moon's node, and compare the difference of these with the lunar ecliptic limits. Should, however, the difference in any instance fall between the two limits, farther calculation will be necessary.

This problem may be solved more expeditiously by means of tables of the sun's mean motion with respect to the moon's node. (See Prob. XXVIII.)

324. Central Eclipse. The magnitude and duration of an eclipse depend upon the proximity of the moon to the node at the time of opposition. In order that the centre of the moon may be on the same right line with the centres of the sun and earth, or, in technical language, that a *central* eclipse may happen, the opposition must take place precisely in the node. A strictly central eclipse, therefore, seldom, if ever, occurs. As the mean semi-diameter of the earth's shadow is 41′ 18″, the mean semi-diameter of the moon 15′ 35″, and the mean hourly motion of the moon with respect to the sun 30′ 29″, the mean duration of a central eclipse would be about $3\frac{3}{4}$h.

325. Particular Facts. Since the moon moves from west to east, an eclipse of the moon must commence on the eastern limb, and end on the western.

In the preceding investigations, we have supposed the cone of the earth's shadow to be formed by lines drawn from the edge of the sun, and touching the earth's surface. This, probably, is not the exact case of nature; for the duration of the eclipse, and thus the apparent diameter of the earth's shadow, is found by observation to be somewhat greater than would result from this supposition. This circumstance is accounted for by supposing those solar rays that, from their direction, would glance by and raze the earth's surface, to be stopped and absorbed by the lower strata of the atmosphere. In such a case the conical boundary of the earth's shadow would be formed by certain rays exterior to the former, and would be larger.

The moon in approaching and receding from the earth's total shadow, or umbra, passes through the penumbra, and thus its light, instead of being extinguished and recovered suddenly, experiences at the beginning of the eclipse a gradual diminution, and at the end a gradual increase. On this account the times of the beginning and end of the eclipse cannot be noted with precision, and in consequence astronomers differ as to the amount of the increase in the size of the earth's shadow from the cause above mentioned. It is the practice, however, in computing an eclipse of the moon, to increase the semi-diameter of the shadow by a $\frac{1}{60}$ part; or, which amounts to the same, to add as many seconds as the semi-diameter contains minutes

It is remarked in total eclipses of the moon, that the moon is not wholly invisible, but appears with a dull reddish light. This phenomenon is doubtless another effect of the earth's atmosphere, though of a totally different nature from the preceding. Certain of the sun's rays, instead of being stopped and absorbed, are bent from their rectilinear course by the refracting power of the atmosphere, so as to form a cone of faint light, interior to that cone which has been mathematically described as the earth's shadow, which falling upon the moon renders it visible.

As an eclipse of the moon is occasioned by a real loss of its light, it must begin and end at the same instant, and present precisely the same appearance to every spectator who sees the moon above his horizon during the eclipse. It will be shown that the case is different with eclipses of the sun.

CALCULATION OF AN ECLIPSE OF THE MOON.

326. The apparent distance of the centre of the moon from the axis of the earth's shadow, and the arcs passed over by the centre of the moon and the axis of the shadow during an eclipse of the moon, being necessarily small, they may, without material error, be considered as right lines. We may also consider the apparent motion of the sun in longitude, and the motions of the moon in longitude and latitude, as uniform during the eclipse. These suppositions being made, the calculation of the circumstances of an eclipse of the moon is very simple.

327. Relative Orbit. Let NF (Fig. 79) be a part of the ecliptic, N the moon's ascending node, NL a part of the moon's orbit, C the centre of a section of the earth's shadow at the moon, CK perpendicular to NF a circle of latitude, and C′ the centre of the moon at the instant of opposition: then CC′, which is the latitude of the moon in opposition, is the distance of the centres of the shadow and moon at that time. The moon and shadow both have a motion, and in the same direction, as from N towards F and L. It is the practice, however, to regard the shadow as stationary, and to attribute to the moon a motion equal

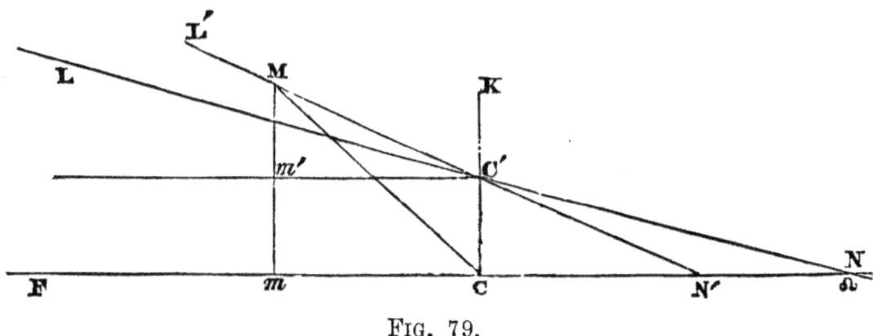

FIG. 79.

to the relative motion of the moon and shadow. The orbit that would be described by the moon's centre if it had such a motion, is called the *Relative Orbit* of the moon. Inasmuch as the circumstances of the eclipse depend altogether upon the relative motion of the moon and shadow, this mode of proceeding is obviously allowable.

As the shadow has no motion in latitude, the relative motion of the moon and shadow in latitude will be equal to the moon's actual motion in latitude: and since the centre of the earth's shadow moves in the plane of the ecliptic at the same rate as the sun, the relative motion of the moon and shadow in longitude will be equal to the difference between the motions of the sun and moon in longitude. We obtain, therefore, the relative position of the centres of the moon and shadow at any interval t, following opposition, by laying off Cm equal to the difference of the motions of the sun and moon in longitude in this interval, through m drawing mM perpendicular to NF, and cutting off mM equal to the latitude at opposition plus the motion in latitude in the interval t: M will be the position of the moon's centre in the relative orbit, the centre of the shadow being supposed to be stationary at C. As the motion of the sun in longitude, and of the moon in longitude and latitude, are considered uniform, the ratio of C'm' (= Cm, the difference between the motions of the sun and moon in longitude) to Mm' the moon's motion in latitude, is the same, whatever may be the length of the interval considered. It follows, therefore, that the relative orbit of the moon N'C'M is a *right line.*

The relative orbit passes through C', the place of the moon's centre at opposition: its position will therefore be known, if its inclination to the ecliptic be found. Now we have

$$\text{tan inclina.} = \frac{Mm'}{C'm'} = \frac{\text{moon's motion in latitude}}{\text{moon's mot. in long.} - \text{sun's mot. in long.}}$$

328. Requisite Data. The following data are requisite in the calculation of the circumstances of a lunar eclipse:

$T =$ time of opposition.
$M =$ moon's hourly motion in longitude.
$n =$ moon's hourly motion in latitude.
$m =$ sun's hourly motion in longitude.
$\lambda =$ moon's latitude at opposition.
$d =$ moon's semi-diameter.
$\delta =$ sun's semi-diameter.
$P =$ moon's horizontal parallax.
$p =$ sun's horizontal parallax.
$s =$ semi-diameter of the earth's shadow.
$I =$ inclination of relative orbit.
$h =$ moon's hourly motion on relative orbit.

T, M, n, m, λ, d, δ, P, and p, are derived from Tables of the sun and moon. (See Problems IX and XIV.)

The quantities s, I, and h, may be determined from these:

$$s = P + p - \delta + \tfrac{1}{60}(P + p - \delta) \text{ (322 and 325)....(57)};$$

$$\text{tang } I = \frac{n}{M - m} \text{ (327)....(58)}.$$

The triangle C'Mm' gives

$$C'M = \frac{C'm'}{\cos MC'm'}, \text{ or, } h = \frac{M - m}{\cos I} \text{....(59)}.$$

329. Process of Calculation. The above quantities being supposed to be known, let N'CF (Fig. 80) represent the ecliptic, and C the stationary centre of the earth's shadow. Let CC' $= \lambda$, and let N'C'L' represent the relative orbit of the

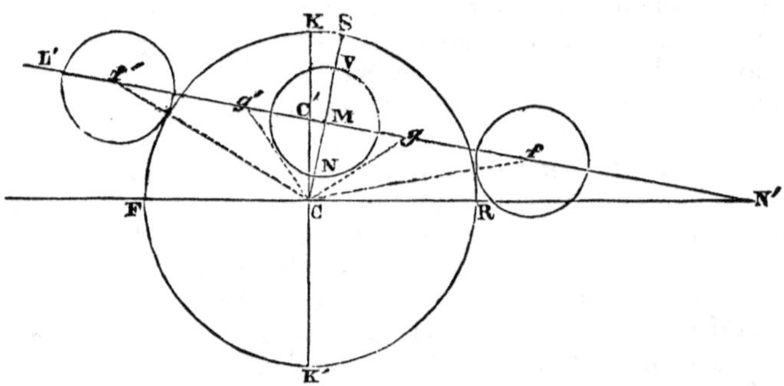

FIG. 80.

moon. We here suppose the moon to be north of the ecliptic at the time of opposition and near its ascending node; when it is south of the ecliptic λ is to be laid off below N'CF, and when it is approaching either node, the relative orbit is inclined to the right. Let the circle KFK'R, described about the centre C, represent the section of the earth's shadow at the moon; and let f, f', and g, g', be the respective places of the moon's centre, at the beginning and end of the eclipse, and at the beginning and end of the total eclipse. $Cf = Cf' = s + d$, and $Cg = Cg' = s - d$. Draw CM perpendicular to N'C'L', and M will represent the place of the moon's centre when nearest the centre of the shadow: it will also be its place at the middle of the eclipse; for since $Cf = Cf'$, and CM is perpendicular to N'Cf' $Mf = Mf'$.

Middle of the eclipse. The time of opposition being known, that of the middle of the eclipse will become known when we have found the interval (x) employed by the moon in passing from M to C'. Now

$$\text{(expressed in parts of an hour) } x = \frac{MC'}{h};$$

and in the right-angled triangle CC′M we have $CC' = \lambda$, and $< C'CM = < C'N'C = I$, and therefore $MC' = \lambda \sin I$; whence, by substitution,

$$x = \frac{\lambda \sin I}{h} = \frac{\lambda \sin I}{\frac{M-m}{\cos I}} \text{ (equ. 59)} = \frac{\lambda \sin I \cos I}{M-m};$$

$$\text{or (expressed in seconds), } x = \frac{3600s. \cos I}{M-m}.\lambda \sin I..(60).$$

Hence, if $M =$ time of middle, we have

$$M = T \mp x = T \mp \frac{3600s. \cos I}{M-m}.\lambda \sin I....(61).$$

It is obvious that the *upper* sign is to be used when the latitude is *increasing*, and the *lower* sign when it is *decreasing.*

The distance of the centre of the moon from the centre of the shadow at the middle of the eclipse,

$$= CM = CC' \cos C'CM = \lambda \cos I....(62).$$

Beginning and end of the eclipse. Let any point l of the relative orbit be the place of the moon's centre at the time of any given phase of the eclipse. Let $t =$ the interval of time between the given phase and the middle; and $k = Cl$, the distance between the centres of the moon and shadow. In the interval t the moon's centre will pass over the distance Ml; hence

$$t = \frac{Ml}{h} = \frac{Ml \cos I}{M-m}:$$

but, $$Ml = \sqrt{\overline{Cl}^2 - \overline{MC}^2} = \sqrt{k^2 - \lambda^2 \cos^2 I} \text{ (equa. 62)},$$

and therefore $$t = \frac{\cos I}{M-m}\sqrt{k^2 - \lambda^2 \cos^2 I};$$

or (in seconds), $$t = \frac{3600s. \cos I}{M-m}\sqrt{(k + \lambda \cos I)(k - \lambda \cos I)}....(63).$$

Let T' denote the time of the supposed phase of the eclipse, and M the time of the middle; and we shall have

$$T' = M + t, \text{ or, } T' = M - t,$$

according as the phase follows or precedes the middle.

Now, at the beginning and end of the eclipse, we have,

$$k = Cf \text{ or } Cf' = s + d:$$

substituting in equation (63) we obtain

$$t' = \frac{3600s. \cos I}{M-m}\sqrt{(s + d + \lambda \cos I)(s + d - \lambda \cos I)}....(64).$$

t' being found, the time of the beginning (B), and the time of the end (E), result from the equations

$$B = M - t', \ E = M + t'.$$

Beginning and end of the total eclipse. At the beginning and end of the total eclipse, $k = Cg = Cg' = s - d$; whence, by equation (63),

$$t'' = \frac{3600s. \cos I}{M-m}\sqrt{(s - d + \lambda \cos I)(s - d - \lambda \cos I)}....(65):$$

and, denoting the time of the beginning by B' and the time of the end by E', we have $B' = M - t''$, $E' = M + t''$.

Quantity of the eclipse. In a partial eclipse of the moon the magnitude or quantity of the eclipse is measured by the relative portion of that diameter of the moon, which, if produced, would pass through the centre of the earth's shadow, that is involved in the shadow. The whole diameter is divided into twelve equal parts, called *Digits*, and the quantity is expressed by the number of digits and fractions of a digit in the part immersed. When the moon passes entirely within the shadow, as in a total eclipse, the quantity of the eclipse is expressed by the number of digits contained in the part of the same diameter prolonged outward, which is comprised between the edge of the shadow and the inner edge of the moon. Thus the number of digits contained in SN (Fig. 80) expresses the quantity of the eclipse represented in the figure. Hence, if Q = the quantity of the eclipse, we shall have

$$Q = \frac{NS}{\frac{1}{12}NV} = \frac{12NS}{NV} = \frac{12\,(NM + MS)}{NV} = \frac{12\,(NM + CS - CM)}{NV} = \frac{12\,(d + s - \lambda \cos I)}{2d};$$

or,

$$Q = \frac{6\,(s + d - \lambda \cos I)}{d} \ldots (66).$$

If $\lambda \cos I$ exceeds $(s + d)$ there will be no eclipse. If it is intermediate between $(s + d)$ and $(s - d)$ there will be a partial eclipse; and if it is less than $(s - d)$ the eclipse will be total.

CONSTRUCTION OF AN ECLIPSE OF THE MOON.

330. The times of the different phases of an eclipse of the moon may easily be determined by a geometrical construction, within a minute or two of the truth. Draw a right line N′F (Fig. 81) to represent the ecliptic; and assume upon it any point

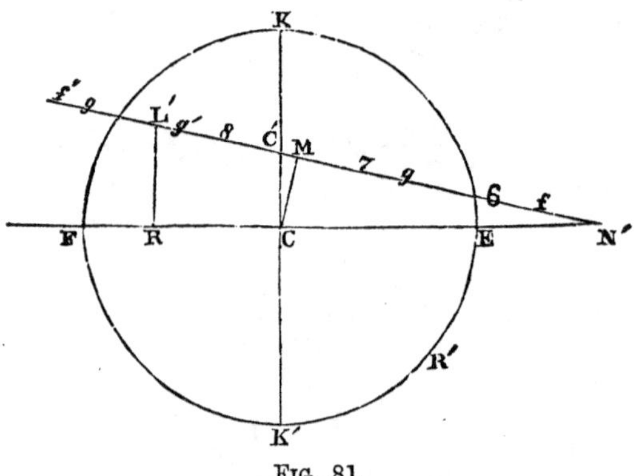

FIG. 81.

C, for the position of the centre of the earth's shadow, at the time of opposition. Then, having fixed upon a scale of equal parts, lay off CR = M — m, the difference of the hourly motions of the sun and moon in longitude; and draw the perpendiculars CC′ = λ the moon's latitude in opposition, and RL′ = $\lambda \pm n$, the moon's latitude an hour after opposition. The right line C′L′, drawn through C′ and L′, will represent the moon's relative orbit. It should be observed, that if the latitudes are south they must be laid off below N′F, and that N′C′L′ will be inclined to the right when the latitude is decreasing. With a radius CE =

s (equation 56) describe the circle EKFK′, which will represent the section of the earth's shadow. With a radius $= s + d$, and another radius $= s - d$, describe about the centre C arcs intersecting N′L′ in f, f', and g, g'; f and f' will be the places of the moon's centre at the beginning and end of the eclipse, and g and g' the places at the beginning and end of the total eclipse. From the point C let fall upon N′C′L′ the perpendicular CM; and M will be the place of the moon's centre at the middle of the eclipse. To render the construction explicit, let us suppose the time of opposition to be 7h. 23m. 15s. At this time the moon's centre will be at C′. To find its place at 7h., state the proportion, 60m. : 23m. 15s. : : moon's hourly motion on the relative orbit : a fourth term. This fourth term will be the distance of the moon's centre from the point C′ at 7 o'clock; and if it be taken in the dividers and laid off on the relative orbit from C′ backward to the point 7, it will give the moon's place at that hour. This being found, take in the dividers the moon's hourly motion on the relative orbit, and lay it off repeatedly, both forward and backward, from the point 7, and the points marked off, 8, 9, 10, 6, 5, will be the moon's places at those hours respectively. Now, the object being to find the times at which the moon's centre is at the points f, f', g, g', and M, let the hour spaces thus found be divided into quarters, and these subdivided into 5-minute or minute spaces, and the times answering to the points of division that fall nearest to these points, will be within a minute or so of the times in question. For example, the point f' falls between 9 and 10, and thus the end of the eclipse will occur somewhere between 9 and 10 o'clock. To find the number of minutes after 9 at which it takes place, we have only to divide the space from 9 to 10 into four equal parts, or 15-minute spaces, subdivide the part which contains f' into three equal parts, or 5-minute spaces, and again that one of these smaller parts within which f' lies, into five equal parts or minute spaces.

ECLIPSES OF THE SUN.

331. Luminous Frustum and Cone. An eclipse of the sun is caused by the interposition of the moon between the sun and earth; whereby the whole, or part of the sun's light, is prevented from falling upon certain parts of the earth's surface.

Let AGB and agb (Fig. 82) be sections of the sun and earth by a plane passing through their centres S and E; Aa, Bb, tangents to the circles AGB and agb on the same side; and Ad, Bc, tangents to the same on opposite sides. The figure AabB will be a section through the axis, of a frustum of a cone formed by rays tangent to the sun and earth on the same side, and the triangular space Fcd will be a section of a cone formed by rays tangent on opposite sides. An eclipse of the sun will take place somewhere upon the earth's surface, whenever the moon comes within the frustum AabB, and a total or an annular eclipse whenever it comes within the cone Fcd.

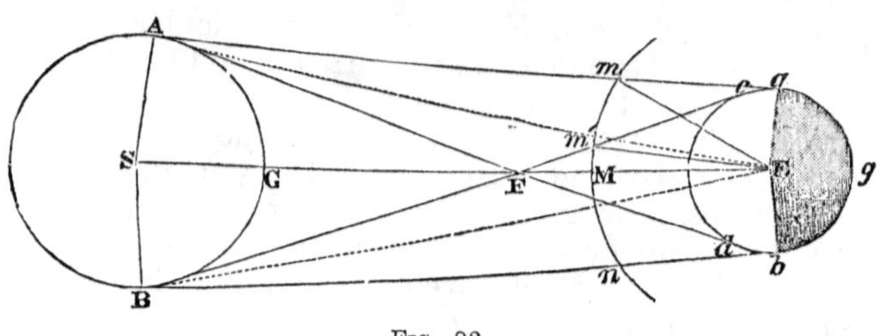

Fig. 82.

332. Semi-diameters of Frustum and Cone. Let mm'M (Fig 82) be a circular arc described about the centre E, and with a radius equal to the distance between the centres of the moon and earth at the time of conjunction. The angle mES is the apparent semi-diameter of a section of the frustum, and m'ES the apparent semi-diameter of a section of the cone, at the distance of the moon. To find expressions for these semi-diameters in terms of determinate quantities, let the first be denoted by A, and the second by A$'$; and let P = the parallax of the moon, p = the parallax of the sun, and δ = the semi-diameter of the sun. Then we have

$$m\mathrm{ES} = \mathrm{A} = m\mathrm{EA} + \mathrm{AES} = \mathrm{E}ma - \mathrm{EA}m + \mathrm{AES};$$

or,
$$\mathrm{A} = \mathrm{P} - p + \delta \ldots . (67):$$

and
$$m'\mathrm{ES} = m'\mathrm{EB} - \mathrm{BES} = \mathrm{E}m'c - \mathrm{EB}m' - \mathrm{BES};$$

or,
$$\mathrm{A}' = \mathrm{P} - p - \delta \ldots . (68).$$

Taking the mean values of P, p, and δ (322), we find for the mean value of A, 1° 13′ 3″; and for the mean value of A$'$, 41′ 1″.

333. Circumstances of Moon's Position in Solar Eclipses. As the plane of the moon's orbit is not coincident with the plane of the ecliptic, an eclipse of the sun can happen only when conjunction or new moon takes place in one of the nodes of the moon's orbit, or so near it that the moon's latitude does not exceed the sum of the semi-diameters of the moon and luminous frustum at the moon's orbit. This may be illustrated by means of Fig. 78, already used for a lunar eclipse, by supposing the sun to be in the directions Es, Es', Es'', and that s, s', s'', are sections of the luminous frustum corresponding to these directions of the sun; also that m, m', m'', represent the moon in the corresponding positions of conjunction. Thus, denoting the moon's semi-diameter by d, and the greatest latitude of the moon in conjunction, at which an eclipse can take place, by L, we have

$$L = P - p + \delta + d \ldots . (69).$$

For a total eclipse, the greatest latitude will be equal to the sum of the semi-diameters of the moon and the luminous cone. Hence, denoting it by L′,

$$L' = P - p - \delta + d \ldots . (70).$$

In order that an *annular eclipse* may take place, the apparent semi-diameter of the moon must be less than that of the sun, and the moon must come at conjunction entirely within the luminous frustum. Whence, if L″= the maximum latitude at which an annular eclipse is possible, we have

$$L'' = P - p + \delta - d \ldots . (71).$$

In the same manner as in the case of an eclipse of the moon, it has been found that when at the time of mean new moon the difference between the mean longitude of the sun or moon and that of the node, exceeds 19° 44′, there cannot be an eclipse of the sun; but when the difference is less than 13° 33′, there must be one. These numbers are called the *Solar Ecliptic Limits*.

334. Prediction of Eclipses:—Period. In order to discover at what new moons in the course of a year an eclipse of the sun will happen, with its approximate time, we have only to find the mean longitudes of the sun and node at each mean new moon throughout the year (301), and take the difference of the longitudes and compare it with the solar ecliptic limits. (For a more direct method of solving this problem, see Prob. XXVIII.)

Eclipses both of the sun and moon recur in nearly the same order and at the same intervals at the expiration of a period of 223 lunations, or 18 years of 365 days, and 15 days;* which for this reason is called the *Period of the Eclipses*. For, the time of a revolution of the sun with respect to the moon's node is 346.619851d., and the time of a synodic revolution of the moon is 29.5305887d. These numbers are very nearly in the ratio of 223 to 19. Thus, in a period of 223 lunations, the sun will have returned 19 times to the same position with respect to the moon's node, and at the expiration of the period will be in the same position with respect to the moon and node as at its commencement. The eclipses which occur during one such period being noted, subsequent eclipses are easily predicted.

This period was known to the Chaldeans and Egyptians, by whom it was called *Saros*.

335. Number of Eclipses in a Year. As the solar ecliptic limits are more extended than the lunar, eclipses of the sun must occur more frequently than eclipses of the moon.

* More exactly, 18 years (of 365 days) plus 15d. 7h. 42m. 29s.

As to the *number of eclipses* of both luminaries, there cannot be fewer than two nor more than seven in one year. The most usual number is four, and it is rare to have more than six. When there are seven eclipses in a year, five are of the sun and two of the moon; and when but two, both are of the sun. The reason is obvious. The sun passes by both nodes of the moon's orbit but once in a year, unless it passes by one of them in the beginning of the year, in which case it will pass by the same again a little before the end of the year, as it returns to the same node in a period of 346 days. Now, if the sun be at a little less distance than 19° 44′ from either node at the time of mean new moon, he may be eclipsed (333), and at the subsequent opposition the moon will be eclipsed near the other node, and come round to the next conjunction before the sun is 13° 33′ from the former node; and when three eclipses happen about either node, the like number commonly happens about the opposite one; as the sun comes to it in 173 days afterwards, and six lunations contain only four days more. Thus there may be two eclipses of the sun and one of the moon about each of the nodes; and the twelfth lunation from the eclipse in the beginning of the year may give a new moon before the year is ended, which, in consequence of the retrogradation of the nodes, may be within the solar ecliptic limit; and hence there may be seven eclipses in a year, five of the sun and two of the moon. But when the moon changes in either of the nodes, it cannot be near enough to the other node, at the next full moon, to be eclipsed; as in the interval the sun will move over an arc of 14° 32′, whereas the greatest lunar ecliptic limit is but 13° 21′, and in six lunar months afterwards it will change near the other node. In this case there cannot be more than two eclipses in a year, both of which will be of the sun. If the moon changes at the distance of a few degrees from either node, then an eclipse both of the sun and moon will probably occur in the passage of that node and also of the other.

Although solar eclipses are more frequent than lunar, when considered with respect to the whole earth, yet at any given place more lunar than solar eclipses are seen. The reason of this circumstance is, that an eclipse of the sun (unlike an eclipse of the moon) is visible only over a part of a hemisphere of the earth. To show this, suppose two lines to be drawn from the centre of the moon tangent to the earth at opposite points: they will make an angle with each other equal to double the moon's horizontal parallax, or of 1° 54′. Therefore, should an observer situated at one of the points of tangency, refer the centre of the moon to the centre of the sun, an observer at the other would see the centres of these bodies distant from each other an angle of 1° 54′, and their nearest limbs separated by an arc of more than 1°.

336. Moon's Shadow Cast upon the Earth. Instead of regarding an eclipse of the sun as produced by an interposition of the moon between the sun and earth, as we have hitherto considered it, we may regard it as occasioned by the moon's shadow falling upon the earth. Fig. 83 represents the moon's shadow, as projected from the sun and covering a portion of the earth's surface. Wherever the umbra falls, there is total eclipse; and wherever the penumbra falls, a partial eclipse.

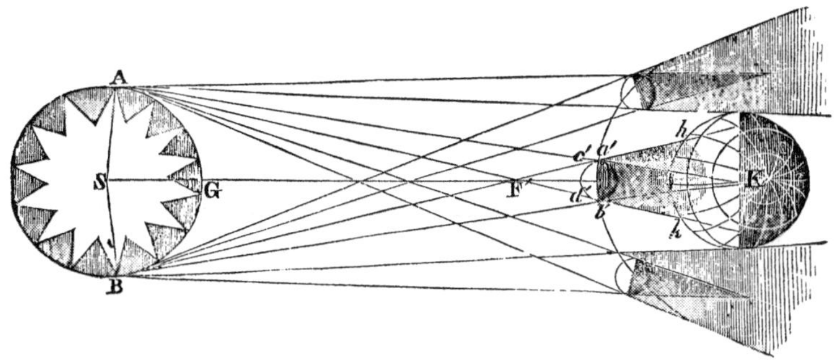

FIG. 83.

In order to discover the extent of the portion of the earth's surface over which the eclipse is visible at any particular time, we have only to find the breadth of the portion of the earth covered by the penumbral shadow of the moon; but we will first ascertain the length of the moon's shadow. As seen at the vertex of the moon's shadow, the apparent diameters of the moon and sun are equal. Now, as seen at the centre of the earth, they are nearly equal, sometimes the one being a little greater and sometimes the other. It follows, therefore, that *the length of the moon's shadow is about equal to the distance of the earth, being sometimes a little greater and at other times a little less.*

When the apparent diameter of the moon is the greater, the shadow will extend beyond the earth's centre; and when the apparent diameter of the sun is the greater, it will fall short of it. If we increase the mean apparent diameter of the moon as seen from the earth's centre, viz. 31′ 7″, by $\frac{1}{60}$, the ratio of the radius of the earth to the distance of the moon, we shall have 31′ 38″ for the mean apparent diameter of the moon as seen from the nearest point of the earth's surface. Comparing this with the mean apparent diameter of the sun as viewed from the same point, which is sensibly the same as at the centre of the earth, or 32′ 3″, we perceive that it is less; from which we conclude, that when the sun and moon are each at their mean distance from the earth, the shadow of the moon does not extend as far as the earth's surface.

337. To find a General Expression for the Length of the Moon's Shadow. Let AGB, $a'g'b'$, and agb (Fig. 84) be sections of the sun, moon, and earth, by a plane passing through

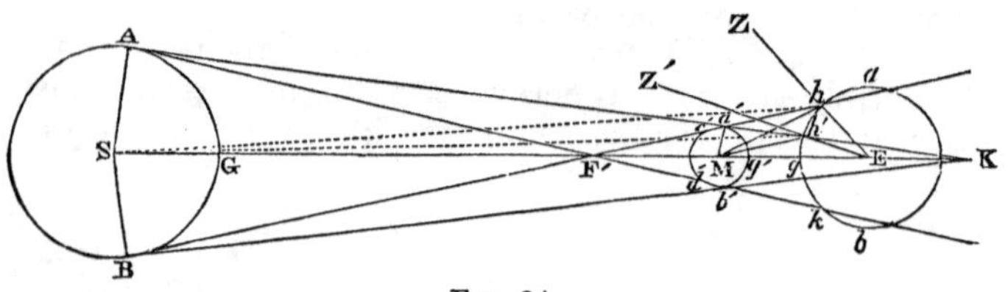

FIG. 84.

their centres S, M, and E, supposed to be in the same right line, and Aa', Bb', tangents to the circles AGB, $a'g'b'$: then $a'Kb'$ will represent the moon's shadow. Let L = the length of the shadow; D = the distance of the moon; D′ = the distance of the sun; d = the apparent semi-diameter of the moon; and δ = the apparent semi-diameter of the sun. At K the vertex of the shadow, MKa' the apparent semi-diameter of the moon, will be equal to SKA the apparent semi-diameter of the sun; and as the distance of this point from the centre of the earth, even when it is the greatest, is small in comparison with the distance of the sun (336), the apparent semi-diameter of the sun will always be very nearly the same to an observer situated at K as to one situated at the centre of the earth. Now, since the apparent semi-diameter of the moon is inversely proportional to its distance,

$$\text{angle } MKa' : d :: ME : MK;$$

and thus, $$\delta : d :: ME : MK :: D : L \text{ (nearly)}:$$

whence, $$L = D\frac{d}{\delta} \ldots. (72).$$

If a more accurate result be desired, we have only to repeat the calculation, after having diminished δ in the ratio of D′ to (D′ + L — D).

338. To find the Breadth of the Penumbral Shadow cast upon the earth, let the lines Ad', Bc' (Fig. 84) be drawn tangent to the circles AGB, $a'g'b'$, on opposite sides, and prolonged to the earth. The space $hc'd'k$ will represent the penumbra of the moon's shadow, and the arc gh one-half the breadth of the portion of the earth's surface covered by it. Let this arc or the angle gEh = S, and denote the semi-diameter of the sun and the semi-diameter and parallax of the moon by the same letters as in previous articles. The triangle MEh gives

$$\text{angle } MEh = S = MhZ - hME.$$

The angle hME is the moon's parallax in altitude at the station h, and MhZ is its zenith distance at the same station. Denote the former by P′ and the latter by Z. Thus,

$$S = Z - P' \ldots. (73).$$

The triangle hMS gives

$$hME = P' = MSh + MhS;$$

$MhS = d + \delta$; and MSh is the sun's parallax in altitude at the station h: let it be denoted by p'. We have, then,

$$P' = d + \delta + p' = d + \delta \text{ (nearly)} \ldots .(74);$$

and to find Z we have (equa. 9, p. 63),

$$P' = P \sin Z, \text{ or } \sin Z = \frac{P'}{P} \ldots .(75).$$

P′ and Z being found by these equations, equa. (73) will then make known the value of S.

If great accuracy be required, the calculation must be repeated, giving now to p' in equa. (74) the value furnished by equa. (9) which expresses the relation between the parallax in altitude of a body and its horizontal parallax, instead of neglecting it as before; and Z must be computed from the following equation:

$$\sin Z = \frac{\sin P'}{\sin P} \ldots .(76).$$

The penumbral shadow will obviously attain to its greatest breadth when the sun is at its least and the moon is at its greatest distance. The values of d, δ, and P under these circumstances are respectively 14′ 24″, 16′ 18″, and 52′ 50″. Performing the calculations, we find that *the breadth of the greatest portion of the earth's surface ever covered by the penumbral shadow is* 70° 17′, *or about* 4,850 *miles.*

339. The Breadth of the Umbra may be found in a similar manner.

The arc gh' (Fig. 84) represents one half of it: denote this arc or the equal angle gEh' by S′.

$$MEh' = S' = Mh'Z' - h'ME;$$

or,

$$S' = Z - P' \ldots .(77).$$

$$h'ME = P' = MSh' + Mh'S;$$

but $Mh'S = d - \delta$, and $MSh' = p'$, the sun's parallax in altitude at h'; whence,

$$P' = d - \delta + p' = d - \delta \text{ (nearly)} \ldots .(78):$$

and we have, as before,

$$P' = P \sin Z, \text{ or } \sin Z = \frac{P'}{P} \ldots .(79).$$

The greatest breadth will obtain when the sun is at its greatest and the moon is at its least distance. We shall then have

$$\delta = 15' \, 45'', \; d = 16' \, 46'', \; P = 61' \, 32''.$$

Making use of these numbers, we deduce for the *maximum breadth of the portion of the earth's surface covered by the moon's shadow*, 1° 54′; or 130 miles.

It should be observed that the deductions of the last two articles answer to the supposition that the moon is in the node, and that the axis of the shadow and penumbra passes through the centre of the earth. In every other case, both the shadow and penumbra will be cut obliquely by the earth's surface, and the sections will be ovals, and very nearly true ellipses, the lengths of which may materially exceed the above determinations.

340. Phases of Eclipse Different at each Place. Parallax not only causes the eclipse to be visible at some places and invisible at others, as shown in Art. 335, but, by making the distance between the centres of the sun and moon unequal, renders the circumstances of the eclipse at those places where it is visible, different at each place. This may also be inferred from the circumstance that the different places, covered at any time by the shadow of the moon, will be differently situated within this shadow. It will be seen, therefore, that an eclipse of the sun has to be considered in two points of view: 1st. *With respect to the whole earth*, or as a *general eclipse;* and, 2d. *With respect to a particular place.*

341. Particular Facts. The following are the principal facts relative to eclipses of the sun that remain to be noticed: 1st. The duration of a general eclipse of the sun cannot exceed about 6 hours. 2d. A solar eclipse does not happen at the same time at all places where it is seen: as the motion of the moon toward the sun, and consequently of its shadow, is from west to east, the eclipse must begin *earlier* at the *western* parts and *later* at the *eastern.* 3d. The moon's shadow being tangent to the earth at the commencement and end of the eclipse, the sun will be just rising at the place where the eclipse is first seen, and just setting at the place where it is last seen. At intermediate places, the sun will at the time of the beginning and end of the eclipse have various altitudes. 4th. An eclipse of the sun begins on the *western* side and ends on the *eastern.* 5th. When the straight line passing through the centres of the sun and moon passes also through the place of the spectator, the eclipse is said to be *central:* a central eclipse may be either annular or total, according as the apparent diameter of the sun is greater than that of the moon, or the reverse. 6th. A total eclipse of the sun cannot last at any one place more than *eight minutes;* and an annular eclipse more than *twelve and a half minutes.* 7th. In most solar eclipses the moon's disc is covered with a faint light, a phenomenon which is attributed to the reflection of the light from the illuminated part of the earth.

CALCULATION OF AN ECLIPSE OF THE SUN.

342. The complete calculation of a solar eclipse involves the solution of two distinct problems, viz.: (1), the determination of all the circumstances of the eclipse for the earth as a whole; (2), the determination of the times of all the phases, and the corresponding apparent relative positions of the moon and sun for a particular place. Different methods of solving these problems have been devised. Processes of calculation, comparatively simple and direct, are given in the Appendix.

OCCULTATIONS.

343. An occultation is an eclipse, or deprivation of the light of a star, resulting from the interposition of the moon between the star and the eye of the observer. At all places on the earth which at a given time have the moon in the horizon, its apparent place will differ from its true place (78), by the amount of the horizontal parallax. It follows, therefore, that a star will be eclipsed by the moon, somewhere upon the earth, in case its true distance from the moon's centre is less than the sum of the moon's semi-diameter and horizontal parallax.

344. Limits of Position of Stars Liable to Occultation. The greatest value of the apparent semi-diameter of the moon is 16′ 46″, and that of its horizontal parallax is 61′ 32″. If we add the sum of these quantities to 5° 20′ 6″, the greatest possible latitude of the moon, we obtain as the result 6° 38′ 24″. It is then only the stars which have a latitude, either north or south, less than 6° 38′ 24″ that can experience an occultation from the moon.

In order that any of the stars situated within this distance from the ecliptic may suffer occultation at some point on the earth, it is necessary that, at the time of the true conjunction (144) of the moon and star, the actual difference of latitude of the two should not exceed the sum of the actual apparent semi-diameter and horizontal parallax of the moon.

CHAPTER XVII.

The Planets, and the Phenomena occasioned by their Motions in Space.

APPARENT MOTIONS OF THE PLANETS WITH RESPECT TO THE SUN

345. The apparent motion of an inferior planet with reference to the sun, is materially different from that of a superior planet. The inferior planets always accompany the sun, being seen alternately on the east and west side of it, and never receding from it beyond a certain moderate distance, while the superior planets are seen, at different times, at every variety of angular distance. This difference of apparent motion arises from the difference of situation of the orbits of an inferior and superior planet, with respect to the orbit of the earth; the one lying within, and the other without the earth's orbit.

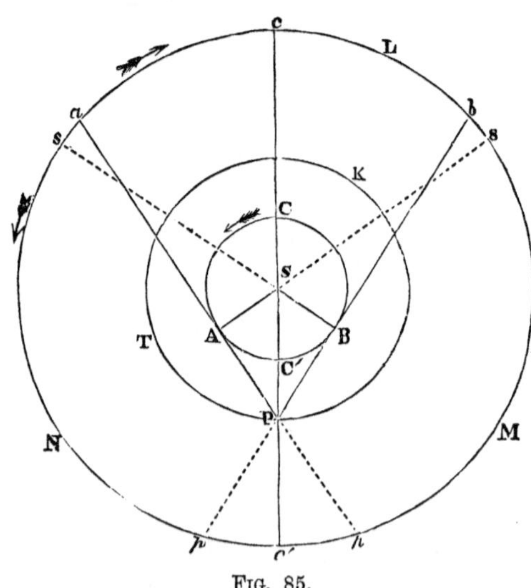

Fig. 85.

346. Apparent Motion of an Inferior Planet. Let CAC'B (Fig. 85) represent the orbit of either one of the inferior planets, Venus for example, and PKT the orbit of the earth—which we will suppose to be circles, and to lie in the same plane,—and let MLN represent the circle of intersection of this plane with the sphere of the heavens, to some point of which the planet will be referred by an observer on the earth. Suppose, for the present, that the earth is stationary in the position P, and through P draw the lines PA, PB, tangent to the orbit of Venus, and prolong them until they intersect the heavens at *a* and *b*. The earth being at P, Venus will be in superior conjunction at C, and in inferior conjunction at C′. Now, by inspecting the figure, it will be seen that in passing from C to C′ she will be seen in the heavens on the east side of the sun, and in passing from C′ to C on the west side of the sun; also, that in passing from C to A she will recede from the sun in the heavens, from A to C′ approach it, from C′ to B recede from it again, and from B to C approach it again. *a* and *b* will be the positions of the planet in the heavens at the times of the greatest eastern and western elongations.

When to the east of the sun, Venus is seen in the evening, and called the *Evening Star;* and when to the west of the sun, is seen in the morning, and called the *Morning Star.*

We have in the foregoing investigation supposed the earth to be stationary, a supposition which is contrary to the fact; but it is plain that the only effect of the earth's motion in the case under consideration, as it is slower than that of the planet, is to cause the points A, C′, B, to advance in the orbit, without altering the nature of the apparent motion of the planet with respect to the sun. The orbits of the earth and planet are also ellipses of small eccentricity, and are slightly inclined to each other, instead of being circles and lying in the same plane: on this account, as the greatest elongations will occur in various parts of the orbits, they will differ in value. The greatest elongation of Venus varies from 45° to 47° 15′. Its mean value is about 46°.

Owing to the circumstance of the orbit of Mercury being within the orbit of Venus, the greatest elongation of this planet is less than that of Venus. It is never so great as 30°.

347. Apparent Motion of a Superior Planet. Suppose PKT (Fig. 85) to be the orbit of a superior planet, and CAC′B that of the earth; and as the velocity of the earth is much greater than that of the planet, let us, for the present, regard the planet as stationary in the position P, while the earth describes the circle CAC′. When the earth is at C, the planet being at P, is in conjunction with the sun. When the earth is at A, SAP, the elongation of the planet, is 90°. When it arrives at C′, the planet is in opposition, or 180° distant from the sun: and when it reaches B, the elongation is again 90°. At intermediate points, the elongation will have intermediate values. If, now, we restore to the planet its orbital motion, we shall manifestly be conducted to the same results relative to the change of elongation, as the only effect of such motion will be to throw the points A, C′, B, forward in the orbit. It appears, then, that in the course of a synodic revolution a superior planet will be seen at all angular distances from the sun, both on the east and west side of him. From conjunction to opposition, that is, while the earth is passing from C to C′, the planet will be to the right, or to the west of the sun; and will therefore be below the horizon at sunset, and rise some time in the course of the night. But, from opposition to conjunction, or while the earth is moving from C′ to C, it will be to the east of the sun, and therefore above the horizon at sunset.

348. To find the Length of the Synodic Revolution of a Planet. Let us first take an inferior planet, Venus for instance. Suppose we assume, at a given instant, the sun, Venus, and the earth to be in the same right line; then, after any elapsed time (a day for instance), Venus will have described an angle m, and the earth an angle M around the sun. Now, the value of m is greater than that of M; therefore, at the end of a day, the separation of the planet from the earth (measuring the separation by an angle formed by two lines drawn from the planet and the earth to the sun), will be $m - M$; at the end of two days (the mean daily motions continuing the same) the angle of separation wil be $2(m - M)$; at the end of three days, $3(m - M)$; at the end of s days, $s(m - M)$. When the angle of separation amounts to 360°, that is, when $s(m - M) = 360°$, the sun, planet, and earth must be again in the same right line, and in that case

$$s = \frac{360°}{m - M} \ldots . (80).$$

In which expression s denotes the mean duration of a synodic revolution, m and M being taken to denote the mean daily motions.

We may obtain from equation (80) *another equation*, in which the synodic revolution is expressed in terms of the sidereal periods of the earth and planet.

Let P and p denote the sidereal periods in question, then, since

$$1\text{d.} : M° :: P : 360°,$$

and

$$1 : m :: p : 360;$$

$$M = \frac{360°}{P}, \text{ and } m = \frac{360°}{p}; \text{ substituting,}$$

$$s = \frac{360°}{360°\left(\frac{1}{p} - \frac{1}{P}\right)} = \frac{Pp}{P - p} \ldots . (81).$$

Equations (80), (81), although investigated for an inferior

planet, will answer equally well for a superior planet, provided we regard m as standing for the mean daily motion of the earth, M for that of the planet, p for the sidereal period of the earth, and P for that of the planet. For the earth holds towards a superior planet the place of an inferior planet, and a synodic revolution of the earth, to an observer on the planet, will obviously be a synodic revolution of the planet to an observer on the earth.

349. Lengths of Synodic Revolutions of Planets. Equa. (80) shows that the length of a mean synodic revolution depends altogether upon the amount of the difference of the mean daily motions of the earth and planet, and is the greater the less is this difference.

It follows, therefore, that the synodic revolution is the longest for the planets nearest the earth.

It appears by equa. (81) that the length of a synodic revolution is, for an inferior planet, greater than the sidereal period of the planet, and, for a superior planet, greater than the sidereal period of the earth. The actual lengths of the synodic revolutions of the different planets are given in Table V.

The mean synodic revolution of a planet being known, and also the time of one conjunction or opposition, we may easily ascertain its mean elongation at any given time, and thus approximately the time of its rising, setting, and meridian passage.

350. Planets as Evening or Morning Stars. A planet will rise and set at the same hours at the end of a synodic revolution; and will be an evening star, that is, above the horizon at sunset, during half of a synodic revolution, and a morning star, that is, above the horizon at sunrise, during an equal interval of time. The inferior planets will be evening stars from superior to inferior conjunction; and the superior planets from opposition to conjunction.

Mercury is an evening star for a period of 2 months; Venus during an interval of 9½ months; Mars for 1 year and 1 month; Jupiter for 6½ months; Saturn and Uranus each a few days more than 6 months.

STATIONS AND RETROGRADATIONS OF THE PLANETS.

351. The apparent motions of the planets in the heavens, as has already been stated (11), are not, like those of the sun and moon, continually from west to east, or direct, but are sometimes also from east to west, or retrograde. The retrograde motion takes place over arcs of but a small number of degrees; and in changing the direction of their motions, the planets are for several days stationary in the heavens. These phenomena are called the *Stations* and *Retrogradations* of the planets. We now propose to inquire theoretically into the particulars of the motions in question, and to show how the phenomena just mentioned result from the motions of the planets in connection with the motion of the earth.

352. Case of an Inferior Planet. Let CAC′B (Fig. 85) represent the orbit of an inferior planet, and PKT the orbit of the earth; both considered as circles, and as situated in the same plane. If the earth were continually stationary in some point P of its orbit, it is plain that while the planet was moving from B the position of greatest western elongation to A the position of greatest eastern elongation, it would advance in the heavens from *b* to *a*; that, while it was moving from A to B, that is, from greatest eastern to greatest western elongation, it would retrograde in the heavens from *a* to *b*; and that, in passing the points A and B, as it would be moving directly towards or from the earth, it would for a time appear stationary in the heavens, in the positions *a* and *b*.

But the earth is in fact in motion, and the actual apparent motion of the planet is in consequence materially different from this. Let A, A′ (Fig. 86) be the positions of the planet and earth, at the time of the greatest eastern elongation, C′, P their positions at inferior conjunction, and B, B′ their positions at the

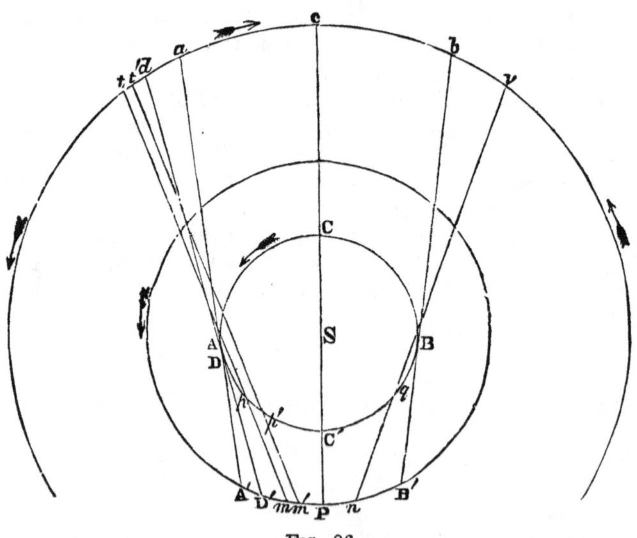

FIG. 86.

greatest western elongation. At the time of the greatest eastern elongation, while the planet describes a certain distance AD on the line of the centres of the earth and planet, the earth moves forward in its orbit a certain distance A′D′; so that, instead of appearing stationary at *a* in the interval, the planet will advance in the heavens from *a* to *d*. From the same cause it will have a direct motion about the time of the greatest western elongation. As it advances from A towards C′, the direct motion will continue; but, as the daily arc described by the planet will make a less and less angle with the daily arc described by the earth, the rate of motion will continually decrease, and finally, when the planet has come into a position with respect to the earth, such that the lines of direction of the planet, *mp*, *m′p′*, at the beginning and end of the day are parallel, it will be stationary in the heavens. As the daily arc of the planet is greater than that of the earth, and becomes parallel to it in inferior conjunction, the planet will be in the position in question before it comes into inferior conjunction.

Subsequent to this, the inclination of the daily arcs still diminishing, the lines of direction of the planet at the beginning and end of the day will diverge, and therefore the motion will be retrograde. After inferior conjunction, the inclination of the arcs will, at corresponding positions of the earth and planet, obviously be the same as before. It follows, therefore, that the planet will be at its western station when it is at the same angular distance from the sun as at its eastern station; that its motion will be retrograde until it has passed inferior conjunction and arrived at its western station; and that after this it will be direct. *q* and *n* represent the positions of the planet and the earth at the time of the western station; C′*q* = C′*p*, and P*n*=P*m*.

The diminution of the elongation of the planet at its two stations is not the only effect of the earth's motion in the case under consideration; it also accelerates the direct, and retards the retrograde motion of the planet, and gives to the planet along with the sun an apparent motion of revolution around the earth.

353. Case of a Superior Planet. Suppose AC′B (Fig. 86) to be the orbit of the earth, and A′PB′ that of the planet. Since the earth is an inferior planet to an observer stationed upon a superior planet, it appears by the foregoing article that it will, to an observer so situated, have a retrograde motion while it is passing over a certain arc *p*C′*q* in the inferior part of its orbit, and a direct motion during the remainder of the synodic revolution. Now, it is plain that the direction of the planet's motion, as seen from the earth, will always be the same as the direction of the earth's motion as seen from the planet. When the earth is at C′, the middle of the arc *p*C′*q*, the planet is in opposition. It follows, therefore, that a superior planet has a retrograde motion during a small portion of its synodic revolution, about the time of opposition. (See Table V.)

PHASES OF THE INFERIOR PLANETS.

354. To the naked sight the disc of the planet Venus appears circular, like that of each of the other planets, but the telescope shows this to be an optical illusion. When Venus is repeatedly observed with a telescope, it is seen to present in its various positions with respect to the sun the same variety of phases as the moon; being a full circle at superior conjunction, a half circle at the greatest eastern and western elongations, and a crescent, with the horns turned from the sun, before and after inferior conjunction.

Mercury exhibits precisely similar phases, but being smaller, at a greater distance from the earth, and much nearer the sun, its phases are not so easily observed as those of Venus.

355. Explanation. The phases of Venus are easily accounted for, by supposing it to be an opake spherical body, and to shine by reflecting the sun's light, and by taking into consideration its motion with respect to the sun and earth. The hemisphere turned towards the sun is illuminated and the other is in the dark, and as the planet revolves around the sun, various portions of the enlightened half are turned towards the earth; in superior conjunction, the whole of it; at the greatest elongations, one half; and near inferior conjunction, but a small part. This will be abundantly evident on inspecting Fig. 87. The phases corresponding to the positions represented are delineated in the figure.

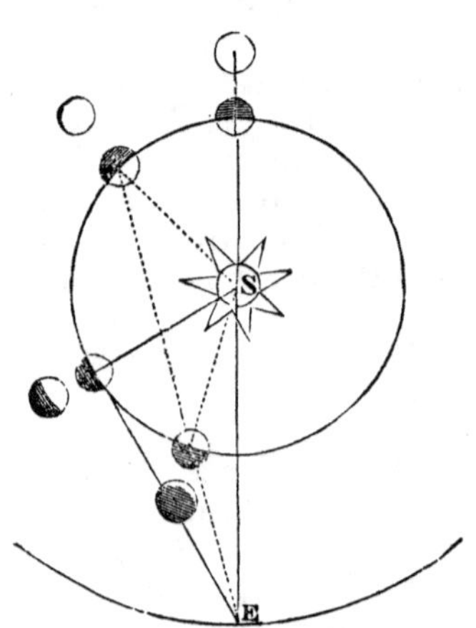

Fig. 87.

The phases of Mercury are obviously suceptible of a similar explanation.

356. Changes of Form of the Disc of Mars. The disc of the planet Mars also undergoes changes of form, but they are of comparatively moderate extent. It is sometimes gibbous, but never has the form of a crescent. Indeed, on the supposition that Mars is an opake body illuminated by the sun, we would not see the whole of the enlightened hemisphere, except in conjunction and opposition, but there would always be more than half of it turned towards the earth, and therefore the disc should always be larger than a half circle.

The discs of the other superior planets do not experience any perceptible variation of form, for the reason, doubtless, that their orbits are so large with respect to the orbit of the earth, that all, or very nearly all of their illuminated hemispheres, is constantly visible from the earth. Jupiter offers the only exception to this general truth; it is slightly gibbous in quadratures.

TRANSITS OF THE INFERIOR PLANETS.

357. The two inferior planets Venus and Mercury, at inferior conjunction, sometimes, though rarely, pass between the sun and earth, and are seen as a dark spot crossing the sun's disc. This phenomenon is called a *Transit.* It will take place, in the case of either planet, whenever, at the time of inferior conjunction, it is so near either node that its geocentric latitude is less than the apparent semi-diameter of the sun.

358. Epochs of Transits:—Periods of Recurrence. The transits of Venus take place alternately at intervals of 8 and 105½ or 121½ years. The last were in the years 1761 and 1769. The next will be in 1874 and 1882; of which the latter will be visible in this country.

In consequence of the greater distance of Mercury from the earth, a greater portion of its orbit is directly interposed between the sun and earth than of the orbit of Venus; moreover, the synodic revolution of Mercury is shorter than that of Venus. On these accounts it happens that the transits of Mercury are much more frequent than those of Venus. The last transit of Mercury was on November 11, 1861. The next two will take place in 1868 and 1878, in November and May. The first, which will occur on the 4th, will be visible in this country.

359. A Transit is Calculated in a precisely similar manner with a solar eclipse; the planet in the one calculation answering to the moon in the other.

360. A Transit is an Important Phenomenon in a practical point of view, as it furnishes an indirect but accurate method of ascertaining the sun's parallax. In order to understand how this phenomenon can be used for this purpose, we have only to consider that, in consequence of the difference of

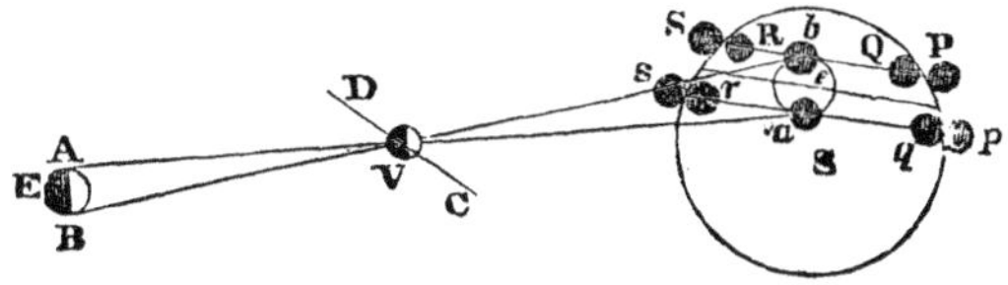

FIG. 88.

the distances or parallaxes of the sun and Venus, observers at different stations upon the earth will refer the planet to different points upon the sun's disc, and that therefore, to such observers, the transit will take place along different chords, and be accomplished in unequal portions of time. This fact is represented to the eye in Fig. 88. It is then to be expected, that, if the durations of the transit at two different places should be noted, the *difference* of the parallaxes of the sun and Venus, upon which alone the difference of the durations depends, could be computed. This computation is in fact possible. Also, the *ratio* of the parallaxes being inversely as that of the distances, could be found by the elliptic theory of the planetary motions, and thus the parallax both of the sun and Venus would become known.

361. The Parallax of the Sun was quite accurately deduced from observations upon the transits of Venus in 1769 and 1761. Expeditions were fitted out on the most efficient scale, by the British, French, Russian, and other governments, and sent to various parts of the earth, remote from each other, to observe the transit of 1769, that the parallax of the sun might be computed from the results of the observations. The sun's parallax, as determined by Professor Encke from the observations made upon the transit in question, and that of 1761, is $8''.5776$. We have already seen that the sun's parallax has recently been more accurately determined (150).

APPEARANCE, DIMENSIONS, ROTATION, AND PHYSICAL CONSTITUTION OF THE PLANETS.

362. Variations of Apparent Diameter. It appears from admeasurement with the telescope and micrometer, that the apparent diameter of a planet is subject to sensible variations. The apparent diameter of Venus, as well as of Mercury, is greatest in inferior conjunction, and least in superior conjunction; while the apparent diameter of each of the other planets is greatest in opposition and least in conjunction. These variations of the apparent diameters of the planets are necessary consequences of the changes that take place in the distances of the planets from the earth. (See Fig. 85.)

363. Absolute and Relative Magnitudes. The real diameter of a planet is deduced from its apparent diameter and horizontal parallax. (See Art. 310.) When the diameters of the planets have been found, their relative surfaces and volumes are easily obtained; for the surfaces are as the squares of the diameters, and the volumes as the cubes.

The order of magnitude of the planets is as follows: 1 Jupiter, 2 Saturn, 3 Neptune, 4 Uranus, 5 the Earth, 6 Venus, 7 Mars, 8 Mercury, 9 Pallas, 10 Vesta, 11 Ceres, 12 Juno, 13 the other planetoids. The range of magnitude, for the principal planets, is from 1 to about 25,000. The relative magnitudes of the principal planets are given in Table IV.

364. Rotation of Planets. Spots more or less dark have

been seen upon the discs of most of the principal planets; and by passing across them from east to west and reappearing at the eastern limbs, have established that the planets upon which they are observed rotate upon axes from west to east. From repeated careful observations upon the situations of these spots, the periods of rotation, and the positions of the axes, have been determined.

The periods of rotation of Mercury, Venus, the Earth, and Mars, are all about 24 hours, and of Jupiter and Saturn about 10 hours. Those of the other planets are not known. The axes of rotation remain continually parallel to themselves, as the planets revolve in their orbits.

365. The Amount of Light and Heat, which the sun bestows upon the planets decreases in the same ratio that the square of the distance increases. (See Table IV.)

It will be seen in the sequel that the planets are all opake bodies, like the earth, and shine wholly by the reflected light of the sun; and that most, if not all of them, are surrounded with atmospheres.

MERCURY.

366. In consequence of its proximity to the sun, Mercury is rarely visible to the naked eye. When seen under the most favorable circumstances, about the time of greatest elongation, and at periods of the year when twilight is of short duration, it presents the appearance of a star of the third or fourth magnitude. Its phases indicate that it is opake and illuminated by the sun. Its apparent diameter varies with its distance from the earth, from 5″ to 13″. Its real diameter is a little less than 3,000 miles, or $\frac{4}{11}$ of that of the earth; and its volume is about $\frac{1}{19}$ of the earth's volume.*

Mercury performs a complete rotation on its axis in 24h. 5½m., and according to Schröter, its axis is inclined to the plane of the ecliptic under as small an angle as 20°.

367. Telescopic Appearances. Owing to the dazzling splendor of its light, and the tremulous motion induced by the ever-varying density of the air and vapors near the earth's surface, through which it is seen, the telescope does not present a well-defined image of the disc of this planet. Schröter is the only observer who has supposed that he discerned distinct spots upon it. Later observers have only noticed on rare occasions, slight inequalities of brightness on its disc.

From the fact that such appearances are only occasionally seen, it has been inferred that the planet is surrounded with a dense atmosphere loaded with clouds, that reflect a strong light, and, except when the atmosphere clears up in an unusual degree, prevent the darker body of the planet from being seen. But the evidence in support of this conclusion needs confirmation.

Schröter, in making observations upon Mercury at the time the disc had the form of a crescent, discovered that one of the horns of the crescent became blunt at

* The exact diameters, volumes, times of rotation, &c., of the different planets, as far as known, may be found in Table IV.

the end of every 24 hours; from which he inferred that the planet turned upon an axis, and had mountains upon its surface, which were brought at the end of every rotation into the same position with respect to his eye and the sun.

VENUS.

368. Venus is the brightest of all the planets, and generally appears larger and brighter than any of the fixed stars. But it is much more conspicuous at some times than at others, during a synodic revolution. It is found by calculation, that the epochs in the course of a synodic revolution at which Venus gives most light to the earth, are those at which, being in the inferior part of its orbit, it has an elongation from the sun of a little less than 40°. The disc is then less than one-quarter of a circle, but the increased proximity to the earth more than compensates for the diminished size of the disc. Venus attains to greater splendor in some revolutions than in others, in consequence of being nearer the earth when in the favorable position just noticed. A combination of the most favorable circumstances recurs every eight years, when Venus becomes visible in full daylight, and casts a sensible shadow at night. This last happened in February, 1862.

As seen through a telescope, Venus presents a disc of nearly uniform brightness, and spots have very rarely been seen upon it. From the regular succession of phases through which the disc passes, as the planet changes its position with respect to the earth and sun, we infer that it is an opake spherical body, shining by the reflected light of the sun. Its apparent diameter varies with its distance from the earth, from 10″ to 66″. Its real diameter is 7,600 miles; and its volume $\frac{1}{8}$ less than that of the earth. The period of its rotation is 23h. 21m. The inclination of its axis to the plane of its orbit is not exactly known, but is supposed to be not far from 18°.

369. Evidences of an Atmosphere surrounding Venus. From the remarkable vivacity of the light of this planet, which far exceeds that of the light reflected from the moon's surface, as well as the transitory nature of the few darkish spots that have been seen upon its disc, it is inferred that it is surrounded by a dense and highly reflective atmosphere, which in general screens the whole of the darker body of the planet from view. The truth of this inference is confirmed by certain delicate observations made by Schröter. This Astronomer distinctly discerned, when the disc was seen as a narrow crescent, a faint light stretching beyond the proper termination of one of the horns of the crescent into the dark part of the face of the planet, as is represented in Fig. 89, where the left extremity of the dotted line represents the natural terminating point of one of the horns.

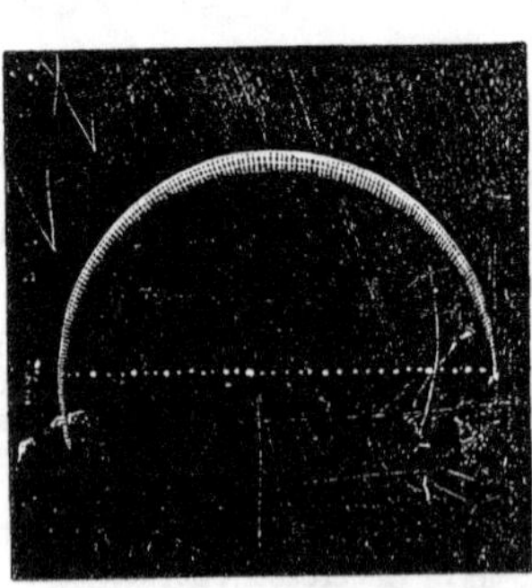

FIG. 89.

The same appearance has since been repeatedly noticed by other observers It was distinctly perceptible before and after the last inferior conjunction of the planet, December 11th, 1866. The planet was watched from day to day by Professor Lyman, with the nine-inch equatorial of the Sheffield Observatory, Yale College, until, on the day before conjunction, its distance from the nearest limb of the sun was only 1° 8′. The very slender crescent which it exhibited, was each day seen more and more extended beyond a semicircle; until at favorable moments, when the sun, but not the planet, was covered by a passing cloud, it was distinctly observed as an entire ring of light, thinnest on the side furthest from the sun. The entire ring was seen also, by the same observer, with a five-foot telescope, so placed as to have the sun covered by a distant chimney. The maximum extent of the crescent observed at Dorpat, at the conjunction in 1849, was 240°; the planet being 3° 26′ from the sun's centre.

This remarkable prolongation of the cusps must be attributed mainly to the refraction of the sun's rays by the atmosphere of the planet. On this assumption, Mädler, from the Dorpat observations of the extent of the cusps, made the horizontal atmospheric refraction of Venus 43′.7. The observations of Professor Lyman, at the late conjunction, give 45′.3. This is about $\frac{1}{4}$ greater than the horizontal refraction produced by the earth's atmosphere; and indicates that the density of the atmosphere of Venus is decidedly greater than that of the earth's atmosphere.

370. *Clouds in the Atmosphere.* Since the transparency of Venus's atmosphere is variable, becoming occasionally such as to admit of the body of the planet's being seen through it, we must suppose that it contains aqueous vapor and clouds, and therefore that there are bodies of water upon the surface of the planet. It is in fact supposed that isolated clouds have actually been seen. The most natural explanation of the bright spots which have sometimes been noticed on the disc, is, that they are clouds, more highly reflective than the atmosphere, or than the clouds in general.

371. *Inequalities on the Surface.* There are great inequalities on the surface of Venus, and, it would seem, mountains, much higher than any upon our globe. Schröter detected these masses by several infallible marks. In the first place, the edge of the enlightened part of the planet is shaded, as seen in Figs. 89, 90, 91, and as the moon appears when in crescent even to the naked eye.

FIG. 90.

FIG. 91.

This appearance is doubtless caused by shadows cast by mountains; which are naturally best seen on that part of the planet to which the sun is rising or setting, where they are longest. In the next place, the edge of the disc shows marked irregularities. Thus it sometimes appears rounded at the corners, as in Fig. 90, owing undoubtedly to part of the disc being rendered invisible there by the shadow or interposition of some line of eminences, and at other times, as in Fig. 91, a single bright point appears detached from the disc—the top of a high mountain, illuminated across a dark valley.

Schröter found that these appearances recurred regularly, at equal intervals of about 23½ hours; the same period as that which Cassini had previously found for the completion of a rotation, by observations upon the spots.

MARS.

372. Mars is of the apparent size of a star of the first or second magnitude, and is distinguished from the other planets by the ruddy color of its light. The observed variation in the form of its disc (356) shows that it derives its light from the sun. Its greatest and least apparent diameters are respectively 4″ and 30″. Its real diameter is somewhat less than 5,000 miles, and its bulk is about $\frac{1}{7}$ of that of the earth.

Mars revolves on its axis in 24h. 37m.; and its axis is inclined to the plane of the ecliptic in an angle of about 60°. It appears, from measurements made with the micrometer, that its polar diameter is less than the equatorial, and thus, that like the earth, it is flattened at its poles. According to the latest determinations its oblateness (105) is $\frac{1}{51}$.

373. Telescopic Appearances :—Inferences. When the disc of Mars is examined with telescopes of great power, it is generally seen to be diversified with large spots of different shades, which, with occasional variations, retain constantly the same size and form.

These are conjectured to be continents and seas. In fact, Sir J. F. W. Herschel has on several occasions, in examining this planet with a good telescope, noticed that some of its spots are of a dull red color, while others have a greenish hue. The former he supposes to be land, and the latter water. Fig. 92 represents Mars in its gibbous state, as seen by Herschel in his twenty-feet reflector, on the 16th of August, 1830. The darker parts are the supposed seas. The bright spot at the top is at one of the poles of Mars. At other times a similar bright spot is seen at the other pole. These brilliant white spots have been conjectured, with a great deal of probability, to be snow; as they are reduced in size, and sometimes disappear when they have been long exposed to the sun, and are greatest when just emerging from the long night of their polar winter.

FIG. 92

The great divisions of the surface of Mars are seen with different degrees of distinctness at different times, and sometimes disappear, either partially or entirely; parts of the disc also appear at times particularly dark or bright. From these facts it is to be inferred that this planet is environed with an atmosphere, and that this contains aqueous vapor, which, by varying in quantity and density, renders its transparency variable.

374. *No mountains have been detected upon Mars.* But this is no good reason for supposing that they are really wanting there; for, if the surface of Mars be actually diversified with mountains and valleys, since its disc never differs much from a full circle, we have no reason to expect that its edge would present that shaded appearance and those irregularities which have been noticed on Venus and Mercury, when of the form of a crescent. The same remarks will apply, with still greater force, to the other superior planets.

375. *The ruddy color of the light of Mars* has generally been attributed to its atmosphere, but Sir John Herschel finds a sufficient cause for this phenomenon in the ochrey tinge of the general soil of the planet.

JUPITER AND ITS SATELLITES.

376. Jupiter is the most brilliant of the planets, except Venus, and sometimes even surpasses Venus in brightness. The general fact and special circumstances of the eclipses of its satellites, and of the transits of the shadows across the disc of their primary (243), indicate that Jupiter, as well as its satellites, are opake bodies, and shine by the reflected light of the sun. Its apparent diameter, when greatest, is 51″, and when least, 31″.

Jupiter is the largest of all the planets; its equatorial diameter is 11 times that of the earth, or 88,000 miles, and its bulk is very nearly 1,300 times that of the earth. It turns on an axis nearly perpendicular to the ecliptic, and completes a rotation in 9h. 55½m. The polar diameter is $\frac{1}{17}$ less than the equatorial.

377. Belts of Jupiter. When Jupiter is examined with a good telescope, its disc is always observed to be crossed by several obscure spaces, which are nearly parallel to each other and to the equator. These are called the *Belts* of Jupiter (see Fig. 93; which represents the appearance of Jupiter as seen by Sir John Herschel, in his twenty-feet reflector, on the 23d of September, 1832.) They vary somewhat in number, breadth, and situation on the disc, but never in direction. Sometimes only one or two are visible; on other occasions as many as eight have been seen at the same time. Sir William Herschel even saw them, on one or two occasions, broken up and distributed over the whole face of the planet; but this phenomenon is extremely rare. Branches running out from the belts, and subdivisions, as represented in the figure, are by no means uncommon. *Dark Spots* of invariable form and size have also been occasionally seen upon them. These have been observed to have a rapid motion across the disc, and to return at equal intervals to the same position on the disc, after the same manner as the sun's spots; which leaves no room to

FIG. 93.

doubt that they are on the body of the planet, and that this turns upon an axis. *Bright Spots* have also recently been detected upon the belts by two observers; Dawes and Lassell. The belts generally retain pretty nearly the same appearance for several months together, but occasionally marked changes of form and size take place in the course of an hour or two. They are even said to change sometimes very sensibly in the course of a few minutes.

Explanation of the Belts. The occasional variations of Jupiter's belts, and the occurrence of spots upon them, which are undoubtedly permanent portions of the mass of the planet, render it extremely probable that they are the body of the planet seen through an atmosphere of variable transparency, but, in general, having extensive tracts of comparatively clear sky in a direction parallel to the equator. These are supposed to be determined by currents analogous to our trade-winds, but of a much more steady and decided character, as would be the necessary consequence of the superior velocity of rotation of this planet. As remarked by Herschel, that it is the comparatively dark body of the planet which appears in the belts, is evident from this, that they do not come up in all their strength to the edge of the disc, but fade away gradually before they reach it. The bright belts, intermediate between the dark ones, are believed to be bands of clouds, or tracts of less pure air.

It is possible that these bright belts may be of the nature of auroral rather than aqueous clouds, and that the dark belts may result from their dispersion along certain tracts, the process being controlled by the varying operation of the sun and planets: after the manner that the planets operate upon the photosphere of the sun, to develop spots upon the sun's disc. Such clouds may have a certain degree of luminosity, and yet at the distance of the earth may shine by the reflected light of the sun. The general prevalence of dark belts on either side of the equator, separated by a bright band at the equator, is analogous to the two spot-belts of the sun, with an intervening region from which the spots are absent.

If, as has been maintained by the author in other publications, the collision of the particles of the earth and planets with the ether of space develops heat, not only directly, but by the origination of electric currents which subsequently pass off in the form of heat, then since a point on the equator of Jupiter has a rotatory velocity 28 times greater than that of a point on the equator of the earth, the temperature at the surface of Jupiter may be much greater than that of the earth, notwithstanding its greater distance from the sun. Upon this idea, it is natural to expect a certain degree of similarity in the photospheric condition of this planet and the sun.

378. The Satellites of Jupiter, as it has been already remarked, are visible with telescopes of very moderate power. With the exception of the second, which is a little smaller, they are a little larger than the moon. The orbits of the satellites lie very nearly in the plane of Jupiter's equator. They are, therefore, viewed nearly edgewise from the earth, and in consequence the satellites always appear nearly in a line with each other.

Sir W. Herschel, in examining the satellites of Jupiter with a telescope, perceived that they underwent periodical variations of brightness. These variations he supposed to proceed from a rotation of the satellites upon axes which caused them to turn different faces towards the earth; and from repeated and careful observations made upon them, he discovered that each satellite made one turn upon its axis in the same time that it accomplished a revolution around the primary, and therefore, like the moon, presented continually the same face to the primary.

SATURN, WITH ITS SATELLITES AND RING.

379. Saturn shines with a pale dull light. Its apparent diameter varies less than 6″, by reason of the change of distance, and is 16″ at the mean distance. The eclipses of its satellites indicate that it is opake, and illuminated by the sun. Saturn is the largest of the planets, next to Jupiter. Its equatorial diameter is 9 times that of the earth, or 72,000 miles; and its volume is 670 times that of the earth. The rotation on its axis is performed in 10h. 29m. The inclination of its axis to the ecliptic is about 62°. Its oblateness is $\frac{1}{10}$.

380. Belts of Saturn. The disc of Saturn, like that of Jupiter, is frequently crossed with dark bands, or belts, in a direction parallel to its equator. But Saturn's belts are far more indistinct than those of Jupiter. Extensive dusky spots are also occasionally seen upon its surface. (See Fig. 94.)

The cause of Saturn's belts is doubtless the same as that of Jupiter's. They accordingly establish the existence of an atmosphere upon the surface of Saturn. The results of Herschel's observations on the occultations of the satellites by the planet, indicate the existence of a dense atmosphere.

381. Saturn's Ring. The planet Saturn is distinguished from all the other planets in being surrounded by a broad, thin, luminous ring, situated in the plane of its equator, and entirely detached from the body of the planet. (See Fig. 94.) This ring sometimes casts a shadow upon the planet, and is, in turn, at times partially obscured by the shadow of the planet; from which we conclude that it is opake, and receives its light from the sun.

FIG. 94.

It is inclined to the plane of the ecliptic in an angle of about 28°, and during the motion of Saturn in its orbit remains continually parallel to itself. The face of the ring is, therefore, never viewed perpendicularly from the earth, and for this reason never appears circular, al though such is its actual form. Its apparent form is that of an ellipse, more or less eccentric, according to the obliquity under which it is viewed, which varies with the position of Saturn in its orbit. When it is seen under the larger angles of obliquity, it appears as a luminous band nearly encircling the planet, and is visible in telescopes of small power. Stars can also be seen between it and the planet in these positions. At other times, when viewed very obliquely, it can be seen only with telescopes of high power. When it is approaching the latter state, it has the appearance of two handles or *ansæ*, one on each side of the planet.

It is also at times invisible. This is the case whenever the earth and sun are on different sides of the plane of the ring, for the reason that the illuminated face is then turned from the earth. When the plane of the ring passes through the centre of the sun, the illuminated edge can be seen only in telescopes of extraordinary power, and appears as a thread of light cutting the disc of the planet.

382. Circumstances of Disappearance of Ring. Since the orbit of Saturn is very large in comparison with the orbit of the earth, the plane of the ring, during the greater part of the revolution of Saturn, will pass without the orbit of the earth; and when this is the case the ring will be visible, as the earth and sun will be on the same side of its plane. During the period, which is about a year, that the plane of the ring is passing by the orbit of the earth, the earth will sometimes be on the same side of it as the sun, and sometimes on opposite sides. In the latter case the ring will be invisible, and in the former will be seen so obliquely as to be visible only in telescopes of considerable or great power. All this will perhaps be better understood on consulting Fig. 95, where *efg* represents the orbit of the earth.

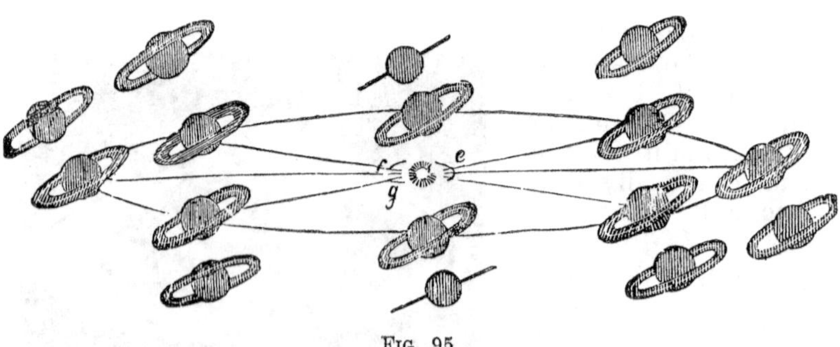

FIG 95.

The appearances of the ring in the different positions of the planet in its orbit are delineated in the figure.

The plane of the ring will pass through the sun every semi-revolution of Saturn, or, at a mean, about every fifteen years; and at the epochs at which the longitude of the planet is respectively 170° and 350°. The ring will then disappear once in about fifteen years; but, owing to the different situations of the earth in its orbit, under varied circumstances: and the disappearance will occur when the longitude of the planet is about 170° or 350°. The ring will be seen to the greatest advantage when the longitude of the planet is not far from 80°, or 260°. The last disappearance took place in 1861; the next will be in 1877. At the present time (1867) the north face of the ring is visible.

383. Rotation of Ring.—Dimensions. From observations made upon bright spots seen on the face of the ring, Herschel discovered that it rotated from west to east about an axis perpendicular to its plane, and passing through the centre of the planet (or very nearly). The period of its rotation is 10h. 32m. It is remarkable that this is almost the exact period in which a satellite assumed to be at a mean distance equal to the mean distance of the particles of the ring, would revolve around the primary, according to the third law of Kepler.

The breadth of the ring is 28,400 miles, which is a little more than one-half greater than its distance from the surface of the planet, and exceeds one-third the equatorial diameter of the planet.

384. Divisions of the Ring. What we have called Saturn's Ring consists in fact of two principal concentric rings; which turn together, although entirely detached from each other. The void space between them is perceived in telescopes of high power, under the form of a black oval line. Calculations from the micrometric measures of Professor Struve give for the breadth of the inner ring 16,500 miles, and of the outer, 10,150 miles. The interval between the rings is 1,700 miles, and the distance from the planet to the inside of the interior ring is 18,300 miles. The thickness of the rings is not well known; the edge subtends an angle less than $\frac{1}{20}''$, which at the distance of the planet, answers to 210 miles.

The division of the ring was discovered as early as the year 1665. The improved telescopes in the hands of modern observers, have revealed the existence of a dark line on the exterior ring, indicative of a subdivision of this ring. It is outside of the middle of the ring, and its breadth is estimated by Dawes at about one-third of that of the principal division of the whole ring.

385. *A new Ring of Saturn*, interior to the other two, was discovered by G. P. Bond, then assistant at the Observatory at Harvard College, on the 11th of November, 1850. It was subsequently observed by the Messrs. Bond on repeated occasions from that date to the 7th of January, 1851. It shone with a pale dusky light. Its inner edge was distinctly defined, but the side next the old ring was not so definite; so that it was impossible to make out with certainty whether the new was connected with the old ring or not.

The same appearances were noticed by Dawes, at his observatory near Maidstone, England, on the 25th and 29th of November, and subsequently by Lassell, with his large reflector, at Starfield, near Liverpool. According to Dawes, the breadth of the new ring is 1''.7, or 7,200 miles; and its distance from the inner edge of the bright ring 0''.3, or 1,270 miles.

386. *Form of Cross Section of the Ring.* Bessel has shown that the double ring is not bounded by parallel plane surfaces. He infers this to be the case from the fact that at almost every disappearance or reappearance of the ring, the two ansæ have not disappeared or reappeared at the same time. He has also found, from a discussion of the observations which have been made upon the disappearances and reappearances of the ring, that they cannot be satisfied by supposing the two faces of the ring to be parallel planes. In view of all the facts, it seems most probable that the cross section of each ring has the approximate form of a very eccentric ellipse instead of a rectangle, and that it varies somewhat in size from one part of the ring to another. It may have irregularities on its surface, as great or greater than those which diversify the surface of the earth.

387. *Centre of Gravity of each Ring—Stability of the Rings.* Whatever may be the form of the rings, their matter is not uniformly distributed; for micrometric measurements of great delicacy made by Struve, have made known the fact, that the rings are not concentric with the planet, but that their centre of gravity revolves in a minute orbit about the centre of the planet. Laplace had previously inferred, from the principle of gravitation, that this circumstance was essential to the stability of the rings. He demonstrated that if the centre of gravity of either ring were once strictly coincident with the centre of gravity of the planet, the slightest disturbing force, such as the attraction of a satellite, would destroy the equilibrium of the ring, and eventually cause the ring to precipitate itself upon the planet.

388. *Physical Constitution of the Ring.* G. P. Bond has propounded a bold and ingenious theory, relative to the physical constitution of Saturn's rings; which is, that "they are in a fluid state, and within certain limits change their form and position in obedience to the laws of equilibrium of rotating bodies." He conceives also, that under peculiar circumstances of disturbance several subdivisions of the two fluid rings may take place, and continue for a short time until the sources of disturbance are removed, when the parts thrown off would again reunite. Professor Pierce has followed up the speculations of Bond, by undertaking to demonstrate from purely mechanical considerations, that Saturn's ring cannot be solid. He maintains that there is no conceivable form of irregularity, and no combination of irregularities, consistent with an actual ring, which would serve to retain it permanently about the primary if it were solid. He is led by his investigations to the curious result, that Saturn's ring is sustained in a position of stable equilibrium about the planet, solely by the attractive power of his satellites; and that no planet can have a ring unless it is surrounded by a sufficient number of properly arranged satellites. Upon the theory of the development of heat by collision with the ether of space (377), the temperature of the mass of Saturn's rings should be much higher than that of the body of the planet, as its actual velocity of rotation is nearly twice as great, and the possibility of a liquid condition of its mass may be admitted.

389. *Origin.* In respect to the origin of Saturn's ring, Sir John Herschel has offered the interesting suggestion, that as the smallest difference of velocity in space between the planet and ring must infallibly precipitate the latter on the former, never more to separate, it follows either that their motions in their common orbit around the sun must have been adjusted by an external power with the minutest precision, or that the ring must have been formed about the planet while subject to their common orbital motion, and under the full influence of all the acting forces. The latter supposition accords with Laplace's theory of the progressive development of the planetary system.

390. The Satellites of Saturn were discovered, the 6th, in the order of distance, by Huyghens, in 1655, with a telescope of 12 feet focus; the 3d, 4th, 5th and 8th, by Dominique Cassini, between the years 1670 and 1685, with refracting telescopes of 100 and 136 feet in length; and the 1st and 2d, by Sir William Herschel, in 1789, with his great reflecting telescope of 40 feet focus. All these but the 1st and 2d, are visible in a telescope of large aperture, with a magnifying power of 200. The 7th satellite, in the order of distance from the primary, was discovered by the Messrs. Bond, with the great refractor of the Cambridge Observatory, on the 16th of September, 1848; and observed two days afterwards by Lassell. It has received the name of Hyperion. The periods of revolution and the mean distances of the satellites of Saturn from their primary, together with the mythological names proposed for them by Sir John Herschel, are given in Table VI.

All of Saturn's satellites, with the exception of the 8th, revolve very nearly in the plane of the ring, and of the equator of the primary. The orbit of the 8th is inclined under a considerable angle to this plane. The 6th satellite is much the largest, and is estimated to be not much inferior to Mars in size. The others interior to this, diminish in size, towards the ring. The 1st and 2d are so small, and so near the ring, that they have never been discerned but with the most powerful telescopes which have yet been constructed, and with these only at the time of the disappearance of the ring (to ordinary telescopes), when they have been seen as minute points of light skirting the narrow line of the luminous edge of the ring. The new satellite (the 7th) is described as fainter than either of these two interior satellites, discovered by Sir William Herschel.

The 8th satellite is subject to periodical variations of lustre, which indicate a rotation about an axis in the period of a sidereal revolution of Saturn.

URANUS AND ITS SATELLITES.

391. The planet now known by the name of Uranus, was discovered by Sir William Herschel. It is not visible to the naked eye, except in opposition, when it becomes barely discernible. In a telescope it appears as a small, round, uniformly illuminated disc. Its apparent diameter is about 4″, from which it never varies much, owing to the small size of the earth's orbit in comparison with its own. Its real diameter is 33,000 miles, and its bulk 73 times that of the earth. Analogy leads us to believe that this planet is opake and turns on an axis, but there is no positive evidence that this is the case.

Of the eight satellites of Uranus, six were discovered by Herschel, one by Lassell, and one by O. Struve.

NEPTUNE.

392. It is a remarkable fact that the existence of this planet was first detected from the disturbances it produced in the motions of Uranus. It having been ascertained that there were outstanding inequalities in the motion of this planet, which could not be referred to the action of the other planets, Le Verrier, the eminent French astronomer, undertook in 1845 the problem of determining the orbit and mass of a planet capable of producing such inequalities. The same problem was independently undertaken and successfully solved by Mr. Adams, of Cambridge, England. Le Verrier, as the final result of his computations, indicated the probable place of the theoretical planet in the heavens; and Dr. Galle, of Berlin, upon directing the great telescope of the Royal Observatory on the region indicated, on the evening of the 23d of September, 1846, descried the new planet within 1° of its most probable place, as assigned by Le Verrier.

The apparent diameter of Neptune is a little less than 3″. Its real diameter is 36,000 miles; and its volume 93 times that of the earth. Neptune, like Uranus, is destitute of visible spots and belts, and the period of its axial rotation is unknown.

Neptune's satellite was discovered by Lassell in 1846. The same observer has since obtained traces of the existence of a second satellite.

THE PLANETOIDS.

393. Vesta is the brightest of the minor planets. In the telescope, it appears as a star of 6th or 7th magnitude. Pallas, Ceres, and Juno appear of the 7th or 8th magnitude. The great majority of the other planetoids are of the 10th or 11th magnitude. Pallas is the largest of this class of bodies. According to Lamont, Director of the Royal Observatory, Munich, its diameter is 670 miles. The diameter of Vesta is believed not to exceed 300 miles; and that of Ceres to be somewhat smaller. Juno is the smallest of the four planetoids first discovered. All of the other minor planets are supposed to be less than 100 miles in diameter.

CHAPTER XVIII.

COMETS.

THEIR GENERAL APPEARANCE:—VARIETIES OF APPEARANCE.

394. The general appearance of comets is that of a mass of some luminous nebulous substance, to which the name *Coma* has been given, condensed towards its centre around a brilliant *nucleus* that is in most cases not very distinctly defined; from which proceeds, in a direction opposite to the sun, a stream of

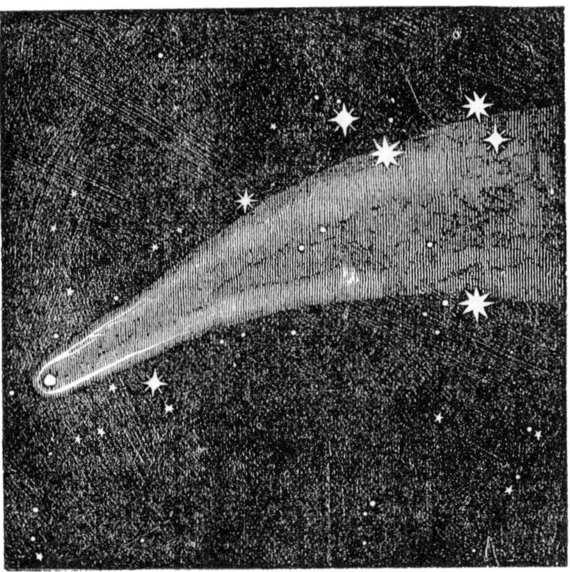

FIG. 96.

similar but less luminous matter, called the *Tail* or Train of the comet (Fig. 96). The nucleus, with the surrounding coma, forms the *Head* of the comet.

The tail gradually increases in width, and at the same time diminishes in distinctness from the head to its extremity, where it is generally many times wider than at the head, and fades away until it is lost in the general light of the sky. It is, in general, less bright along its middle than at the borders. From this cause the tail sometimes seems to be divided, along a greater or less portion of its length, into two separate tails or streams of light, with a comparative dark space between them. Ordinarily it is not straight, that is, coincident with a great circle of the heavens, but concave towards that part of the heavens which the comet has just left. This curvature of the tail is most observable near its extremity. The most remarkable example is that of the comet of 1744, which was bent so as to form nearly a quarter of a circle. Nor does the general direction of the tail usually coincide exactly with the great circle passing through the sun and the head of the comet, but deviates more or less from this, the position of exact opposition to the sun in the heavens, on the side towards the quarter of the heavens just traversed by the comet. This deviation is quite different for different comets, and varies materially for the same comet while it continues visible. It has even amounted in some instances to a right angle.

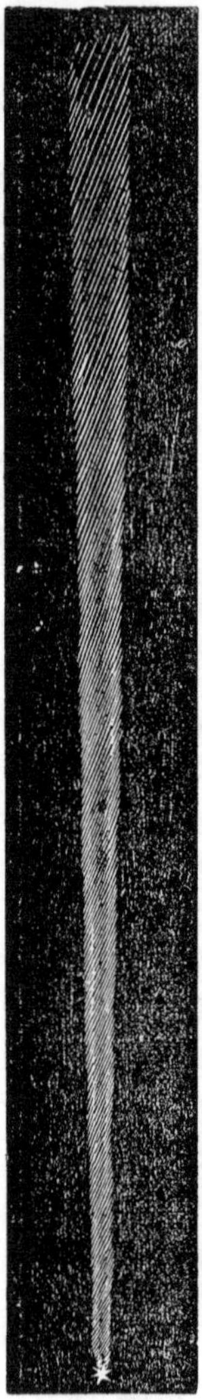

395. Variations of Length of Tail. The apparent length of the tail varies, from one comet to another, from zero to 100° and more; and ordinarily the tail of the same comet increases and diminishes very much in length during the period of its visibility. When a comet first appears, in general no tail is perceptible, and its light is very faint. As it approaches the sun, it becomes brighter; the tail also, after a time, shoots out from the coma, and increases from day to day in extent and distinctness. As the comet recedes from the sun, the train precedes the head, being still on the opposite side from the sun, and grows less and less at the same time that, along with the head, it decreases in brightness, till at length the comet resumes nearly its first appearance, and finally disappears. (See Fig. 98.) It sometimes happens that, owing to peculiar circumstances, a comet does not make its appearance in the firmament until after it has passed the sun in the heavens, and not until it has attained to more or less distinctness, and is furnished with a train of considerable or even great length. This was remarkably the case with the great comet of 1843. (See Art. 237; also Fig. 97.)

FIG. 97.
Great Comet of 1843.

396. Effects of the Position of the Earth, on the apparent size and brightness of a comet. The tail of a comet is the longest, and the whole comet is intrinsically the most luminous, not long after it has passed its perihelion. Its apparent size and lustre will not, however, necessarily be the greatest at this time, as they will depend upon the distance and position of the earth, as well as the actual size and intrinsic brightness of the comet. To illustrate

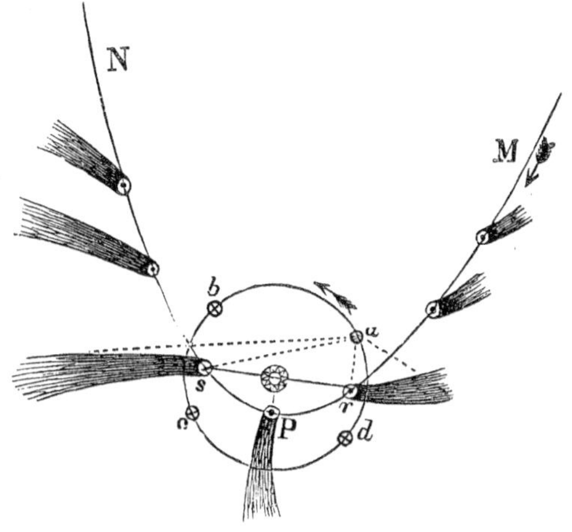

FIG. 98.

this, let *abcd* (Fig. 98) represent the orbit of the earth, and MPN the orbit of a comet, having its perihelion at P. Now, if the earth should chance to be at *a* when the comet, moving towards its perihelion, is at *r*, it might very well happen that the comet would appear larger and more distinct than when it had reached the more remote point *s*, although when at the latter point it would in reality be larger and brighter than when at *r*. It would be the most conspicuous possible if the earth should be in the vicinity of *c* or *b* soon after the perihelion passage; and it would be the least conspicuous possible if the comet be supposed to be moving in the direction NPM, and to pass from N around to M, while the earth is moving around from *a* to *b* or *c*; so as to be continually comparatively remote from the comet, and so that the comet will be in conjunction with the sun at the time after the perihelion passage when its actual size and intrinsic lustre are the greatest. It is to be observed that the apparent lustre of a comet is sometimes very much enhanced by the great obliquity of the tail, in some of its positions, to the line of sight. This seems to have been the case with the comet of 1843, on February 28 (see Fig. 63), and, it has been already intimated, was one reason of its being so very bright as to be seen in open day in the immediate vicinity of the sun.

Since the earth may have every variety of position in its orbit at the successive returns of the same comet to its perihelion, it will be seen, on examining Fig. 98, that the circumstances of the appearance and disappearance of the comet, as well as its size and distinctness, may be very different at its different returns. This has been strikingly true in the case of Halley's Comet. Biela's Comet was also invisible on its return to its perihelion in 1839, by reason of its continual proximity to the line of direction of the sun as seen from the earth, and its great distance from the earth.

397. Varieties of Aspect. Individual comets offer considerable varieties of aspect. Some comets have been seen which were wholly destitute of a tail: such, among others, was the comet of 1682, which Cassini describes as being as round and as bright as Jupiter. Others have had more than one luminous train. The comet of 1744 was provided with six, which were spread out like an immense fan, through an angle of 117°; and that of 1823 with two, one directed from the sun in the heavens, and, what is very remarkable, another smaller and fainter one directed towards the sun. Others still have had no perceptible nucleus, as the comets of 1795 and 1804.

The comets that are visible only in telescopes, which are very numerous, have generally no distinct nucleus, and are often entirely destitute of every vestige of a tail. They have the appearance of round masses of luminous vapor, somewhat more dense towards the centre. Such are Encke's and Biela's comets. (Fig. 99.) The point of greatest condensation is often more or less

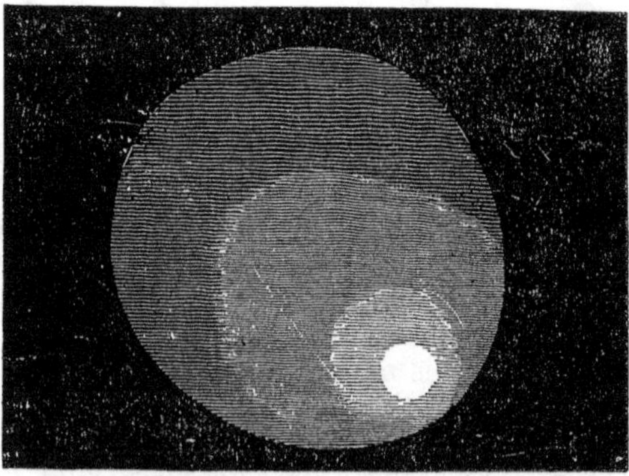

FIG. 99.—*Encke's Comet.*

removed from the centre of figure on the side towards the sun; and sometimes also on the opposite side.

398. The Comets which have had the Longest Trains, are those of 1680, 1769, and 1618. The tail of the great comet of 1680, when apparently the longest, extended to a distance of 70° from the head; that of the comet of 1769, a distance of 97°: and that of the comet of 1618, 104°. These are the apparent lengths as seen at certain places. By reason of the different degrees of purity and density of the air through which it is seen, the tail of the same comet often appears of a very different length to observers at different places. Thus, the comet of 1769, which at the Isle of Bourbon seemed to have a tail of 97° in length, at Paris was seen with a tail of only 60°. From this general fact we may infer that the actual train extends an unknown distance beyond the extremity of the apparent train.

FORM, STRUCTURE, AND DIMENSIONS OF COMETS.

399. The general form and structure of comets, so far as they can be ascertained from the study of the details of their appearance, may be described as follows: The head of a comet consists of a central nucleus, or mass of matter brighter and denser than the other portions of the comet, enveloped on the side towards the sun, and ordinarily at a great distance from its surface in comparison with its own dimensions, by a globular nebulous mass of great thickness, called the *Nebulosity*, or nebulous *Envelope*. This, it is said, never completely surrounds the nucleus, except in the case of comets which have no tails. It forms a sort of hemispherical cap to the nucleus on the side towards the sun. Its form, however, is not truly spherical, but varies between this and that of a paraboloid having the nucleus in its focus and its vertex turned towards the sun. The tail begins where the nebulosity terminates, and seems, in general, to be merely the continuation of this in nearly a straight line beyond the nucleus. There is ordinarily, as has been already intimated, a distinct space containing less luminous matter between the nucleus and the nebulosity, but this is not always the case. The tail of a comet has the shape of a hollow truncated cone, with its smaller base in the nebulosity of the head; with this difference, however, that the sides are usually more or less curved. That the tail is hollow is evident from the fact, already noticed, that on whichever side it is viewed it appears less bright along the middle than at the borders. There can be less luminous matter on a line of sight passing through the middle, than on one passing near one of the edges, only on the supposition that the tail is hollow. The whole tail is generally bent so as to be concave towards the regions of space which the comet has just left.

400. Multiple Tails. In some instances the nucleus is furnished with several envelopes concentric with it: which are formed in succession as the comet approaches the sun, and then recedes from it again. For example, the comet of 1744, eight days after the perihelion passage, had three envelopes. Sometimes each of them is provided with a tail. Each of these several tails being hollow, may in consequence appear so faint along its middle as to have the aspect of two distinct tails. A comet which has in reality three separate trains, might thus appear to be supplied with six, as was the comet of 1744. If the different envelopes were not distinctly separate from each other, then all the trains would appear to proceed from the same nebulous mass.

Supernumerary tails, shorter and less distinct than the principal one, are by no means uncommon; but they generally appear quite suddenly, and as suddenly disappear in a few days, as if the stock of materials from which they were supplied had become exhausted.

401. The general Position of the Tail of a Comet is nearly but not exactly in the prolongation of the line of the centres of the sun and head, or of the radius-vector of the comet. (See Fig. 98.) It deviates from this line on the side of the regions of space which the comet has just left; and the angle of deviation, which, when the comet is first seen at a distance from the sun, is very small or not at all perceptible, increases as the comet approaches the sun, and attains to its maximum value soon after the perihelion passage; after which it decreases, and finally, at a distance from the sun, becomes insensible. For example, the angle of deviation of the tail of the great comet of 1811 attained to its maximum about ten days after the perihelion passage, and was then about 11°. In the case of the comet of 1664, the same angle about two weeks after the perihelion passage was 43°, and was then decreasing at the rate of 8° per day.

The comet of 1823 might seem to present an exception to the general fact that the tail of a comet is nearly opposite to the sun; but Arago has suggested that the probable cause of the singular phenomenon of a secondary tail, apparently directed towards the sun in the heavens, was that the earth was in such a position that the two tails, although in fact inclined to each other under a small angle, were directed towards different sides of the earth, and thus were referred to the heavens so as to appear nearly opposite.

The same principle will serve to show that the deviation of the train of a comet from the position of exact opposition to the sun may appear to be much greater than it actually is, by reason of the earth's happening to be within the angle formed by the direction of the train with the radius-vector prolonged.

402. Vast Size of Comets. Comets are the most voluminous bodies in the solar system. The tail of the great comet of 1680 was found by Newton to have been, when longest, no less than 123,000,000 miles in length. The remarkable comet of

1843, about three weeks after its perihelion passage, had a tail of over 108,000,000 miles in length. Other comets have had trains from fifty to one hundred million miles long. The heads of comets are generally tens, and often hundreds, of thousands of miles in diameter. That of the great comet of 1811 had a diameter of over 1,000,000 miles; that of Halley's comet, in 1836, a diameter of 350,000 miles, and that of Encke's comet, in 1828, a diameter of over 300,000 miles. The head of the great comet of 1843 was about 30,000 miles in diameter.

403. The Nuclei of comets, so far as they have been accurately determined, do not exceed a few hundred miles in diameter. For example, the great comet of 1811 had a nucleus of 428 miles, and that of 1798 one of 125 miles in diameter. Instances are cited of comets with nuclei of several thousand miles in diameter (e. g., the third comet of 1845, and the fourth comet of 1825); but there is little reason to doubt that in these cases, the apparent telescopic nucleus ordinarily observed was measured, instead of the true nucleus, which is only occasionally seen. When a comet is viewed with the naked eye, it usually offers the appearance of a star-like nucleus at the centre of the head. Telescopes resolve this, more or less, into a bright nebulous mass, which is the ordinary telescopic nucleus. But occasionally they show, in the case of a bright comet, within this a stellar point, distinguished by its brightness and appearance of solidity from the nebulosity about it. This is the true nucleus. The nucleus, so-called, of Donati's comet, is stated to have been 5,600 miles in diameter, but according to Bond, the true nucleus that was occasionally discernible in his telescope, was less than 500 miles in diameter.

404. Variation of Dimensions. The dimensions of comets are subject to continual variations. The tail increases in actual length as the comet approaches the sun, and attains its greatest size a certain time after the perihelion passage; after which it gradually decreases. The head, on the contrary, diminishes in size during the approach to the sun, and augments during the recess from him. These changes of dimension, both in the case of the head and of the tail of the comet, are often very great, and sometimes quite sudden and rapid. Encke's comet, at its return in 1828, in the course of two months, while its distance from the sun was diminished in the ratio of 1 to 3, underwent an apparent diminution of volume in the ratio of 16,000 to 1. The apparent nucleus of Donati's comet was 1,000 times less soon after the perihelion passage than when it was previously seen at a distance two or three times greater. The tail of the great comet of 1843 increased in length after the perihelion passage, at the rate of 5,000,000 miles per day; and that of Donati's comet increased in length for ten days after the perihelion passage, at the average rate of 2,500,000 miles per day.

PHYSICAL CONSTITUTION OF COMETS.

405. Small Mass and Density. The quantity of matter which enters into the constitution of a comet is exceedingly small. This is proved by the fact that comets have had no influence upon the motions of the planets or satellites, although they have, in many instances, passed near these bodies. The comet of 1770, which was quite large and bright, passed in close proximity to Jupiter's satellites, without deranging their motions in the least perceptible degree. Moreover, since this small quantity of matter is dispersed over a space of tens of thousands or millions of miles (if we include the tail), in linear extent, the nebulous matter of comets must be incalculably less dense than the solid matter of the planets. In fact, the cometic matter, with the exception perhaps of that of the nucleus, is inconceivably more rare and subtile than the lightest known gas, or the most evanescent film of vapor that ever makes its appearance in our sky; for faint telescopic stars are distinctly visible through all parts of the comet, with, it may be, the exception of the nucleus, notwithstanding the great space occupied by the matter of the comet, which the light of the star traverses. The matter of the tail of a comet is even more attenuated than that of the general mass of the nebulosity of the head, but is apparently of the same nature, and derived from the head.

406. Nucleus and Nebulosity. The nucleus is supposed by some astronomers to be, in some instances, a solid, partially or wholly convertible into vapor under the influence of the sun; by others, to be in all cases the same species of matter as the nebulosity, only in a more condensed state; and by others still, to be a solid of permanent dimensions, with a thick stratum of condensed vapors resting upon its surface. Whichever of these views be adopted, it is a matter of observation that the nebulosity frequently receives fresh supplies of matter from the nucleus. It was the opinion of Sir William Herschel, and it has been the more generally received notion since his time, that the nucleus of a comet is surrounded with a transparent atmosphere of vast extent, within which the nebulous envelope floats, as do clouds in the earth's atmosphere. But Olbers, and after him Bessel, conceives the nebulous matter of the head to be either in the act of flowing away into the tail under the influence of a repulsion from the nucleus and the sun, or in a state of equilibrium under the action of these forces and the attraction of the nucleus.

407. Luminosity of Comets. Observations with the polariscope have established that comets shine in a great degree by reflected light. This is especially true of the tail of the comet; the nucleus and nebulosity present feeble traces of polarization, and we must therefore conclude, emit a strong light of their own, or shine wholly by light radiantly reflected. If the head of a comet shone entirely by reflected light, and the amount of reflecting surface remained constantly the same, its apparent brightness would be inversely proportional to the product of the squares of the distances from the sun and earth. By this rule, the head of Donati's comet should have been 188 times brighter on the 2d of October than on June 15th; whereas it was actually 6,300 times brighter. From which we may infer that the quantity of light emanating from it had increased in the proportion of 33 to 1. This increase of actual light was confined chiefly to the nebulosity of the head, and is probably attributable, in a great degree, to an augmentation of the quantity of nebulous matter received from the nucleus.

CONSTITUTION AND MODE OF FORMATION OF THE TAILS OF COMETS.

408. Upon this topic we may lay down the following *postulates:* 1. The general situation of the tail of a comet with respect to the sun, shows that the sun is concerned, either directly or indirectly, in its formation. The changes which take place in the dimensions of a comet, both in approaching the sun and receding from it, conduct to the same inference. 2. Since the tail lies in the direction of the radius-vector prolonged beyond the head, the particles of matter of which it is made up must have been driven off by some force exerted in a direction from the sun. 3. This force cannot emanate from the nucleus, for such a force would expel the nebulous matter surrounding the nucleus in all directions, instead of one direction only. It is, however, conceivable that, as Olbers supposes, the nebulous matter is, in the first instance, expelled from the nucleus by its repulsive action, taking effect chiefly on the side towards the sun, and afterwards driven past the nucleus into the tail by a repulsion from the sun. 4. There seems, then, to be little room to doubt that the matter of the tail is driven from the head by some force foreign to the comet, and taking effect from the sun outwards. 5. This force, whatever may be its nature, extends far beyond the earth's orbit; for comets have been seen provided with tails of great length, though their perihelion distance exceeded the radius of the earth's orbit (*e.g.*, the great comet of 1811). 6. It is natural to suppose that, like all central forces, *the repulsive force exerted by the sun upon cometic matter varies inversely as the square of the distance.* This law of variation has in fact been established by the investigations of Bessel and Professor Pierce, and confirmed by the author's determination of the form and dimensions of the tail of Donati's comet, upon the theory that it was made up of particles individually repelled by the sun with an intensity of force varying according to this law.*

409. Explanation of Situation and Curved Form of Tail. Let PCA (Fig. 100) be a portion of a comet's orbit, the

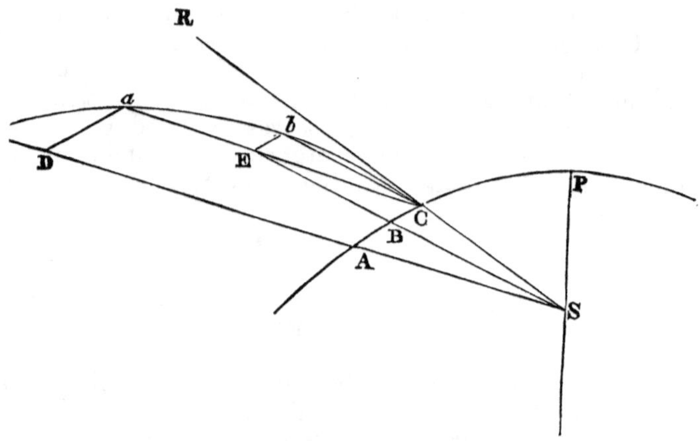

Fig. 100.

sun being at S; and suppose a particle to be expelled in the direction SAD, when the head is at A, and another particle to be driven off in the direction SBE, when the head is at B. Each particle will retain the orbital motion which obtained at the time of its departure, as it moves away from the sun; and thus, when the comet has reached the point C, instead of being at any points, D and E, on the lines SAD and SBE, will be respectively at certain points, a and b, farther forward. The line Cba, which, when the comet is at C, is the locus of all the particles that have been emitted during the interval of time in which the comet has been moving over the arc AC, is the tail. We here suppose the head to be a mere point. If we conceive the particles to be continually emitted from the marginal parts of the head, we shall have the hollow conical tail actually observed. It is easy to see that Cba, the line of the tail, must be a curved line concave towards the regions of space which the comet has left. Supposing the arc AC to be so small, or its curvature so slight, that it may be considered as a straight line, and neglecting the change of velocity in the orbit, Ca will be parallel to AD, and Cb parallel to BE; whence RCa = CSA, and RCb = CSB. Thus the line joining any particle with the nucleus always makes an angle with the prolongation of the radius-vector, approximately equal

* See American Journal of Science, Vol. xxix., pp. 79 and 383, and Vol. xxxii. p. 54, etc.

to the motion in anomaly during the interval that has elapsed since the particle left the head. It follows from this, that if we suppose the velocity of the particles to be continually the same, and the motion in anomaly uniform, the deviations of the particles *a* and *b* from the line of the radius-vector SCR will be in the ratio of the distances C*a* and C*b*. But in point of fact the velocity increases with the distance, so that the curvature of the tail will be less than on the supposition just made.

As to the amount of the deviation of the tail from the line of the radius-vector, it must depend upon the proportion between the velocity of the particles and the velocity of the head in its orbit; and it follows from the principle just established that unless the velocities of emission augment as rapidly as the velocity of revolution, the deviation in question will increase to the perihelion, and afterwards decrease, as it is in fact known to do.

410. Dispersion of the Cometic Matter in the Plane of the Orbit. Observations made upon Donati's comet, have established that the nebulous matter was much more widely dispersed in the direction of the plane of the comet's orbit than in the direction perpendicular to the orbit; so that the transverse sections of the tail were approximately elliptical in form, and more elongated in proportion as their distance from the head was greater. The same fact was still more conspicuous in the case of the great comet of 1861, and is probably a general law. It is shown in the memoir above referred to, that this phenomenon had its origin in the case of Donati's comet, in an *inequality in the force of repulsion* exerted by the sun upon different portions of nebulous matter expelled from the nucleus. The limits between which the repulsive force varied were 0 and 1.21 (the intensity of the sun's ordinary force of attraction at the same distance being the unit). It is shown also that nearly one-half of the tail, on the concave side, was made up of matter that was not actually repelled by the sun, but became widely separated from the head of the comet, after being expelled by a projectile force beyond the sphere of attraction of the nucleus, simply because it was subject to a diminished intensity of solar attraction. The concave edge of the tail consisted of matter subject to an attractive force equal to $\frac{45}{100}$ of the full force of the sun's attraction. The greatest intensity of repulsive action (1.21) obtained at the convex, or preceding side of the tail.

If we assume that the escaping particles did not receive any initial lateral velocity from a repulsive or projectile force ex erted by the nucleus, the limits of the effective solar repulsion and attraction, for the two edges of the tail, become 1.5 and 0.6 (instead of 1.21 and 0.45).

In Fig. 101, the train of the comet as theoretically determined is compared with that actually observed. The full curve runs through the positions of the particles that left the head at several assumed dates, calculated for October, 5d. 7h. mean time at Greenwich, and is accordingly the outline of the train as theoretically determined for that instant. The dotted curved line is the

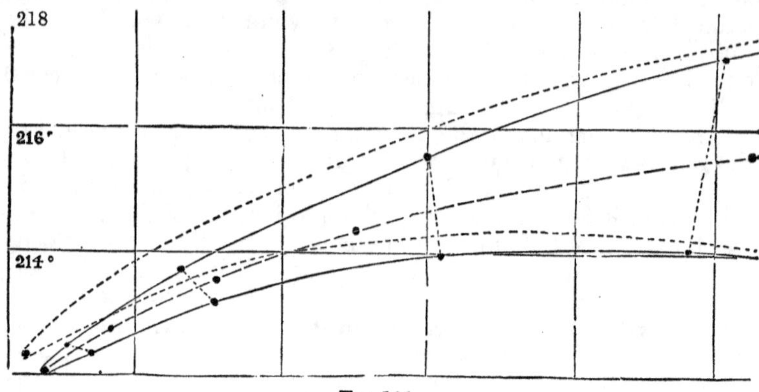

FIG 101.

outline of the actual train as observed 1½ hours later, when its form and dimensions were sensibly the same as at 7h. The broken line nearly in the middle of the theoretical train, runs through the calculated positions of several particles that left the head of the comet at different dates, and were neither attracted nor repelled by the sun, and therefore proceeded on in tangents to the orbit.

The single straight streamers seen in connection with this and other comets (See Plate III.), must have been urged by a force of repulsion many times greater than the maximum limit of repulsion for the principal tail (1.21).

411. Columnar Structure of the Tail of Donati's Comet. The tail of the comet of Donati, was seen on certain occasions to be traversed, for a part of its length, by *bands of unequal brightness*, diverging from the vicinity of the head (See Plate III.). This proves to have been a consequence of frequent alternations in the ejection of nebulous matter from the head of the comet; for it appears, as a result of the calculations above mentioned, that all the matter variously repelled which issued at any instant, must, at any subsequent date, have been arranged nearly in a straight line that produced would pass near the head. Fig. 102, shows, for the date of the calculations (October 5th), the lines made up of the particles that proceeded from the head of the comet, at the dates given, viz.: September 29th, September 26th, &c. The train may accordingly be considered as having been composed of a series of diverging bands, or columns of nebulous matter emanating from the head on successive days, or other equal intervals of time; which alternated in brightness when there were alternations in the quantity of matter discharged.

412. *The Source of the Nebulous Stream*, called the tail of the comet, has been generally supposed to be the envelope, or envelopes of the head; but at the present day the preponderating weight

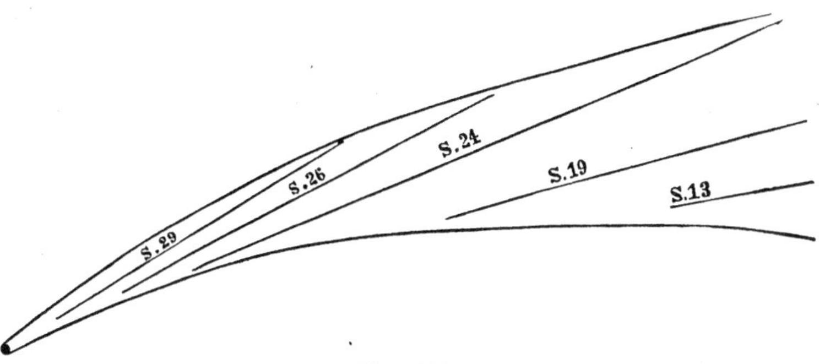

FIG. 102.

of evidence is opposed to this view, and in favor of the theory that the envelope and tail are but different portions of one continuous stream of cometic matter emanating from the nucleus, or from the bright nebulosity contiguous to the nucleus proper. It appears to be an insuperable objection to the former hypothesis, that a small extent of the nebulous stream, in the immediate vicinity of the envelope from which it proceeds, contains as much luminous matter as the envelope itself, and yet the envelope usually continues in existence for many days. Some of the envelopes that were seen to rise apparently from the nucleus of Donati's Comet, did not become dissipated until two weeks after their first appearance. Besides it is certain that a considerable portion of the matter detached from the nucleus, does move in a continuous stream through the apparent envelope into the tail; for jets, or single streams, are frequently seen to proceed from the nucleus, on the side toward the sun, and after being bent back by the solar repulsion, to become merged in the general stream that seems to issue from the envelope. It is also possible to deduce the actual form and dimensions of both the envelope and tail, on the hypothesis of a single continuous stream proceeding from a certain portion of the nucleus exposed to the action of the sun.*

CONDITION AND ORIGIN OF THE NEBULOUS ENVELOPES.

413. Successive Envelopes. As already intimated there are frequently two or more envelopes that appear to be indefinitely continued into the train (See Plate III.). These are detached in succession from the nucleus, and while receding continually

* (See the American Journal of Science, Vol. XXVII., January, 1859.)

from it and expanding, decline in lustre, and finally disappear; according to one of the above-mentioned hypotheses because they are dissipated by the repulsive action of the sun upon their particles, and according to the other because the supply of outstreaming matter at the nucleus falls off. The late Director of the Observatory of Harvard College, in his great work on the comet of Donati, states that no less than seven envelopes were detached in succession from the nucleus of the comet, at intervals of from four to seven days. Their rate of recess from the nucleus was about 1,000 miles per day. The great comet of 1861 presented a succession of eleven envelopes, rising at regular intervals on every second day. Their evolution and final dissipation were accomplished with much greater rapidity than the corresponding phenomena of the comet of 1858.

414. Expelling Force. Since the cometic particles which were distributed along the concave side of the tail of Donati's comet were not repelled by the sun (410), we must infer that they were not expelled from the nucleus by a force of repulsion, but were in all probability detached by some projectile force in operation at or near its surface. On the other hand, the cometic particles that were in a condition to be repelled by the sun, may have become detached from the nucleus under the operation of a force of repulsion exerted by its mass, or from its surface. We may conceive a repulsive force, exerted by both the nucleus and the sun, to be a consequence of the particles being more nearly in the condition of the ultimate molecules, in which there is reason to believe that they become subject to both a molecular and heat repulsion, operating at indefinitely great distances.*

If we conceive the bright nebulous mass adjacent to the nucleus, which appears to be the fountain head of the nebulous stream that constitutes both the envelope and train of a comet, to be in a magnetic condition similar to that which has been attributed to the photosphere of the sun, it is to be observed that particles may become detached from the tops of magnetic columns simply in consequence of a diminution in the magnetic intensity of the nucleus and its photosphere; and such diminution of magnetic intensity should continually occur, from day to day, as the comet recedes from the sun, and consequently has a decreasing velocity in its orbit. For, according to the *theory of cosmical magnetization*, the intensity of the magnetic currents developed should be directly proportional both to the orbital velocity and the velocity of rotation.† A statical force of electric repulsion might also operate to detach particles, whether magnetic or not, in directions normal to the surface of the nucleus.

The Projectile Force, whose existence we have here recognized, may have its origin in electric discharges along magnetic vaporous columns, like the similar force supposed to be in action upon the surface of the sun's photosphere (293). In support of this view it may be urged that, if we assume the hypothesis that the nebulous matter at the nucleus of a comet is made up of particles susceptible of magnetization, and capable of being expelled by discharges along lines of magnetic polarization, we are enabled to give an adequate explanation of diverse luminous phenomena presented by comets, that are wholly inexplicable upon all previous hypotheses.

415. Theoretical Process of Evolution of an Envelope. We would first remark that a rotatory motion of the nucleus, in conjunction with its orbital motion, should, by the collision of the molecules with the ether of space, bring it into a magnetized state, with the poles in the vicinity of 90° from the plane of the orbit. Now if we conceive the matter, disposed in magnetic columns, to be expelled in the lines of direction of the columns, and subsequently to be repelled by the sun, we have to observe that the lines of discharge will be nearly parallel to the surface of the nucleus near the magnetic equator, and that their angle of incli-

* See American Journal of Science, Vol. XXXVIII., p. 70.
† See American Journal of Science, Vol. XLI., p. 62.

nation to the surface will increase with the distance of the columns from the magnetic equator, or approximately from the plane of the orbit.

The envelope should therefore consist of two portions proceeding from parts of the nucleus that lie on opposite sides of the magnetic equator. The nebulous streams issuing from the points on either side of the equator will pass to the other side, intersecting its plane at points more and more distant from the nucleus, until the initial directions of the streams become, at points at a certain distance from the equator, parallel to its plane, or to the plane of the orbit nearly. If we conceive the magnetic equator to lie in the plane of the orbit, and confine our attention to the streams proceeding from points on the meridian, whose plane contains the radius-vector, then at a point on this meridian about 35° from the orbit the nebulous stream would issue in a direction parallel to the radius-vector, and at points that have a higher magnetic latitude its direction would diverge more and more from this line. All such streams would be bent back by the force of the sun's repulsion, and would form, collectively, an apparent envelope on the side towards the sun. This would have a parabolic form, if the discharge extend beyond 35° of magnetic latitude, and the expelling force and the solar repulsion have each a constant intensity for all latitudes. But if the latter should increase, or the former decrease, with the latitude, the outline of the envelope would approach more nearly to the circular form, as was observed in the case of Donati's comet.

416. Phenomena confirmatory of the Theory. Various peculiarities of form, and diversities of brightness presented by Donati's comet, and several others, seem to indicate that each envelope does in fact consist of two portions, that do not in general originate simultaneously; and which in part pass from the one side to the other of the nucleus. The following are some of the peculiar features referred to:

(1.) The *spiral form*, or *awry position* of each of the successive envelopes, when first seen distinctly separate from the nucleus. The explanation is that the discharge of cometic matter began from the one magnetic hemisphere sooner than from the other; from that which is most exposed to the sun's action, we may suppose.

(2.) The *depression*, or *deficiency of cometic matter* about the vertex of the envelope, frequently noticed, especially in the later stages of the envelope. This has been in some instances represented as a notch in the envelope. This deficiency of light at the vertex is obviously what should result if the discharge should relatively fall off at the magnetic latitudes (about 35°) from which the nebulous streams issue in directions parallel to the radius-vector; and we shall soon see that it is reasonable to expect that the discharge should begin to decline at these sooner than at higher latitudes.

(3.) The remarkably *dark band* seen to extend nearly along the axis of the tail of Donati's comet, for a certain distance from the nucleus. This band was too dark to be explained by the supposition that the tail was hollow. Upon the present theory, the brightest portions of the tail, near the head, should have been in the plane perpendicular to the orbit, and through the radius-vector; that is, in the plane through the sun and the magnetic axis of the nucleus. A section of the tail and envelope in this plane would show the brightest parts of the two branches streaming away from the two magnetic hemispheres, on the side towards the sun, bending around past the nucleus, and separated there by a dark space. In the earlier and later stages of an envelope, the dark shade would be enhanced by the deficiency of the streams that would return along the axis (415).

(4.) *The great difference noticed in the brightness of the two branches of the train of Donati's comet*, near the head. This may reasonably be ascribed to an inequality in the discharge of nebulous matter from the two magnetic hemispheres. This inequality of brightness was not changed by the earth's passage through the plane of the comet's orbit (on Sept. 8). G. P. Bond infers from this, that "the initial plane passing through the two branches would seem to have a strong inclination to that of the orbit." Theory, as we have seen, assigns it a position nearly perpendicular to the plane of the orbit.

(5.) *The remarkable shifting of the superior brightness and eccentric position from one branch of the tail of Donati's comet, near the head, to the other, about October 10th.* At about that date, the plane through the sun, comet, and earth, was perpendicular to the plane of the comet's orbit, and the earth should therefore have passed from one side to the other of the initial plane of the branches of the train. Cotemporaneously with these changes, the dark, axial stripe nearly disappeared, and reappeared at later dates.

417. Explanation of the Rise and Recess of Successive Envelopes. To understand how one envelope after another may rise and recede to a certain distance from the nucleus, we have to consider that masses of vaporous magnetic matter may rise, at certain intervals, from the nucleus to a certain height in its atmosphere, under the operation of the sun's rays; and that such matter should ascend most abundantly from the equatorial regions, where the sun is supposed to act most directly, and flow off towards the poles. It will be seen, if we consider the diverse directions that would be assumed by the magnetic columns in different magnetic latitudes, that, as a necessary consequence, the nebulous streams proceeding from them would rise to a greater and greater height towards the sun, until the process of discharge reached the magnetic latitude of 35°. The combination of all the nebulous streams thus originating would present the appearance of a luminous envelope on the side of the nucleus towards the sun, the outer boundary of which would recede steadily from the nucleus.

418. Diversities in the Brightness of an Envelope. The great diversity often observed in the brightness of different parts of the same envelope, may be ascribed to intersections, on the line of sight, of the separate streams of cometic particles, and to varying velocities in different parts of the same stream. Besides the ordinary diversities which are thus satisfactorily explained, sudden interruptions of brightness are often observed at certain parts of an envelope; these may result from sudden variations in the intensity of the expelling force, or in the quantity of matter discharged.

The *dark spots* sometimes seen are probably due to a deficiency of nebulous matter near the nucleus, on certain magnetic parallels. Such deficiency may result primarily from an intermission in the ascent of nebulous matter from the equatorial regions of the nucleus. As the ascended matter flows off towards the poles, any vacuity thus arising will gradually pass from one latitude to another, and the spot answering to it in the envelope will rise and expand with the envelope.

CHAPTER XIX.

THE FIXED STARS.

CONSTELLATIONS.—DIVISION INTO MAGNITUDES.

419. In order to distinguish the fixed stars from each other, they are arranged into groups, called *Constellations*, which are imagined to form the outlines of figures of men, animals, or other objects, from which they are named. Thus, one group is conceived to form the figure of a Bear, another of a Lion, a third of a Dragon, and a fourth of a Lyre. The division of the stars into constellations is of very remote antiquity; and the names given by the ancients to individual constellations are still retained.

The resemblance of the figure of a constellation to that of the animal or other object from which it is named, is in most instances altogether fanciful. Still, the prominent stars hold certain definite positions in the figure conceived to be drawn on the sphere of the heavens. Thus, the brightest star in the constellation Leo is placed in the heart of the Lion, and hence it has sometimes been called *Cor Leonis* or the *Lion's Heart:* and the brightest star in the constellation Taurus is situated in the eye of the Bull, and therefore sometimes called the *Bull's Eye;* while that conspicuous cluster of seven stars in this constellation, known by the name of the Pleiades, is placed in the neck of the figure. Again, the line of three bright stars noticed by every observer of the heavens in the beautiful constellation of Orion, is in the belt of the imaginary figure of this bold hunter drawn in the skies. The three larger stars of this constellation are, respectively, in the right shoulder, in the left shoulder, and in the left foot.

420. Different Classes of Constellations. The constellations are divided into three classes: *Northern Constellations*, *Southern Constellations*, and *Constellations of the Zodiac*. Their whole number is 91: Northern 34, Southern 45, and Zodiacal 12. The number of the ancient constellations was but 48. The rest have been formed by modern astronomers from southern stars not visible to the ancient observers, and others variously situated, which escaped their notice, or were not attentively observed.

The zodiacal constellations have the same names as the signs of the zodiac (Def. 25, p. 17): but it is important to observe that the individual signs and constellations do not occupy the same places in the heavens. The signs of the zodiac coincided with the zodiacal constellations of the same name, as now defined, about the year 140 B. C. Since then the equinoctial and solstitial points have retrograded nearly one sign: so that now the vernal equinox, or first point of the sign Aries, is near the beginning of the constellation Pisces; the summer solstice, or first point of Cancer, near the beginning of the constellation Gemini; the autumnal equinox, or first point of Libra, at the beginning of Virgo; and the winter solstice, or first point of Capricornus, at the beginning of Sagittarius.

It follows from this, that when the sun is in the sign Aries, he is in the constellation Pisces, and when in the sign Taurus, in the constellation Aries, &c. For the rest, it should be observed that the constellations and signs of the zodiac have not precisely the same extent.

421. Modes of Designation of Individual Stars. The stars of a constellation are distinguished from each other by the letters of the Greek alphabet, and in addition to these, if necessary, the Roman letters, and the numbers 1, 2, 3, &c.; the characters, according to their order, denoting the relative magnitude of the stars. Thus α Arietis designates the largest star in the constellation Aries; β Draconis, the second star of the Dragon, &c.

Some of the fixed stars have particular names, as *Sirius*, *Aldebaran*, *Arcturus*, *Regulus*, *&c.*

422. Magnitudes. The stars are also divided into classes, or *magnitudes*, according to the degrees of their apparent brightness. The largest or brightest are said to be of the *first magnitude;* the next in order of brightness, of the *second magnitude*; and so on to stars of the *sixth magnitude*, which includes all those that are barely perceptible to the naked eye. All of a smaller kind are called *telescopic* stars, being invisible without the assistance of the telescope. The classification according to apparent magnitude is continued with the telescopic stars down to stars of the twentieth magnitude (according to Sir John Herschel), and the twelfth according to Struve.

The following are all the stars of the first magnitude that occur in the heavens, viz.: *Sirius* or the *Dog-star*, *Betelgeux*, *Rigel*, *Aldebaran*, *Capella*, *Procyon*, *Regulus*, *Denebola*, *Cor. Hydræ*, *Spica Virginis*, *Arcturus*, *Antares*, *Altair*, *Vega*, *Deneb* or *Alpha Cygni*, *Dubhe* or *Alpha Ursæ Majoris*, *Alpherat* or *Alpha Andromedæ*, *Fomalhaut*, *Achernar*, *Canopus*, *Alpha Crucis*, and *Alpha Centauri.* It is the practice of Astronomers to mark more or less of these stars as intermediate between the first and the second magnitude; and in some catalogues some of them are assigned to the second magnitude. All of these stars, with the exception of the last four, come above the horizon in all parts of the United States.

423. Celestial Globe. There are two principal modes of

representing the relative positions of the stars; the one by delineating them on a globe, where each star occupies the spot in which it would appear to an eye placed in the centre of the globe, and where the situations are reversed when we look down upon them; the other is by a chart or map, where the stars are generally so arranged as to be represented in positions similar to their natural ones, or as they would appear on the internal concave surface of the globe. The construction of a globe or chart, is effected by means of the right ascensions and declinations of the stars. Two points diametrically opposite to each other on the surface of an artificial globe are taken to represent the poles of the heavens, and a circle traced 90° distant from these for the equator: another point 23½° from one of the poles is then fixed upon for one of the poles of the ecliptic, and with this point as a geometrical pole a great circle described; the points of intersection of the two circles will represent the equinoctial points. The point which represents the place of a star is found by marking off the right ascension and declination of the star upon the globe.

All the fixed stars visible to the naked eye, together with some of the telescopic stars, are represented on celestial globes of 12 or 18 inches in diameter.

424. Catalogue of Stars. The places of the fixed stars are generally expressed by their right ascensions and declinations, but sometimes also by their longitudes and latitudes. A table containing a list of fixed stars designated by their proper characters, and giving their mean right ascensions and declinations, or their mean longitudes and latitudes, is called a *Catalogue* of those stars. (See Table XC (*a*)).

NUMBER AND DISTRIBUTION OVER THE HEAVENS.

425. The number of stars visible to the naked eye, in the entire sphere of the heavens, is from 6,000 to 7,000; of which nearly 4,000 are in the northern hemisphere; but not more than 2,000 can be seen with the naked eye at any one hour of the night at a given place. The telescope brings into view many millions, and every material augmentation of its space-penetrating power greatly increases the number.

As to the number of stars belonging to each different magnitude, astronomers assign from 20 to 24 to the first magnitude, from 50 to 60 to the second, about 200 to the third, and so on; the numbers increasing very rapidly as we descend in the scale of brightness; the whole number of stars already registered down to the seventh magnitude, inclusive, amounting to 12,000 or 15,000.

The reason of this increase in the number of the stars, as we descend from one magnitude to another, is undoubtedly that in general the stars are less bright in proportion as their distance is greater; while the average distance between contiguous stars is about the same for one magnitude as for another. It is easy to see that upon these suppositions the number of stars posited at any given distance, and having therefore the same apparent magnitude, will be greater in proportion as this distance is greater, and thus as the apparent magnitude is lower.

426. Comparative Brightness. It is not to be understood that the classification of the stars into different magnitudes, is made according to any fixed definite proportion subsisting between the degrees of apparent brightness of the stars belonging to different classes. Stars of almost every gradation of brightness, between the highest and the lowest, are met with. Those which offer marked differences of lustre, form the basis of the classification; others, which do not differ very widely from these, are united to them. As a necessary consequence, there are some stars of intermediate lustre, which cannot be assigned with certainty to either magnitude. Thus, in the catalogue published by the Astronomical Society of London, 3 stars are marked as intermediate between the first and second magnitudes, and 29 between the second and third.

Different astronomers also not unfrequently assign the same star to different magnitudes.

As to the proportions of light emitted from the average stars of the different magnitudes, according to the experimental comparisons of Sir Wm. Herschel, they are, from the first to the sixth magnitude, approximately in the ratio of the numbers, 100, 25, 12, 6, 2, 1.

427. Distribution of the Stars. With the exception of the three or four brightest classes, the stars are not distributed indiscriminately over the sphere of the heavens, but are accumulated in far greater numbers on the borders of that belt of cloudy light in the heavens, which is called the milky way, and in the milky way itself, which the telescope shows to consist of an immense number of stars of small magnitude in close proximity. According to Struve, the total number of stars visible in the Herschelian telescope of 20 feet focus and 19 inches aperture, is a little over 20,000,000.

428. Stratum of the Milky Way. The great accumulation of stars in a zone of the heavens, encompassing the earth in the direction of a great circle, suggested to the mind of Herschel the idea that the stars of our firmament are not disseminated indifferently throughout the surrounding regions of space, but are for the most part arranged in a stratum, the thickness of which is very small in comparison with its breadth; the sun and solar system being near the middle of the thickness. If S (Fig. 103) represents the place of the sun, it will be seen that

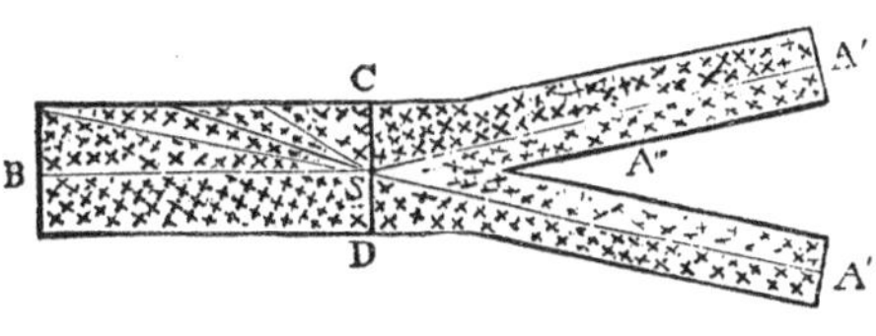

FIG. 103.

upon this supposition the number of stars in the direction SC of the thickness of the stratum, will be less than in any other direction, and that the greatest number will lie in the direction of the breadth, as SB. On one side of the point S, the stratum is supposed to be divided for a certain distance into two laminæ, as shown in the figure, which represents a section of the supposed stratum. This supposition is necessary to account for the two branches, with a dark space between them, into which the milky way is divided for about one-third of its course.

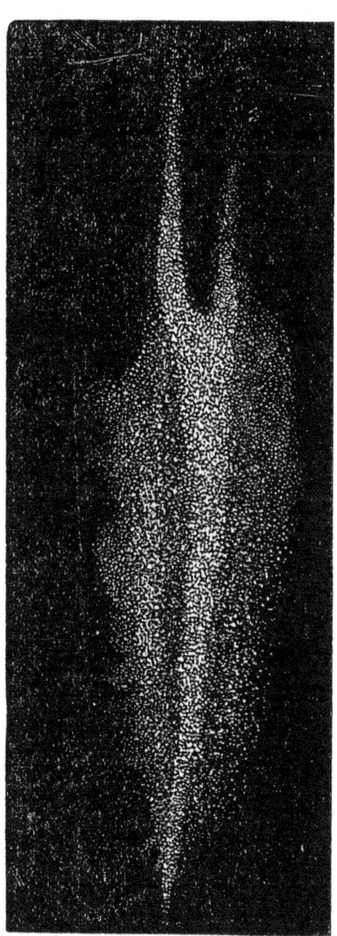
FIG. 104.

Herschel undertook to gauge this stratum in various directions, on the principle that the distance through to its borders in any direction was greater in proportion as the number of stars seen in that direction was greater. He thus found that its actual form was very irregular; its section, instead of being truly that of a segment of a sphere divided for a certain distance into two laminæ, as represented in Fig. 103, having the form represented in Fig. 104. He estimated the thickness of the stratum to be less than 160 times the interval between the stars, and the breadth to be nowhere greater than 1,000 times the same distance. These are his first results; we shall see in the sequel that they were materially modified by his subsequent investigations.

Sir John Herschel conceives that the superior brilliancy and larger development of the milky way in the southern hemisphere, from the constellation Orion to that of Antinous, indicate that the sun and his system are at a distance from the centre of the stratum in the direction of the Southern Cross, and that the central parts are so vacant of stars that the whole approximates to the form of an annulus.

ANNUAL PARALLAX AND DISTANCE OF THE STARS.

429. The *Annual Parallax* of a fixed star is the angle made by two lines conceived to be drawn, the one from the sun and the other from the earth, and meeting at the star, at the time the earth is in such part of its orbit that its radius-vector is perpendicular to the latter line; or, in other words, it is the greatest angle that can be subtended at the star by the radius of the earth's orbit. Thus, let S (Fig. 105) be the sun, *s* a fixed star, and E the earth, in such a position that the radius-vector SE is perpendicular to E*s* the line of direction of the star, then the angle S*s*E is the annual parallax of the star *s*.

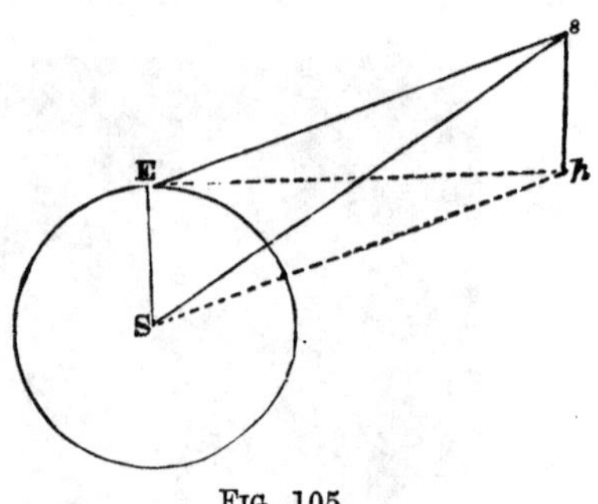

Fig. 105.

430. Least Distance of the Stars. If the annual parallax of a star were known, we might easily find its distance from the earth; for in the right-angled triangle SE*s* we would know the angle S*s*E and the side SE, and we should only have to compute the side E*s*. Now, if any of the fixed stars have a sensible parallax, it could be detected by a comparison of the places of the star, as observed from two positions of the earth in its orbit, diametrically opposite to each other; and accordingly, the attention of astronomers furnished with the most perfect instruments, has long been directed to such observations upon the places of some of the fixed stars, in order to determine their annual parallax. But, after exhausting every refinement of observation, they have not been able to establish, until quite recently, that any of them have a measurable parallax. Now, such is the nicety to which the observations have been carried, that, did the angle in question amount to as much as 1″, it could not possibly have escaped detection by the methods of observation employed. We may then conclude that the *annual parallax of the nearest fixed star is less than* 1″.

Taking the parallax at 1″, the distance of the star comes out 206,265 times the distance of the sun from the earth, or about 20 millions of millions of miles. The distance of the nearest fixed star must therefore be greater than this. A juster notion of the immense distance of the fixed stars, than can be conveyed by figures, may be gained from the consideration that light, which traverses the distance between the sun and earth in 8m. 18s., and would perform the circuit of our globe in $\frac{1}{8}$ of a second, employs $3\frac{1}{4}$ years in coming from the nearest fixed star to the earth.

431. Determination of the Parallax of a Fixed Star. The long continued endeavor to detect an annual parallax of a fixed star, by the direct method of comparing the places of the star, determined at an interval of half a year, has at last been crowned with success. The parallax of α Centauri has been thus determined by Professor Henderson, from observations made in 1832 and 1833, with a large mural circle. Subsequent observations with a more efficient instrument by Maclear have furnished an angle of parallax differing but little from that obtained by Henderson. Its value is 0″.913, which answers to a distance about $\frac{1}{10}$ less than the least limit of distance of the stars, just determined. The parallax and distance of Sirius and of the pole-star, have since been determined in a similar manner, but with less certainty. The result obtained for the parallax of the pole-star is 0″.11, and for that of Sirius an angle a little greater. A parallax of 0″.11 answers to a distance that light would require nearly 30 years to traverse.

432. Parallax of a Star found by the Differential Method. The honor of being the first to determine with certainty the parallax and distance of a fixed star belongs to Bessel. The star observed by him is that designated as 61 Cygni. It is a star of about the 6th magnitude, barely visible to the naked eye. When viewed through a telescope it is seen to consist of two stars of nearly equal brightness, at a distance from each other of about 16″. These stars have a motion of revolution around each other, and the two move together at the same rate of 5″.3 per year, as one star, along the sphere of the heavens. It is hence inferred that they are bound together into one system by the principle of gravitation, and are at pretty nearly the same distance from the earth. The great proper motion of this double star, as compared with other stars, led to the suspicion that it was nearer than any other; and thus to attempts to determine its parallax. The principle of Bessel's method is to find the *difference* between the parallaxes of the star 61 Cygni, and some other star of much smaller magnitude, and therefore supposed to be at a much greater distance, seen in as nearly the same direction as possible. This difference will differ from the absolute parallax of the double star by only a small fraction of its whole amount. It was found by measuring with a position micrometer (62) the annual changes in the distance of the two stars, and in the position of the line joining them. To make it evident that such changes will be an inevitable consequence of any difference of parallax in the two stars, conceive two cones having the earth's orbit for a common base, and their vertices respectively at the two stars, and imagine their surfaces to be produced past the stars until they intersect the heavens. The intersections will be ellipses, but, by reason of the different distances of the two stars, of different sizes, as represented in Fig. 106; and they will be apparently described by the stars in the course of one revolution of the earth in its orbit. The two stars will always be similarly situated in their parallactic ellipses: thus, if one is at A the other will be at *a*; and after the earth has made one-quarter of a revolution, they will be at B and *b*; and after another quarter of a revolution at C and *c*, &c. Now it will be manifest, on inspecting the figure, the ellipses being of unequal size, that the lines of the stars will be of unequal lengths, and have different directions in the different situations of the stars.

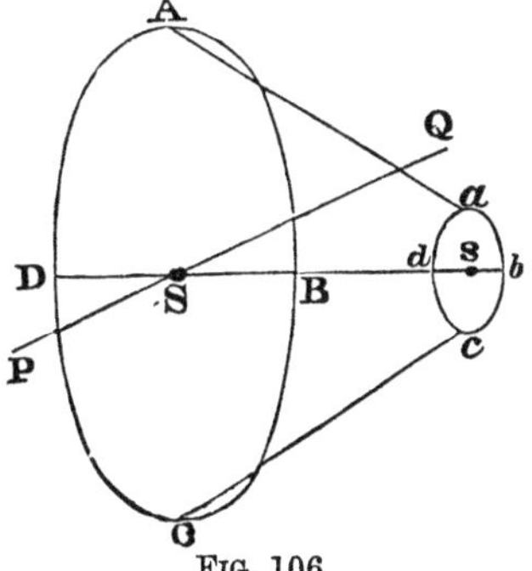

FIG. 106.

A much smaller angle of parallax may be found, with the same degree of certainty, by this indirect method, than by the direct process explained in Art. 430; for since the two stars are seen in pretty nearly the same direction, they will be equally affected by refraction and aberration; and since it is only the relative situations of the two stars that are measured, no allowance has to be made for precession and nutation, or for errors in the construction or adjustment of the instrument. It is therefore independent of the errors that are inevitably committed in the determination of these several corrections, when it is attempted to find directly the absolute parallax, by observing the right ascension and declination at opposite seasons of the year. The measurements made with the micrometer in the hands of the most accurate observers, may be relied on as exact to within a small fraction of 1″.

For the sake of greater certainty Bessel made the measurements of parallactic changes of relative situation between the star 61 Cygni and two small stars instead of one,—the middle point between the two members of the double star being taken for the situation of this star. He found the difference of parallax to be for the one star 0″.3584, and for the other star 0″.3289: and assuming the absolute parallax of the two stars to be equal, found for the most probable value of the difference of parallax 0″.3483. Whence he calculated the distance of the star 61 Cygni to be 592,200 times the mean distance of the earth from the sun; a distance which would be traversed by light in 9⅓ years.

The number of stars whose parallax and distance have been determined, more or less accurately, by both methods, now amounts to 12. The least parallax obtained is that of Capella, which is 0″.05; but it must be regarded as quite uncertain.

433. Comparative Distances of Stars of Different Magnitudes. According to Peters, the mean parallax of stars of the second magnitude is 0″.116, which answers to a distance that light would traverse in 28 years. From this result the mean parallax and distance of stars of each of the different magnitudes have been approximately deduced by means of the relative distances of stars of the different magnitudes, as determined by Struve on the assumption that the stars are uniformly distributed through space (at least in certain directions), and that the light from the stars of the different magnitudes varies according to a certain admitted law. The mean distance of stars of the first magnitude, as computed, is traversed by light in 15·5 years; and that of a star of the sixth magnitude in 120 years. Light requires 138 years to come from the most remote star visible to the naked eye. The same principle of computation of distances being extended to the telescopic stars, it appears that the stars just visible in the Herschelian telescope of 20 ft. focus are separated from us by a distance that light takes 3,500 years to journey over. This is on the supposition that the rays of light do not experience any sensible degree of extinction in traversing the regions of space.

NATURE AND MAGNITUDE OF THE STARS.

434. The vast distance at which the fixed stars are visible, and shine with a light not much inferior to the planets, leaves no room to doubt that they are all suns, or self-luminous bodies.

If it should be conjectured that some of the fainter stars might be bodies shining by reflected light, like the planets, the answer is, that if we were to suppose the existence of opake bodies, at the distance of the stars, so inconceivably vast in their dimensions as to send a sensible light to the eye, if illuminated to the same degree as the planets, the stars of the smaller magnitudes are too remote from the brighter ones to receive sufficient light from them; for, the smallest measurable space in the field of the larger telescopes is, at the distance of the nearer stars, nearly as large as the earth's orbit. It is perhaps possible, that some of the faintest members of some of the double stars, as surmised by Sir John Herschel, may shine by reflected light.

435. Magnitude of the Stars. To be able to determine the magnitude of a star, we must know its distance, and also its apparent diameter. Now the distances of but few stars have as yet been found; and the discs of all the stars, even in the most powerful telescopes, are altogether spurious; so that in no instance have we the data, nor have we reason to expect that they will be hereafter obtained, for determining with certainty the magnitude of a fixed star.

But we may infer from the quantity of their light as compared with that of the sun, and the mean distances of stars of the different magnitudes, as approximately determined (433), that the stars are in general much larger than the sun. According to the mean result of recent photometrical comparisons made by Messrs. G. P. Bond and Alvan Clark, between the bright star α Lyræ and the sun, if the sun were removed to 133,500 times its present distance it would send us the same quantity of light as this star. From this we may infer that if it were removed to the distance of the nearest star (430), it would appear as a star of the second magnitude; and that if it were removed to the mean distance of stars of the first magnitude, it would appear as a star of the sixth magnitude, and be just visible to the naked eye. It would seem then that the sun is much smaller than most, if not all, of the stars of the first magnitude; and is presumably also smaller than most of the stars of the other magnitudes.

VARIABLE STARS.

436. There are many stars which exhibit periodical changes of brightness; these are termed *Variable Stars.* More than a hundred stars are now known to belong to this class. One of the most remarkable of the variable stars is *o* Ceti, or *Mira.* From being as bright as a star of the second magnitude, it gradually decreases until it entirely disappears; and after remaining for a time invisible, reappears, and gradually increasing in lustre, finally recovers its original appearance. The mean period of these changes is 331½ days. The star remains at its greatest brightness about two weeks, employs about three months in waning to its disappearance, continues invisible for about five months, and during the remaining three months of its period increases to its original lustre. Such has been the general course of its phases. But it does not always recover the same degree of brightness, nor increase and diminish by the same gradations. It is even related by Hevelius, that in one instance it remained invisible for a period of four years. A similar phenomenon has been noticed in the case of the star χ Cygni. According to the testimony of Cassini, it was scarcely visible throughout the years 1699, 1700, and 1701, at those times when it should have been most conspicuous. On the other hand a variable star situated in the Northern Crown, sometimes fluctuates in its brightness very slightly for several years, and then suddenly resumes its regular variations, in the course of which it entirely disappears.

The greater number of variable stars undergo a regular increase and diminution of lustre without ever becoming entirely invisible. Algol, or β Persei, is a remarkable variable star of this description. For a period of 2d. 14h., it appears as a star of the second magnitude, after which it suddenly begins to diminish in splendor, and in about 3½ hours is reduced to a star of the fourth magnitude. It then begins again to increase, and in 3½ hours more is restored to its usual brightness, going through all its changes in 2d. 20h. 49m.

Besides the single variable stars, there are also a number of double stars, one or both the members of which are variable; as γ Virginis, ϵ Arietis, ζ Bootis, &c.

437. General Facts. Two general facts have been noticed with respect to the variable stars which are worthy of remark, viz.: that the color of their light is red, and that their period of increase of lustre is shorter than that of the decrease. The star Algol, offers an exception to both of these general facts. The ruddy color is especially observable in the case of the smaller variable stars. It is a curious and suggestive fact that a number of the variable stars present a hazy appearance at their minimum, as if some form of nebulous matter were interposed between them and the eye.

438. Temporary Stars. There are also some instances on record of temporary stars having made their appearance in the heavens; breaking forth suddenly in great splendor, and without changing their positions among the other stars, after a time entirely disappearing. One of the most noted of these is the star which suddenly shone forth with great brilliancy on the 11th of November, 1572, between the constellations Cepheus and Cassiopeia, and was attentively observed by Tycho Brahé. It was then as bright as any of the permanent stars, and continued to increase in splendor till it surpassed Jupiter when brightest, and was visible at mid-day. It began to diminish in December of the same year, and in March, 1574, entirely disappeared, after having remained visible for sixteen months, and has not since been seen.

It was noticed that while visible the color of its light changed from white to yellow, and then to a very distinct red; after which it became pale, like Saturn.

In the years 945 and 1264, brilliant stars appeared in the same region of the heavens. It is conjectured from the tolerably near agreement of the intervals of the appearance of these stars, and that of 1572, that the three may be one and the same star, with a period of about 300 years. The places of the stars of 945 and 1264 are, however, too imperfectly known to establish this with any degree of certainty.

Besides these three temporary stars, several others have made their appearance, viz.: one in the year 134 B. C., seen by Hipparchus; another in 389 A. D., in the constellation Aquila; a third in the 9th century, in Scorpio; a fourth in 1604, in Serpentarius, seen by Kepler; a fifth in 1670, in the Swan; and a sixth in 1848, in Ophiuchus.

What is no less remarkable than the changes we have noticed, several stars, which are mentioned by the ancient astronomers, have now ceased to be visible, and some are now visible to the naked eye which are not in the ancient catalogues.

439. Explanation of Variable Stars. The most probable explanation of the phenomenon of variable stars is that they are self-luminous bodies rotating upon axes, and having, like the sun, spots developed periodically on their surface, under the varying action of revolving planets upon their photospheres. The range of the planetary action must be regarded as much greater than in the case of the sun. The fluctuations generally observable in the periods and in the maxima and minima of brightness of the variable stars, are analogous to the fluctuations that occur in the periods and maxima and minima of the sun's spots. Prof. Wolf has minutely investigated this correspondence of phenomena, in the case of certain stars, by constructing curves showing their variations of light in detail. The hazy appearance often presented by variable stars at their minimum, may result from the interposition of nebulous matter expelled from the star in the process of formation of the spots on its surface (293). The ruddy color frequently noticed may be ascribed to a lower temperature consequent upon a greater prevalence of spots, or to more intense electric discharges within the photosphere.

In the case of the star Algol the phenomena are precisely such as would result from the periodical interposition of an opake body. In those cases in which the period of the diminution of the light is a large fraction of the entire period of the star, as well as those in which there are occasional interruptions in the regular recurrence of the phenomena, the supposition of the interposition of an opake body will not answer.

Temporary stars may be supposed to be suns which have entirely omitted the evolution of light for a long period of time, and then burst forth anew with a sudden and peculiar splendor, under the influence of a planetary action returning to its maximum at the end of a long period. Or they may possibly result from an encounter of two stars at the point of intersection of the vast orbits which they pursue in the regions of space. The remarkable fact, noticed by Sir John Herschel, that all the temporary stars on record, of which the places are distinctly indicated, have occurred in or close upon the borders of the Milky Way, where, as we shall see, the stars are most condensed, lends some support to the latter hypothesis.

DOUBLE STARS.

440. Many of the stars which to the naked eye appear single, when examined with telescopes are found to consist of two (in some instances three or more) stars in close proximity to each other. These are called *Double Stars*, or *Multiple Stars*. (See Fig. 107.) This class of bodies was first attentively observed by Sir William Herschel, who, in the years 1782 and 1785, published

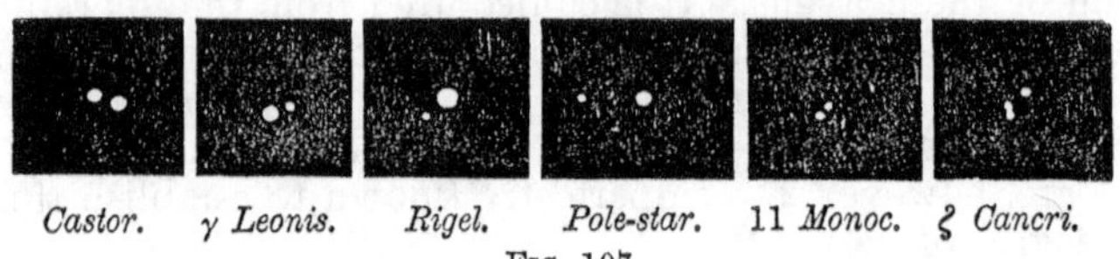

Castor. *γ Leonis.* *Rigel.* *Pole-star.* 11 *Monoc.* *ξ Cancri.*

Fig. 107.

catalogues of a large number of them which he had observed. The list has since been greatly increased by Professor Struve, of Dorpat, Sir J. F. W. Herschel, and other observers, and now amounts to several thousand.

441. Degree of Proximity. Double stars are of various degrees of proximity. In a great number of instances, the angular distance of the individual stars is less than 1″, and the two can only be separated by very powerful telescopes. In other instances, the distance is ½′ and more, and the separation can be effected with telescopes of very moderate power. They are divided into different classes or orders, according to their distances; those in which the proximity is the closest forming the first class.

442. Comparative Size. The two members of a double star are generally of quite unequal size (See Fig. 107). But in some instances, as that of the star Castor, they are of nearly the same apparent magnitude. Double stars occur of every variety of magnitude. Sirius is the largest of the double stars. It is attended by a minute companion star, at a distance of 10″. This was first discovered by Clark, with his great telescope of 18½in. aperture.

In some instances one of the constituents of a double star is itself double. ε Lyræ offers the remarkable combination of a double-double star.

443. *Different Colors.* It is a curious fact, that the two constituents of a double star in numerous instances shine with different colors; and it is still more curious that these colors are in general complementary to each other. Thus, the larger star is usually of a ruddy or orange hue, while the smaller one appears blue or green. This phenomenon has been supposed to be in some cases the effect of contrast; the larger star inducing the accidental color in the feebler light of the other. Sir John Herschel cites as probable examples of this effect the two stars ι Cancri, and γ Andromedæ. But it is maintained by Nichol that this explanation cannot be admitted; for, if true, it ought to be universal, whereas there are many systems similar in relative magnitudes to the contrasted ones, in which both stars are yellow, or otherwise belong to the red end of the spectrum. Again, if the blue or violet color were the effect of contrast, it ought to disappear when the yellow star is hid from the eye; which, however, it does not do. Thus, the star β Cygni consists of two stars, of which one is yellow, and the other shines with an intensely blue light; and when one of them is concealed from view by an interposed slip of darkened copper, the other preserves its color unchanged. The color, then, of neither of the stars can be accidental.

It may be remarked in this connection, that the isolated stars also shine with various colors. For example, among stars of the first magnitude, Sirius, Vega, Altair, Spica are white, Aldebaran, Arcturus, Betelgeux red, Capella and Procyon yellow. In smaller stars the same difference is seen, and with equal distinctness when they are viewed through telescopes. According to Herschel, insulated stars of a deep red color, occur in many parts of the heavens, but no decidedly green or blue star has ever been noticed unassociated with a companion brighter than itself.

444. Discovery of Binary Stars. Sir William Herschel instituted a series of observations upon several of the double stars, with the view of ascertaining whether the apparent relative situation of the individual stars experienced any change in consequence of the annual variation of the parallax of the star. With a micrometer adapted to the purpose, (62), he measured from time to time the apparent distance of the two stars, and the angle formed by their line of junction with the meridian at the time of the meridian passage, called the *Angle of Position.* Instead, however, of finding that annual variation of these angles, which the parallax of the earth's annual motion would produce, he observed that, in many instances, they were subject to regular progressive changes which seemed to indicate a real motion of the stars with respect to each other. After continuing his observations for a period of twenty-five years, he satisfactorily ascertained that the changes in question were in reality produced by a motion of revolution of one star around the other, or of both around their common centre of gravity; and in two papers, published in the Philosophical Transactions for the years 1803 and 1804, he announced the important discovery that there exist sidereal systems composed of two stars revolving about each other in regular orbits. These stars have received the appellation of *Binary Stars*, to distinguish them from other double stars which are not thus physically connected, and whose apparent proximity may be occasioned by the circumstance of their being situated on nearly the same line of direction from the earth, though at very different distances from it. Similar stars, consisting of more than two constituents, are called *Ternary*, *Quaternary*, &c.

Since the time of Sir W. Herschel, the observations upon the binary stars have been continued by Sir John Herschel, Sir James South, Struve, Bessel, Mädler, and other astronomers. According to Mädler the number of known binary and ternary stars is now about 600. Every year materially increases the list; and will probably continue to do so for some time to come: for, while the changes of relative situation are in some instances exceedingly slow, the actual number of such systems is probably a large fraction of the whole number of double stars; at least, if we confine our attention to double stars whose constituents are within $\frac{1}{2}'$ of each other. This may be inferred from the fact, that the number of such double and multiple stars actually observed, which amounts to over 3000, is at least ten times greater than the number of instances of fortuitous juxtaposition that would obtain on the supposition of a uniform distribution of the stars. Besides, there are a number of double stars not yet discovered to have a motion of revolution, which still give indications of a physical connection. Thus, their constituents are found to have constantly the same proper motion in the same direction; showing that they are in all probability moving as one system through space.

445. Periods and Orbits of Binary Stars. From the observations made upon some of the binary stars, astronomers have been enabled to deduce the form of their orbits, and approximately the lengths of their periods. The orbits are ellipses of considerable eccentricity. The periods are of various lengths, as will be seen from the following enumeration of some of those considered as best ascertained: μ^2 Bootis 650 years, γ Virginis 171 years, p Ophiuchi 92 years, α Centauri 77 years, ζ Cancri 58 years, ζ Herculis 36 years. Fig. 108 represents a portion of the

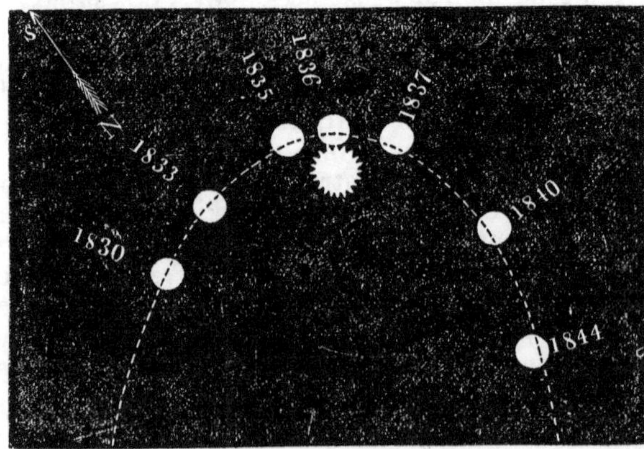

Fig. 108.

apparent orbit of the double star γ Virginis, and shows the relative positions of the two members of the double star in various years. At the time of their nearest approach, in 1836, the interval between them was a fraction of 1″, and they could not be separated by the best telescopes, with a magnifying power of 1000. Since then their distance has been continually increasing. In 1844 it amounted to 2″, and a power of from 200 to 300 was sufficient to separate them. The orbit represented in the figure is the stereographic projection of the true orbit on a plane perpendicular to the line of sight.

The actual distance between the members of a binary star has been found for 61 Cygni, and α Centauri. Bessel makes it for the first about two and a half times the distance of Uranus from the sun.

It is important to observe, that the revolution of one star around another is a different phenomenon from the revolution of a planet around the sun. It is the revolution of one sun around another sun; of one solar system around another solar system; or rather of both around their common centre of gravity. We learn from it the important fact, that the fixed stars are endued with the same property of attraction that belongs to the sun and planets.

PROPER MOTIONS OF THE STARS.

446. It has already been stated that the fixed stars, so called, are not all of them rigorously stationary. By a careful comparison of their places, found at different times with the accurate instruments and refined processes of modern observation, it has been found that great numbers of them have a progressive motion along the sphere of the heavens, from year to year. The velocity and direction of this motion are uniformly the same for the same star, but different for different stars. One of the stars which has the greatest proper motion, is the double star 61 Cygni. During the last fifty years it has shifted its position in the heavens 4′ 21″; the annual proper motion of each of the individual stars being 5″.2. An isolated star, called ε Indi, has a still greater proper motion. It changes its place 7″.7 every year. The proper motions of some of the stars are either partially or entirely attributable to a motion of the sun and the whole solar system in space; but the motions of others cannot be reconciled with this hypothesis, and must be regarded as indicative of a real motion of these bodies in space.

447. Motion of the Solar System through Space. The first successful attempt to explain the proper motions of the fixed stars on the hypothesis of a motion of the solar system through space, was made by Sir William Herschel. After a careful examination of these motions, he conceived that the majority of them could be explained on the supposition of a general recess of the stars from a point near that occupied by the star λ Herculis towards a point diametrically opposite. Whence he inferred that the sun, with its attendant system of planets, was moving rapidly through space in a direction towards this constellation. Doubt has since been thrown upon these conclusions by Bessel and other astronomers; but they have recently been decisively reëstablished by M. Argelander, of Abo. The investigations of

Argelander, which were communicated to the Academy of St. Petersburgh in 1837, have since been confirmed by M. Otto Struve, of the Pulkowa Observatory, and other eminent observers.

Taking the mean of all the more recent determinations, we find the most probable situation of the point towards which the sun's motion is directed to be as follows: R. A. 260° 14′, N. Dec. 35° 10′. This point is a little to the east and north of the star *u* in the constellation Hercules, and about 9° distant from the point first supposed by Herschel.

448. Velocity of Sun's Motion through Space. O. Struve finds that for a star situated at right angles to the direction of the sun's motion, and placed at the mean distance of the stars of the first magnitude, the annual angular displacement due to the sun's motion is 0″.339 (with a probable error of 0″.025). So that, if we assume, according to the best determinations, 0″.209 for the hypothetical value of the parallax of a star of the first magnitude, it follows that at the distance of the star supposed the annual motion of the sun subtends an angle 1.623 times greater than the radius of the earth's orbit: which makes it nearly 150,000,000 of miles. This is at the rate of 4.7 miles per second.

449. *Velocity of the proper motions of the stars.* The above angle of 0″.339 is about the greatest annual displacement which a star can experience in consequence of the sun's motion. Whence it appears that the whole of the proper motion of any star which is over and above this amount must certainly be due to a real motion in space. Thus, in the case of the star 61 Cygni, nearly 5″ of its annual proper motion (5″.23) must result from an actual motion in space. This is 14.37 times greater than the parallax of this star (0″.35). Accordingly, if we suppose the direction of its motion to be perpendicular to its line of direction from the sun or earth, its annual motion is 14.37 times greater than the radius of the earth's orbit, or at the rate of nearly 42 miles per second. As we have no means of ascertaining the actual direction of its motion, it is impossible to discover how much the velocity exceeds this determination.

450. *Sun's motion comparatively slow.* By comparing the particular motions presented by stars of different classes with the motion of the solar system, viewed perpendicularly at the distance of a star of the first magnitude, as above given, it is found that the former, at the mean, are 2.4 times greater than that of the sun; whence it follows that this luminary may be ranked among those stars which have a comparatively slow motion in space.

CLUSTERS OF STARS.

451. In many parts of the heavens stars are seen crowded together into clusters, often in numbers too great to be counted. Some of these clusters are visible to the naked eye. One of the most conspicuous is that called the *Pleiades*. To the unaided sight it appears to consist of six or seven stars, but with a telescope of moderate power 50 or 60 conspicuous stars are seen grouped together within the same space, and more than 100 smaller ones are distinctly discernible.

In the constellation Cancer is a luminous spot called *Præsepe*, or the bee-hive, which a telescope of moderate power resolves entirely into stars. Within a space about $\frac{1}{2}^\circ$ square, more than 40 conspicuous stars are seen, besides many smaller ones. In the sword-handle of Perseus is another cloudy spot thickly crowded with stars, which become separately visible with a telescope of low power.

One of the richest clusters in the northern hemisphere occurs in the constellation Hercules, between the stars η and ϵ. It is visible to the naked eye, on clear nights, as a hazy mass of light; which is readily resolved into stars by a good telescope. Viewed through a telescope of high power it presents the magnificent aspect of an innumerable host of stars crowded together towards the centre into a perfect blaze of light.

The richest and largest cluster in the whole heavens is seen in the constellation Centaurus, in the southern hemisphere. It is visible to the naked eye as a nebulous star, and is designated ω *Centauri*. The telescope shows it to consist of an immense multitude of stars congregated together in the form of a magnificent globular cluster (see Fig. 1, Plate IV.). In the field of view of a large telescope, it has an apparent diameter nearly equal to that of the moon.

NEBULÆ.

452. With the aid of the telescope, a great number of faintly luminous spots, or patches, are seen scattered here and there over the sphere of the heavens. These are called *Nebulæ*. Some of these nebulous objects are partially visible to the naked eye, but the great majority of them cannot be discerned without the assistance of a good telescope, and very many are beyond the reach of any but the most powerful instruments.

453. Number and Distribution of Nebulæ. The number of nebulæ hitherto discovered, is over 5,000. They are very unequally distributed over the heavens, especially in the northern hemisphere. They are most abundant in the constellations Virgo, Leo, Coma Berenices, Canes Venatici, and Ursa Major;

and occur in astonishing profusion in certain regions in this quarter of the heavens, as in the northern wing of Virgo. When the telescope is directed towards these regions it is observed that the nebulæ follow each other in rapid succession, from the diurnal motion of the heavens; while, in some parts of the heavens, hours elapse after one of them has passed through the field before another enters. In the southern hemisphere there are two detached spaces of considerable extent, visible to the naked eye, called the *Magellanic Clouds*, that shine with a nebulous light like the milky way, which are thickly sown with nebulæ.

454. Diversity of Form and Appearance. As seen through telescopes of moderate power, the nebulæ are, for the most part, round or oval in form; but, when carefully examined with the larger telescopes, they are found to present a great variety of aspects and forms. A large number are found to consist of a multitude of minute stars distinctly separate, and condensed about one or more points within the mass. Many others take on the appearance of incipient resolvability, characterized by the phrase *star-dust*, and are doubtless real clusters too distant, or too much condensed, to show their individual stars. Others still offer no appearance of stars, and remain the same cloud-like objects when the highest telescopic power is applied to them. These *Irresolvable Nebulæ* were supposed by Sir William Herschel to be masses of actual nebulous matter disseminated through space, but are now generally believed to be clusters, or beds of stars, like the rest; only too vastly remote to be revealed as such by any optical means yet employed.

455. Classification of Nebulæ. The nebulæ are classified according to their aspects and forms, as seen through the larger telescopes, as follows: (1) *Globular Clusters*, (2) *Irregular Clusters*, (3) *Oval Nebulæ*, (4) *Annular Nebulæ*, (5) *Planetary Nebulæ*, (6) *Stellar Nebulæ*, (7) *Spiral Nebulæ*, (8) *Irregular Nebulæ*, (9) *Double Nebulæ*.

456. Globular Clusters take their name from their supposed actual form. Their component stars are so crowded together as to form an almost definite outline, and they run up to a blaze of light towards the centre, where their condensation is the greatest. The number of stars congregated in a single cluster is to be told only by thousands and tens of thousands; although their apparent size does not in any instance exceed the $\frac{1}{10}$ part of the moon's disc. They are, in general, difficult of resolution, and appear in telescopes of moderate power as small, round, nebulous specks, resembling a comet without a tail. Fig. 3, Plate IV., represents a globular cluster to be seen in the constellation Pegasus.

457. Irregular Clusters. These are more or less irregular and indefinite in their outline. They are generally less rich in stars, and less condensed towards the centre than the globular clusters. Fig 2, Plate IV., represents an irregular cluster situated in the constellation Capricornus. The *Pleiades*, and *Coma Berenices*, are instances of irregular clusters whose individual stars are seen in telescopes of low power.

Irregular clusters occur of almost every degree of condensation, from a cluster which seems to be only a space of an irregular and ill-defined outline, somewhat more rich in stars than the surrounding regions, to the perfectly defined globular cluster highly condensed at the centre.

458. Oval Nebulæ. Nebulæ having a distinct elliptic outline occur of various degrees of eccentricity, from moderately oval to an elongation almost linear (see Figs. 5, 6, and 7, Plate IV.). They are more condensed, though in very different degrees, in their central parts, and often present great and sudden variations of brightness from one portion of their mass to another.

This is very observable in Fig. 9, Plate V. Such nebulæ, which retain their oval form in the field of the most powerful telescope, are doubtless spheroidal clusters, in their general form, though more or less complex in their internal structure. Many of them are either wholly or partially resolvable into individual stars. Others afford to the eye only indistinct intimations of their stellar structure. In general the spheroidal clusters are far more difficult of resolution than globular clusters.

459. Dynamical Equilibrium of Sidereal Systems. It cannot be doubted that the systematic organization of sidereal systems has been determined under the operation of the principle of universal gravitation; and it is plain that in the instance of globular and spheroidal clusters, a general state of equilibrium would be possible only upon the supposition that the individual stars of each cluster revolve around some central point. Such a general dynamical equilibrium of a cluster may however exist, and the internal structural condition be subject at the same time to secular changes, from the varying combinations of individual orbital positions, and the disturbing actions of some of the component stars on one another.

460. Annular Nebulæ. A very small number of observed nebulæ have the annular form (Fig. 12, Plate V.). A conspicuous example of this singular class of nebulæ may be seen with a telescope of moderate power, midway between the stars β and γ Lyræ. The central vacuity is not perfectly dark, but filled with a faint nebulous light. The telescope of Lord Rosse, has resolved it into minute stars, and shown it to be fringed on its outer edge with filaments of stars (Fig. 11, Plate V.). Chacornac, of the Paris observatory, describes it as presenting, in Foucault's great telescope of plated glass, the appearance of a hollow cylindrical bed of very small stars, with a thin stratum of minute stars stretching across the centre.

461. Planetary Nebulæ have a round planet-like disc of

an equable light throughout, or only slightly mottled, and often perfectly definite in outline. As many as 25 of these curious objects have been discovered. A large planetary nebula occurs near β Ursæ Majoris. It is nearly 3′ in diameter. There is a still larger planetary nebula in the constellation Bootes. If we suppose the former nebula to be at no greater distance than α Centauri, the nearest fixed star, its linear diameter must still be more than three times the diameter of the orbit of Neptune. Its actual distance must be vastly greater than here supposed, and its dimensions correspondingly greater, unless its individual stars are very minute in comparison with the most distant isolated stars. If we suppose them to be of the same size as the more distant stars, its distance should be equally great, and its dimensions more than 1,000 times greater than the above determination.

One of the planetary nebulæ has been resolved by Lord Rosse's telescope, and another shown to be an annular nebula This class of nebulæ are generally supposed to be either cylindrical beds of stars, or assemblages of stars in the form of hollow spherical shells.

462. Stellar Nebulæ are those in which one or more stars are seen apparently connected with a nebulosity. This class of nebulæ comprises several varieties, the most important of which is that of the *Nebulous Stars.* Nebulous stars are stars encircled by a faint nebulosity; in some cases terminating in a distinct outline, in others shading off gradually into the general light of the sky (Fig. 14, Plate V.). Fig. 16, Plate V., shows the appearance of a nebulous star in Gemini, as seen through Lord Rosse's telescope. The stars surrounded by these nebulous atmospheres have the same appearance as other stars; and their atmospheres offer no indication of resolvability into stars with the most powerful telescopes.

Fig. 15, Plate V., is a remarkable stellar nebula in the constellation Cygnus. It consists of a star of the 11th magnitude, surrounded by a very bright and perfectly round planetary nebula of uniform light, nearly 15′ in diameter. Herschel regards it as constituting a connecting link between planetary nebulæ and nebulous stars.

In the other varieties of stellar nebulæ stars are seen occupying various positions, in apparent connection with nebulous masses which are generally of an oval form. Sometimes the nebulosity is spindle-shaped, with a star at each end. One variety has received the name of *Cometic Nebulæ,* from their close resemblance to a comet with a spreading tail. Fig. 18, Plate V., represents a cometic nebula in the tail of Scorpio.

463. Spiral Nebulæ. The great telescope of Lord Rosse has revealed the remarkable fact that some of the nebulæ are made up of spiral convolutions proceeding from a common nucleus, or from two nuclei. The most conspicuous example of this curious form is represented in Fig. 10, Plate V. It is situated near the star η, at the extremity of the tail of the Great Bear. The spiral nebulous coils diverge from two bright centres, about 5′ apart. As seen in the field of his great reflecting telescope, they are described by Lord Rosse as "breaking up into stars." Another beautiful spiral nebula is situated in the northern wing of Virgo. In some of the instances cited by Lord Rosse, the spiral arrangement was only partially made out.

464. Irregular Nebulæ. Under this head are classed all the remaining single nebulæ that, as seen through the best telescopes, have no simple geometrical form. The majority of these are of great extent in comparison with other nebulæ, and are devoid of all symmetry of form. They are also remarkable for the great irregularities observable in the distribution of their light, indicating a singular complexity of internal structure.

The Great Nebula in the sword handle of Orion is the most conspicuous example of this class of nebulæ. It consists of irregular nebulous patches extending over a surface about 40′ square, or about twice the size of the moon's disc. From its great magnitude and beauty, singularly grotesque form, and the wonderful variety of its light, it is the most remarkable of all the nebulæ. One portion, near the trapezium or sextuple star θ, is uncommonly bright, and is visible to the naked eye. Other portions are quite hazy and dim; and still other intervening parts are dark, and even absolutely black. Sir John Herschel describes the brightest portions as resembling the head and yawning jaws of some monstrous animal, with a sort of proboscis running out from the snout. The constitution of this singular nebula remained enveloped in mystery from the time of its first discovery by Huyghens, in 1656, until the telescope of Lord Rosse was directed upon it; when the brighter portion near the trapezium was distinctly perceived to consist of clustering stars. The elder Bond, with the great Cambridge refractor, subsequently succeeded in resolving the same part of the nebula. More recently G. P. Bond has detected indications of an arrangement of the separated stars in spiral lines.

The Great Nebula in Andromeda is another remarkable irregular nebula. In the field of the Cambridge telescope it has the irregular outline and peculiar appearance represented in Fig. 8, Plate IV. Its extreme length is 2½°, and breadth over 1°. It is traversed, for a considerable portion of its length, by "two dark bands or canals." Certain parts offered, in the same telescope, decided indications of a stellar constitution. The brighter portion of this nebula is distinctly visible to the naked eye. As viewed with a telescope of moderate power, it has an elongated oval form, similar to Fig. 7, Plate IV.

The Crab Nebula. Fig. 4, Plate IV., represents the appearance of this curious nebula as seen through Lord Rosse's telescope. It is described as studded with stars, mixed with a nebulosity probably consisting of stars too minute to be recognised, and exhibiting filaments extending out from the southern portion of the nebula. In ordinary telescopes these outlying branches, which have suggested the name of crab nebula, are invisible, and the part seen has an oval form.

The Dumb-bell Nebula is so named from the fact that as seen through a telescope of moderate size, in which the brighter portion alone is visible, it has the apparent form of a dumb-bell. In Lord Rosse's telescope the nebula appears less regular in its form; and it is at the same time seen to consist of innumerable stars mixed with irresolvable nebulosity. When its fainter portions are included, its outer limit has an oval form (see Fig. 9, Plate V.), which shows the nebula as viewed through the smaller telescope of 3 feet aperture, constructed by Lord Rosse.

465. Double Nebulæ. A considerable number of double nebulæ occur in different parts of the heavens. M. D'Arrest, of Copenhagen, enumerates fifty whose constituents are not over 5′ apart, and estimates that there may be as many as 200 such double nebulæ. The two constituents are most commonly circular in their apparent form, and are probably real globular clusters. (Fig. 17, Plate V.)

The individual members of most of these nebulæ are probably physically connected. In one instance considerable changes have been recognised in the distance and relative position of the nebulæ in the interval from 1785 to 1862, which seem to indicate a motion of revolution of the one around the other.

466. Variability of Nebulæ. Systematic observations have been made by Struve, D'Arrest, and other astronomers, with the view of ascertaining whether any of the nebulæ were subject to variations of brightness. The result is that in a small number of cases some degree of variability has been positively ascertained. One case is that of the nebula in Orion, in certain parts of which material changes of brightness have been observed. But the most marked case is that of a small and faint nebula, discovered by Hind, in 1852, in the constellation Taurus. It has since gradually faded from year to year, and in 1862 was barely discernible in the great Pulkowa refractor. It is now entirely invisible in the telescope with which it was first detected. It is an interesting fact that this diminution of brightness has proceeded pari passu with that of a small star which presented itself almost in contact with the nebula. It has been observed also that there are many variable stars in a part of the nebula in Orion that is subject to change. Corresponding changes have been observed in the faint nebulous haze noticed around some of the variable stars; for instance, the new star that suddenly burst forth in May, 1866, in Corona Borealis, and then rapidly declined in brightness.

DISTANCE AND MAGNITUDE OF NEBULÆ.

467. Resolved Nebulæ. Herschel undertook to determine the distance of resolved nebulæ, by noting the space-penetrating power of the telescope which first succeeded in revealing their distinct stars. According to his determinations, therefore, the most remote of the resolved nebulæ are at the same distance as the most remote of the isolated stars discerned in his large telescope. The theoretical space-penetrating power of his telescope was 2,080 times the mean distance of stars of the first magnitude. This should accordingly be the limiting distance of the resolved nebulæ seen in Herschel's telescope. The corresponding limit for stars and nebulæ, as seen in Lord Rosse's telescope, should be 3,120. But Struve, after determining the comparative distances of stars of the different photometric magnitudes, by comparing the actual number of stars of the different magnitudes, has been enabled to ascertain the actual space-penetrating power of any telescope in which all the stars up to any particular magnitude could be seen. According to his determinations, the actual space-penetrating power of Herschel's telescope of 20 feet focus was 183; that of the 40 feet reflector was 368, instead of 2080 as deduced upon optical principles; and that of Lord Rosse's great telescope is 422, instead of 3,120, the theoretical determination.

The unit of distance in these numerical values is the mean distance of stars of the first magnitude. According to Peters, this corresponds to a parallax of 0″.21, and is traversed by light in 15.5 years. We may therefore conclude that light employs about 6,540 years in coming from the most remote telescopic stars hitherto discerned to the earth. It traverses the distance from the nearest star (α Centauri) to the earth in 3½ years.

The resolvable nebulæ require telescopes of various powers to reveal their individual stars, and must therefore be distributed at the same variety of distance as the isolated telescopic stars of similar magnitudes.

468. Irresolvable Nebulæ. Herschel also undertook to determine the probable distance of the more remote irresolvable nebulæ. He estimated that a certain cluster of stars (75 of Messier's catalogue), which at one-fourth of its distance would be visible to the naked eye, would be visible as a faint irresolvable nebula, in his great reflector, if it were removed to 48 times its actual distance, or to more than 35,000 times the distance of Sirius. Struve's investigation reduces this determination to 787 times the mean distance of stars of the first magnitude (467). The corresponding result for Lord Rosse's telescope would be only a small fraction greater.

469. Extinction of the Light of the Stars, in its passage through space. The course of investigation followed up by Struve, at the same time that it affixed a much lower limit to the power of telescopes to pierce into the depths of space, conducted in explanation of this fact, to an important theoretical conclusion, viz., *that the light of the stars is partially extinguished in its transit through space.* He estimated the amount of this extinction to be such that light, in its passage through a distance equal to that of a star of the first magnitude, loses $\frac{1}{107}$ of its intensity. Sir John Herschel controverts this theory of the distinguished Pulkowa astronomer, but makes no attempt to overthrow the principal argument upon which it rests. If we reject, with Herschel, the testimony of the stars relative to the power of telescopes to penetrate the depths of space in which they lie, we must then adopt the determinations obtained upon optical principles alone as the exponents of telescopic power; we must accordingly conclude that stars can be discerned with the most powerful telescopes when separated from us by a distance so vast that light requires 48,000 years to traverse it; and that nebulæ might still be visible at a distance which light would require 500,000 years to pass over. At that distance, the united impression of the light of 10,000 stars upon the eye would only equal that from 100 single stars, so remote as to be just discernible in the most powerful telescope; and therefore clusters containing hundreds of thousands of stars should be visible at a much greater distance.

470. Magnitude of Nebulæ. At the distance of 422 stellar intervals (the utmost actual reach of Lord Rosse's telescope) a linear extent of 10′, in the heavens, answers to 1.23 times one of these intervals (467). Some of the planetary nebulæ have an apparent diameter as great as 10′, and as they are probably more remote than the most distant telescopic stars, their actual diameters are probably greater than 1.23 stellar units. The irregular nebulæ have a much greater extent. For example, the more conspicuous portion of the nebula in Orion extends to 30′, or 3.7 stellar intervals, in the east and west direction, and nearly as far in the north and south direction. The outlying branches run out much further. The extreme length of the nebula in Andromeda is no less than 18 times the same unit or the mean distance of stars of the first magnitude. Its extreme breadth is 7½ units. We here suppose these two nebulæ to be at the distance of the most remote telescopic stars. As they are barely resolvable by the most powerful telescopes, their distance cannot be less than this, unless their component stars are smaller, or intrinsically less luminous than the more remote isolated stars.

If the space-penetrating power of telescopes, as obtained upon optical principles, be adopted, the above numerical results must be increased seven-fold.

NUMBER, MUTUAL DISTANCE, AND COMPARATIVE BRIGHTNESS OF THE COMPONENT STARS OF CLUSTERS.

471. Possible Number of Stars in a Nebula. We may obtain an approximate estimate of the number of stars that may be congregated together in a nebula that is completely resolvable by a powerful telescope, by considering that if the telescope just shows them distinctly separate, the apparent distance between two contiguous stars may be assumed to be less than 1″. A space of one square minute should then contain more than 3,600 stars. The planetary nebula near the star β, in the constellation of the Great Bear (461), has an apparent extent of 7 square minutes. If it were just resolvable it should then contain more than 25,000 stars. As it is really irresolvable, the number of its individual stars must be still greater. Upon the same basis of calculation, the more conspicuous portion of the nebula in Orion, occupying, according to Sir John Herschel, $\frac{1}{25}$ of a square degree, should contain more than 500,000 stars; and the similar portion of the nebula in Andromeda (90′ long by 15′ broad) not less than 4,000,000 stars. If we suppose this vast nebula to be one continuous bed of stars, of different sizes, for its entire extent, it must comprise the enormous number of 30,000,000 stars.

It is true that these great nebulæ where resolved, in their brighter portions, show distinct stars in numbers that can be counted; but the space intervening between them is full of a nebulosity that is probably composed of smaller stars too closely compacted to be separated by the telescope.

472. Limit of Distance between Stars in a Resolved Nebula. An angular space of 1″, at a distance equal to 422 stellar intervals, corresponds to a linear distance 2,019 times the distance of the earth from the sun, or about 67 times the radius of Neptune's orbit. The distance between two contiguous stars of a nebula, that are just separated by a powerful telescope, cannot exceed this amount.

If the light of the stars suffers no sensible extinction in its passage, and therefore telescopes really penetrate as far into space as the optical theory requires, this determination is only $\frac{1}{4}$ of the actual value.

Clusters whose individual stars are separated by the distance just determined, would, if posited at a less distance than the furthest reach of telescopes, be more readily resolved; while any that might be at a greater distance would be wholly irresolvable by any telescope yet constructed.

473. Explanation of Inequalities of Brightness in a Nebula. Globular and irregular clusters (456–7,) are brighter and more difficult of resolution at the central than at the outer portions of the cluster. This is what should result if they were composed of stars of equal size and equally spaced. But in some instances the increase of brightness towards the centre is too great to admit of this supposition; and we infer that the stars are there condensed into a smaller space.

Oval and irregular nebulæ are more difficult of resolution at the fainter than at the brighter parts. From this we may infer that the stars are larger or more luminous in the brightest portions of such nebulæ; or that instances of close juxtaposition more frequently occur, in groups of two or three, which appear united as one, as suggested by Sir John Herschel.

STRUCTURE OF THE SIDEREAL UNIVERSE.

474. System of the Milky Way. We have already seen (428) that Sir William Herschel made the grand discovery that the sun is one of the individual stars of a vast bed, or organized system of stars, called the system of the milky way; that the sun is posited near its middle plane, and that its innumerable stars constitute the starry host which diversify our firmament. He at first conceived that his telescope penetrated to the outermost limits of the stratum, but later investigations, recently confirmed by the observations and researches of Bessel, Argelander, and Struve, have fully established that it extends in all directions beyond the reach of the most powerful telescopes; and that we can obtain no definite knowledge of its exterior form.

Herschel's star-gauges afford positive information only with regard to the comparative *densities* of the fathomless starry stratum in different directions, within the range of telescopic vision. From these we learn that the individual stars are not uniformly distributed throughout the system, but are greatly condensed towards the medial plane. Struve, by an elaborate discussion, has established that the distance between neighboring stars decreases, according to a regular law, on both sides of this plane as the distance from it increases; the decrease being much more rapid at first, and the rate gradually declining with the increasing distance. Within this plane of greatest condensation there is also a line of greatest density, from both sides of which the density gradually decreases. A corresponding line of superior density exists in each plane of the starry stratum parallel to the principal plane. The axis of greatest condensation is nearly coincident with the line passing through the points of intersection of the galactic circle, or middle line of the milky way in the heavens, with the equator. These points lie in R. Asc. 6h. 40m., and R. Asc. 18h. 40m., between the constellations Orion and Canis Minor, and between Serpentarius and Antinous. According to Struve the sun is on the north side of the plane of greatest condensation, and at an estimated distance from it equal to the distance of α Centauri from the sun and earth. It is also to one side of the axis of greatest density in the direction of the constellation Virgo, and at a distance nearly equal to the distance of the nearest stars of the second magnitude from the earth. The galactic circle, and therefore, also, the principal plane of the milky way, passes through the points on the equator above-mentioned, and within about 30° of the north and south poles of the heavens; through points in the constellations Cassiopeia and the Southern Cross. The north pole of the galactic circle, or of the whole system, lies in R. Asc. 12h. 38m., and Dec. 31°.5, between the constellations Coma Berenices and Canes Venatici.

475. The Galaxy, or Belt of the Milky Way. The luminous belt in the heavens called the milky way, as seen by the naked eye, varies in breadth at different points between the limits 5° and 16°, and has an average breadth of about 10°. It presents a succession of luminous patches, unequally condensed, intermingled with others of a fainter shade. From the bright star α Cygni, in the northern hemisphere, it runs towards the southwest in two clustering streams, which reunite beyond the southern constellation Scorpio, at a distance of 120° from the point of separation. Near the place in which it crosses the equator, between Antinous and Serpentarius, the double stream attains its greatest width of 22°. The middle point of crossing is the ascending node, on the equator, of the galactic circle.

To give a more accurate idea of the system of the milky way, we must add that its principal plane, so called, is not strictly a single plane, but a broken plane, or two planes differing about 10° in their direction, and separating at the line of the nodes in the equator. The two condensed branches answering to the two separate streams in the heavens just noticed, lie on opposite sides of this broken plane. The line of greatest density before referred to (474) also is not truly a right line, but has sensible inflexions; and there occur in its vicinity remarkable alternations of starry condensations and vacant spaces. Similar interruptions of continuity are observed in various directions through the mass. In some directions dark intervening spaces are seen, in which, according to Sir John Herschel, the telescope seems to penetrate to the very confines of the starry stratum. In other directions, there appear to be vast starless regions lying between the more remote portions and outlying branches of the milky way, or other systems entirely detached from it.

476. Relations of Clusters and Nebulæ to the System of the Milky Way. Globular and irregular clusters are far more abundant in the denser portions of the milky way than in other portions of equal extent. The irregular nebulæ, some of which have been resolved, are, for the most part, either portions or outlying branches of the system. Some of those which have not been resolved, may possibly be independent systems exterior to that of the milky way.

Oval nebulæ, and the irresolvable nebulæ generally, do not hold the same relations to our starry firmament. They are mostly absent from that great belt in which the stars are so numerous and condensed, and the conspicuous clusters abound, and are congregated towards its poles. The region richest in nebulæ lies around its north pole. They are more uniformly disseminated and more widely dispersed over the zone which surrounds its south pole; and are at the same time less numerous. But on the other hand, as already intimated, there are two luminous tracts of the southern heavens, called the *Magellanic Clouds*, in which they occur in large numbers. In these they are found associated with groups and clusters of stars of every form, and must be presumed to be no more remote than these resolved clusters. In the northern hemisphere they in general occur dissociated from resolved clusters, and may be much more remote. According to the estimate already obtained (468) their extreme limit of distance does not exceed twice that of the most distant isolated stars visible in telescopes.

477. Theoretical Inferences. The peculiarity that has just been noticed in the position of most of the oval and irresolvable nebulæ of the northern hemisphere, leads to the supposition that they may have originated in a different manner from the clusters and nebulæ that are chiefly accumulated in the denser portions of the system of the milky way, and undoubtedly are component parts of it; and that they may differ from these in some of the features of their physical constitution. The latter supposition acquires additional probability from a recent discovery that the character of the light received from some of the nebulæ is in certain respects different from that of the light received from the sun and the stars. A spectral analysis of the light from some of these nebulæ, by two eminent physicists, has disclosed the remarkable fact, that it is not made up of rays of widely different refrangibilities, but is, the greater part of it, monochromatic; and that the spectrum is not crossed by dark lines, like that obtained from the light of the sun, or of a star. From this, the experimenters draw the conclusion that the nebulæ in question can no longer be regarded as clusters of suns, similar in constitution to the centre of our planetary system, but as objects having quite a different and peculiar composition; and that instead of being considered as made up of bodies having a solid nucleus, they must be regarded as enormous masses of luminous gas or vapor. The latter conclusion does not follow of necessity from the results of the experiments; they only show that the light from these nebulæ comes from masses of pure gas or vapor, rendered luminous either by ignition or electric discharges, but afford no certain knowledge with regard to the existence of a solid nucleus.

478. General Motion of Revolution of the Stars. Mädler, after an elaborate discussion of the proper motions of a large number of stars, has arrived at the conclusion that the collective body of stars visible to us has, together with the sun, a common movement of revolution around a centre situated in the group of the Pleiades. He estimates the period of revolution to be about 27 millions of years. A general circulation of the sun and the stars of our firmament around a common centre of attraction, must also be regarded as highly probable upon physical grounds, but it cannot be doubted that the centre of attraction would lie in the principal plane of the milky way. The group of the Pleiades lies considerably to the south of this plane, and therefore in all probability the actual centre is situated to the north of the Pleiades, in the constellation Perseus, as suggested by Argelander.

479. Hypotheses respecting the Milky Way. Mädler supposes that the stars of the milky way are arranged in several concentric rings of unequal thickness, and of varying dimensions in different directions, but lying nearly in the same plane. He conceives the sun to be eccentrically situated in the system, and at a short distance from the general plane of the rings; so that on one side the rings are seen distinctly separate.

Professor Stephen Alexander, of Princeton College, has advanced the hypothesis that the milky way, and the stars within it, together constitute a spiral with several branches, and a central spheroidal cluster.

The hypothesis of Sir William Herschel has already been considered (428 and 474). Another conception of the probable structure, and present dynamical condition of the system of the milky way, is briefly presented in a Note in the Appendix.

GENERAL DYNAMICAL CONDITION OF SIDEREAL SYSTEMS.

480. Three different general conceptions may be formed of the possible nature of the motions of the individual members of a cluster or system of stars.

(1.) They may all be in the act of falling in right lines towards their common centre of attraction.

(2.) They may be in the act of receding from a centre about which they were originally collected, under the influence of some dispersing force.

(3.) They may be revolving in separate orbits around their common centre of attraction, or possibly around different centres.

First Hypothesis.—This was proposed by Sir William Herschel. It accords with the different aspects presented by clusters condensed towards a centre, but cannot be applied to annular nebulæ, some of which are known to consist of stars, nor to spiral formed clusters. It involves also the highly improbable supposition that there is in the condition of the system no provision for stability, but only for its inevitable destruction, in the final collision of all its constituent stars at its centre.

Second Hypothesis.—The second supposition is advocated by Professor Alexander, who has propounded a systematic theory of the evolution of sidereal systems, under the operation of a certain supposed process of dispersion.

Third Hypothesis.—The supposition that the individual stars of a system are moving in separate orbits about a common centre of attraction, is that which is suggested by the analogy of our planetary system, as well as that of the revolution of binary and triple stars around their common centre of gravity. It is supported also by the results of Mädler's investigations with respect to a general revolution of the system of the milky way about a centre (478). It implies the existence of the only causes of stability that can be conceived to be in operation; viz., a centre of attraction, and a motion of revolution around that centre. For the rotation of a cluster of separate stars around an axis, as one single body of matter, is mechanically impossible. In the history of such an organized system, from its beginning, there may be epochs of collision among its individual members, but when all such cases, inevitably resulting from correspondences of original position, have occurred, the motions which remain outstanding may ultimately tend to a permanent stability.

CHAPTER XX.

THEORIES OF THE EVOLUTION OF SIDEREAL AND PLANETARY SYSTEMS.

NEBULAR HYPOTHESIS.

481. Primitive Nebulous Condition of all Systems. Although the telescope, by revealing the stellar constitution of many of the nebulæ regarded by Sir William Herschel as giving no intimations of resolvability, has removed the supposed direct evidence of the existence of detached masses of nebulous matter disseminated through space, there still remains strong indirect evidence of a *primitive nebulous condition of all worlds and systems of worlds.* Numerous correspondences of structural and dynamical features, and intimations of a progressive creation, lead to this conception as the only ground upon which they can reasonably be explained. Thus Laplace adduces five general phenomona as indications of a common origin of the system of planets circulating around the sun; and infers that they must all have originally formed portions of one vast nebulous body rotating about an axis. These are:

1. The planets all revolve in the same direction around the sun; viz.: from west to east.
2. Their orbits lie nearly in the plane of the sun's equator.
3. Their orbits are ellipses of small eccentricity.
4. The sun and all the planets, so far as the circumstances of their rotation are known, rotate about axes in the same direction that the planets revolve around the sun.
5. The satellites revolve around their primaries in the same direction that these revolve around the sun, and turn about their axes. They also revolve, as far as known, approximately in the plane of the equator of each primary; and describe ellipses of small eccentricity.

The only known exception to the general direction of revolution occurs in the case of the satellites of Uranus, which have a common retrograde motion. Their orbits are also inclined to the plane of the ecliptic under a large angle (79°); but their common plane may still coincide with the plane of Uranus's equator, and the direction of their motion of revolution may be the same as that of the rotation of the primary. (See Note III.).

The hypothesis proposed by Herschel in explanation of sidereal systems, and since extended by Laplace to the explanation of the solar system, is called the *Nebular Hypothesis.* It is, comprehensively stated, that all worlds and systems of worlds have been slowly evolved from primordial nebulous masses, under the operation of the general forces and properties which the Creator has either permanently imparted to matter, or is incessantly renewing in it.

DEVELOPMENT OF THE SOLAR SYSTEM.

482. Origin of the Planets and Satellites. The mechanical theory of the formation of the solar system propounded by Laplace, is briefly this: The rotating nebulous body from which the system has been evolved, in the progress of ages slowly contracted and condensed, by the gravitation of its parts towards the centre, and by the process of cooling at its surface. This contraction of necessity accelerated the rotation of the body, and augmented the centrifugal force: until finally the increasing centrifugal force at the equator balanced the gravity. When this mechanical condition was reached at the surface, and for a certain depth where the influence of the cooling had especially prevailed, a vaporous zone became detached, and revolved independently of the interior mass. This zone, by concentration at special points, eventually separated into fragments; which, from the preponderating attraction of the larger fragment, or because of slight differences of initial velocity, became incorporated into one revolving body. This body would take up a motion of rotation in the same direction that it revolves; since the parts most remote from the sun would have the most rapid motion of revolution. By an indefinite continuation of the same process a succession of zones would become detached, and a system of vaporous bodies revolving around a central condensed mass would be formed. Each of these revolving bodies being also in the same condition of rotation as the original nebulous mass, might pass through a similar succession of changes, and thus a system of satellites circulating around a primary, in the direction of the rotation, be developed.

The solar system presents one instance, that of Saturn's ring, in which the detached vaporous zone condensed uniformly without separating into parts. The planetoids appear to afford an instance of the opposite extreme, in which the ring broke up into a great number of small fragments that continued to revolve separately.

483. Origin of Cometary Bodies. Laplace supposed the comets did not belong, originally, to the solar system, but wandered into its precincts from other systems, and so became permanently united with it by the bond of gravitation. But, with the evidence now afforded by accumulated facts, several considerations may be urged which tend to show that comets have been derived from the same nebulous body as the planets and satellites. The principal of these are the following:

1. The comets of short period form a class but little distinguished, in their orbital motions, from the planetoids. They revolve in the same direction, and in orbits having about the same average inclination to the ecliptic, as those of the planetoids. Their orbits are only somewhat more eccentric.

2. All the known comets that describe orbits whose aphelia lie within the limits of the solar system, or do not fall more than fifty millions of miles beyond the orbit of Neptune, revolve in the same direction as the planets.

3. If we compare all the comets whose elliptic orbits have been determined with more or less accuracy, among themselves, we find that the more eccentric orbits of the comets of long period are more inclined to the plane of the ecliptic than the less eccentric orbits of the comets of short period.

If we consider the class of comets which recede to a distance of more than fifty millions of miles beyond the limits of the solar system, it appears that as many among them have a retrograde as a direct motion; while the majority move in orbits inclined under large angles to the ecliptic. These exceptional facts do not necessarily imply that this class of comets have an origin extraneous to the system; but rather that the mode of their evolution from the primary nebulous body was different from that of the planets and comets of short period. Now, besides the process of evolution supposed to have been in operation in the case of the planets, we may conceive,

(1.) That certain portions of the body, near its surface, became, by mutual attraction of their parts and by cooling, condensed upon particular points into masses of sufficient density to revolve independently. Such masses, as they would have less initial velocities in proportion as they were more remote from the equator, would, in general, describe orbits more eccentric in proportion as they are more inclined to the ecliptic. Besides, the masses which became detached at the equator in the manner here supposed, must have separated from the general mass in the intervals between the epochs of the separation of the equatorial planetary rings, during which the velocity of rotation at the equator was less than that answering to a motion of revolution in a circle. The comets of the first two classes may have thus originated. If so, as they must have performed many revolutions within the attenuated mass of the nebulous body, they are now doubtless moving in orbits much more eccentric than those which they first described.

(2.) That fragments may have been suddenly detached from the general nebulous mass, by the operation of some expelling force. If we adopt the most probable hypothesis, that this force acted indifferently in all directions outward from the surface, and assume it to have been of sufficient intensity to impart, when exerted under certain obliquities to the surface, a velocity in the direction of the parallel of latitude considerably greater than the velocity of rotation at the place of discharge, then among the comets thus originating that come within our firmament, a retrograde may be as frequent as a direct motion. For, those which were detached with the higher velocities, either obliquely in the direction of the rotation or in the opposite direction, would move in too large orbits to become visible from the earth. If all the comets detached, however, could be seen, there should be a preponderance in the number of those having a direct motion. (See Note III. in Appendix.)

PART II.

PHYSICAL ASTRONOMY.

CHAPTER XXI.

Principle of Universal Gravitation.

484. Force of Gravity. It is demonstrated in treatises on Mechanics, that if a body move in a curve in such a manner that the areas traced by the radius-vector about a fixed point, increase proportionally to the times, it is solicited by an incessant force constantly directed towards this point.

The following is a geometrical proof of this principle. Conceive the orbit to be a polygon of an infinite number of sides. Let ABCD (Fig. 109) be a portion of it; and S the fixed point about which the radius-vector describes areas proportional to the times, or equal areas in equal times. Since the impulses are only communicated at the angular points A, B, C, D, &c., of the polygon, the motion will be uniform along each of the sides AB, BC, CD, &c.: and since we may suppose the times of describing these sides to be equal, we shall have the triangular area SAB equal to the triangular area SBC, and SBC equal to SCD, &c. Produce AB and make B*c* equal to AB, which may be taken to represent the velocity along AB; and join C*c*. C*c* will be parallel to the line of direction of the impulse that takes effect at B. Upon SB let fall the perpendiculars A*m*, *cn*, C*r*. Then, since AB = B*c*, A*m* = *cn*; and since the equivalent triangles SAB, SBC, have a common base SB, A*m* = C*r*. It follows, therefore, that *cn* = C*r*, and consequently, that C*c* is parallel to BS. The impulse which the body receives at B is therefore directed from B towards S. In the same manner it may be shown that the impulse which it receives at C is directed from C towards S. The line of direction of the force passes, therefore, in every position of the body, through the point S.

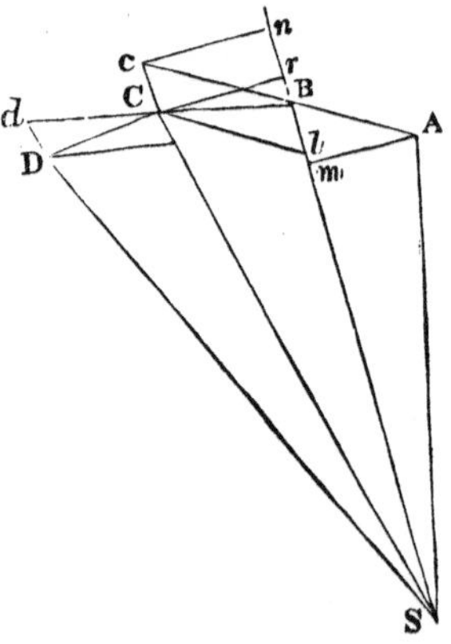

Fig. 109.

Now, by Kepler's first law, the areas described by the radius-vectors of the planets about the sun, are proportional to the times. It follows therefore from this law, that each planet is acted upon by a force which urges it continually towards the sun.

This fact is technically expressed by saying that the planets *gravitate* towards the sun, and the force which urges each planet towards the sun is called its *Gravity*, or Force of Gravity, towards the sun.

485. Its Law of Variation. It is also proved by the principles of Mechanics, that if a body, continually urged by a force directed to some point, describe an ellipse of which that point is a focus, the force by which it is urged must vary inversely as the square of the distance.

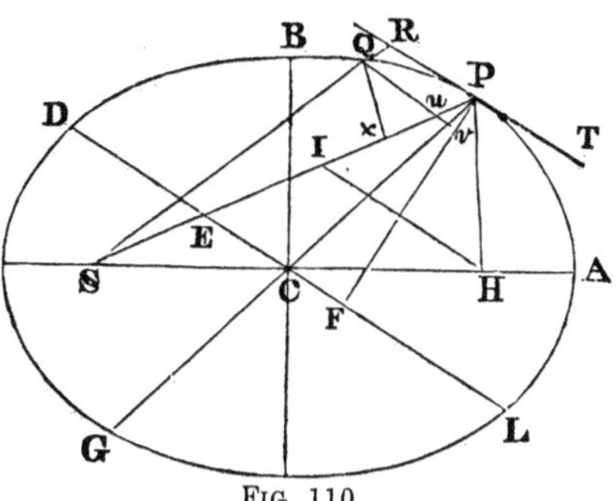

FIG. 110.

Thus, let ABG (Fig. 110) be the supposed elliptic orbit of the body, CA and CB its semi-axes, and S the focus towards wh ch the force is constantly directed. Also let P be one position of the body, PR a tangent to the orbit at P; and draw RQ parallel to PS, Quv, HI, and CD, parallel to PR, Qx perpendicular to SP, and join S and Q. CP and CD are semi-conjugate diameters. Denote them, respectively, by A' and B'; and denote the semi-axes, CA and CB, by A and B Since HI is parallel to PR, and, by a well-known property of the ellipse, the angle RPS is equal to the angle HPT, PH is equal to PI: and since HC = SC, and CE is parallel to HI, E is the middle of SI. We have, therefore,

$$PE=\frac{PS+PI}{2}=\frac{PS+PH}{2}=CA=A.$$

Now the force at P is measured by $2Pu$; and we may state the proportion

$$Pu : Pv :: PE : PC :: A : A'; \text{ which gives } Pv = Pu\frac{A'}{A}.$$

By the equation of the ellipse referred to its centre and conjugate diameters, PG and DL,

$$\overline{Qv}^2=\frac{B'^2}{A'^2}(Pv\times Gv)=\frac{B'^2}{A'^2}\left(Pu\frac{A'}{A}\times Gv\right).$$

If we regard Q as indefinitely near to P, then $Qu = Qv$, and $Gv = 2CP = 2A'$; and therefore

$$\overline{Qu}^2=\frac{B'^2}{A'^2}\left(Pu\frac{A'}{A}.2A'\right)=\frac{B'^2}{A}.2Pu.. \; (a.)$$

But $$Qu : Qx :: PE : PF :: CA : PF:$$

and, by Analytical Geometry,

$$CD\times PF=CA\times CB, \text{ or, } CA : PF :: CD : CB :: B' : B.$$

Hence $$Qu : Qx :: B' : B, \; \overline{Qu}^2 : \overline{Qx}^2 :: B'^2 : B^2, \text{ and } \overline{Qu}^2=\overline{Qx}^2\frac{B'^2}{B^2}.$$

Substituting in equation (a), $\overline{Qx}^2\frac{B'^2}{B^2}=\frac{B'^2}{A}.2Pu$; whence $\overline{Qx}^2=\frac{B^2}{A}.2Pu.$

Now triangular area $SQP=k=SP\times\frac{Qx}{2}$; whence $\overline{Qx}^2=\frac{4k^2}{\overline{SP}^2}$. Substituting, there results

$$\frac{4k^2}{\overline{SP}^2}=\frac{B^2}{A}.2Pu; \text{ or } 2Pu=\frac{A}{B^2}.4k^2.\frac{1}{\overline{SP}^2}\ldots. \; (I).$$

To compare the intensities of the force at different points of the orbit, we must take the values of $2Pu$, by which they are measured, for the same interval of time. On this supposition k is constant, and therefore the force is inversely proportional to the square of the distance SP.

It therefore follows from Kepler's second law, viz.: that the planets describe ellipses having the centre of the sun at one of their foci; that the force of gravity of each planet towards the sun varies inversely as the square of the distance from the sun's centre.

486. It operates on all the Planets alike. By taking into view Kepler's third law, it is proved that it is one and the same force, modified only by distance from the sun, which causes all the planets to gravitate towards him, and retains them in their orbits. This force is conceived to be an attraction of the matter of the sun for the matter of the planets, and is called the *Solar Attraction.*

To deduce this consequence from Kepler's third law, let t, t', denote the periodic times of any two planets; r, r', their distances from the sun at any assumed point of time; k, k', the areas described by their radius-vectors in any supposed unit of time; and A, B, and A$'$, B$'$, the semi-axes of their elliptic orbits. Then kt, $k't$, will be equal to the areas of the entire orbits; which are also measured by πAB, $\pi A'B'$.

Thus $kt : k't' :: AB : A'B'$, and $k^2t^2 : k'^2t'^2 :: A^2B^2 : A'^2B'^2$.

But, by Kepler's third law, $t^2 : t'^2 :: A^3 : A'^3$.

Dividing, and reducing, $k^2 : k'^2 :: \frac{B^2}{A} : \frac{B'^2}{A'}$:

that is, the squares of the areas described in equal times are as the parameters of the orbits.

Now, let f, f', denote the forces soliciting the two planets. Then, by equation (I), Art. 485,

$$f = \frac{A}{B^2}.4k^2.\frac{1}{r^2}, \text{ and } f' = \frac{A'^2}{B'^2}.4k'^2.\frac{1}{r'^2};$$

whence $$f : f' :: \frac{A}{B^2}.k^2.\frac{1}{r^2} : \frac{A'}{B'^2}.k'^2.\frac{1}{r'^2} :: \frac{A}{B^2}.\frac{B^2}{A}.\frac{1}{r^2} : \frac{A'}{B'^2}.\frac{B'^2}{A'}.\frac{1}{r'^2},$$

or $$f : f' :: \frac{1}{r^2} : \frac{1}{r'^2}.$$

From which it appears that the planets are solicited by a force of gravitation towards the sun, which varies from one planet to another according to the law of the inverse square of their distance.

487. Planets Endued with an Attractive Force. The motions of the satellites are in conformity with Kepler's laws; hence, the planets which have satellites are endued with an attractive force of the same nature with that of the sun.

The existence of a similar attractive power in each of the planets that are devoid of satellites, is proved by the fact that the observed inequalities of their motions, and of those of the other planets, may be shown upon this supposition to be necessary consequences of the attractions of the planets for each other.

In like manner the inequalities in the motions of the satellites and their primaries, show that the satellites possess the same property of attraction as the sun.

488. The Constituent Particles Attract each other. We learn from the motions produced by the action of the sun and planets upon each other, that the intensities of their attractive forces are, at the same distance, proportional to their masses, and that the whole attraction of the same body for different bodies, is, at the same distance, proportional to the masses of these bodies. From which we may infer that a mutual attraction exists between the particles of bodies, and that the whole force of attraction of one body for another, is the result of the attractions of its individual particles. Moreover, analysis shows, that in order that the law of attraction of the whole body may be that of the inverse ratio of the square of the distance, this must also be the law of attraction of the particles. The fact, as well as the law of the mutual attraction of particles, is also revealed by the tides and other phenomena referable to such attraction.

489. Theory of Universal Gravitation. The celestial phenomena compared with the general laws of motion, conduct us therefore to this great principle of nature; namely, *that all particles of matter mutually attract each other in the direct ratio of their masses, and in the inverse ratio of the squares of their distances.* This is called the principle of *Universal Gravitation.* The theory of its existence was first promulgated by Sir Isaac Newton, and is hence often called *Newton's Theory of Universal Gravitation.* The force which urges the particles of matter towards each other is called the *Force of Gravitation*, or the *Attraction of Gravitation.*

In the following chapters our object will be to develop the most important effects of the principle of gravitation thus arrived at by induction. The perfect accordance that will be observed to obtain between the deductions from the theory of universal gravitation and the results of observation, will afford additional confirmation of the truth of the theory.

CHAPTER XXII.

THEORY OF THE ELLIPTIC MOTION OF THE PLANETS.

490. Accelerating Force due to Sun's Attraction. Let the attraction of the unit of mass of the sun for the unit of mass of a planet, at the unit of distance, be designated by 1. The whole attraction exerted by the sun upon the unit of mass, at the same distance, will then be expressed by the mass of the sun (M); or, in other words, by the number of units which its mass contains. And the attraction F, at any distance r, will result from the proportion $M : F :: r^2 : 1^2$, which gives $F = \frac{M}{r^2}$. This, in the language of Dynamics, is the *Accelerating Force* of the planet, due to the attraction of the sun.

As $\frac{M}{r^2}$ expresses the attraction of the sun for a unit of mass of the planet, its attraction for the entire mass m of the planet will be expressed by $m\frac{M}{r^2}$. This is the *moving force* of the planet, and since it is, at the same distance, proportional to the mass of the planet, the velocity due to its action is the same, whatever may be the mass.

Attractive Force of Planet. The planet has also an attraction for the sun, as well as the sun for the planet, and the expression for its attractive force, or for the accelerating force animating the sun, will obviously be $\frac{m}{r^2}$. The sun will then tend towards the planet, as the planet towards the sun. But if the two bodies were to set out from a state of rest, the velocity of the planet would be as many times greater than the velocity of the sun, as the mass of the sun is greater than that of the planet. For the velocity of the planet would be to that of the sun as the attractive force of the sun is to the attractive force of the planet, that is, as $\frac{M}{r^2} : \frac{m}{r^2}$, or as $M : m$.

As the attractions of the particles of the sun and planet are mutual and equal, the attraction of the planet for the entire mass of the sun must be equal to the attraction of the sun for the entire mass of the planet.

491. The Sun and any Planet revolve about their Common Centre of Gravity.

To show this, we would remark, in the first place, that it is a principle of Mechanics that the mutual actions of the different members of a system of bodies cannot affect the state of the centre of gravity of the system. This is called the *Principle of the Preservation of the Centre of Gravity.* It follows from it that the common centre of gravity of the sun and any planet is at rest, unless it has a motion of translation in common with the two bodies, imparted by a force extraneous to the system. As we are concerned at present only with the relative motion of the sun and planet, such motion of translation, if it does exist, may be left out of account. Now, let S (Fig. 111) be the sun, and P any planet, supposed for the moment to be at rest. If neither of the two bodies should receive a velocity in a direction inclined to PS, the line of their centres, they would move towards each other by virtue of their mutual attraction, and meet at C their common centre of gravity.* But, if the body P have a projectile velocity given to it in any direction P*t*, inclined to the line PS, it is susceptible of proof that its motion relative to the sun may be in an ellipse, as is observed to be the case with the planets.

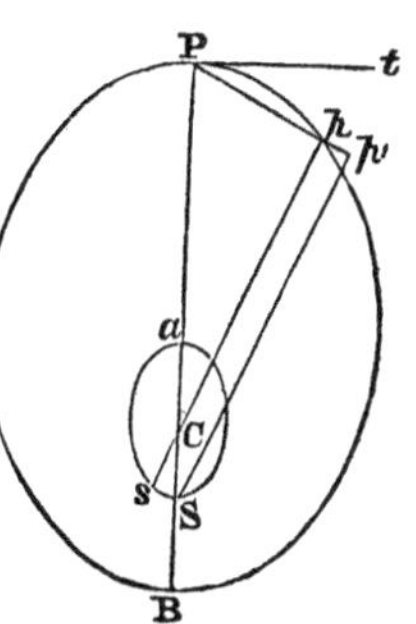

FIG. 111.

Now, while the planet moves in space, the line of the centres of the planet and sun must continually pass through the stationary position of the centre of gravity; and therefore, when the planet has advanced to any point *p*, the sun will have shifted its position to some point *s* on the line *p*C prolonged. Moreover, as the two bodies mutually gravitate towards each other, the path of each in space will be continually concave towards the other body, and therefore also towards the centre of gravity C, which is constantly in the same direction as the other body. Since the planet performs a revolution around the sun, the sun and planet must each continue to move about the point C until they have accomplished a revolution and returned to the line PCS. Also as the distance PS of the two bodies will be the same at the end as at the beginning of the revolution, as well as the ratio of their distances PC and SC from the centre of gravity, they will return to the positions, P, S, from which they set out, and will therefore move in continuous curves.

Moreover, these curves are similar to the apparent orbit described by P around S. For, draw Sp' parallel and equal to *sp*, and join P*p* and S*s*. Then, since *s*C : C*p* :: SC : CP, P*p* is parallel to S*s*; and therefore P*p* produced passes through p'. Whence, CP : C*p* :: SP : Sp'. Moreover, the angle PCp = PSp'. It follows,

* The common centre of gravity of two bodies lies on the line joining their centres, and divides this line into parts inversely proportional to the masses of the bodies.

therefore, that the area PCp is similar to the area PSp'; and thus that the orbit of P around C is similar to the apparent orbit of P around S. The latter is known from observation to be an ellipse. The former is therefore also an ellipse.

As the distances of the sun and planet from their common centre of gravity are constantly reciprocally proportional to their masses, the orbit of the sun will be exceedingly small in comparison with the orbit of the planet.

492. Entire Accelerating Force of Planet. If to both the sun and planet there should be applied a force equal to the accelerating force of the sun, $\frac{m}{r^2}$, (49°), but in an opposite direction, the sun would be solicited by two forces that would destroy each other, but the planet would now be urged towards the sun remaining stationary, with the accelerating force $\frac{M+m}{r^2}$, or a force the intensity of which was equal to the sum of the intensities of the attractive forces of the sun and planet, at the distance of the planet. Now, the application of a common force will not alter the relative motion of the two bodies. Hence, in investigating this motion, we are at liberty to conceive the sun to be stationary, if we suppose the planet to be solicited by the accelerating force $\frac{M+m}{r^2}$. As the mass of the sun is very much greater than that of any planet, but little error will be committed in neglecting the attraction of the planet, and taking into account only the sun's action $\frac{M}{r^2}$.

493. General Theoretical Results. Analysis makes known the general laws of the motion of a body, when impelled by a projectile force, and afterwards continually attracted towards the sun's centre by a force varying inversely as the square of the distance. We learn by it that the body will necessarily describe some one of the conic sections around the sun situated at one of its foci. We learn, also, that the nature of the orbit, as well as the length of the major axis, is wholly dependent, for any given distance of the planet, upon the *intensity* of the projectile force; but that the position of the axis, and the eccentricity of the orbit, depend also upon the *angle* of projection (that is, the angle included, at the commencement of the motion, between the line of direction of the projectile force and the radius-vector). As to the relative intensity of projectile force necessary to the production of each one of the conic sections, a certain intensity of force will produce a parabola; any less intensity, an ellipse or circle; and any greater, a hyperbola.

494. Theoretical Determination of Orbit of Planet. If the velocity that would at a given distance be imparted by the sun's attraction in a second of time, which is the measure of its intensity at the given distance, be found, and also the distance of a planet at any time, as well as its velocity and the angle made by the direction of its motion with the radius-vector, the form, dimensions, and position of the planet's orbit can be computed. This is to determine the orbit *à priori.* The practice has been, however, to determine the various elements of a planet's orbit by observation (as already described, Chap. IX.).

The elements being known, the equations of the elliptic motion, investigated on the principles of Mechanics, serve to make known the position and velocity of the planet at any time.

The physical theory of the motion of a satellite around its primary is obviously the same as that of the motion of a planet around the sun.

495. Centre of Gravity of the Solar System. According to the principle of the preservation of the centre of gravity (491), the centre of gravity of the whole solar system must either be at rest, or have a motion of translation in space in common with the system, resulting from the action of a foreign force. We have already seen (447) that it has been ascertained from observation, that it is in fact in motion.

The sun and planets revolve around their common centre of gravity. The path of the sun's centre results from the joint action of all the planets, and is a complicated curve. As the quantity of matter in all the planets taken together is very small, compared with that in the sun (less than $\frac{1}{700}$), the extent of the curve described by the centre of the sun cannot be very great. It is found by computation, that the distance between the sun's centre and the centre of gravity of the system can never be equal to the sun's diameter.

496. Centre of Gravity of a Planet and its Satellites. It is demonstrated in treatises on Mechanics, that if foreign forces act upon a system of bodies, the centre of gravity of the system will move just as the whole mass of the system concentrated at the centre of gravity would move, under the action of the same forces. It follows from this principle, that from the attraction of the sun for a primary planet and its satellites, their common centre of gravity will revolve around the sun, just as the whole quantity of matter in the planet and its satellites concentrated at this point would, under the influence of the same attraction. Moreover, the same considerations which show that the sun and planets revolve about their common centre of gravity, will also show that a primary planet and its satellites revolve about their common centre of gravity. It appears, therefore, that in the case of a planet which has satellites, it is not, strictly speaking, the centre of the planet that moves agreeably to the first and second laws of Kepler, but the common centre of gravity of the planet and its satellites; the planet and satellites revolving around the centre of gravity, as it describes its orbit about the sun.

The mass of the earth is to that of the moon as 82 to 1, while the distance of the moon is to the radius of the earth as 60 to 1: it follows, therefore, that the common centre of gravity of the earth and moon lies within the body of the earth.

497. Kepler's third Law not rigorously true. It appears from the physical investigation of the elliptic motion of the planets, that Kepler's third law is not strictly true. In consequence of the action of the planets upon the sun, the ratio of the periodic times of the different planets depends upon the masses of the planets, as well as their distances from the sun. If p and p' be the periodic times of any two of the planets, a and a' their mean distances from the sun's centre, and m and m' their quantities of matter, that of the sun being denoted by 1, then, disregarding the actions of the other planets, theory gives

$$p^2 : p'^2 :: \frac{a^3}{1+m} : \frac{a'^3}{1+m'}.$$

As m and m' are very small fractions, the error resulting from their omission will be very small. If we omit them, we shall have

$$p^2 : p'^2 :: a^3 : a'^3;$$

which is Kepler's third law.

CHAPTER XXIII.

THEORY OF THE PERTURBATIONS OF THE ELLIPTIC MOTION OF THE PLANETS AND THE MOON.

498. WE have, in a previous chapter, given a general idea of the mode of determining, from theory and observation combined, the law and amount of the perturbations or inequalities of the lunar and planetary motions. We propose now to give some insight into the nature and manner of operation of the disturbing forces, and will commence with the perturbations of the moon produced by the action of the sun.

499. Components of Disturbing Force. We have already shown (209) how the intensity and direction of the disturbing force of the sun, in any given position of the moon in its orbit, may be determined. Let us now derive the disturbing forces that take effect in the three directions in which the motion of the moon can be changed; namely, in the direction of the radius-vector, of the tangent to the orbit, and of the perpendicular to its plane. Let E (Fig. 112) be the earth, M the moon, and S the sun. Let the force exerted by the sun upon the moon be decomposed into two forces, one acting along the line MS′ parallel to ES, and the other from M towards E. If the component along MS′ were equal to the force exerted by the sun upon the earth, the motion of the moon about the earth would not be changed by the action of these two forces. Hence, the difference between them will be the disturbing force in the direction MS′. The component along ME is another disturbing force. It is called the *Addititious Force*, because it tends to increase the gravity of the moon towards the earth. The disturbing force along MS′ will generally be inclined to the plane of the orbit, and may be decomposed into three forces, one in the direction of the tangent, another in the direction of the radius-vector, and a third in the direction of the perpendicular to the plane. The first mentioned component is called the *Tangential Force;* the second is called the *Ablatitious Force;* and the third we shall call the *Perpendicular Force.*

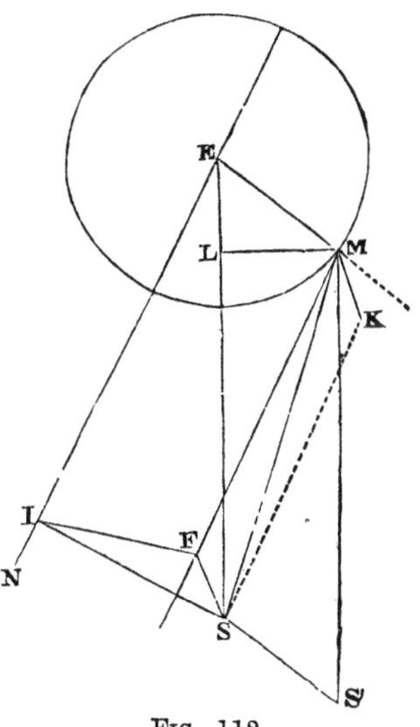

FIG. 112.

The actual disturbing force in the direction of the radius-vector is equal to the difference between the addititious and ablatitious forces, and is called the *Radial Force.* This and the tangential and perpendicular forces constitute the disturbing forces, the direct operation of which is to be considered.

500. To obtain General Analytical Expressions for these Forces, let the distance of the sun from the earth (which for the present we shall suppose to be constant) be denoted by a, and the distances of the moon from the earth and sun, respectively, by y and z. Also let F $=$ the force exerted by the earth upon the moon, P $=$ the force exerted by the sun upon the earth, and Q $=$ the force exerted by the sun upon the moon. Then, if we denote the mass of the earth by 1, and take m to stand for the mass of the sun, we shall have (490),

$$F=\frac{1}{y^2},\ P=\frac{m}{a^2},\ Q=\frac{m}{z^2}.$$

Let the force Q be represented by the line MS (Fig. 112); and let its component parallel to ES, or MS′=R, and its component along the radius-vector, or ME=T.

$$Q : T :: MS : ME;\ \text{or},\ \frac{m}{z^2} : T :: z : y.$$

Whence, additious force $T=\frac{my}{z^3}$....(82).

In a similar manner we obtain

$$R=\frac{ma}{z^3}....(83).$$

The disturbing force in the direction of the sun

$$=R-P=\frac{ma}{z^3}-\frac{m}{a^2}=ma\left(\frac{1}{z^3}-\frac{1}{a^3}\right).$$

Now, let α, β, γ, denote the angles made by the line MS′ with the tangent, radius-vector, and perpendicular to the plane of the orbit, and we shall have for the approximate components of the disturbing force R — P, along these lines:

$$\text{tangential force}=ma\left(\frac{1}{z^3}-\frac{1}{a^3}\right)\cos\alpha....(84);$$

$$\text{ablatitious force}=ma\left(\frac{1}{z^3}-\frac{1}{a^3}\right)\cos\beta....(85);$$

$$\text{perpendicular force}=ma\left(\frac{1}{z^3}-\frac{1}{a^3}\right)\cos\gamma....(86).$$

Combining equ. (85) with equ. (82), we obtain for the radial force,

$$\text{radial force}=my\frac{1}{z^3}-ma\left(\frac{1}{z^3}-\frac{1}{a^3}\right)\cos\beta.$$

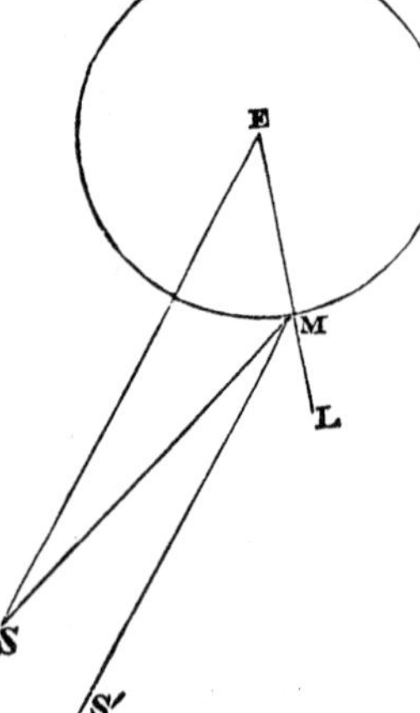

Fig. 113.

The obliquity of the orbit of the moon to the plane of the ecliptic, affects but very slightly the value of the tangential and radial forces. If we leave it out of account, or suppose the moon's orbit to lie in the plane of the ecliptic, we shall have (Fig. 113) β = S′ML = SEM, the elongation of the moon = ϕ, and α = complement of ϕ, which gives

$$\text{tang. force}=ma\left(\frac{1}{z^3}-\frac{1}{a^3}\right)\sin\phi....(87);$$

$$\text{rad. force}=my\frac{1}{z^3}-ma\left(\frac{1}{z^3}-\frac{1}{a^3}\right)\cos\phi\ (88).$$

Equation (86) *may be transformed* into another which is better adapted to the purposes we have in view. Let MK (Fig. 112) represent the perpendicular to the plane of the moon's orbit, MF the intersection of the plane SMK with the plane of the moon's orbit, and SI, IF, the intersections of a plane passing through S and perpendicular to EN, the line of nodes, with the plane of the ecliptic and the plane of the orbit. SF will be perpendicular to both IF and MF. Denote SIF, the inclination of the orbit to the ecliptic, by I, SEN the angular distance of the sun from the node by N, and SE and SM by a and z, as before.

Now, in equ. (86) γ stands for the angle S′MK, but S′MK = SMK (nearly), and

$$\cos SMK=\sin SMF=\frac{SF}{SM}.$$

$$SF = SI \sin SIF, \text{ and } SI = SE \sin SEI;$$

whence $$SF = SE \sin SEI \sin SIF = a \sin N \sin I:$$

substituting,

$$\cos \gamma = \cos SMK = \frac{a \sin N \sin I}{SM} = \frac{a \sin N \sin I}{z}.$$

Thus we have

$$\text{perpen. force} = ma\left(\frac{1}{z^3} - \frac{1}{a^3}\right)\frac{a \sin N \sin I}{z} \ldots .(89).$$

The variable z may be eliminated from equations (87), (88), and (89), and other equations obtained, involving only the variables y and ϕ. Let ML (Fig. 112) be drawn through the place of the moon perpendicular to ES. Then, using the same notation as before,

$$LS = z \text{ (nearly)}, \; EL = EM \cos LEM = y \cos \phi.$$

But $$LS = SE - EL;$$

whence $$z = a - y \cos \phi, \text{ and } z^3 = a^3 - 3a^2 y \cos \phi:$$

neglecting the terms containing the higher powers of y than the first, as they are very minute, y being only about $\frac{1}{400}$ a.

Thus, $$\frac{1}{z^3} = \frac{1}{a^3 - 3a^2 y \cos \phi} = \frac{1}{a^3} + \frac{3y \cos \phi}{a^4};$$

neglecting all the terms of the quotient that involve higher powers of y than the first. Substituting this value of $\frac{1}{z^3}$ in equ. (87), we obtain,

$$\text{tangential force} = \frac{3my \cos \phi \sin \phi}{a^3};$$

or (App. For. 13),

$$\text{tangential force} = \frac{3my \sin 2\phi}{2\,a^3} \ldots .(90).$$

Making the same substitution in equ. (88), and neglecting the term containing y^2, there results,

$$\text{radial force} = \frac{my\,(1 - 3\cos^2 \phi)}{a^3};$$

or (App. For. 9),

$$\text{radial force} = -\frac{my\,(1 + 3 \cos 2\phi)}{2\,a^3} \ldots .(91).$$

In equ. (89) we have to substitute, besides, the value of z, viz. $a - y \cos \phi$; then dividing and neglecting as before, we have

$$\text{perpen. force} = \frac{3my \cos \phi}{a^3} \sin N \sin I \ldots .(92).$$

501. *Variations of disturbing forces.* If the disturbing forces retained constantly the same intensity and direction, the result would be a continual progressive departure from the elliptic place; but, in point of fact, these forces are subject to periodical changes of intensity and direction from several causes, from which results a compensation of effects, and an eventual return to the elliptic place. The causes of the variation of the disturbing forces are:

(1.) The revolution of the moon around the earth.
(2.) The elliptic form of the apparent orbit of the sun.
(3.) The elliptic form of the orbit of the moon.
(4.) The inclination of the two orbits.

As the variations of the radial and tangential forces, resulting from the inclination of the orbits, are very minute, we shall leave them out of account, and in the consideration of the effects of these forces shall, for the sake of simplicity, regard the orbits as lying in the same plane.

The first mentioned circumstance is the most prominent cause of variation, and gives rise to the more conspicuous perturbations. The other two serve to modify the variations of the forces resulting from the first, and occasion each a distinct set of periodical perturbations.

502. Tangential Force. Let us now investigate, in succession, the effects of each of the disturbing forces, commencing with the tangential force. The tangential force takes effect directly upon the velocity of the moon in its orbit; and as its line of direction does not pass through the earth, it disturbs the equable description of areas. It also affects the radius-vector indirectly, by changing the centrifugal force. To understand the detail of its action we must inquire into the variations which it undergoes.

If we regard y as constant in the expression for the tangential force, (equa. 90), which amounts to considering the moon's orbit as circular, the expression will become equal to zero when $\sin 2\phi=0$, and will have its maximum value when $\sin 2\phi=1$. It will also change its sign with $\sin 2\phi$. It appears, therefore, that the tangential force is zero in the syzigies and quadratures, where it also changes its direction, and that it attains its maximum value in the octants. It will be seen, on inspecting Fig. 114, that it will be a retarding force in the first quadrant (AB). Accordingly, it will be an accelerating force in the second, a retarding force again in the third, and an accelerating force again in the fourth.

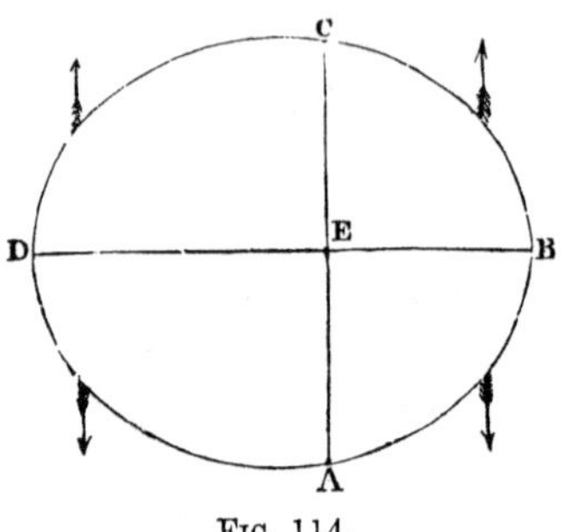

FIG. 114.

This will also appear upon considering the direction of the disturbing force parallel to the line of the centres of the sun and earth, in the various quadrants. In the nearer half of the orbit the sun tends to draw the moon away from the earth, and the force in question is directed towards the sun. In the more remote half the sun tends to draw the earth away from the moon, but we may regard it, instead, as urging the moon from the earth by the same force; for the relative motion will be the same on this supposition. In the part of the orbit supposed, then, the disturbing force under consideration will be directed from the sun, as represented in Fig. 114.

It appears, then, that the tangential force will alternately retard and accelerate the motion of the moon during its passage through the different quadrants, and that the maximum of velocity will occur in the syzigies, A, C, where the accelerating force becomes zero, and the minimum of velocity in the quadratures, B, D, where the retarding force becomes zero. On the supposition that the orbit is a circle, the arcs AB, BC, CD, and DA, would be equal, and the retardation of the velocity in one quadrant would be compensated for by an equal acceleration in the next, and at the close of a synodic revolution the velocity of the moon would be the same as at its commencement. As the velocity is greatest in the syzigies and least in the quadratures, and as the degree of retardation is the same as that of acceleration, the mean motion* must have place in the octants. Now, as the moon moves from the syzigy A with a motion greater than the mean motion, its true place will be in advance of its mean place, and will become more and more so till it reaches the octant, where the true motion is equal to the mean. The difference between the true and mean place will then be the greatest; for after that, the true motion becoming less than the mean, the mean place will approach nearer to the true, till at the quadrature they coincide. Beyond B, the true motion still continuing less than the mean, the mean place will be in advance of the true, and the separation will increase till at the octant the true motion has attained to an equality with the mean motion, after which, the mean motion being the slowest, the true place will approach the mean till at the syzigy C they again coincide. Corresponding effects will take place in the two remaining quadrants. We perceive, therefore, that the tangential force produces an inequality of longitude, which attains to its maximum positive and negative value in the octants, and is zero in the syzigies.

* The expressions, mean motion, true motion, mean place, true place, are here to be understood only in relation to the perturbation under consideration.

This is the inequality known in Spherical Astronomy by the name of *Variation* (217).

503. *Modifications of the effects of the tangential force, that result from the elliptic form of the sun's orbit.* Suppose that at the moment when the moon sets out from conjunction the sun is in the apogee of its orbit: then it is plain that, during the whole revolution of the moon, the sun's disturbing force would be on the increase by reason of the diminution of the sun's distance, and that, in consequence, the retardation in the first quadrant would be less than the acceleration in the second, and the retardation in the third less than the acceleration in the fourth. So that, when the moon has again come round into conjunction, the acceleration will have overcompensated the retardation. This kind of action would go on so long as the sun approaches the earth; but when it has passed the perigee of its orbit, and begun to recede from the earth, the reverse effect would take place, and a retardation of the moon's orbital motion would happen each revolution. If the anomalistic revolution of the sun were an exact multiple of the synodic revolution of the moon, the acceleration in each revolution of the moon during the passage of the sun from the apogee to the perigee of its orbit, would be compensated for by an equivalent retardation in the revolution of the moon, answering to the same distance of the sun in its passage from the perigee to the apogee; and the velocity of the moon would be the same at the close of an anomalistic revolution of the sun as at its commencement. But as this relation does not, in fact, subsist between the anomalistic revolution of the sun and the synodic revolution of the moon, a compensation between the accelerations and retardations, answering to the different revolutions of the moon, will not be effected until conjunctions shall have occurred at every variety of distance of the sun in each half of its orbit. Since the anomalistic and synodic revolutions are incommensurable, the sun will be, in reality, in every variety of position in its orbit at the time of conjunction, in process of time, so that eventually the original velocity in conjunction will be regained. It appears, therefore, that the variation of the moon's motion from one revolution to another, occasioned by the elliptic form of the sun's orbit, is periodic. Its period will be the interval of time in which the moon will perform a certain number of synodic revolutions, while the sun performs a certain number of anomalistic revolutions. Avoiding unnecessary precision, we find it to consist of but a moderate number of years.

504. *Consequences of the elliptic form of the moon's orbit.* We remark, in the first place, that the orbit being an ellipse, the areas AEB, BEC, CED, and DEA (Fig. 114), will be unequal, and therefore, by the laws of elliptic motion, the arcs AB, BC, CD, and DA, will be described in unequal times. It follows from this, that the retardation in the first quadrant will not be exactly compensated by the acceleration in the second, and that the retardation in the third will not be exactly compensated by the acceleration in the fourth. Therefore, at the end of the synodic revolution the moon will have an excess or deficiency of velocity. Its mean motion will then vary from one revolution to another, by reason of the ellipticity of its orbit. This variation will be periodic, like that just considered, and for similar reasons. The excess or deficiency of velocity at the close of any one revolution, will in time be compensated by an equal deficiency or excess occurring at the close of another revolution, when the sun has a certain different position with respect to the perigee of the moon's orbit.

505. Radial Force. We pass now to the consideration of the action of the radial force. The direct general effect of the radial force, is an alteration in the intensity of the moon's gravity towards the earth, and in its law of variation. Its specific effects are periodical variations in the magnitude, eccentricity, and position of the orbit. As it is directed towards the earth, it will not disturb the equable description of areas. To discover the variations of this force we have only to discuss the general analytical expression for it, already investigated. It is,

$$\text{radial force} = \frac{my\,(1-3\cos^2\phi)}{a^3}.$$

We shall have radial force $= 0$, when $1 - 3\cos^2\phi = 0$, or when $\cos\phi = \pm\sqrt{\tfrac{1}{3}}$. This value of $\cos\phi$ answers to four points lying on either side of the quadratures, and about 35° distant from them. When $\cos\phi$ is numerically greater than $\sqrt{\tfrac{1}{3}}$ the result will be negative, and when it is less than $\sqrt{\tfrac{1}{3}}$ the result will be positive. It follows, therefore, that the radial force increases the gravity of the moon in the quadratures, and for about 35° on each side of them, and that during the remainder of a synodic revolution it diminishes it.

When the moon is in quadratures, $\cos\phi = 0$, and

$$\text{radial force} = \frac{my}{a^3} \quad . \quad . \quad . \quad . \quad (93).$$

In the syzigies, we have $\cos\phi = \pm 1$, which gives

$$\text{radial force} = -\frac{2my}{a^3} \quad . \quad . \quad . \quad (94).$$

It appears, then, that the diminution of the moon's gravity in the syzigies is double of its increase in the quadratures.

We learn also from equations (93) and (94), that the radial force in the quadratures and syzigies varies directly as the distance; from which we conclude that the gravity of the moon varies at these points by a different law from that of the inverse squares. In the quadratures the gravity will be increased most at the greatest distance, where it is the least; and thus it will vary in a less rapid ratio than the square of the distance. In the syzigies it will be diminished most at the greatest distance, or where it is the least; and accordingly, at these points it will vary in a more rapid ratio than the square of the distance.

506. *Moon's distance increased by radial force.* With the aid of the Differential Calculus, we readily find that the mean diminution of the moon's gravity from the sun's action is $\frac{mr}{2a^3}$; r representing in this case the mean distance of the moon from the earth. The value of this expression is equal to about the 360th part of the whole gravity of the moon to the earth.

In consequence of this diminution, the moon must describe its orbit at a greater distance from the earth, with a less angular velocity, and in a longer time, than if it were acted on only by the attraction of the earth.

507. *The radial force of the sun alters the eccentricity of the moon's orbit* and differently in different revolutions of the moon, according to the position of the line of syzigies with respect to the line of apsides. When these lines are coincident the eccentricity is increased. For suppose PMAN (Fig. 115) to be the elliptic orbit of the moon that would be described under the influence of a force varying inversely as the square of the distance. In going from the apogee to the perigee, the gravity will increase in a greater ratio than that of the inverse square of the distance; the true orbit will therefore fall within the ellipse, and the perigean distance (EP′) will be less than for the ellipse. Consequently, the eccentricity will increase so much the more as the major axis diminishes. On the other hand, in going from the perigee to the apogee, the gravity will decrease in a greater ratio than the inverse square of the distance, and the moon will consequently recede further from the earth than if the orbit described was an ellipse. Therefore, in this half of the orbit the eccentricity will also be increased. When the apsides are in quadratures the eccentricity will be diminished; for the gravity will then vary from the apogee to the perigee, and from the perigee to the apogee, in a less ratio than that of the inverse squares; and therefore the results will be contrary to those just obtained. The eccentricity will have its maximum value when the apsides are in syzigies, and its minimum when they are in quadratures; for, in every other position of the line of apsides with respect to the line of syzigies, the radial force in the apogee and perigee will be less than in these positions (equa. 91), and therefore alter less the proportional gravity of the moon in the apogee and perigee. It is evident, from the gradual decrease of the radial force as we recede from the syzigies and quadratures, that the eccentricity will continually diminish in the progress of the apsides from the syzigies to the quadratures, and that it will continually increase from the quadratures to the syzigies.

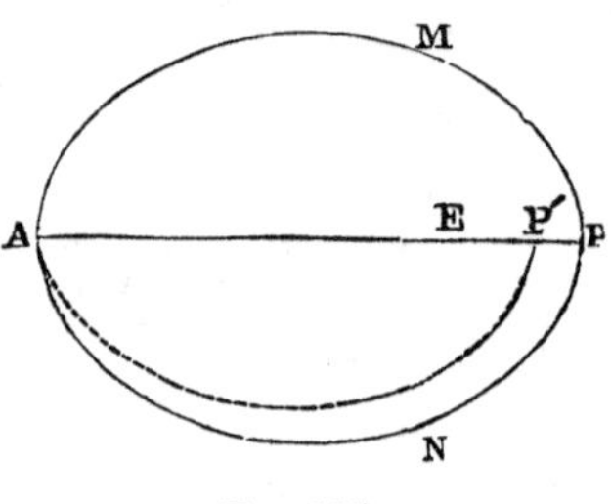

FIG. 115.

The change in the eccentricity of the moon's orbit, thus produced, will be attended with a corresponding change in the equation of the centre, and thus of the longitude. And this change is the conspicuous inequality of the moon, known by the name of Evection (217).

508. *The radial force also produces a motion of the line of apsides.* If the moon were only acted upon by the attraction of the earth its orbit would be an ellipse, and the motion from one apsis to another, or, in other words, from one point where the orbit cuts the radius-vector at right angles to the other, would be 180°. In point of fact, however, the gravity due to the earth's attraction is constantly either diminished or increased by the radial disturbing force of the sun, and therefore its true orbit must continually deviate from the ellipse that would be described under the sole action of the earth's attraction. When from the action of this force there is a diminution of the force of gravity, the moon will continually recede from the ellipse in question, its path will be less bent, and it must therefore move through a greater angular distance before the central force will have deflected its course into a direction at right angles to the radius-vector. Accordingly, it will move through a greater angular distance than 180° in going from one apsis to another, and thus the apsides will advance. On the other hand, when the same force increases the force of gravity, the moon's path will fall within the ellipse, its curvature will be increased, and therefore it will be brought to intersect the radius-vector at right angles at a less angular distance. In this case, therefore, the apsides will move backward. Now, we have shown (505) that the radial disturbing force of the sun alternately diminishes and increases the moon's gravity to the earth. It follows, therefore, that the motion of the apsides will be alternately direct and retrograde; but since, as has been shown (505), the diminution subsists during a longer part of the moon's revolution, and is moreover greater than the increase, the direct motion will exceed the retrograde, and therefore in an entire revolution the apsides will advance.

The observed motion of the apsides of the moon's orbit is not, however, wholly produced by the radial disturbing force. It is in part due to the action of the tangential force. This force alters the centrifugal force of the moon, and thus changes its gravity towards the earth, at the same time with the radial force.

509. *Explanation of the Annual Equation.* The elliptic form of the sun's orbit is the occasion of a change in the radial force, from which results a perturbation of longitude called the *Annual Equation* (217). The mean diminution of the moon's gravity, arising from the action of the sun, or the mean radial force, is equal to $\frac{mr}{2a^3}$ (506). Hence this diminution is inversely proportional to the cube of the sun's distance from the earth. Therefore, as the sun approaches the perigee of its orbit, its distance from the earth diminishing, the mean diminution of the moon's gravity to the earth will increase, and consequently the moon's distance from the earth will become greater, and its motion slower, than it otherwise would be. The contrary will take place while the sun is moving from the perigee to the apogee.

510. Perpendicular Force. The disturbing force perpendicular to the plane of the moon's orbit, produces a tendency in the moon to quit that plane, from which there results a change in the position of the line of the nodes, and a change in the inclination of the plane of the orbit to that of the ecliptic. If we examine the general expression for this force, viz.:

$$\text{perpen. force} = \frac{3my\cos\phi}{a^3}\sin N \sin I,$$

we see that for any given values of N and I, it will be zero in the quadratures, and have its greatest value in the syzigies; and that it will change its direction in the quadratures, lying, in the nearer half of the orbit, on the same side of its plane as the sun, and in the more remote half, on the opposite side. We perceive also that it will be zero for every value of ϕ, or for every elongation of the moon, when the angle N is zero, that is, when the sun is in the plane of the orbit; and will attain its maximum, for any given elongation, when the line of direction of the sun is perpendicular to the line of nodes. It will also be the less, other things being the same, the smaller is the inclination I.

511. *Retrograde Motion of the Nodes.* Let NM'R (Fig. 116) represent the orbit of the moon, and S the sun, supposed stationary, the line of the nodes being in quadratures; and let L, L′ be the points of the orbit 90° distant from the nodes. The direction of the force, in the various points of the orbit, is indicated by the arrows drawn in the figure. When the moon is at any point M′ between L and the descending node N′, it will be drawn out of the plane in which it is moving by the disturbing force M′K′, and compelled to move in such a line as M′t′. The node N′ will therefore retrograde to some point n′. When the moon is at any point M, moving from the ascending node N towards L, its course will be changed to the line Mt, lying, like the line M′t′, below the orbit, which being produced backward, meets the plane of the ecliptic in some point n behind N. The nodes, therefore, retrograde in this position of the moon, as well as in the former. When the moon is in the half N′L′N of the orbit, lying below the ecliptic, the absolute direction of the disturbing force will be reversed, and thus its tendency will be the same as before, namely, to draw the moon towards the ecliptic. It follows, therefore, that throughout this half of the orbit, as in the other, the motion of the nodes will be retrograde. Accordingly, when the nodes are in quadratures, or 90° distant from the sun, they will retrograde during every part of the revolution of the moon.

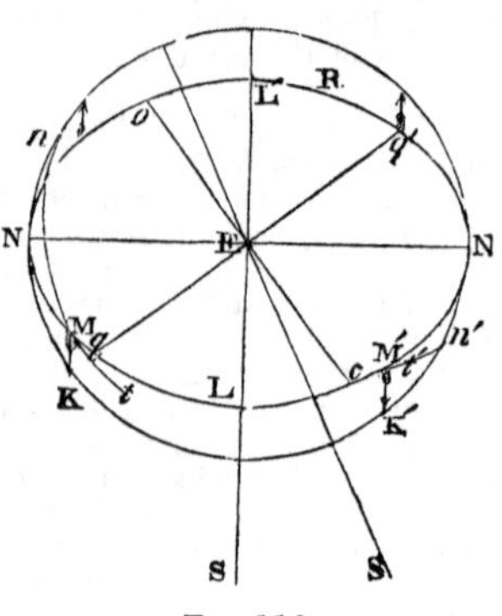

FIG. 116.

Suppose the sun now to be fixed on the line of nodes, or the nodes to be in syzigies. In this case the perpendicular force will be zero (510), and therefore there will be no disturbance of the plane of the moon's orbit.

Next, let the situation of the sun be intermediate between the two just considered, as represented in Figs. 116 and 117. The effect of the disturbing force will be the same as in the first situation from the quadrature q (Fig. 116) to the node N′, and from the quadrature q' to the node N. But throughout the arcs Nq, N′q', the direction of the force, and therefore the effects, will be reversed. The node will then retrograde, as before, while the moon moves over the arcs qN′ and q'N, and advance while it is in the arcs Nq, N′q'. But as the force is greatest over the arcs qN′, q'N, which contain the syzigies (510), and as these arcs are also longer than the arcs Nq, N′q', the node will, on the whole, retrograde each revolution. The velocity of retrogradation will, however, be less than when the nodes are in quadratures, and proportionably less as the distance of the sun from this position is greater.

In the position represented in Fig. 117, a direct motion will take place over the arcs q'N′ and qN: but as Nq' and N′q, the arcs of retrograde motion, are of greater extent than q'N′ and qN, and moreover contain the syzigies, the retrograde motion in each revolution must exceed the direct, as before.

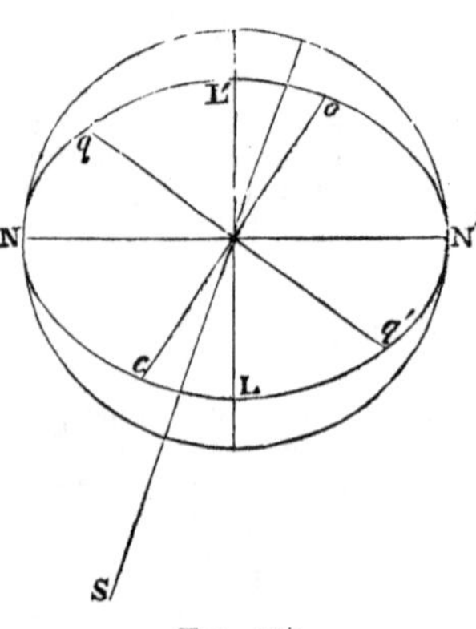

FIG. 117.

If we suppose the sun to be situated on the other side of the line of nodes, the effect of the disturbing force will obviously be the same in any one position of the sun, as in the position diametrically opposite to it. It appears, then, that the line of the nodes has a retrograde motion in every possible position of the sun.

512. *Effect of Sun's Motion.* We have thus far supposed the sun to remainstationary in the various positions in which we have considered it, during the revolution of the moon. It remains, then, to consider the effect of the sun's motion in this interval. And first, it is plain, that, as the sun advances from S towards N′ (Fig. 116), the arcs Nq, N$'q'$ will increase, and the arcs qN$'$ and q'N diminish; from which it appears, that during the advance of the sun from the point 90° behind the descending node to this node, its motion in the course of each revolution of the moon will cause the retrograde motion of the node to be slower than it otherwise would be. While the sun moves from the ascending node to the point 90° from it, the effect of its motion will obviously be just the reverse of this. During its passage from the descending to the ascending node, the effect will be the same in either quadrant as in that diametrically opposite.

The variation in the intensity of the perpendicular force, conspires with the difference of situation of the sun and its motion during a revolution of the moon in diminishing or increasing, as the case may be, the velocity of retrogradation of the nodes.

513. *Change of the inclination of the orbit.* If we refer to Fig. 116 we shall see that when the nodes are in quadrature the inclination will diminish while the moon is moving from the ascending node N to the point L 90° distant from it, and increase while it is moving from L to the other node N$'$. In the other half of the orbit the tendency of the disturbing force is the same (511), and therefore while the moon is moving from N$'$ to L$'$ the inclination will diminish, and while it is moving from L$'$ to N the inclination will increase. The diminutions and increments will compensate each other, and the original inclination will be regained at the close of the revolution.

When the nodes are in syzigies there will be no change of inclination (510).

In the situations of the sun represented in Figs. 116 and 117, the inclination will decrease from q to L and from q' to L$'$, and increase from L to q' and from L$'$ to q; the effects being the same as when the nodes are in quadratures over the arcs qL and LN$'$ in Fig. 116, and NL and Lq' in Fig. 117, and being reversed over

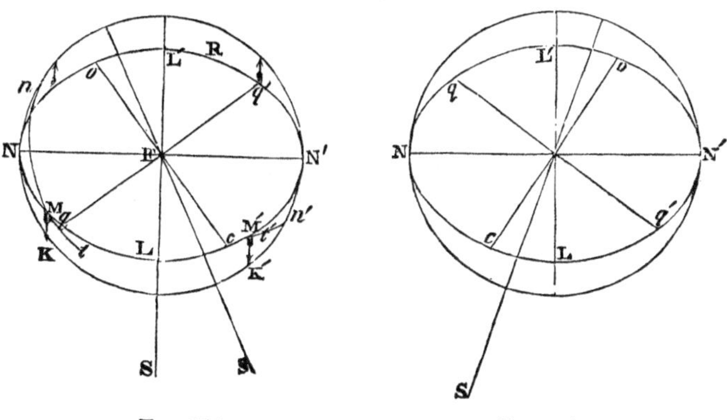

FIG. 116. FIG. 117.

the arcs Nq and N$'q'$ in Fig. 116, and qN and q'N$'$ in Fig. 117. When the sun has the position represented in Fig. 116, the arcs of increase Lq' and L$'q$ will be greater than the arcs of diminution qL and q'L$'$. The disturbing force will also be greater in the former arcs than in the latter. In the position supposed, therefore, there will be, on the whole, an increase of inclination every revolution. When the sun is in the position represented in Fig. 117, the arcs of diminution qL and q'L$'$ will be the greater; and the force in them will also be the greater. In this case, therefore, there will be a diminution of the inclination each revolution of the moon.

When the sun is on the other side of the line of nodes, the results will be the same as in the positions diametrically opposite.

514. *Consequences of the sun's motion* during the revolution of the moon. As the sun moves from S towards N$'$ (Fig. 116) the arcs Lq', L$'q$, over which there is an increase of the inclination, will increase; and the arcs qL, q'L$'$, over which there is a diminution, will diminish. The motion of the sun will, therefore, in approaching the descending node, render the increase of the inclination each revolution of the moon greater than it otherwise would be. When the sun is receding from the ascending node, the corresponding arcs will experience corresponding changes, and therefore the diminution will now be less than if the sun were stationary.

The results will be similar for the opposite quadrants on the other side of the line of nodes

515. *Epochs of greatest and least Inclinations.* Since the inclination diminishes as the sun recedes from either node, and increases as it approaches either node, it will be the least when the nodes are in quadratures, and the greatest when they are in syzigies.

It is important to observe that the change of inclination which we have been considering is modified by the retrograde motion of the node; and thus, that, besides the variations of this element connected with the motions of the moon and sun, there is another extending through the period employed by the node in completing a revolution with respect to both the sun and moon.

516. Perturbations Periodic. The perturbations of the elliptic motion of the moon, comprising inequalities of orbit longitude, and variations in the form and position of the orbit, which have now been under consideration, depend upon the configurations of the sun and moon, with respect to each other, the perigee of each orbit, and the node of the moon's orbit. Their effects will disappear when the configurations upon which they depend become the same. They are therefore *periodical.*

517. The Perturbations of the Motions of a Planet, produced by the action of another planet, are precisely analogous to the perturbations of the motions of the moon, produced by the action of the sun. The disturbing forces are obviously of the same kind, and they are subject to variations from precisely similar causes. But, owing to the smallness of the masses of the planets and their great distances, their disturbing forces are much more minute than the disturbing force of the sun. From this cause, together with the slow relative motion of the disturbing and disturbed body, the motion of the apsides and nodes, and the accompanying variations of eccentricity and inclination, are very much more gradual in the case of the planets than in the case of the moon. Their periods comprise many thousands of years, and on this account they are called *Secular Motions* or *Variations.* In consequence of the greater feebleness of the disturbing forces, the periodical inequalities are also much less in amount. Moreover, as the motion of a planet is much slower than that of the moon, and as the variations of its orbit are more gradual than those of the lunar orbit, the compensations produced by a change of configurations are much more slowly effected, and thus the periods of the inequalities are much longer.

518. Acceleration of the Moon. The motions of the moon would be subject to no secular variations if the apparent orbit of the sun were unchangeable; but the secular variation of the eccentricity of the sun's orbit, which answers to an equal variation of the eccentricity of the earth's orbit, that is produced by the action of the planets, gives rise to a secular inequality in the motion of the moon, called the *Acceleration of the Moon.* This inequality was discovered from observation. Its physical cause was first made known by Laplace.

CHAPTER XXIV.

RELATIVE MASSES AND DENSITIES OF THE SUN, MOON, AND PLANETS:—RELATIVE INTENSITY OF THE FORCE OF GRAVITY AT THEIR SURFACE.

519. Determination of the Masses of the Planets. The perturbations which a planet produces in the motions of the other planets, depend for their amount chiefly upon the ratio of the mass of the planet to the mass of the sun, and the ratio of the distance of the planet from the sun to the distance of the planet disturbed from the same body. Now, the ratio of the distances is known by the methods of Spherical Astronomy; consequently, the observed amount of the perturbations ought to make known the ratio of the masses, the only unknown element upon which it depends.

This is one method of determining the masses of the planets. The masses of those planets which have satellites may be found by another and simpler method, viz.: by comparing the attractive force of the planet for either one of its satellites with the attractive force of the sun for the planet. These forces are to each other directly as the masses of the planet and sun, and inversely as the squares of the distances of the satellite from the primary and of the primary from the sun. Thus calling the forces f, F, the masses m, M, and the distances d, D, we have

$$f : \mathrm{F} :: \frac{m}{d^2} : \frac{\mathrm{M}}{\mathrm{D}^2};$$

whence we obtain $m : \mathrm{M} :: fd^2 : \mathrm{FD}^2$. If we regard the orbits as circles, then d and D will be the mean distances, respectively, of the satellite from the primary, and of the primary from the sun, and are given in Tables II, III, and VI. The ratio of f to F is equal to the ratio of the versed sines of the arcs actually described by the satellite and primary, in some short interval of time; since these are sensibly equal to the distances that the two bodies are deflected in this interval from the tangents to their orbits, towards the centres about which they are revolving: and since the rates of motion and dimensions of the orbits of the planet and satellite are known, these arcs and their versed sines are easily determined.

Table IV exhibits the relative masses of the sun, moon, and

planets, according to the most received determinations, that of the sun being denoted by 1.

520. Computation of the Densities of the Planets. The quantities of matter of the sun, moon, and planets, as well as their bulks, being known, their densities may be easily computed; for, the densities of bodies are proportional to their quantities of matter divided by their bulks.

Table IV contains the densities of the sun, moon, and planets, that of the earth being denoted by 1. It will be seen on inspecting it, that the densities of the planets decrease from Mercury to Saturn; and that the four planets most distant from the sun are much less dense than the four which are nearest the sun.

521. The Comparative Forces of Gravity at the surface of the sun, moon, and planets, may also readily be found, when the masses and bulks of these bodies are known. For supposing them to be spherical, and not to rotate on their axes, the force of gravity at their surface will be directly as their masses and inversely as the squares of their radii, or, in other words, proportional to their masses divided by the squares of their radii. The centrifugal force at the surface of a planet, generated by its rotation on its axis, diminishes the gravity due to the attraction of the matter of the planet. The diminution thus produced on any of the planets is not, however, very considerable. The method of determining the centrifugal force at the surface of a body in rotation, is given in treatises on Mechanics. (See Table IV.)

CHAPTER XXV.

FORM AND DENSITY OF THE EARTH:—CHANGES OF ITS PERIOD OF ROTATION.—PRECESSION OF THE EQUINOXES, AND NUTATION.

522. WE have already seen (105) that measurements made upon the earth's surface establish that the figure of the earth is that of an oblate spheroid, and that the oblateness at the poles is $\frac{1}{299}$.

523. Density of the Earth. From the amount and law of variation of the force of gravity upon the earth's surface, ascertained by observations upon the length of the seconds' pendulum, it is proved that the matter of the earth is not homogeneous, but denser towards the centre, and that it is arranged in concentric strata of nearly an elliptical form and uniform density.

The fact of the greater density of the earth towards its centre has also been established by observations upon the deviation of a plumb-line from the vertical, produced by the attraction of a mountain; the amount of the deviation being ascertained by observing the difference in the zenith distances of the same star, as measured with a zenith-sector on opposite sides of the mountain. To the north of the mountain the plummet was drawn towards the south, and the zenith distance of a star to the north of the zenith was diminished; while to the south of the mountain the plummet was drawn towards the north, and the zenith distance of the same star was increased by an equal amount: and thus the difference of the two measured zenith distances was equal to twice the deviation of the plumb-line from the true vertical in either of the positions of the instrument (allowance being made for the difference of latitude of the two stations, as determined from the distance between them and the known length of a degree).

Such observations were made for the purpose of determining the mean density of the earth by Dr. Maskelyne, in 1774, on the sides of the mountain Schehallien in Scotland. The observed deviation of the plumb-line made known the ratio of the attraction of the mountain to that of the whole earth, and thus the relative quantities of matter in the mountain and earth. These being ascertained, and the figure and bulk of the mountain having been determined by a survey, the relative density of the earth and mountain became known by the principle mentioned in Art. 520, and thence the actual density of the earth; the density of the mountain having been found by experiment. The result was, that the mean density of the earth is 4.95. Later determinations make it 5.44.

524. Explanation of Spheroidal Form of Earth. The spheroidal form of the surface of the earth and of its internal strata is easily accounted for, if we suppose the earth to have been originally in a fluid state. The tendency of the mutual attraction of its particles would be to give it a spherical form; but by virtue of its rotation, all its particles, except those lying immediately on the axis, would be animated by a centrifugal force increasing with their distance from the axis. If, therefore, we conceive of two columns of fluid extending to the earth's centre, one from near the equator, and the other from near either pole, the weight of the former would by reason of the centrifugal force be less than that of the latter. In order, then, that they may sustain each other in equilibrio, that near the equator must increase in length, and that near the pole diminish. As this would be true at the same time for every pair of columns situated as we have supposed, the surface of the whole body of fluid about the poles must fall, and that of the fluid about the equator rise. In this manner the earth would become flattened at the poles and protuberant at the equator.

Upon a strict investigation it appears that a homogeneous fluid of the same mean density with the earth, and rotating on its axis at the same rate that the earth does, would be in equilibrium, if it had the figure of an oblate spheroid, of which the axis was to the equatorial diameter as 229 to 230, or of which the oblateness was $\frac{1}{230}$. If the fluid mass supposed to rotate on its axis be not homogeneous, but be composed of strata that increase in density from the surface to the centre, the solid of equilibrium will still be an elliptic spheroid, but the oblateness will be less than when the fluid is homogeneous.

525. Possible Changes of Period of Rotation. The time of the earth's rotation, as well as the position of its axis, would change if any variation should take place in the distribution of the matter of the earth, or in case of the impact of a foreign body.

If any portion of matter be, from any cause, made to approach the axis, its velocity will be diminished, and the velocity lost being imparted to the mass, will tend to accelerate the rotation. If any portion of matter be made to recede from the axis, the opposite effect will be produced, or the rotation will be retarded. In point of fact, the changes that take place in the position of the matter of the earth, whether from the washing of rains upon the sides of mountains, or evaporation, or any other known cause, are not sufficient ever to produce any sensible alteration in the circumstances of the earth's rotation on its axis.

526. Earth's Dimensions and Axis Invariable. It is ascertained from direct observation, that there has in reality been no perceptible change in the period of the earth's rotation since the time of Hipparchus, 120 years before the beginning of the present era. We may therefore conclude, *à posteriori*, that there has been no material change in the form and dimensions of the earth in this interval.

Were the axis of the earth to experience any change of position with respect to the matter of the earth, the latitudes of places would be altered. A motion of 100 feet might increase or diminish the latitude of a place to the amount of 1″, an angle which can be measured by modern instruments. Now, in point of fact, the latitudes of places have not sensibly varied since their first determination with accurate instruments; therefore, in this interval the axis of the earth cannot have materially changed. Indeed, since the earth's surface and its internal strata are arranged symmetrically with respect to the present axis of rotation, it is to be inferred that this axis is the same as that which obtained at the epoch when the matter of the earth changed from a fluid to a solid state.

527. Physical Theory of Precession and Nutation. The motions of the earth's axis, along with the whole body of the earth, which give rise to the Precession of the Equinoxes and Nutation, are consequences of the spheroidal form of the earth, inasmuch as they are produced by the actions of the sun and moon upon that portion of the matter of the earth which lies on the outside of a sphere conceived to be described about the earth's axis. The physical theory of the phenomena in question is analogous to that of the retrogradation of the moon's nodes. The sun produces a retrograde movement of the points in which the circle described by each particle of the protuberant mass cuts the plane of the ecliptic, as it does of the moon's nodes; the effect produced is, however, exceedingly small, by reason of the inertia of the interior spherical mass connected with the external mass upon which the action takes place. The moon, in like manner, occasions a retrograde movement of the nodes of the same particles on the plane of its orbit. The actions of the sun and moon will not be the same each revolution of a particle. That of the sun will vary during the year with the angular distance of the sun from the node (510); and that of the moon will vary during each month with the distance of the moon from the node, and also during a revolution of the nodes of the moon's orbit by reason of the change in the inclination of the orbit to the equator. The mean effect of both bodies is the *precession;* the inequality resulting from the change in the sun's action during the year is the *solar nutation;* and the inequality consequent upon the retrogradation of the moon's nodes is the *lunar nutation*, or the chief part of it.

CHAPTER XXVI.

THE TIDES.

528. THE alternate rise and fall of the surface of the ocean twice in the course of a lunar day, or about 25 hours, is the phenomenon known by the name of the *Tides.* The rise of the water is called the *Flood Tide*, and the fall the *Ebb Tide.*

529. Times of High and Low Water. The interval between one high water and the next is, at a mean, half a mean lunar day, or 12h. 25m. Low water has place nearly, but not exactly, at the middle of this interval; the tide, in general, employing nine or ten minutes more in ebbing than in flowing. As the interval between one period of high water and the second following one is a lunar day, or 1d. 0h. 50m., the *retardation* in the time of high water from one day to another is 50m., in its mean state.

The time of high water is mainly dependent upon the position of the moon, being always, at any given place, about the same length of time after the moon's passage over the superior or inferior meridian. As to the length of the interval between the two periods, at different places, in the open sea it is only from two to three hours; but on the shores of continents, and in rivers, where the water meets with obstructions, it is very different at different places, and in some instances is of such length that the time of high water seems to precede the moon's passage.

530. Height of Tide. The height of the tide at high water is not always the same, but varies from day to day; and these variations have an evident relation to *the phases of the moon.* It is greatest soon after the syzigies; after which it diminishes and becomes the least soon after the quadratures.

The tides which occur near the syzigies, are called the *Spring Tides;* and those which occur near the quadratures are called the *Neap Tides.*

The highest of the spring tides is not that which has place nearest to new or full moon, but is in general the third following tide. In like manner the lowest of the neap tides is the third or fourth tide after the quadrature.

The spring tides are, in general, from once and a half to twice the height of the neap tides. At Brest, in France, the former rise to the height of 19.3 feet, and the latter only to 9.2 feet.

On the Atlantic coast of the United States the spring tides exceed the neap tides in the proportion of 3 to 2.

The tides are also affected by the declinations of the sun and moon: thus, the highest spring tides in the course of the year are those which occur near the equinoxes. The extraordinarily high tides which frequently occur at the equinoxes are, however, in part attributable to the equinoctial gales. Also, when the moon or the sun is out of the equator, the evening and morning tides differ somewhat in height. At Brest, in the syzigies of the summer solstice, the tides of the morning of the first and second day after the syzigy are smaller than those of the evening by 6.6 inches. They are greater by the same quantity in the syzigies of the winter solstice.

The distance of the moon from the earth has also a sensible influence upon the tides. In general, they increase and diminish as the distance diminishes and increases, but in a more rapid ratio.

531. Daily Retardation of High Water. The daily retardation of the time of high water varies with the phases of the moon. It is at its minimum towards the syzigies, when the tides are at their maximum; and at its maximum towards the quadratures, when the tides are at their minimum. It also varies with the distances of the sun and moon from the earth, and with their declinations.

532. Physical Theory of the Tides. The facts which have been detailed indicate that the tides are produced by the actions of the sun and moon upon the waters of the ocean; but in a greater degree by the action of the moon. To explain them, let us suppose at first that the whole surface of the earth is covered with water. We remark, in the first place, that it is not the whole attractive force of the moon or sun which is effective in raising the waters of the ocean, but the difference in the actions of each body upon the different parts of the earth; or, more precisely, that the phenomenon of the tides is a consequence of the inequality and non-parallelism of the attractive forces exerted by the moon, as well as by the sun, upon the different particles of the earth's mass. From this cause there results a diminution in the gravity of the particles of water at the surface, for a certain distance about the point immediately under the moon, and the point diametrically opposite to this, and an augmentation for a certain distance on the one side and the other of the circle 90° distant from these points, or of which they are the geometrical poles: in consequence of which the water falls about this circle and rises about these points. That the actions of the moon upon the different parts of the earth's mass are really unequal, is evident from the fact that these parts are at different distances from the moon. To show that the inequality will give rise to the results just noted, let us suppose that the circle *acbd* (Fig.118) represents the earth, and M the place of the moon; then *a* will be the point of the earth's surface directly under the moon, *b* the point diametrically opposite to this, and the right line *dc*, perpendicular to MO, will represent the circle traced on the earth's surface 90° distant from *a* and *b*. Now, the attraction of the moon for the general mass of the earth is the same as if the whole mass were concentrated at the centre O. But the centre of the earth is more distant from the moon than the point *a* at the surface. It follows, therefore, that a particle of matter situated at *a* will be drawn towards the moon with a proportionally greater force than the centre, or than the general mass of the earth. Its gravity or tendency towards the earth's centre will therefore be diminished by the amount of this excess. On the other hand, the centre is nearer to the moon than the point *b*. It is therefore attracted more strongly than a particle at *b*. The excess will be a force tending to draw the centre away from the particle; and the effect will be the same as if the particle were drawn away from the centre by the same force acting in the opposite direction. The result then is, that this particle has its gravity towards the earth's centre diminished, as well as the particle at *a*. If now we consider a particle at some point *t* near to *a*, the moon's action upon it (*tr*) may be considered as taking effect partially in the direction *tk* parallel to OM, and partially in the direction of the tangent or horizontal line *ts*. The component (*ts*) in the latter direction, will have no tendency to alter the gravity of the particle towards the earth's centre. The component (*sr*) in the direction *tk*, will obviously be less than the actual force of attraction *tr*; and the difference will be greater in proportion as the particle is more remote from *a*. This component will decrease gradually from *a*, where it is equal to the attractive force, while the attraction for the centre is less than for *a* by a certain finite difference: it is plain, therefore, that the component in question will be greater than the attraction for the centre, in the vicinity of the point *a*, and for a certain distance from it in all directions. The gravity of the particles will therefore be diminished for a certain distance from this point. In a similar manner it may be shown that it will also be diminished for a certain distance from the point *b*. Let us now

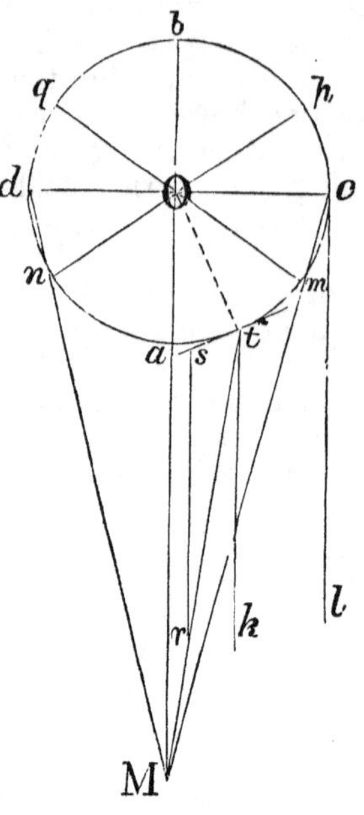

Fig. 118.

consider a particle at c, 90° from the points a and b. The attraction of the moon for it will take effect in the two directions cl and cO. The force in the latter direction alone will alter the gravity of the particle; and this, it is plain, will increase it. The same effect will extend to a certain distance from c in both directions.

A strict mathematical investigation would show that the gravity is diminished for a distance of 55° from a and b in all directions; and is augumented for a distance of 35° on each side of the circle dc, 90° distant from the points a and b. These distances are represented in the Figure.

This may be easily made out by means of the expression for the radial disturbing force of the sun in its action upon the moon (505), viz. $\frac{m}{a^3} \times y\,(1 - 3\cos^2\phi)$. If we consider m as denoting the mass of the moon, a the moon's distance from the earth's centre, y the distance of a particle of matter at some point t of the earth's surface from the earth's centre, and ϕ the angular distance or elongation (M$\mathrm{O}t$) of the same particle from the moon, as seen from the centre of the earth, it will express the change in the gravity of a particle at the earth's surface, produced by the moon's action. The points a and b will answer to conjunction and opposition, and the points c and d to the quadratures. Now we have already seen (505) that the gravity of the moon is increased at the quadratures, and for 35° on each side of them; and diminished at the syzigies, and 55° from them in both directions. It follows, therefore, that the same is true for particles of matter at the earth's surface.

In consequence of the earth's diurnal rotation, the parts of the surface, at which the rise and fall of the water will take place, will be continually changing. Were the entire rise and fall produced instantaneously, the points of highest water would constantly be the precise points in which the line of the centres of the moon and earth intersects the surface, and it would always be high water on the meridian passing through these points, both in the hemisphere where the moon is, and in the opposite one. On the west side of this meridian, the tide would be flowing; on the east side of it, it would be ebbing; and on the meridian at right angles to the same, it would be low water. But it is plain that the effects of the moon's action would not be instantaneously produced, and therefore that the points of highest water will fall behind the moon.

533. Comparative Effects of Sun and Moon. It is evident that the sun will produce precisely similar effects with the moon, and will raise a tide wave similar to the lunar tide wave, which will follow it in its diurnal motion.

To show that the effects of the sun are less in degree than those of the moon, let us take the general expression for the change of the moon's gravity, arising from the action of the sun, namely,

$$\frac{m}{a^3} \times y\,(1 - 3\cos^2\phi)\ \ldots\ (a).$$

From what we have seen in the previous article, this formula will serve to express the change in the gravity of a particle of matter upon the earth's surface, produced by the sun's action, if we take $m =$ the mass of the sun, as before, a $=$ its distance expressed in terms of the radius of the earth as unity, $y =$ the distance of the particle from the centre of the earth, and ϕ = its elongation from the sun, as seen from the earth's centre. If we designate the corresponding quantities for the moon by m' a' y, ϕ, we shall have for the change of the gravity of a particle, produced by the moon's action,

$$\frac{m'}{a'^3} \times y\,(1 - 3\cos^2\phi)\;.\;.\;.\;(b).$$

For particles at equal elongations from the sun and moon, we shall have ϕ the same in expressions (a) and (b), and y may be regarded as the same without material error. For such particles, then, the alterations of the gravity, produced by the sun and moon, will bear the same ratio to each other as the quantities $\frac{m}{a^3}$ and $\frac{m'}{a'^3}$. Now, if we give to m, m', a, a', their values, we shall find that the latter quantity is about $2\frac{1}{3}$ times greater than the former. Accordingly, the effect of the moon's action, at corresponding elongations of the particles, and therefore generally, is about $2\frac{1}{3}$ times greater than that of the sun.

534. Combined Effects of Sun and Moon. The actual tide will be produced by the joint action of the sun and moon, or it may be regarded as the result of the combination of the lunar and solar tide waves.

At the time of the syzigies, the action of the sun and moon will be combined in producing the tides, both bodies tending to produce high as well as low water at the same places. But at the quadratures they will be in opposition to each other, the one tending to raise the surface of the water where the other tends to depress it, and *vice versâ*. The tides should, therefore, be much higher at the syzigies than at the quadratures.

Between the syzigies and the quadratures the two bodies will neither directly conspire with each other, nor directly oppose each other, and tides of intermediate height will have place. The points of highest water will also, in the configuration supposed, neither be the vertices of the lunar nor of the solar tide wave, but certain points between them. This circumstance will occasion a variation in the length of the interval between the time of the moon's passage and the time of high water.

Spring and Neap Tides. The effect of the moon's action being to that of the sun's nearly as $2\frac{1}{3}$ to 1 (533), the spring tides will be to the neap tides nearly as $2\frac{1}{2}$ to 1. For, let x = the effect of the moon, and y = the effect of the sun: then the ratio of $x + y$ to $x - y$ will be the ratio of the heights of the spring and neap tides. Now,

$$x = 2.34y, \text{ and thus } \frac{x + y}{x - y} = \frac{2.34y + y}{2.34y - y} = 2.5.$$

We have already seen that the ratio obtained from observation is less than this.

The height of the joint tide, as well as the interval between the time of high water and that of the moon's meridian passage, will vary not only with the elongation of the moon from the sun, but also with *the distance and declination of the moon.* For, expressions (*a*) and (*b*) above given, show that the intensities of the moon's and sun's actions vary inversely as the cube of their distance; and the changes of the declinations of the two bodies must be attended with a change both in the absolute and relative situation of the vertices of the lunar and solar tide-waves.

COMPARISON OF THE THEORY OF THE TIDES WITH THE RESULTS OF OBSERVATION.

535. The laws of the tides, which should obtain on the hypothesis of the earth being entirely covered with water, are found to correspond only partially with those of the actual tides. The continents have a material influence upon the formation and propagation of the tide-wave. The actual phenomena of the tides have been carefully observed, for many years, at numerous points along the coast lines of continents and on the shores of islands; and the results of the observations have been subjected to a thorough discussion by several distinguished astronomers and physicists. As one result of the discussion the determination has been effected of a system of *Cotidal Lines;* that is, a set of lines connecting those places at which high tide occurs at the same instant, from hour to hour. A chart has been constructed showing, at intervals of 1^h, 2^h, 3^h, &c., after the meridian transit of the moon at Greenwich, the cotidal lines of the Southern, Atlantic, and Pacific Oceans. These lines show the varying form of the ridge of the tide-wave as it proceeds on its course, and by the greater or less distance between them the rate of propagation of the wave in different oceans and in different parts of the same ocean. Along the coasts they are, for the most part, constructed from actual observations, but their extensions into the open sea are mostly inferential.

536. Tide-wave of the Atlantic Ocean. By examining the chart of cotidal lines we learn that the floodtide of the Atlantic Ocean is, for the most part, produced by a *derivative tide-wave*, sent off from the great wave which, in the Southern Ocean, follows the moon in its diurnal motion around the earth. At 6 hours after the meridian transit of the moon at Greenwich, the derivative tide-wave stretches from the coast of Upper Guinea to the coast of Brazil, a little to the south of the narrowest part of the Atlantic. Three hours later it has advanced, by estimation, in mid-ocean, to about 24° of north latitude; and in 3 hours more, or 12 hours after the meridian transit of the moon at Greenwich, it has reached the Atlantic coast of the United States. It advances more rapidly in the open sea than along the coasts, where the depth of the water is less. It is therefore convex towards the north. Thus, at the hour just mentioned, it stretches nearly parallel to the general trend of our Atlantic coast, along its whole extent into the northern Atlantic, and there curves around to the south-east, so as to strike, at its eastern end, the N. W. coast of Africa (lat. 23°). The same wave does not reach the coast of Spain until more than two hours later.

537. Nature and Velocity of the Tide-wave. The tidal-wave is of the nature of a *wave of translation.* In this form of wave there is no oscillation proper; but the particles of the fluid, in a cross section perpendicular to the line of propagation, by the transit of the wave are raised, transferred forward, and brought to rest in the direction of the motion in a new place; with the same extent of transference of each particle throughout the whole depth of the wave. Whereas in ordinary *oscillatory waves*, such as those caused by the wind, the individual particles oscillate in vertical circles, or ellipses, and return to their original position. A wave of translation travels with a velocity equal to that acquired by a heavy body in falling freely by gravity through a height equal to half the mean depth of the fluid, reckoned from the top of the wave to the bottom of the channel. Its velocity is therefore directly proportional to the square root of the depth of the fluid. The rate of propagation of an oscillatory wave, on the other hand, is independent of the depth, and varies only with the breadth of the wave.

The moon tends to draw the wave which it raises along with it in its diurnal course, at the rate of 1,000 miles per hour at the equator; but it appears that the tidal wave actually travels at a much less rapid rate. Setting out from the Eastern Pacific, where it lags about 2 hours behind the moon, it travels westward in about 12 hours to New Zealand. From thence to the Cape of Good Hope, passing south of Australia, it occupies 17 hours, and has an average velocity of about 470 miles per hour. From the Cape of Good Hope the portion of the wave that passes northward into the Atlantic traverses the distance to the coast of the United States in about 11 hours; which is at the average rate of 565 miles per hour. The tide-wave accordingly does not reach our Atlantic coast until about 40 hours after it originated in the South-eastern Pacific. The average velocity of 565 miles in the South and North Atlantic, answers to a depth of 21,500 feet, or about 4 miles. The average velocity in mid-ocean is greater than this, and answers to a greater depth.

The velocity of the tide-wave becomes rapidly reduced after the wave strikes the shallow waters of the coast, to 100 miles per hour; or even less than 50 miles per hour in bays and sounds. As a necessary consequence the breadth of the wave diminishes with its velocity. At a velocity of 565 miles per hour it has a breadth of 7,000 miles. When the velocity is reduced to 100 miles per hour the breadth is only 1,240 miles.

TIDES OF THE ATLANTIC COAST OF THE UNITED STATES.

538. General Phenomena. The phenomena of the tides as they occur along the entire coast line of the United States, have been carefully deduced by the late Superintendent of the Coast Survey, from the systematic tidal observations carried on in connection with the Survey. The following are the more important general results obtained from the discussion:

1. The cotidal lines, in the vicinity of the Atlantic coast, are nearly parallel to the general trend of the coast. The ridge of the tide-wave, as it approaches the coast, is therefore nearly parallel to the coast line. This wave, when it reaches the most prominent points of the coast, has a mean height of about 2 feet above the lowest point of the ebb-wave, or mean low-water level.

2. The coast is physically divided, by projecting headlands, into three great bays, each of which has its particular system of cotidal lines, running nearly parallel to the shore. These bays may be designated as the Southern, Middle, and Eastern Bays. The Southern Bay lies between Cape Florida and Cape Hatteras; the Middle Bay between Cape Hatteras and Nantucket (eastern end); and the Eastern Bay between Nantucket and Cape Sable (Nova Scotia). The latter is supposed to be a portion of a greater bay, from Nantucket to Cape Race (Newfoundland). In the Southern Bay, the mean rise and fall, or range of the tides along the shores, increases from about 2 feet at the capes to 7 feet at Port Royal, at the head of the bay. In the Middle Bay, the range increases from 2 feet to nearly 5 feet at Sandy Hook and Cape May. In the Eastern Bay the tides are more complex, owing to greater irregularities in the shore line, and the influence of shoals. The heights increase rapidly from Nantucket to Cape Cod; the mean range being 2 feet at Nantucket and 9.2 feet at Provincetown. At Cape Ann (the northern cape of Massachusetts Bay) it is about the same. From Cape Ann northward to Portsmouth there is a decrease of about half a foot in the mean range of the tides. From thence, following the shore line towards the northeast, it increases at an augmenting rate until at the entrance of the Bay of Fundy the tide rises, on the average, 18 feet above low water.

539. Tides of Inner Bays. The tides of Delaware Bay, New York Bay, and Narragansett and Buzzard Bays, present, on a smaller scale, the same phenomena of increase in the height of the tides in ascending, as the three great bays, or undulations of the coast. On the contrary, in Chesapeake Bay, which widens and changes direction at right angles immediately from the entrance, the tides diminish in height, as a general rule, in going up the bay.

The tide-wave, on entering Massachusetts Bay, increases somewhat, viz., from 9 feet above low water at the entrance to 10 feet at Plymouth and Boston.

In the *Bay of Fundy*, the tides rise to a much greater height than on any other part of the Atlantic Coast. At St. Johns, N. B., the mean rise and fall of the tide is 19.3 feet; and at Shadwood Point, at the head of the bay, no less than 36 feet. The ordinary spring tides attain, at the latter place, to the height of 50 feet. Special tides have been known to rise 20 feet higher. This remarkable accumulation of the tidal waters results from the great contraction in the width of the bay or channel into which the ascending wave is forced.

450. Tides of Channels. In channels peculiar tides occur, in consequence of the meeting of the waves which enter the channels at their two extremities. Where the two flood waves meet in the same state, a tide equal to the sum of their two heights is produced by their superposition. At other points the tides are variously modified by the interference of the waves.

Tides in Sounds present similar peculiarities.

541. Tides of Long Island Sound. The great tidal wave from the Atlantic enters the Sound between Point Judith and Montauk Point; and another portion of this wave enters New York Bay, and passing through Hell Gate, meets the wave propagated through the Sound from the eastward. The point of meeting of the crests of the two waves is off Sands' Point, at the head of the Sound. At Montauk Point the mean height of the tide-wave, above low water, is 2 feet, and at Sandy Hook 4.8 feet. At Sands' Point it is 7.7 feet; exceeding the sum of the heights at the two entrances by nearly 1 foot, owing to the narrowing of the Sound. The mean range of the tide declines in both directions from Sands' Point. At Bridgeport it is 6.6 feet; at New Haven 5.8 feet; at New London and Stonington between 2½ feet, and 2¾ feet; and at Point Judith 3 feet.

The tide is propagated from Montauk Point to the head of the Sound in 3 hours. It travels from Fisher's Island to Sands' Point, 95 miles, in 2h. 1m.; or at the average rate of 47 miles per hour. This agrees approximately with the velocity as theoretically computed from the soundings taken by the Coast Survey, according to the law of propagation of a wave of translation (537). At Fisher's Island it is about 60 miles, and becomes reduced to 30 miles at the head of the Sound, where the depth of the water is less. In the East River the rate of propagation of the tide is only about 7½ miles per hour.

Owing to the retardation of the tide-wave in the shallow waters near the Connecticut shore, it is nearly parallel to the shore, from the head of the Sound to a distance of some 20 miles east of New Haven harbor. Accordingly high water occurs at about the same hour along this extent of the shore. Farther to the east, the line of the tide-wave is inclined to the shore line, and the tides occur earlier.

542. *Tidal Currents.* The currents produced by the tides in the shallow waters of bays, sounds, and rivers, are not to be confounded with the transmission of the tide-wave. Their velocity is but a few miles per hour; and the turn of the current, or tide-stream, does not in general correspond to the turn of the tide, and may occur at quite a different hour. For example, at Montauk Point the ebb-stream does not begin until half ebb-tide, and in New York Bay it begins at one-sixth of the ebb tide.

Tidal currents owe their origin to the resistance opposed by shallow waters, and contracted channels, to the free propagation of the tide-wave, and to differences of hydrostatic level. They have the greatest velocity in narrow channels, as in the Race off Fisher's Sound, and in Hell Gate. About the time of the turn of the tide, at the head of the Sound, there is a certain interval of *slackwater* there. After the tide-wave begins to move in the opposite direction, the accumulative effect of the resistances determines, in a certain interval of time, a sensible current, which shows itself first at the surface and in-shore, but soon becomes general. In mid-channel, throughout the Sound, the outward motion of the water commences shortly after high water at the head of the Sound, and evidently depends upon it.

A similar, but still more striking fact is observed in the *Irish Channel.* The turn of the stream, whether flood or ebb, is *simultaneous* throughout the entire length of the channel. It is coincident with the time of high or low water at Morecambe Bay, north of Liverpool, where the tides coming round the extremities of Ireland finally meet. The times of slackwater throughout the channel, therefore, correspond with the times of high and low water at Morecambe Bay. In the Irish Channel there are two spots, in one of which the stream runs with considerable velocity without the tide either rising or falling, while in the other the water rises and falls from sixteen to twenty feet without having any visible horizontal motion at its surface.

The average maximum *drift of the current* in Long Island Sound, is 2.2 knots per hour. The average maximum current velocity opposite the west end of Fisher's Island is nearly 4¾ knots per hour; and at Hell Gate nearly 6 knots. In New York harbor it is 3.7 knots, and in the Bay 3 knots. The point of meeting of the two flood streams in the East River, is a little to the east of Throgs' Neck. To the east and west of that point, both the flood and ebb streams run in opposite directions.

The mean duration of the flood stream at different points of Long Island Sound varies between 4¼ hours and 7¾ hours. The corresponding limits for the ebb stream are 5h. and 8¼h. The mean duration of slackwater varies between 0m. and 45m. It is at most places less than 10m. The duration of the ebb or flood stream, differs as much as ¾ of an hour in successive tides; but commonly not more than 10m. The set of the currents is ordinarily nearly parallel to the shore.

543. Tides of Rivers. The tide-wave that enters the mouth of a river is propagated according to the same laws as a wave that comes in at the entrance of a sound, or channel. The velocity varies with the depth of water; and the height of the tide increases where the river contracts, and decreases where it widens. Thus, in a tidal river of considerable length, the tide may have various heights at different points. The ascending flood tide may also be encountered by the descending ebb tide. On the Hudson the tide rises at West Point, 55 miles from New York, 2.7 feet; at Tivoli, nearly 100 miles from New York, 4 feet; and at Albany, 2.3 feet.

In the shallow parts of rivers, the tide-wave becomes converted into a tidal current, by which alone the tide is transmitted. In rivers the duration of the ebb tide is considerably longer than that of the flood. Thus, at Philadelphia and Richmond, the ebb continues 2½ hours longer than the flood tide.

TIDES OF THE PACIFIC COAST.

544. Cotidal Lines. The cotidal lines of the Pacific coast of the United States are approximately parallel to the coast. Thus, high tide occurs at about the same hour from San Francisco to Vancouver's Island. South of San Francisco the tide-wave arrives at an earlier hour; at the southern extremity of California, about 2½ hours earlier.

545. Diurnal Inequality. The tides of the Pacific coast

are remarkable for *the great inequality that prevails between the heights of two successive tides*, as measured from the high water mark of each tide to the next succeeding low water mark. The difference of level of the two successive high tides is less conspicuous, but quite marked. The differences are greater for the ebb than for the flood tides. These diurnal inequalities increase with the moon's declination, north or south; and vanish entirely when the moon is in the equator. When the moon's declination is north, the highest of the two high tides of the twenty-four hours occurs at San Francisco about 11½ hours after the moon's superior transit; and when the declination is south, the lowest of the two high tides occurs about this interval after the transit. When the moon has its greatest declination the mean range of the highest tide is nearly 7 feet, and of the lowest tide from 1½ ft. to 3 ft. The lowest tide sometimes amounts to only two or three inches.

According to Professor Bache, the tides that occur on the western coast, near the maximum of the moon's declination and for several days on each side of it, result from the interference of a semi-diurnal and diurnal wave, which at the maximum of each are nearly equal in magnitude, the crest of the diurnal wave being at that period about eight hours in advance of that of the semi-diurnal wave. This diurnal wave exists only when the moon has a considerable declination.

On the *Atlantic coast* the corresponding inequality at the time of the moon's greatest declination, is a small fraction of the height of the tide, and is generally not more than one foot. A similar remark may be made of the tides of the coast of Europe.

TIDES OF THE GULF OF MEXICO.

546. On the northern coast of the Gulf of Mexico, from Florida westward, there is but one tide in the 24 hours, and the mean range of this tide is only from 1 foot to 1½ feet. The second tide is doubtless obliterated by the interference of the semi-diurnal flood-tide with a diurnal ebb-tide; as happens approximately on the Pacific coast (545). For some three to five days, about the time when the moon is crossing the equator, when the diurnal inequality should vanish, from the absence of the diurnal wave (545), there are generally two tides at the same places on the coast, the rise and fall being quite small. The greatest rise and fall of the single day-tide occurs when the moon's declination is the greatest.

The small height of the tides in the Gulf of Mexico is attributable chiefly to the fact that the width of the gulf is three or four times greater than that of the two channels through which the tide-wave enters it.

TIDES OF THE MEDITERRANEAN.

547. The average height of the tide in the Mediterranean is said not to exceed 1½ feet, though at some ports, as Tunis and Venice, it sometimes amounts to 3 or 4 feet. The Mediterranean is of sufficient extent for the sun and moon to produce a sensible tide by their direct action. A derivative tide-wave, from the Atlantic Ocean, should also enter the Straits of Gibraltar, and spread out laterally as it advances; but the ebb and flow from this cause is said to be slight.

TIDES OF INLAND SEAS AND LAKES.

548. Lakes and inland seas have no perceptible tides, or only very small tides, for the reason that their extent is not sufficient to admit of any sensible inequality of gravity, as the result of the action of the moon (532). A tide of nearly 2 inches has been detected at Chicago, on the southwestern shore of Lake Michigan.

TIDES OF THE COAST OF EUROPE.

549. The tide-wave advancing from the south, makes a considerable angle with the coast of Europe, and thus the tide occurs continually later in following the coast from the Straits of Gibraltar northward; and along its entire extent from two to twelve hours later than the corresponding tides on the coast of the United States. Similar varieties of tidal phenomena occur on either coast.

The highest tides prevail in the Bristol Channel, and the Bay of St. Malo, on the northwest coast of France. At the head of the Bristol Channel, and of the Bay of St. Malo, the spring tides sometimes rise to the height of 50 feet. The mean range of spring tides is 26 feet at Liverpool, nearly 13 feet at Portsmouth, and about 20 feet at London Docks. On the coast of France, the height of the tides at different ports falls approximately between the same limits as on the coast of England.

The lowest tides occur on the eastern coast of Ireland, to the north of the entrance to St. George's Channel. At Courtown, about 30 miles north of Tuskar, there is scarcely any rise or fall of the water. From that point the height of the tide increases about equally in every direction, from 0 to 15 feet on the opposite coast. The remarkably low tides at that locality result from the fact that the tide stream is diverted by a promontory at the entrance of the channel to the opposite shore.

ESTABLISHMENT OF THE PORT.—TIDE-TABLES.

550. The interval between the time of the moon's crossing the meridian and the time of high water at a given place is nearly constant. It varies only between moderate assignable limits. The mean interval on the days of new and full moon is called the *establishment* of the port. The average of the intervals during a month's tides, is called the *mean*, or *correct establishment*. The mean establishment of Boston is 11h. 27m.; of New Haven 11h. 16m.; of New York 8h. 13m.; of Charleston, S. C., 7h. and 26m.; of San Francisco 12h. 12m.

551. Calculation of Time of High Water. When the mean establishment of a port is known, the time of high water on any day may be approximately determined. The hour of transit of the moon on the given day is to be taken from the Nautical Almanac and added to the mean establishment; the result will be the time of high water. If the time thus determined falls in the succeeding day, half a lunar day (12h. 25m.) is to be subtracted, as this is the mean interval between two successive tides.

On the day of new or full moon, the time of the next high water after noon, will be approximately equal to the establishment of the port.

In the annual Coast Survey Reports a table is published, giving the interval between the time of the moon's transit and the time of high water for different hours of transit, and for the principal ports on the U. S. coast. If the time of the moon's transit on any day be obtained from the Nautical Almanac, the interval corresponding to this time in the table, added to the time of transit, will give more accurately the time of high water.

552. *A tide table for the coast of the United States*, is published in the same Reports, giving for numerous points of the coast the mean values of the interval between the time of the moon's transit and time of high water, the rise and fall of the tides, the rise and fall of the spring and neap tides, the duration of flood and of ebb tide, and the duration of the stand, or the period of time during which the surface of the water neither rises nor falls. A table is also given showing, for various ports, the rise and fall of tides corresponding to different hours of the moon's transit; from which, by taking the time of transit for any day from the Almanac, the corresponding rise and fall of the tide may be obtained for any of the ports mentioned in the table

PART III.

ASTRONOMICAL PROBLEMS.

EXPLANATIONS OF THE TABLES.

THE Tables which form a part of this work, and which are employed in the resolution of the following Problems, consist of Tables of the Sun, Tables of the Moon, Tables of the Mean Places of some of the Fixed Stars, Tables of Corrections for Refraction, Aberration, and Nutation, and Auxiliary Tables.

The Tables of the Sun, which are from XVII to XXXIV, inclusive, are, for the most part, abridged from Delambre's Solar Tables. The mean longitudes of the sun and of his perigee for the beginning of each year, found in Table XVIII, have been computed from the formulæ of Prof. Bessel, given in the Nautical Almanac of 1837. The Table of the Equation of Time was reduced from the table in the Connaissance des Tems of 1810, which is more accurate than Delambre's Table, this being in some instances liable to an error of 2 seconds. The Table of Nutation (Table XXVII) was extracted from Francœur's Practical Astronomy. The maximum of nutation of obliquity is taken at 9″.25. The Tables of the Sun will give the sun's longitude within a fraction of a second of the result obtained immediately from Delambre's Tables, as corrected by Bessel. The Tables of the Moon, which are from XXXIV to LXXXV, inclusive, are abridged and computed from Burckhardt's Tables of the Moon. To facilitate the determination of the hourly motions in longitude and latitude, the equations of the hourly motions have all been rendered positive, like those of the longitude. Some few new tables have been computed for the same purpose. The longitude and hourly motion in longitude will very rarely differ from the results of Burckhardt's Tables more than 0″.5, and never as much as 1″. The error of the latitude and hourly motion in latitude will be still less. The other tables have been taken from some of the most approved modern Astronomical Works. (For the principles of the construction of the Tables, see Note 1., Appendix.)

Before entering upon the explanation of each of the tables, it will be proper to define a few terms that will be made use of in the sequel.

The given quantity with which a quantity is taken from a table, is called the *Argument* of this quantity.

The angular arguments are expressed in some of the tables according to the sexagesimal division of the circle. In others, they are given in parts of the circle supposed to be divided into 100 1000, or 10000, &c., parts.

Tables are of *Single* or *Double Entry*, according as they contain one or two arguments. The *Epoch* of a table is the instant of time for which the quantities given by the table are computed. By the *Epoch* of a quantity, is meant the value of the quantity found for some chosen epoch, from which its value at other epochs is to be computed by means of its known rate of variation.

Table I, contains the latitudes and longitudes from the meridian of Greenwich, of various conspicuous places in different parts of the earth. The longitudes serve to make known the time at any one of the places in the table, when that at any of the others is given. The latitude of a place is an important element in various astronomical calculations.

Table II, is a table of the Elements of the Orbits of the Planets with their secular variations, which serve to make known the elements at any given epoch different from that of the table. From these the elliptic places of the planets at the given epoch may be computed. Table III., is a similar table for the Moon.

Table II. (*a*) gives the mean distances, &c., of the Planetoids.

Tables IV, V, VI, VII, require no explanation.

Table VIII, gives the mean Astronomical Refractions; that is, the refractions which have place when the barometer stands at 30 inches, and the thermometer of Fahrenheit at 50°.

Table IX, contains the corrections of the Mean Refractions for + 1 inch in the barometer, and — 1° in the thermometer, from which the corrections to be applied, at any observed height of the barometer and thermometer, are easily derived.

Table X, gives the Parallax of the Sun for any given altitude on a given day of the year; for reducing a solar observation made at the surface of the earth to what it would have been, if made at the centre.

Table XI, is designed to make known the Sun's Semi-diurnal Arc, answering to any given latitude and to any given declination of the sun; and thus the time of the sun's rising and setting, and the length of the day.

Table XII, serves to make known the value of the Equation of Time, with its essential sign, which is to be applied to the apparent time to convert it into the mean. If the sign of the equation taken from the table be changed, it will serve for the conversion of mean time into apparent. This table is constructed for the year 1840.

Table XIII, is to be used in connection with Table XII, when the given date is in any other year than 1840. It furnishes the Secular Variation of the Equation of Time, from which the proportional part of its variation in the interval between the given date and the epoch of Table XII is easily derived.

Table XIV, contains certain other Corrections to be applied to the equation of time taken from Table XII, when its exact value to within a small fraction of a second, is desired.

Table XV, gives the Fraction of the Year corresponding to each date. This table is useful when quantities vary by known and uniform degrees, in deducing their values at any assumed time from their values at any other time.

Table XVI, is for converting Hours, Minutes, and Seconds into decimal parts of a Day.

Table XVII, is for converting Minutes and Seconds of a degree into the decimal division of the same. It will also serve for the conversion of minutes and seconds of time into decimal parts of an hour.

The last two tables will be found frequently useful in arithmetical operations

Table XVIII, is a table of Epochs of the Sun's Mean Longitude, of the Longitude of the Perigee, and of the Arguments for finding the small equations of the Sun's place. They are all calculated for the first of January of each year, at mean noon on the meridian of Greenwich. Argument I. is the mean longitude of the Moon minus that of the Sun; Argument II. is the heliocentric longitude of the Earth; Argument III. is the heliocentric longitude of Venus; Argument IV. is the heliocentric longitude of Mars; Argument V. is the heliocentric longitude of Jupiter, Argument VI. is the mean anomaly of the Moon; Argument VII. is the heliocentric longitude of Saturn; and Argument N is the supplement of the longitude of the Moon's Ascending Node. Argument I. is for the first part of the equation depending on the action of the Moon. Arguments I. and VI. are the arguments for the remaining part of the lunar equation. Arguments II. and III. are for the equation depending on the action of Venus; Arguments II. and IV. for the equation depending on the action of Mars; Arguments II. and V. for the equation depending on the action of Jupiter; and Arguments II. and VII. for the equation depending on the action of Saturn. Argument N is the argument for the Nutation in longitude: it is also the argument for the Nutation in right ascension, and of the obliquity of the ecliptic.

Table XIX, shows the Motions of the Sun and Perigee, and the variations of the arguments, in the interval between the beginning of the year and the first of each month.

Table XX, shows the Motions of the Sun and Perigee, and the variations of the arguments from the beginning of any month to the beginning of any day of the month; also the same for Hours.

Table XXI, gives the Sun's Motions for Minutes and Seconds. Tables XVIII to XXI, inclusive, make known the mean longitude of the Sun from the mean equinox, at any moment of time.

Table XXII, Mean Obliquity of the Ecliptic for the beginning of each year contained in the table. It is found for any intermediate time by simple proportion.

Tables XXIII, and XXIV, furnish the Sun's Hourly Motion and Semi-diameter.

Table XXV, is designed to make known the Equation of the Sun's Centre. When the equation has the negative sign, its supplement to 12s. is given: this is to be added along with the other equations of longitude, and 12s. are to be subtracted from the sum.

The numbers in the table are the values of the equation of the centre, or of its supplement, diminished by 46″.1. This constant is subtracted from each value, to balance the different quantities added to the other equations of the longitude, in order to render them affirmative. The epoch of this table is the year 1840.

Table XXVI, gives the Secular Variation of the Equation of the Sun's Centre, from which the proportional part of the variation in the interval between the given date and the year 1840, may be derived.

Table XXVII, is for the Nutation in Longitude, Nutation in Right Ascension, and Nutation of the Obliquity of the Ecliptic. The nutation in longitude and nutation in right ascension, serve to transfer the origin of the longitude and right ascension from the mean to the true equinox. And the nutation of obliquity serves to change the mean into the true obliquity.

Tables XXVIII to XXXIII, inclusive, give the Equations of the Sun's Longitude, due respectively to the attractions of the Moon, Venus, Jupiter, Mars, and Saturn.

Table XXXIV, is for the variable part of the Sun's Aberration. The numbers have all been rendered positive by the addition of the constant 0″.3.

Table XXXV, contains the Epochs of the Moon's Mean Longitude, and of the Arguments of the equations used in determining the True Longitude and Latitude of the Moon. They are all calculated for the first of January of each year, at mean noon on the meridian of Greenwich. The Argument for the Evection is diminished by 30′; the Anomaly by 2°; the Argument for the Variation by 9°, and the mean longitude by 9° 45′; and the Supplement of the Node is increased by 7′. This is done to balance the quantities which are added to the different equations in order to render them affirmative.

Tables XXXVI to XL, inclusive, give the Motions of the Moon, and the variations of the arguments, for Months, Days, Hours, Minutes, and Seconds; and, together with Table XXXV, are for finding the Moon's Mean Longitude and the Arguments, at any assumed moment of time.

Tables XLI to LIII, inclusive, give the various Equations of the Moon's Longitude. It is to be observed with respect to Table XLI, that the right hand figure of the argument is supposed to be dropped. But when the greatest attainable accuracy is desired, it can be retained, and a cipher conceived to be written after the numbers in the columns of Arguments in the table. In Tables L, LI, LII, and LV, the degrees will be found by referring to the head or foot of the column. (See Problem II., note 2.)

Table LIV is for the Nutation of the Moon's Longitude.

Tables LV to LIX, inclusive, are for finding the Latitude of the Moon.

Tables LX to LXIII, inclusive, are for the Equatorial Paral lax of the Moon.

Table LXIV furnishes the Reductions of Parallax and of the Latitude of a Place. The reduction of parallax is for obtaining the parallax at any given place from the equatorial parallax. The reduction of latitude is foı reducing the true latitude of a place, as determined by observation, to the corresponding latitude on the supposition of the earth being a sphere. The ellipticity to which the numbers in the table correspond is $\frac{1}{300}$.

Tables LXV and LXVI, Moon's Semi-diameter, and the Augmentation of the Semi-diameter depending on the altitude.

Tables LXVII to LXXXV, inclusive, are for finding the Hourly Motions of the Moon in Longitude and Latitude.

Table LXXXVI, Mean New Moons, and the Arguments for the Equations for New and Full Moon, in January. The time of mean new moon in January of each year has been diminished by 15 hours, the sum of the quantities which have been added to the equations in Table LXXXIX. Thus, 4h. 20m. has been added to equation I.; 10h. 10m. to equation II.; 10m. to equation III.; and 20m. to equation IV.

Tables LXXXVII and LXXXVIII, are used with the preceding in finding the Approximate Time of Mean New or Full Moon in any given month of the year.

Table LXXXIX furnishes the Equations for finding the Approximate Time of New or Full Moon.

Table XC contains the Mean Right Ascensions and Declinations of 50 principal Fixed Stars, for the beginning of the year 1840, with their Annual Variations.

Table XCI is for finding the Aberration and Nutation of the Stars in the preceding catalogue.

Table XCII contains the Mean Longitudes and Latitudes of some of the principal Fixed Stars, for the beginning of the year 1840, with their Annual Variations.

Tables XCIII, XCIV, XCV, Second, Third, and Fourth Differences. These tables are given to facilitate the determination, from the Nautical Almanac, of the moon's longitude or latitude for any time between noon and midnight.

Table XCVI, Logistical Logarithms. This table is convenient in working proportions, when the terms are minutes and seconds, or degrees and minutes, or hours and minutes,—especially when the first term is 1h. or 60m

To find the logistical logarithm of a number composed of minutes and seconds, or degrees and minutes, of an arc; or of minutes and seconds, or hours and minutes, of time.

1. If the number consists of minutes and seconds, at the top of the table seek for the minutes, and in the same column opposite the seconds in the left-hand column will be found the logistical logarithm.

2. If the number is composed of hours and minutes, the hours must be used as if they were minutes, and the minutes as if they were seconds.

3. If the number is composed of degrees and minutes, the degrees must be used as if they were minutes, and the minutes as if they were seconds.

To find the logistical logarithm of a number less than 3600.

Seek in the second line of the table from the top the number next less than the given number, and the remainder, or the complement to the given number, in the first column on the left: then in the column of the first number, and opposite the complement, will be found the logistical logarithm of the sum. Thus, to obtain the logarithm of 1531, we seek for the column of 1500, and opposite 31 we find 3713.

PROBLEM I.

To work, by logistical logarithms, a proportion the terms of which are degrees and minutes, or minutes and seconds, of an arc; or hours and minutes, or minutes and seconds, of time.

With the degrees or minutes at the top, and minutes or seconds at the side, or if a term consists of hours and minutes, or minutes and seconds, with the hours or minutes at the top, and minutes or seconds at the side, take from Table XCVI. the logistical logarithms of the three given terms; add together the logistical logarithms of the second and third terms and the arithmetical complement of that of the first term, rejecting 10 from the index.* The result will be the logistical logarithm of the fourth term, with which take it from the table.

Note 1. The logistical logarithm of 60′ is 0.

Note 2. If the second or third term contains tenths of seconds, (or tenths of minutes, when it consists of degrees and minutes,) and is less than 6′, or 6°, multiply it by 10, and employ the logarithm of the product in place of that of the term itself. The

---

* Instead of adding the arithmetical complement of the logarithm of the first term, the logarithm itself may be subtracted from the sum of the logarithms of the other two terms.

result obtained by the table, divided by 10, will be the fourth term of the proportion, and will be exact to tenths.

Note 3. If none of the terms contain tenths of minutes or seconds, and it is desired to obtain a result exact to tenths, diminish the index of the logistical logarithm of the fourth term by 1, and cut off the right-hand figure of the number found from the table, for tenths.

Exam. 1. When the moon's hourly motion is 30′ 12″, what is its motion in 16m. 24s.?

As	60m.	0
:	30′ 12″	2981
: :	16m. 24s.	5633
:	8′ 15″	8614

2. If the moon's declination change 1° 31′ in 12 hours, what will be the change in 7h. 42m.?

As	12h.	ar. co. 9.3010
:	1° 31′	1.5973
: :	7h. 42m.	8917
:	0° 58′	1.7900

3. When the moon's hourly motion in latitude is 2′ 26″.8, what is its motion in 36m. 22s.?

2′ 26″.8
60

146″.8
10

	As	60m.	0
1468	:	1468″	3896
	: :	36m. 22s.	2174
	:	890″	6070

Ans. 1′ 29″ 0.

4. When the sun's hourly motion in longitude is 2′ 28″, what is its motion in 49m. 11s.? Ans. 2′ 1″.

5. If the sun's declination change 16′ 33″ in 24 hours, what will be the change in 14h. 18m.? Ans. 9′ 52″.

6. If the moon's declination change 54″.7 in one hour, what will be the change in 52m. 18s.? Ans. 47″.7.

PROBLEM II.

To take from a table the quantity corresponding to a given value of the argument, or to given values of the arguments of the table

Case 1. *When quantities are given in the table for each sign and degree of the argument.*

With the signs of the given argument at the top or bottom, and the degrees at the side, (at the left side, if the signs are found at the top; at the right side, if they are found at the bottom,) take out the corresponding quantity. Also take the difference between this quantity and the next following one in the table, and say, 60′ : this difference : : odd minutes and seconds of given argument : a fourth term. This fourth term, added to the quantity taken out, when the quantities in the table are increasing, but subtracted when they are decreasing, will give the required quantity.

Note 1. When the quantities change but little from degree to degree of the argument, the required quantity may often be estimated without the trouble of stating a proportion.

Note 2. In some of the tables the degrees or signs of the quantity sought, are to be had by referring to the head or foot of the column in which the minutes and seconds are found. (See Tables L, LI, LII, and LV.) The degrees there found are to be taken, if no horizontal mark intervenes; otherwise, they are to be increased or diminished by 1°, or 2°, according as one or two marks intervene. They are to be increased, or diminished, according as their number is less or greater than the number of degrees at the other end of the column.

Note 3. If, as is the case with some of the tables, the quantities in the table have an algebraic sign prefixed to them, neglect the consideration of the sign in determining the correction to be applied to the quantity first taken out, and proceed according to the rule above given. The result will have the sign of the quantity first taken out. It is to be observed, however, that if the two consecutive quantities chance to have opposite signs, their numerical sum is to be taken instead of their difference; also that the quantity sought will, in every such instance, be the numerical difference between the correction and the quantity first taken out, and, according as the correction is less or greater than this quantity, is to be affected with the same or the opposite sign.

Exam. 1. Given the argument $7^{s.}$ 6° 24′ 36″, to find the corresponding quantity in Table L.

$7^{s.}$ 6° gives 0° 43′ 17″.4.

The difference between 0° 43′ 17″.4 and the next following quantity in the table is 1′ 7″.3.

60′ : 1′ 7″.3 : : 24′ 36″ : 27″.6.*

* The student can work the proportion, either by the common method, or by logistical logarithms, as he may prefer. In working this and all similar proportions by the arithmetical method, the seconds of the argument may be converted into the equivalent decimal part of a minute by means of Table XVII, (using the seconds as if they were minutes.) It will be sufficient to take the fraction to the nearest tenth

From	0° 43′ 17″.4
Take	27 .6
	0 42 49 .8

2. Given the argument $2^{s.}$ 18° 41′ 20″, to find the corresponding quantity in Table XXV.

$2^{s.}$ 18° gives 1° 52′ 32″.5.

The difference between 1° 52′ 32″.5 and the next following quantity in the table is 21″.8.

60′ : 21″.8 : : 41′ 20″ : 15″.0.

To	1° 52′ 32″.5
Add	15 .0
	1 52 47 .5

3. Given the argument $9^{s.}$ 2° 13′ 33″, to find the corresponding quantity in Table XII.

$9^{s.}$ 2° gives 29.8s.

The arithmetical sum of 29.8s. and the next following quantity in the table is 30.4s.

60′ : 30.4s. : : 13° 33′ : 6.9s.

From	29.8s.
Take	6.9
	22.9s.

Ans. — 22.9s

4. Given the argument $5^{s.}$ 8° 14′ 52″, to find the corresponding quantity in Table LII. Ans. 12′ 36″.0.

5. Given the argument $11^{s.}$ 11° 23′ 10″, to find the corresponding quantity in Table LVI. Ans. 11′ 48′ .0.

6. Given the argument $0^{s.}$ 26° 20′, to find the corresponding quantity in Table XII. Ans. — 41^{s}.0.

Case 2. *When the argument changes in the table by more or less than 1°; or when it is given in lower denominations than signs.*

Take out of the table the quantity answering to the number in the column of arguments next less than the given argument. Take the difference between this quantity and the next following one, and also the difference of the consecutive values of the argument inserted in the table, and say, difference of arguments : difference of quantities : : excess of the given argument over the value next less in the table : a fourth term. This fourth term applied to the quantity first taken out, according to the rule given in the preceding case, will give the quantity sought.

Note. In some of the tables the columns entitled Diff. are made up of the differences answering to a difference of 10′ in the argument. In obtaining quantities from these tables, it will be found more convenient to take for the first and second terms of the pro portion, respectively, 10′, and the difference furnished by the table and work the proportion by the arithmetical method. (See note at bottom of page 268.)

Exam. 1. Given the argument 0ˢ· 24° 42′ 15″, to find the corresponding quantity in Table LI.

0ˢ· 24° 30′ gives 9° 47′ 14″.3.

The difference between 9° 47′ 14″.3 and the next following quantity = 3 × 63″.0 = 189″.0. The argument changes by 30′. And the excess of 0ˢ· 24° 42′ 15″ over 0ˢ· 24° 30′, is 12 15″. Thus,

30′ : 189″.0 : : 12′ 15″ : 77″.2.

But the correction may be found more readily by the following proportion:

10′ : 63″.0 : : 12′.25 : 77″.2.

To	9° 47′ 14″.3
Add	77 .2
	9 48 31 .5

2. Given the argument 1° 12′, to find the corresponding quantity in Table VIII.

1° 10′ gives 23′ 13″,

and 5′ : 33″ : : 2′ : 13″ the correction.

From	23′ 13″
Take	13
	23 0

3. Given the argument 6ˢ· 6° 7′ 23″, to find the corresponding quantity in Table LV. Ans. 90° 20′ 53″.5.

4. Given the argument 49° 27′, to find the corresponding quantity in Table LXIV. Ans. 11′ 19″.8.

Case 3. *When the argument is given in the table in hundredth, thousandth, or ten thousandth parts of a circle.*

The required quantity can be found in this case by the same rule as in the preceding; but it can be had more expeditiously by observing the following rules. If the argument varies by 10, multiply the difference of the quantities between which the required quantity lies by the excess of the given argument over the next less value in the table, and remove the decimal point one figure to the left; the result will be the correction to be applied to the quantity taken out of the table. The same rule will apply in taking quantities from tables in which the differences answering to a change of 10 in the argument are given, although the argument should actually change by 50 or 100. If the argument changes by 100, multiply as above, and remove the decimal point two figures to the left. When the common difference of the arguments is 5, proceed as if it were 10, and double the result. In like manner, when the common difference is 50, proceed as if it were 100, and double the result.

Exam. 1. Given the argument 973, to find the corresponding quantity in Table XLV column headed 13.

970 gives 23″.5.

The difference is 1″.2, and the excess 3.

	1″.2	From	23″.5
	3	Take	.4
Corr.	.36		23 .1

2. Given the argument 4834, to find the corresponding quantity in Table XLII, column headed 5.

4800 gives 2′ 3″.7.

The difference is 6″.8, and the excess 34.

6″.8		
34		
	From	2′ 3″.7
2.312 . . .	Take	2 .3
		2 1 .4

3. Given the argument 5444, to find the corresponding quantity in Table XLI. Ans. 15′ 37″.7.

4. Given the argument 4225, to find the corresponding quantity in Table XLIII, column headed 8. Ans. 0′ 47″.2.

Case 4. *When the table is one of double entry, or quantities are taken from it by means of two arguments.*

Take out of the table the quantity answering to the values of the arguments of the table next less than the given values; and find the respective corrections to be applied to it, due to the excess of the given value of each argument over the next less value in the table, by the general rule in the preceding case. These corrections are to be added to the quantity taken out, or subtracted from it, according as the quantities increase or decrease with the arguments.

Note 1. If the tenths of seconds be omitted, the corrections above mentioned can be estimated without the trouble of stating a proportion, or performing multiplications.

Note 2. The rule above given may, in some rare instances, give a result differing a few tenths of a second from the truth. The following rule will furnish more exact results. Find the quantities corresponding, respectively, to the value of the argument at the top next less than its given value and the other given argument, and to the value next greater and the other given argument. Take the difference of the quantities found, and also the difference of the corresponding arguments at top, and say, difference of arguments : difference of quantities : : excess of given value of the argument at the top over its next less value in the table : a fourth term. This fourth term added to the quantity first found, if it is less than the other, but subtracted from it, if it is greater, will give the required quantity. The error of the first rule may be diminished without any extra calculation, by attending to the difference of the quantities answering to the value of the argument at the side next *greater* than its given value and the values of the other argument between which its given value lies.

Exam. 1. Given the argument 64 at the top and 77 at the side to find the corresponding quantity in Table LXXXI.

50 and 70 give 47″.7.

The difference between 47″.7 and the next quantity below it is 1″.4. The excess of 77 over 70 is 7, and the argument at the side changes by 10.

```
                      1″.4
                         7
                      ----         From  47″.7
Corr. due excess 7,  .98, or 1″.0.  Take   1 .0
                                          -----
   Quantity corresponding to 50 and 77,   46 .7
```

The difference between 47″.7 and the adjacent quantity in the next column on the right is 3″.3. The excess of 64 over 50 is 14, and the argument at the top changes by 50.

```
                      3″.3
                        14
                      ----
                      .462
                         2
                      ----         From  46″.7
Corr. due excess 14,  .924         Take   0 .9
                                         -----
                                          45 .8
```

2. Given the argument 223 at the top and 448 at the side, to find the corresponding quantity in Table XXX.

220 and 440 give 16″.0.

The difference between 16″.0 and the quantity next below it is 2″.2.

```
                       2″.2
                          8
                     ------
                   2 ) 1.76
                       ----          From  16″.0
Corr. for excess 8,    .88, or 0″.9.  Take   0 .9
                                           -----
   Quantity corresponding to 220 and 448,  15 .1
```

The difference between 16″.0 and the adjacent quantity in the next column on the right is 0″.7.

```
                       0″.7
                          3
                       ----          To  15″.1
Corr for excess 3,      .21          Add     .2
                                         -----
                                         15 .3
```

3. Given the argument 472 at the top and 786 at the side, to find the corresponding quantity in Table XXXI.

Ans. 9″.7.

4. Given the argument 620 at the top and 367 at the side, to find the corresponding quantity in Table LXXXI.

Ans. 55″.2.

5 Given the argument 348 at the top and 932 at the side, to find (by the rule given in Note 2) the corresponding quantity in Table XXXII. Ans. 15″.4.

PROBLEM III.

To convert Degrees, Minutes, and Seconds of the Equator into Hours, Minutes, &c., of Time.

Multiply the quantity by 4, and call the product of the seconds, thirds; of the minutes, seconds; and of the degrees, minutes.

Exam. 1. Convert 83° 11′ 52″ into time.

$$\begin{array}{r} 83^\circ\ 11'\ 52'' \\ 4 \\ \hline 5^{h.}\ 32^{m.}\ 47^{s.}\ 28''' \end{array}$$

2. Convert 34° 57′ 46″ into time.

Ans. 2h. 19m. 51sec. 4‴.

PROBLEM IV.

To convert Hours, Minutes, and Seconds of Time into Degrees, Minutes, and Seconds of the Equator.

Reduce the hours and minutes to minutes: divide by 4, and call the quotient of the minutes, degrees; of the seconds, minutes; and multiply the remainder by 15, for the seconds.

Exam. 1. Convert 7h. 9m. 34sec. into degrees, &c.

$$\begin{array}{l} 7^{h.}\ 9^{m.}\ 34^{s.} \\ 60 \\ \hline 4\ )\ 429\ \ 34 \\ \hline 107^\circ\ 23'\ 30'' \end{array}$$

2. Convert 11h. 24m. 45s. into degrees, &c.

Ans. 171° 11′ 15″

PROBLEM V.

The Longitudes of two Places, and the Time at one of them being given, to find the corresponding Time at the other.

When the given time is in the morning, change it to astronomical time, by adding 12 hours, and diminishing the number of the day by a unit. When the given time is in the evening, it is already in astronomical time.

Find the difference of longitude of the two places, by taking the numerical difference of their longitudes, when these are of the same name, that is, both east or both west; and the sum, when they are of different names, that is, one west and the other east. When one of the places is Greenwich, the longitude of the other is the difference of longitude.

Then, if the place at which the time is required is to the *east* of the place at which the time is given, *add* the difference of longitude, in time, to the given time; but, if it is to the *west*, *subtract* the difference of longitude from the given time. The sum or remainder will be the required time.

Note. The longitudes used in the following examples, are given in Table I.

Exam. 1. When it is October 25th, 3h. 13m. 22sec. A. M. at Greenwich, what is the time as reckoned at New York?

Time at Greenwich, October,	24d.	15h.	13m.	22s.	
Diff. of Long. . . .		4	56	4	
Time at New York . .	24	10	17	18	P. M.

2. When it is June 9th, 5h. 25m. 10sec. P. M. at Washington, what is the corresponding time at Greenwich?

Time at Washington, June,	9d.	5h.	25m.	10s.	
Diff. of Long. . . .		5	8	6	
Time at Greenwich . .	9	10	33	16	P. M.

3. When it is January 15th, 2h. 44m. 23sec. P. M. at Paris what is the time at Philadelphia?

Longitude of Paris . .		0h.	9m.	21s..6	E.
Do. of Philadelphia, .		5	0	39 .6	W.
		5	10	1.2	
Time at Paris, January, .	15d.	2h.	44m.	23s.	
Diff. of Long. . . .		5	10	1	
Time at Philadelphia, .	14	21	34	22	

Or January 15th, 9h. 34m. 22sec. A. M.

4. When it is March 31st, 8h. 4m. 21sec. P. M. at New Haven, what is the corresponding time at Berlin?

Ans. April 1st, 1h. 49m. 43sec. A. M.

5. When it is August 10th, 10h. 32m. 14sec. A. M. at Boston, what is the time at New Orleans?

Ans. Aug. 10th, 9h. 16m. 4sec. A. M.

6. When it is noon of the 23d of December at Greenwich, what is the time at New York?

Ans. Dec. 23d, 7h. 3m. 55sec. A. M.

PROBLEM VI.

The Apparent Time being given, to find the corresponding Mean Time; or the Mean Time being given to find the Apparent.

When the given time is not for the meridian of Greenwich, reduce it to that meridian by the last problem. Then find by the tables the sun's mean longitude corresponding to this time. Thus, from Table XVIII take out the longitude answering to the given year, and from Tables XIX, XX, and XXI, take out the motions in longitude for the given month, days, hours, and minutes, neglecting the seconds. The sum of the quantities taken from the tables, rejecting 12 signs, when it exceeds that quantity, will be the sun's mean longitude for the given time.

With the sun's mean longitude thus found, take the Equation of Time from Table XII. Then, when Apparent Time is given to find the Mean, apply the equation with the sign it has in the table; but when Mean Time is given to find the Apparent, apply it with the contrary sign; the result will be the Mean or Apparent Time required.

This rule will be sufficiently exact for ordinary purposes, for several years before and after the year 1840. When the given date is a number of years distant from this epoch, take also with the sun's mean longitude the Secular Variation of the Equation of Time from Table XIII, and find by simple proportion the variation in the interval between the given year and 1840. The result, applied to the equation of time taken from Table XII, according to its sign, if the given time is subsequent to the year 1840, but with the opposite sign if it is prior to 1840, will give the equation of time at the given date, which apply to the given time as above directed.

Note 1. When the exact mean or apparent time to within a small fraction of a second is demanded, take the numbers in the columns entitled I, II, III, IV, V, N, in Tables, XVIII, XIX, XX, answering respectively to the year, month, days, and hours, of the given time. With the respective sums of the numbers taken from each column, as arguments, enter Table XIV, and take out the corresponding quantities. These quantities added to the equation of time as given by Tables XII and XIII, and the constant 3.0s. subtracted, will give the true Equation of Time, if the given time is Mean Time. When Apparent Time is given, it will be farther necessary to correct the equation of time as given by the tables, by stating the proportion, 24 hours : change of equation for 1° of longitude : : equation of time : correction.

Note 2. The Equation of Time is given in the Nautical Almanac for each day of the year, at apparent, and also at mean noon, on the meridian of Greenwich, and can easily be found for any intermediate time by a proportion. Directions for applying it to the given time are placed at the head of the column. The Equation is given on the first and second pages of each month.

Exam. 1. On the 16th of July, 1840, when it is 9h. 35m. 22s. P M., mean time at New York, what is the apparent time at the same place?

Time at New York, July, 1840,	$16^{d.}$	$9^{h.}$	$35^{m.}$	$22^{s.}$
Diff. of Long. . . .		4	56	4
Time at Greenwich, July, 1840,	16	14	31	26

	M. Long.			
1840	$9^{s.}$	10°	12′	49′
July	5	29	23	16
16d.		14	47	5
14h.			34	30
31m.			1	16
M. Long.	3	24	58	56

The equation of time in Table XII, corresponding to $3^{s.}$ 24° 58′ 56″, is + $5^{m.}$ $44^{s.}$

Mean Time at New York, July, 1840,	$16^{d.}$	$9^{h.}$	$35^{m.}$	$22^{s.}$
Equation of time, sign changed, .			—5	44
Apparent Time, ...	16	9	29	38 P.M.

2. On the 9th of May, 1842, when it is 4h. 15m. 21sec. A. M. apparent time at New York, what is the mean time at the same place, and also at Greenwich?

Time at New York, May, 1842,	$8^{d.}$	$16^{h.}$	$15^{m.}$	$21^{s.}$
Diff. of Long. . . .		4	56	4
Time at Greenwich, . . .	8	21	11	25

	M. Long.			
1842 . .	$9^{s.}$	10°	43′	18″
May . .	3	28	16	40
8d. . .		6	53	58
21h. . .			51	45
11m. . .				27
M. Long. .	1	16	46	8. Equa. of time,—3m.45s.

Apparent Time at Greenwich, May, 1842,	8d.	21h.	11m.	25s.
Equation of Time, ...			—3	45
Mean Time at Greenwich, . . .	8	21	7	40
Diff. of Long.		4	56	4
Mean Time at New York, . . .	8	16	11	36

Or, May 9th, 4h. 11m. 36s. A. M.

3. On the 3d of February, 1855, when it is 2h. 43m 36s. apparent time at Greenwich, what is the exact mean time at the same place?

Appar. Time at Greenwich, Feb., 1855, 3d. 2h. 43m. 36s

	M. Long.				I.	II.	III.	IV.	V.	N.
1855 . .	9s.	10°	34′	30″	433	279	806	889	866	863
Feb. . .	1	0	33	18	47	85	138	45	7	5
3d. . .		1	58	17	68	5	9	3	0	0
2h. . .			4	56	3					
43m. . .			1	46						
	10	13	12	47	551	369	953	937	873	868

Appar. Time at Greenwich, Feb., 1855,	3d.	2h.	43m.	36s.
Equation of time by Table XII, .			+14	8.6
100yrs. : 13s. (Sec. Var., Table XIII) : : 15yrs. : 1.9s. . . .				—1.9
Approx. Mean Time at Greenwich, .	3	2	57	42.7
24h. : 6s. (change of equa. for 1° of long.) : : 14m. : 0.1s. . .				+0.1
II. III.				0.8
II. IV.				0.4
II. V.				1.0
I.				0.3
N.				0.1
Constant.				—3.0
Mean Time at Greenwich,	3	2	57	42.4

4. On the 18th of November, 1841, when it is 2h. 12m. 26sec. A. M. mean time at Greenwich, what is the apparent time at Philadelphia? Ans. Nov. 17th, 9h. 26m. 28s. P. M.

5. On the 2d of February, 1839, when it is 6h. 32m. 35sec. P. M., apparent time at New Haven, what is the mean time at the same place? Ans. 6h. 46m. 39s. P. M.

6. On the 23d of September, 1850, when it is 9h. 10m. 12sec. mean time at Boston, what is the *exact* apparent time at the same place? Ans. 9h. 18m. 1.0s.

PROBLEM VII.

To correct the Observed Altitude of a Heavenly Body for Refraction.

With the given altitude take the corresponding refraction from Table VIII. Subtract the refraction from the given altitude, and the result will be the true altitude of the body at the given station.

This rule will give exact results if the barometer stands at 30 inches, and Fahrenheit's thermometer at 50°, and results sufficiently exact for ordinary purposes in any state of the atmosphere. When there is occasion for greater precision, take from Table IX the corrections for + 1 inch in the height of the barometer, and −1° in the height of Fahrenheit's thermometer, and compute the corrections for the difference between the observed height of the barometer and 30in. and for the difference between the observed height of the thermometer and 50°. Add these to the mean refraction taken from Table VIII, if the barometer stands higher than 30in. and the thermometer lower than 50°; but in the opposite case subtract them, and the result will be the true refraction, which subtract from the observed altitude.

Exam. 1. The observed altitude of the sun being 32° 10′ 25″, what is its true altitude at the place of observation?

Observed alt. . . .	32° 10′ 25″
Refraction (Table VIII) .	—1 32
True alt. at the station, .	32° 8 53

2. The observed altitude of Sirius being 20° 42′ 11″, the barometer 29.5 inches, and the thermometer of Fahrenheit 70°, required the true altitude at the place of observation. The difference between 29.5 inches and 30 inches is 0.5 inches, and the difference between 70° and 50° is 20°.

Obs. alt. .	20° 42′ 11″.0		
Refrac. (Table VIII),	2′ 33″.0;	Bar. +1 in., 5″.12;	ther. −1°, 0″.310
Corr. for −0.5 in., bar.	−2 .6	.5	20
Corr. for +20°, ther.	−6 .2	2.560	6.20
True refrac. .	2 24 .2		
True alt.	20 39 46 .8		

3. The observed altitude of the moon on the 11th of April, 1838, being 14° 17′ 20″, required the true altitude at the place of observation. Ans. 14° 13′ 35″.

4. Let the observed altitude of Aldebaran be 48° 35′ 52″, the barometer at the same time standing at 30 7 inches, and the thermometer at 42°, required the true altitude. Ans. 48° 34′ 58″.8.

PROBLEM VIII.

The Apparent Altitude of a Heavenly Body being given, to find its True Altitude.

Correct the observed altitude for refraction by the foregoing problem. Then,

1. If the sun is the body whose altitude is taken, find its parallax in altitude by Table X, and add it to the observed altitude corrected for refraction. The result will be the true altitude sought.

2. If it is the altitude of the moon that is taken, and the horizontal parallax at the time of the observation is known, find the parallax in altitude by the following formula:

$$\log.\sin(\text{par. in alt.}) = \log.\sin(\text{hor.par.}) + \log.\cos(\text{app.alt.}) - 10;$$

and add it, as before, to the apparent altitude corrected for refraction.

3. If one of the planets is the body observed, the following formula will serve for the determination of the parallax in altitude when the horizontal parallax is known:

$$\log.(\text{par. in alt.}) = \log.(\text{hor. par.}) + \log.\cos(\text{appar. alt}) - 10.$$

Note 1. The equatorial horizontal parallax of the moon at any given time may be obtained from the tables appended to the work. (See Problem XIV.) But it can be had much more readily from the Nautical Almanac. The equatorial horizontal parallax being known, the horizontal parallax at any given latitude may be obtained by subtracting the Reduction of Parallax, to be found in Table LXIV. The horizontal parallax of any planet, the altitude of which is measured, may also be derived from the Nautical Almanac.

Note 2. The fixed stars have no sensible parallax, and thus the observed altitude of a star, corrected for refraction, will be its true altitude at the centre of the earth as well as at the station of the observer.

Note 3. If the true altitude of a heavenly body is given, and it is required to find the apparent, the rules for finding the parallax in altitude and the refraction are the same as when the apparent altitude is given; the true altitude being used in place of the apparent. But these corrections are to be applied with the opposite signs from those used in the determination of the true altitude from the apparent; that is, the parallax is to be subtracted, and the refraction added. It will also be more accurate to make use of equa. (a), p. 422, in the case of the moon.

Exam. 1. The observed altitude of the sun on the 1st of May, 1837, being 26° 40′ 20″, what is its true altitude?

Obs. alt.	26° 40′ 20″
Refraction	—1 56
True alt. at the station, .	26 38 24
Parallax in alt. (Table X), .	+8
True altitude . . .	26 38 32

2. Let the apparent altitude of the moon at New York on the 17th of March, 1837, 8h. P. M., be 66° 10′ 44″; the barometer 30.4in. and the thermometer 62°; required the true altitude.

			logarithms
Appar. alt. . .		66° 10′ 44″	
Mean refrac. .		0 25.7	
Corr. for + 0.4in., bar.		+ 0.3	
Corr. for + 12°, ther.		—0.6	
True refrac. . .		0 25.4	
True alt. at N. York,		66 10 18.6 .	cos. 9.60637
Equa. par. by N. Almanac,	54′ 13″		
Reduc. for lat. 40°,	4		
Hor. par. at New York,	54 9	. . .	sin. 8.19731
Par. in alt. . .		21 52 .	sin. 7.80368
True altitude .		. 66 32 11	

3. On the 18th of February, 1837, the true meridian altitude of the planet Jupiter at Greenwich was 56° 54′ 57″, what was its apparent altitude at the time of the meridian passage, the horizontal parallax being taken at 1″.9, as given by the Nautical Almanac?

True alt. . .	56° 54′ 57″ .	cos. 9.7371
Hor. par. 1″.9	.	log. 0.2787
Par. in alt.	—1.0 .	log. 0.0158
Refraction	+ 37.9	
Appar. alt. . .	56 55 34	

4. What will be the true altitude of the sun on the 22d of September, 1840, at the time its apparent altitude is 39° 17′ 50″?

Ans. 39° 16′ 46″.

5. Given 29° 33′ 30″ the apparent altitude of the moon at Philadelphia on the 15th of June, 1837, at 9h. 30m. P. M., and 58′ 33″ the equatorial parallax of the moon at the same time, to find the true altitude. Ans. 30° 22′ 41″.

6. Given 15° 24′ 23″ the true altitude of Venus, and 8″ its horizontal parallax, to find the apparent altitude Ans. 15° 27′ 41″.

PROBLEM IX.

To find the Sun's Longitude, Hourly Motion, and Semi-diameter, for a given time, from the Tables.

For the Longitude.

When the given time is not for the meridian of Greenwich, reduce it to that meridian by Problem V; and when it is apparent time, convert it into mean time by the last problem.

With the mean time at Greenwich, take from Tables XVIII, XIX, XX, and XXI, the quantities corresponding to the year, month, day, hour, minute, and second, (omitting those in the last two columns,) and place them in separate columns headed as in Table XVIII, and take their sums.* The sum in the column entitled *M. Long.* will be the tabular mean longitude of the sun; the sum in the column entitled *Long. Perigee* will be the tabular longitude of the sun's perigee; and the sums in the columns I, II, III, IV, V, N, will be the arguments for the small equations of the sun's longitude, including the equation of the equinoxes in longitude.

Subtract the longitude of the perigee from the sun's mean longitude, adding 12 signs when necessary to render the subtraction possible; the remainder will be the sun's mean anomaly. With the mean anomaly take the equation of the sun's centre from Table XXV, and correct it by estimation for the proportional part of the secular variation in the interval between the given year and 1840; also with the arguments I, II, III, IV, V, take the corresponding equations from Tables XXVIII, XXX, XXXI, and XXXII. The equation of the centre and the four other equations, together with the constant 3″, added to the mean longitude, will give the sun's True Longitude, reckoned from the Mean Equinox.

With the argument N take the equation of the equinoxes or Lunar Nutation in Longitude from Table XXVII. Also take the Solar Nutation in longitude, answering to the given date, from the same table. Apply these equations according to their signs to the true longitude from the mean equinox, already found; the result will be the True Longitude from the Apparent Equinox.

For the Semi-diameter and Hourly Motion.

With the sun's mean anomaly, take the hourly motion and semidiameter from Tables XXIII and XXIV.

---

* In adding quantities that are expressed in signs, degrees, &c., reject 12 or 24 signs whenever the sum exceeds either of these quantities. In adding arguments expressed in 100 or 1000, &c. parts of the circle, when they consist of two figures reject the hundreds from the sum; when of three figures, the thousands; and when of four figures, the ten thousands.

Notes.

1. If the tenths of seconds be omitted in taking the equations from the tables of double entry, the error cannot exceed 2″; in case the precaution is taken to add a unit, whenever the tenths exceed .5.

2. The longitude of the sun, obtained by the foregoing rule, may differ about 3″ from the same as derived from the most accurate solar tables now in use. When there is occasion for greater precision, take from Tables XVIII, XIX, and XX, the quantities in the columns entitled VI and VII, along with those in the other columns. With the sums in these columns, and those in the columns I, II, as arguments, take the corresponding equations from Tables XXIX and XXXIII. Also with the sun's mean anomaly take the equation for the variable part of the aberration from Table XXXIV. Add these three equations along with the others to the mean longitude, and omit the addition of the constant 3″. The result will be exact to within a fraction of a second.

Exam. 1. Required the sun's longitude, hourly motion, and semi-diameter, on the 25th October, 1837, at 11h. 27m. 38s. A. M mean time at New York.

	d.	h.	m.	s.
Mean time at N. York, Oct. 1837,	24	23	27	38
Diff. of Long.		4	56	4
Mean time at Greenwich, . .	25	4	23	42

	M. Long.	Long. Perigee.	I.	II.	III.	IV.	V.	N.
	s ° ′ ″	s ° ′ ″						
1837 . . .	9 10 55 47.2	9 10 8 5	816	280	549	321	348	895
October . .	8 29 4 54.1	46	250	748	215	397	63	40
25d. . . .	23 39 19.9	4	810	66	107	35	5	4
4h. . . .	9 51.4		6	0	1			
23m. . . .	56.7							
42s. . . .	1.7	9 10 8 55	882	94	872	753	416	939
		7 3 50 51						
	7 3 50 51.0							
Eq. Sun's Cent.	11 28 12 43.5	9 23 41 56 Mean Anomaly.						
I. . .	2.5							
II. III. . .	9.0	Sun's Hourly Motion, . . 2′ 29″.7						
II. IV. . .	7.7							
II. V. . .	19.3	Sun's Semi-diameter . 16′ 17″.2						
Const. . . .	3.0							
	7 2 4 16.0							
Lunar Nutation	—6.3							
Solar Nutation	—1.2							
Sun's true long.	7 2 4 8.5							

2. Required the sun's longitude, hourly motion, and semi-diameter, on the 15th of July, 1837, at 8h. 20m. 40s. P. M mean time at Greenwich.

	M. Long.	Long. Peri.	I.	II.	III	IV.	V.	N.	VI.	VII.
	s ° ′ ″	s ° ′ ″								
1837 . . .	9 10 55 47.2	9 10 8 5	816	280	549	321	348	895	787	600
July . . .	5 28 24 7.8	31	129	496	806	263	41	27	569	17
15d. . . .	13 47 56.6	2	473	38	62	20	3	2	508	2
8h. . . .	19 42.8		11	1	1				11	
20m. . . .	49.3									
40s.	1.6	9 10 8 38	429	815	418	604	392	924	875	619
		3 23 28 25								
	3 23 28 25.3									
Eq. Sun's Cent.	11 29 33 10.3	6 13 19 47	Mean Anomaly.							
I. . .	10.7									
II. III. . .	6.6	Sun's Hourly Motion, .								2′ 23′ 1
II. IV. . .	5.0									
II. V. . .	7.7	Sun's Semi-diameter,								15′ 45″.4
I. VI. .	1.8									
II. VII. . .	0.2									
Aber. . . .	0.6									
	3 23 2 8.2									
Lunar Nutation	— 7.8									
Solar Nutation	+ 0.8									
Sun's true long.	3 23 2 1.2									

3. Required the sun's longitude, hourly motion, and semi-diameter, on the 10th of June, 1838, at 9h. 45m. 26s. A. M. mean time at Philadelphia, (omitting the three smallest equations of longitude.)

Ans. Sun's longitude, 2s. 19° 11′ 57″; hourly motion, 2′ 23″.3; semi-diameter, 15′ 46″.1.

4. Required the sun's longitude, hourly motion, and semi-diameter, on the 1st of February, 1837, at 12h. 30m. 15s. mean astronomical time at Greenwich.

Ans. Sun's longitude, 10s. 13° 1′ 44″.6; hourly motion, 2′ 32″ 1 · semi-diameter, 16′ 14″.7.

PROBLEM X.

To find the Apparent Obliquity of the Ecliptic, for a given time, from the Tables.

Take the mean obliquity for the given year from Table XXII. Then with the argument N, found as in the foregoing problem, and the given date, take from Table XXVII the lunar and solar nutations of obliquity. Apply these according to their signs to the mean obliquity, and the result will be the apparent obliquity.

Exam. 1. Required the apparent obliquity of the ecliptic on the 15th of March, 1839.

	N.		
1839, .	3		
March,	9		
15d. .	2		
		M. Obliquity,	23° 27′ 36″.9
	14	. . . .	+9 .1
Solar Nutation for March 15th,			+0 .5
Apparent Obliquity, . .			23 27 46 .5

2. Required the apparent obliquity of the ecliptic on the 12th of July, 1845. Ans. 23° 27′ 28″.2.

PROBLEM XI.

*Given the Sun's Longitude and the Obliquity of the Ecliptic, to find his Right Ascension and Declination.**

Let ω = obliquity of the ecliptic; L = sun's longitude; R = sun's right ascension; and D = sun's declination; then to find R and D, we have

$$\log.\tan R = \log.\tan L + \log.\cos \omega - 10,$$
$$\log.\sin D = \log.\sin L + \log.\sin \omega - 10.$$

The right ascension must always be taken in the same quadrant as the longitude. The declination must be taken less than 90°; and it will be north or south according as its trigonometrical sine comes out positive or negative.

Note. The sun's right ascension and declination are given in the Nautical Almanac for each day in the year at noon on the meridian of Greenwich, and may be found at any intermediate time by a proportion.

Exam. 1. Given the sun's longitude 205° 23′ 50″, and the obliquity of the ecliptic 23° 27′ 36″, to find his right ascension and declination.

L = 205°	23′	50″	. . .	tan.	9.67649
ω = 23	27	36	. . .	cos.	9.96253
R = 203	32	5	. . .	tan.	9.63902
L = 205	23	50	. . .	sin.	9.63235—
ω = 23	27	36	. . .	sin.	9.60000
D = 9	49	52 S.	. . .	sin.	9.23235—

2. The obliquity of the ecliptic being 23° 27′ 30″, required

* The obliquity of the ecliptic at any given time for which the sun's longitude is known, is found by the foregoing Problem.

the sun's right ascension and declination when his longitude is 44° 18′ 25″.

Ans. Right ascension 41° 50′ 30″, and declination 16° 8′ 40″ N.

PROBLEM XII.

Given the Sun's Right Ascension and the Obliquity of the Ecliptic, to find his Longitude and Declination.

Using the same notation as in the last problem, we have, to find the longitude and declination,

$$\text{log. tang L} = \text{log. tang R} + \text{ar. co. log. cos } \omega,$$
$$\text{log. tang D} = \text{log. sin R} + \text{log. tang } \omega - 10.$$

Exam. 1. What is the longitude and declination of the sun, when his right ascension is 142° 11′ 34″, and the obliquity of the ecliptic 23° 27′ 40″?

R = 142°	11′	34″	. . .	tan.	9.88979 −
ω = 23	27	40	. . ar. co.	cos.	0.03747
L = 139	46	30	. . .	tan.	9.92726 −
R = 142	11	34	. . .	sin.	9.78746
ω = 23	27	40	. . .	tan.	9.63750
D = 14	53	55 N.	. . .	tan.	9.42496

2. Given the sun's right ascension 310° 25′ 11″, and the obliquity of the ecliptic 23° 27′ 35″, to find the longitude and declination.

Ans. Longitude 307° 59′ 57″, and declination 18° 17′ 0″ S.

PROBLEM XIII.

The Sun's Longitude and the Obliquity of the Ecliptic being given, to find the Angle of Position.

Let p = angle of position; ω = obliquity of the ecliptic; and L = sun's longitude. Then,

$$\text{log. tang } p = \text{log. cos L} + \text{log. tang } \omega - 10.$$

The angle of position is always less than 90°. The northern part of the circle of latitude will lie on the *west* or *east* side of the northern part of the circle of declination, according as the sign of the tangent of the angle of position is *positive* or *negative*.

Exam. 1. Given the sun's longitude 24° 15′ 20″, and the obliquity of the ecliptic 23° 27′ 32″, required the angle of position.

$L=24°$	$15'$	$20''$	. .	cos. 9.95986
$\omega=23$	27	32	. .	tan. 9.63745
$p=21$	35	10	. .	tan. 9.59731

The northern part of the circle of latitude is to the west of the circle of declination.

2. When the sun's longitude is 120° 18′ 55″, and the obliquity of the ecliptic 23° 27′ 30″, what is the angle of position?

Ans. 12° 21′ 17″; and the northern part of the circle of latitude lies to the east of the circle of declination.

PROBLEM XIV.

To find from the Tables, the Moon's Longitude, Latitude, Equatorial Parallax, Semi-diameter, and Hourly Motion in Longitude and Latitude, for a given time.

When the given time is not for the meridian of Greenwich, reduce it to that meridian, and when it is apparent time convert it into mean time.

Take from Table XXXV, and the following tables, the arguments numbered 1, 2, 3, &c., to 20, for the given year, and their variations for the given month, days, &c., and find the sums of the numbers for the different arguments respectively; rejecting the hundred thousands and also the units in the first, the ten thousands in the next eight, and the thousands in the others. The resulting quantities will be the arguments for the first twenty equations of longitude.

With the same time, take from the same tables the remaining arguments with their variations, entitled Evection, Anomaly, Variation, Longitude, Supplement of the Node, II, V, VI, VII, VIII, IX, and X; and add the quantities in the column for the Supplement of the Node.

For the Longitude.

With the first twenty arguments of longitude, take from Tables XLI to XLVI, inclusive, the corresponding equations; and with the Supplement of the Node for another argument, take the corresponding equation from Table XLIX. Place these twenty-one equations in a single column, entitled *Eqs. of Long.;* and write beneath them the constant 55″. Find the sum of the whole, and place it in the column of Evection. Then the sum of the quantities in this column will be the corrected argument of Evection.

With the corrected argument of Evection, take the Evection from Table L, and add it to the sum in the column of Eqs. of Long. Place this in the column of Anomaly. Then the sum of the quantities in this column will be the corrected Anomaly.

With the corrected Anomaly, take the Equation of the Centre from Table LI, and add it to the last sum in the column of Eqs. of Long. Place the resulting sum in the column of Variation. Then the sum of the quantities in this column will be the corrected argument of Variation.

With the corrected argument of Variation, take the variation from Table LII, and add it to the last sum in the column of Eqs. of Long.; the result will be the sum of the principal equations of the Orbit Longitude, amounting in all to twenty four, and the constants subtracted for the other equations. Place this sum in the column of Longitude. Then the sum of the quantities in this column will be the Orbit Longitude of the Moon, reckoned from the mean equinox.

Add the orbit longitude to the supplement of the node, and the resulting sum will be the argument of Reduction.

With the argument of Reduction, take the Reduction from Table LIII, and add it to the Orbit Longitude. The sum will be the Longitude as reckoned from the mean equinox. With the Supplement of the Node, take the Nutation in Longitude from Table LIV, and apply it, according to its sign, to the longitude from the mean equinox. The result will be the Moon's True Longitude from the Apparent Equinox.

For the Latitude.

The argument of the Reduction is also the 1st argument of Latitude. Place the sum of the first twenty-four equations of Longitude, taken to the nearest minute, in the column of Arg. II. Find the sum of the quantities in this column, and it will be the Arg. II of Latitude, corrected. The Moon's true Longitude is the 3d argument of Latitude. The 20th argument of Longitude is the 4th argument of Latitude. Take from Table LVIII the thousandth parts of the circle, answering to the degrees and minutes in the sum of the first twenty-four equations of longitude, and place it in the columns V, VI, VII, VIII, and IX; but not in the column X. Then the sums of the quantities in columns V, VI, VII, VIII, IX, and X, rejecting the thousands, will be the 5th, 6th, 7th, 8th, 9th, and 10th arguments of Latitude.

With the Arg. I of Latitude, take the moon's distance from the North Pole of the Ecliptic, from Table LV; and with the remaining nine arguments of latitude, take the corresponding equations from Tables LVI, LVII, and LIX. The sum of these quantities, increased by 10″, will be the moon's true distance from the North Pole of the Ecliptic. The difference between this distance and 90° will be the Moon's true Latitude; which will be *North* or *South*, according as the distance is less or greater than 90°.

For the Equatorial Parallax.

With the corrected arguments, Evection, Anomaly, and Variation, take out the corresponding quantities from Tables LXI, LXII, and LXIII. Their sum, increased by 7″, will be the Equatorial Parallax

For the Semi-diameter.

With the Equatorial Parallax as an argument, take out the moon's semi-diameter from Table LXV.

For the Hourly Motion in Longitude.

With the arguments 2, 3, 4, 5, and 6 of Longitude, rejecting the two right-hand figures in each, take the corresponding equations of the hourly motion in longitude from Table LXVII. Find the sum of these equations and the constant 3″, and with this sum at the top, and the corrected argument of the Evection at the side, take the corresponding equation from Table LXIX; also with the corrected argument of the Evection take the corresponding equation from Table LXVIII.

Add these equations to the sum just found, and with the resulting sum at the top, and the corrected anomaly at the side, take the corresponding equation from Table LXX; also with the corrected anomaly take the corresponding equation from Table LXXI.

Add these two equations to the sum last found, and with the resulting sum at the top, and the corrected argument of the Variation at the side, take the corresponding equation from Table LXXII. With the corrected argument of the Variation, take the corresponding equation from Table LXXIII.

Add these two equations to the sum last found, and with the resulting sum at the top, and the argument of the Reduction at the side, take the corresponding equation from Table LXXIV. Also, with the argument of the Reduction take the corresponding equation from Table LXXV. These two equations, added to the last sum, will give the sum of the principal equations of the hourly motion in longitude, and the constants subtracted for the others To this add the constant 27′ 24″.0, and the result will be the Moon's Hourly Motion in Longitude.

For the Hourly Motion in Latitude.

With the argument I of Latitude, take the corresponding equation from Table LXXIX. With this equation at the side, and the sum of all the equations of the hourly motion in longitude, except the last two, at the top, take the corresponding equation from Table LXXXI. With the argument II of Latitude, take the corresponding equation from Table LXXXII. And with this equation at the side, and the sum of all the equations of the hourly motion in longitude, except the last two, at the top, take the equation from Table LXXXIII. Find the sum of these four equations and the constant 1″. To the resulting sum apply the constant – 237″.2. The difference will be the Moon's true Hourly Motion in Latitude. The moon will be tending *North* or *South*, according as the sign is *positive* or *negative*

Note. The errors of the results obtained by the foregoing rules, occasioned by the neglect of the smaller equations, cannot exceed for the longitude 15″, for the latitude 8″, for the parallax 7″, for the hourly motion in longitude 5″, and for the hourly motion in latitude 3″; and they will generally be very much less. When greater accuracy is required, take from Tables XXXV to XXXIX the arguments from 21 to 31, along with those from 1 to 20, and their variations. The sums of the numbers for these different arguments, respectively, will be the arguments of eleven small additional equations of longitude. Also, take from the same tables the arguments entitled XI and XII, along with those in the preceding columns. Retain the right-hand figure of the sum in column 1 of arguments, and conceive a cipher to be annexed to each number in the columns of arguments of Table XLI. The numbers in the columns entitled *Diff. for* 10, will then be the differences for a variation of 100 in the argument.

For the Longitude. With the arguments 21 to 31, take the cor responding equations from Tables XLVII and XLVIII, and place them in the same column with the equations taken out with the arguments 1, 2, &c. to 20. Take also equation 32 from Table XLIX, as before. Find the sum of the whole, (omitting the constant 55″,) and then continue on as above. The longitude from the mean equinox being found, take the lunar nutation in longitude from Table LIV, and the solar nutation answering to the given date from Table XXVII. Apply them both, according to their sign, to the longitude from the mean equinox, and the result will be the more exact longitude from the apparent equinox, required.

For the Latitude. With the arguments XI and XII, take the corresponding equations from Table LIX. Add these with the other equations, and omit the constant 10″. The difference between the sum and 90° will be the more exact latitude.

For the Equatorial Parallax. With the arguments 1, 2, 4, 5, 6, 8, 9, 12, 13, take the corresponding equations from Table LX. Find the sum of these and the other equations, omitting the constant 7″, and it will be the more exact value of the Parallax.

For the Hourly Motion in Longitude. With the arguments 1, 7, 8, 9, 10, 11, 12, 13, 14, 15, 16, 17, and 18, of longitude, along with the arguments 2, 3, 4, 5, and 6, heretofore used, take the corresponding equations from Table LXVII. Find the sum of the whole, omitting the constant 3″, and proceed as in the rule already given.

To obtain the motion in longitude for the hour which precedes or follows the given time, with the arguments of Tables LXX, LXXII, and LXXIV, take the equations from Tables LXXVI and LXXVII. Also, with the arguments of Evection, Anomaly, Variation, and Reduction, take the equations from Table LXXVIII. Find the sum of all these equations. Then, for the hour which follows the given time, add this sum to the hourly motion at the given time already found, and subtract 2″.0; for the hour which precedes, subtract it from the same quantity, and add 2″.0.

It will expedite the calculation to take the equations of the second order from the tables at the same time with those of the first order which have the same arguments.

For the Hourly Motion in Latitude. The moon's hourly motion in latitude may be had more exactly by taking with the arguments of Latitude V, VI, &c. to XII, the corresponding equations from Table LXXX, and finding the sum of these and the other equations of the hourly motion in latitude.

To obtain the moon's motion in latitude for the hour which precedes or follows the given time, with the Argument I of Latitude, take the equation from Table LXXXIV, and with this equation and the sum of all the equations of the hourly motion in longitude except the last two, take the equation from Table LXXXV. Find the sum of these two equations. Then, for the hour which follows the given time, add this sum to the Hourly Motion in Latitude already found, taken with its sign, and subtract 1″.3; and for the hour which precedes, subtract it from the same quantity, and add 1″.3.

It will also be more exact to enter Table LXXXI with the sum of all the equations of Tables LXXIX and LXXX, diminished by 1″, instead of the equation of Table LXXIX, for the argument at the side. The numbers over the tops of the columns in Table LXXXI are the common differences of the consecutive numbers in the columns. The numbers in the last column are the common differences of the consecutive numbers in the same horizontal line.

Exam. 1. Required the moon's longitude, latitude, equatorial parallax, semi-diameter, and hourly motions in longitude and latitude, on the 14th of October, 1838, at 6h. 54m. 34s. P. M. mean time at New York.

Mean time at New York, October,	$14^{d.}$	$6^{h.}$	$54^{m.}$	$34^{s.}$
Diff. of Long.		4	56	4
Mean time at Greenwich, October,	14	11	50	38

	1	2	3	4	5	6	7	8	9	10	11	12	13	14	15	16	17	18	19	20
1838 . .	00153	3508	3868	3163	0329	7757	4579	0360	8583	211	175	354	319	670	576	178	492	315	715	870
October .	74741	7419	3969	8343	1602	1569	5752	6550	6630	152	497	237	329	71	333	992	483	087	578	125
14d. . .	03559	8449	3522	3731	4362	4837	0750	5074	0316	912	405	916	444	289	397	476	547	337	028	006
11h. . .	125	298	477	131	154	171	26	179	11	32	14	32	16	45	14	17	19	12	1	0
50m. 38sec.	9	23	36	10	11	13	2	13	1	2	1	2	1	3	1	1	1	1		
	78587	9697	1872	5378	6458	4347	1109	2176	5541	309	092	541	109	078	321	664	542	752	322	001

	Evection.	Anomaly.	Variation.	Longitude.	Sup. of Node.	II.	V.	VI.	VII.	VIII.	IX.	X.
	s ° ′ ″	s ° ′ ″	s ° ′ ″	s ° ′ ″	s ° ′ ″	s ° ′						
1838	3 23 54 9	0 10 25 33	1 24 24 51	11 4 21 29.8	11 11 51 10	5 1 21	876	880	419	423	137	588
October . .	6 29 24 24	10 26 44 34	2 28 4 28	11 27 9 22.4	14 27 24	5 14 32	285	780	710	204	783	451
14d.	4 27 6 53	5 19 50 42	5 8 28 47	5 21 17 35.4	41 18	4 24 59	442	513	367	438	466	069
11h.	5 11 12	5 59 17	5 35 15	6 2 21.0	1 27	5 7	16	18	13	15	16	2
50m.	23 34	27 13	25 24	27 27.0	7	23	1	1	1	1	1	0
38sec. . . .	18	21	19	20.9	0	0	0	0	0	0	0	0
Sum of Equa. .	34 47	3 16 14	12 36 47	12 40 50.8		12 41	35	35	35	35	35	
	3 26 35 17	5 6 43 54	10 19 35 51	5 11 59 27.3	11 27 1 26	3 29 3	655	227	545	116	438	110
			Reduction . . .	11 35.6	5 11 59 27							
				5 12 11 2.9	5 9 0 53	Arg. I of Latitude.						
			Nutation in Longitude	—0.9								
			Moon's True Longitude . . .	5 12 11 2.0								

Arg.	Eqs. ☽'s Long. ° ′ ″
1	0 23 25.7
2	2 2.1
3	0 6.8
4	2 35.3
5	0 23.3
6	0 56.6
7	0 22.0
8	0 10.8
9	1 48.5
10	10.7
11	18.0
12	24.5
13	8.4
14	14.2
15	1.0
16	8.2
17	11.7
18	3.3
19	0.8
20	10.1
32	10.3
Const.	55.0
Sum	0 34 47.3
Ev.	2 41 26.6
Sum	3 16 13.9
An.	9 20 33.5
Sum	12 36 47.4
Var.	4 3.4
Sum	12 40 50.8

Arg.	Eqs. ☽'s Lat. ° ′ ″
I.	87 57 33.1
II.	1 20.0
☽'s long.	10.5
20 long.	14.9
V.	8.5
VI.	6.4
VII.	31.3
VIII.	21.7
IX.	19.8
X.	6.6
Const.	10.0
	88 1 2.8
	90 0 0.0
Moon's Lat.	1 58 57.2

Arguments.	☽'s Eq. Par. ′ ″
Evection . .	0 26.0
Anomaly . .	52 55.1
Variation . .	33.3
Constant . .	7.0
Moon's Eq. Par.	54 1.4

Moon's Semi-diameter, 14′43″

Hourly Motion in Longitude.

Arguments.	Equa. ″
2 of long. . .	5.0
3 do. . .	1.0
4 do. . .	0.0
5 do. . .	0.3
6 do. . .	1.5
Constant . .	3.0
Sum	10.8
Evec. & Sum Eqs.	0.2
Evection .	21.8
Sum	32.8
An. & Sum Eqs.	11.6
Anomaly . .	21.7
Sum	66.1
Var. & Sum Eqs.	9.4
Variation . .	44.9
Sum	120.4
Red. & Sum Eqs.	2.6
Reduction . .	4.1
Sum	127.1
	2′ 7″.1
Constant . .	27 24 .0
Moon's Hourly Mot. in Long.	29 31 .1

Hourly Motion in Latitude.

Arguments.	Equa. ″
I.	33.8
Pre. Eq. & Sum Eq.	46.9
II.	2.9
Pre. Eq. & Sum Eq.	1.4
Constant	1.0
	86.0
	—237.2
	—151.2
	—2′ 31″.2

Moon's Hourly Motion in Latitude, tending S., 2′ 31′.2

	1	2	3	4	5	6	7	8	9	10	11	12	13	14	15	16	17	18	19	20	21	22	23	24	25	26	27	28	29	30	31
1838 . .	00153	3508	3868	3163	0329	7757	4579	0360	8583	211	175	354	319	670	576	178	492	315	715	870	72	46	38	60	75	32	07	04	04	89	84
October .	74741	7419	3969	8343	1602	1569	5752	6550	6630	152	497	237	329	71	333	992	483	087	578	125	07	32	65	26	23	08	07	40	41	47	68
14d. . .	03559	8449	3522	3731	4362	4837	0750	5074	0316	912	405	916	444	289	397	476	547	337	028	006	10	11	32	34	53	48	91	54	40	02	03
11h. . .	125	298	477	131	154	171	26	179	11	32	14	32	16	45	14	17	19	12	1	0	0	0	5	1	2	2	3	2	1	0	0
50m. 38sec.	9	23	36	10	11	13	2	13	1	2	1	2	1	3	1	1	1	1													
	78587	9697	1872	5378	6458	4347	1109	2176	5541	309	092	541	109	078	321	664	542	752	322	001	89	89	40	21	53	90	08	00	86	38	55

	Evection.	Anomaly.	Variation.	Longitude.	Supp. of Node.	II.	V.	VI.	VII.	VIII.	IX.	X.	XI.	XII.
	s ° ′ ″	s ° ′ ″	s ° ′ ″	s ° ′ ″	s ° ′ ″	s ° ′								
1838 . . .	3 23 54 9	0 10 25 32.9	1 24 24 51	11 4 21 29.8	11 11 51 10.4	5 1 21	876	880	419	423	137	588	595	714
October . .	6 29 24 24	10 26 44 33.7	2 28 4 28	11 27 9 22.4	0 14 27 23.8	5 14 32	285	780	710	204	783	451	358	182
14d. . . .	4 27 6 53	5 19 50 41.6	5 8 28 47	5 21 17 35.4	41 18.3	4 24 59	442	513	367	438	466	069	541	437
11h. . . .	5 11 12	5 59 17.2	5 35 15	6 2 21.0	1 27.4	5 7	16	18	13	15	16	2	19	15
50m. . . .	23 34	27 13.1	25 24	27 27.0	6.6	23	1	1	1	1	1	0	1	1
38sec. . .	18	20.7	19	20.9	0.0	0	0	0	0	0	0	0	0	0
Sum of Equa.	34 47	3 16 14.0	12 36 47	12 40 50.9		12 41	35	35	35	35	35		35	35
	3 26 35 17	5 6 43 53.2	10 19 35 51	5 11 59 27.4	11 27 1 26.5	3 29 3	655	227	545	116	438	110	549	384
			Reduction . . .	11 35.6	5 11 59 27.4									
				5 12 11 3.0	5 9 0 53.9	Arg. I of Latitude.								
			Lunar Nutation . . .	—0.8										
			Solar Nutation . . .	—0.8										
			Moon's True Longitude . . .	5 12 11 14										

Arg.	Eqs. ☽'s Lon.		
	°	′	″
1	0	23	25.7
2		2	2.1
3		0	6.8
4		2	35.3
5		0	23.3
6		0	56.6
7		0	22.0
8		0	10.8
9		1	48.5
10			10.7
11			18.0
12			24.5
13			8.4
14			14.2
15			1.0
16			8.2
17			11.7
18			3.3
19			0.8
20			10.1
21			3.2
22			6.1
23			6.2
24			6.1
25			4.8
26			5.5
27			5.4
28			5.0
29			4.4
30			3.8
31			4.6
32			10.3
Sum	0	34	47.4

	°	′	″
Sum	0	34	47.4
Ev.	2	41	26.6
Sum	3	16	14.0
An.	9	20	33.5
Sum	12	36	47.5
Var.		4	3.4
Sum	12	40	50.9

Arg.	Eqs. ☽'s Lat.		
	°	′	″
I.	87	57	33.1
II.		1	20.0
☽'s long.			10.5
20 long.			14.9
V.			8.5
VI.			6.4
VII.			31.3
VIII.			21.7
IX.			19.8
X.			6.6
XI.			7.6
XII.			6.5
Sum .	88	1	6.9
	90	0	0.0
Moon's Lat.	1	58	53.1

Hourly Motion in Longitude.

Arguments.	Equa.	Eq. 2d ord.
	″	″
1 of long. . .	0.40	
2 do. . .	5.0	
3 do. . .	1.0	
4 do. . .	0.0	
5 do. . .	0.3	
6 do. . .	1.5	
7 do. . .	0.02	
8 do. . .	0.58	
9 do. . .	0.12	
10 do. . .	0.33	
11 do. . .	0.19	
12 do. . .	0.65	
13 do. . .	0.04	
14 do. . .	0.03	
15 do. . .	0.11	
16 do. . .	0.06	
17 do. . .	0.14	
18 do. . .	0.16	
Sum	10.6	
Evec. & Sum Eqs.	0.2	
Evection . . .	21.8	0.02
Sum	32.6	
An. & Sum Eqs.	11.6	0.06
Anomaly . . .	21.7	0.74
Sum	65.9	
Var. & Sum Eqs.	9.4	0.06
Variation . . .	44.9	0.67
Sum	120.2	1.55

Arguments.	Equa.	Eq. 2d ord.
	″	″
Sum	120.2	1.55
Red. & Sum Eqs.	2.6	0.04
Reduction . .	4.1	0.12
Sum . . .	2′ 6.9	±1.7
Constant . .	27 24.0	∓2.0
Moon's Hourly Mot. in Long.	29 30.9	29′ 30.9
For the hour following,	29 30.6	
For the hour preceding,	29 31.2	

Arguments.	☽'s Eq. Par.	
	′	″
1 of long. . .	0	0.3
2 do. . . .		1.6
4 do. . . .		0.0
5 do. . . .		0.3
6 do. . . .		0.4
8 do. . . .		0.5
9 do. . . .		3.2
12 do. . . .		0.0
13 do. . . .		1.7
Evection, . . .		26.0
Anomaly, . . .	52	55.1
Variation, . . .		33.3
Moon's Eq. Par.	54	2.4

Moon's Semi-diameter, 14′ 43″.5

Hourly Motion in Latitude.

Args.	Equa.	Eq. 2d ord
	″	″
I.	33.8	0.59
Pre. Eq. &c.	46.9	0.44
II.	2.9	
Pre. Eq. &c.	1.4	
V.	0.36	
VI.	0.29	
VII.	0.01	
VIII.	0.03	
IX.	0.02	
X.	0.04	
XI.	0.00	
XII.	0.04	
Sum .	85.79	1.03
Const.	—237.2	
	—151.4	
	2′ 31″.4 tending S.	
		′ ″
Moon's Hourly Motion in Lat.,		—2 31.4
		± 1.0
		∓ 1.3
For the hour following,		2 31.7
For the hour preceding,		2 31.1

Exam. 2. Required the moon's longitude, latitude, equatorial parallax, semi-diameter, and hourly motions in longitude and latitude, on the 9th of April, 1838, at 8h. 58m. 19s. P. M. mean time at Washington.

Ans. Long. 6ˢ· 19° 45′ 31″.2; lat. 36′ 21″.9 S.; equat. par. 54′ 36″.3; semi-diameter 14′ 52″.7; hor. mot. in long. 30′ 15″ 2; and hor. mot. in lat. 2′ 47″.0, tending south.*

PROBLEM XV.

The Moon's Equatorial Parallax, and the Latitude of a Place, being given, to find the Reduced Parallax and Latitude.

With the latitude of the place, take the reductions from Table LXIV, and subtract them from the Parallax and Latitude.

Exam. 1. Given the equatorial parallax 55′ 15″, and the latitude of New York 40° 42 40″ N., to find the reduced parallax and latitude.

Equatorial parallax, . . .	55′ 15″
Reduction, ...	5
Reduced parallax, . . .	55 10
Latitude of New York, .	40° 42′ 40″ N.
Reduction, ...	11 20
Reduced Lat. of New York,	40 31 20

2. Given the equatorial parallax 60′ 36″, and the latitude of Baltimore 39° 17′ 23″ N., to find the reduced parallax and latitude.

Ans. Reduced par. 60′ 32″, and reduced lat. 39° 6′ 9″.

3. Given the equatorial parallax 57′ 22″, and the latitude of New Orleans 29° 57′ 45″ N., to find the reduced parallax and latitude.

Ans. Reduced par. 57′ 19″, and reduced lat. 29° 47′ 50″.

PROBLEM XVI.

To find the Longitude and Altitude of the Nonagesimal Degree of the Ecliptic, for a given time and place.

For the given time, reduced to mean time at Greenwich, find the sun's mean longitude and the argument N from Tables XVIII, XIX, XX, and XXI. To the sun's mean longitude, apply according to its sign the nutation in right ascension, taken from Table

* The smaller equations were omitted in working this example.

XXVII with argument N; and the result will be the right ascension of the mean sun, (see Art. 127,) reckoned from the true equinox.

Reduce the mean time of day at the given place, expressed astronomically, to degrees, &c., and add it to the right ascension of the mean sun from the true equinox. The sum, rejecting 360° when it exceeds that quantity, will be *the right ascension of the midheaven*, or *the sidereal time in degrees.*

Next, find the reduced latitude of the place by Problem XV; and when it is *north, subtract* it from 90°; but when it is *south, add* it to 90°. The sum or difference will be *the reduced distance of the place from the north pole.*

Also, take the obliquity of the ecliptic for the given year from Table XXII.*

These three quantities having been found, the longitude and altitude of the nonagesimal degree may be computed from the following formulæ:

$$\log. \cos \tfrac{1}{2}(H-\omega) - \log. \cos \tfrac{1}{2}(H+\omega) = A \ . \ . \ . \ (1);$$

$$\log. \text{tang} \tfrac{1}{2}(H-\omega) + 10 - \log. \text{tang} \tfrac{1}{2}(H+\omega) = B \ . \ . \ . \ (2);$$

$$\log. \text{tang } E = A + \log. \text{tang} \tfrac{1}{2}(S - 90°) \ . \ . \ . \ (3);$$

$$\log. \text{tang } F = \log. \text{tang } E + B \ . \ . \ . \ (4);$$

$$N = E + F + 90° \ . \ . \ . \ (5);$$

$$\log. \text{tang} \tfrac{1}{2}h = \log. \cos E + \log. \text{tang} \tfrac{1}{2}(H+\omega) + \text{ar. co. log} \cos F - 20 \ . \ . \ . \ (6).$$

in which

H = the reduced distance of the place from the north pole;

ω = the Obliquity of the Ecliptic;

S = the Sidereal Time converted into degrees;

N = the required Longitude of the Nonagesimal;

h = the required Altitude of the Nonagesimal;

E and F are auxiliary angles.

We first find the logarithmic sums A and B. With these we determine the angles E and F by formulæ (3) and (4), and with these again N and h by formulæ (5) and (6).

The angles E, F, are to be taken less than 180°; and less or greater than 90°, according as the sign of their tangent proves to be positive or negative.

Note 1. In case the given place lies within the arctic circle, we must take, in place of formula (5), the following:

$$N = E - F + 90°.$$

---

* If great precision is required, the apparent obliquity is to be used in place of the mean. (See Prob. X.)

Note 2. As the obliquity of the ecliptic varies but slowly from year to year, the values which have once been found for the logarithms A, B, and log. tang $\frac{1}{2}$ (H + ω) (C), will answer for several years from the date of their determination, unless very great accuracy is required.

Note 3. The angle *h* derived from formula (6), is the distance of the zenith of the given place from the north pole of the ecliptic. This is not always equal to the altitude of the nonagesimal Throughout the southern hemisphere, and frequently in the northern near the equator, it is the supplement of the altitude. In employing this angle in the following Problem, it is, however, for the sake of simplicity, called the altitude of the nonagesimal in all cases.

Exam. 1. Required the longitude and altitude of the nonagesimal degree of the ecliptic at New York, on the 18th of September, 1838, at 3h. 52m. 56s. P. M. mean time.

The sun's mean longitude taken from the tables, for the given time, is 5ˢ· 27° 19′ 17″, and the argument N is 987. The nutation taken from Table XXVII with argument N is − 1″. Hence, the right ascension of the mean sun, reckoned from the true equinox, is 5ˢ· 27° 19′ 16″. The given time of day, expressed astronomically, is 3h. 52m. 56sec.; which in degrees is 58° 14′ 0″.

The reduced latitude of New York, found by Problem XV, is 40° 31′ 20″, and this taken from 90° leaves the polar distance 49° 28′ 40″. The obliquity of the ecliptic, derived from Table XXII, is 23° 27′ 37″.

Given time in degrees, . . .	58° 14′ 0″
R. Asc. of mean sun, . . .	177 19 16
Sidereal time in degrees (S), . .	235 33 16
	90
	2) 145 33 16
$\frac{1}{2}$ (S − 90)	72 46 38

H . .	49° 28′ 40″				
ω . .	23 27 37				
Diff . .	26 1 3				
Sum . .	72 56 17				
$\frac{1}{2}$ diff. . .	13 0 31	cos.	9.98870	tan. + 10,	19.36366
$\frac{1}{2}$ sum . .	36 28 8	cos.	9.90535	tan. C.	9.86871
		A.	0.08335	B.	9.49495
$\frac{1}{2}$ (S − 90°)	72 46 38	tan.	0.50866		
E .	75 38 55	tan.	0.59201	cos.	9.39422
		B.	9.49495	C.	9.86871
F .	50 41 55	tan	0.08696	ar. co. cos.	0.19832
	90 0 0				
		$\frac{1}{2}$ alt. non.	16° 7′ 54″	tan.	9.46125
long. non.	216 20 50				
		alt. non.	32 15 48		

2. Required the longitude and altitude of the nonagesimal degree of the ecliptic at New York, on the 10th of May, 1838, at 11h. 33m. 56sec. P. M. mean time.

Ans. Long. 200° 12′ 23″, and alt. 37° 0′ 34″.

PROBLEM XVII.

To find the Apparent Longitude and Latitude, as affected by Parallax, and the Augmented Semi-diameter of the Moon; the Moon's True Longitude, Latitude, Horizontal Semi-diameter, and Equatorial Parallax, and the Longitude and Altitude of the Nonagesimal Degree of the Ecliptic, being given.

We have for the resolution of this Problem the following formulæ:

$$\log. x = \log. P + \log. \cos h + \text{ar. co.} \log. \cos \lambda - 10 \ldots (1);$$

$$c = \log. x + \log. \text{tang } h - 10 \ldots (2);$$

$$\log. u = c + \log. \sin K - 10 \ldots (3);$$

$$\log. u' = c + \log. \sin (K + u) - 10 \ldots (4);$$

$$\log. p = c + \log. \sin (K + u') - 10 \ldots (5);$$

$$\text{Appar. long.} = \text{true long.} + p \ldots (6);$$

$$\log. \text{tang } \lambda' = \log. p + \text{ar. co.} \log. \cos \lambda + \text{ar. co.} \log. u + \log. \sin (\lambda - x) - 10^{*} \ldots (7);$$

$$\log. v = \log. P + \log. \cos h + \log. \cos \lambda' - 10 \ldots (8);$$

$$\log. z = \log. v + \log. \text{tang } h + \log. \text{tang } \lambda' + \log. \cos (K + \tfrac{1}{2}p) - 30 \ldots (9);$$

$$\pi = v - z \ldots (10);$$

$$\text{Appar. lat.} = \text{true lat.} - \pi \ldots (11);$$

$$\log. R' = \log. p + \text{ar. co.} \log. \cos \lambda + \text{ar. co.} \log. u + \log \cos \lambda' + \log. R - 10 \ldots (12):$$

in which

P = the Reduced Parallax of the Moon;
h = the Altitude of the Nonagesimal;
λ = the True Latitude of the Moon (minus when south);
K = the Longitude of the Moon, minus the longitude of the Nonagesimal;
p = the required Parallax in Longitude;
λ' = the *approximate* Apparent Latitude of the Moon;
π = the required Parallax in Latitude;
R = the True Semi-diameter of the Moon;
R' = the Augmented Semi-diameter of the Moon;
x, u, u', v, z, are auxiliary arcs.

* Formula (7) will be rendered more accurate by adding to it the ar. co. cos x — 10, and will generally give the apparent latitude with sufficient accuracy; thus rendering formulæ (8), (9), (10), and (11) unnecessary.

Formulæ (1), (2), (3), (4), and (5), being resolved in succession, we derive the apparent longitude from formula (6); then the apparent latitude from equations (7), (8), (9), (10), (11); and lastly, the augmented semi-diameter from equation (12.)

The latitude of the moon must be affected with the *negative* sign when *south;* and the apparent latitude will be *south* when it comes out *negative*. In performing the operations, it is to be remembered that the *cosine* of a negative arc has the *same* sign as the cosine of a positive arc of an equal number of degrees; but that the *sine* or *tangent* of a negative arc has the *opposite* sign from the sine or tangent of an equal positive arc. Attention must also be paid to the signs in the addition and subtraction of arcs. Thus, two arcs affected with essential signs, which are to be added to each other, are to be added *arithmetically* when they have like signs, but subtracted if they have unlike signs; and when one arc is to be taken from another, its sign is to be changed, and the two united according to their signs. An arithmetical sum, when taken, will have the same sign as each of the arcs: and an arithmetical difference the same sign as the greater arc.

The use of negative arcs may be avoided, though the calculation would be somewhat longer, by using the true polar distance d, and the approximate apparent polar distance d', in place of λ and λ', substituting sin d for cos λ, cos $(d+x)$ for sin $(\lambda-x)$, sin d' for cos λ', log. co-tang d' for log. tang λ'; and observing that p is to be subtracted from the true longitude in case the longitude of the nonagesimal exceeds the longitude of the moon; that z, when it comes out negative, is to be added to v, which is always positive to the north of the tropic, otherwise subtracted; and that the parallax in latitude is to be applied according to its sign to the true polar distance.

In seeking for the logarithms of the trigonometrical lines, it will be sufficient to take those answering to the nearest tens of seconds.

Note 1. When great accuracy is not desired, u' may be taken for p, from which it can never differ more than a fraction of a second.

Note 2. In solar eclipses the moon's latitude is very small, and formula (7) may be changed into the following:

$$\log. \lambda' = \log. p + \text{ar. co. log. } \cos\lambda + \text{ar. co. log. } u + \log. (\lambda - x) - 10$$

and cos λ' omitted in formula (12) without material error.

Formulæ (8), (9), (10), and (11), may also now be dispensed with, unless very great precision is desired, and the value of λ' given by the above formula taken for the apparent latitude.

It is to be observed also, that in eclipses of the sun P is taken equal to the reduced parallax of the moon minus the sun's horizontal parallax. By this the parallax of the sun in longitude and latitude is referred to the moon, and the relative apparent places of the sun and moon are correctly obtained, without the necessity of a separate computation of the sun's parallax in longitude and latitude.

Exam. 1. About the time of the middle of the occultation of the star Antares, on the 10th of May, 1838, the moon's longitude, by the Connaissance des Tems, was 247° 37′ 6″.7; latitude 4° 14 14′.7 S.; semi-diameter 15′ 24″.2; and equatorial parallax 56 31″.7; and the longitude of the nonagesimal at New York was 200° 12′ 23″; the altitude 37° 0′ 34″; required the apparent lon gitude and latitude, and the augmented semi-diameter of the moon at New York, at the time in question.

Equat. par.	56′ 31″.7	Moon's long.	247° 37′ 7″
Reduction	4 .6	Long. nonag.	200 12 23
	P = 56 27 .1	K =	47 24 44
		h =	37 0 34
		λ =	− 4 14 14.7

P	3387″.1		log. 3.52983
h	37° 0′ 34″		cos. 9.90230
			a. 3.43213
λ	− 4 14 15		ar. co. cos. 0.00119
x	45 12	2712″	log. 3.43332
h	37 0 34		tan. 9.87725
			c. 3.31057
K	47 24 44		sin. 9.86701
u	25 5	1505″	log. 3.17758
			c. 3.31057
K + u	47 49 49		sin. 9.86991
u'	25 15	1515″.2	log. 3.18048
			c. 3.31057
K + u'	47 49 59		sin. 9.86993
p	25 15.3	1515″.3	log. 3.18050
True long.	247 37 6.7		
Appar. long.	248 2 22.0		
p			log. 3.18050
λ−x	−4 59 27		sin. 8.93957−
λ			ar. co. cos. 0.00119
u			ar. co. log. 6.82242
λ′	−5 1 10		tan. 8.94368−

λ' . .	5° 1′ 10″ .	. .	cos.	9.99833
			a.	3.43213
v	44 54.4 .	2694″.4 .	log.	3.43046
h	. . .	. .	tan.	9.87725
λ'	. . .	. .	tan.	8.94368—
$K+\frac{1}{2}p$.	47 37 22 .	. .	cos.	9.82867
z	—2 0.2 .	120″.2 .	log.	2.08006 —
$v-z$ . . .	46 54.6			
$v-z$ (sign changed)	—46 54.6			
True lat. . .	—4 14 14.7			
Appar. lat. . .	5 1 9.3 S.			
p	. . .	. .	log.	3.18050
λ	. . .	. ar. co.	cos.	0.00119
u	. . .	. ar. co.	log.	6.82242
λ'	. . .	. .	cos.	9.99833
R . . .	15 24.2 .	924″.2 .	log.	2.96577
Augm. semi-diam.	15 29.4 .	929″.4 .	log.	2.96821

Exam. 2. About the middle of the eclipse of the sun on the 18th of September, 1838, the moon's longitude was 175° 29′ 19″.0, latitude 47′ 47″.5, equatorial parallax 53′ 53″.5, and semi-diameter 14′ 41″.1 ; and the longitude of the nonagesimal at New York was 216° 20′ 50″, the altitude 32° 15′ 48″: required the apparent longitude and latitude, and the augmented semi-diameter of the moon.

Equat. paral.	53′ 53″.5	Moon's long.	175° 29′ 19″
Reduction,	4 .4	Long. nonag.	216 20 50
	53 49 .1	K =	—40 51 31
Sun's paral.	8 .6	h =	32 15 48
P =	53 40 .5	λ =	0 47 47.5

P	. 3220″.5 .	.	log.	3.50792
h . .	32° 15′ 48″ .	. .	cos.	9.92716
λ . .	47 47.5 .	. ar. co.	cos.	0.00004
x . .	45 23.5 .	2723″.5 .	log.	3.43512
h . .	32 15 48 .	. .	tan.	9.80023
			c.	3.23535
K . .	—40 51 31 .	. .	sin.	9.81570—
u . .	—18 45 .	1125″ .	log.	3.05105—

			c. 3.23535
K+u	−41° 10′ 16″		sin. 9.81844−
u'	−18 52.9	1132″.9	log. 3.05379−
			c. 3.23535
K+u'	−41 10 24		sin. 9.81844−
p	−18 52.9	1132″.9	log. 3.05379−
True long.	175 29 19.0		
Appar. long.	175 10 26.1		
p			log. 3.05379
λ			ar. co. cos. 0.00004
u			ar. co. log. 6.94895
$\lambda - x$	2′ 24″.0	144″.0	log. 2.15836
Appar. latitude	2′ 24″.9 N.	144″.9	log. 2.16114
p			log. 3.05379
λ			ar. co. cos. 0.00004
u			ar. co. log. 6.94895
R	14′ 41″.1	881″.1	log. 2.94502
Augm. semi-diam.	14 46 .7	886″.7	log. 2.94780

PROBLEM XVIII.

To find the Mean Right Ascension and Declination, or Longitude and Latitude of a Star, for a given time, from the Tables.

Take the difference between the given year and 1840. Then seek in Table XV for the fraction of the year answering to the given month and days, and add it to this difference, if the given time is after the beginning of the year 1840; but if it is before, subtract it. Multiply the sum or difference by the annual variation given in the catalogue, (Table XC, or XCII,) and the product will be the variation in the interval between the given time and the epoch of the catalogue. Apply this product to the quantity given in the catalogue, according to its sign, if the given time is after the beginning of the year 1840, but with the opposite sign if it is before, and the result will be the quantity sought. (*See Prob.* XXI. *Note.*)

Exam. 1. Required the mean right ascension and declination of the star Sirius on the 15th of August, 1842.

Interval between given time and beginn. of 1840, (t,)	2.619yrs.
Annual variation of right ascension, . . .	2.646s
Variation of right ascension for interval t, . .	6.93s.

A similar operation gives for the variation of declination in the same interval, 11″.65.

Mean right ascen., beginning of 1840, Table XC,	6h. 38m. 5.76s.
Variation for interval t, ...	+ 6.93
Mean right ascension required, . . .	6 38 12.69
Mean declination, beginning of 1840, . . .	16° 30′ 4″.79 S
Variation for interval t, ...	+ 11 .65
Mean declination required, ...	16 30 16 .44 S.

2. Required the mean longitude and latitude of Aldebaran on the 20th of October, 1838.

Interval between given time and begin. of 1840, (t)	1.200yrs.
Annual variation of longitude, ...	50″.210
Variation of longitude for interval t, . . .	60″.2

A similar operation gives for the variation of latitude in the same interval 0″.4.

Mean longitude, beginning of 1840, .	2s. 7° 33′ 5″.9
Variation for interval t, . . .	−1 0 .2
Mean longitude required, . . .	2 7 32 5 .7
Mean latitude, beginning of 1840, . .	5° 28′ 38″.0 S.
Variation for interval t, ...	+0 .4
Mean latitude required, ...	5 28 38 .4 S.

3. Required the mean right ascension and declination of Capella on the 9th of February, 1839?

Ans. Mean right ascension 5h. 4m. 48.74s., and mean declination 45° 49′ 38″.53 N.

4. Required the mean longitude and latitude of Aldebaran on the 16th of April, 1845?

Ans. Mean longitude 2s. 7° 37′ 31″.4, and mean latitude 5° 28′ 36″.2.

PROBLEM XIX.

To find the Aberrations of a Star in Right Ascension and Declination, for a given Day. (*See Prob.* XXI. *Note.*)

This problem may be resolved for any of the stars in the catalogue of Table XC by means of the following formulæ :

log. (aber. in right ascen.) $= M + \log. \sin (\odot + \varphi) - 10.$
log. (aber. in declin.) $= N + \log. \sin (\odot + \theta) - 10,$

in which M, N, are constant logarithms, ⊙ the longitude of the sun on the given day, and φ, θ, auxiliary angles. M, N, and the angles φ, θ, are given for each of the stars in the catalogue, in Table XCI. ⊙ may be derived from an ephemeris of the sun, or it may be computed from the solar tables by Problem IX.

Exam. 1. What was the amount of aberration, in right ascension and declination, of α Orionis on the 20th of December, 1837, the sun's longitude on that day being 8ˢ· 28° 28′?

Right Ascension.

Table XCI, φ .	6ˢ· 3° 13′	M .	.	0.1361
⊙ .	8 28 28			
$\odot + \varphi$.	3 1 41	.	. sin.	9.9998
Aberration = 1″.37 .	.	.	. log.	0.1359

Declination.

Table XCI, θ .	8ˢ· 28° 23′	N .	.	0.7521
⊙ .	8 28 28			
$\odot + \theta$.	5 26 51	.	. sin.	8.7399
Aberration = 0″.31 .	.	.	. log.	$\bar{1}.4920$

2. Required the aberrations in right ascension and declination of α Andromedæ on the 1st of May, 1838, the sun's longitude being 1ˢ· 10° 38′.

Ans. Aberr. in right ascension − 1″.07, and aberr. in declination − 11″.69.

PROBLEM XX.

To find the Nutations of a Star in Right Ascension and Declination, for a given Day.

This Problem may be solved by means of the formulæ,

log. (nuta. in right asc.) $= M' + \log. \sin (☊ + \varphi') - 10;$
log. (nuta. in declin.) $= N' + \log. \sin (☊ + \theta') - 10;$

in which M′, N′, are constant logarithms, ☊ the mean longitude of the moon's ascending node, and φ', θ', auxiliary angles. M′, N′, and the angles φ', θ', are given for each of the stars in the catalogue, in Table XCI. The mean longitude of the moon's ascending node is given for every tenth day of the year in the Nautical Almanac, page 242, and may be easily found for any intermediate day from the daily motion inserted at the foot of the column of longitudes. It may also be had by finding the supplement of the moon's node, for the given time, from the lunar tables, and subtracting it from 12ˢ· 0° 7′. (*See Note to Prob.* XXI.)

Exam. 1. What was the amount of the nutation, in right ascension and declination, of α Orionis on the 20th of December, 1837, the mean longitude of the moon's node on that day being 18° 54′ ?

Right Ascension.

Table XCI, φ′	. 6ˢ· 0° 15′	M′ .	. 0.0481
☊	. 0 18 54		
☊ + φ′.	6 19 9	.	. sin. 9.5159 —
Nutation =	— 0″.37	.	log. $\bar{1}$.5640 —

Declination.

Table XCI, θ′	. 3ˢ· 2° 37′	N′ .	. 0.9657
☊	. 0 18 54		
☊ + θ′.	3 21 31	.	. sin. 9.9686
Nutation =	8″.60	.	log. 0.9343

2. Required the nutations in right ascension and declination of α Andromedæ on the 1st of May, 1838.

Ans. Nutation in right ascension — 0″.54, and nutation in declination — 1″.43.

Note. When the apparent place of a star is desired with great accuracy, the *solar* nutations must also be estimated and allowed for. These may be determined by repeating the process for finding the lunar nutations, only using twice the sun's longitude in place of the longitude of the moon's node, and multiplying the results by the decimal .075.

The calculation of the solar nutations in Example 1st, is as follows:

Right Ascension.

Table XCI, φ′ .	. 6ˢ· 0° 15′	M′ .	. 0.0481
2 ☉ .	. 5 26 56		
2 ☉ + φ′ .	11 27 11	.	. sin. 8.6914 —
	— 0″.05	.	log. $\bar{2}$.7395 —
	.075		
Solar Nutat. =	— 0″.00		

	Declination.		
Table XCI, θ' . .	3^s. 2° 37′ N′ .		0.9657
$2\odot$.	5 26 56		
$2\odot+\theta'$.	8 29 33 . .		sin. 10.0000—
		— 9″.24 . .	0.9657—
		.075	
Solar Nutat. =		— 0″.69	

In Example 2d, we find for the solar nutation in right ascension, — 0″.08, and for the solar nutation in declination, — 0″.51.

PROBLEM XXI.

To find the Apparent Right Ascension and Declination of a Star, on a given Day.

Find the mean right ascension and declination for the given day by Problem XVIII; then compute the aberrations in right ascension and declination by Problem XIX, and the lunar and solar nutations in right ascension and declination by Problem XX. Apply the aberrations and nutations according to their signs, to the mean right ascension and declination on the given day, observing that the declination when south is to be marked negative, and the results will be the apparent right ascension and declination sought.

Exam. 1. What was the apparent right ascension and declination of α Orionis on the 20th of December, 1837?

	h. m. s.		° ′ ″
Table XC, M. right ascen.	5 46 30.71	M. dec.	7 22 17.14 N
Variations .	— 6.59	. .	—2.42
	5 46 24.12		7 22 14.72
Aberr. . .	+1.37	. .	+0.31
Lun. nutat. .	—0.37	. .	+8.60
Sol. nutat. .	0.00	. .	—0.69
App. right asc.	5 46 25.12	App. dec.	7 22 22.94 N.

2. Required the apparent right ascension and declination of α Andromedæ on the 1st of May, 1838.

Ans. Appar. right ascen. 0h. 0m. 0.90s., and appar. dec. 28° 11′ 39″.92.

NOTE.—In Prob. XVIII. use Table XC. (*a*) for calculations after 1860. Table XCI. will not give accurate results for dates after 1860. The method now adopted in solving Prob. XXI. is by means of tables published annually in the N. Almanac.

PROBLEM XXII.

To find the Aberrations of a Star in Longitude and Latitude, for a given Day.

The formulæ for the computation are,

log. (aber. in long.) = 1.30880 + log. cos (6s. + ⊙ − L) + ar. co. log. cos λ − 10;

log. (aber. in lat.) = 1.30880 + log. sin (6s. + ⊙ − L) + log. sin λ — 20;

in which ⊙ = longitude of the sun on the given day; L = mean longitude of the star; and λ = mean latitude of the star.

Exam. 1. Required the aberrations in longitude and latitude of Antares on the 26th of February, 1838, the sun's longitude on that day being 11s. 7° 29′.

By Prob. XVIII, L =	8s. 7° 30′,	and λ =	4° 32′ S
6s. + ⊙ .	17 7 29	Const. log.	1.3088
6s. + ⊙ − L	8 29 59 .	. cos.	6.4637 −
λ . .	4 32 .	ar. co. cos.	0.0014
Aberr. in long. =	−0″.00 .	log.	$\bar{3}$.7739 −
		Const. log.	1.3088
6s. + ⊙ − L	8s. 29° 59′ .	. sin.	10.0000 −
λ . . .	4 32 .	. sin.	8.8978
Aberr. in lat. =	− 1″.61 .	log.	0.2066 −

2. Required the aberrations in longitude and latitude of Arcturus on the 5th of October, 1838, the sun's longitude being 6s. 11° 47′.

Ans. Aberr. in long. — 23″.34, and aberr. in lat. 1″.85.

Note. The *nutation* in longitude of a fixed star may be found after the same manner as the nutation in longitude of the sun. See Problem IX.)

PROBLEM XXIII.

To find the Apparent Longitude and Latitude of a Star, for a given Day.

Find the mean longitude and latitude on the given day by Problem XVIII. Find also the aberrations in longitude and latitude by Problem XXII, and the nutation in longitude, as in Problem IX. Apply the aberration and nutation in longitude, according to their signs, to the mean longitude, and the result will be the apparent longitude; and apply the aberration in latitude according to its sign, to the mean latitude, and the result will be the apparent latitude.

Exam. 1. Required the apparent longitude and latitude of Antares on the 26th of February, 1838.

Table XC, M. long.	8s. 7° 31′ 45″.2	M. lat. 4° 32′ 51″.6 S.
Var. . .	— 1 32 .57	. . 0 .78
	8 7 30 12 .63	. . 4 32 50 .82
Aberr. .	0 .00	. . — 1 .61
Nutat. .	— 4 .40	
App. long.	8 7 30 8 .23	App. lat. 4 32 49 .21 S

2. Required the apparent longitude and latitude of Arcturus on the 5th of October, 1838.

Ans. Appar. long. 6s. 21° 58′ 37″.4, and appar. lat. 30° 51′ 19″.1.

PROBLEM XXIV.

To compute the Longitude and Latitude of a Heavenly Body from its Right Ascension and Declination, the Obliquity of the Ecliptic being given.

This Problem may be solved by means of the following formulæ:

$$\log.\ \text{tang}\ x = \log.\ \text{tang}\ D + \text{ar. co.}\ \log.\ \sin R;$$
$$\log.\ \text{tang}\ L = \log.\ \cos(x - \omega) + \log.\ \text{tang}\ R + \text{ar. co.}\ \log\ \cos x - 10;$$
$$\log.\ \text{tang}\ \lambda = \log.\ \text{tang}\ (x - \omega) + \log.\ \sin L - 10;$$

in which

R = the Right Ascension;
D = the Declination (minus when South);
L = the Longitude;
λ = the Latitude;
ω = the Obliquity of the ecliptic;

x is an auxiliary arc. It must be taken according to the sign of its tangent, but always less than 180°. The longitude is generally in the same quadrant as the right ascension. The latitude must be taken less than 90°, and will be *north* or *south*, according as the sign is *positive* or *negative*.

Note. When the mean longitude and latitude are to be derived from the mean right ascension and declination, the mean obliquity of the ecliptic is taken. When the apparent longitude and latitude are to be derived from the apparent right ascension and declination, found as in Problem XXI, the apparent obliquity is taken.

The mean obliquity of the ecliptic at any assumed time is easily deduced from Table XXII. The apparent obliquity is found by Problem X.

Exam. 1. On the 20th of June, 1838, the right ascension of Capella was 76° 11′ 29″, the declination 45° 49′ 35″ N., and the obliquity of the ecliptic 23° 27′ 37″; required the longitude and latitude.

D = 45° 49′ 35″	. . .	tan.	0.0125295
R = 76 11 29	. .	ar. co. sin.	0.0127367
x = 46 39 56	. . .	tan.	0.0252662
ω = 23 27 37			
$x - \omega$ = 23 12 19	. . .	cos.	9.9633623
R = 76 11 29	. . .	tan.	0.6094483
x = 46 39 56	. .	ar. co. cos.	0.1635240
Long. = 79 36 4	. . .	tan.	0.7363346
L = 79 36 4	. . .	sin.	9.9928075
$x - \omega$ = 23 12 19	. . .	tan.	9.6321632
Lat. = 22 51 49	. . .	tan.	9.6249707

2. Given the right ascension of *Spica* 199° 11′ 35″, and declination 10° 19′ 24″ S., and the obliquity of the ecliptic 23° 27′ 36″, on the 1st of January, 1840, to find the longitude and latitude.

Ans. Long. 201° 36′ 32″, and lat. 2° 2′ 30″ S.

PROBLEM XXV.

To compute the Right Ascension and Declination of a Heavenly Body from its Longitude and Latitude, the Obliquity of the Ecliptic being given.

The formulæ for the solution of this problem are,

$$\log. \text{tang } y = \log. \text{tang } \lambda + \text{ar. co. log. sin L};$$
$$\log. \text{tang R} = \log. \cos(y+\omega) + \log. \text{tang L} + \text{ar. co. log. cos } y - 10;$$
$$\log. \text{tang D} = \log. \text{tang } (y+\omega) + \log. \sin \text{R} - 10;$$

in which

L = the Longitude;
λ = the Latitude (minus when South);
R = the Right Ascension;
D = the Declination;
ω = the Obliquity of the ecliptic;

y is an auxiliary arc. It must be taken according to the sign of its tangent, but always less than 180°. The right ascension is generally in the same quadrant with the longitude. The declination must be taken less than 90°, and will be *north* or *south*, according as the sign is *positive* or *negative*.

Note. The mean or apparent obliquity of the ecliptic is taken, according as the given and required elements are mean or apparent.

Exam. 1. On the 1st of January, 1830, the longitude of *Sirius* was $3^{s.}$ 11° 44′ 18″, the latitude 39° 34′ 1″ S., and the obliquity of the ecliptic 23° 27′ 41″ : required the right ascension and declination.

λ =	− 39° 34′ 1″	. .	tan. 9.9171381 −
L =	101 44 18	. ar. co.	sin. 0.0091788
y =	139 50 14	. .	tan. 9.9263169 −
ω =	23 27 41		
$y + \omega$ =	163 17 55	. .	cos. 9.9812819 −
L =	101 44 18	. .	tan. 0.6823798 −
y =	139 50 14	. ar. co.	cos. 0.1167843 −
Right ascen =	99 24 48		tan. 0.7804460 −
R =	99 24 48	. .	sin. 9.9941121
$y + \omega$ =	163 17 55	. .	tan. 9.4771803 −
Dec. =	16 29 20 S.	. .	tan. 9.4712924 −

2. Given the longitude of Aldebaran 67° 33′ 5″, and latitude 5° 28′ 38″ S., and the obliquity of the ecliptic 23° 27′ 36″, on the 1st of January, 1840, to find the right ascension and declination.

Ans. Right ascension 66° 41′ 4″, and declination 16° 10′ 57″ N.

PROBLEM XXVI.

The Longitude and Declination of a Body being given, and also the Obliquity of the Ecliptic, to find the Angle of Position.

The formula is

log. sin p = log. sin ω + log. cos L + ar. co. log. cos D − 10 :

p = Angle of Position (required);

L = Longitude ;

D = Declination ;

ω = Obliquity of the ecliptic.

The angle of position p must be taken less than 90°. It is to be observed also that when the longitude is less than 90°, or more than 270°, the northern part of the circle of latitude lies to the *west* of the circle of declination, but that when the longitude is between 90° and 270°, it lies to the *east*.

Note. The angle of position may also be computed from the

right ascension and latitude, by means of a formula similar to that just given, namely,

log. sin p = log. sin ω + log. cos R + ar. co. log. cos λ — 10;

Exam. 1. Given the longitude of *Regulus* 147° 27′ 54″, and declination 12° 47′ 45″ N., and the obliquity of the ecliptic 23° 27′ 41″, to find the angle of position.

ω =	23° 27′ 41″	. .	sin. 9.6000260
L =	147 27 54	. .	cos. 9.9258601
D =	12 47 45	. ar. co.	cos. 0.0109217
Angle of pos. =	20 7 58	. .	sin. 9.5368078

The circle of latitude lies to the east of the circle of declination.

2. Given the longitude of Fomalhaut 331° 27′ 56″, and declination 30° 31′ 14″ S., and the obliquity of the ecliptic 23° 27′ 41″, to find the angle of position. Ans. 23° 57′ 20″.

The circle of latitude lies to the west of the circle of declination

PROBLEM XXVII.

To find from the Tables the Time of New or Full Moon, for a given Year and Month.

For New Moon.

Take from Table LXXXVI, the time of mean new moon in January, and the Arguments I, II, III, and IV, for the given year. Take from Table LXXXVII, as many lunations with the corresponding variations of Arguments I, II, III, and IV, as the given month is months past January, and add these quantities to the former, rejecting the ten thousands from the sums in the columns of the first two arguments, and the hundreds from the sums in the columns of the other two. Seek the number of days from the first of January to the first of the given month, in the *second* or *third* column of Table LXXXVIII, according as the given year is a *common* or *bissextile* year, and subtract it from the sum in the column of mean new moon: the remainder will be tabular time of mean new moon for the given month. It will sometimes happen that the number of days taken from Table LXXXVIII, will exceed the number of days of the sum in the column of mean new moon: in this case one lunation more, with the corresponding arguments, must be added.

With the sums in the columns I, II, III, and IV, as arguments, take the corresponding equations from Table LXXXIX, and add them to the time of mean new moon: the sum will be the *Approximate* time of new moon for the given month, expressed in mean time at Greenwich.

Next, for the approximate time of new moon calculate the true longitudes and hourly motions in longitude of the sun and moon subtract the less longitude from the greater, and the hourly motion of the sun from the hourly motion of the moon ; and say, as the difference between the hourly motions : the difference between the longitudes : : 60 minutes : the correction of the approximate time. The correction *added* to the approximate time, when the sun's longitude is *greater* than the moon's, but *subtracted*, when it is *less*, will give the true time of new moon required, in mean time at Greenwich. This time may be reduced to the meridian of any given place by Problem V.

For Full Moon.

Take from Table LXXXVI, the time of mean new moon, and the corresponding Arguments I, II, III, and IV, for January of the given year, and from Table LXXXVII, a half lunation with the corresponding changes of the arguments. Then, when the time of mean new moon for January is on or after the 16th, subtract the latter quantities from the former, increasing, when necessary to render the subtraction possible, either or both of the first two arguments by 10,000, and of the last two by 100 ; but add them when the time is before the 16th. The result will be the tabular time of mean full moon and the corresponding arguments, for January. Proceed to find the approximate time of full moon after the same manner as directed for the new moon.

For the approximate time of full moon calculate the true longitudes and hourly motions in longitude of the sun and moon. Subtract the sun's longitude from the moon's, adding 360° to the latter if necessary. Take the difference between the remainder and VI signs, and call the result R. Also subtract the hourly motion of the sun from the hourly motion of the moon. Then say, as the difference between the hourly motions : R : : 60m. : the correction of the approximate time. The correction *added* to the approximate time of full moon, when the excess of the moon's longitude over the sun's is *less* than VI signs, but *subtracted* when it is *greater*, will give the true time of full moon.

Exam. 1. Required the time of new moon in September, 1838 expressed in mean time at New York.

	M. New Moon.			I.	II.	III.	IV.
	d.	h.	m.				
1838,	24	16	53	0681	9175	99	85
8 lun.	236	5	52	6468	5737	22	93
	260	22	45	7149	4912	21	78
Days,	243						
Sept'r,	17	22	45				
I.		0	16				
II.		9	35				
III.			3				
IV.			10				
Sept'r.	18	8	49	Approximate time.			

Moon's true long. found for approx. time, is	5ˢ·	25°	29′	19′
Sun's do. do. do.	5	25	27	27
Difference, ...			1	52
Moon's hourly motion in long. is . .			29	28
Sun's do. do. . . .			2	27
Difference, ...			27	1

As 27′ 1″ : 1′ 52″ : : 60ᵐ· : 4ᵐ· 9ˢ·, the correction.

Approx. time of new moon, September, .	18ᵈ·	8ʰ·	49ᵐ·	0ˢ
Correction, ...			− 4	9
True time, in mean time at Greenwich, .	18	8	44	51
Diff. of meridians, ...		4	56	4
True time, in mean time at New York, .	18	3	48	47

Exam. 2. Required the time of full moon in April, 1838, expressed in mean time at New York.

	M. Full Moon.			I.	II.	III.	IV.
	d.	h.	m.				
1838,	24	16	53	0681	9175	99	85
½ lun.	14	18	22	404	5359	58	50
	9	22	31	0277	3816	41	35
3 lun.	88	14	12	2425	2151	46	97
	98	12	43	2702	5967	87	32
Days,	90						
April,	8	12	43				
I.		8	29				
II.		16	7				
III.			15				
IV.			30				
April,	9	14	4	Approximate time.			

Moon's true long. found for approx. time, is	6ˢ·	19°	44′	17″
Sun's do. do. do.	0	19	45	22
	5	29	58	55
	6	0	0	0
	R.	.	1	5
Moon's hourly motion in long. is . .			30	15
Sun's do. do. . .			2	27
Difference			27	48

As 27′ 48″ : 1′ 5″ : : 60ᵐ· : 2ᵐ· 20ˢ·, the correction.

Approximate time of full moon, April, .	$9^{d.}$	14^{h}	$4^{m.}$	0^{s}
Correction, ...			+2	20
True time, in mean time at Greenwich, .	9	14	6	20
Diff. of meridians, ...		4	56	4
True time, in mean time at New York, .	9	9	10	16

3. Required the time of new moon in September, 1837, expressed in mean time at Philadelphia; taking the longitudes for the approximate time from the Nautical Almanac.

Ans. 29d. 3h. 0m. 5s.

4. Required the time of full moon, in October, 1837, expressed in mean time at Boston. Ans. 13d. 6h. 30m. 25s.

PROBLEM XXVIII.

To determine the number of Eclipses of the Sun and Moon that may be expected to occur in any given Year, and the Times nearly at which they will take place.

For the Eclipses of the Sun.

Take, for the given year, from Table LXXXVI the time of mean new moon in January, the arguments and the number N. If the number N differs less than 37 from either 0, 500, or 1000, an eclipse must occur at that new moon. If the difference is between 37 and 53, there may be an eclipse, but it is doubtful, and the doubt can only be removed by a calculation of the true places of the moon and sun. If the difference exceeds 53, an eclipse is impossible.

If an eclipse may or must occur at the new moon in January, calculate the approximate time of new moon by Problem XXVII, and it will be the time nearly of the middle of the eclipse, expressed in mean time at Greenwich. This may be reduced to the meridian of any other place by Problem V.

To find the first new moon after January, at which an eclipse of the sun may be expected, seek in column N of Table LXXXVII the first number after that answering to the half lunation, that, added to the number N for the given year, will make the sum come within 53 of 0, 500, or 1000. Take the corresponding lunations, changes of the arguments, and the number N, and add them, respectively, to the mean new moon in January, the arguments, and the number N, for the given year. Take from the *second* or *third* column of Table LXXXVIII, according as the given year is a *common* or *bissextile* year, the number of days next less than the days of the sum in the column of mean new moon, and subtract it from this sum; the remainder will be the tabular time of mean new moon in the month corresponding to the days taken from Table LXXXVIII. At this new moon there may be an eclipse of the sun; and if the sum in the column N is within 37 of the numbers mentioned above, there must be one. Find the approximate time of new moon, and it will be the time nearly of the middle of the eclipse.

If any of the other numbers in the last column of Table LXXXVII are found, when added to the number N of the given year, to give a sum that falls within the limit 53, proceed in a similar manner to find the approximate times of the eclipses.

Note. When the sum of the numbers N, or the number N itself, in case the eclipse happens in January, is a little above 0, or a little less than 500, the moon will be to the north of the sun, and there is a *probability* that the eclipse will be visible at any given place in north latitude at which the approximate time of the eclipse, found as just explained and reduced to the meridian of the place, comes during the day-time. When the number N found for the eclipse is more than 500, the moon will be to the south of the sun, and the eclipse will seldom be visible in the northern hemisphere, except near the equator.

For the Eclipses of the Moon.

Find the time of full moon and the corresponding arguments and number N, for January of the given year, as explained in Problem XXVII. Then proceed to find the times at which eclipses of the moon may or must occur, after the same manner as for eclipses of the sun, only making use of the limits 35 and 25, instead of 53 and 37.*

Note. An eclipse of the moon will be visible at a given place, if the time of the eclipse thus found nearly, and reduced to the meridian of the place, comes in the night.

Exam. 1. Required the eclipses that may be expected in the year 1840, and the times nearly at which they will take place.

For the Eclipses of the Sun.

	M. New Moon. d.	h.	m.	I.	II.	III.	IV.	N.
1840,	3	10	30	0085	6386	65	63	844
2 lun.	59	1	28	1617	1434	31	98	170
	62	11	58	1702	7820	96	61	014
	60							
March,	2	11	58	As the sum of the numbers N comes within 37 of 0, there must be an eclipse.				
I.		8	3					
II.		19	38					
III.			12					
IV.			13					
March,	3	16	4	Mean time at Greenwich.				

* The numbers 53, 37, and 35, 25, are the lunar and solar ecliptic limits, as determined by Delambre. The limits given in the text, converted into thousandth parts of the circle, are 55, 37, and 37, 21.

	M. New Moon.			I.	II.	III.	IV.	N.
	d.	h.	m.					
1840,	3	10	30	0085	6386	65	63	844
8 lun.	236	5	52	6468	5737	22	93	682
	239	16	22	6553	2123	87	56	526
	213							
August,	26	16	22	As the sum of the numbers N comes within 37 of 500, there must be an eclipse.				
I.		0	54					
II.		0	49					
III.			15					
IV.			16					
August,	26	18	36	Mean time at Greenwich				

For the Eclipses of the Moon.

	M. Full Moon.			I.	II.	III.	IV.	N.
	d.	h.	m.					
1840,	3	10	30	0085	6386	65	63	844
½ lun.	14	18	22	404	5359	58	50	43
	18	4	52	489	1745	23	13	887
1 lun.	29	12	44	808	717	15	99	85
	47	17	36	1297	2462	38	12	972
	31							
Febr.	16	17	36	As the sum of the numbers N, although it comes within 35 of 1000, does not come within 25, the eclipse may be considered doubtful.				
I.		7	27					
II.		0	23					
III.			5					
IV.			27					
Febr.	17	1	58	Mean time at Greenwich.				

	M. Full Moon.			I.	II.	III.	IV.	N.
	d.	h.	m.					
1840,	18	4	52	489	1745	23	13	887
7 lun.	206	17	8	5659	5020	7	94	596
	224	22	0	6148	6765	30	07	483
	213							
August,	11	22	0	As the sum of the numbers N comes within 25 of 500, there must be an eclipse.				
I.		1	37					
II.		19	16					
III.			3					
IV.			25					
August,	12	19	21	Mean time at Greenwich.				

2. Required the eclipses that may be expected in the year 1839, and the times nearly at which they will take place, expressed in mean civil time at New York.

Ans. One of the sun on the 15th of March, at 9h. 20m A. M.; and one of the sun on the 7th of September, at 5h. 24m. P. M.

3. Required the eclipses that may be expected in the year 1841 and the times nearly at which they will take place, expressed in mean civil time at New York.

Ans. Four of the sun, namely, one on the 22d of January, at 12h. 18m. P. M.; one on the 21st of February, at 6h. 17m. A. M , one on the 18th of July, at 9h. 24m. A. M.; and one on the 16th of August, at 4h. 28m. P. M.: and two of the moon, namely, one on the 5th of February, at 9h. 10m. P. M.; and one on the 2d of August, at 5h. 5m. A. M.

The eclipses of the sun in January and August may be considered as doubtful.

PROBLEM XXIX.

To calculate an Eclipse of the Moon.

The calculation of the circumstances of a lunar eclipse is effected with the following fundamental data, derived from the tables of the sun and moon:

Approximate Time of Full Moon (at Greenwich),	T
Sun's Longitude at that time, ...	L
Do. Hourly Motion, ...	s
Do. Semi-diameter, ...	δ
Do. Parallax, ...	p
Moon's Longitude, ...	l
Do. Latitude, ...	λ
Do. Equatorial Parallax, ...	P
Do. Semi-diameter, ...	d
Do. Hourly Motion in longitude, ...	m
Do. Hourly Motion in latitude, . .	n

We obtain the time T by Problem XXVII; the quantities appertaining to the sun, namely, L, s, and δ, by Problem IX;* and those which have relation to the moon, namely, l, λ, P, d, m, and n, by Problem XIV.

From these quantities we derive the following:

True Time of Full Moon, (at given place,) .	T′
Moon's Latitude at that time, ...	λ'
Semi-diameter of earth's shadow, ...	S
Inclination of Moon's relative orbit, ...	I

T being known, T′ is found as explained in Problem XXVII. To obtain λ', we state the following proportion,

1 hour : correction for the time of full moon : : $n : x$;

* p may be taken $= 9''$.

from this we deduce the value of x; and thence find λ by the equation

$$\lambda' = \lambda \pm x.$$

When the true time of full moon, expressed in mean time at Greenwich, is *later* than the approximate time, the *upper* sign is to be used, if the latitude is *increasing*, the *lower* if it is *decreasing*; but when the true time is *earlier* than the approximate time, the *lower* sign is to be used if the latitude is *increasing*; the *upper* if it is *decreasing*.

The value of S is derived from the equation

$$S = (P + p - \delta) + \tfrac{1}{60}(P + p - \delta);$$

and the angle I from the formula

$$\text{log. tang I} = \text{log. } n + \text{ar. co. log. } (m - s).$$

The foregoing quantities having all been determined, the various circumstances of the eclipse may be calculated by the following formulæ:

For the Time of the Middle of the Eclipse.

$$3.55630 + \text{log. cos I} + \text{ar. co. log. } (m - s) - 20 = R;$$
$$\text{log. } t = R + \text{log. } \lambda' + \text{log. sin I} - 10;$$
$$M = T' \pm t:$$

t = interval between time of middle of eclipse and time of full moon; M = time of middle of the eclipse.

The *upper* sign is to be taken in the last equation when the latitude is *decreasing*; the *lower*, when it is *increasing*.

For the Times of Beginning and End.

$$\text{log. } c = \text{log } \lambda' + \text{log. cos I} - 10;$$
$$\text{log. } v = \frac{\text{log. } (S + d + c) + \text{log. } (S + d - c)}{2} + R;$$
$$B = M - v, \text{ and } E = M + v:$$

v = half duration of the eclipse; B = time of beginning; and E = time of end.

Note. If c is equal to or greater than $S + d$, there cannot be an eclipse.

For the Times of Beginning and End of the Total Eclipse.

$$\text{log. } v' = \frac{\text{log. } (S - d + c) + \text{log. } (S - d - c)}{2} + R;$$
$$B' = M - v', \text{ and } E' = M + v':$$

v' = half duration of the total eclipse; B′ = time of beginning of total eclipse; and E′ = time of end of total eclipse.

Note. When c is greater than $S - d$, the eclipse cannot be total.

For the Quantity of the Eclipse.

$$\text{log. } Q = 0.77815 + \text{log. } (S + d - c) + \text{ar. co. log. } d - 10:$$

Q = the quantity of the eclipse in digits.

Note 1. An eclipse of the moon begins on the eastern limb, and ends on the western. In partial eclipses the southern part of the moon is eclipsed when the latitude is north, and the northern part when the latitude is south.

Note 2. When the eclipse commences before sunset, and ends after sunset, the moon will rise more or less eclipsed. To obtain the quantity of the eclipse at the time of the moon's rising, find the moon's hourly motion on the relative orbit by the equation

$$\log. h = \log. (m - s) + \text{ar. co. log. cos I};$$

in which h = hourly motion on relative orbit. Also find the interval between the time of sunset and the time of the middle of the eclipse, which call i. Then,

$$1 \text{ hour} : i :: h : x.$$

Deduce the value of x from this proportion, and substitute it in he equation

$$c' = \sqrt{c^2 + x^2};$$

in which c designates the same quantity as in previous formulæ. Find the value of c', and use it in place of c in the above formula for the quantity of the eclipse, and it will give the quantity of the eclipse at the time of the moon's rising. When the eclipse begins before and ends after sunrise, the quantity of the eclipse at the time of the moon's setting may be found in the same manner, only using sunrise instead of sunset.

Example. Required to calculate, for the meridian of New York, the eclipse of the moon in October, 1837.

Elements.

Approximate time of full moon, .	$T =$	11h. 10m. (Oct. 13)
Sun's longitude at that time, . .	$L =$	6s. 20° 24′ 28″
Do. hourly motion, ...	$s =$	2 29
Do. semi-diameter, ...	$\delta =$	16 4
Do. parallax, ...	$p =$	9
Moon's longitude, ...	$l =$	0 20 21 51
Do. latitude, ...	$\lambda =$	11 28 S.
Do. equatorial parallax, . .	$P =$	59 32
Do. semi-diameter, ...	$d =$	16 13
Do. hourly motion in long. . .	$m =$	35 54
Do. hourly motion in lat. (tending north),	$n =$	3 19

	d.	h.	m.	s.
Approx. time of full moon, October, .	13	11	10	00
Correction found by Prob. XXVII, .			+4	42
True time, in mean time at Greenwich, .	13	11	14	42
Diff. of meridians, ...		4	56	4
True time, in mean time at New York, $T' =$	13	6	18	38

$60^{m.} : 4^{m.}\ 42^{s.} :: 3'\ 19'' : x = 16''$.

Moon's lat. at approx. time, ... $\lambda = 11'\ 28''$ N
Correction, ... $x = \ -16$

Moon's lat. at true time, ... $\lambda' = 11\ \ 12$

Moon's equatorial parallax, ... $P = 59'\ 32''$
Sun's do . . . . . . $p = \ \ 9$

Sum, ... $59\ \ 41$
Sun's semi-diameter, ... $\delta = 16\ \ 4$

Diff. . . . . $P + p - \delta = 43\ \ 37$
Add . . . . . . $\frac{1}{60}(P + p - \delta) = \ \ 44$

Semi-diameter of earth's shadow, ... $S = 44\ \ 21$

Moon's hor. mot. less sun's $(m - s) = 2005''$. ar. co. log. 6.69789
Moon's hor. motion in latitude, $n = 199$ . . log. 2.29885

Inclination of rel. orbit, $I = 5°\ 40'$ . . . . tan. 8.99674

Time of Middle.

3.5563C
I . . . . . . $5°\ 40'$ . . cos. 9.99787
$m - s$ . . . . . $2005''$ ar. co. log. 6.69789

R. 0.25206
λ' . . . . . . $672''$. log. 2.82737
I . . . . . . $5°\ 40'$ . . . sin. 8.99450

t . . . . $0^{h.}\ 1^{m.}\ 58^{s.} = 118^{s.}$ . . . log. 2.07393
T′ . . . . 6 18 38 P. M.

Middle, . . . 6 20 36 P. M.

Times of Beginning and End.

λ' . . . . log. 2.82737
I . . . . cos. 9.99787

c . . . . . $11'\ 9'' = 669''$. log. 2.82524

$S + d + c$ . . . $4303''$. log. 3.63377
$S + d - c$ . . . 2965 . log. 3.47202

2) 7.10579

3.55289
R. 0.25206

v . . . . $1^{h.}\ 46^{m.}\ 22^{s.} = 6382^{s.}$. log. 3.80495

v . . .	$1^{h.}\ 46^{m.}\ 22^{s.} = 6382^{s.}$	.	log. 3.80495
Middle, . .	6 20 36		
Beginning, . .	4 34 14 P. M.		
End, . .	8 6 58 P. M.		
$S - d + c$	2357″	.	log. 3.37236
$S - d - c$	1019		log. 3.00817
			2) 6.38053
			3.19026
			R 0.25206
v' . . .	$0^{h.}\ 46^{m.}\ 9^{s.} = 2769^{s.}$	.	log. 3.44232
Middle, . .	6 20 36		
Beg. of total eclipse,	5 34 27 P. M.		
End of total eclipse,	7 6 45 P. M		
			0.77815
$S + d - c$ . . .	. .		log. 3.47202
d	973″ .	ar. co.	log. 7.01189
Quantity, . .	18.3 digits, . . .		log. 1.26206

PROBLEM XXX.

To calculate an Eclipse of the Sun, for a given Place.

Having found by the rule given in the note to Problem XXVIII, that there is a probability that the eclipse will be visible at the given place, and calculated the approximate time of new moon by Problem XXVII, find from the tables, for this time or for the nearest whole or half hour, the sun's longitude, hourly motion, and semi-diameter; and the moon's longitude, latitude, equatorial parallax, semi-diameter, and hourly motions in longitude and latitude. Find also by Problem XVI, the longitude and altitude of the nonagesimal degree; and thence compute by Problem XVII, the apparent longitude, latitude, and augmented semi-diameter of the moon, (using the relative horizontal parallax.) With these data compute the apparent distance of the centres of the sun and moon, at the time in question, by means of the following formulæ:

$$\text{log. tang } \theta = \text{log. } \lambda' + \text{ar. co. log. } \alpha;$$
$$\text{log. } \Delta = \text{log. } \alpha + \text{ar. co. log. cos } \theta:$$

in which

Δ = appar. distance of centres ;
λ' = appar. Lat. of Moon ;
a = Diff. of appar. Long. of Moon and Sun = diff. of appar long. of Moon (found as above) and true long. of Sun.

θ is an auxiliary arc. The value of θ being derived from the first equation, the second will then make known the value of Δ.

a and λ' are in every instance to be affected with the positive sign.*

For the Approximate Times of Beginning, Greatest Obscuration, and End.

Let the time for which the above calculations are made, be denoted by T. If the apparent distance of the centres of the sun and moon, found for the time T, is less than the sum of their apparent semi-diameters, there is an eclipse at this time. But if it is greater, either the eclipse has not yet commenced, or it has already terminated. It has not commenced if the apparent longitude of the moon is less than the longitude of the sun; and has terminated, if the apparent longitude of the moon is greater than the longitude of the sun.

1. If there should be an eclipse at the time T, from the sun's longitude and hourly motion in longitude, and the moon's longitude and latitude, and hourly motions in longitude and latitude, found for this time, calculate the longitudes and the moon's latitude for two instants respectively an hour before, and an hour after the time T. The semi-diameter of the sun, and the equatorial parallax and semi-diameter of the moon, may, in our present inquiry, be regarded as remaining the same during the eclipse. Find the apparent longitude and latitude, and the augmented semi-diameter of the moon, (using in all cases the relative parallax,) and thence compute by the formulæ already given, the apparent distance of the centres of the sun and moon at the two instants in question.

Observe for each result, whether it is less or greater than the sum of the apparent semi-diameters of the two bodies. If the moon is apparently on the same side of the sun at the times T and T + 1h., take the difference of the distances of the two bodies in apparent longitude at these times, but, if it is on opposite sides, take their sum, and it will be the variation of this distance in the

---

* Δ, the apparent distance of the centres, may be found without the aid of logarithms by means of the following equation:

$$\Delta = \sqrt{a^2 + \lambda'^2}.$$

If the logarithmic formulæ are used, it will be sufficient here to take out the angle θ to the nearest minute. When we have occasion to obtain the distance of the centres exact to within a small fraction of a second, θ must be taken to the nearest tens of seconds, if it exceeds 20° or 30°.

hour following T. Find in like manner the variation of the distance during the hour preceding T. Then, if the apparent distance of the centres at the times (T − 1h.), (T + 1h.) is less than the sum of the apparent semi-diameters, deduce from these results the variations of the distance in apparent longitude during the preceding and following hours, allowing for the second difference, and observing whether the two bodies are approaching each other, or receding from each other. Thence, find the distance in apparent longitude at the times (T − 2h.), (T + 2h.) Find by the same method the apparent latitude of the moon at the instants (T − 2h.), (T + 2h.), observing that the variation of the apparent latitude in any given interval is the difference between the latitudes at the beginning and end of it, if they are both of the same name; their sum, if they are of opposite names.

From these results derive the apparent distance of the centres of the sun and moon at the two instants in question.

If there should still be an eclipse at the time (T + 2h.) or (T − 2h.), find by the same method the distance of the centres at the time (T + 3h.) or (T − 3h.) These calculations being effected, the times of the beginning, greatest obscuration, and end of the eclipse, will fall between some of the instants T, (T − 1h.), (T + 1h.), &c., for which the apparent distance of the centres is computed.

2. If the eclipse occurs after the time T, the different phases will happen between the instants T, (T + 1h.), (T + 2h.), &c. Find the apparent distance of the centres of the sun and moon for the times (T + 1h.), (T + 2h.), by the same method as that by which it is found for the times (T + 1h.), (T − 1h.), in the case just considered. Then, if the eclipse has not terminated, deduce the distance of the moon from the sun in apparent longitude, and the moon's apparent latitude, for the time (T + 3h.), from these distances and latitudes at the times T, (T + 1h.), (T + 2h.); as in the preceding case the distance and latitude for the time (T + 2h.) were deduced from the same at the times (T − 1h.), T, (T + 1h.) With the results obtained compute the apparent distance of the centres of the two bodies at the time (T + 3h.)

3. In case the eclipse occurs before the time T, the apparent distance of the centres must be found by similar methods for the times (T − 1h.), (T − 2h.), &c.

The calculation is to be continued until the distance, from being less, becomes greater than the sum of the semi-diameters.

Now, let h = variation of apparent distance of centres in the interval of one hour comprised between the first two of the instants for which the distance is computed; d = difference between the sum of the semi-diameters of the sun and moon and the apparent distance of their centres at the first instant; and t = interval between first instant and the time of the beginning of the eclipse Then,

$$h : d : : 60^{m.} \cdot t \text{ (nearly.)}$$

Find the value of t given by this proportion, and add it to the time at the first instant, and the result will be a first approximation to the time of the beginning of the eclipse, which call b. Find, by interpolation,* the distance of the moon from the sun in apparent longitude (a), and the moon's apparent latitude (λ'), for this time, and thence compute the apparent distance of the centres. Take h = variation of apparent distance in the interval between the time b and the nearest of the two instants above mentioned, between which the beginning falls, and d = difference between the apparent distance of the centres at the time b and the sum of the semi-diameters, and compute again the value of t. Add this to the time b, or subtract it from it, according as b is before or after the beginning, and the result will be a second approximation to the time of the beginning, which call B. A result still more approximate may be had, by taking h = variation of apparent distance of centres in the interval B — b, d = difference between apparent distance at the time B and sum of semi-diameters, finding anew the value of t given by the preceding proportion, and adding it to, or subtracting it from, as the case may be, the time B. But preparatory to the calculation of the exact times, it will suffice, in general, to take the first approximation.

The end of the eclipse will fall between the last two of the several instants for which the apparent distance of the centres of the moon and sun have been computed. The approximate time of the end is found by the same method as that of the beginning.†

* The second differences may easily be taken into the account in finding the quantities a and λ' for the time b. Thus, let k = variation of a for the interval of an hour comprised between the instants above mentioned, k' = same for the succeeding hour, and i = interval between b and the nearer of the two instants, (in minutes.) Then, if we put $f = \frac{k}{6}$, $c = \frac{k - k'}{36}$, and v = var. of a in interval i,

$$v = \frac{i\left\{f \pm \left(c + \frac{c}{2}\right)\right\}}{10}$$

The upper sign is to be used when the time b is nearer the first than the second instant, the lower when it is nearer the second than the first. c is to be used with its sign. The error by this method will not exceed the number c, (supposing the changes of k, k', from 10m. to 10m. to increase or decrease by equal degrees.)

The general formula for interpolation is $Q = q + \frac{t}{h}d' + \frac{t(t-h)}{2h^2}d'' + \frac{t(t-h)(t-2h)}{2.3.h^3}d''' +$ &c., in which q is the first of a series of values, found at equal intervals, of the quantity whose value Q at the time t is sought. t is reckoned from the time for which q is found. h is one of the equal intervals. d', d'', d''', &c., are the first, second, third, &c., differences. If we make $h = 1$, we have

$$Q = q + td' + \frac{t(t-1)}{2}d'' + \frac{t(t-1)(t-2)}{2.3}d''' + \text{\&c.}$$

† In effecting the reductions of the quantities a and λ' to the first approximate time of end, k' must stand for the variation of a during the hour preceding that comprised between the last two instants, and the last instant must be substituted for the first. (See Note above.)

The middle of the interval between the approximate times of the beginning and end of the eclipse, will be a first approximation to the time of greatest obscuration.

Note. When the object is merely to prepare for an ooservation results sufficiently near the truth may be obtained by a graphical construction. The elements of the construction are the difference of the apparent longitudes of the moon and sun, and the apparent latitude of the moon, found as above, for two or more instants during the continuance of the eclipse. Draw a right line EF, (Fig. 119,) to represent the ecliptic, assume on it some point C for the

Fig. 119.

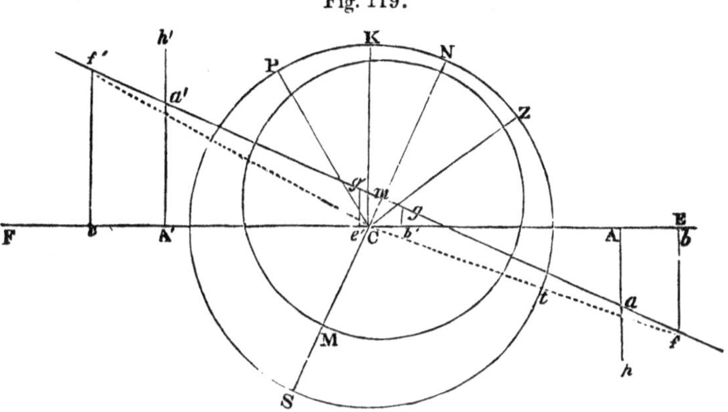

position of the sun at the instant of apparent conjunction, and lay off CA, CA′, equal to the two differences of apparent longitude; and to the right or left, according as the moon is to the west or east of the sun at the instants for which the calculations have been made. Erect the perpendiculars A*p*, A′*p*′, and mark off A*a*, A′*a*′ equal to the two apparent latitudes. Through *a*, *a*′, draw a right line, and it will be the apparent relative orbit of the moon, or will differ but little from it. From C let fall the perpendicular C*m* upon the relative orbit, *m* will be the apparent place of the moon at the instant of greatest obscuration. Take a distance in the dividers equal to the sum of the apparent semi-diameters of the moon and sun, and placing one foot of it at C, mark off with the other the points *f*, *f*′, for the beginning and end of the eclipse, and by means of a square mark on EF the points *b*, *e*, which answer to the beginning and end. If the eclipse be total or annular, mark the points of immersion and emersion, *g*, *g*′, with an opening in the dividers equal to the difference of the semi-diameters, and find the corresponding points *b*′, *e*′ on the line EF.

If the calculations are made from hour to hour, the distance AA′ is the apparent relative hourly motion of the sun and moon in longitude. This distance laid off repeatedly to the right and left will determine the points 1, 2, &c., answering to 1h., 2h., &c. before and after the times for which the calculations are made. If the spaces in which the points b, e, answering to the beginning and end of the eclipse, occur, be divided into quarters, and then subdivided into three equal parts or five-minute spaces, the approximate times of the beginning and end of the eclipse will become known.

From the point m, as a centre, describe the lunar disc; and from the point C, as a centre, describe the sun's disc, and we shall have the figure of the greatest eclipse. The quantity of the eclipse will result from the proportion

SN : MN : : 12 : number of digits eclipsed.

Draw from the centre C to the place of commencement f, the line Cf; and through the same point C raise a perpendicular to the ecliptic. With the longitude of the sun at the time of the beginning, calculate its angle of position by Problem XIII, and lay it off in the figure, placing the circle of declination CP to the left if the tangent of the angle of position be positive, to the right if it be negative.

Compute also for the time of beginning the angle of the vertical circle of the sun with the circle of declination, that is, the angle PSZ in Fig. 6 p. 13, for which we have in the triangle PSZ the side PS = co-declination, the side PZ = co-latitude, and the included angle ZPS. (The requisite formulæ are given in the Appendix.) Form this angle in the figure at the point C, placing CZ to the right or left of CP, according as the time is in the forenoon or afternoon; CZ will be the vertical, and Z the *vertex*, or highest point of the sun. The arc Zt on the limb of the sun will be the angular distance from the vertex of the point on the limb at which the eclipse commences.

For the True Times of Beginning, Greatest Obscuration, and End.

The approximate times of beginning, greatest obscuration, and end of the eclipse, being calculated by the rules which have been given, find from the tables, or from the Nautical Almanac, (see Problem XXXI,) the moon's longitude, latitude, equatorial parallax, semi-diameter, and hourly motions in longitude and latitude, for the approximate time of greatest obscuration.* With the moon's longitude and latitude, and hourly motions in longitude and latitude, found for this time, calculate the longitude and latitude for the approximate times of beginning and end. The parallax and semi-diameter may, without material error, be considered the same during the eclipse. With the moon's true longitude, latitude, and semi-diameter at the approximate times of beginning, greatest obscuration, and end, calculate its apparent longitude and latitude,

* It will, in general, suffice to calculate the moon's longitude and latitude from the elements already found for the approximate time of full moon, if these have been accurately determined. The equatorial parallax and semi-diameter may be found by interpolation from the Nautical Almanac.

and augmented semi-diameter, for these several times, (making use of the relative parallax.) With the sun's longitude and hourly motion previously found for the approximate time of new moon, find his longitude at the approximate times of beginning, greatest obscuration, and end. The sun's semi-diameter found for the approximate time of new moon, will serve also for any time during the eclipse. With the data thus obtained, calculate by the formulæ given on page 375 the apparent distance of the centres of the sun and moon at the approximate times of the three phases.

Note. When very great accuracy is required, the moon's longitude, latitude, equatorial parallax, semi-diameter, and hourly motions in longitude and latitude, must be calculated directly from the tables, or from the Nautical Almanac, for the approximate times of the beginning and end, as well as for that of the greatest obscuration.

For the Beginning.

Subtract the apparent longitude of the moon at the approximate time of beginning from the true longitude of the sun at the same time, and denote the difference by a. Do the same for the approximate time of greatest obscuration. Subtract the latter result from the former, paying attention to the signs, and call the remainder k. Next, take the difference between the apparent latitudes of the moon at the approximate times of beginning and greatest obscuration, if they are of the same name; their sum, if they are of opposite names; and denote the difference or sum, as the case may be, by n. This done, compute the correction to be applied to the approximate time of beginning by means of the following formulæ:

$$\log. b = \log. a + \log. k + \text{ar. co.} \log. n - 10;$$

$$c = \lambda' - b, \text{S} = d + \delta - 5'';$$

$$\log t = \log. (\text{S} + \Delta) + \log. (\text{S} - \Delta) + \text{ar. co.} \log. n + \text{ar. co.} \log. c + \log. \text{L} + 1.47712 - 20:$$

in which

t = Correction of approx. time of beginn. (required);
a = Diff. of appar. long. of Moon and Sun at approx. time;
L = Half duration of eclipse in minutes (known approximately);
k = Appar. relative motion of Sun and Moon in long. in the interval L;
n = Moon's appar. motion in lat. in same interval;
λ' = Moon's appar. lat.;
d = Augmented semi-diameter of the Moon;
δ = Semi-diam. of Sun;
Δ = Appar. distance of centres of Sun and Moon.
b and c are auxiliary quantities.

First find the value of b by the first equation, and substitute it in the second. Then derive the values of c and S from the second and third equations, and substitute them in the fourth, and it will make known the value of t, which is to be applied to the approximate time of the beginning of the eclipse according to its sign.

The quantities a, k, n, &c., are all to be expressed in seconds. The apparent latitude λ' must be affected with the *negative* sign when it is *south.* The motion in latitude, n, must also have the negative sign in case the moon is apparently receding from the north pole. a and k are always positive.*

The result may be verified, and corrected, by computing the apparent distance of the centres at the time found, and comparing it with the sum of the semi-diameters minus 5″.

Note. When great precision is desired, the quantities k and n must be found for some shorter interval than the half duration of the eclipse. Let some instant be fixed upon, some five or ten minutes before or after the approximate time of the beginning of the eclipse, according as the contact takes place before or after. For this time deduce the longitude and latitude of the moon, from the longitude and latitude at the approximate time of beginning, by means of their hourly variations; and thence calculate the apparent longitude and latitude, and the augmented semi-diameter. Find the longitude of the sun for the time in question, from its longitude and hourly motion already known for the approximate time of beginning. Then proceed according to the rule given above, only using the quantities thus found for the time assumed, in place of the corresponding quantities answering to the approximate time of greatest obscuration. L will always represent the interval for which k and n are determined.

For the End.

Subtract the longitude of the sun at the approximate time of the end from the apparent longitude of the moon at the same time. Do the same for the approximate time of greatest obscuration. Then proceed according to the rule for the beginning, only substituting everywhere the approximate time of the end for the approximate time of the beginning, and taking in place of the formula $c=\lambda'-b$, the following:

$$c=\lambda'+b.$$

* It will be somewhat more accurate to use in place of k and n, as above defined, the values of the following expressions: $\frac{k}{6}-2\frac{1}{2}\frac{k'-k}{36}$ or $\frac{k}{6}-3\frac{1}{2}\frac{k'-k}{36}$; $\frac{n}{6}-2\frac{1}{2}\frac{n'-n}{36}$ or $\frac{n}{6}-3\frac{1}{2}\frac{n'-n}{36}$. The first of each of these pairs of expressions is to be used in case the true time of beginning is after the approximate time;—the second in the other case. k' and n' are the apparent relative motions in longitude and latitude during the last half of L. In case these expressions are used the following constant logarithm is to be employed instead of that above given, viz. 0.69897.

In the calculation of the end of the eclipse, k and n will answer to the last half of L, and k' and n' to the first half.

For the Greatest Obscuration.

Take the sum of the distances of the moon from the sun in apparent longitude at the approximate times of the beginning and end of the eclipse, and call it k. Take the difference of the apparent latitudes of the moon at the same times, if the two are of the same name; but if they are of different names, take their sum. Denote the difference or sum by n. Let $a' =$ the distance of the moon from the sun in apparent longitude at the true time of greatest obscuration; $\lambda' =$ the apparent latitude of the moon at the approximate time of greatest obscuration.

$$k : n :: \lambda' : a'.$$

Find the value of a' by this proportion, affecting λ', n, k, always with the positive sign.

Ascertain whether the greatest obscuration has place before or after the apparent conjunction, by observing whether the apparent latitude of the moon is increasing or decreasing about this time, the rule being, that when it is *increasing*, the greatest obscuration will occur *before* apparent conjunction; when it is *decreasing*, *after*. If the approximate and true times of greatest obscuration are both before or both after apparent conjunction, from the value found for a' subtract the distance of the moon from the sun in apparent longitude at the approximate time; but if one of the times is before and the other after apparent conjunction, take the sum of the same quantities. Denote the difference or sum by m. Also, let $D =$ duration of eclipse, and $t =$ correction to be applied to the approximate time of greatest obscuration. Then to find t, we have the proportion

$$k : m :: D : t.$$

If the apparent latitude of the moon is decreasing, t is to be applied according to the sign of m; but if the apparent latitude is increasing, it is to be applied according to the opposite sign.

A still more exact result may be had by repeating the foregoing calculations, making use now of the apparent latitude at the time just found. When the greatest accuracy is required, the values of k and n may be found more exactly after the same manner as for the beginning or end.

For the Quantity of the Eclipse.

Find by interpolation the apparent latitude of the moon at the true time of greatest obscuration. With this, and the distance in longitude a' obtained by the proportion above given, compute by the formulæ on page 375, the apparent distance of the centres of the sun and moon at the time of greatest obscuration. Subtract this distance from the sum of the apparent semi-diameters of the two bodies, diminished by 5″, and denote the remainder by R Then,

Sun's semi-diam. (diminished by 3″) : R : : 6 digits : number of digits eclipsed.

When the apparent distance of the centres of the sun and moon at the time of greatest obscuration is less than the difference between the sun's semi-diameter and the augmented semi-diameter of the moon, the eclipse is either *annular* or *total; annular*, when the sun's semi-diameter is the *greater* of the two; *total,* when it is the *less.*

For the Beginning and End of the Annular or Total Eclipse.

The times of the beginning and end of the annular or total eclipse may be found as follows: the greatest obscuration will take place very nearly at the middle of the eclipse in question, and will not differ, at most, more than five or eight minutes (according as the eclipse is total or annular) from the beginning and end: to obtain the half duration of the eclipse, and thence the times of the beginning and end, we have the formulæ

$$\log. \tan g\, \theta = \log. \lambda' + \text{ar. co. } \log. a, \quad \log. k' = \log. k + \text{ar. co. } \log. \sin \theta :$$

$$S = \delta - d - 1'', \text{ or } S = d - \delta + 1'';$$

$$\log. c = \frac{\log. (S + \Delta) + \log. (S - \Delta)}{2};$$

$$\log. t = \text{ar. co. } \log. k' + \log. c + \log. D + 1.77815 - 10;$$

$$\text{Time of Begin.} = M - t, \text{ Time of End} = M + t:$$

in which

M = Time of greatest obscuration;
λ' = Moon's apparent latitude at that time;
a = Distance of moon from sun in appar. long.;
k = Variation of this distance during the whole eclipse, or relative mot. in appar. long. during this interval;
k' = Moon's appar. mot. on relative orbit for same interval;
θ = Inclination of relative orbit;
δ = Semi-diameter of sun;
d = Augm. semi-diam. of moon;
Δ = Appar. distance of centres;
D = Duration of eclipse, (partial *and* annular or total;)
t = Half duration of annular or total eclipse.

The *first* value of S is used when the eclipse is *annular*, the *second* when it is *total*. The quantities may all be regarded as positive. The results may be verified and corrected by finding directly the apparent distance of the centres for the times obtained, and comparing it with the value of S.

For the Point of the Sun's Limb at which the Eclipse commences.

Find the angle of position of the sun, and the angle between its vertical circle and circle of declination, at the beginning of the eclipse, as explained at page 380. Let the former be denoted by p, and the latter by v. Give to each the negative sign, if laid off towards the right; the positive sign if laid off towards the left. Let a = distance of the moon from the sun in apparent longitude at the beginning of the eclipse; λ' = the moon's apparent latitude at the same time; and θ = angular distance of the point of contact from the ecliptic. Compute the angle θ by the formula

$$\log.\ \text{tang}\ \theta = \log.\ \lambda' + \text{ar. co. log.}\ a\ ;$$

taking it always less than 90°, and positive or negative according to the sign of its tangent. λ' is negative when south; a is always positive.

Let A = distance on the limb of the point of contact from the vertex. The above operations being performed, the value of A results from the equation

$$A = p + v + 90° - \theta\ ;$$

p, v, and θ being taken with their signs.

If the result is affected with the positive sign, the point first touched will lie to the right of the vertex. If with the negative sign, it will lie to the left of the vertex.

Note. The circumstances of an occultation of a fixed star by the moon may be calculated in nearly the same manner as those of a solar eclipse. The star in the occultation holds the place of the sun in the eclipse. The immersion and emersion of the star correspond to the beginning and end of the eclipse. The elements which ascertain the relative apparent place and motion of the moon and star, take the place of those which ascertain the relative apparent place and motion of the moon and sun. Thus the star's longitude, corrected for aberration and nutation, (see Problem XXIII,) must be used instead of the sun's longitudes; the apparent distances of the moon from the star in latitude, instead of the moon's apparent latitudes; and the moon's augmented semi-diameter, instead of the sum of the semi-diameters of the sun and moon. The difference of the longitudes, and the relative motion in longitude. must also now be reduced to a parallel to the ecliptic passing through the star, (see Appendix, page 431.) If λ = apparent latitude of star, a = diff. of appar. longitudes of moon and star, and k = relative motion in longitude, we must substitute in the formulæ for the eclipse, for $\lambda', \lambda' - \lambda$; for $a, a \cos \lambda$; and for $k, k \cos \lambda$. n will stand for the relative motion in latitude, or for the variation of $\lambda' - \lambda$.

Example. Required to calculate an eclipse of the sun, for the latitude and meridian of New York, that will occur on the 18th of September, 1838.

For the Approximate Times of the Phases.

Approximate time of New Moon.

Sept. 18^d. 8^h. 49^m.

Sun's longitude, ...	175° 27′ 31″.4
Do. hourly motion, ...	2 26 .7
Do. semi-diameter, ...	15 57 .0
Moon's longitude, ...	175 29 19
Do. latitude, ...	47 47
Do. equatorial parallax, . . .	53 53
Do. semi-diameter, ...	14 41
Do. hor. mot. in long.	29 29
Do. hor. mot. in lat.	2 41
Do. appar. long. (Prob. XVII), . .	175 10 26
Do. appar. lat. (λ'), ...	2 25 N.
Do. augm. semi-diameter, . . .	14 47
Diff. of appar. long. (a), ...	17 5
Appar. dist. of cen. (Δ), ...	17 15
Sum of semi-diameters, ...	30 44

7^h. 49^m.

Sun's longitude, ...	175° 25′ 4″
Moon's appar. long.	174 47 3
Do. appar. lat. (λ')	8 12 N.
Do. augm. semi-diameter, . . .	14 49
Diff. of appar. long. (a), ...	38 1
Appar. dist. of cen. (Δ), ...	38 53
Sum of semi-diameters, ...	30 46

9^h. 49^m.

Sun's longitude, ...	175° 29′ 58″
Moon's appar. long.	175 36 15
Do. appar. lat. (λ'), ...	2 18 S.
Do. augm. semi-diameter, . . .	14 44
Diff. of appar. long. (a), ...	6 17
Appar. dist. of cen. (Δ), ...	6 42
Sum of semi-diameters, ...	30 41

	a	diff. or k.	λ'	diff. or n.	Δ	diff.	sum semi-d.
7^h. 49^m.	2281″		492″ N		2333″		1846″
		1256″		347″		1298″	
8 49	1025		145 N		1035		1844
		1402		283			
9 49	377		138 S		402		1841
		1548		219		1556	
10 49	1925		357 S		1958		1839

For the Approximate Time of Beginning.

$h = 1298''$, $d = 2333'' - 1846'' = 487''$;

$1298'' : 487'' : : 60^{m.} : t = 22^{m.}.5$

$7^{h.}\ 49^{m.}$
22

1st Approxi. $8^{h.}\ 11^{m.}$

$7^{h.}\ 49^{m.}$. $a = 2281''$. $\lambda' = 492''$ N.
Corrections for $22^{m.}$ 447 . 133 (See Note, p. 324)

$8^{h.}\ 11^{m.}$. $a = 1834$. $\lambda' = 359$ N.

$a = 1834''$ ar. co. log. 6.73660 . . log. 3.26340
$\lambda' = 359$. log. 2.55509

$\theta = 11°\ 4'\ 30''$. tan. 9.29169 . ar. co. cos. 0.00817

Appar. dist. of cen. $\Delta = 1869''$ . . log. 3.27157
Sum of semi-diam. . 1846

$487'' : 23'' : : 22^{m.} : t = 1^{m.}\ 2^{s.}$

$8^{h.}\ 11^{m.}$
$+1$

2d Approxi. $8^{h.}\ 12^{m.}$

For the Approximate Time of the End.

$h = 1556''$, $d = 1958'' - 1839'' = 119''$.

$1556'' : 119'' : : 60^{m.} : t = 4^{m.}.6$.

$10^{h.}\ 49^{m.}$
-5

1st Approxi. $10^{h.}\ 44^{m.}$

$10^{h.}\ 49^{m.}$. $a = 1925''$ . . . $\lambda' = 357''$ S.
Corrections for $5^{m.}$ 132 . . . 17

$10^{h.}\ 44^{m.}$. $a = 1793$ . . . $\lambda' = 340$ S.

$a = 1793''$. ar. co. log. 6.74642 . log. 3.25358
$\lambda' = 340$ . . log. 2.53148

$\theta =$ . . . tan. 9.27790 . ar. co. cos. 0.00767

Appar. dist. of cen. $\Delta = 1825''$. 3.26125
1839

$133'' : 14'' : : 5^{m.} : t = 0^{m.}.5$.

	$10^{h.}\ 44^{m.}$
	0 .5
2d Approxi.	$10^{h.}\ 44^{m.}.5$

For the Approximate Time of Greatest Obscuration

	h.	m.
Approx. time of begin. .	8	12
Approx. time of end, .	10	44
2)	18	56
1st Approxi. .	9	28

For the True Times of the Phases.

	Approx. time of Beginning. $8^{h.}\ 12^{m.}$	Approx. time of Greatest Obscur. $9^{h.}\ 28^{m.}$	Approx. time of End. $10^{h.}\ 44^{m.}$
Sun's longitude,	175° 26′ 1″.0	175° 29′ 6″.8	175° 32′ 12″.6
Do. semi-diam.,	15 57 .0	15 57 .0	15 57 .0
Moon's app. lon.	174 55 36 .7	175 27 7 .7	176 2 17 .2
Do. app. lat.	5 45 .3 N.	0 43 .5 S.	5 32 .4 S
Do. augm. semid.	14 48 .0	14 45 .1	14 41 .7

	a	*k*	λ′	*n*	Δ	S
$8^{h.}\ 12^{m.}$	1824″.3	1705″.2	345″.3 N	388″.8	1856″.7	1840″.0
9 28	119 .1	1923 .7	43 .5 S	288 .9		
10 44	1804 .6		332 .4 S		1835 .0	1833 .7

For the True Time of Beginning.

a .	1824″.3	log. 3.26109
k .	1705 .2	log. 3.23178
n .	388 .8	ar. co. log. 7.41028—
b = —	8001 .1	log. 3.90315—
λ′ .	345 .3	
λ′ — *b* = *c* =	8346 .4	ar. co. log. 6.07850
S + Δ .	3696 .7	log. 3.56781
S — Δ .	—16 .7	log. 1.22272—
n . . .	. . . .	ar. co. log. 7.41028—
L . .	76m.	log. 1.88081
		Const. log. 1.47712
Corr. of approx time,	+ 43ˢ.4 .	log. 1.63724 +

Corr. of approx. time,				+ $43^{s.}$.4		
Approx. time, .	$8^{h.}$	$12^{m.}$	0	.0		
True time of begin.	8	12	43	.4,	in Greenwich time	
Diff of merid. .	4	56	4			
True time of begin.	3	16	39	.4,	in New York time.	

For the True Time of End.

a . .	1804″.6	log. 3.25638
k . .	1923 .7	log. 3.28414
n . .	288 .9 . .	ar. co. log. 7.53925—
b =	− 12016 .3	log. 4.07977—
λ′ .	− 332 .4	
λ′ + *b* = *c* =	−12348 .7 . .	ar. co. log. 5.90838—
S + Δ . .	3668 .7	log. 3.56451
S − Δ . .	—1 .3	log. 0.11394—
n	. . .	ar. co. log. 7.53925—
L . . .	76m.	log. 1.88081
		Const. log. 1.47712

Corr. of approx. time,			−$3^{s.}$	0		. log. 0.48401—
Approx. time, .	$10^{h.}$	$44^{m.}$	0	.0		
True time of end, .	10	43	57	.0,	in Greenwich time.	
Diff. of merid. .	4	56	4			
True time of end, .	5	47	53,		in New York time.	

For the True Time of Greatest Obscuration.

True time of beginning, ...	$8^{h.}$	$12^{m.}$	$43^{s.}$	.4
Do. of end, ...	10	43	57	.0
2)	18	56	40	.4
2d Approx.	9	28	20	.2

$9^{h.}$ $49^{m.}$ . .	λ′ = 138″	S.
9 28 . .	λ′ = 43 .5	S.
Diff. 21	Diff. 94 .5	

$21^{m.}$: $20^{s.}$: : 94″.5 : 1″.5
43 .5

$9^{h.}$ $28^{m.}$ $20^{s.}$ . . λ′ = 45. 0

1705″.2	388″.8
1923 .7	288 .9

k = 3628 .9 : n = 677 .7 : : λ′ = 45″.0 : a' = 8″.4

Time of beginn. 8h. 12m. 43s. .4, at 9h. 28m. a = 119″.1
Time of end, 10 43 57 .0 a' = 8 .4

D = 2 31 13 .6 m = − 110 .7

3628″.9 : 110″.7 : : 2h. 31m. 13s. .6 : 4m. 36s. .8
9h. 28 0 .0

True time (nearly) 9 32 36 .8

21m. : 4m. 37s. : : 94″.5 : 20″.8
43 .5

At 9h. 32m. 37s., λ′ = 64 .3

3628″.9 : 677″.7 : : 64″.4 : 12″.0 ; at 9h. 32m. 37s., a = 8″.4
a' = 12 .0

m = 3 .6

3628″.9 : 3″.6 : : 2h. 31m. 13s..6 : 9s..0
9h. 32m. 36 .8

9 32 27 .8

True time of greatest obscur. . 9h. 32m. 27s..8, in Greenw. time
Diff. of merid. 4 56 4

True time of greatest obscur. . 4 36 23 .8, in N. Y. time.

For the Quantity of the Eclipse.

9h. 32m. 37s. . λ′ = 64″.3
21m. : 9s. : : 94″.5 : 0 .6

At nearest approach of centres, . λ′ = 63 .7
" " " . . a = 12 .0

a . 12″.0 . ar. co. log. 8.92082, . . log. 1.07918
λ′ . 63 .7 . . . 1.80414

θ . . . . . tan. 0.72496, . ar. co. cos. 0.73253

Shortest distance of centres, 64″.8 . . log. 1.81171
Sum of semi-diameters, 1837 .0

1772 .2

15′ 54″ : 1772″.2 : : 6 : 11.14 digits eclipsed.

For the Situation of the Point at which the Obscuration commences.

$8^{h.}\ 12^{m.}$ . . a = 1824″, . . λ' = 345″.3 N
$76^{m.} : 43^{s.}$: : 1705″ : 16, $76^{m.} : 43^{s.}$: : 389″ : 3 .7

At the beginn. . a = 1808, . . λ' = 341 .6

a . 1808 . ar. co. log. 6.74280
λ' . 341.6 . . log. 2.53352

θ = 10° 41′ 57″ . . tan. 9.27632
Obliq eclip. (Prob. X), 23° 27′ 47″ . sin. 9.60005 . tan. 9.63753
Sun's longitude, 175 26 3 . sin. 8.90093 . cos. 9.99862 –

sin. 8.50098, tan. 9.63615 –
Sun's declination, 1° 49′ 0″; Angle of pos. 23° 23′ 50″.
Mean time of begin. $3^{h.}\ 16^{m.}\ 39^{s.}$, Lat. 40° 42′ 40″, Dec. 1° 49′ 0″
Equa. of time, . 5 58 90 90

Appar. time, . 3 22 37, PZ = 49 17 20, PS = 88 11
60

4) 202 37

Hour angle P = 50° 39′ 15″ . cos. 9.80210
Co. lat. PZ = 49 17 20 . tan. 0.06526

m = 36° 23′ 0″ . . tan. 9.86736
Co. dec. PS = 88 11 0

m' = 51 48 0 . ar. co. sin. 0.10466
m = 36 23 0 . . sin. 9.77320
P = 50 39 15 . . tan. 0.08627

S = 42 38 10 . . tan. 9.96413
Angle of position, . . – 23° 23′ 50″
Angle from eclip. (θ), . . – 10 41 50
Angle of dec. circle from vertex (S), 42 38 10
90

Angular dist. of point first touched from vertex, 98 32, to the right

For the Beginning and End of the Annular Eclipse.

Approx. time, $9^{h.}\ 32^{m.}\ 27^{s.}.8$ = true time of greatest obscur.
At this time, a = 12″.2, λ' = 63″.7.

a = 12″.2 . ar. co. log. 8.91364 . . log. 1.08636
λ' = 63 .7 . . log. 1.80414

θ = 79° 9′ 30″ . tan. 0.71778 . ar. co. cos. 0.72564

Δ = 64″.9 . . log. 1.81200

S + Δ = 135″.8 . log. 2.13290, θ = 79° 9′ 30″ . ar. co. sin. 0.00783
S − Δ = 6 .2 . log. 0.79239, k = 3628″.9 . log. 3.55977

2) 2.92529, k′ . . ar. co. log. 6.43240

1.46264 . . . . 1.46264
D = 152ᵐ· . . . . log. 2.18184
Const. log. 1.77815

t = 0ʰ· 1ᵐ· 11ˢ·.6 . log. 1.85503

Time of greatest obscur. .	4	36	23 .8	
Formation of ring, . .	4	35	12 .2,	New York time.
Rupture of do. . .	4	37	35 .4	" "

PROBLEM XXXI.

To find the Moon's Longitude, Latitude, Hourly Motions, Equatorial Parallax, and Semi-diameter, for a given time, from the Nautical Almanac.

Reduce the given time to mean time at Greenwich; then,

For the Longitude.

Take from the Nautical Almanac the calculated longitudes answering to the noon and midnight, or midnight and noon, next preceding and next following the given time. Commencing with the longitude answering to the first noon or midnight, subtract each longitude from the next following one: the three remainders will be the *first differences*. Also subtract each first difference from the following for the *second differences*, which will have the plus or minus sign, according as the first differences increase or decrease.

Find the quantity to be added to the second longitude by reason of the first differences, by the proportion, 12ʰ· : excess of given time above time of second longitude : : second first difference : *fourth term*.

With the given time from noon or midnight at the side, take from Table XCIII the quantities corresponding to the minutes, tens of seconds, and seconds, of the mean or half sum of the two second differences, at the top: the sum of these will be the *correction for second differences*, which must have the *contrary* sign to the mean.

The sum of the second longitude, the fourth term, and the correction for second differences, will be the longitude required.

For the Latitude.

Prefix to *north* latitudes the *positive* sign, but to *south* latitudes the *negative* sign, and proceed according to the rules for the longitude, only that attention must now be paid to the signs of the first differences, which may either be plus or minus.

The sign of the resulting latitude will ascertain whether it is *north* or *south*.

For the Hourly Motion in Longitude.

Solve the proportion, $12^{h.}$: given time from noon or midnight . : half sum of second differences : a fourth term ; which must have the same sign as the half sum of the second differences.

Take the sum of the second first difference, half the mean of the second differences, with its sign changed, and this fourth term, and divide it by 12 : the quotient will be the required hourly motion in longitude.

For the Hourly Motion in Latitude.

With the given time from noon or midnight, the second first difference of latitude, and the mean of the second differences, find the hourly motion in latitude in the same manner as directed for finding the hourly motion in longitude. When the hourly motion is *positive*, the moon is tending *north;* and when it is *negative*, she is tending *south.*

For the Semi-diameter and Equatorial Parallax.

The moon's semi-diameter and equatorial parallax may be taken from the Nautical Almanac, with sufficient accuracy, by simple proportion, the correction for second differences being too small to be taken into account, unless great precision is required.

Corrections for Third and Fourth Differences.

When the moon's longitude and latitude are required with great precision, corrections must also be applied for the third and fourth differences. To determine these, take from the Almanac the three longitudes or latitudes immediately preceding the given time, and the three immediately following it, and find the first, second, third, and fourth differences, subtracting always each number from the following one, and paying attention to the signs. With the given time from noon or midnight at the side, and the middle third difference at the top, take from Table XCIV the correction for third differences, which must have the same sign as the middle third difference when the given time from noon or midnight is less than 6 hours ; the contrary sign, when the given time is more than 6 hours.

With the given time, and half sum of fourth differences, take from Table XCV the correction for fourth differences, giving it always the same sign as the half sum.

The sum of the third longitude or latitude, the proportional part of the middle first difference answering to the given time from noon or midnight, and the corrections for second, third, and fourth differences, having regard to the signs of all the quantities, will be the longitude or latitude required.

APPENDIX.

TRIGONOMETRICAL FORMULÆ.*

I. Relative to a Single Arc or Angle a.

1. $\sin^2 a + \cos^2 a = 1$
2. $\sin a = \tan a \cos a$
3. $\sin a = \dfrac{\tan a}{\sqrt{1 + \tan^2 a}}$
4. $\cos a = \dfrac{1}{\sqrt{1 + \tan^2 a}}$
5. $\tan a = \dfrac{\sin a}{\cos a}$
6. $\cot a = \dfrac{1}{\tan a} = \dfrac{\cos a}{\sin a}$
7. $\sin a = 2 \sin \frac{1}{2} a \cos \frac{1}{2} a$
8. $\cos a = 1 - 2 \sin^2 \frac{1}{2} a$
9. $\cos a = 2 \cos^2 \frac{1}{2} a - 1$
10. $\tan \frac{1}{2} a = \dfrac{\sin a}{1 + \cos a}$
11. $\cot \frac{1}{2} a = \dfrac{\sin a}{1 - \cos a}$
12. $\tan^2 \frac{1}{2} a = \dfrac{1 - \cos a}{1 + \cos a}$
13. $\sin 2 a = 2 \sin a \cos a$
14. $\cos 2 a = 2 \cos^2 a - 1 = 1 - 2 \sin^2 a$

II. Relative to Two Arcs a and b, of which a is supposed to be the greater.

15. $\sin (a + b) = \sin a \cos b + \sin b \cos a$
16. $\sin (a - b) = \sin a \cos b - \sin b \cos a$
17. $\cos (a + b) = \cos a \cos b - \sin a \sin b$

---

* The radius is supposed to be equal to unity in all of the formulæ.

18. $\cos(a-b)=\cos a\cos b+\sin a\sin b$

19. $\tan(a+b)=\dfrac{\tan a+\tan b}{1-\tan a\tan b}$

20. $\tan(a-b)=\dfrac{\tan a-\tan b}{1+\tan a\tan b}$

21. $\sin a+\sin b=2\sin\tfrac{1}{2}(a+b)\cos\tfrac{1}{2}(a-b)$

22. $\sin a-\sin b=2\sin\tfrac{1}{2}(a-b)\cos\tfrac{1}{2}(a+b)$

23. $\cos a+\cos b=2\cos\tfrac{1}{2}(a+b)\cos\tfrac{1}{2}(a-b)$

24. $\cos b-\cos a=2\sin\tfrac{1}{2}(a+b)\sin\tfrac{1}{2}(a-b)$

25. $\tan a+\tan b=\dfrac{\sin(a+b)}{\cos a\cos b}$

26. $\tan a-\tan b=\dfrac{\sin(a-b)}{\cos a\cos b}$

27. $\cot a+\cot b=\dfrac{\sin(a+b)}{\sin a\sin b}$

28. $\cot b-\cot a=\dfrac{\sin(a-b)}{\sin a\sin b}$

29. $\dfrac{\sin a+\sin b}{\sin a-\sin b}=\dfrac{\tan\tfrac{1}{2}(a+b)}{\tan\tfrac{1}{2}(a-b)}$

30. $\dfrac{\cos b+\cos a}{\cos b-\cos a}=\dfrac{\cot\tfrac{1}{2}(a+b)}{\tan\tfrac{1}{2}(a-b)}$

31. $\dfrac{\tan a+\tan b}{\tan a-\tan b}=\dfrac{\cot b+\cot a}{\cot b-\cot a}=\dfrac{\sin(a+b)}{\sin(a-b)}$

32. $\dfrac{\cot b-\tan a}{\cot b+\tan a}=\dfrac{\cot a-\tan b}{\cot a+\tan b}=\dfrac{\cos(a+b)}{\cos(a-b)}$

33. $\sin^2 a-\sin^2 b=\sin(a+b)\sin(a-b)$

34. $\cos^2 a-\sin^2 b=\cos(a+b)\cos(a-b)$

35. $1\pm\sin a=2\sin^2(45^\circ\pm\tfrac{1}{2}a)$

36. $\dfrac{1\pm\sin a}{1\mp\sin a}=\tan^2(45^\circ\pm\tfrac{1}{2}a)$

37 $\dfrac{1\pm\sin a}{\cos a}=\tan(45^\circ\pm\tfrac{1}{2}a)$

38. $\dfrac{1-\sin a}{1-\cos a}=\dfrac{\sin^2(45^\circ-\tfrac{1}{2}a)}{\sin^2\tfrac{1}{2}a}$

39. $\dfrac{1+\sin b}{1+\cos a}=\dfrac{\sin^2(45^\circ+\tfrac{1}{2}b)}{\cos^2\tfrac{1}{2}a}$

40. $\dfrac{1+\tan b}{1-\tan b}=\tan(45^\circ+b)$

41. $\dfrac{1-\tan b}{1+\tan b}=\tan(45^\circ-b)$

42. $\sin a \cos b = \frac{1}{2} \sin (a + b) + \frac{1}{2} \sin (a - b)$
43. $\cos a \sin b = \frac{1}{2} \sin (a + b) - \frac{1}{2} \sin (a - b)$
44. $\sin a \sin b = \frac{1}{2} \cos (a - b) - \frac{1}{2} \cos (a + b)$
45. $\cos a \cos b = \frac{1}{2} \cos (a + b) + \frac{1}{2} \cos (a - b)$

III. Trigonometrical Series.

46.
$$\begin{cases} \sin a = a - \frac{a^3}{2.3} + \frac{a^5}{2.3.4.5} - \&c. \\ \cos a = 1 - \frac{a^2}{2} + \frac{a^4}{2.3.4} - \frac{a^6}{2.3.4.5.6} + \&c. \\ \tan a = a + \frac{a^3}{3} + \frac{2a^5}{3.5} + \frac{17a^7}{3^2.5.7} + \&c. \\ \cot a = \frac{1}{a} - \frac{a}{3} - \frac{a^3}{3^2.5} - \frac{2a^5}{3^3.5.7} - \&c. \end{cases}$$

Let a = length of an arc of a circle of which the radius is 1, and (a'') = number of seconds in this arc, then to replace an arc expressed by its length, by the number of seconds contained in it, we have the formula

47. $a = (a'') \sin 1''$; $\log. \sin 1'' = \overline{6}.685574867$.

IV. Differences of Trigonometrical Lines.

48. $\Delta \sin x = + 2 \sin \frac{1}{2} \Delta x . \cos (x + \frac{1}{2} \Delta x)$
49. $\Delta \cos x = - 2 \sin \frac{1}{2} \Delta x . \sin (x + \frac{1}{2} \Delta x)$
50. $\Delta \tan x = + \frac{\sin \Delta x}{\cos x . \cos (x + \Delta x)}$
51. $\Delta \cot x = - \frac{\sin \Delta x}{\sin x . \sin (x + \Delta x)}$

V. Resolution of Right-angled Spherical Triangles *

Table of Solutions.

Given.	Required.		Solution.
Hypothen. and an angle	side op. giv. ang.	52	$\sin x = \sin h . \sin a$
	side adj. giv. ang.	53	$\tan x = \tan h . \cos a$
	the other angle	54	$\cot x = \cos h . \tan a$
Hypothen. and a side	the other side	55	$\cos x = \frac{\cos h}{\cos s}$
	ang. adj. giv. side	56	$\cos x = \tan s . \cot h$
	ang. op. giv. side	57	$\sin x = \frac{\sin s}{\sin h}$

* Baily's Astronomical Tables and Formulæ.

<table>
<tr><td rowspan="3">A side and the angle opposite</td><td>the hypothen.</td><td>58</td><td>$\sin x = \dfrac{\sin s}{\sin a}$</td><td rowspan="3">the ambiguous cases.</td></tr>
<tr><td>the other side</td><td>59</td><td>$\sin x = \tan s \,.\, \cot a$</td></tr>
<tr><td>the other angle</td><td>60</td><td>$\sin x = \dfrac{\cos a}{\cos s}$</td></tr>
<tr><td rowspan="3">A side and the angle adjacent</td><td>the hypothen.</td><td>61</td><td>$\cot x = \cos a \,.\, \cot s$</td><td></td></tr>
<tr><td>the other side</td><td>62</td><td>$\tan x = \tan a \,.\, \sin s$</td><td></td></tr>
<tr><td>the other angle</td><td>63</td><td>$\cos x = \sin a \,.\, \cos s$</td><td></td></tr>
<tr><td rowspan="2">The two sides</td><td>the hypothen.</td><td>64</td><td>$\cos x =$ rectang. cos. of the giv. sides</td><td></td></tr>
<tr><td>an angle</td><td>65</td><td>$\cot x =$ sin adj. side × cot. op. side</td><td></td></tr>
<tr><td rowspan="2">The two angles</td><td>the hypothen.</td><td>66</td><td>$\cos x =$ rectang. cot. of the given angles</td><td></td></tr>
<tr><td>a side</td><td>67</td><td>$\cos x = \dfrac{\text{cos. opp. ang.}}{\text{sin. adj. ang.}}$</td><td></td></tr>
</table>

In these formulæ, x denotes the quantity sought.
a = the *given* angle
s = the *given* side
h = the hypothenuse.

NAPIER'S RULES.

The formulæ for the resolution of right-angled spherical triangles are all embraced in two rules discovered by Lord Napier, and called *Napier's Rules for the Circular Parts.* The circular parts, so called, are the two legs of the triangle, or sides which form the right angle, the complement of the hypothenuse, and the complements of the acute angles. The right angle is omitted. In resolving a right-angled spherical triangle, there are always three of the circular parts under consideration, namely, the two given parts and the required part. When the three parts in question are contiguous to each other, the middle one is called the *middle part,* and the others the *adjacent parts.* When two of them are contiguous, and the third is separated from these by a part on each side, the part thus separated is called the middle part, and the other two the *opposite parts.* The rules for the use of the circular parts are (the radius being taken = 1),

1. Sine of the middle part = the rectangle of the tangents of the adjacent parts.
2. Sine of the middle part = the rectangle of the cosines of the opposite parts.

PARTICULAR CASES OF RIGHT-ANGLED SPHERICAL TRIANGLES.

Equations 52 to 67, or Napier's rules, are sufficient to resolve all the cases of right-angled spherical triangles; but they lack precision if the unknown quantity is very small and determined by means of its cosine or cotangent; or, if the unknown quantity is near 90°, and given by a sine or a tangent: in these cases the following formulæ may be used:

68. $\tan^2 \frac{1}{2}a = -\dfrac{\cos(B+C)}{\cos(B-C)}$

69. $\tan^2 \frac{1}{2}B = \dfrac{\sin(a-c)}{\sin(a+c)}$

70. $\tan^2 \frac{1}{2}c = \tan\frac{1}{2}(a+b)\tan\frac{1}{2}(a-b)$

71. $\tan(45° - \frac{1}{2}b) = \sqrt{\tan(45° - x)}$, $\tan x = \sin a \sin B$

72. $\tan^2 \frac{1}{2}b = \tan\left(\dfrac{B-C}{2} + 45°\right)\tan\left(\dfrac{B+C}{2} - 45°\right)$.

a is the hypothenuse, B, C, the acute angles, and b, c, the sides opposite the acute angles.

VI. Resolution of Oblique-Angled Spherical Triangles.

General Formulæ.

Let A, B, C, denote the three angles of a spherical triangle, and a, b, c, the sides which are opposite to them respectively.

73. $\dfrac{\sin A}{\sin a} = \dfrac{\sin B}{\sin b} = \dfrac{\sin C}{\sin c}$

or, *the sines of the angles are proportional to the sines of the opposite sides.*

74. $\cos c = \cos a \cos b + \sin a \sin b \cos C$

75. $\cos c = \cos(a-b) - 2\sin a \sin b \sin^2 \frac{1}{2}C$

76. $\cos C = \sin A \sin B \cos c - \cos A \cos B$

77. $\sin a \cos c = \sin c \cos a \cos B + \sin b \cos C$

78. $\sin a \cot c = \cos a \cos B + \sin B \cot C$

79. $\sin a \cos B = \sin c \cos b - \sin b \cos c \cos A$

Case I. *Given the three sides,* a, b, c.

To find one of the angles.

80. $\sin^2 \frac{1}{2}A = \dfrac{\sin(k-b)\sin(k-c)}{\sin b \sin c}$

or,

81. $\cos^2 \frac{1}{2}A = \dfrac{\sin k \sin(k-a)}{\sin b \sin c}$

82. $k = \dfrac{a+b+c}{2}$

Case II. *Given the three angles,* A, B, C

To find one of the sides.

83. $\sin^2 \frac{1}{2}a = \dfrac{-\cos K \cos(K-A)}{\sin B \sin C}$

or,

84. $\cos^2 \frac{1}{2}a = \frac{\cos(K - B)\cos(K - C)}{\sin B \sin C}$

85. $K = \frac{A + B + C}{2}$

Case III. *Given two sides a and b, and the included angle* C.

1°. To find the two other angles A and B.

86. $\tan \frac{1}{2}(A + B) = \cot \frac{1}{2}C . \frac{\cos \frac{1}{2}(a - b)}{\cos \frac{1}{2}(a + b)}$ } Napier's Analogies.

87. $\tan \frac{1}{2}(A - B) = \cot \frac{1}{2}C . \frac{\sin \frac{1}{2}(a - b)}{\sin \frac{1}{2}(a + b)}$ }

2°. To find the third side *c*.

88. $\tan \frac{1}{2}c = \tan \frac{1}{2}(a - b) . \frac{\sin \frac{1}{2}(A + B)}{\sin \frac{1}{2}(A - B)}$

or,

$\tan \frac{1}{2}c = \tan \frac{1}{2}(a + b) . \frac{\cos \frac{1}{2}(A + B)}{\cos \frac{1}{2}(A - B)}$

or equa. 73.

Case IV. *Given two angles* A *and* B, *and the adjacent side* **c**.

1°. To find the other two sides, *a* and *b*.

89. $\tan \frac{1}{2}(a + b) = \tan \frac{1}{2}c . \frac{\cos \frac{1}{2}(A - B)}{\cos \frac{1}{2}(A + B)}$ } Napier's Analogies.

90. $\tan \frac{1}{2}(a - b) = \tan \frac{1}{2}c . \frac{\sin \frac{1}{2}(A - B)}{\sin \frac{1}{2}(A + B)}$ }

2°. To find the third angle C.

91. $\cot \frac{1}{2}C = \tan \frac{1}{2}(A - B) . \frac{\sin \frac{1}{2}(a + b)}{\sin \frac{1}{2}(a - b)}$

or,

$\cot \frac{1}{2}C = \tan \frac{1}{2}(A + B) . \frac{\cos \frac{1}{2}(a + b)}{\cos \frac{1}{2}(a - b)}$

or equa. 73.

Case V. *Given two sides a, b, and an opposite angle* A.

To find the other opposite angle B; take equation 73, or the proportion; sines of the angles are as sines of the opposite sides. (For the methods of determining the remaining angle and side, see page 402, Case 3.)

Case VI. *Given two angles* A, B, *and an opposite side a.*

To find the other opposite side *b*; sines of the angle are proportional to the sines of the opposite sides. (For the methods of determining the remaining side and angle, see page 402, Case 4.)

OTHER METHODS OF RESOLVING OBLIQUE-ANGLED SPHERICAL TRIANGLES.*

Except when three sides or three angles are given, the data always include an angle A, and the adjacent side b, besides a third part. The required parts in the different cases may be found by the following formulæ, and formula 73.

92. $\tan m = \tan b \cos A$ 93. $\cot n = \tan A \cos b$

94. $c = m + m'$ 95. $C = n + n'$

96. $\frac{\cos a}{\cos b} = \frac{\cos m'}{\cos m}$ 97. $\frac{\cos A}{\cos B} = \frac{\sin n}{\sin n'}$

98. $\frac{\tan A}{\tan B} = \frac{\sin m'}{\sin m}$ 99. $\frac{\tan a}{\tan b} = \frac{\cos n}{\cos n'}$

100. $\sin k = \sin A \sin b.$

From the angle C (Fig. 120) a perpendicular CD is let fall upon the opposite side c, which divides the triangle into two right-angled triangles, that are resolved separately. In the one, ACD, A and b are known, and it is easy to find the other parts, which, joined to the third given part, serve to resolve the second right-angled triangle BCD, and determine the unknown quantity required. m, m' denote the two segments of the base; n, n' the two parts of the angle C; and k the perpendicular arc CD.

Fig. 120.

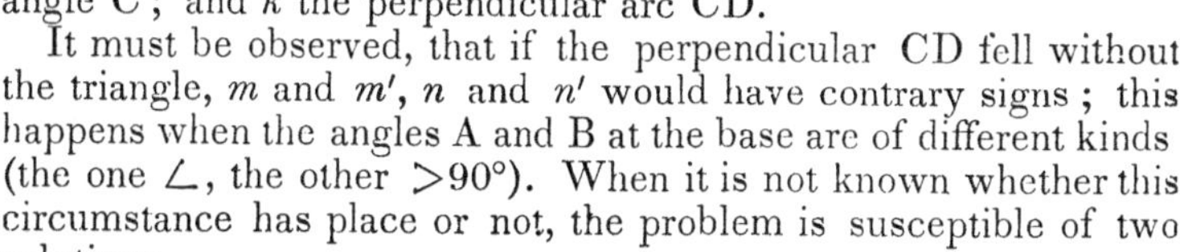

It must be observed, that if the perpendicular CD fell without the triangle, m and m', n and n' would have contrary signs; this happens when the angles A and B at the base are of different kinds (the one $\angle$, the other $>90°$). When it is not known whether this circumstance has place or not, the problem is susceptible of two solutions.

The detail of the different cases is as follows: the data are A, b, and another arc or angle.

Case 1. *Given two sides and the included angle; or* b, c, A.

Equation 92 makes known m, 94 m', which may be negative (what the calculation shows,) 96 a, 98 B, and equation 73, (page 399,) C, which is known in kind.

Case 2. *Given two angles and the adjacent side; or* A, C, b.

Equation 93 makes known n, 95 n', which may be negative (what the calculation shows,) 97 B, 99 a; finally, equation 73 (page 399) gives c, which is known in kind.

* Francœur's Practical Astronomy.

Case 3. *Given two sides and an opposite angle; or b, a, A.*

Equation 92 gives m, 96 m', 94 c, 98 and 73 B and C;

or else, 93 gives n, 99 n', 95 C, 97 and 73 B and c.

This problem admits in general of two solutions. In effect, the arc m' or angle n' being given by its cos., may have either the sign + or −; there are then two values for c, and also for C. m' and n' enter into equations 97 and 98 by their sines, whence result therefore also two values of B.

Case 4. *Given two angles, and an opposite side; or* A, B, b.

Equation 92 gives m, 98 m', 94 c, 96 a, and equation 73 makes known C;

or else 93 gives n, 97 n', 95 C, 99 and 73 a and c.

There are also two solutions in this case; for, m' or n' is given by a sin., and therefore two supplementary arcs satisfy the question. Thus c in 94, and a in 96, receive two values; same for C in 95, and a in 99, &c.

Instead of solving the two right-angled triangles, into which the oblique-angled triangle is divided, by equations 92 to 99, we may employ Napier's rules, from which these equations have been obtained.

Isosceles Triangles.

When the triangle is *isosceles*, B = C, $b = c$, the perpendicular arc must be let fall from the vertex A, and the equations furnished by Napier's rules, become very simple. We find

101. $\sin \frac{1}{2} a = \sin \frac{1}{2} A \sin b$

102. $\tan \frac{1}{2} a = \tan b \cos B$

103. $\cos b = \cot B \cot \frac{1}{2} A$

104. $\cos \frac{1}{2} A = \cos \frac{1}{2} a \sin B$

The knowledge of two of the four elements A, B, a, b, which form the isosceles triangle, is sufficient for the determination of the two others.

INVESTIGATION OF ASTRONOMICAL FORMULÆ.

Formulæ for the Parallax in Right Ascension and Declination, and in Longitude and Latitude. (See Article 93, page 65.)

Fig. 121

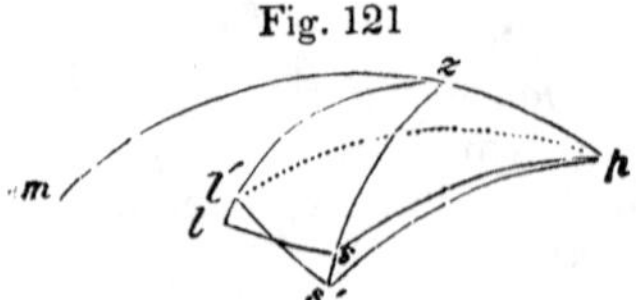

Let s (Fig. 121) be the *true place* of a star seen from the centre of the earth, s' the *apparent place*, seen from a point on the surface of which z is the zenith, the latitude being l. The displacement $ss' = p$ is the parallax in altitude, which takes effect in the vertical circle zs'; p is the pole; the hour angle $zps = q$ is changed into zps', and $sps' = \alpha$ is the *variation of the hour angle, or the parallax in right ascension*; the polar distance $ps = d$ is changed into ps'; the difference δ of these arcs is the *parallax in declination* or of *polar distance*.* We have, (For. 73, p. 399,)

$$\sin s' : \sin ps\ (d) :: \sin sps'\ (\alpha) : \sin ss'\ (p),$$
$$\sin zps'\ (q + \alpha) : \sin zs'\ (Z) :: \sin s' : \sin pz\ (90^\circ - l).$$

Multiplying, term by term, we obtain

$$\sin s' \sin (q + \alpha) : \sin d \sin Z :: \sin \alpha \sin s' : \sin p \cos l;$$

whence, $$\sin \alpha = \frac{\sin p \cos l}{\sin d \sin Z} \sin (q + \alpha).$$

Or, substituting for p its value given by equa. (8,) p. 62, and replacing H by P,

$$\sin \alpha = \frac{\sin P \cos l}{\sin d} \sin (q + \alpha) \ldots (A).$$

This equation makes known α when the apparent hour angle $zps = q + \alpha$, seen from the earth's surface, is given; but if we know the true hour angle $zps = q$, seen from the centre of the earth, developing $\sin (q + \alpha)$, (For. 15, p. 395), and putting $\frac{\sin P \cos l}{\sin d} = m$,

$$\sin \alpha = m (\sin q \cos \alpha + \sin \alpha \cos q),$$

or, dividing by $\sin \alpha$,

$$1 = m (\sin q \cot \alpha + \cos q);$$

whence, by transformation,

$$\tan \alpha = \frac{m \sin q}{1 - m \cos q} = m \sin q + m^2 \sin q \cos q \text{ (very nearly.)}$$

Restoring the value of m,

$$\tan \alpha = \frac{\sin P \cos l}{\sin d} \sin q + \left(\frac{\sin P \cos l}{\sin d}\right)^2 \sin q \cos q.$$

Putting the arc α in place of its tangent, and P in place of sin P, and expressing these arcs in seconds, (For. 47, p. 397,) there results,

$$\alpha = \frac{P \cos l}{\sin d} \sin q + \left(\frac{P \cos l}{\sin d}\right)^2 \sin q \cos q \sin 1'' \ldots (B).$$

The *parallax in declination* (δ) is the difference of the arcs ps ($= d$) and ps' ($= d + \delta$.) Let $zs = z$, and $zs' = Z$. The triangles zps and zps' give (For. 74 and 73),

$$1^\circ.\ \cos pzs = \frac{\cos d - \sin l \cos z}{\cos l \sin z} = \frac{\cos (d + \delta) - \sin l \cos Z}{\cos l \sin Z},$$

* Francœur's Uranography, p. 418.

2°. $\sin pzs = \frac{\sin d \sin q}{\sin z} = \frac{\sin (d+\delta) \sin (q+\alpha)}{\sin Z}.$

From the first equation we derive

$$\cos (d+\delta) = \frac{\cos d \sin Z - \sin l \cos z \sin Z}{\sin z} + \sin l \cos Z$$

$$= \frac{\cos d \sin Z - \sin l (\cos z \sin Z - \sin z \cos Z)}{\sin z}$$

$$= \frac{\cos d \sin Z - \sin l \sin (Z - z)}{\sin z},$$

or, (equ. 8, p. 62,)

$$= \frac{\sin Z}{\sin z} (\cos d - \sin P \sin l);$$

from the second,

$$\frac{\sin Z}{\sin z} = \frac{\sin (d+\delta)}{\sin d} \cdot \frac{\sin (q+\alpha)}{\sin q}:$$

substituting,

$$\cos (d+\delta) = \frac{\sin (d+\delta)}{\sin d} \cdot \frac{\sin (q+\alpha)}{\sin q} (\cos d - \sin P \sin l)$$

$$\frac{\cos (d+\delta)}{\sin (d+\delta)} = \frac{\sin (q+\alpha)}{\sin q} \left(\frac{\cos d}{\sin d} - \frac{\sin P \sin l}{\sin d}\right)$$

$$\cot (d+\delta) = \frac{\sin (q+\alpha)}{\sin q} \left(\cot d - \frac{\sin P \sin l}{\sin d}\right) \ldots \text{(C)}.$$

Put $\tan x = \frac{\sin P \sin l}{\sin d}$;

$$\text{then, } \cot (d+\delta) = \frac{\sin (q+\alpha)}{\sin q} (\cot d - \tan x)$$

$$= \frac{\sin (q+\alpha)}{\sin q} \left(\frac{\cos d}{\sin d} - \frac{\sin x}{\cos x}\right)$$

$$= \frac{\sin (q+\alpha)}{\sin q} \cdot \frac{\cos d \cos x - \sin d \sin x}{\sin d \cos x}$$

$$= \frac{\sin (q+\alpha) \cos (d+x)}{\sin q \sin d \cos x} \ldots \text{(D)}.$$

The apparent polar distance $(d+\delta)$ being computed by either of the formulæ (C) and (D), we have $\delta = (d+\delta) - d$.

Formulæ may be obtained that will give the parallax in declination without first finding the apparent declination, (except approximately.)

From equa. (C) we obtain

$$\frac{\sin P \sin l}{\sin d} = \cot d - \frac{\sin q \cot (d+\delta)}{\sin (q+\alpha)},$$

and we also have

$$\cot d - \cot(d+\delta) = \frac{\cos d}{\sin d} - \frac{\cos(d+\delta)}{\sin(d+\delta)} = \frac{\sin\delta}{\sin d \sin(d+\delta)};$$

the sum of these equations gives

$$\frac{\sin P \sin l}{\sin d} = \cot(d+\delta)\left(1 - \frac{\sin q}{\sin(q+\alpha)}\right) + \frac{\sin\delta}{\sin d \sin(d+\delta)}.$$

Now, $$1 - \frac{\sin q}{\sin(q+a)} = \frac{\sin(q+\alpha) - \sin q}{\sin(q+\alpha)}$$

$$= \frac{2\sin\frac{1}{2}\alpha\cos(q+\frac{1}{2}\alpha)}{\sin(q+\alpha)} = \frac{\sin\alpha\cos(q+\frac{1}{2}\alpha)}{\sin(q+\alpha)\cos\frac{1}{2}\alpha} \text{ (For. 22, 13)}$$

$$= \frac{\cos(q+\frac{1}{2}\alpha)\sin P\cos l}{\sin d\cos\frac{1}{2}\alpha}, \text{ by equa. (A).}$$

Substituting,

$$\frac{\sin P \sin l}{\sin d} = \cot(d+\delta)\frac{\cos(q+\frac{1}{2}\alpha)\sin P\cos l}{\sin d\cos\frac{1}{2}\alpha} + \frac{\sin\delta}{\sin d\sin(d+\delta)},$$

or, $$\sin\delta = \sin P\sin l\sin(d+\delta) - \frac{\cos(d+\delta)\cos(q+\frac{1}{2}\alpha)\sin P\cos l}{\cos\frac{1}{2}\alpha} \ldots \text{(E)}$$

$$= \sin P\sin l\,[\sin(d+\delta) - \tan y\cos(d+\delta)],$$

making $$\tan y = \frac{\cot l\cos(q+\frac{1}{2}a)}{\cos\frac{1}{2}a};$$

whence, $$\sin\delta = \frac{\sin P\sin l}{\cos y}\sin(d+\delta-y) \;\ldots\; \text{(F)}.$$

To facilitate the calculation, the sines of δ and P in eqs. (E) and (F), may be replaced by the arcs.

To obtain an expression for the parallax in declination in terms of the *true declination*, develope $\sin(d+\delta-y)$ in equation (F) which gives

$$\sin\delta = \frac{\sin P\sin l}{\cos y}[\sin(d+\delta)\cos y - \sin y\cos(d+\delta)];$$

developing $\sin(d+\delta)$ and $\cos(d+\delta)$, and reducing, we have

$$\sin\delta = \frac{\sin P\sin l}{\cos y}[\sin(d-y)\cos\delta + \cos(d-y)\sin\delta];$$

dividing by $\cos\delta$,

$$\tan\delta = \frac{\sin P\sin l}{\cos y}[\sin(d-y) + \cos(d-y)\tan\delta],$$

$$\text{whence } \tan \delta = \frac{\dfrac{\sin P \sin l}{\cos y} \sin (d-y)}{1 - \dfrac{\sin P \sin l}{\cos y} \cos (d-y)}$$

$$= \frac{\sin P \sin l}{\cos y} \sin (d-y) + \left(\frac{\sin P \sin l}{\cos y}\right)^2 \times$$

$$\sin (d-y) \cos (d-y) \text{ (very nearly ;)}$$

or, replacing tan δ and sin P by δ and P, expressing these **arcs in** seconds, (For. 47, p. 397), and reducing by For. 13, p. 395,

$$\delta = \frac{P \sin l}{\cos y} \sin (d-y) + \left(\frac{P \sin l}{\cos y}\right)^2 \frac{\sin 1''}{2} \sin 2(d-y) \quad \ldots \text{(G.)}$$

If the place of a body be referred to the ecliptic, similar formulæ will give the *parallax in latitude and longitude*, but as the ecliptic and its pole are continually in motion by virtue of the diurnal rotation of the heavens, it is necessary, in order to be able to determine the parallax in longitude at any given instant, to know the situation of the ecliptic at the same instant.

This is ascertained by finding the situation of the point of the ecliptic 90° distant from the points in which it cuts the horizon, and which are respectively just rising and setting, called the *Nonagesimal Degree*, or the *Nonagesimal*.

Fig. 122.

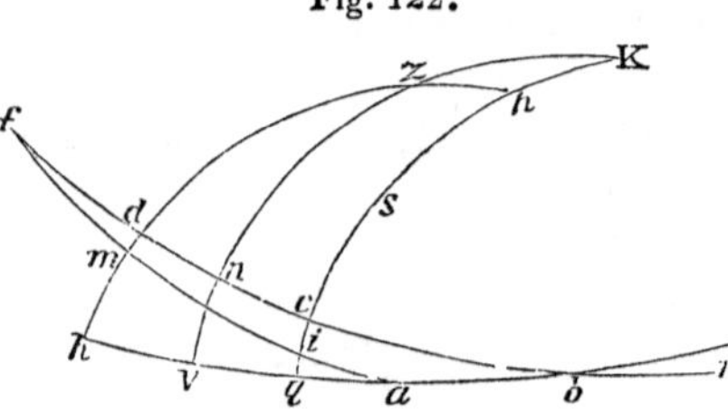

Let K (Fig. 122) be the pole of the ecliptic fb, p the pole of the equator fa ; f is the vernal equinox, the origin of longitudes and of right ascensions; hbs is the eastern horizon, b the *horoscope*, or the point of the ecliptic which is just rising; $pz = 90° - l$ (the latitude of given place) ; $Kp = \omega$ the obliquity of the ecliptic. The circle $Kznv$ is at the same time perpendicular at n to the ecliptic fb, and at v to the horizon hb ; it is a circle of latitude and a vertical circle, since it passes through the pole K and the zenith z: b is 90° from all the points of the circle Knv ; zn is the latitude of the zenith, fn its longitude ; the point n is the nonagesimal, since $bn =$ 90° ; nv is the altitude of this point, and the complement of zn ; nv measures the inclination of the ecliptic to the horizon at the given instant, or the angle b, so that $b = nv = Kz$; thus $fn = N$ *the longitude of the nonagesimal*, and $nv = h$ *the altitude of the nonagesimal*, designate the situation of this point, and consequently ascertain the position of the ecliptic and its pole **at the** moment of observation.*

* Francœur's Uranography, p. 421

The points m and d are those of the equator and ecliptic which are on the meridian; the arc fm, in time, is the sidereal time s. which is known; the arc $fi = 90°$, since the plane Kpi, passing through the poles K and p, is at the same time perpendicular to the ecliptic and to the equator; the arc $mi = fi - fm = 90° - s$; then the angle $zpK = 180° - zpi = 180° - mi = 90° + s$.*

Now, in the spherical triangle pKz we know the sides $Kp = \omega$, $zp = 90° - l = H$, and the included angle $zpK = 90° + s$; and may therefore find $Kz = h$ the altitude of the nonagesimal, and the angle $pKz = nc = fc - fn = 90° - N =$ complement of the longitude N of the nonagesimal. Let S = sum of the angles Kzp and zKp, then, (For. 86, page 400,)

$$\tan \tfrac{1}{2}S = \frac{\cos \frac{1}{2}(H - \omega)}{\cos \frac{1}{2}(H + \omega)} \cdot \cot \tfrac{1}{2}(90° + s),$$

or,

$$\tan \tfrac{1}{2}S = \frac{\cos \frac{1}{2}(H - \omega)}{\cos \frac{1}{2}(H + \omega)} \cdot \tan \tfrac{1}{2}(90° - s):$$

but,

$$\tan \tfrac{1}{2}S = -\tan(180° - \tfrac{1}{2}S), \text{ and } \tan \tfrac{1}{2}(90° - s) = -\tan \tfrac{1}{2}(s - 90°);$$

substituting, and denoting $(180° - \frac{1}{2}S)$ by E, we have

$$\tan E = \frac{\cos \frac{1}{2}(H - \omega)}{\cos \frac{1}{2}(H + \omega)} \cdot \tan \tfrac{1}{2}(s - 90°) \quad . \; . \; . \quad (H).$$

Again, let $D = zKp - Kzp$, then, (For. 87,)

$$\tan \tfrac{1}{2}D = \frac{\sin \frac{1}{2}(H - \omega)}{\sin \frac{1}{2}(H + \omega)} \cdot \cot \tfrac{1}{2}(90° + s);$$

whence, by transforming as above, and denoting $(180° - \frac{1}{2}D)$ by F we have

$$\tan F = \frac{\sin \frac{1}{2}(H - \omega)}{\sin \frac{1}{2}(H + \omega)} \cdot \tan \tfrac{1}{2}(s - 90°) \; . \; . \; . \; (I).$$

Now, $\quad \frac{1}{2}S + \frac{1}{2}D = pKz = 90° - N;$

whence, $\quad N = 90° - (\frac{1}{2}S + \frac{1}{2}D),$

or,

$$N = 360° + 90° - (\tfrac{1}{2}S + \tfrac{1}{2}D) = 180° - \tfrac{1}{2}S + 180° - \tfrac{1}{2}D + 90°;$$

consequently, $\quad N = E + F + 90° \; . \; . \; . \; (J),$

rejecting 360° when the sum exceeds that number.

Next, for the altitude of the nonagesimal, we have, (For. 88,)

$$\tan \tfrac{1}{2}h = \frac{\cos \frac{1}{2}S}{\cos \frac{1}{2}D} \cdot \tan \tfrac{1}{2}(H + \omega),$$

$$= \frac{\cos E}{\cos F} \cdot \tan \tfrac{1}{2}(H + \omega) \; . \; . \; . \; (K).$$

N and h being known, to obtain the *formulæ for the parallax in longitude and latitude*, we have only to replace in the formulæ

* Francœur's Uranography, p. 421.

for the parallax in right ascension and declination, the altitude l of the pole of the equator by that $90°-h$ of the pole K of the ecliptic, and the distance im of the star s from the meridian by the distance nc to the vertical through the nonagesimal. Let us change then in formulæ (A), (B), (C), (D), (E), (F), and (G), l into $90°-h$, and q into $fc-fn=\text{L}-\text{N}$, L being the longitude fc of the star s. Besides, d will become the distance sK to the pole of the ecliptic, or complement of the latitude $\lambda = sc$. Making these substitutions, and denoting the parallax in longitude by Π, and the parallax in latitude by π, we obtain in terms of the apparent longitude and latitude,

$$\sin \Pi = \frac{\sin \text{P} \sin h}{\sin d} (\sin \overline{\text{L}-\text{N}+\Pi}) \quad \ldots \text{(L)},$$

$$\cot (d+\pi) = \frac{\sin (\text{L}-\text{N}+\Pi)}{\sin (\text{L}-\text{N})} \left(\cot d - \frac{\sin \text{P} \cos h}{\sin d}\right) \quad \ldots \text{(M)},$$

$$\tan x = \frac{\sin \text{P} \cos h}{\sin d} \quad \ldots \text{(N)},$$

$$\cot (d+\pi) = \frac{\sin (\text{L}-\text{N}+\Pi) \cos (d+x)}{\sin (\text{L}-\text{N}) \sin d \cos x} \quad \ldots \text{(O)},$$

$$\sin \pi = \sin \text{P} \cos h \sin (d+\pi) - \frac{\cos (d+\pi) \cos (\text{L}-\text{N}+\frac{1}{2}\Pi) \sin \text{P} \sin h}{\cos \frac{1}{2}\Pi} \quad \ldots \text{(P)},$$

$$\tan y = \frac{\tan h \cos (\text{L}-\text{N}+\frac{1}{2}\Pi)}{\cos \frac{1}{2}\Pi} \quad \ldots \text{(Q)},$$

$$\sin \pi = \frac{\sin \text{P} \cos h}{\cos y} \sin (d+\pi-y) \quad \ldots \text{(R)};$$

and in terms of the true longitude and latitude,

$$\Pi = \frac{\text{P} \sin h}{\sin d} \sin (\text{L}-\text{N}) + \left(\frac{\text{P} \sin h}{\sin d}\right)^2 \times \sin (\text{L}-\text{N}) \cos (\text{L}-\text{N}) \sin 1'' \quad \ldots \text{(S)},$$

$$\pi = \frac{\text{P} \cos h}{\cos y} \sin (d-y) + \tfrac{1}{2} \left(\frac{\text{P} \cos h}{\cos y}\right)^2 \times \sin 2(d-y) \sin 1'' \quad \ldots \text{(T)},$$

$$\tan y = \frac{\tan h \cos (\text{L}-\text{N}+\frac{1}{2}\Pi)}{\cos \frac{1}{2}\Pi}.$$

To facilitate the computation, $\sin \Pi$, $\sin \pi$, and $\sin \text{P}$, in formulæ (L), (P), and (R), may be replaced by the arcs themselves.

The distance d of the star from the pole of the ecliptic enters into these formulæ in place of the latitude λ.

To find the apparent distance d', we have

$$d' = d + \pi;$$

for the apparent latitude λ',

$$\lambda' = \lambda - \pi;$$

for the apparent longitude L′,

$$L' = L + \Pi.$$

The logarithmic formulæ given on page 352, were derived from equations (L), (O), and (P), and the logarithmic formula on page 353 from equa. (O).

To determine now the effect of parallax upon the apparent diameter of the moon.

Let ACB (Fig. 71, p. 163) represent the moon, and E the station of an observer; also let R = apparent semi-diameter of the moon, and D = its distance. The triangle AES gives

$$\sin AES = \frac{AS}{ES}, \text{ or } \sin R = \frac{AS}{D}.$$

At any other distance D′ we should have for the apparent semi-diameter R′,

$$\sin R' = \frac{AS}{D'};$$

whence,

$$\frac{\sin R'}{\sin R} = \frac{D}{D'}.$$

Thus, if R′ = moon's apparent semi-diameter to an observer at the earth's surface, as at O (Fig. 34, p. 61), R = the same as it would be seen from the centre C, and S represents the situation of the moon,

$$\frac{\sin R'}{\sin R} = \frac{CS}{OS} = \frac{\sin ZOS}{\sin ZCS} = \frac{\sin Z}{\sin z}.$$

But we have, (see page 404,)

$$\frac{\sin Z}{\sin z} = \frac{(\sin d + \delta)}{\sin d} \cdot \frac{\sin (q + a)}{\sin q},$$

or, in terms of the apparent longitude and latitude, (see page 408,)

$$\frac{\sin Z}{\sin z} = \frac{\sin (d + \pi)}{\sin d} \cdot \frac{\sin (L - N + \Pi)}{\sin (L - N)}.$$

Hence, $$\sin R' = \frac{\sin R \sin (d + \pi) \sin (L - N + \Pi)}{\sin d \sin (L - N)} \dots (U)$$

*Aberration in Longitude and Latitude, and in Right Ascension and Declination.** (See Art. 100, page 70.)

Aberration is caused by the motion of light in conjunction with the motion of the earth. Light comes to us from the sun in 8^{m}. $17^{s}.8$, during which time the earth describes an arc $a = 20''.44$,

* Francœur's Uranography, p. 442, &c.

of its orbit *pbdin* (Fig. 123,) supposed circular: p is the place of the earth. Let us take any plane whatsoever, which we will call *relative*, passing through the star and the sun, and let dd' be the intersection of this plane and the ecliptic, with which it makes an angle k: let us seek the quantity φ by which the aberration displaces the star in the direction perpendicular to this plane. The question is to project on to a line perpendicular to the relative plane, the small constant arc a which the earth describes, this being the quantity that the star is displaced from its line of direction in a direction parallel to the line of the earth's motion, (see Art. 196 of the text:) this projection is φ, variable according to the position of the *relative plane* in relation to which it is estimated. The velocity along the tangent at p, makes with ph an angle $\theta = pch =$ the arc pd'; $a \cos \theta$ is then the projection of this velocity on the line ph. The angle of our two planes being k, this projection will be reduced to $a \cos \theta \sin k$, when it is taken perpendicularly to the relative plane. Thus,

Fig. 123.

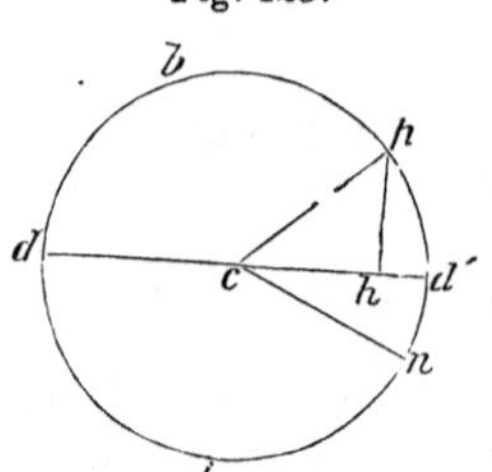

$$\varphi = a \sin k \cos \theta \; . \; . \; . \; (V).$$

The aberration displaces the star from the relative plane by this quantity φ, k designating the inclination of this plane to the ecliptic, and θ the arc pd', reckoned from p the place of the earth to d' the point of intersection of these two planes. Let us give to the relative plane the positions which are met with in applications.

Let us suppose at first that $k = 90°$, or $\sin k = 1$; the relative plane will then be perpendicular to the ecliptic. Let n be the vernal equinox; we have $pd' = np - nd'$; np is the longitude of the earth, or $180° +$ that ⊙ of the sun; nd' is the longitude l of the star; whence

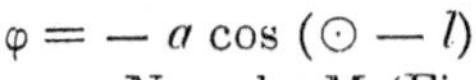

$$\varphi = -a \cos (\odot - l).$$

Now, let M (Fig. 124) be the true place of the star, M′ the star as displaced by aberration, KM is the circle of true latitude, KM′ the circle of apparent latitude, and MM′ $= \varphi$: this arc has its centre C on the axis which passes through the pole K of the ecliptic; the longitude of the star is then altered by the part OO′ of the ecliptic comprised between these two planes; and since OO′ is to the arc MM′ as the radius 1 is to the radius CM $=$ sin KM $=$ cos latitude λ of the star, we have

Fig. 124.

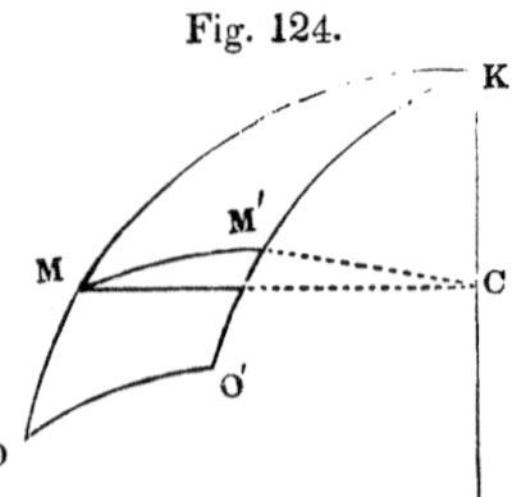

$$\text{aberr. in long.} = -\frac{a}{\cos \lambda} \cos (\odot - l) \; . \; . \; . \; (W).$$

If the relative plane is kc, (Fig. 125,) perpendicular to the circle of latitude Kcd, the aberration φ perpendicularly to it, will be the aberration in latitude. Let kd be the ecliptic, and o the earth; the angle k is measured by the arc cd $=\lambda$; the arc $ok = \theta = \odot -$ long. of k; and as $kd = 90°$, long. of point $k = l - 90°$: substituting in equation (V), we find

Fig. 125.

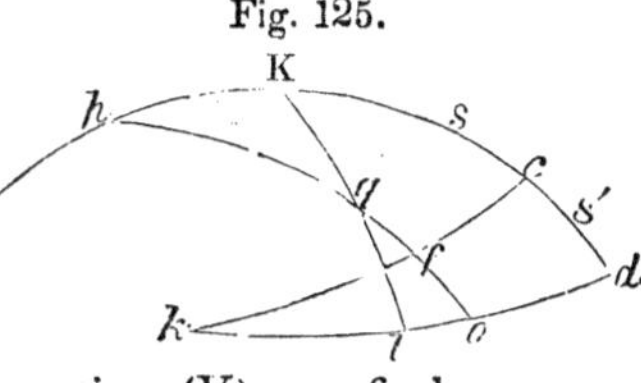

$$\text{aberr. in lat.} = -a \sin\lambda \sin(\odot - l) \quad . \quad . \quad . \quad \text{(X)}.$$

These aberrations of the star produce a small apparent orbit, which is confounded with its projection on the tangent plane to the celestial sphere. Let us suppose the orbit to be referred to two co-ordinate axes passing through the true place of the star and lying in the tangent plane, of which one is parallel to the plane of the ecliptic, and the other perpendicular to this, or tangent to the circle of latitude at the star; and let $\frac{x}{\cos\lambda}$ = aberr. in long., and y = aberr. in lat.; y will be the ordinate, and x (the aberr. in long., reduced to the parallel through the star) the abscissa: we have

$$\frac{x}{\cos\lambda} = -\frac{a}{\cos\lambda}\cos(\odot - l),$$

$$y = -a \sin\lambda \sin(\odot - l);$$

or,

$$\frac{x}{a} = -\cos(\odot - l),$$

$$\frac{y}{a\sin\lambda} = -\sin(\odot - l).$$

Squaring the last two equations, and adding them together, $\odot$ disappears, and we find

$$y^2 + x^2 \sin^2\lambda = a^2 \sin^2\lambda \quad . \quad . \quad . \quad \text{(Y)},$$

whatever may be the place of the earth. Such is the equation of the *apparent orbit*, which, as we perceive, is an ellipse of which the semi-axes are a and $a \sin\lambda$, and whose centre is the true place of the star. When the star is at the pole of the ecliptic, $\lambda = 90°$, and the ellipse becomes a circle of which the radius is a. When $\lambda = 0$, this ellipse is reduced to an arc $2a$ of the ecliptic.

To find the aberration in right ascension, the relative plane must be perpendicular to the equator. Let kc be the equator, (Fig. 125,) p its pole, psd the relative plane, which is the circle of declination of the star s; kd the ecliptic, o the earth, k the vernal equinox, kc = R, sc = D. Aberration carries the star s out of the plane pcd a distance φ, which it is the question to determine. Equa. (V) is here

$$\varphi = a \sin d \cos do = a \sin d \cos(kd - ko)$$
$$= a \sin d (\cos kd \cos ko + \sin kd \sin ko)$$
$$= a \sin d \cos kd \cos ko + a \sin d \sin kd \sin ko$$

but ko = long. of earth $= 180° + \odot$; we have also the angle $k =$ the obliquity ω of the ecliptic, and the right-angled spherical triangle kcd gives, by Napier's rules,

$$\cot kd = \cot R \cos \omega, \quad \sin d \sin kd = \sin R.$$

The 1st equa. multiplied by the 2d, gives

$$\sin d \cos kd = \cos R \cos \omega,$$

whence $\quad \varphi = - a(\cos R \cos \omega \cos \odot + \sin R \sin \odot).$

The displacement from M to M′ (Fig. 124) conducts, as before, to the division of φ by cos D, to have the corresponding arc of the equator: thus the *aberration in right ascension is,*

$$u = - a \sin R \sec D \sin \odot - a \cos \omega \cos R \sec D \cos \odot \ (Z).$$

Taking the relative plane perpendicular to the circle of declination, we find for the *aberration in declination,*

$$v = - a \sin D \cos R \sin \odot - a \cos \omega (\tan \omega \cos D - \sin R \sin D) \cos \odot \ . \ . \ . \ (a).$$

These formulæ may easily be adapted to logarithmic computation:

In formula (Z) let $a \sin R \sec D = A$, and $a \cos \omega \cos R \sec D = B$; then,

$$u = - A\left(\sin \odot + \frac{B}{A} \cos \odot\right) \ . \ . \ . \ (Z').$$

$$\text{Put } \tan \varphi = \frac{B}{A} = \frac{a \cos \omega \cos R \sec D}{a \sin R \sec D} = \cos \omega \cot R \ . \ . \ . \ (b)$$

and we shall have

$$u = - A\left(\sin \odot + \frac{\sin \varphi}{\cos \varphi} \cos \odot\right)$$

$$= - A \frac{\sin \odot \cos \varphi + \sin \varphi \cos \odot}{\cos \varphi}$$

$$= - \frac{A}{\cos \varphi} \sin (\odot + \varphi).$$

Restoring the value of A, and taking $\frac{1}{\cos D}$ for sec D, we obtain

$$u = - \frac{a \sin R}{\cos D \cos \varphi} \sin (\odot + \varphi) \ . \ . \ . \ (c).$$

The auxiliary arc φ is given by equation (b); it must be substituted in equation (c), with its sign, and we then obtain u. Tan φ, and the co-efficient of $\sin (\odot + \varphi)$ are constant, for the same star, for a long period of time, since these quantities vary very slowly with ω and the precession. Moreover, the co-efficient of $\sin (\odot + \varphi)$ is the maximum value of u, since it answers to $\sin (\odot + \varphi) = 1$. Thus we shall be able to calculate in advance, for any designated star, the values of φ and of the *maximum* of the aber ration in right ascension, or of the logarithm of this maximum.

The results of these calculations for 50 principal stars are given in Table XCI, columns entitled M and φ.

If in equation (a) we make $a \sin D \cos R = A'$, and $a \cos \omega (\tan \omega \cos D - \sin R \sin D) = B'$, we shall have the equation

$$v = -A' (\sin \odot + \frac{B'}{A'} \cos \odot),$$

in which A′ and B′ are constants. This equation is of the same form with equa.(Z′). We therefore have, in the same manner as for the right ascension,

$$\tan \theta = \frac{B'}{A'} = \frac{a \cos \omega (\tan \omega \cos D - \sin R \sin D)}{a \sin D \cos R}$$

$$= \frac{a \sin \omega \cos D - a \cos \omega \sin R \sin D}{a \sin D \cos R}$$

$$= \frac{\sin \omega \cot D}{\cos R} - \cos \omega \tan R \quad . \; . \; . \; (d).$$

$$v = -\frac{A'}{\cos \theta} \sin (\odot + \theta) = -\frac{a \sin D \cos R}{\cos \theta} \times \sin (\odot + \theta) \quad . \; . \; . \; (e).$$

θ is given by equation (d), and being substituted in equation (e), we shall have v. θ and the co-efficient of $\sin (\odot + \theta)$ are constant for the same star, and we can therefore calculate in advance the value of this arc, and of the co-efficient, which is the *maximum* of the aberration in declination. Columns entitled θ and N, Table XCI, contain the quantities θ and the logarithms of the maxima of the aberration in declination for 50 principal stars.

For convenience in calculation, the angles φ, θ, and the maxima, M, N, in Table XCI, have been rendered positive in all cases. This has been accomplished by adding $12^{s.}$ to φ and θ whenever the calculation conducted to a negative value, and by adding $6^{s.}$ to $\odot + \varphi$, or $\odot + \theta$, whenever the co-efficient had the sign —, (this sign being changed to +;) in this manner the sign of each of the two factors is changed, which does not alter the sign of the pro duct.

*Formulæ for the Nutation in Right Ascension and Declination.**
(See Article 124, p. 90.)

In deriving these formulæ, we must begin with borrowing certain results established by Physical Astronomy. It has been proved, in confirmation of Bradley's conjectures, that the phenomena of nutation are explicable on the hypothesis of the pole of the earth describing around its mean place (that place which, see page

* Woodhouse's Astronomy, p. 357, &c.

87, it would hold in the small circle described around the pole of the ecliptic, were there no *inequality* of precession) an ellipse, in a period equal to the revolution of the moon's nodes. The major axis of this ellipse is situated in the solstitial colure and equal to 18″.50; it bears that proportion to the minor axis (such are the results of theory) which the cosine of the obliquity bears to the cosine of twice the obliquity: consequently, the minor axis will be 13″.77.

Let CdA (Fig. 126) represent such an ellipse, P being the mean place of the pole, K the pole of the ecliptic. CDOA is a circle

Fig. 126.

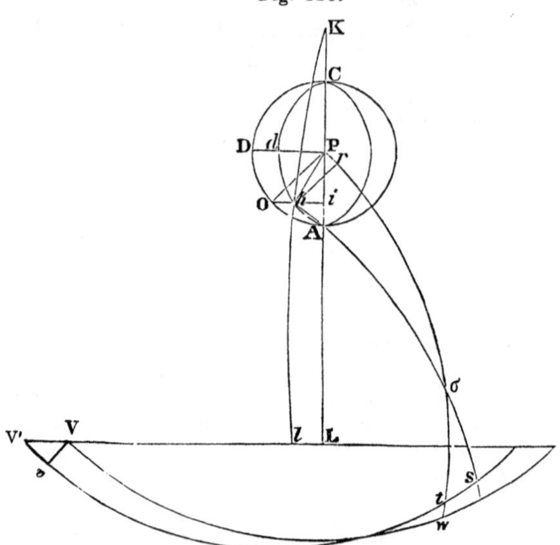

described with the centre P and radius CP. VL is the ecliptic, Vw the equator, KPL the solstitial colure. In order to determine the true place of the pole, take the angle APO equal to the retrogradation of the moon's ascending node from V: draw Oi perpendicular to PA, and the point in the ellipse, through which Oi passes, is the true place of the pole. This construction being admitted, the *nutations* in right ascension and north polar distance may, Pp being very small, be thus easily computed.

Nutation in North Polar Distance.

$$\begin{aligned}\text{Nutation in N. P. D.} &= P\sigma - p\sigma. = Pr = Pp \cos pP\sigma, \text{ nearly,}\\ &= Pp \cos (APp + AP\sigma)\\ &= Pp \cos (APp + R - 90^\circ)\\ &= Pp \sin (APp + R),\end{aligned}$$

R denoting the right ascension.

Nutation in Right Ascension.

The right ascension of the star σ is, by the effect of nutation, changed from Vw into $V'ts$. Now,

$$V'ts = V'v + Vw + ts, \text{ nearly,}$$

whence,
$$Vw - V'ts = -V'v - ts$$
$$= -VV' \cos VV'v - Pp \sin Pp\sigma \frac{\sin \sigma s}{\sin P\sigma};$$

in which expression $V'v\,(=VV' \cos VV'v)$ is, as in the case of precession, common to all stars.

In order to reduce farther the above expression, we have

$$pP\sigma = APp + AP\sigma = APp + R - 90°,$$

$$\text{and } VV' = Ll = Pp \frac{\sin APp}{\sin PK};$$

whence,
$$-V'v - ts = -Pp \sin APp \cot \omega$$
$$- Pp \sin (APp + R - 90°) \cot \text{N. P. D.}$$
$$= -Pp \sin APp \cot \omega + Pp \cos (APp + R) \cot \delta,$$

δ representing the north polar distance, and ω the obliquity of the ecliptic.

But these forms are not convenient for computation. In order to render them convenient, we must, from the properties of the ellipse, deduce the values of Pp, and of the tangent of APp, and then substitute such values in the above expressions: thus,

$$\frac{Pp}{PO} = \frac{\sec APp}{\sec APO} = \frac{\cos APO}{\cos APp} = \frac{\cos (12^s - \Omega)}{\cos APp} = \frac{\cos \Omega}{\cos APp},$$

Ω designating the longitude of the moon's ascending node;

whence
$$Pp = \frac{PO \cos \Omega}{\cos APp}.$$

Again,
$$\frac{\tan APp}{\tan APO} = \frac{pi}{Oi} = \frac{Pd}{PD} = \frac{Pd}{PO};$$

hence,
$$\tan APp = \frac{Pd}{PO} \tan APO = \frac{Pd}{PO} \tan (12^s - \Omega)$$
$$= -\frac{Pd}{PO} \tan \Omega.$$

Now substitute, and there will result

The Nutation in North Polar Distance

$$= \frac{PO \cos \Omega}{\cos APp} (\sin APp \cos R + \cos APp \sin R)$$
$$= PO (\tan APp \cos R \cos \Omega + \cos \Omega \sin R)$$
$$= -Pd \cos R \sin \Omega + PO \cos \Omega \sin R$$
$$= -6''.887 \cos R \sin \Omega + 9''.250 \cos \Omega \sin R \qquad (f)$$

which is the difference, as far as nutation is concerned, between the *mean* and *apparent* north polar distance. The *apparent* north polar distance, therefore, must be had by adding the preceding quantity, with its sign changed, to the mean.

$$\text{Nutation in right ascension} = \text{P}d \sin \Omega \cot \omega$$
$$+ \text{PO} \cos \Omega \cos \text{R} \cot \delta + \text{P}d \sin \Omega \sin \text{R} \cot \delta,$$

which, as far as nutation is concerned, is the difference of the mean and apparent right ascensions: and, consequently, the above expression must be subtracted from the mean, in order to obtain the apparent right ascension; or, which is the same, must be added after a negative sign has been prefixed; in which case, we have, substituting for PO, Pd their numerical values,

The Nutation in Right Ascension

$$= - 6''.887 \sin \Omega \cot \omega$$
$$-9''.250 \cos \Omega \cos \text{R} \cot \delta - 6''.887 \sin \Omega \sin \text{R} \cot \delta \quad . \; . \; . \; (g).$$

Formulæ (f) and (g) are of the same form with (Z) and (a) for the aberrations in right ascension and declination, and therefore formulæ may be derived from them similar to (c) and (e), adapted to logarithmic computation. The quantities corresponding to φ, M, θ, N, have been calculated for the stars in the catalogue of Table XC, and inserted in Table XCI, in the columns entitled φ', M$'$, θ', N$'$.

The *Solar Nutation* arises from like causes as the Lunar, and admits of similar formulæ. As an ellipse, made the locus of the true place of the pole, served to exhibit the effects of the lunar nutation, so an ellipse, of different, and much smaller dimensions, may be made to represent the path which the true pole of the equator would, by reason of the sun's inequality of force in causing precession, describe about the mean place of the pole. Thus, in Figure 130, the ellipse AdC will serve to represent the locus of the pole, when AP $= 0''.545$, P$d = 0''.500$, and APO, instead of being $= \Omega$, is equal to $2 \odot$, or twice the sun's longitude, taken in the order of the signs; the equations, therefore, for the solar nutation in north polar distance, and right ascension, analogous to eqs. f and g will be

The Solar Nutation in North Polar Distance

$$= - 0''.500 \cos \text{R} \sin 2 \odot + 0''.545 \sin \text{R} \cos 2 \odot \quad . \; . \; . \; (h).$$

The Solar Nutation in Right Ascension

$$= - 0''.500 \sin 2 \odot \cot \omega$$
$$\cdot\, 0''.545 \cos 2 \odot \cos \text{R} \cot \delta - 0''.500 \sin 2 \odot \sin \text{R} \cot \delta \; . \; . \; (i).$$

If the apparent place of a star should be required with great precision, it would be necessary to compute the solar nutations from these formulæ, and apply them as corrections to the mean right ascension and declination. The calculation would be performed after the same manner as for the lunar nutation; but it is much abridged by remarking that the form of the equations is the same as that of the equations for the lunar nutation, and that the co-efficients are very nearly the 0.075 of those of the latter equations. Thus we can make use of the same arcs φ', θ', and log. *maxima*, M′, N′, repeat the calculation for the lunar nutation, taking 2 ☉ instead of ☊, and multiply the nutations in right ascension and declination thus obtained by 0.075. The results will be the solar nutations required. (See Prob. XX.)

*Formulæ for computing the effects of the Oblateness of the Earth's Surface upon the Apparent Zenith Distance and Azimuth of a Star.**

From the centre of the earth, an observer would see a star at I, (Fig. 127,) and would have V for his zenith: from the surface his zenith is Z, and he sees this star at B; IB $= p$ is the parallax in altitude; the azimuth VZI is changed into VZB. If for a given time, we wish to calculate the apparent zenith distance BZ, and the apparent azimuth VZB, we have first to resolve the spherical triangle IZP, in which we know the two sides ZP = co-latitude and IP = co-declination, and the included hour angle P; the azimuth VZI (= A), and the arc IZ (= n) will thus be known. But from the earth's surface, the star is seen at B: the azimuth VZB = VZI + IZB = A + α; the zenith distance BZ $= n + p$, since, VZ (= i) being very small, we have sensibly IB + IZ = BZ. *By reason of the want of sphericity of the earth, parallax then increases the true azimuth and zenith distance of a star* by small quantities, α and p, which it is necessary to calculate. In the triangle VIZ we have

Fig. 127.

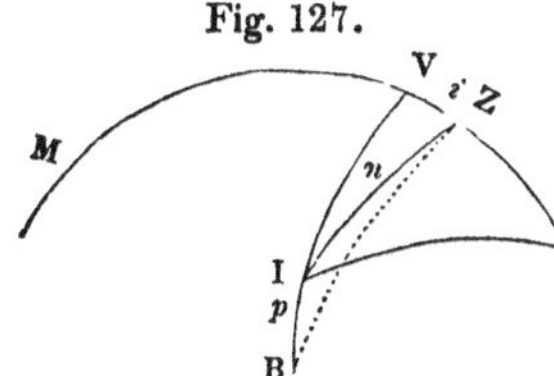

$$\cos IV = \cos i \cos n + \sin i \sin n \cos A = \cos n + k \sin n;$$

making $\cos i = 1$, $\sin i = i$, and $i \cos A = k$. Now, $k \angle i$, and *à fortiori* $\cos k = 1$, $\sin k = k$; whence

$$\cos IV = \cos n \cos k + \sin n \sin k = \cos (n - k),$$

and $$IV = n - k = n - i \cos A.$$

Thus we correct the calculated arc n by the quantity $- i \cos A$, to have

$$IV = z = n - i \cos A \ldots (j).$$

If this value of z be introduced into equation (a), page 422, we

---

* Francœur's Uranography, p. 426, &c.

shall have p, and thence the apparent zenith distance $Z = n + p = BZ$.

Afterwards, to obtain $IZB = \alpha$, or the *parallax in azimuth*, the triangles ZBV, ZBI give

$$\frac{\sin ZBV}{\sin i} = \frac{\sin (A + \alpha)}{\sin (z + p)}, \quad \frac{\sin ZBV}{\sin n} = \frac{\sin \alpha}{\sin p};$$

whence, by equating the values of sin ZBV,

$$\frac{\sin n \sin \alpha}{\sin p} = \frac{\sin i \sin (A + \alpha)}{\sin (z + p)}:$$

substituting for $\sin p$ its value $\sin H \sin (z + p) = \sin H \sin Z$, (equa. 8, page 51,) and reducing, we have

$$\frac{\sin \alpha}{\sin H \sin i} = \frac{\sin (A + \alpha)}{\sin n},$$

and as i is very small, $\sin i \sin (A + \alpha)$ does not differ sensibly from $i \sin A$, and we thus have *in seconds*, (For. 47, page 397,)

$$\alpha = \frac{Hi \sin A \sin 1''}{\sin n} \ldots (k).$$

*Solution of Kepler's Problem, by which a Body's Place is found in an Elliptical Orbit.** (See Art. 199, p. 127.)

Let APB (Fig. 128) be an ellipse, E the focus occupied by the sun, round which P the earth or any other planet is supposed to revolve. Let the time and planet's motion be dated from the ap-

Fig. 128.

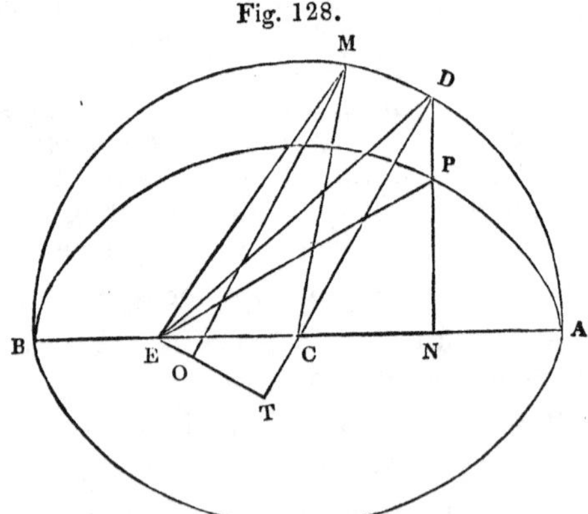

side or aphelion A. The *condition given* is the time elapsed from the planet's quitting A; the *result sought* is the place P; to be determined either by finding the value of the angle AEP, or by

* Woodhouse's Astronomy, p. 457, &c.

cutting off, from the whole ellipse, an area AEP bearing the same proportion to the area of the ellipse which the given time bears to the periodic time.

There are some technical terms used in this problem which we will now explain.

Let a circle AMB be described on AB as its diameter, and suppose a point to describe this circle uniformly, and the whole of it in the same time as the planet describes the ellipse; let also t denote the time elapsed during P's motion from A to P; then if AM $= \frac{t}{\text{period}} \times 2$ AMB, M will be the place of the point that moves uniformly, while P is that of the planet; the angle ACM is called the *Mean Anomaly*, and the angle AEP is called the *True Anomaly*.

Hence, since the time (t) being given, the angle ACM can always be immediately found, (see Art. 198, p. 127,) we may vary the enunciation of Kepler's problem, and state its object to be *the finding of the true anomaly in terms of the mean.*

Besides the mean and true anomalies, there is a third called the *Eccentric Anomaly*, which is expounded by the angle DCA, and which is always to be found (geometrically) by producing the ordinate NP of the ellipse to the circumference of the circle. This eccentric anomaly has been devised by mathematicians for the purposes of expediting calculation. It holds a mean place between the two other anomalies, and mathematically connects them. There is one equation by which the mean anomaly is expressed in terms of the eccentric; and another equation by which the true anomaly is expressed in terms of the eccentric.

We will now deduce the two equations by which the *eccentric* is expressed, respectively, in terms of the *true* and *mean* anomalies.

Let t = time of describing, AP,
P = periodic time in the ellipse,
a = CA,
ae = EC,
v = $\angle$ PEA,
u = $\angle$ DCA; (whence, ET, perpendicular to DT, = EC $\times \sin u$,)
ρ = PE,
π = 3.14159, &c.;

then, by Kepler's law of the equable description of areas,

$$t = \text{P} \times \frac{\text{area PEA}}{\text{area of ellip.}} = \text{P} \times \frac{\text{area DEA}}{\text{area circle}} = \frac{\text{P}}{\pi a^2}(\text{DEC} + \text{DCA})$$

$$= \frac{\text{P}}{\pi a^2}\left(\frac{\text{ET . DC}}{2} + \frac{\text{AD . DC}}{2}\right) = \frac{\text{P}a}{2\pi a^2}(\text{EC} . \sin u + \text{DC} . u)$$

$$= \frac{\text{P}}{2\pi}(e \sin u + u): \text{ hence, if we put } \frac{\text{P}}{2\pi} = \frac{1}{n},$$

we have

$$nt = e \sin u + u \ldots (l),$$

an equation connecting the mean anomaly nt, and the eccentric u.

In order to find the other equation, that subsists between the true and eccentric anomaly, we must investigate, and equate, two values of the radius-vector ρ, or EP.

First value of ρ, in terms of v the true anomaly,

$$\rho = \frac{a(1-e^2)}{1 - e \cos v} \ldots (1).$$

Second, in terms of u the eccentric anomaly,

$$\rho = a \ (1 + e \cos u) \ldots (2).$$

$$\begin{aligned}
\text{For, } \rho^2 &= \text{EN}^2 + \text{PN}^2 \\
&= \text{EN}^2 + \text{DN}^2 \times (1 - e^2) \\
&= (ae + a \cos u)^2 + a^2 \sin^2 u \, (1 - e^2) \\
&= a^2 \{e^2 + 2e \cos u + \cos^2 u\} + a^2 (1 - e^2) \sin^2 u \\
&= a^2 \{1 + 2e \cos u + e^2 \cos^2 u\}.
\end{aligned}$$

Hence, extracting the square root,

$$\rho = a \, (1 + e \cos u).$$

Equating the expressions (1), (2), we have

$$(1 - e^2) = (1 - e \cos v)\,(1 + e \cos u), \text{ whence,}$$

$$\cos v = \frac{e + \cos u}{1 + e \cos u}, \text{ an expression for } v \text{ in terms of } u;$$

but, in order to obtain a formula fitted to logarithmic computation, we must find an expression for $\tan \frac{v}{2}$: now, (see For. 12, p. 397,)

$$\tan \frac{v}{2} = \sqrt{\left(\frac{1 - \cos v}{1 + \cos v}\right)} = \sqrt{\left(\frac{(1 - e)\,(1 - \cos u)}{(1 + e)\,(1 + \cos u)}\right)}$$

$$= \sqrt{\left(\frac{1 - e}{1 + e}\right)} \tan \frac{u}{2} \ldots (m).$$

These two expressions (l) and (m), that is,

$$nt = e \sin u + u,$$

$$\tan \frac{v}{2} = \sqrt{\left(\frac{1 - e}{1 + e}\right)} \tan \frac{u}{2},$$

analytically resolve the problem, and, from such expressions, by certain formulæ belonging to the higher branches of analysis, may v be expressed in the terms of a series involving nt.

Instead, however, of this exact but operose and abstruse method of solution, we shall now give an approximate method of expressing the true anomaly in terms of the mean.

MO is drawn parallel to DC. (1.) Find the half difference of the angles at the base EM of the triangle ECM, from this expression,

$$\tan \tfrac{1}{2} (\text{CEM} - \text{CME}) = \tan \tfrac{1}{2} (\text{CEM} + \text{CME}) \times \frac{1-e}{1+e},$$

in which CEM + CME = ACM, the mean anomaly.

(2.) Find CEM by adding $\frac{1}{2}$ (CEM + CME) and $\frac{1}{2}$ (CEM — CME) and use this angle as an approximate value to the eccentric anomaly DCA, from which, however, it really differs by ∠ EMO.

(3.) Use this approximate value of ∠DCA = ∠ECT in computing ET which equals the arc DM; for, since (see p. 419),

$t = \dfrac{\text{P}}{\text{area circle}} \times \text{DEA}$, and (the body being supposed to revolve in the circle ADM) $= \dfrac{\text{P}}{\text{area circle}} \times \text{ACM}$, area AED = area ACM, or, area DEC + area ACD = area DCM + area ACD; consequently the area DEC = the area DCM, and, expressing their values,

$$\frac{\text{ET} \times \text{DC}}{2} = \frac{\text{DM} \times \text{DC}}{2}, \text{ and thus, ET} = \text{DM}.$$

Having then computed ET = DM, find the sine of the resulting arc DM, which sine = OT; the difference of the arc and sine (ET — OT) gives EO.

(4.) Use EO in computing the angle EMO, the real difference between the eccentric anomaly DCA and the ∠MEC; add the computed ∠EMO to ∠MEC, in order to obtain ∠DCA. The result, however, is not the exact value of ∠DCA, since ∠EMO has been computed only approximately; that is, by a process which commenced by assuming ∠MEC for the value of the ∠DCA.

For the purpose of finding the eccentric anomaly, this is the entire description of the process, which, if greater accuracy be required, must be repeated; that is, from the last found value of ∠DCA = ∠ECT, ET, EO, and ∠EMO must be again computed.

Formula for calculating the Parallax in Altitude of a Heavenly Body from its True Zenith Distance. (See Art. 88, p. 62.)

In the actual state of astronomy, the true co-ordinates of the places of the heavenly bodies are generally known, or may be obtained by computation from the results of observations already made, and from these there is often occasion to deduce the apparent co-ordinates. For this purpose there is required an expression for the parallax in altitude in terms of the true zenith distance.

If we make $Z = z + p$ in equation (8) p. 62, we shall have

$$\sin p = \sin \text{H} \sin (z+p), \text{ or } \sin \text{H} = \frac{\sin p}{\sin (z+p)};$$

whence,

$$1 + \sin \text{H} = 1 + \frac{\sin p}{\sin (z+p)} = \frac{\sin (z+p) + \sin p}{\sin (z+p)},$$

and

$$1-\sin H = 1 - \frac{\sin p}{\sin (z+p)} = \frac{\sin (z+p) - \sin p}{\sin (z+p)}.$$

Dividing,

$$\frac{1+\sin H}{1-\sin H} = \frac{\sin (z+p)+\sin p}{\sin (z+p)-\sin p};$$

or,

$$\text{tang}^2 (45° + \tfrac{1}{2} H) = \frac{\text{tang} (\frac{1}{2} z + p)}{\text{tang} \frac{1}{2} z} \text{ (App. For. 36, 29)};$$

whence,

$$\text{tang} (\tfrac{1}{2} z + p) = \text{tang} \tfrac{1}{2} z \text{ tang}^2 (45° + \tfrac{1}{2} H) \ldots (a).$$

This equation makes known $\frac{1}{2} z + p$, from which we may obtain p by subtracting $\frac{1}{2} z$.

Formulæ for computing the Annual Variations in the Right Ascension and Declination of a Heavenly Body. (See Art. 119, p. 88.)

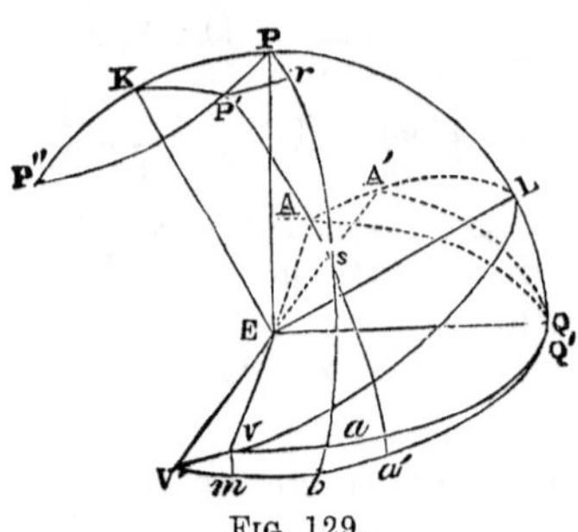

FIG. 129.

Let VLA (Fig. 129) be the ecliptic, K its pole, PP'P'' the circle described by the mean pole, P the mean pole, and VQA the mean equator at any given time, P' the mean pole and V'Q'A' the mean equator a year afterwards, and s a star. Draw P'r perpendicular to the declination circle Psa. We have

an. var. in dec.$=sa'-sa=$P$s-$P'$s=$Pr;

but since PP'r may be considered as a right-angled plane triangle,

P$r=$PP' cos P'P$r=$PP' sin QPa....(a).

Regarding KPP' as a right-angled isosceles triangle, we obtain

sin KPP' or 1 : sin KP' :: sin PKP' : sin PP';

whence,

sin PP'=sin PKP' sin KP', or PP'=PKP' sin KP' (nearly)......(b):

substituting in equation (a), there results,

Pr=PKP' sin KP' sin QPa.

PKP' = 50''.24; KP' = obliquity of the ecliptic = ω;

QPa=VQ$-$Va=90°$-$R (R designating the right ascension of the star s). Thus, finally,

an. var. in dec.=50''.24 sin ω cos R..... (c).

Next, we have

an. var. in r. asc.=V'$a'-$Va=V'$a'-mb$=V'$m+ba'$.......(d);

but,

V'm=VV' cos VV'm=50''.24 cos ω;

and since the right-angled triangles sP'r and sba' are similar,

sin sr or sin sP' (nearly) : sin P'r :: sin sa' : sin ba';

whence,

$$\sin ba' = \sin P'r \frac{\sin sa'}{\sin P's}, \text{ or } ba' = P'r\frac{\sin sa'}{\sin P's} \text{ (nearly).}$$

The triangle PP'r gives P'r=PP' sin P'Pr=PP' cos QPa=PKP' sin KP' cos QPa (equa. b); and sin P's=cos sa'. Substituting, we obtain

$$ba = \text{PKP}' \sin \text{KP}' \cos \text{QP}a\frac{\sin sa'}{\cos sa'} = \text{PKP}' \sin \text{KP}' \cos \text{QP}a \text{ tang } sa'.$$

Replacing PKP′, KP′, and QPa by their values, as above, and taking the declination sa for sa' and denoting it by D, there results,

$$ba'=50''.24 \sin\omega \sin R \text{ tang } D.$$

Now, substituting in equation (d) the values of V′m, and ba', we have

$$\text{an. var. in r. asc.}=50''.24 \cos\omega+50''.24 \sin\omega \sin R \text{ tang } D\ldots. \quad (e)$$

The results of formulæ (c, e,) are to be used with their algebraic signs, if the reduction is from an earlier to a later epoch, otherwise with the contrary signs. The declination is always to be considered *positive* if *North*, and *negative* if *South.*

$$V'm=50''.24 \cos\omega=50''.24 \cos 23^\circ 27'=46''.0,$$

is the annual retrograde motion of the equinoctial points along the equator.

Formulæ for computing the Heliocentric Longitude and Latitude, and Radius-vector of a Planet, from its Geocentric Longitude and Latitude. (Referred to in Art. 177, p. 119.)

The longitude of the node and the inclination of the orbit are supposed to be known. Let NP (Fig. 130) be part of the orbit of a planet, SNC the plane of the

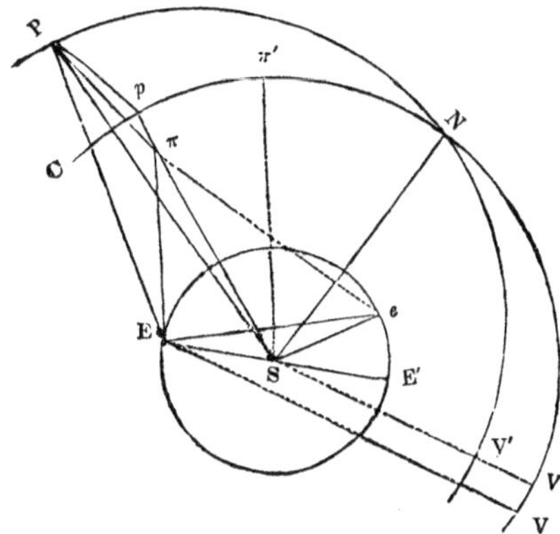

FIG. 130.

ecliptic, N the ascending node, S the sun, E the earth, and P the planet; also, let Pπ be a perpendicular let fall from P upon the plane of the ecliptic, and EV, SV, the direction of the vernal equinox. Let λ = PEπ the geocentric latitude of the planet; l = PSπ its heliocentric latitude; G = VEπ its geocentric longitude; L = VSπ its heliocentric longitude; S = VES the longitude of the sun; N = VSN the heliocentric longitude of the node; I = PNC the inclination of the orbit; r = SE the radius-vector of the earth; and v = SP the radius-vector of the planet.

The point π is called the *reduced place* of the planet, and Sπ its *curtate distance.* All the angles of the triangle SEπ have also received particular appellations; SπE the angle subtended at the reduced place of the planet by the radius of the earth's orbit, is called the *Annual Parallax*, SEπ the *Elongation*, and ESπ the *Commutation.* Let A = SπE, E = SEπ, and C = ESπ. Draw Sπ' parallel to Eπ: then A = πSπ' = VSπ — VSπ' = VSπ — VEπ = L — G; E = VEπ — VES = G — S; C = VSE — VSπ = 180° + VSE′ — VSπ = 180° + VES — VSπ = 180° + S — L = T — L (putting T = 180° + S).

(1.) *For the latitude.*—The triangles EPπ, SPπ, give

$$E\pi \text{ tang } \lambda = P\pi = S\pi \text{ tang } l, \text{ whence } \frac{\text{tang }\lambda}{\text{tang } l} = \frac{S\pi}{E\pi};$$

but,

$$S\pi : E\pi :: \sin E : \sin C, \text{ or, } \frac{S\pi}{E\pi} = \frac{\sin E}{\sin C};$$

substituting,
$$\frac{\text{tang}\,\lambda}{\text{tang}\,l} = \frac{\sin E}{\sin C},$$

whence, $\text{tang}\,\lambda \sin C = \text{tang}\,l \sin E \ldots .(a).$

or, $\text{tang}\,\lambda \sin (T - L) = \text{tang}\,l \sin (G - S) \ldots .(b).$

Again, the triangle N*Pp* gives, by Napier's first rule,

$$\sin Np = \cot PNp \tan Pp, \text{ or, } \sin (L - N) = \cot I \tan l \ldots .(c).$$

Either of the equations (*b*) and (*c*) will give the value of l, when the longitude L is known.

(2.) *For the Longitude.*—If we substitute in equation (*b*) the value of tang l, given by equation (*c*), and replace (G — S) by E, we have

$$\text{tang}\,\lambda \sin (T - L) = \sin (L - N) \text{ tang I} \sin E;$$

but $T - L = (T - N) - (L - N) = D - (L - N)$, (denoting (T — N) by D); substituting, and designating L — N by x,

$$\text{tang}\,\lambda \sin (D - x) = \sin x \text{ tang I} \sin E;$$

whence,

$$\text{tang}\,\lambda \sin D \cos x - \text{tang}\,\lambda \cos D \sin x = \text{tang I} \sin E \sin x,$$

or, $\text{tang}\,\lambda \sin D - \text{tang}\,\lambda \cos D \text{ tang}\,x = \text{tang I} \sin E \text{ tang}\,x,$

which gives

$$\text{tang}\,x = \frac{\text{tang}\,\lambda \sin D}{\text{tang}\,\lambda \cos D + \text{tang I} \sin E} \ldots .(d).$$

Substituting the values of x, D, and E, we have, finally,

$$\text{tang}\,(L - N) = \frac{\text{tang}\,\lambda \sin (T - N)}{\text{tang}\,\lambda \cos (T - N) + \text{tang I} \sin (G - S)} \ldots .(e).$$

As N is known, the value of L will result from this equation.

The co-ordinates employed to fix the position of a planet in the plane of its orbit, are its orbit longitude and its radius-vector, both of which result from the heliocentric longitude and latitude, the longitude of the node and the inclination of the orbit being known.

In Fig. 130, V′NP represents the orbit longitude, and SP ($= v$) the radius-vector, for the position P. Now, the triangle PSπ gives

$$SP = \frac{S\pi}{\cos PS\pi}, \text{ or, } v = \frac{S\pi}{\cos l};$$

and the triangle ESπ gives

$$\sin A : \sin E :: SE : S\pi = \frac{SE \sin E}{\sin A} = \frac{r \sin E}{\sin A};$$

whence, by substitution,

$$v = \frac{r \sin E}{\sin A \cos l} = \frac{r \sin (G - S)}{\sin (L - G) \cos l} \ldots .(f).$$

The orbit longitude $L' = NP + \text{long. of node} \ldots .(g)$:

and to find NP, the triangle N*Pp* gives

$$\cos PNp = \cot NP \text{ tang } Np, \text{ or tang } NP = \frac{\text{tang } Np}{\cos I};$$

and $Np = \text{long. of planet} - \text{long. of node}.$

Formulæ for computing the Geocentric Longitude and Latitude of a Planet from its Heliocentric Longitude and Latitude and Radius-Vector.

Let S (Fig. 130) be the sun, E the earth, P the planet, π its reduced place, and V the vernal equinox. Denote the heliocentric longitude VSπ by L, the heliocentric latitude PSπ by l, and the radius-vector SP by v; and denote the geocentric longitude by G, and the geocentric latitude by λ. Also let E = SEπ the elongation; C = ESπ the commutation; A = SπE the annual parallax; and r = SE the radius-vector of the earth. Now,

$$VE\pi = SE\pi + VES,$$

or $G = E + \text{long. of sun}.$

This equation will make known the geocentric longitude when the value of E is found. In the triangle $PS\pi$ the side $S\pi = SP \cos PS\pi = v \cos l$, and is therefore known, the side ES is given by the elliptical theory, and the angle C may be derived from the following equation: $C = VSE - VS\pi =$ long. of earth — long. of planet; and to find E we have, by Trigonometry,

$$ES + S\pi : ES - S\pi :: \tan \tfrac{1}{2}(E\pi S + SE\pi) : \tan \tfrac{1}{2}(E\pi S - SE\pi),$$

or, $$r + v \cos l : r - v \cos l :: \text{tang} \tfrac{1}{2}(A + E) : \text{tang} \tfrac{1}{2}(A - E);$$

whence,

$$\text{tang} \tfrac{1}{2}(A - E) = \frac{r - v \cos l}{r + v \cos l} \text{tang} \tfrac{1}{2}(A + E) = \frac{1 - \dfrac{v \cos l}{r}}{1 + \dfrac{v \cos l}{r}} \text{tang} \tfrac{1}{2}(A + E).$$

Let $\text{tang}\,\theta = \dfrac{v \cos l}{r}$: then,

$$\text{tang} \tfrac{1}{2}(A - E) = \frac{1 - \text{tang}\,\theta}{1 + \text{tang}\,\theta} \text{tang} \tfrac{1}{2}(A + E);$$

or, $$\text{tang} \tfrac{1}{2}(A - E) = \text{tang}(45^\circ - \theta) \text{tang} \tfrac{1}{2}(A + E) \dots (a)$$

But, $$A + E = 180^\circ - C, \text{ and } E = \tfrac{1}{2}(A + E) - \tfrac{1}{2}(A - E).$$

Next, to find the *geocentric latitude.*

$$S\pi \,\text{tang}\, l = P\pi = E\pi \,\text{tang}\, \lambda$$

whence, $$\frac{S\pi}{E\pi} = \frac{\text{tang}\,\lambda}{\text{tang}\,l};$$

but, $$S\pi : E\pi :: \sin E : \sin C, \text{ or } \frac{S\pi}{E\pi} = \frac{\sin E}{\sin C},$$

and therefore $$\frac{\sin E}{\sin C} = \frac{\text{tang}\,\lambda}{\text{tang}\,l},$$

or $$\text{tang}\,\lambda = \frac{\sin E \,\text{tang}\, l}{\sin C} \dots (b).$$

When a planet is in conjunction or opposition, the sines of the angles of elongation and commutation are each nothing. In these cases, then, the geocentric latitude cannot be found by the preceding formula; it may, however, be easily determined in a different manner. Suppose the planet to be in conjunction at P, (Fig. 56, p. 120;) then,

$$\text{tang}\,\lambda = \frac{P\pi}{E\pi} = \frac{P\pi}{ES + S\pi}.$$

But the triangle $SP\pi$ gives

$$P\pi = v \sin l \text{ and } S\pi = v \cos l, \text{ and } ES = r;$$

hence, $$\text{tang}\,\lambda = \frac{v \sin l}{r \div v \cos l} \dots (c).$$

To find *the distance* of the planet from the earth, represent the distance by D; then, from the triangles $P\pi S$ and $EP\pi$, we have

$$P\pi = EP \sin PE\pi = D \sin \lambda,$$

and $$P\pi = SP \sin PS\pi = v \sin l;$$

whence, $$D = \frac{v \sin l}{\sin \lambda} \dots (d.)$$

The distance of a planet being known, its *horizontal parallax* may be computed from the equation

$$\sin H = \frac{R}{D} \dots (e.) \quad \text{(Art. 88).}$$

CALCULATION OF AN ECLIPSE OF THE SUN..

(1). *Of the circumstances of the general eclipse.*

It is a simple inference from what has been established in Art. 333, that an eclipse of the sun will begin and end upon the earth, at the times before and after conjunction, when the distance of the centres of the moon and sun is equal to $P-p+\delta+d$; that the total eclipse will begin and end when this distance is equal to $P-p-\delta+d$; and the annular eclipse when the distance is equal to $P-p+\delta-d$.

The times of the various phases of the general eclipse of the sun may be obtained by a process precisely analogous to that by which the times of the phases of an eclipse of the moon are found. Let C (Fig. 131) be the centre of the sun, and C′ the centre of the moon, at the time of conjunction. We may suppose the sun to remain stationary at C, if we attribute to the moon a motion equal to its motion relative to the sun; for, on this supposition, the distance of the centres of the two bodies will, at any given period during the eclipse, be the same as that which obtains in the actual state of the case. Let N′C′L′ represent the orbit that would be described by the moon if it had such a motion, which is called the *Relative Orbit* Let CM be drawn perpendicular to it; and let $Cf = Cf' = P-p+\delta+d$, and $Cg = Cg' = P-p-\delta+d$, or $P-p+\delta-d$, according as the eclipse is total or annular. Then, M will be the place of the moon's centre at the middle of the eclipse; f and f' the places at the beginning and end of the eclipse; and g and g' the places at the beginning and end of the total, or of the annular eclipse. We shall thus have, as in eclipses of the moon,

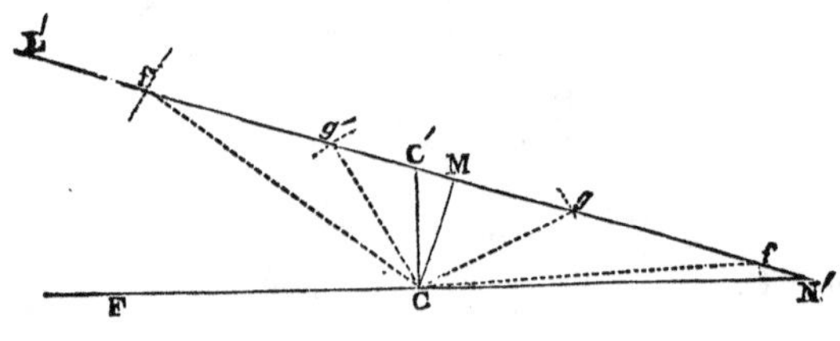

FIG. 131.

$$\text{tang}\, I = \frac{n}{M-m},\quad CM = \lambda \cos I,\ C'M = \lambda \sin I$$

$$\text{Interval from con. to mid.} = \frac{3600s.\ \lambda \sin I \cos I}{M-m} \quad \ldots (a).$$

Interval from middle to beginning or end

$$= \frac{3600s.\ \cos I}{M-m}\sqrt{(k'+\lambda \cos I)\,(k'-\lambda \cos I)} \quad \ldots (b).$$

Interval for total eclipse

$$= \frac{3600s.\ \cos I}{M-m}\sqrt{(k''+\lambda \cos I)\,(k''-\lambda \cos I)} \quad \ldots (c).$$

Interval for annular eclipse

$$= \frac{3600s.\ \cos I}{M-m}\sqrt{(k'''+\lambda \cos I)\,(k'''-\lambda \cos I)} \quad \ldots (d).$$

$$\text{Quantity} = \frac{6\,(k'-\lambda \cos I)}{d} \quad \ldots (e).$$

$$k' = P-p+\delta+d,\ k'' = P-p-\delta+d,\ k''' = P-p+\delta-d \quad \ldots (f).$$

The letters λ, M, m, &c., represent quantities of the same name as in the formulæ for a lunar eclipse; but they designate the values of these quantities at the time of *conjunction*, instead of opposition. These values are in practice obtained from tables of the sun and moon, as in a lunar eclipse.

The times of the different circumstances of a general eclipse of the sun may also be found within a minute or two of the truth, by *construction*, in a precisely similar manner with those of an eclipse of the moon (330).

(2.) *Of the phases of the eclipse at a particular place.*

The phase of the eclipse, which obtains at any instant at a given place, is indicated by the relation between the apparent distance of the centres of the sun and moon, and the sum, or difference, of their apparent semi-diameters; and the calculation of the time of any given phase of the eclipse, consists in the calculation of the time when the apparent distance of the centres has the value relative to the sum or difference of the semi-diameters, answering to the given phase. Thus, if we wish to find the time of the beginning of the eclipse, we have to seek the time when the apparent distance of the centres of the sun and moon first becomes equal to the sum of their apparent semi-diameters.

The calculation of the different phases of an eclipse of the sun, for a particular place, involves, then, the determination of the apparent distance of the centres of the sun and moon, and of the apparent semi-diameters of the two bodies at certain stated periods.

The true semi-diameter of the sun, as given by the tables, may be taken for the apparent without material error. For the method of computing the apparent semi-diameter of the moon, for any given time and place, see Problem XVII.

According to the celebrated astronomer Duséjour, in order to make the observations agree with theory, it is necessary to diminish the sun's semi-diameter, as it is given by the tables, 3″.5. This circumstance is explained by supposing that the apparent diameter of the sun is amplified by reason of the very lively impression which its light makes upon the eye. This amplification is called *Irradiation*. He also thinks that the semi-diameter of the moon ought to be diminished 2″, to make allowance for an *Inflexion* of the light which passes near the border of this luminary, supposed to be produced by its atmosphere. It must be observed, however, that the astronomers of the present day do not agree either as to the necessity or the amount of the diminutions just spoken of.

The determination of the apparent distance of the centres of the sun and moon may easily be accomplished, as will be shown in the sequel, when the apparent longitude and latitude of the two bodies have been found. Now, the true longitude of the sun, and the true longitude and latitude of the moon, may be found from the tables, (Probs. IX. and XIV.); and from these the apparent longitudes and latitudes may be deduced by correcting for the parallax. But the customary mode of proceeding is a little different from this: the true longitude and latitude of the sun are employed instead of the apparent, and the parallax of the sun is referred to the moon; that is, the difference between the parallax of the moon and that of the sun is, by fiction, taken as the parallax of the moon. This supposititious parallax is called the moon's *Relative Parallax*. (See Prob. XVII.)

We will first show how to find the *approximate times* of the different phases of the eclipse. Put T = the time of new moon, known to within 5 or 10 minutes. (Prob. XXVII.) For the time T calculate by the tables the sun's longitude, hourly motion, and semi-diameter, and the moon's longitude, latitude, horizontal parallax, semi-diameter, and hourly motions in longitude, and latitude. Subtract the sun's horizontal parallax from the reduced horizontal parallax of the moon,* and calculate the apparent longitude and latitude, and the apparent semi-diameter of the moon. From a comparison of the apparent longitude of the moon with the true longitude of the sun, we shall know whether apparent ecliptic conjunction occurs before or after the time T. Let T′ denote the time an hour earlier or later than the time T, according as the apparent conjunction is earlier or later. With the sun and moon's longitudes, the moon's latitude, and the hourly motions in longitude and latitude, at the time T, calculate the longitudes and the moon's latitude for the time T′; and for this time also calculate the moon's apparent longitude and latitude. Take the difference between the apparent longitude of the moon and the true longitude of the sun at the time T, and it will be the apparent distance of the moon from the sun in longitude, at this time. Let it be denoted by n. Find, in like manner, the apparent distance of the moon from the sun in longitude at the time T′, and denote

* The reduced horizontal parallax of the moon is its horizontal parallax as reduced from the equator to the given place. (See Prob. XV.)

it by n'. In the same manner as at the time T, we find whether apparent conjunction occurs before or after the time T'. If it occurs between the times T and T', the sum of n and n', otherwise their difference, will be the apparent relative motion of the sun and moon in longitude in the interval T'—T, or T—T'; from which the relative hourly motion will become known. The difference of the apparent latitudes of the moon, at the times T and T', will make known the apparent relative hourly motion in latitude. With the relative hourly motion in longitude and the difference of the apparent longitudes at the time T, find by simple proportion the interval between the time T and the time of apparent ecliptic conjunction; and then, with the apparent latitude of the moon at the time T and its hourly motion in latitude, find the apparent latitude at the time of apparent conjunction thus determined. Then, knowing the relative hourly motion of the sun and moon in longitude and latitude, together with the time of apparent conjunction, and the apparent latitude at that time, and regarding the apparent relative orbit of the moon as a right line (which it is nearly), it is plain that the time of beginning, greatest obscuration, and end, as well as the quantity of the eclipse, may be calculated after the same manner as in the general eclipse; the disc of the sun answering to the section of the luminous frustum mentioned in Art. 331, and the apparent elements answering to the true. Let C (Fig. 132) represent the centre of the sun supposed stationary, CC' the apparent latitude of the moon at apparent conjunction, N'C' the apparent relative orbit of the moon, determined by its passing through the point C' and making a determinate angle with the ecliptic N'F, or by its passing through the situations of the moon at the times T and T'. Also, let RKFK' be the border of the sun's disc; f, f' the positions of the moon's centre at the beginning and end of the eclipse, determined by describing a circle around C as a centre, with a radius equal to the sum of the apparent semi-diameters of the sun and moon; and M (the foot of the perpendicular let fall from C upon N'C') its position at the time of greatest obscuration.

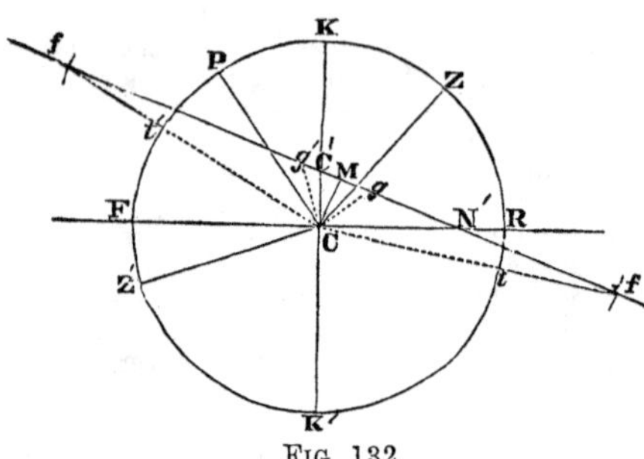

FIG. 132.

If the eclipse should be total or annular, then g, g' will be the positions of the moon's centre at the beginning and end of the total or annular eclipse; these points being determined by describing a circle around C as a centre, and with a radius equal to the difference of the apparent semi-diameter of the sun and moon.

The results will be a closer approximation to the truth, if the same calculations that are made for the time T' be made also for another time T''.

The various circumstances of the eclipse may also be had by construction, after the same manner as in a lunar eclipse (330).

In order to be able to observe the beginning or end of a solar eclipse, it is necessary to know the position of the point on the sun's limb where the first or last contact takes place. The situation of these points is designated by the distance on the limb, intercepted between them and the highest point of the limb, called the *Vertex*. The contacts will take place at the points t, t', (Fig. 132,) on the lines Cf, Cf'. To find the position of the vertex, with the sun's longitude found for the beginning of the eclipse, calculate the angle of position of the sun at that time, (see Prob. XIII.), and lay it off to the right of the circle of latitude CK, when the sun's longitude is between 90° and 270°, and to the left when the longitude is less than 90° or more than 270°. Suppose CP to be the circle of declination thus determined. Next, let Z (Fig. 6, p. 13) be the zenith, P the elevated pole, and S the sun; then in the triangle ZPS we shall know ZP the co-latitude, ZPS the hour angle of the sun, and we may deduce PS, the co-declination of the sun, from the longitude of the sun as derived from the tables (equ. 24). These three quantities being known, ZSP, the angle made by the vertical through the sun with its circle of declination, may be computed; and being laid off in the figure to the right or left of CP (Fig. 132), according as the time of beginning is before or after noon, the point Z or Z', as the case may be, in which the vertical intersects the limb RKK', will be the vertex, and the arc Zt, $Z't$, on the limb, will ascertain the situation of t, the first point of contact, with respect to it.

The situation of the last point of contact may be found by the same mode of proceeding.

Let us now show how *to find the exact times of the beginning, greatest obscuration, and end of the eclipse*, the approximate times being known. Let B designate the approximate time of beginning, taken to the nearest minute. Calculate for the time B by means of the tables, the sun's longitude, hourly motion, and semi-diameter; also the moon's longitude, latitude, horizontal parallax, semi-diameter, and hourly motions in longitude and latitude. Then, making use of the relative parallax, calculate the apparent longitude, latitude, and semi-diameter of the moon. Subtract the apparent longitude of the moon from the true longitude of the sun; the difference will be the apparent distance of the moon from the sun in longitude: let it be denoted by a. Denote the apparent latitude of the moon by λ.

Now, let EC (Fig 133) represent an arc of the ecliptic, and K its pole; and let S be the situation of the sun, and M the apparent situation of the moon at the time B. Then MS is the apparent distance of the centres of the two bodies at this time. Denote it by Δ. $Sm=a$, and $Mm=\lambda$. The right-angled triangle MSm being very small, may be considered as a plane triangle, and we therefore have, to determine Δ, the equation

$$\Delta^2 = a^2 + \lambda^2 \ldots . (g).*$$

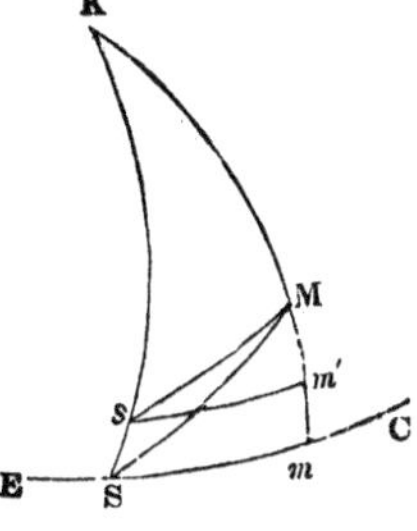

FIG. 133.

Having computed the value of Δ, we find, by comparing it with the sum of the apparent semi-diameters of the sun and moon, whether the beginning of the eclipse occurs before or after the approximate time B. Fix upon a time some 4 or 5 minutes before or after B, according as the beginning is before or after, and call it B'. With the sun and moon's longitudes, the moon's latitude, and the hourly motions in longitude and latitude, at the time B, find the longitudes and the moon's latitude at the time B', and compute for this time the apparent longitude, latitude, and semi-diameter of the moon. Subtract the apparent longitude of the moon from the true longitude of the sun, and we shall have the apparent distance of the moon from the sun at the time B'. Take the difference between this and the same distance a at the time B, and we shall have the apparent relative motion of the sun and moon in longitude during the interval of time between B and B'. Then find, by simple proportion, the apparent relative hourly motion in longitude, and denote it by k. Take the difference between the apparent latitudes of the moon at the times B and B', and it will be the apparent relative motion of the sun and moon in latitude, in the interval; from which deduce the apparent relative hourly motion in latitude, and call it n. Now, put $t=$ the interval between the approximate and true times of the beginning of the eclipse, and suppose S and M (Fig. 133) to be the situations of the sun and moon at the true time of beginning. In the time t, the apparent relative motions in longitude and latitude will be, respectively, equal to kt and nt, and accordingly we shall have

$$Sm = a - kt, \quad Mm = \lambda + nt.$$

The small right-angled triangle SMm may be considered as a plane triangle; the hypothenuse $SM=\psi=$ the sum of the apparent semi-diameters of the sun and moon, minus 5″.5 (*p.* 427). We have then, to find t, the equation

$$(a - kt)^2 + (\lambda + nt)^2 = \psi^2,$$

or, developing and transposing,

$$(n^2 + k^2)\, t^2 - 2\,(ak - \lambda n)\, t = \psi^2 - (a^2 + \lambda^2) = \psi^2 - \Delta^2;$$

* In place of equation (g) the following equations may be employed in logarithmic computation:

$$\text{tang}\,\theta = \frac{\lambda}{a}, \quad \Delta = \frac{a}{\cos\theta};$$

where θ is an auxiliary arc.

making $A = \psi^2 - \Delta^2$, and $B = ak - \lambda n$, we have $(n^2 + k^2)\, t^2 - 2Bt = A$,

and
$$t = \frac{B - \sqrt{B^2 + A\,(n^2 + k^2)}}{n^2 + k^2} \quad \ldots \ (h).$$

The negative sign must be prefixed to the radical, for, if we suppose A to be equal to zero, t must be equal to zero. Multiplying the numerator and denominator by $B + \sqrt{B^2 + A\,(n^2 + k^2)}$, and restoring the value of A we obtain

(in seconds),
$$t = \frac{3600\text{s.}\ (\Delta^2 - \psi^2)}{B + \sqrt{B^2 + (\psi^2 - \Delta^2)\,(n^2 + k^2)}} \quad \ldots \ (i).$$

Although this equation has been investigated for the beginning of the eclipse, it is plain that it will answer equally well for the determination of the other phases, if we give the proper values and signs to ψ, a, λ, n, and k. k is positive before conjunction and negative, after it, and the radical quantity is negative after conjunction; n is negative, when the moon appears to recede from the north pole of the ecliptic; λ has the sign —, when it is south; a is always positive.*

The value of t taken with its sign is to be added to the time B.

The values of the quantities a, λ, n, and k, are found for the other phases after the same manner as for the beginning.

To obtain the value of ψ at the time of greatest obscuration, find the relative motions in longitude and latitude (k and n), during some short interval near the middle of the eclipse, which is the approximate time of greatest obscuration; then compute the inclination of the relative orbit by the equation

$$\text{tang I} = \frac{n}{k} \quad \ldots \ (j).$$

after which ψ will result from the equation

$$\psi = \lambda \cos \text{I} \quad \ldots \ (k).$$

λ is the moon's latitude at the time of apparent conjunction, which is easily calculated, by means of the values of k and n, and the apparent longitude and latitude of the moon, found for some instant near the time of apparent conjunction.

For the beginning and end of the total eclipse, we have, ψ = appar. semi-diam. of moon — appar. semi-diam. of sun + 1″.5; and for the beginning and end of the annular eclipse, ψ = appar. semi-diam. of sun — appar. semi-diam. of moon — 1″.5.

If the value of ψ, given by equation (k), be substituted in equation (i), this equation will make known the time of greatest obscuration; but this may be found more conveniently by a different process. Let NCF (Fig. 134) represent a portion of the ecliptic, EML a portion of the relative orbit passed over about the time of greatest obscuration, C the stationary position of the sun's centre, and M the place of the moon's centre at the instant of its nearest approach to C. Also, let a = CR the apparent distance of the moon from the sun in longitude at the time of the nearest approach of the centres, λ' = RM the moon's apparent latitude at the same time, k = Mk the apparent relative motion in longitude in some short interval about this time, and n = kn the moon's apparent motion in latitude during the same interval. The right-angled triangles Mnk and CMR are similar, for their sides are respectively perpendicular to each other; whence

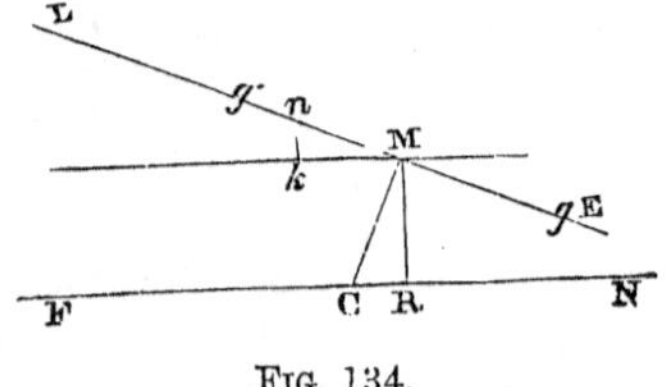

FIG. 134.

$$Mk : MR :: kn : CR;$$

and
$$CR = MR\,\frac{kn}{Mk},\ \text{or},\ a = \lambda'\,\frac{n}{k} \quad \ldots \ (l).$$

* Developing the radical in equation (h), and neglecting all the terms after the second, as being very small, we obtain for the beginning and end of the eclipse the more convenient formula

$$t = \frac{1800\text{s.}\ (\Delta^2 - \psi^2)}{B.}$$

If the moon's apparent latitude be found for the approximate time of greatest obscuration, and substituted for λ' in equation (l), this equation will give very nearly the apparent distance (ι) of the two bodies in longitude at the true time of greatest obscuration. With this, and the apparent distance at the approximate time of greatest obscuration, together with the relative apparent motion in longitude, the true time of greatest obscuration may be found nearly by simple proportion. A more accurate result may then be had by finding the moon's apparent latitude for the time obtained, substituting it for λ' in equation (l) and then repeating the calculations.

A simpler, though less accurate method than that already given, of finding the times of beginning and end of the total or annular eclipse, is to compute the half duration of the total or annular eclipse, and add it to, and subtract it from, the time of greatest obscuration. This interval may easily be determined, if we can find the rate of motion on the relative orbit, and the distance passed over by the moon's centre during the interval. Let g, g' (Fig. 134) be the places of the moon's centre at the instants of the two interior contacts, and Mn, the distance passed over in some short interval (L). Let $\theta = <$ Mnk the complement of the inclination of the relative orbit, $k =$ Mk, $k' =$ Mn, and $t =$ half duration of total or annular eclipse. The triangles Mnk, CRM, give

$$\text{M}n = \frac{\text{M}k}{\sin \text{M}nk}, \text{ or } k' = \frac{k}{\sin \theta} \quad . \; . \; . \; (m):$$

and
$$\text{tang RCM} = \text{tang M}nk = \frac{\text{RM}}{\text{CR}}, \text{ or, tang } \theta = \frac{\lambda'}{a} \quad . \; . \; . \; (n).$$

Finding the value of θ by the last equation, and substituting it in equation (m), we obtain the value of k'; and then, to find t, we have

$$k' : \text{L} :: \text{M}g : t, \text{ or } t = \frac{\text{L} \times \text{M}g}{k'}.$$

$$\text{M}g = \sqrt{\overline{\text{C}g}^2 - \overline{\text{CM}}^2} = \sqrt{\psi^2 - \triangle^2} \quad . \; . \; . \; . \; (p);$$

whence,
$$t = \frac{\text{L}\sqrt{\psi^2 - \triangle^2}}{k'} = \frac{\text{L}\sqrt{(\psi + \triangle)(\psi - \triangle)}}{k'} \quad . \; . \; . \; (o)$$

The apparent distance of the centres of the two bodies at the time of greatest obscuration being known, the quantity of the eclipse may be readily found. We have but to subtract the apparent distance from the sum of the apparent semi-diameters, and state the proportion, as the sun's apparent diameter : the remainder :: 12 digits : the digits eclipsed. (For a more particular description of the method of calculating a solar eclipse, see Prob. XXX.)

CALCULATION OF AN OCCULTATION.

The calculation of an occultation is very nearly the same as that of a solar eclipse. The only difference is in the data. The star has no diameter, parallax, or motion in longitude; and as it is situated without the ecliptic, we have, in place of the latitude of the moon, employed in solar eclipses, the difference between the latitude of the moon and that of the star, and in place of the difference between the longitudes of the two bodies and their relative hourly motion in longitude, these quantities referred to an arc passing through the star and parallel to the ecliptic. Thus, if EC (Fig. 133) represent the ecliptic, K its pole, s the situation of the star, M that of the moon, and sm' an arc passing through s and parallel to the arc EC, we have in place of mM, m'M $= m$M $- mm'$, and in place of Sm, sm'. The hourly variation of Sm must also be reduced to the arc sm'.

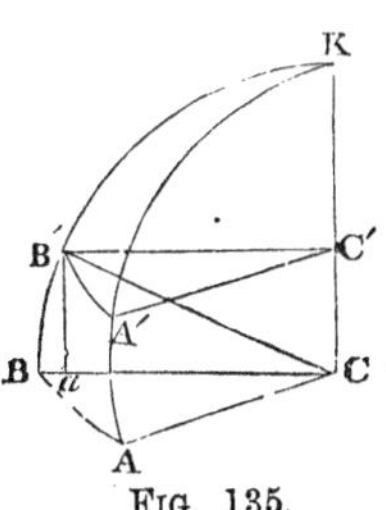

FIG. 135.

The reduction of the difference of longitude of the moon and star, to the parallel to the ecliptic, passing through the star, is effected by multiplying the difference by the cosine of the latitude of the star. For, let AB (Fig. 135) be an arc of the ecliptic, and A'B' the corresponding arc of a circle parallel to it, then, since similar arcs of circles are proportional to their radii we have

$$BC : B'C' :: AB : A'B' = \frac{AB.B'C'}{BC}.$$

But, $$B'C' = Ca = B'C \cos BCB' = BC \cos BB':$$

hence, $$A'B' = \frac{AB.BC \cos BB'}{BC} = AB \cos BB'.$$

The reduction of the relative hourly motion in longitude to the parallel in question, is obviously effected in the same manner.

NOTE I.

CONSTRUCTION OF TABLES.

The determination of the place of the sun or moon, or of a planet, may be greatly facilitated by the use of tables. The principle and mode of construction of tables adapted to this purpose are nearly the same for each body.

We will first explain the mode of constructing tables for facilitating the computation of the sun's longitude. We have the equation

True long. = mean long. + equa. of centre + inequalities + nutation.

If, then, tables can be constructed which will furnish by inspection the mean longitude, the equation of the centre, the amounts of the various inequalities in longitude, and the nutation in longitude, at any assumed time, we may easily find the true longitude at the same time.

(1.) *For the mean longitude.*—The sun's mean motion in longitude in a mean tropical year, is 360°. From this we may find by proportion the mean motions in a common year of 365 days, and a bissextile year of 366 days.

With these results, and the mean longitude for the epoch of Jan. 1, 1850 (see Table II.), we may easily derive the mean longitude at the beginning of each of the years prior and subsequent to the year 1850. The second column of Table XVIII. contains the mean longitude of the sun at the beginning of each of the years inserted in the first column. The third column of this table contains the mean longitude of the perigee at the same epochs: it was constructed by means of the mean longitude of the perigee found for the beginning of the year 1800, and its mean yearly motion in longitude.*

Having the sun's mean daily motion in longitude, we obtain by proportion the motion in any proposed number of months, days, hours, minutes, or seconds. Table XIX contains the respective amounts of the sun's motion from the commencement of the year to the beginning of each month; Table XX, the sun's mean motion from the beginning of any month to the beginning of any day of the month, and his motion for hours from 1 to 24; and Table XXI, the same for minutes and seconds from 1 to 60. With these tables the sun's mean motion in longitude in the interval between any given time in any year and the beginning of the year may be had: and if this be added to the epoch for the given year, taken out from Table XVIII, the result will be the mean longitude at the given time. (See Problem IX.)

Tables XIX and XX also contain the motions of the sun's perigee, from which and the epoch given by Table XVIII results the longitude of the perigee at any proposed time. The longitude of the perigee is given in the Solar Tables for the purpose of making known *the mean anomaly*, the mean anomaly being equal to the mean longitude minus the longitude of the perigee.

(2.) *For the equation of the centre.*—To find the equation of the centre of an orbit we have the following equation:

$$\text{Equa. of centre} = A \sin \theta + B \sin 2\theta + C \sin 3\theta + \&c.;$$

* The quantities in Table XVIII are called *Epochs.* The *Epoch* of a quantity is its value at some chosen epoch.

in which A, B, C, &c., are constants that rapidly decrease in value, and which may be determined for any particular orbit, and θ the mean anomaly. Now, by giving to the mean anomaly θ in this equation a series of values increasing by small equal differences (of 1°, for instance,) from zero to 360°, and computing the corresponding values of the equation of the centre, then registering in a column the different values assigned to θ, and in another column to the right of this the computed values of the equation of the centre, we shall obtain a table which will give on inspection the equation of the centre corresponding to any particular mean anomaly. In this manner was constructed Table XXV. In this table, however, for the sake of compactness, the values of the equation, instead of being registered in one column, are put in as many different columns as there may be different numbers of signs in the value of the mean anomaly; each column answering to the particular number of signs placed at the head of it.

If the equation of the centre at an assumed time be required, find the mean anomaly by the tables, and with the value found for it take out the equation of the centre from Table XXV.

The given quantity with which a quantity is taken from a table is called the *Argument* of that quantity. Accordingly the mean anomaly is the argument of the equation of the centre in Table XXV.

(3.) *For the inequalities.*—The equations of the inequalities, as we have already stated, are of the form C sin A, the argument A being the difference between the longitude of the disturbing planet and that of the earth, or some multiple of this difference. With the equations of the inequalities a table of each inequality may be constructed, upon the same principles as Table XXV. But, as the expression for the whole perturbation in longitude (Art. 212), produced by any one planet, involves only two variables, the longitude of the earth and the longitude of the planet, it is thought to be more convenient to have a table of *double entry*, which will give the amount of the perturbation by means of the two variables as arguments. Such a table may be constructed, by assigning to the longitude of the earth and the longitude of the disturbing planet a series of values increasing by a common difference, and computing with each set of the values of these quantities, the corresponding amount of the perturbation.

In connection with the tables of the perturbations, we must have tables that make known the values of the arguments at any given time. Now, the mean longitude of the sun may be found by the solar tables (XVIII to XXI), and thence the mean heliocentric longitude of the earth by subtracting 180°; and the mean longitude of the disturbing planet may be had from similar tables. The columns of Table XVIII, marked I, II, III, IV, V, VI, VII, contain the arguments of all the perturbations for the beginning of each of the years registered in the first column, expressed in thousandth parts of a circle. Tables XIX and XX contain the variations of the arguments for months, days, and hours. Their variations for minutes and seconds are too small to be taken into account. With these tables, and Table XVIII, the values of the arguments at any given time may be found, and by means of the arguments the perturbations may be taken from Tables, XXVIII, XXIX, XXXII, XXXI, XXX, and XXXIII.

(4.) *For the nutation.*—The formula for the lunar nutation in longitude, is 17″.3 sin N—0″.2 sin 2 N, in which N denotes the supplement (to 360°) of the longitude of the moon's ascending node. With this formula the second column of the Table XXVII was constructed. The value of N, in thousandth parts of a circle, results from Tables XVIII, XIX, and XX. The solar nutation is also given by Table XXVII.

Tables may also be constructed that will facilitate the computation of the radius-vector. We have

$$\text{True rad.-vector} = \text{elliptic rad.-vector} + \text{perturbations.}$$

A table of the elliptic radius-vector may be formed by means of the polar equation of the orbit, and tables of the perturbations from their analytical expressions (Art. 214). The tables of the perturbations will have the same arguments as the tables of the perturbations of longitude.

Lunar and planetary tables are constructed upon the same principles as the solar tables we have been describing, which serve to make known the orbit longitude and radius-vector. But other tables are necessary in the case of these bodies, for the computation of the ecliptic longitude and the latitude.

The difference between the orbit longitude and the ecliptic longitude is called the *Reduction* to the ecliptic. A formula for the reduction has been investigated, in which the variable is the difference between the orbit longitude and the longitude of the node (or, what amounts to the same, the orbit longitude plus the supplement of the longitude of the node to 360°). If this formula be reduced to a table, by taking the reduction from the table and adding it to the orbit longitude, we shall have the ecliptic longitude. Table LIII is a table of reduction for the moon.

For the latitude, we have the equation

$$\text{True lat.} = \text{lat. in orbit} + \text{perturbations.}$$

We have already seen (Art. 201) that

$$\sin (\text{lat. in orbit}) = \sin (\text{orbit long.} - \text{long. of node}) \sin (\text{inclina.})$$

A table constructed from this formula will have for its argument the orbit longitude minus the longitude of the node, which is also the argument of the reduction. (See Table LV.)

The tables of the perturbations in latitude are constructed upon the same principles as the tables of the perturbations in longitude and radius-vector.

NOTE II.

(*Referred to on p.* 175.)

The fact stated in the text (Art. 273) that the penumbra of the solar spot does not begin to be formed until the spot, which at first has an umbral blackness, has attained a measurable size, is cited by astronomical writers as a circumstance strongly favoring the hypothesis of the origination of the spot in a disturbance from below. But this fact may be reconciled with the opposite hypothesis advocated in the text, if we reflect that the penumbral lies at a considerable depth below the luminous envelope, and that the process of discharge and ascent of a column of photospheric matter (Art. 293), which results in disclosing to view a portion of the penumbral envelope, should *occupy a certain interval of time in passing down to it.* During this interval this envelope may have an umbral blackness, and it may owe its subsequent visibility, as distinct from the umbra, entirely to the fact that it acquires a luminosity in consequence of the electric discharges that attend the process of spot evolution, which has penetrated to its depth in the atmosphere of the sun. This view is supported by the fact that it furnishes a simple explanation of the decrease in the brightness of the penumbra from the umbra to its outer margin. We have only to observe that the process of expulsion and ascent of vaporous matter, which begins at a certain point of the photosphere, at the same time that it proceeds downward, spreads laterally, and that when it has penetrated to the depth of the penumbral envelope at the point below that where it originated, an opening of a certain size will have been formed in the luminous envelope, and except below the central portions of this opening, the lower envelope will still be in a comparatively quiescent condition, and retain its natural depth of shade. Subsequently the same process of evolution is repeated at this envelope; an opening is made in it that has the umbral depth of shade, and this is surrounded by a region of luminous activity, which is the penumbra of the spot, and is fringed by a dark border consisting of the part which the descending action has not yet reached. In the case of the larger spots and of long continuance, the same process penetrates to the third envelope, and the former umbra shows, in its turn, a black central nucleus, surrounded by a fringe of a shade perceptibly less dark.

Upon the present hypothesis with regard to the mode of development of the spots, the principal varieties, consisting of a spot without a penumbra, a spot without an umbra, a spot without the central black nucleus at the centre of the umbra, and a spot with this nucleus, are but the varieties that present themselves according as the process of discharge, beginning at the surface of the photosphere, penetrates only through the first envelope, or through the first and as far as the second, or through the first and second, or through the first, second, and third. Upon the old hypothesis, it is necessary, in order to explain these diverse phenomena, to assume that there are three possible centres of explosive action, posited below the successive envelopes.

So long as the active evolution continues at the lower envelopes, the ascending vaporous column, expanding as it rises, should check any eventual tendency of the opening in the luminous envelope to close. When the activity subsides at the penumbral stratum, and the opening in it begins to close, this should be followed by a similar collapse in the regions above it; and so the contraction of a spot should generally begin in accordance with observation at the umbra, and be followed by the encroachment of the luminous margin upon the penumbra. But it is conceivable, also, that the closing up of a spot may result from a condensation into luminous clouds of portions of the expanded matter ascending within the region of the spot; and that the luminous veil that is often seen to form over a spot, and the bridges of light which suddenly span the umbra, are the first indications of the beginning of such condensations.

To give a more complete exposition of the author's theory of the development of solar spots, the following general conclusions are added to those given in the text.

1. The spots are, for the most part, due to diminutions occurring in the magnetic intensity that obtains in the photosphere of the sun.

2. Each planet operates on the photosphere by repulsive impulses, and develops electro-magnetic currents circulating in a direction opposite to that of the rotation. The effective currents thus originating are differential, and result from the fact, that on the left or east side of the line from the planet to the sun's centre the motion of the surface is in a direction opposite to that in which the impulses are propagated, and on the other side in the same direction.

3. The general tendency of such new currents should be to increase the magnetic intensity at the surface of the photosphere; but it is possible that in peculiar conditions of the photospheric matter, as to density and other qualities, the superficial currents developed by planetary action may prevail over those set in motion below the surface, and the opposite magnetic effect be produced.

4. The motion of the sun through space also develops similar magnetic currents, which should have a similar effect. These currents may be considered as originating on that side of the sun where the absolute motion of a point of its surface is the most rapid.

5. If the sun were stationary the motion of revolution of a single planet would have but little direct effect to change its magnetic action on the sun as a whole; except so far as this may vary by reason of the varying distance of the planet. But in point of fact the effective action of a planet will change with its distance in longitude from the point towards which the sun is moving in space. It will be at its maximum when the planet is in heliocentric conjunction with this point, and at a minimum when it is in opposition to it. In the former case its heliocentric longitude will be 253°, and in the latter 73°.

6. The joint magnetic action of two planets varies with their relative position; it has its maximum value when the planets are in conjunction, and its minimum when they are in opposition.

7. The epochs of the conjunction of one planet with another, or of a planet with the point towards which the sun is moving, are, in general, the epochs of minimum of spots, since the magnetic intensity at the surface of the photosphere is on the increase before every such epoch. The approximate coincidence of planetary conjunctions with special epochs of minimum of spots is a recognized fact in the case of Jupiter and the earth, and Venus and the earth. On the other hand, the opposition of two planets tends to determine a maximum of spots.

8. Jupiter and Venus are the most influential planets. The period of the spots is determined mainly by the motion of revolution of Jupiter, but appears to be modified by the varying planetary configurations, and also by changes occurring in the physical condition of the sun's photosphere. The varying action of Jupiter, in the course of a revolution, has been attributed to its variations of distance from the sun, but it seems improbable that effects so marked should result from so slight a cause. The mean period of the spots is, according to Wolf, nearly one year, and according to Schwabe, nearly two years less than Jupiter's period of revolution. This difference cannot be explained by any mere recurrence of planetary configurations. In fact,

epochs of maximum and minimum of spots have occurred under every variety of configurations; and also when Jupiter has had every variety of position in its orbit. The following Table shows the mean positions of Jupiter and Saturn at various such epochs, from 1750 to 1860; together with the relative numbers showing, according to Wolf, the spot-condition of the sun.

Epochs of Maxima.	Relative Numbers.	Mean Heliocentric Longitude.		Epochs of Minima.	Obs'd Min. -Mean Min.	Mean Heliocentric Longitude.	
		Jupiter.	Saturn.			Jupiter.	Saturn.
1750.0	68.2	4°	231°	1744.5	+0.558	197°	164°
1761.5	75.0	353	12	1755.7	+0.639	177	301
1770.0	79.4	251	116	1766.5	+0.320	145	73
1779.5	99.2	179	232	1775.8	—1.499	67	187
1788.5	90.6	93	342	1784.8	—3.618	340	297
1804.0	70.0	203	172	1798.5	—1.037	36	105
1816.8	45.5	232	329	1810.5	—0.156	41	251
1829.5	53.5	258	124	1823.2	+1.425	66	47
1837.2	111.0	131	218	1833.8	+0.906	28	177
1848.6	100.4	117	358	1844.0	—0.013	338	301
1860.5	98.6	119	143	1856.2	+1.068	348	91

It will be seen that since 1761 the maxima have occurred when Jupiter was in the second or third quadrant of longitude; and that since 1755, the minima have occurred when he was in the other two quadrants. But previous to those dates the condition of things was reversed, and the transition from the one condition to the other was gradual. This fact seems to necessitate the supposition that the physical condition of the sun's photosphere is liable to changes, by reason of which the ordinary effect of the planets is, at intervals, wholly reversed.

9. The highest maxima of spots have occurred when Jupiter was in the second quadrant of longitude; that is, after he has passed a certain distance beyond the point (long. 73°) where his magnetic action on the sun is most directly opposed to the effect of the sun's motion through space, and advanced towards the aphelion of the orbit, where the direct magnetizing tendency of the planet is the least.

It should here be remarked that the author has undertaken, in other publications, to show that the earth derives its magnetic condition from the collision of the molecules of its revolving and rotating mass with the ether of space, and that the sun and the planets should each be in a magnetized condition from a similar cause; also that in the new terrestrial magnetic currents being continually developed by the earth's motions of revolution and rotation combined, in connection with those generated in the photosphere (upper atmosphere) of the earth by ethereal impulses, and streams of nebulous matter, proceeding from the sun, we have the principal operative causes of the disturbances experienced by the magnetic needle on the earth's surface.

NOTE III.

(*Referred to on page* 275.)

A remarkable analogy in the periods of rotation of the primary planets was discovered a few years since (1848) by Daniel Kirkwood, of Pottsville, Pennsylvania. This analogy is now generally known by the name of *Kirkwood's Law*, and is as follows:

"Let P be the point of equal attraction between any planet and the one next interior, the two being in conjunction: P′ that between the same and the one next exterior.

Let also D = the sum of the distances of the points P, P′ from the orbit of the planet; which I shall call the diameter of the sphere of the planet's attraction;

D′ = the diameter of any other planet's sphere of attraction found in like manner;

n = the number of sidereal rotations performed by the former during one sidereal revolution around the sun;

n = the number performed by the latter; then it will be found that

$$n^2 : n'^2 :: D^3 : D'^3; \text{ or } n = n' \left(\frac{D}{D'}\right)^{\frac{3}{2}}.$$

That is, *the square of the number of rotations made by a planet during one revolution round the sun, is proportional to the cube of the diameter of its sphere of attraction;* or $\frac{n}{D^{\frac{3}{2}}}$ is a constant quantity for all the planets of the solar system.

The analogy thus announced has been subjected to a rigid mathematical examination by Sears C. Walker, with the following result: "We may therefore conclude," says he, "that whether Kirkwood's Analogy is or is not the expression of a physical law, it is at least that of a physical fact in the mechanism of the universe." (See the American Journal of Science, New Series, vol. x. pp. 19-26.)

There are but three planets, viz., Venus, the Earth, and Saturn, for which all the elements embraced in this law are known. The diameters of the spheres of attraction of Mercury and Neptune are, from the nature of the case, incapable of determination. The mass of the one planet into which the planetods are supposed once to have been united is not known with certainty, as there may be planetoids yet undiscovered, and its period of rotation is hypothetical only. The diameters of the spheres of attraction of Mars and Jupiter can only be approximately determined; and the period of rotation of Uranus is unknown.

The interest naturally awakened by the announcement of so important a discovery was heightened by the fact, that it was at once perceived that it furnished a new and powerful argument in support of the nebular hypothesis (or cosmogony) devised by Laplace. (See a paper on this subject by Dr. B. A. Gould, Jr., in the Journal of Science, New Series, vol. x. p. 26, etc.)

NOTE IV.

(Referred to on page 277.)

It remains to deduce, if possible, the known law of the distribution of the inclinations of the cometary orbits. This law is, that the number of orbital inclinations of different values increases with the angle from 0° to 50° or 60°, and then decreases; as appears from the following tabular statement, from which the comets of short period have been deducted.

0° to 10°	10° to 20°	20° to 30°	30° to 40°	40° to 50°	50° to 60°	60° to 70°	70° to 80°	80° to 90°
11	15	16	28	34	32	23	22	14

Let us first consider the case of a discharge from the equator, and conceive, for the present, the direction of discharge to be tangent to the surface. Since the orbits of the class of comets under consideration are very eccentric, the initial velocity must be very much less than the velocity of rotation at the equator. If we fix upon a maximum limit (V) to this velocity, at any assumed epoch, and from this and the velocity of rotation at the equator (v) deduce the direction of the expelling force, and the velocity ($v°$,) due to its action, it will be seen:

(1.) That this force will take effect in directions opposed to that of the rotation, and inclined to it under angles differing but little from 180°, whatever direction may be assumed for V, the resultant or effective velocity.

(2.) That the velocity, v', due to the expelling force, will be either equal to the velocity of rotation, v, or a little greater, or a little less.

Under these circumstances, nebulous masses may be projected in every variety of direction in the tangential plane, which would move in planes having any angle of inclination, and describe orbits of every variety of eccentricity greater than that answering to the assumed maximum initial velocity, V.

Now, if we conceive the point at which a discharge occurs to lie in any latitude (l), the velocity of rotation will be less in the proportion of cos l to 1; and making use of the parallelogram of velocities as before, and retaining the same effective velocity V, we find that for any assumed direction of V, the line of direction of the expelling force will deviate more from that of direct opposition to the motion of rotation than at the equator; and that the deviation will increase with the latitude. For any latitude (l) the orbits described by the masses detached may have every variety of inclination from l° to 90°; but the larger inclinations will result from an action of the expelling force exerted in a direction inclined under a smaller angle to the meridian in proportion as l is greater.

If we conceive this force to act in some direction oblique to the surface, instead of tangentially, the velocity due to its action will be replaced by its tangential component, which is now to be taken equal to V. The aphelion of the orbit will also now be removed to a certain distance from the nebulous body, instead of being within its surface at the point of discharge.

In view of what has now been stated, it may be seen that *the actual law of distribution of the inclinations might result if the frequency of discharge were to decrease with the angle included between the line of direction of the operating force and the meridian.*

This theoretical result suggests, as the possible origin of the separation of fragments from the surface of the nebulous body, *the flow of electric currents in all directions from points near the equator;* similar to the currents we have conceived to be developed by planetary action on the sun's photospheric surface, and to give rise to the solar spots (293). Such currents, in proportion to the resistance they experience, would develop statical electricity, the repulsive action of which might occasion discharges in the direction of the radial currents and obliquely upwards. Upon this conception, if we consider the disengagements that may occur from any point of any one meridian, and bear in mind that the electric currents supposed may proceed, indifferently, from all points near the equator, it will be observed that the frequency of the detachment of fragments from the point in question will decrease in proportion as the radial current, or direction of the expelling force, makes a less angle with the meridian. For an arc of the equator (say 10°) will subtend, at the point considered, a greater angle in proportion as it is nearer to the meridian. At each point of the meridian, therefore, the liability to expulsive action should be greatest in directions nearly perpendicular to the meridian; for which directions the resulting orbital inclinations would be nearly equal to the latitude of the point. The effective directions of expulsion would fall, for each latitude, l, on the meridian, between l° and 90°. The complete result, for all points on the meridian, should then be that the number of inclinations would augment with the angle of inclination up to a certain large value of this angle, and then decrease; in accordance with the observed law. The agreement would apparently be more exact if, as would naturally be supposed, the frequency of discharge became less at the higher latitudes.

We must suppose that the masses discharged which have since become permanent members of the solar system, must, after being projected into space, have become condensed sufficiently before returning within the attenuated mass of the nebulous body, to have pursued their course unaffected by the resistance of the medium traversed, except within certain limits.

It appears that whether we consider the movements of the magnetic needle upon the earth, resulting from solar action, or the development of spots on the sun's surface by planetary influence, or the rise of nebulous envelopes from the nucleus of a large comet, under the operation of the sun, or the origination of cometary bodies, we are conducted, in each instance, to the primary conception of electric currents radiating from points near the equator of the body subject to external influence, as playing an important part in the production of the phenomena observed. Analogy would then lead us to infer that the exciting cause of the electric currents supposed to have furnished the operating cause of the detachment of cometary masses from the same nebulous body from which the planets have been derived, has been an action of the planets upon the surface of this body, similar to that which has been operative in the other cases. The same process may have continued in operation down to the present epoch, originating, in the later ages, meteorites rather than true cometary bodies. In fact we have seen that a continual process of discharge of nebulous magnetic matter from the sun is in operation in the present age. (Art. 293.)

NOTE V.

ORIGIN OF SIDEREAL SYSTEMS.

We propose, in the present note, to develop very briefly a theory of the possible evolution of all sidereal systems from primordial, rotating, nebulous masses.

Fundamental hypothesis, and circumstances of evolution. Let us assume that the component stars of every cluster were originally integrant portions of a vast nebulous body, and that this body had a motion of rotation about an axis. It is obvious that every portion of the rotating mass that might become detached would thenceforward tend to revolve independently about its centre in a certain orbit. Every such orbit would cross the plane of the equator in two points, or nodes, unless the detachment occurred in that plane, in which event the orbit would lie within it. Again, if we conceive the separation to have taken place without bringing disturbing forces into play, the detached portion of the mass should have an initial motion in a direction parallel to the equatorial plane, and an initial velocity proportional to the cosine of the angular distance from that plane. If we suppose impulsive or repulsive forces to have been in energetic operation, we may approximately determine the nature and amount of their influence upon these initial circumstances, and thence upon the orbit subsequently described.

Case I. *A simultaneous disruption of the whole mass of the nebulous body.* We will regard the original nebulous body as sensibly spherical in form, and first conceive the disruption to have occurred at the same epoch in all its parts; or at epochs separated by small intervals in comparison with the vast duration of one rotation period.

General results. The first result will be the formation of a globular cluster of stars, separated either by equal or unequal intervals of space. We will confine our attention to the case of equal intervals. All the stars of each spherical layer will then set out on their various courses at the same epoch. If we consider those which lie on any one meridian of the outer layer, their initial velocities will decrease proportionally to their angular distance from the equator, and they will therefore set out in elliptic courses that will be more eccentric in proportion as this distance is greater. In case the disruption occurs at the period when the centrifugal force of rotation is equal to the force of gravity, at the equator of the nebulous body, the equatorial stars will move in circles, and the others in orbits of every degree of eccentricity, from a circle at the equator to a right line at the poles. The stars in question will all pass, at the end of one quadrant of a revolution, through the plane of the equator at various points of the line perpendicular to the plane of the meridian from which they set out. Another meridian of the outer surface would give a new set of stars, with a new common line of nodes for all their orbits. As a general result then of the orbital motions communicated to the stars of the outer spherical layer, *this layer will assume the form of an oblate spheroid, with its shorter axis coincident with the line of the original axis of rotation.* If we consider the next spherical layer of stars, these will all have taken up their independent movements simultaneously with those at the surface, and as a general result the whole layer will assume the oblate spheroidal form, like the first. The same will be true of each successive layer, and the contractions will proceed simultaneously, while the order of the layers will remain unchanged.

Periods of revolution. Since the initial velocities, from the surface inward, are proportional to the distance from the axis of rotation, and the attractive forces by which the stars are solicited at the outset, are proportional to the distance from the centre, it follows that all stars proceeding from points at the same angular distance from the equator will accomplish their revolutions in the same period of time.

This will be true whether the accelerating force soliciting each star, in its orbital motion, varies inversely as the square of the distance from the centre, or directly as the distance; and must be approximately true if the actual law of variation lies between these two. Now if in the contraction of the starry layers they all retained their spherical form, each star would be subject to the attraction of the same spherical mass, during an entire revolution, and therefore the law of variation of its accelerating force would be that of the inverse squares; but in point of fact the layers in question contract into spheroids continually increasing in oblateness, and hence the accelerating force must vary according to a less rapid law.

Upon the supposition that the law of the inverse squares obtains, the period of revolution of an equatorial star in a circle, would be to that of a star from very near the pole, in a very eccentric ellipse, as 2.8 to 1. On the other hand, if the force varied directly as the distance, the two periods would be equal. The actual ratio should then lie between 2 8 and 1; and may be assumed to be not far from 2. The equatorial star would complete a half revolution in an interval of time equal to the duration of one oscillation of an ideal celestial pendulum, having a length equal to the radius of the globular cluster at the epoch of its formation, and solicited by the force of gravitation in operation at the surface of the cluster. Either pole of the cluster would contract to the equatorial plane, and attain to its limit of expansion on the opposite side, while such an ideal pendulum is completing a half oscillation, or thereabouts. Every such dynamical cluster has its *vast cycle*, at the close of which all its stars return approximately to their original positions. Such a state of things would be realized in about the interval of two periods of revolution of the equatorial stars, supposed to move in circles, or in four oscillations of the representative pendulum.

Increase of density. As the contraction of the original globular cluster proceeds, the density continually increases, and attains, at any part of the equatorial plane, its greatest value at the epoch when the stars are crossing the plane in that region. The condensations not only augment, as the contraction goes on, in directions towards the equatorial plane, but also towards the centre; since all the stars, except those moving in the plane of the equator, tend towards the centre, in their orbital motions. The greatest density will be attained at the centre; and at the epoch when the polar stars have reached its vicinity.

Possible collisions. In all the movements of the stars of the supposed cluster, the only collisions that could directly ensue would be in those cases in which two stars set out from two points of a meridian, at the same distance from the equator, at the same epoch. Each of the two stars should, at the end of a quarter revolution, reach the plane of the equator at the same point. Such exact coincidences of position and date of origin should, however, rarely occur. But frequent close approximations of two or more stars may well occur, and eventuate in the formation of *double* or *triple stars*, revolving around each other.

Spheroidal clusters. We are led, by the theoretical views that have now been presented, to regard the oval nebulæ, or spheroidal clusters, seen in the heavens, as *original globular clusters in some of their different stages of spheroidal condensation.* Upon this hypothesis they should in general be more condensed and more difficult of resolution than existing globular clusters (as is observed to be the case). The amount of oblateness, with the attendant condensation, would be an index of the age of the cluster. Present globular clusters would be just at the beginning of their first, or of a new grand cycle; and destined in future ages to pass through the continued series of spheroidal forms that we have signalized.

Inequalities of brightness in different parts of spheroidal clusters. Instances of such inequalities, not resulting from a mere central condensation, are observable in Fig. 6, Plate IV, and in Fig. 9, Plate V. They may be attributed to inequalities of density existing at certain stages of the contraction of the original globular cluster. For example, at the end of a half revolution of the polar stars they would be condensed about the contracted poles, which would have passed to the opposite sides of the equatorial plane, and the other stars of each original spherical layer would be condensed in the resulting spheroidal layer, but in a decreasing degree towards the equator. This theoretical condition answers to the dumb-bell nebula (Fig. 9, Plate V). The differences observable in the distribution of the light on opposite sides of the equatorial plane (or of the larger axis of the faint elliptic outline) are such as might result from a want of entire correspondence in the epochs of the initial orbital motions of the stars on opposite sides of that plane. Fig. 6,

Plate IV, answers to a similar period in the structural history of a cluster; but the inner layers have experienced a more marked condensation towards the centre. This would be the result if the initial velocities of the stars of these layers were, from some special cause, materially less than the normal values. Under the same circumstances these layers would be more nearly spherical in their form than in the normal type above considered.

Case II. *A simultaneous disruption of the nebulous body along a limited number of meridians.* We will now suppose the disruptive evolution of the starry masses to be confined to certain meridians, and the regions contiguous to them on either side, and enquire into the subsequent form and internal condition of the cluster. It will readily be seen that, since the initial velocities and the periods of revolution decrease from the equatorial to the polar stars, the stars proceeding from any one of these meridians, will, as they follow their natural orbits, take on collectively *a spiral arrangement.* This will be best seen when viewed perpendicularly to the equatorial plane. At the close of a half revolution of the polar stars, the spiral, thus viewed, would occupy the second quadrant of revolution of the entire set of stars considered. As the revolution proceeded, the angular extent of the spiral would continually increase. The stars proceeding from the meridian contiguous to that first supposed, would form a similar spiral contiguous to that just considered; and the entire collection of stars setting out from the one meridional region of disruption would form a spiral band, increasing in width from its inner to its outer end. The similarly situated stars of the successive layers, proceeding inwards, would form shorter spirals, that would be combined with the others and add to the width of the spiral band. A similar spiral band would result from each of the other collections of detached stars.

Spiral nebulæ. According to what has just been shown, the spiral formation is a natural consequence of a cotemporaneous evolution of stars from various points of the same meridian. The spiral arrangement of stars should therefore exist in every cluster, and be more or less discernible, unless the disruption was general and nearly simultaneous throughout the original nebulous mass. Accordingly spiral lines and fringes of stars are in many instances observable on the borders of spheroidal and irregular clusters.

The theory of the origin of the true spiral nebulæ has been sufficiently indicated. The length of the spiral coils in Fig. 10, Plate V, indicates that the equatorial stars of the cluster have completed three-quarters of a revolution, and the polar stars one revolution and a half. The secondary condensation, on the right of the figure, may be ascribed to the circumstance of the stars that set out from points on the meridian near the equator, being at the present epoch in the vicinity of one of the nodes; three-quarters of a revolution having been completed. The nebula seems to consist of two spiral bands or coils, made up of an indefinite number of spiral filaments, or smaller bands. The line of sight is probably nearly perpendicular to the line of the two centres of condensation, and oblique to the equatorial plane. The great comparative dispersion of the filaments of the lower coil may be in part attributable to the detachment of stars beginning on one meridian, and extending gradually around to successive meridians.

Case III. *An irregular disruption.* If irregular deviations from the normal type of evolution that has been considered occur, the result should be the formation of *irregular clusters* differing more or less from true globular or spheroidal clusters. By reason of the want of correspondence in the epochs of detachment on different meridians of the nebulous body, such clusters should be less condensed, and with less regularity than the regular clusters.

Case IV. *A disruption beginning at the equator, and extending gradually towards the poles.* When this deviation from the normal type occurs, the obvious result will be that the arrival of the stars at the equatorial plane will be delayed, in proportion to their angular distance from it, at the outset; and therefore that the contraction of each of the original spherical layers will take place more slowly. The condensation towards the equatorial plane will also go on more slowly. The law of retardation of the dates of initial movement, for stars at increasing distances from the equator, may theoretically be such as to determine any known law of decrease of density along each spheroidal layer, from the equator to the axis.

System of the Milky Way. It is accordingly conceivable that the observed law of decrease of the density of the system of the milky way, as the distance from its principal plane increases, may have resulted from the operative cause just considered. If the same law of evolution prevailed cotemporaneously, or approximately so, throughout the mass of the nebulous body from which this system is supposed to have been derived, the result would be the formation of spheroidal layers of stars in which the density would vary according to a common law. The density would therefore decrease outwardly from the principal plane of the milky way at the same rate for stars at all distances. Struve found this to be nearly true for all stars beyond the 6th or 7th magnitude.

Motion of revolution of the sun. Since the sun is now near the equatorial plane, and moving away from it (Art. 474), we must suppose that it has at least completed either one-quarter or three-quarters of a revolution in its vast orbit. The position of the centre of the system is too imperfectly known (Art. 478) to make it possible to determine with certainty which of the two nodes it is now leaving. Taking the average velocity of the sun at $4\frac{2}{7}$ miles per second, as given in the text, and supposing the distance of the centre of the system to be equal to the exterior limit of stars of the 6th magnitude, the period of revolution of the sun should be 35 millions of years. Mädler's estimate of the distance of the centre of the system of the milky way places it a little beyond the exterior limit of stars of the 5th magnitude. The above larger estimate answers nearly to Struve's determination of the sun's orbital velocity (Art. 448).

Particular features of the Milky Way. The variable breadth of the belt of the milky way, its bifurcation, and alternations of bright and dark patches, may have proceeded from a want of correspondence in the dates of evolution of stars on different meridians of the original nebulous mass, as well as abnormal interruptions of the process of separation at certain parts of particular meridians. Thus, if on a certain meridian, the separation into stars should not have occurred for a certain distance from the equator, on either side, it would follow that just before or after $\frac{1}{4}$ of a revolution of the stars that set out from the points nearest the equator, the stars from greater latitudes would be concentrated upon two points at a short distance from the equator, on either side. If the same initial circumstances prevailed over a series of meridians, there would be formed two bands of greatest condensation at a certain distance from the equator, on opposite sides. The same bands would manifest themselves before and after $\frac{3}{4}$ of a revolution.

Primary condition, and subsequent processes of change, of fragments disunited from the primitive nebulous mass. Since the parts of any such fragment unequally distant from the axis must have had, at the time of separation, unequal velocities of rotation, it must have taken up a rotation about an axis of its own, and tended to assume the form of a sphere, or spheroid somewhat flattened at the poles. It should therefore, at some subsequent date, have broken up into a cluster of stars, or into a planetary system revolving around a central sun. If the mass detached should have been of comparatively great extent, it may have separated into a combination of stars and clusters of stars, as in the Magellanic Clouds; or into irregular beds of stars, as in the irregular nebulæ.

The concentration attending the formation and subsequent contraction of such systems should have occasioned a vacuity of stars in the spaces contiguous to them.

Irresolvable nebulæ. If, as we have already been led to suppose, the process of evolution of the system of the milky way from a primitive nebulous mass, extended gradually from the equator towards the poles, the vast nebulous mass left detached at the outer polar regions, and subject to peculiar conditions, may have become disunited into large masses, from which clusters have been derived that are now at an earlier stage of development, and at a greater distance than the telescopic stars and clusters.

Annular nebulæ, which are the rarest objects in the heavens, may have resulted from the matter of the polar regions of a nebulous body being mostly drawn to surrounding points of condensation, or not having yet condensed into true stars, or into stars comparatively minute.

Planetary nebulæ probably belong to the same type of development as annular nebulæ; since some of them have been resolved into annular nebulæ. Their equable light may result from the process of star-formation having penetrated only a certain depth into the original nebulous mass.

General considerations. We may conclude from the previous theoretical discussion, that if we assume all systems of stars to have been derived, by separation, from rotating nebulous bodies of vast extent, according to one or the other of a certain small number of types of evolution, the forms and internal conditions that would be inevitably passed through, in the progress of ages, would be the counterparts of the various forms and apparent structural conditions of the clusters and nebulæ actually observed. We may suppose, it is true, in explanation of the single case of the spheroidal clusters, that the rotating bodies took on a decidedly spheroidal form before disintegration, and that the derived clusters have now, sensibly, their original form. Upon this view we must still admit that these clusters are destined, in the lapse of future ages, to pass through changes similar to those we have deduced for globular clusters. But spiral, and some other forms of nebulæ, cannot but have passed through vast ages of development, and in the light of this indication of the age of the heavens, it seems improbable that spheroidal clusters should be of comparatively recent origin.

As to the real nature of the process of separation of starry fragments from rotating nebulous masses, we are led, on physical grounds, to conceive that it must have consisted in a concentration upon special points of the mass at certain favorable epochs, rather than in a violent separation. It would seem that such a process beginning at any part of the nebulous body, should tend to extend indefinitely through it. But it is to be observed that the propagation of a force through such a mass would of itself require a vast period of time. It is possible that the want of correspondence in the epochs of separation, in different parts of the body, we have inferred existed in the development of the system of the milky way, and to a less extent also in spiral nebulæ, may have resulted from the far greater comparative size, in these instances, of the original nebulous mass.

TABLE I.

Latitudes and Longitudes of Places.

Names of Places.	Latitude.	Longitude from Greenwich in Arc.	Long. from Greenwich in Time.	Longitude from Washington in Time.
			h. m. s.	h. m. s.
Abo, *Obs*.	60° 26′ 56″.8 N.	22° 17′ 11″.4 E.	—1 29 8.8	— 6 37 20.0
Albany, *Obs*.	42 39 50 .0 N.	73 44 39 .0 W.	+4 54 56.6	— 0 13 12.6
Altona, *Obs*.	53 32 45 .3 N.	9 56 32 .3 E.	—0 39 46.1	— 5 47 57.4
Ann Arbor, *Obs*.	42 16 48 .0 N.	83 43 3 .0 W.	+5 34 52.2	+ 0 26 41.0
Astor Point, *Oregon*.	46 11 27 .6 N.	123 49 31 .7 W.	+8 15 18.1	+ 3 7 6.9
Athens, *Obs*	37 58 20 .0 N.	23 43 47 .8 E.	—1 34 55.2	— 6 43 6.4
Baltimore, *Wash. Mt*.	39 17 47 .8 N.	76 36 38 .6 W.	+5 6 26.6	— 0 1 45.4
Berlin, *Obs*.	52 30 16 .7 N.	13 23 52 .8 E.	—0 53 35.5	— 6 1 46.7
Boston, *State House*.	42 21 27 .6 N.	71 3 30 .0 W.	+4 44 14.0	— 0 23 57.2
Cambridge, *Obs*.	42 22 48 .6 N.	71 7 24 .9 W.	+4 44 29.7	— 0 23 41.5
Cape of Good Hope, *Obs*	33 56 3 .0 S.	18 28 45 .0 E.	—1 13 55.0	— 6 22 6.2
Cape Horn.	55 58 41 S.	67 10 53 .0 W.	+4 28 43.5	— 0 39 27.7
Charleston, *St. Mich.'s Ch*	32 46 33 .1 N.	79 55 37 .6 W.	+5 19 42.5	+ 0 11 31.3
Copenhagen, *Obs*.	55 40 53 .0 N.	12 34 57 .0 E.	—0 50 19.8	— 5 58 31.0
Dorpat, *Obs*.	58 22 47 .1 N.	26 43 23 .4 E.	—1 46 53.6	— 6 55 4.8
Dublin, *Obs*.	53 23 13 .0 N.	6 20 30 .0 W.	+0 25 22.0	— 4 42 49.2
Edinburgh, *Obs*.	55 57 23 .2 N.	3 10 45 .0 W.	+0 12 43.0	— 4 55 28.2
Galveston, *Cath*.	29 18 17 .3 N.	101 33 33 .0 W.	+6 46 14.2	+ 1 38 3.0
Gotha, *Obs*.	50 56 5 .2 N.	10 43 54 .9 E.	—0 42 55.7	— 5 51 6.9
Gottingen, *Obs*.	51 31 47 .9 N.	9 56 31 .5 E.	—0 39 46.1	— 5 47 57.3
Greenwich, *Obs*.	51 28 38 .0 N.	0 0 0 .0	0 0 0.0	— 5 8 11.2
Hamburg, *Obs*.	53 33 5 .0 N.	9 58 23 .4 E.	—0 39 53.6	— 5 48 4.8
Hamilton Coll., *Obs*.	43 3 16 .5 N.	75 24 16 .8 W.	+5 1 37.1	— 0 6 34.1
Kazan, *Obs*.	55 47 23 .1 N.	49 6 36 .6 E.	—3 16 26.3	— 8 24 37.5
Königsberg, *Obs*.	54 42 50 .7 N.	20 30 5 .4 E.	—3 22 0.4	— 6 30 11.2
Liverpool, *Lassell Obs*.	53 25 28 .0 N.	2 54 40 .5 W.	+0 11 38.7	— 4 56 32.5
London, *Obs*.	51 31 29 .8 N.	0 9 16 .5 W.	+0 0 37.1	— 5 7 34.1
Madras, *Obs*.	13 4 8 .1 N.	80 14 15 .0 E.	—5 20 57.0	—10 29 8.2
Madrid, *Obs*.	40 24 27 .7 N.	3 41 21 .0 W.	+0 14 45.4	— 4 53 25.8
Marseilles, *Obs*.	43 17 49 .0 N.	5 22 14 .8 E.	—0 21 29.0	— 5 29 40.2
Milan, *Obs*.	45 28 0 .7 N.	9 11 39 .6 E.	—0 36 46.6	— 5 44 57.8
Moscow, *Obs*.	55 45 19 .8 N.	37 34 14 .4 E.	—2 30 17.0	— 7 38 28.2
Mount Desert, Maine.	44 21 3 .9 N.	68 13 15 .5 W.	+4 32 53.3	— 0 35 17.9
Naples, *Obs*.	40 51 46 .6 N.	14 14 42 .9 E.	—0 56 58.9	— 6 5 10.1
New Haven, *Sheff. Obs*.	41 18 36 .5 N.	72 55 30 .0 W.	+4 51 42.0	— 0 16 29.2
New Orleans.	29 57 45 .0 N.	90 6 49 .0 W.	+6 0 27.0	+ 0 52 15.8
New York, *City Hall*.	40 42 43 .2 N.	74 0 3 .1 W.	+4 56 0.2	— 0 12 11.0
Palermo, *Obs*	38 6 44 .0 N.	13 21 2 .6 E.	—0 53 24.2	— 6 1 35.4
Paramatta, *Obs*.	33 48 49 .8 S.	151 1 33 .7 E.	—3 39 31.4	— 8 47 42.6
Paris, *Obs*.	48 50 13 .2 N.	2 20 9 .4 E.	—0 9 20.6	— 5 17 31.8
St. Petersburg, *Obs*.	59 56 29 .7 N.	30 18 22 .2 E.	—2 1 13.5	— 7 9 24.7
Philadelphia, *Obs*.	39 57 7 .5 N.	75 9 23 .4 W.	—5 15 44.8	— 0 7 33.6
Point Venus, Otaheite.	17 29 21 .0 S.	149 28 55 .0 W.	+9 57 56.0	+ 4 49 45.0
Princeton, *Seminary*.	40 20 40 .0 N.	74 39 34 .3 W.	+4 58 38.3	— 0 9 32.9
Pulkowa, *Obs*.	59 46 18 .7 N.	30 19 39 .9 E.	—2 1 18.7	— 7 9 29.9
Quebec, *Obs*.	46 48 30 .0 N.	71 12 15 .0 W.	+4 44 49.0	— 0 23 22.2
Rio de Janeiro, *Obs*.	22 53 51 .0 S.	43 3 39 .0 W.	+2 52 14.6	— 2 15 56.6
Rome, *Obs*.	41 53 52 .2 N.	12 28 40 .5 E.	—0 49 54.7	— 5 58 5.9
San Francisco, *Tel. Hill*.	37 47 59 .2 N.	122 23 19 .4 W.	+8 9 32.5	+ 3 1 21.3
Santiago de Chile, *Obs*.	33 26 25 .4 S.	70 38 14 .5 W.	+4 42 33.0	— 0 25 38.2
Savannah, *Exch*.	32 4 53 .4 N.	81 5 14 .3 W.	+5 24 20.9	+ 0 16 9.7
Stockholm, *Obs*.	59 20 31 .0 N.	18 3 42 .0 E.	—1 12 14.8	— 6 20 26.0
Vienna, *Obs*.	48 12 35 .5 N.	16 23 7 .9 E.	—1 5 32.5	— 6 13 43.7
Washington, *Obs*.	38 53 39 .3 N.	77 2 48 .0 W.	+5 8 11.2	— 0 0 0.0

TABLE II.

Elements of the Planetary Orbits.

Epoch, January 1, 1850, mean noon at Paris.

Planet's Name.	Inclination to the Ecliptic.	Sec. Var	Longitude of Ascending Node.	Sec. Var.	Longitude of Perihelion.	Sec. Var.
Mercury	7° 0′ 8″	+ 6″.3	46° 33′ 9″	71′ 4″	75° 7′ 14″	93′ 11″
Venus	3 23 31	+ 4 .5	75 19 52	54 49	129 27 14	82 26
Earth					100 21 21	102 50
Mars	1 51 5	— 2 .4	48 23 43	46 39	333 17 54	110 24
Jupiter	1 18 40	—23	98 54 20	57 14	11 54 53	94 49
Saturn	2 29 28	—15	112 21 44	51 10	90 6 12	115 55
Uranus	0 46 30	+ 3	73 14 14	23 39	168 16 45	87 32
Neptune	1 46 59		130 6 52		47 14 37	

	M. Distance from Sun, or Semi-Axis.	Mean Distance from Sun in Miles.	Eccentricity	Sec. Variation.
Mercury	0.3870987	35,353,000	0.2056048	+0.000020294
Venus	0.7233322	66,060,000	0.0068433	—0.000053843
Earth	1.0000000	91,328,000	0.0167711	—0.000042582
Mars	1.523691	139,156,000	0.0932611	+0.000095284
Jupiter	5.202798	475,161,000	0.0482388	+0.000159350
Saturn	9.538852	871,164,000	0.0559956	—0.000312402
Uranus	19.182639	1,751,912,000	0.0465775	—0.000025072
Neptune	30.03697	2,743,216,000	0.0087195	

	Mean Longitude at the Epoch.	M. Sidereal Period in Mean Solar Days.	Motion in Mean Long. in 1 yr. of 365 days.	Mean Daily Motion in Longitude.
		d.		
Mercury	327° 15′ 20″	87.9692580	53° 43′ 3″	4° 5′ 32″.6
Venus	245 33 15	224.7007869	224 47 30	1 36 7 .8
Earth	280 46 43 .5	365.2563744	359 45 41	0 59 8 .33
Mars	83 40 31	686.9796458	191 17 9	0 31 26 .7
Jupiter	160 1 20	4332.5848212	30 20 32	0 4 59 .3
Saturn	14 50 41	10759.2198174	12 13 36	0 2 0 .6
Uranus	28 26 41	30686.8208296	4 17 45	0 0 42 .4
Neptune	335 8 58	60126.72	2 11 58	0 0 21 .7

TABLE III.—*Elements of Moon's Orbit.*

Epoch, January 1, 1801, mean noon at Paris.

Mean inclination of orbit		5° 8′ 40″
Mean longitude of node at epoch		13 53 17. 7
Mean longitude of perigee at epoch		266 10 7 .5
Mean longitude of moon at epoch		118 12 8 .3
Mean distance from earth		60.2665590
Eccentricity		0.05490807
	d. h. m. s.	d.
Mean sidereal revolution	27 7 43 11.5	= 27.32166142
Mean tropical "	27 7 43 4.7	= 27.32158242
Mean synodical "	29 12 44 2.9	= 29.53058872
Mean anomalistic "	27 13 18 37.4	= 27.55459950
Mean nodical "	27 5 5 36.0	= 27.21222222

Mean revolution of nodes: sidereal.... = 6793.432 d.; tropical.... = 6798.33557 d.
Mean revolution of perigee: sidereal... = 3232.57534; tropical.. = 3231.4751

TABLE IV.

Diameters, Volumes, Masses, etc., of Sun, Moon, and Planets.

	Apparent Diameter.			Equatorial Diam ter.	Equatorial Diameter in Miles.	Volume.
	Least.	At Mean Distance.	Greatest.			
Mercury . .	4″.5	6″.7	12″.9	0.3732	2,958	0.0518
Venus	9 .7	17 .0	66 .3	0.9525	7,549	0.8686
Earth				1.0000	7,925.6	1.0000
Mars	4 .1	11 .1	30 .1	0.6201	4,915	0.2345
Jupiter . . .	30 .8	37 .2	50 .6	11.1401	88,294	1303.91
Saturn . . .	14 .6	16 .1	20 .3	9.0621	71,823	667.54
Uranus. . . .	3 .5	3 .9	4 .3	4.1864	33,124	73.369
Neptune. . .	2 .6	2 .7	2 .9	4.5383	35,910	93.470
Sun	31′ 32	32′ 3 .4*	32′ 36	107.263	850,123	1,240,285.0
Moon	28 48	31 7 .0	33 32	0.2725	2,160†	0.0203

	Mass.	Density.	Gravity at Equator.	Sidereal Rotation.	Light and Heat.
				h. m. s.	
Mercury	$\frac{1}{3000000}$‡	2.020	0.751	24 5 28	6.680
Venus.	$\frac{1}{400246}$	0.903	0.865	23 21 24	1.911
Earth	$\frac{1}{314000}$	1.000	1.000	23 56 4	1.000
Mars.	$\frac{1}{2994790}$	0.447	0.273	24 37 20	.431
Jupiter.	$\frac{1}{1050}$	0.229	2.410	9 55 26	.037
Saturn	$\frac{1}{3512}$	0.134	1.089	10 29 17	.011
Uranus.	$\frac{1}{24000}$	0.178	0.746		.003
Neptune.	$\frac{1}{18780}$	0.179	0.812	d. h. m	.001
Sun.	1	0.253	27.292	25 4 29§	
Moon	$\frac{1}{25697760}$	0.602	0.164	27 7 43	

TABLE V.

Elements of the Retrograde Motion of the Planets.

Planets.	Arc of Retrogradation.	Duration of Retrogradation.	Elongation at the Stations.	Synodic Revolution.
		d. h. d. h.		d.
Mercury .	9° 22′ to 15° 44′	23 12 to 21 12	14° 49′ to 20° 51′	115.8775
Venus. . . .	14 35 to 17 12	40 21 to 43 12	27 40 to 29 41	583.9214
Mars	10 6 to 19 35	60 18 to 80 15	128 44 to 146 37	779.9364
Jupiter. . .	9 51 to 9 59	116 18 to 122 12	113 35 to 116 42	398.8841
Saturn . . .	6 41 to 6 55	138 18 to 135 9	107 25 to 110 46	378.0919
Uranus . .	3 36	151	103 30	369.6563
Neptune .				367.4888

* The value 32′ 0″ used in the text (269) is the above value corrected for irradiation, and agrees with that obtained from observations of the transits of Mercury over the sun's disc.

† The value of the moon's diameter (2161.6 m.) obtained in the text, is probably about 2 miles too large, as, according to Airy, a correction of 2″ should be applied to the moon's measured diameter for the effect of irradiation, and 1″ answers to about 1 mile.

‡ This is the mass of Mercury adopted by Le Verrier. Many astronomers still retain Encke's determination, which is $\frac{1}{4865751}$. This gives for the density of Mercury 1.246.

§ This is Faye's recent determination. According to Carrington, the most probable value is 24d. 23h. 30m. Spörer's value is 25d. 5h. 37m.

Elements of the Orbits of the Satellites.

The distances are expressed in equatorial radii of the primaries.

I. *Satellites of Jupiter.*

Satellites.	Mean Distance.	Sidereal Revolution.	Inclination to Orbit of Jupiter.	Epoch of El'ts.	Mass; that of Jupiter = 1.
		d. h. m. s.			
1..............	6.04853	1 18 27 33.506	3° 5' 30"	Jan. 1,	0.000017328
2..............	9.62347	3 13 14 36.393	Variable.	1801.	0.000023235
3..............	15.35024	7 3 42 33.362	Variable.	G. T.	0.000088497
4..............	26.99835	16 16 31 49.702	2 58 48		0.000042659

II. *Satellites of Saturn.*

Name and Order.	Mean Distance.	Sidereal Revolution.	M. Long. at the Epoch.	Eccentricity and Perisaturnium.	Epoch of El'ts.
		d. h. m. s.			
1. Mimas.....	3.3607	0 22 37 22.9	256° 58' 48"		1790.0
2. Enceladus..	4.3125	1 8 53 6.7	67 41 36		1836 0
3. Tethys	5.3396	1 21 18 25.7	313 43 48	0.04 (?); 54° (?)	Ditto.
4. Dione.....	6.8398	2 17 41 8.9	327 40 48	0.02 (?); 42 (?)	Ditto.
5. Rhea......	9.5528	4 12 25 10.8	353 44 0	0.02 (?); 95 (?)	Ditto.
6. Titan......	22.1450	15 22 41 25.2	137 21 24	0.0293; 256 38'	1830.0
7. Hyperion ..	28	21 7 7		Apsides of Titan have a motion in long. of 30' 28" per an.	
8. Iapetus....	64.3590	79 7 53 40.4	269 37 48		1790.0

The longitudes are reckoned in the plane of the ring from its descending node with the ecliptic. The first seven satellites move in or very nearly in its plane; the orbit of the 8th lies about half way between the plane of the ring and that of the planet's orbit.

III. *Satellites of Uranus.*

Satellites.	M. Distance.	Sidereal Revolution.	Passage through Asc. Node, G. T.	Inclination to Ecliptic.
		d. h. m.		The orbits are in-
1. Ariel	7.44	2 12 28		clined at an angle of
2. Umbriel......	10.37	4 3 27		about 79° to the eclip-
3...............	13.12	5 21 25	d. h. m.	tic in a plane whose as-
4. Titania.......	17.01	8 16 56	1787, Feb. 16 0 10	cending node is in long.
5...............	19.85	10 23 3		165° 30' (Equinox of
6. Oberon	22.75	13 11 7	1787, Jan. 7 0 28	1798). Their motion is
7...............	45.51	38 1 48		retrograde. Their orbits
8	91.01	107 16 39		are nearly circular.

IV. *Satellite o Neptune.* Period, 5.877 d.; M. Distance, 12 radii of Planet.

TABLE VII. *Saturn's Ring.*

Outer diameter of outer ring.................	40".095.......	169,341	miles.
Inner diameter of outer ring.....................		149,060	"
Breadth of outer ring..................		10,149	"
Breadth of inner ring....................		16,484	"
Interval between rings.....................		1,723	"
Breadth of double ring.....................		28,356	"
Distance of ring from planet		18,327	"
Thickness of the rings not exceeding............		210	"

TABLE II (*a*).

The Planetoids.

Names, particulars of discovery, mean distances, etc.

No.	Name.	Date of Discovery.	Discoverer.	Mean Distance.	Sidereal Period. Yrs.	Eccentricity.
1.	Ceres......	1801, Jan. 1.	Piazzi.	2.7660	4.600	0.08024
2.	Pallas	1802, March 28.	Olbers.	2.7700	4.610	0.23969
3.	Juno.......	1804, Sept. 1.	Harding.	2.6687	4.362	0.25590
4.	Vesta	1807, March 29.	Olbers.	2.3607	3.627	0.09012
5.	Astræa	1845, Dec. 8.	Hencke.	2.5775	4.136	0.18999
6.	Hebe	1847, July 1.	Hencke.	2.4254	3.777	0.20115
7.	Iris........	" Aug. 13.	Hind.	2.3862	3.686	0.23125
8.	Flora	" Oct. 18.	Hind.	2.2014	3.266	0.15670
9.	Metis	1848, April 25.	Graham.	2.3862	3.686	0.12320
10.	Hegeia	1849, April 12.	De Gasparis.	3.1494	5.589	0.10056
11.	Parthenope.	1850, May 11.	De Gasparis.	2.4526	3.841	0.09888
12.	Victoria	" Sept. 13.	Hind.	2.3344	5.567	0.21890
13.	Egeria	" Nov. 2.	De Gasparis.	2.5756	4.133	0.08775
14.	Irene	1851, May 19.	Hind.	2.5895	4.167	0.16525
15.	Eunomia ...	" July 29.	De Gasparis.	2.6429	4.297	0.18801
16.	Psyche.....	1852, March 17.	De Gasparis.	2.9263	5.006	0.13575
17.	Thetis	" April 17.	Luther.	2.4737	3.890	0.12686
18.	Melpomene.	" June 24.	Hind.	2.2958	3.479	0.21723
19.	Fortuna....	" Aug. 22.	Hind.	2.4414	3.815	0.15792
20.	Massilia....	" Sept. 19.	De Gasparis.	2.4093	3.740	0.14383
21.	Lutetia	" Nov. 15.	Goldschmidt.	2.4354	3.081	0.16204
22.	Calliope....	" Nov. 16.	Hind.	2.9091	4.962	0.10361
23	Thalia	" Dec. 15.	Hind.	2.6250	4.263	0.23180
24.	Themis	1853, April 5.	De Gasparis.	3.1420	5.570	0.11701
25.	Phocea.....	" April 6.	Chacornac.	2.4023	3.723	0.25335
26.	Proserpine..	" May 5.	Luther.	2.6556	4.329	0.08752
27.	Euterpe....	" Nov. 8.	Hind.	2.3473	3.596	0.17290
28.	Bellona.....	1854, March 1.	Luther.	2.7784	4.631	0.15039
29.	Amphitrite..	" March 1.	Marth.	2.5548	4.084	0.07238
30.	Urania	" July 22.	Hind.	2.3642	3.635	0.12718
31.	Euphrosyne	" Sept. 1.	Ferguson.	3.1561	5.607	0.21601
32.	Pomona....	" Oct. 26.	Goldschmidt.	2.5831	4.160	0.08240
33.	Polyhymnia.	" Oct. 28.	Chacornac.	2.8646	4.848	0.33769
34.	Circe	1855, April 6.	Chacornac.	2.6839	4.397	0.10961
35.	Leucothea..	" April 19.	Luther.	3.0060	5.215	0.21363
36.	Atlanta	" Oct. 5.	Goldschmidt.	2.7487	4.557	0.29788
37.	Fides	" Oct. 5.	Luther.	2.6422	4.295	0.17489
38.	Leda	1856, Jan. 12.	Chacornac.	2.7399	4.535	0.15552
39.	Lætitia.....	" Feb. 8.	Chacornac.	2.7710	4.613	0.11081
40.	Harmonia ..	" March 31.	Luther.	2.2679	3.415	0.04608
41.	Daphne	" May 22.	Goldschmidt.	2.7674	4.605	0.27034
42.	Isis	" May 23.	Pogson.	2.4401	3.812	0.22566
43.	Ariadne....	1857, April 15.	Pogson.	2.2038	3.272	0.16756
44.	Nysa	" May 27.	Goldschmidt.	2.4242	3.774	0.14933
45.	Eugenia ...	" June 28.	Goldschmidt.	2.7159	4.476	0.08200

No.	Name.	Date of Discovery.	Discoverer.	Mean Distance.	Sidereal Period. Yrs.	Eccentricity.
46.	Hestia	1857, Aug. 16.	Pogson.	2.5178	3.995	0.16184
47.	Melete.....	" Sept. 9.	Goldschmidt.	2.5976	4.189	0.23686
48.	Aglaia.....	" Sept. 15.	Luther.	2.8831	4.896	0.12788
49.	Doris	" Sept. 19.	Goldschmidt.	3.1044	5.470	0.07580
50.	Pales	" Sept. 19.	Goldschmidt.	3.0861	5.421	0.23783
51.	Virginia	" Oct. 4.	Ferguson.	2.6486	4 310	0.28695
52.	Nemausa...	1858, Jan. 22.	Laurent.	2.3779	3.667	0.06285
53.	Europa	" Feb. 6.	Goldschmidt.	3.0999	5.458	0.00450
54.	Calypso	" April 4.	Luther.	2.6102	4.217	0.21263
55.	Alexandra..	" Sept. 10.	Goldschmidt.	2.7076	4.553	0.19941
56.	Pandora ...	" Sept. 10.	Searle.	2.7692	4.608	0.13895
57.	Mnemosyne	1859, Sept. 22.	Luther.	3.1597	5.616	0.10752
58.	Concordia...	1860, March 24.	Luther.	2.6979	4.431	0.04103
59.	Danae	" Sept. 9.	Goldschmidt.	2.9746	5.131	0.16308
60.	Olympia ...	" Sept. 12.	Chacornac.	2.7147	4.472	0.11883
61.	Erato......	" Sept. 14.	Forster.	3.1296	5.537	0.16964
62.	Echo	" Sept. 14.	Ferguson.	2.3939	3.729	0.18543
63.	Ausonia	1861, Feb. 10.	De Gasparis	2.3972	3.712	0.12732
64.	Angelina....	" March 4.	Tempel.	2.6783	4.385	0.12482
65.	Cybele.....	" March 8.	Tempel.	3.4205	6.658	0.12030
66.	Maia	" April 9.	H. P. Tuttle.	2.6539	4.322	0.15422
67.	Asia.......	" April 17.	Pogson.	2.4209	3.769	0.18443
68.	Hesperia ...	" April 29.	Schiaparelli.	2.9949	5.186	0.17452
69.	Leto.......	" April 29.	Luther.	2.7748	4.622	0.18566
70.	Panopea ...	" May 5.	Goldschmidt.	2 6132	4.224	0.18309
71.	Feronia	" May 29.	C. H. F. Peters	2.2660	3.411	0.11977
72.	Niöbe	" Aug. 13.	Luther.	2.7555	4.574	0.17374
73.	Clytie	1862, April 7.	Tuttle.	2.6648	4.350	0.04424
74.	Galatea	" Aug. 30.	Temple.	2.7777	4.629	0.23820
75.	Eurydice ...	" Sept. 22.	C. H. F. Peters	2.6698	4.362	0.30690
76	Freia	" Oct. 21.	D'Arrest.	3.3877	6 235	0.18772
77.	Frigga	" Nov. 15.	C. H. F. Peters	2.6719	4.368	0.13582
78.	Diana	1863, March 15.	Luther.	2.6228	4.248	0 20549
79.	Eurynome..	" Sept. 14.	Watson.	2.4433	3 819	0.19509
80.	Sappho.....	1864, May 2.	Pogson.	2.2963	3.480	0.20022
81.	Terpsichore.	" Sept 30.	Tempel.	2.8563	4.827	0.21175
82.	Alemene ...	" Nov. 27.	Luther.	2.7603	4.586	0.22599
83.	Beatrix	1865, April 26.	De Gasparis.	2.4287	3.785	0.08410
84.	Clio	" Aug. 25.	Luther	2.3674	3.643	0.23754
85.	Io	" Sept. 19.	C. H. F. Peters	2.6594	4.337	0.19395
86.	Semele	1866, Jan. 4.	Tietjen.	3 0908	5.434	0.20493
87.	Sylvia	" May 17.	Pogson.	......	....	....
88.	Thisbe.....	" June 15.	C. H. F. Peters	2.7503	4.561	0.16670
89.		" Aug. 6.		2.5841	4.032	0.20499

Mean Astronomical Refractions.

Barometer 30 in. Thermometer, Fah. 50°.

Ap. Alt.	Refr.	Ap. Alt.	Refr.	Ap. Alt.	Refr.	Alt.	Refr.
0° 0′	33′ 51″	4° 0′	11′ 52″	12° 0′	4′ 28.1″	42°	1′ .4.6″
5	32 53	10	11 30	10	4 24.4	43	1 2.4
10	31 58	20	11 10	20	4 20.8	44	1 0.3
15	31 5	30	10 50	30	4 17.3	45	0.58.1
20	30 13	40	10 32	40	4 13.9	46	56.1
25	29 24	50	10 15	50	4 10.7	47	54.2
30	28 37	5 0	9 58	13 0	4 7.5	48	52.3
35	27 51	10	9 42	10	4 4.4	49	50.5
40	27 6	20	9 27	20	4 1.4	50	48.8
45	26 24	30	9 11	30	3 58.4	51	47.1
50	25 43	40	8 58	40	3 55.5	52	45.4
55	25 3	50	8 45	50	3 52.6	53	43.8
1 0	24 25	6 0	8 32	14 0	3 49.9	54	42.2
5	23 48	10	8 20	10	3 47.1	55	40.8
10	23 13	20	8 9	20	3 44.4	56	39.3
15	22 40	30	7 58	30	3 41.8	57	37.8
20	22 8	40	7 47	40	3 39.2	58	36.4
25	21 37	50	7 37	50	3 36.7	59	35.0
30	21 7	7 0	7 27	15 0	3 34.3	60	33.6
35	20 38	10	7 17	15 30	3 27.3	61	32.3
40	20 10	20	7 8	16 0	3 20.6	62	31.0
45	19 43	30	6 59	16 30	3 14.4	63	29.7
50	19 17	40	6 51	17 0	3 8.5	64	28.4
55	18 52	50	6 43	17 30	3 2.9	65	27.2
2 0	18 29	8 0	6 35	18 0	2 57.6	66	25.9
5	18 5	10	6 28	19	2 47.7	67	24.7
10	17 43	20	6 21	20	2 38.7	68	23.5
15	17 21	30	6 14	21	2 30.5	69	22.4
20	17 0	40	6 7	22	2 23.2	70	21.2
25	16 40	50	6 0	23	2 16.5	71	19.9
30	16 21	9 0	5 54	24	2 10.1	72	18.8
35	16 2	10	5 47	25	2 4.2	73	17.7
40	15 43	20	5 41	26	1 58.8	74	16.6
45	15 25	30	5 36	27	1 53.8	75	15.5
50	15 8	40	5 30	28	1 49.1	76	14.4
55	14 51	50	5 25	29	1 44.7	77	13.4
3 0	14 35	10 0	5 20	30	1 40.5	78	12.3
5	14 19	10	5 15	31	1 36.6	79	11.2
10	14 4	20	5 10	32	1 33.0	80	10.2
15	13 50	30	5 5	33	1 29.5	81	9.2
20	13 35	40	5 0	34	1 26.1	82	8.2
25	13 21	50	4 56	35	1 23.0	83	7.1
30	13 7	11 0	4 51	36	1 20.0	84	6.1
35	12 53	10	4 47	37	1 17.1	85	5.1
40	12 41	20	4 43	38	1 14.4	86	4.1
45	12 28	30	4 39	39	1 11.8	87	3.1
50	12 16	40	4 35	40	1 9.3	88	2.0
55	12 3	50	4 31	41	1 6.9	89	1.0

Corrections of Mean Refractions.

Ap.Alt.	dif for +1 B.	dif. for −1° F.	Ap.Alt.	Dif. for +1 B.	Dif. for −1° F.	Ap. Alt.	Dif. for +1 B.	Dif. for −1° F.	Alt.	Dif. for +1 B.	Dif. for -1° F.
° ′	″	″	° ′	″	″	° ′	″	″	°	″	″
0 0	74	8.1	4 0	24.1	1.70	12 0	9.00	0.556	42	2.16	0.130
5	71	7.6	10	23.4	1.64	10	8.86	.548	43	2.09	.125
10	69	7.3	20	22.7	1.58	20	8.74	.541	44	2.02	.120
15	67	7.0	30	22.0	1.53	30	8.63	.533	45	1.95	.116
20	65	6.7	40	21.3	1.48	40	8.51	.524	46	1.88	.112
25	63	6.4	50	20.7	1.43	50	8.41	.517	47	1.81	.108
30	61	6.1	5 0	20.1	1.38	13 0	8.30	.509	48	1.75	.104
35	59	5.9	10	19.6	1.34	10	8.20	.503	49	1.69	.101
40	58	5.6	20	19.1	1.30	20	8.10	.496	50	1.63	.097
45	56	5.4	30	18.6	1.26	30	8.00	.490	51	1.58	.094
50	55	5.1	40	18.1	1.22	40	7.89	.482	52	1.52	.090
55	53	4.9	50	17.6	1.19	50	7.79	.476	53	1.47	.088
1 0	52	4.7	6 0	17.2	1.15	14 0	7.70	.469	54	1.41	.085
5	50	4.6	10	16.8	1.11	10	7.61	.464	55	1.36	.082
10	49	4.5	20	16.4	1.09	20	7.52	.458	56	1.31	.079
15	48	4.4	30	16.0	1.06	30	7.43	.453	57	1.26	.076
20	46	4.2	40	15.7	1.03	40	7.34	.448	58	1.22	.073
25	45	4.0	50	15.3	1.00	50	7.26	.444	59	1.17	.070
30	44	3.9	7 0	150	0.98	15 0	7.18	.439	60	1.12	.067
35	43	3.8	10	14.6	.95	15 30	6.95	.424	61	1.08	.065
40	42	3.6	20	14.3	.93	16 0	6.73	.411	62	1.04	.062
45	40	3.5	30	14.1	.91	16 30	6.51	.399	63	.99	.060
50	39	3.4	40	13.8	.89	17 0	6.31	.386	64	.95	.057
55	39	3.3	50	13.5	.87	17 30	6.12	.374	65	.91	.055
2 0	38	3.2	8 0	13.3	.85	18 0	5.94	.362	66	.87	.052
5	37	3.1	10	13.1	.83	19	5.61	.340	67	.83	.050
10	36	3.0	20	12.8	.82	20	5.31	.322	68	.79	.047
15	36	2.9	30	12.6	.80	21	5.04	.305	69	.75	.045
20	35	2.8	40	12.3	.79	22	4.79	.290	70	.71	.043
25	34	2.8	50	12.1	.77	23	4.57	.276	71	.67	.040
30	33	2.7	9 0	11.9	.76	24	4.35	.264	72	.63	.038
35	33	2.7	10	11.7	.74	25	4.16	.252	73	.59	.036
40	32	2.6	20	11.5	.73	26	3.97	.241	74	.56	033
45	32	2.5	30	11.3	.72	27	3.81	.230	75	.52	.031
50	31	2.4	40	11.1	.71	28	3.65	.219	76	.48	.029
55	30	2.3	50	11.0	.70	29	3.50	.209	77	.45	.027
3 0	30	2.3	10 0	10.8	.69	30	3.36	.201	78	.41	.025
5	29	2.2	10	10.6	.67	31	3.23	.193	79	.38	.023
10	29	2.2	20	10.4	.65	32	3.11	.186	80	.34	.021
15	28	2.1	30	10.2	.64	33	2.99	.179	81	.31	.018
20	28	2.1	40	10.1	.63	34	2.88	.173	82	.27	.016
25	27	2.0	50	9.9	.62	35	2.78	.167	83	.24	.014
30	27	2.0	11 0	9.8	.60	36	2.68	.161	84	.20	.012
35	26	2.0	10	9.6	.59	37	2.58	.155	85	.17	.010
40	26	1.9	20	9.5	.58	38	2.49	.149	86	.14	.008
45	25	1.9	30	9.4	.57	39	2.40	.144	87	.10	.006
50	25	1.9	40	9.2	.56	40	2.32	.139	88	.07	.004
55	25	1.8	50	9.1	.55	41	2.24	.134	89	.03	.002

TABLE X.

Parallax of the Sun, on the first day of each Month: the mean horizontal Parallax being assumed = 8″.60.

Altitude.	Jan.	Feb. Dec.	March. Nov.	April. Oct.	May. Sept.	June. Aug.	July.
°	″	″	″	″	″	″	″
0	8.75	8.73	8.67	8.60	8.53	8.48	8.46
5	8.73	8.69	8.64	8.56	8.50	8.44	8.42
10	8.62	8.59	8.54	8.47	8.40	8.35	8.33
15	8.45	8.43	8.38	8.30	8.24	8.19	8.17
20	8.22	8.20	8.15	8.08	8.01	7.97	7.95
25	7.93	7.91	7.86	7.79	7.73	7.68	7.67
30	7.58	7.56	7.51	7.45	7.39	7.34	7.33
35	7.17	7.15	7.11	7.04	6.99	6.94	6.93
40	6.70	6.68	6.64	6.59	6.53	6.49	6.48
45	6.19	6.17	6.13	6.08	6.03	5.99	5.98
50	5.62	5.61	5.58	5.53	5.48	5.45	5.44
55	5.02	5.01	4.98	4.93	4.89	4.86	4.85
60	4.37	4.36	4.34	4.30	4.26	4.24	4.23
65	3.70	3.69	3.67	3.63	3.60	3.58	3.57
70	2.99	2.98	2.97	2.94	2.92	2.90	2.89
75	2.26	2.26	2.25	2.23	2.21	2.19	2.19
80	1.52	1.52	1.51	1.49	1.48	1.47	1.47
85	0.76	0.76	0.76	0.75	0.74	0.74	0.74
90	0.00	0.00	0.00	0.00	0·00	0.00	0.00

TABLE XI.

Semi-diurnal Arcs.

Lat.	Declination.						
	1°	5°	10°	15°	20°	25°	30°
°	h m	h m	h m	h m	h m	h m	h m
5	6 0	6 2	6 4	6 5	6 7	6 9	6 12
10	6 1	6 4	6 7	6 11	6 15	6 19	6 24
15	6 1	6 5	6 11	6 16	6 22	6 29	6 36
20	6 1	6 7	6 15	6 22	6 30	6 39	6 49
25	6 2	6 9	6 19	6 29	6 39	6 50	7 2
30	6 2	6 12	6 23	6 36	6 49	7 2	7 18
35	6 3	6 14	6 28	6 43	6 59	7 16	7 35
40	6 3	6 17	6 34	6 52	7 11	7 32	7 56
45	6 4	6 20	6 41	7 2	7 25	7 51	8 21
50	6 5	6 24	6 49	7 14	7 43	8 15	8 54
55	6 6	6 29	6 58	7 30	8 5	8 47	9 42
60	6 7	6 35	7 11	7 51	8 36	9 35	12 0
65	6 9	6 43	7 29	8 20	9 25	12 0	

Equation of Time, to convert Apparent Time into Mean Time

Argument, Mean Longitude of the Sun.

°	Os	Is	IIs	IIIs	IVs	Vs
°	min. sec.	min. sec.	min. sec.	min. sec.	min. sec.	min. sec.
0	+6 58.4	−1 29.7	−3 38.7	+1 27.0	+6 4.1	+2 49.7
1	6 39.7	1 42.0	3 34.2	1 40.1	6 6.3	2 34.5
2	6 20.9	1 53.8	3 29.1	1 53.1	6 8.0	2 18.9
3	6 2.1	2 5.2	3 23.5	2 6.0	6 9.1	2 2.8
4	5 43.3	2 15.9	3 17.3	2 18.9	6 9.5	1 46.4
5	5 24.5	2 26.1	3 10.7	2 31.7	6 9.3	1 29.5
6	5 5.7	2 35.9	3 3.5	2 44.3	6 8.5	1 12.3
7	4 46.9	2 45.0	2 56.0	2 56.7	6 7.2	0 54.6
8	4 28.2	2 53.6	2 47.9	3 8.9	6 5.2	0 36.6
9	4 9.6	3 1.8	2 39.5	3 20.8	6 2.5	+0 18.2
10	3 51.1	3 9.3	2 30.5	3 32.5	5 59.3	−0 0.4
11	3 32.6	3 16.3	2 21.2	3 43.9	5 55.4	0 19.5
12	3 14.3	3 22.8	2 11.5	3 55.0	5 51.0	0 38.8
13	2 56.2	3 28.6	2 1.4	4 5.8	5 45.8	0 58.4
14	2 38.3	3 33.9	1 51.0	4 16.3	5 40.1	1 18.2
15	2 20.5	3 38.6	1 40.1	4 26.5	5 33.7	1 38.3
16	2 3.0	3 42.7	1 29.0	4 36.3	5 26.7	1 58.5
17	1 45.7	3 46.3	1 17.6	4 45.7	5 19.2	2 19.1
18	1 28.6	3 49.2	1 5.9	4 54.7	5 11.1	2 39.8
19	1 11.7	3 51.5	0 54.1	5 3.3	5 2.3	3 0.7
20	0 55.2	3 53.3	0 42.0	5 11.3	4 53.0	3 21.6
21	0 39.1	3 54.4	0 29.6	5 18.9	4 43.1	3 42.8
22	0 23.3	3 55.0	0 17.1	5 26.0	4 32.7	4 4.0
23	+0 7.8	3 55.0	−0 4.4	5 32.6	4 21.6	4 25.3
24	−0 7.3	3 54.5	+0 8.4	5 38.6	4 10.1	4 46.6
25	0 22.0	3 53.3	0 21.5	5 44.2	3 57.9	5 8.1
26	0 36.3	3 51.5	0 34.5	5 49.3	3 45.3	5 29.5
27	0 50.3	3 49.2	0 47.6	5 53.9	3 32.1	5 51.0
28	1 3.8	3 46.2	1 0.7	5 57.8	3 18.5	6 12.3
29	1 16.9	3 42.8	1 13.8	6 1.2	3 4.3	6 33.7
30	−1 29.7	−3 38.7	+1 27.0	+6 4.1	+2 49.7	−6 54.9

TABLE XIII.

Secular Variation of Equation of Time.

Argument, Sun's Mean Longitude.

	Os	Is	IIs	IIIs	IVs	Vs
sec.	sec.	sec.	sec.	sec.	sec.	sec.
0	−3	+4	+11	+14	+13	+9
3	2	5	11	14	13	8
6	1	6	12	14	12	8
9	−1	6	12	15	12	7
12	0	7	12	14	12	7
15	+1	8	13	14	11	6
18	2	8	13	14	11	6
21	2	9	14	14	10	5
24	3	9	14	14	10	5
27	4	10	14	14	9	4
30	+4	+11	+14	+13	+9	+4

Equation of Time, to convert Apparent Time into Mean Time.

Argument, Mean Longitude of the Sun.

°	VI^s	VII^s	VIII^s	IX^s	X^s	XI^s
	min. sec.	min. sec.	min. sec.	min. sec.	min. sec.	min. sec.
0	— 6 54.9	—15 18.9	—13 58.7	— 1 30.6	+11 30.0	+14 3.1
1	7 16.1	15 27.9	13 43.0	1 0.2	11 47.0	13 56.0
2	7 37.2	15 36.1	13 26.3	— 0 29.8	12 3.3	13 48.4
3	7 58.3	15 43.7	13 8.9	+ 0 0.6	12 18.7	13 40.1
4	8 19.1	15 50.5	12 50.5	0 31.0	12 33.4	13 31.1
5	8 39.8	15 56.5	12 31.4	1 1.3	12 47.2	13 21.6
6	9 0.2	16 1.8	12 11.6	1 31.4	13 0.1	13 11.4
7	9 20.5	16 6.3	11 51.1	2 1.3	13 12.2	13 0.7
8	9 40.6	16 9.9	11 29.9	2 31.0	13 23.5	12 49.4
9	10 0.3	16 12.9	11 7.9	3 0.5	13 33.9	12 37.4
10	10 19.8	16 15.1	10 45.4	3 29.7	13 43.6	12 25.0
11	10 38.9	16 16.5	10 22.0	3 58.6	13 52.3	12 12.2
12	10 57.8	16 17.0	9 58.1	4 27.1	14 0.2	11 58.9
13	11 16.2	16 16.6	9 33.5	4 55.2	14 7.3	11 45.1
14	11 34.4	16 15.4	9 8.4	5 22.9	14 13.5	11 30.9
15	11 52.1	16 13.4	8 42.6	5 50.2	14 18.9	11 16.3
16	12 9.5	16 10.4	8 16.4	6 17.1	14 23.4	11 1.1
17	12 26.5	16 6.7	7 49.6	6 43.5	14 27.2	10 45.6
18	12 42.9	16 2.1	7 22.5	7 9.3	14 30.0	10 29.7
19	12 58.9	15 56.6	6 54.9	7 34.6	14 32.1	10 13.5
20	13 14.4	15 50.1	6 27.0	7 59.3	14 33.3	9 56.9
21	13 29.5	15 42.9	5 58.5	8 23.4	14 33.7	9 40.1
22	13 44.1	15 34.8	5 29.7	8 46.9	14 33.3	9 23.0
23	13 58.0	15 25.8	5 0.5	9 9.8	14 32.2	9 5.7
24	14 11.4	15 16.0	4 31.0	9 32.0	14 30.2	8 48.0
25	14 24.1	15 5.2	4 1.4	9 53.5	14 27.5	8 30.2
26	14 36.3	14 53.6	3 31.6	10 14.3	14 24.0	8 12.2
27	14 47.9	14 41.1	3 1.5	10 34.4	14 19.9	7 54.0
28	14 58.8	14 27.7	2 31.3	10 53.8	14 15.0	7 35.5
29	15 9.2	14 13.6	2 1.0	11 12.3	14 9.4	7 17.0
30	—15 18.9	—13 58.7	— 1 30.6	+11 30.0	+ 14 3.1	6 58.4

TABLE XIII.

Secular Variation of Equation of Time.

Argument, Sun's Mean Longitude.

°	VI^s	VII^s	VIII^s	IX^s	X^s	XI^s
	sec.	sec.	sec.	sec.	sec.	sec.
0	+4	— 2	—10	—15	—15	—10
3	3	3	10	15	14	10
6	3	4	11	15	14	9
9	2	4	12	15	14	8
12	1	5	12	15	13	8
15	+1	6	13	15	13	7
18	0	7	13	15	12	6
21	0	7	14	15	12	5
24	—1	8	14	15	11	5
27	2	9	15	15	11	4
30	--2	—10	—15	—15	—10	— 3

Perturbations of Equation of Time.

III.

II.	0	100	200	300	400	500	600	700	800	900	1000
	sec.	sec.	sec.	sec.	sec.	sec.	sec.	sec.	sec.	sec.	sec.
0	1.4	0.8	1.0	1.7	1.7	1.2	0.7	0.4	0.6	1.4	1.4
100	1.2	1.4	1.1	1.0	1.6	1.8	1.1	0.7	0.6	0.7	1.2
200	0.9	1.0	1.2	1.2	1.2	1.5	1.7	1.1	0.5	0.7	0.9
300	0.7	1.1	1.1	0.9	1.2	1.4	1.5	1.6	1.2	0.5	0.7
400	0.5	0.6	1.2	1.2	0.8	1.0	1.6	1.7	1.5	1.2	0.5
500	1.0	0.5	0.6	1.2	1.4	0.8	0.8	1.5	1.9	1.5	1.0
600	1.7	1.0	0.4	0.5	1.2	1.4	0.9	0.6	1.3	2.0	1.7
700	1.9	1.8	1.1	0.4	0 4	1.1	1.6	1.1	0.7	1.2	1.9
800	1.2	1.8	1.8	1.2	0.4	0.3	1.0	1.6	1.2	0.7	1.2
900	0.7	1.1	1.7	1.8	1.2	0.6	0.2	0.8	1.6	1.3	0.7
1000	1.4	0.8	1.0	1.7	1.7	1.2	0.7	0.4	0.6	1.4	1.4

IV.

II.	0	100	200	300	400	500	600	700	800	900	1000
	sec.	sec.	sec.	sec.	sec.	sec.	sec.	sec.	sec.	sec.	sec.
0	0.6	0.7	0.5	0.3	0.2	0.6	0.7	0.5	0.2	0.1	0.6
100	0.2	0.7	0.6	0.5	0.2	0.3	0.6	0.9	0.5	0.2	0.2
200	0.2	0.4	0.6	0.5	0.4	0.3	0.4	0.6	0.5	0.5	0.2
300	0.4	0.2	0.5	0.5	0.5	0.4	0.4	0.4	0.5	0.5	0.4
400	0.5	0.4	0.4	0.4	0.4	0.4	0.5	0.5	0.4	0.4	0.5
500	0.4	0.5	0.5	0.5	0.4	0.4	0.3	0.4	0.5	0.3	0.4
600	0.3	0.3	0.5	0.6	0.4	0.4	0.3	0.5	0.7	0.4	0.3
700	0.4	0.2	0.3	0.6	0.6	0.4	0.2	0.2	0.7	0.7	0.4
800	0.6	0.3	0.2	0.3	0.7	0.6	0.3	0.2	0.3	0.8	0.6
900	0.8	0.5	0.3	0.1	0.4	0.7	0.5	0.3	0.1	0.5	0.8
1000	0.6	0.7	0.5	0.3	0.2	0.6	0.7	0.5	0.2	0.1	0.6

V.

II.	0	100	200	300	400	500	600	700	800	900	1000
	sec.	sec.	sec.	sec.	sec.	sec.	sec.	sec.	sec.	sec.	sec.
0	1.0	1.0	1.1	1.2	1.1	1.0	0.7	0.4	0.6	0.9	1.0
100	0.9	0.9	0.8	1.0	1.3	1.3	1.0	0.7	0.4	0.5	0.9
200	0.5	0.7	0.7	0.8	1.0	1.0	1.1	1.2	0.9	0.3	0.5
300	0.2	0.5	0.7	0.7	0.8	1.2	1.5	1.5	1.1	0.5	0.2
400	0.3	0.2	0.5	0.7	0.7	0.9	1.3	1.4	1.4	1.0	0.3
500	0.8	0.3	0.2	0.5	0.7	0.7	1.0	1.4	1.4	1.4	0.8
600	1.3	0.7	0.3	0.3	0.5	0.7	0.9	1.1	1.4	1.6	1.3
700	1.5	1.1	0.7	0.3	0.4	0.5	0.8	1.0	1.2	1.4	1.5
800	1.3	1.3	1.0	0.7	0.4	0.4	0.6	0.8	1.0	1.2	1.3
900	1.1	1.2	1.2	1.0	0.8	0.6	0.5	0.6	0.9	1.1	1.1
1000	1.0	1.0	1.1	1.2	1.1	1.0	0.7	0.4	0.6	0.9	1.0

Moon and Nutation.

	sec.	sec.	sec.	sec.	sec.	sec.	sec.	sec.	sec.	sec.	sec.
I.	0.5	0.8	1.0	1.0	0 8	0.5	0.2	0.0	0.0	0.2	0.5
N.	0.1	0.1	0.2	0.2	0.2	0.2	0.2	0.2	0.2	0.1	0.1

Constant 3s.0

For converting any given day into the decimal part of a year of 365 days.

Day	Jan.	Feb.	March	April	May	June
1	.000	.085	.162	.247	.329	.414
2	.003	.088	.164	.249	.331	.416
3	.006	.090	.167	.252	.334	.419
4	.008	.093	.170	.255	.337	.422
5	.011	.096	.173	.258	.340	.425
6	.014	.099	.175	.260	.342	.427
7	.016	.101	.178	.263	.345	.430
8	.019	.104	.181	.266	.348	.433
9	.022	.107	.184	.268	.351	.436
10	.025	.110	.186	.271	.353	.438
11	.027	.112	.189	274	.356	.441
12	.030	.115	.192	.277	.359	.444
13	.033	.118	.195	.279	.362	.446
14	.036	.121	.197	.282	.364	.449
15	.038	.123	.200	.285	.367	.452
16	.041	.126	.203	.288	.370	.455
17	.044	.129	.205	.290	.373	.458
18	·046	.132	.208	.293	.375	.460
19	.049	.134	.211	.296	.378	.463
20	.052	.137	.214	.299	.381	.466
21	.055	.140	.216	.301	.384	.468
22	.058	.142	.219	.304	.386	.471
23	.060	.145	.222	.307	.389	.474
24	.063	.148	.225	.310	.392	.477
25	.066	.151	.227	.312	.395	.479
26	.068	.153	.230	.315	.397	.482
27	.071	.156	.233	.318	.400	.485
28	.074	.159	.236	.321	.403	.488
29	.077		.238	.323	.405	.490
30	.079		.241	.326	.408	.493
31	.082		.244		.411	

For converting any given day into the decimal part of a year of 365 days.

Day	July	August	Sept.	Oct.	Nov.	Dec.
1	.496	.581	.666	.748	.833	.915
2	.499	.584	.668	.751	.836	.918
3	.501	.586	671	.753	.838	.921
4	.504	.589	.674	.756	.841	.923
5	.507	.592	.677	.759	.844	.926
6	.510	.595	.679	.762	846	.929
7	.512	.597	.682	.764	.849	.931
8	.515	.600	.685	.767	.852	.934
9	.518	.603	.688	.770	855	.937
10	.521	.605	.690	.773	.858	.940
11	.523	.608	.693	.775	.860	.942
12	526	611	.696	.778	.863	.945
13	.529	614	.699	.781	.866	.948
14	.532	.616	.701	.784	.868	.951
15	.534	.619	.704	.786	.871	.953
16	.537	.622	.707	.789	.874	.956
17	.540	.625	.710	.792	877	.959
18	.542	.627	.712	.795	879	.962
19	.545	.630	.715	.797	.882	.964
20	.548	.633	.718	.800	885	.967
21	.551	.636	.721	.803	888	.970
22	.553	.638	.723	805	890	.973
23	.556	.641	.726	.808	893	.975
24	.559	.644	.729	.811	896	.978
25	.562	.647	.731	.814	.899	.981
26	564	.649	.734	.816	.901	.984
27	.567	.652	.737	.819	.904	.986
28	.570	.655	.740	.822	.907	.989
29	.573	.658	.742	.825	.910	.992
30	.575	.660	.745	.827	.912	.995
31	.578	.663		.830		.997

For converting time into decimal parts of a day.

Hours		Minutes				Seconds			
h.		m.		m.		s.		s.	
1	.04167	1	.00069	31	.02153	1	.00001	31	.00036
2	.08333	2	.00139	32	.02222	2	.00002	32	.00037
3	.12500	3	.00208	33	.02292	3	.00003	33	.00038
4	.16667	4	.00278	34	.02361	4	.00005	34	.00039
5	.20833	5	.00347	35	.02430	5	.00006	35	.00040
6	.25000	6	.00417	36	.02500	6	.00007	36	.00042
7	.29167	7	.00486	37	.02569	7	.00008	37	.00043
8	.33333	8	.00556	38	.02639	8	.00009	38	.00044
9	.37500	9	.00625	39	.02708	9	.00010	39	.00045
10	.41667	10	.00694	40	.02778	10	.00012	40	.00046
11	.45833	11	.00764	41	.02847	11	.00013	41	.00047
12	.50000	12	.00833	42	.02917	12	.00014	42	.00049
13	.54167	13	.00903	43	.02986	13	.00015	43	.00050
14	.58333	14	.00972	44	.03056	14	.00016	44	.00051
15	.62500	15	01042	45	.03125	15	.00017	45	.00052
16	.66667	16	.01111	46	.03194	16	.00018	46	.00053
17	.70833	17	.01180	47	.03264	17	.00020	47	.00054
18	.75000	18	.01250	48	.03333	18	.00021	48	.00056
19	.79167	19	.01319	49	.03403	19	.00022	49	.00057
20	.83333	20	.01389	50	.03472	20	.00023	50	.00058
21	.87500	21	01458	51	.03542	21	.00024	51	.00059
22	.91667	22	.01528	52	.03611	22	.00025	52	.00060
23	.95833	23	01597	53	.03680	23	.00027	53	.00061
24	1.00000	24	.01667	54	.03750	24	.00028	54	.00062
		25	.01736	55	.03819	25	.00029	55	.00064
		26	.01805	56	.03889	26	.00030	56	.00065
		27	.01875	57	.03958	27	.00031	57	.00066
		28	.01944	58	.04028	28	.00032	58	.00067
		29	.02014	59	.04097	29	.00034	59	.00068
		30	.02083	60	.04167	30	.00035	60	.00069

For converting Minutes and Seconds of a degree, into the decimal division of the same.

Minutes				Seconds			
′		′		″		″	
1	.01667	31	.51667	1	.00028	31	.00861
2	.03333	32	.53333	2	.00056	32	.00889
3	.05000	33	.55000	3	.00083	33	.00917
4	.06667	34	.56667	4	.00111	34	.00944
5	.08333	35	.58333	5	.00139	35	.00972
6	.10000	36	.60000	6	.00167	36	.01000
7	.11667	37	.61667	7	.00194	37	.01028
8	.13333	38	.63333	8	.00222	38	.01056
9	.15000	39	.65000	9	.00250	39	.01083
10	.16667	40	.66667	10	.00278	40	.01111
11	.18333	41	.68333	11	.00306	41	.01139
12	.20000	42	.70000	12	.00333	42	.01167
13	.21667	43	.71667	13	.00361	43	.01194
14	.23333	44	.73333	14	.00389	44	.01222
15	.25000	45	.75000	15	.00417	45	.01250
16	.26667	46	.76667	16	.00444	46	.01278
17	.28333	47	.78333	17	.00472	47	.01306
18	.30000	48	.80000	18	.00500	48	.01333
19	.31667	49	.81667	19	.00528	49	.01361
20	.33333	50	.83333	20	.00556	50	.01389
21	.35000	51	.85000	21	.00583	51	.01417
22	.36667	52	.86667	22	.00611	52	.01444
23	.38333	53	.88333	23	.00639	53	.01472
24	.40000	54	.90000	24	.00667	54	.01500
25	.41667	55	.91667	25	.00694	55	01528
26	.43333	56	.93333	26	.00722	56	.01556
27	.45000	57	.95000	27	.00750	57	.01583
28	.46667	58	.96667	28	.00778	58	.01611
29	.48333	59	.98333	29	.00806	59	.01639
30	.50000	60	1.00000	30	.00833	60	.01667

TABLE XVIII.

Sun's Epochs.

Years.	M. Long.	Long. Peri.	I.	II.	III.	IV.	V.	N.	VI.	VII.
	s ° ′ ″	s ° ′ ″								
1830	9 10 37 46.9	9 10 0 54	228	279	169	598	758	519	989	362
1831	9 10 23 27.4	9 10 1 55	588	278	793	130	842	573	235	396
1832 B.	9 10 9 7.9	9 10 2 57	948	278	418	661	926	627	482	430
1833	9 10 53 56.8	9 10 3 59	342	280	47	194	11	681	764	464
1834	9 10 39 37.3	9 10 5 0	702	279	671	725	95	734	11	498
1835	9 10 25 17.8	9 10 6 2	62	279	296	256	179	788	257	532
1836 B.	9 10 10 58.4	9 10 7 3	422	278	920	788	264	842	504	566
1837	9 10 55 47.2	9 10 8 5	816	280	549	321	348	895	787	600
1838	9 10 41 27.8	9 10 9 6	176	279	173	852	432	949	33	634
1839	9 10 27 8.3	9 10 10 8	536	279	798	383	517	3	279	668
1840 B.	9 10 12 48.8	9 10 11 9	896	278	422	915	601	56	526	702
1841	9 10 57 37.7	9 10 12 11	290	280	51	447	685	110	809	736
1842	9 10 43 18.2	9 10 13 12	650	279	676	979	770	164	55	770
1843	9 10 28 58.8	9 10 14 14	10	279	300	510	854	218	301	804
1844 B.	9 10 14 39.3	9 10 15 15	370	278	924	41	938	272	548	838
1845	9 10 59 28.2	9 10 16 17	764	280	553	574	23	325	831	872
1846	9 10 45 8.7	9 10 17 19	124	280	177	106	107	379	77	906
1847	9 10 30 49.2	9 10 18 20	484	279	802	637	191	433	324	940
1848 B.	9 10 16 29.8	9 10 19 22	844	278	427	168	276	487	570	974
1849	9 11 1 18.6	9 10 20 23	238	280	55	700	360	540	853	8
1850	9 10 46 59.2	9 10 21 25	598	280	680	231	444	594	99	41
1851	9 10 32 39.7	9 10 22 26	958	279	304	762	529	648	346	75
1852 B.	9 10 18 20.2	9 10 23 28	319	278	929	294	613	701	592	109
1853	9 11 3 9.1	9 10 24 29	713	280	557	827	697	755	875	143
1854	9 10 48 49.6	9 10 25 31	73	280	182	358	782	809	121	177
1855	9 10 34 30.2	9 10 26 32	433	279	806	889	866	863	368	211
1856 B.	9 10 20 10.7	9 10 27 34	793	279	430	421	950	916	614	245
1857	9 11 4 59.6	9 10 28 35	187	281	60	953	35	970	897	279
1858	9 10 50 40.1	9 10 29 37	547	280	684	485	119	24	144	313
1859	9 10 36 20.7	9 10 30 39	907	279	308	16	203	78	390	347
1860 B.	9 10 22 1.2	9 10 31 40	267	279	933	547	288	131	636	381
1861	9 11 6 50.1	9 10 32 42	661	281	562	80	372	185	919	415
1862	9 10 52 30.6	9 10 33 43	21	280	186	612	456	239	166	449
1863	9 10 38 11.1	9 10 34 45	381	280	810	143	541	292	412	483
1864 B.	9 10 23 51.7	9 10 35 46	741	279	435	674	625	346	659	517
1865	9 11 8 40.5	9 10 36 48	135	281	64	207	709	400	941	551
1866	9 10 54 21.1	9 10 37 49	495	280	688	738	794	453	188	585
1867	9 10 40 1.6	9 10 38 51	855	280	313	270	878	507	434	619
1868 B.	9 10 25 42.2	9 10 39 52	215	279	937	801	962	561	681	653
1869	9 11 10 31.0	9 10 40 54	609	281	566	334	47	615	963	687
1870	9 10 56 11.6	9 10 41 56	969	280	190	865	131	668	210	721
1871	9 10 41 52.1	9 10 42 57	329	279	814	396	216	742	457	755
1872 B.	9 10 27 32.6	9 10 43 59	690	278	439	928	300	775	703	789
1873	9 11 12 21.5	9 10 45 0	84	280	67	461	384	829	986	823
1874	9 10 58 2.0	9 10 46 2	444	280	692	992	469	883	232	857
1875	9 10 43 42.6	9 10 47 3	804	279	316	523	553	937	479	891
1876 B.	9 10 29 23.1	9 10 48 5	164	279	940	55	637	990	725	925
1877	9 11 14 12.0	9 10 49 6	558	281	570	587	722	44	8	959
1878	9 10 59 52.5	9 10 50 8	918	280	194	119	806	98	255	993
1879	9 10 45 33.1	9 10 51 10	278	279	818	650	890	152	501	27
1880 B.	9 10 31 13.6	9 10 52 11	638	279	443	181	975	205	747	61
1881	9 11 16 2.5	9 10 53 13	32	281	72	714	59	259	30	95
1882	9 11 1 43.0	9 10 54 14	392	280	696	246	143	313	277	129
1883	9 10 47 23.5	9 10 55 16	752	280	320	777	228	366	523	163
1884 B.	9 10 33 4.1	9 10 56 17	112	279	945	308	312	420	770	197

TABLE XIX.

Sun's Motions for Months.

Months		M. Long.	Per.	I	II	III	IV	V	N	VI	VII
		s ° ′ ″	″								
January		0 0 0 0.0	0	0	0	0	0	0	0	0	0
February		1 0 33 18.2	5	47	85	138	45	7	5	125	3
March	Com.	1 28 9 11.4	10	993	162	263	86	14	9	141	6
	Bis.	1 29 8 19.8	10	27	164	267	87	14	9	178	6
April	Com.	2 28 42 29.7	15	42	246	401	131	21	13	266	8
	Bis.	2 29 41 38.0	15	76	249	405	132	21	13	302	8
May	Com.	3 28 16 39.6	20	59	329	534	175	28	18	355	11
	Bis.	3 29 15 47.9	20	92	331	538	176	28	18	391	11
June	Com.	4 28 49 57.9	26	110	414	672	220	35	22	480	14
	Bis.	4 29 49 6.2	26	144	416	676	221	35	23	516	14
July	Com.	5 28 24 7.8	31	129	496	806	263	41	27	569	17
	Bis.	5 29 23 16.1	31	163	499	810	265	42	27	605	17
Aug.	Com.	6 28 57 26.1	36	182	580	943	309	49	31	694	20
	Bis.	6 29 56 34.4	36	216	583	948	310	49	31	730	20
Sep.	Com.	7 29 30 44.2	41	233	665	81	354	56	36	819	23
	Bis.	8 0 29 52.6	41	268	668	86	355	56	36	855	23
Oct.	Com.	8 29 4 54.1	46	250	748	215	397	63	40	908	25
	Bis.	9 0 4 2.5	46	284	750	219	399	63	40	944	25
Nov.	Com.	9 29 38 12.5	51	300	832	353	443	70	45	33	28
	Bis.	10 0 37 20.7	51	333	835	357	444	70	45	69	28
Dec.	Com.	10 29 12 22.3	56	313	915	486	486	77	49	121	31
	Bis.	11 0 11 30.6	56	347	917	491	488	77	49	158	31

TABLE XX.

Sun's Motions for Days and Hours.

Days	M. Long.	Per.	I	II	III	IV	V	N	VI	VII	Hrs.	Long.	I / VI	II	III
	° ′ ″	″										′ ″			
1	0 0 0.0	0	0	0	0	0	0	0	0	0	1	2 27.8	1	0	0
2	0 59 8.3	0	34	3	4	1	0	0	36	0	2	4 55.7	3	0	0
3	1 58 16.7	0	68	5	9	3	0	0	73	0	3	7 23.5	4	0	1
4	2 57 25.0	0	101	8	13	4	1	0	109	0	4	9 51.4	6	0	1
5	3 56 33.3	1	135	11	18	6	1	1	145	0	5	12 19.2	7	1	1
6	4 55 41.6	1	169	14	22	7	1	1	181	0	6	14 47.1	8	1	1
7	5 54 50.0	1	203	16	27	9	1	1	218	1	7	17 14.9	10	1	1
8	6 53 58.3	1	236	19	31	10	2	1	254	1	8	19 42.8	11	1	1
9	7 53 6.6	1	270	22	36	12	2	1	290	1	9	22 10.6	13	1	2
10	8 52 15.0	1	304	25	40	13	2	1	327	1	10	24 38.5	14	1	2
11	9 51 23.3	2	338	27	44	15	2	1	363	1	11	27 6.3	16	1	2
12	10 50 31.6	2	371	30	49	16	2	2	399	1	12	29 34.2	17	1	2
13	11 49 40.0	2	405	33	53	17	3	2	435	1	13	32 2.0	18	1	2
14	12 48 48.3	2	439	36	58	19	3	2	472	1	14	34 29.9	20	2	3
15	13 47 56.6	2	473	38	62	20	3	2	508	2	15	36 57.7	21	2	3
16	14 47 4.9	2	506	41	67	22	3	2	544	2	16	39 25.6	23	2	3
17	15 46 13.3	3	540	44	71	23	4	2	581	2	17	41 53.4	24	2	3
18	16 45 21.6	3	574	47	76	25	4	2	617	2	18	44 21.2	25	2	3
19	17 44 29.9	3	608	49	80	26	4	3	653	2	19	46 49.1	27	2	4
20	18 43 38.3	3	641	52	85	28	4	3	690	2	20	49 16.9	28	2	4
21	19 42 46.6	3	675	55	89	29	5	3	726	2	21	51 44.8	30	2	4
22	20 41 54.9	4	709	58	93	31	5	3	762	2	22	54 12.6	31	2	4
23	21 41 3.3	4	743	60	98	32	5	3	798	2	23	56 40.5	32	3	4
24	22 40 11.6	4	777	63	102	33	5	3	835	2	24	59 8.3	34	3	4
25	23 39 19.9	4	810	66	107	35	5	4	871	2					
26	24 38 28.2	4	844	68	111	36	6	4	907	2					
27	25 37 36.6	4	878	71	116	38	6	4	943	2					
28	26 36 44.9	5	912	74	120	39	6	4	980	2					
29	27 35 53.2	5	945	77	125	41	6	4	16	3					
30	28 35 1.6	5	979	79	129	42	7	4	52	3					
31	29 34 9.9	5	13	82	134	44	7	4	89	3					

TABLE XXI.

Sun's Motions for Minutes and Seconds.

Min.	Long.	Min.	Long.	Sec.	Lon.	Sec.	Lon.
	′ ″		′ ″		″		″
1	0 2.5	31	1 16.4	1	0.0	31	1.3
2	4.9	32	1 18.8	2	0.1	32	1.3
3	7.4	33	1 21.3	3	0.1	33	1.4
4	9.9	34	1 23.8	4	0.2	34	1.4
5	12.3	35	1 26.2	5	0.2	35	1.4
6	14.8	36	1 28.7	6	0.2	36	1.5
7	17.2	37	1 31.2	7	0.3	37	1.5
8	19.7	38	1 33.6	8	0.3	38	1.6
9	22.2	39	1 36.1	9	0.4	39	1.6
10	24.6	40	1 38.6	10	0.4	40	1.6
11	27.1	41	1 41.0	11	0.5	41	1.7
12	29.6	42	1 43.5	12	0.5	42	1.7
13	32.0	43	1 46.0	13	0.5	43	1.8
14	34.5	44	1 48.4	14	0.6	44	1.8
15	37.0	45	1 50.9	15	0.6	45	1.8
16	39.4	46	1 53.3	16	0.7	46	1.9
17	41.9	47	1 55.8	17	0.7	47	1.9
18	44.4	48	1 58.3	18	0.7	48	2.0
19	46.8	49	2 0.7	19	0.8	49	2.0
20	49.3	50	2 3.2	20	0.8	50	2.0
21	51.7	51	2 5.7	21	0.9	51	2.1
22	54.2	52	2 8.1	22	0.9	52	2.1
23	56.7	53	2 10.6	23	0.9	53	2.2
24	59.1	54	2 13.1	24	1.0	54	2.2
25	1 1.6	55	2 15.5	25	1.0	55	2.3
26	1 4.1	56	2 18.0	26	1.1	56	2.3
27	1 6.5	57	2 20.5	27	1.1	57	2.3
28	1 9.0	58	2 22.9	28	1.1	58	2.4
29	1 11.5	59	2 25.4	29	1.2	59	2.4
30	1 13.9	60	2 27.8	30	1.2	60	2.5

TABLE XXII.

Mean Obliquity of the Ecliptic.

Years	23° 27′
	″
1835	38 80
1836	38.35
1837	37.89
1838	37.43
1839	36.98
1840	36.52
1841	36.06
1842	35.61
1843	35.15
1844	34.69
1845	34.23
1846	33.78
1847	33.32
1848	32.86
1849	32.41
1850	31.95
1851	31.49
1852	31.04
1853	30.58
1854	30.12
1855	29.66
1856	29.21
1857	28 75
1858	28.29
1859	27.84
1860	27.38
1861	26 92
1862	26.47
1863	26.01
1864	25.55

TABLE XXIII.

Sun's Hourly Motion.

Argument. Sun's Mean Anomaly.

	0^s	I^s	II^s	III^s	IV^s	V^s	
°	′ ″	′ ″	′ ″	′ ″	′ ″	′ ″	°
0	2 32.92	2 32.20	2 30.28	2 27.74	2 25.32	2 23.60	30
10	2 32.84	2 31.67	2 29.46	2 26.89	2 24.64	2 23.26	20
20	2 32.59	2 31.02	2 28.61	2 26.07	2 24.06	2 23.05	10
30	2 32.20	2 30.28	2 27.74	2 25.32	2 23.60	2 22.99	0
	XI^s	X^s	IX^s	VIII^s	VII^s	VI^s	

TABLE XXIV.

Sun's Semi-diameter.

Argument. Sun's Mean Anomaly.

	0^s	I^s	II^s	III^s	IV^s	V^s	
°	′ ″	′ ″	′ ″	′ ″	′ ″	′ ″	°
0	16 17.3	16 15.0	16 8.8	16 0.6	15 52.7	15 47.0	30
10	16 17.0	16 13.3	16 6.2	15 57.8	15 50.5	15 45.9	20
20	16 16.2	16 11.2	16 3.4	15 55.1	15 48.6	15 45.2	10
30	16 15.0	16 8.8	16 0.6	15 52.7	15 47.0	15 45.0	0
	XI^s	X^s	IX^s	VIII^s	VII^s	VI^s	

TABLE XXV.

Equation of the Sun's Centre.

Argument. Sun's Mean Anomaly.

	Os	Is	IIs	IIIs	IVs	Vs
°	s ° ′ ″	° ′ ″	° ′ ″	° ′ ″	° ′ ″	° ′ ″
0	11 29 59 13.9	0 57 58.5	1 40 10.7	1 54 34.1	1 38 4.8	0 55 52.6
1	0 0 1 17.3	0 59 43.9	1 41 8.9	1 54 30.5	1 37 2.4	0 54 8.7
2	0 3 20.6	1 1 28.0	1 42 5.1	1 54 24.8	1 35 58.1	0 52 24.0
3	0 5 23.9	1 3 10.9	1 42 59.3	1 54 17.0	1 34 52.2	0 50 38.2
4	0 7 27.0	1 4 52.6	1 43 51.8	1 54 7.1	1 33 44.6	0 48 51.6
5	0 9 30.0	1 6 33.0	1 44 42.1	1 53 55.2	1 32 35.4	0 47 4.2
6	0 11 32.8	1 8 12.3	1 45 30.4	1 53 41.0	1 31 24.4	0 45 16.0
7	0 13 35.4	1 9 50.1	1 46 16.8	1 53 24.9	1 30 11.9	0 43 26.9
8	0 15 37.7	1 11 26.5	1 47 1.2	1 53 6.7	1 28 57.7	0 41 37.0
9	0 17 39.6	1 13 1.7	1 47 43.5	1 52 46.5	1 27 42.0	0 39 46.5
10	0 19 41.2	1 14 35.3	1 48 23.9	1 52 24.2	1 26 24.8	0 37 55.3
11	0 21 42.4	1 16 7.5	1 49 2.2	1 51 59.8	1 25 5.9	0 36 3.3
12	0 23 43.1	1 17 38.2	1 49 38.4	1 51 33.4	1 23 45.7	0 34 10.8
13	0 25 43.4	1 19 7.5	1 50 12.6	1 51 5.0	1 22 23.8	0 32 17.7
14	0 27 43.2	1 20 35.2	1 50 44.7	1 50 34.5	1 21 0.6	0 30 23.8
15	0 29 42.3	1 22 1.5	1 51 14.9	1 50 2.2	1 19 36.0	0 28 29.6
16	0 31 40.9	1 23 26.0	1 51 42.9	1 49 27.7	1 18 9.9	0 26 34.8
17	0 33 38.9	1 24 48.9	1 52 8.7	1 48 51.3	1 16 42.4	0 24 39.6
18	0 35 36.2	1 26 10.3	1 52 32.5	1 48 13.0	1 15 13.7	0 22 43.9
19	0 37 32.9	1 27 30.0	1 52 54.3	1 47 32.7	1 13 43.5	0 20 47.9
20	0 39 28.8	1 28 48.0	1 53 13.9	1 46 50.4	1 12 12.1	0 18 51.4
21	0 41 23.9	1 30 4.2	1 53 31.4	1 46 6.3	1 10 39.3	0 16 54.6
22	0 43 18.1	1 31 18.8	1 53 46.8	1 45 20.3	1 9 5.4	0 14 57.5
23	0 45 11.5	1 32 31.7	1 54 0.1	1 44 32.2	1 7 30.3	0 13 0.1
24	0 47 4.0	1 33 42.7	1 54 11.2	1 43 42.4	1 5 54.0	0 11 2.6
25	0 48 55.6	1 34 52.0	1 54 20.4	1 42 50.7	1 4 16.5	0 9 4.8
26	0 50 46.3	1 35 59.4	1 54 27.2	1 41 57.1	1 2 37.8	0 7 6.9
27	0 52 36.0	1 37 5.1	1 54 32.1	1 41 1.7	1 0 58.0	0 5 8 7
28	0 54 24.6	1 38 8.8	1 54 34.9	1 40 4.5	0 59 17.3	0 3 10 5
29	0 56 12.1	1 39 10.8	1 54 35.4	1 39 5.6	0 57 35.4	0 1 12.2
30	0 57 58.5	1 40 10.7	1 54 34.1	1 38 4.8	0 55 52.6	

TABLE XXVI.

Secular Variation of Equation of Sun's Centre.

Argument. Sun's Mean Anomaly.

	Os	Is	IIs	IIIs	IVs	Vs
°	″	″	″	″	″	″
0	— 0	— 9	— 15	— 17	— 15	— 8
2	1	9	15	17	14	8
4	1	10	16	17	14	7
6	2	10	16	17	14	7
8	2	11	16	17	13	6
10	3	11	16	17	13	6
12	4	12	17	17	12	5
14	4	12	17	16	12	5
16	5	13	17	16	12	4
18	5	13	17	16	11	3
20	6	13	17	16	11	3
22	7	14	17	16	10	2
24	7	14	17	15	10	2
26	8	15	17	15	9	1
28	8	15	17	15	9	1
30	— 9	— 15	— 17	— 15	— 8	— 0

TABLE XXV.

Equation of the Sun's Centre.

Argument. Sun's Mean Anomaly.

	VIs	VIIs	VIIIs	IXs	Xs	XIs
	11s	11s	11s	11s	11s	11s
°	° ′ ″	° ′ ″	° ′ ″	° ′ ″	° ′ ″	° ′ ″
0	29 59 13.9	29 2 35.2	28 20 23.0	28 3 53.7	28 18 17.1	29 0 29.3
1	29 57 15.6	29 0 52.4	28 19 22.2	28 3 52.3	28 19 17.0	29 2 15.7
2	29 55 17.3	28 59 10.5	28 18 23.3	28 3 52.8	28 20 19.0	29 4 3.2
3	29 53 19.1	28 57 29.8	28 17 26.1	28 3 55.6	28 21 22.7	29 5 51.8
4	29 51 20.9	28 55 50.0	28 16 30.7	28 4 0.5	28 22 28.4	29 7 41.5
5	29 49 23.0	28 54 11.4	28 15 37.1	28 4 7.4	28 23 35.8	29 9 32.2
6	29 47 25.2	28 52 33.8	28 14 45.4	28 4 16.6	28 24 45.1	29 11 23.8
7	29 45 27.7	28 50 57.5	28 13 55.6	28 4 27.7	28 25 56.1	29 13 16.3
8	29 43 30.3	28 49 22.4	28 13 7.5	28 4 41.0	28 27 9.0	29 15 9.7
9	29 41 33.2	28 47 48.5	28 12 21.5	28 4 56.4	28 28 23.6	29 17 3.9
10	29 39 36.4	28 46 15.7	28 11 37.4	28 5 13.9	28 29 39.8	29 18 59.0
11	29 37 39.9	28 44 44.3	28 10 55.1	28 5 33.5	28 30 57.8	29 20 54.9
12	29 35 43 9	28 43 14.1	28 10 14.8	28 5 55.3	28 32 17.5	29 22 51.6
13	29 33 48.2	28 41 45.4	28 9 36.5	28 6 19.1	28 33 38.9	29 24 48.9
14	29 31 53.0	28 40 17.9	28 9 0.0	28 6 44.9	28 35 1.8	29 26 46.9
15	29 29 58.2	28 38 51.8	28 8 25.6	28 7 12.9	28 36 26.3	29 28 45.5
16	29 28 4.0	28 37 27.2	28 7 53.2	28 7 43.1	28 37 52.6	29 30 44.6
17	29 26 10.1	28 36 4.0	28 7 22.8	28 8 15.2	28 39 20.3	29 32 44·4
18	29 24 17.0	28 34 42.1	28 6 54.4	28 8 49.4	28 40 49.6	29 34 44.7
19	29 22 24.5	28 33 21.9	28 6 28.0	28 9 25.6	28 42 20.3	29 36 45.4
20	29 20 32.5	28 32 3.0	28 6 3.6	28 10 3.9	28 43 52.5	29 38 46.6
21	29 18 41.3	28 30 45.8	28 5 41.4	28 10 44.3	28 45 26.1	29 40 48.2
22	29 16 50.8	28 29 30.1	28 5 21.1	28 11 26.6	28 47 1.3	29 42 50.1
23	29 15 0.9	28 28 15.9	28 5 2.9	28 12 11.0	28 48 37.7	29 44 52.5
24	29 13 11.8	28 27 3.4	28 4 46.8	28 12 57.4	28 50 15.5	29 46 55.0
25	29 11 23.6	28 25 52.4	28 4 32.6	28 13 45.7	28 51 54.8	29 48 57.8
26	29 9 36.2	28 24 43.2	28 4 20.7	28 14 36.0	28 53 35.2	29 51 0.8
27	29 7 49.5	28 23 35.6	28 4 10.8	28 15 28.5	28 55 16.9	29 53 3.9
28	29 6 3.8	28 22 29.7	28 4 3.0	28 16 22.7	28 56 59.8	29 55 7.2
29	29 4 19.1	28 21 25.4	28 3 57.3	28 17 18.9	28 58 43.9	29 57 10.5
30	29 2 35.2	28 20 23.0	28 3 53.7	28 18 17.1	29 0 29.3	29 59 13.9

TABLE XXVI.

Secular Variation of Equation of Sun's Centre.

Argument. Sun's Mean Anomaly.

	VIs	VIIs	VIIIs	IXs	Xs	XIs
°	″	″	″	″	″	″
0	+ 0	+ 8	+ 15	+ 17	+ 15	+ 9
2	1	9	15	17	15	8
4	1	9	15	17	15	8
6	2	10	15	17	14	7
8	2	10	16	17	14	7
10	3	11	16	17	14	6
12	3	11	16	17	13	6
14	4	12	16	17	13	5
16	5	12	16	17	12	4
18	5	12	17	17	12	4
20	6	13	17	16	11	3
22	6	13	17	16	11	2
24	7	14	17	16	10	2
26	7	14	17	16	10	1
28	8	14	17	15	9	1
30	+ 8	+ 15	+ 17	+ 15	+ 9	+ 0

Nutations.

Argument. Supplement of the Node, or N. *Solar Nutation.*

N.	Long.	R. Asc.	Obliq.	N.	Long.	R. Asc.	Obliq.
	"	"	"		"	"	"
0	+ 0.0	+0.0	+ 9.2	500	— 0.0	— 0.0	— 9.3
10	1.0	1.0	9.1	510	1.1	1.0	9.3
20	2.1	2.1	9.1	520	2.2	2.0	9.3
30	3.2	3.0	9.0	530	3.3	2.9	9.2
40	4.2	4.0	8.9	540	4.4	3.9	9.0
50	+ 5.2	+ 4.9	+ 8.7	550	— 5.5	— 4.8	— 8.9
60	6.2	6.0	8.5	560	6.5	5.7	8.7
70	7.2	6.9	8.3	570	7.5	6.6	8.4
80	8.2	7.8	8.1	580	8.5	7.5	8.1
90	9.1	8.7	7.8	590	9.5	8.4	7.8
100	+ 10.0	+ 9.4	+ 7.5	600	— 10.4	— 9.1	— 7.5
110	10.8	10.3	7.1	610	11.2	9.9	7.1
120	11.6	11.1	6.7	620	12.0	10.6	6.7
130	12.4	11.7	6.3	630	12.8	11.4	6.3
140	13.1	12.4	5.9	640	13.5	12.0	5.9
150	+ 13.8	+ 13.0	+ 5.5	650	— 14.2	— 12.6	— 5.4
160	14.4	13.6	5.0	660	14.8	13.2	4.9
170	15.0	14.1	4.5	670	15.3	13.8	4.4
180	15.5	14.5	4.0	680	15.8	14.2	3.9
190	15.9	14.8	3.5	690	16.2	14.7	3.3
200	+ 16.3	+ 15.1	+ 2.9	700	— 16.6	— 15.0	— 2.8
210	16.6	15.4	2.4	710	16.9	15.3	2.2
220	16.9	15.6	1.8	720	17.1	15.4	1.6
230	17.1	15.7	1.2	730	17.2	15.7	1.1
240	17.2	15.9	0.7	740	17.3	15.9	— 0.5
250	+ 17.3	+ 15.9	+ 0.1	750	— 17.3	— 15.9	+ 0.1
260	17.3	15.9	— 0.5	760	17.2	15.9	0.7
270	17.2	15.7	1.1	770	17.1	15.7	1.2
280	17.1	15.6	1.6	780	16.9	15.4	1.8
290	16.9	15.4	2.2	790	16.6	15.3	2.4
300	+ 16.6	+ 15.1	— 2.8	800	— 16.3	— 15.0	+ 2.9
310	16.2	14.8	3.3	810	15.9	14.7	3.5
320	15.8	14.5	3.9	820	15.5	14.2	4.0
330	15.3	14.1	4.4	830	15.0	13.8	4.5
340	14.8	13.6	4.9	840	14.4	13.2	5.0
350	+ 14.2	+ 13.0	— 5.4	850	— 13.8	— 12.6	+ 5.5
360	13.5	12.4	5.9	860	13.1	12.0	5.9
370	12.8	11.7	6.3	870	12.4	11.4	6.3
380	12.0	11.1	6.7	880	11.6	10.6	6.7
390	11.2	10.3	7.1	890	10.8	9.9	7.1
400	+ 10.4	+ 9.4	— 7.5	900	— 10.0	— 9.1	+ 7.5
410	9.5	8.7	7.8	910	9.1	8.4	7.8
420	8.5	7.8	8.1	920	8.2	7.5	8.1
430	7.5	6.9	8.4	930	7.2	6.6	8.3
440	6.5	6.0	8.7	940	6.2	5.7	8.5
450	+ 5.5	+ 4.9	— 8.9	950	— 5.2	— 4.8	+ 8.7
460	4.4	4.0	9.0	960	4.2	3.9	8.9
470	3.3	3.0	9.2	970	3.2	2.9	9.0
480	2.2	2.1	9.3	980	2.1	2.0	9.1
490	1.1	1.0	9.3	990	1.0	1.0	9.1
500	+ 0.0	+ 0.0	— 9.3	1000	— 0.0	— 0.0	+ 9.2

	Long.	Obliq.
Jan.	"	"
1	+ 0.5	— 0.5
11	0.8	0.4
21	1.1	0.2
31	1.2	— 0.1
Feb.		
10	1.2	+ 0.1
20	1.0	0.3
March.		
2	0.7	0.4
12	+ 0.3	0.5
22	— 0.1	0.5
April.		
1	0.5	0.5
11	0.8	0.2
21	1.1	0.2
May.		
1	1.2	+ 0.1
11	1.2	— 0.1
21	1.1	0.3
31	0.8	0.4
June.		
10	0.4	0.5
20	— 0.0	0.5
30	+ 0.4	0.5
July.		
10	0.7	0.4
20	1.0	0.3
30	1.2	— 0.1
Aug.		
9	1.3	+ 0.0
19	1.2	0.4
29	0.9	0.4
Sept.		
8	0.6	0.5
18	+ 0.2	0.5
28	— 0.2	0.5
Oct.		
8	0.6	0.5
18	1.0	0.3
28	1.2	0.2
Nov.		
7	1.2	+ 0.0
17	1.2	0.2
27	1.0	0.4
Dec.		
7	0.6	0.5
17	— 0.2	0.5
27	+ 0.3	0.5
37	+ 0.6	— 0.5

TABLE XXVIII.

Lunar Equation, 1st part.

Argument I.

I	Equa	I	Equ
	″		″
0	7.5	500	7.5
10	8.0	510	7.0
20	8.4	520	6.6
30	8.9	530	6.1
40	9.4	540	5.6
50	9.8	550	5.2
60	10.3	560	4.7
70	10.7	570	4.3
80	11.1	580	3.9
90	11.5	590	3.5
100	11.9	600	3.1
110	12.3	610	2.7
120	12.6	620	2.4
130	13.0	630	2.0
140	13.3	640	1.7
150	13.6	650	1.4
160	13.8	660	1.2
170	14.1	670	0.9
180	14.3	680	0.7
190	14.5	690	0.5
200	14.6	700	0.4
210	14.8	710	0.2
220	14.9	720	0.1
230	14.9	730	0.1
240	15.0	740	0.0
250	15.0	750	0.0
260	15.0	760	0.0
270	14.9	770	0.1
280	14.9	780	0.1
290	14.8	790	0.2
300	14.6	800	0.4
310	14.5	810	0.5
320	14.2	820	0.7
330	14.1	830	0.9
340	13.8	840	1.2
350	13.6	850	1.4
360	13.3	860	1.7
370	13.0	870	2.0
380	12.6	880	2.4
390	12.3	890	2.7
400	11.9	900	3.1
410	11.5	910	3.5
420	11.1	920	3.9
430	10.7	930	4.3
440	10.3	940	4.7
450	9.8	950	5.2
460	9.4	960	5.6
470	8.9	970	6.1
480	8.4	980	6.6
490	8.0	990	7.0
500	7.5	1000	7.5

TABLE XXIX

Lunar Equation, 2d part.

Arguments I. and VI.

I.

VI	0	50	100	150	200	250	300	350	400	450	500
	″	″	″	″	″	″	″	″	″	″	″
0	1.3	1.2	1.2	1.1	1.0	1.0	1.0	1.1	1.2	1.2	1.3
50	1.5	1.5	1.5	1.3	1.1	1.0	0.9	1.0	1.1	1.1	1.1
100	1.7	1.8	1.7	1.4	1.2	1.1	1.0	0.9	0.9	0.9	0.9
150	1.9	1.9	1.8	1.6	1.4	1.3	1.0	0.8	0.8	0.8	0.7
200	1.9	2.0	2.0	1.7	1.5	1.4	1.0	0.8	0.8	0.8	0.7
250	2.0	2.0	2.0	1.8	1.6	1.5	1.1	0.9	0.7	0.7	0.6
300	1.9	1.9	1.9	1.9	1.7	1.6	1.2	1.0	0.8	0.7	0.7
350	1.8	1.9	1.9	1.9	1.7	1.6	1.4	1.0	1.0	0.9	0.8
400	1.6	1.7	1.8	1.9	1.7	1.6	1.4	1.2	1.1	1.0	1.0
450	1.5	1.5	1.6	1.7	1.7	1.7	1.6	1.4	1.2	1.2	1.1
500	1.3	1.4	1.4	1.5	1.7	1.7	1.7	1.5	1.4	1.4	1.3
550	1.1	1.2	1.2	1.4	1.6	1.7	1.7	1.7	1.6	1.5	1.5
600	1.0	1.0	1.1	1.2	1.4	1.6	1.8	1.8	1.8	1.7	1.6
650	0.8	0.9	1.0	1.1	1.3	1.5	1.7	1.8	1.9	1.9	1.8
700	0.7	0.7	0.8	1.1	1.2	1.4	1.7	1.9	1.9	1.9	1.9
750	0.6	0.6	0.7	1.0	1.1	1.3	1.6	1.9	1.9	2.0	2.0
800	0.7	0.7	0.7	0.9	1.1	1.2	1.5	1.8	2.0	1.9	1.9
850	0.7	0.8	0.8	0.9	0.9	1.1	1.4	1.7	1.8	1.8	1.9
900	0.9	0.9	0.9	0.9	1.0	1.1	1.2	1.5	1.7	1.7	1.7
950	1.1	1.0	1.1	1.0	1.0	1.0	1.1	1.3	1.4	1.6	1.5
0	1.3	1.2	1.2	1.1	1.0	1.0	1.0	1.1	1.2	1.2	1.3

I.

VI	500	550	600	650	700	750	800	850	900	950	1000
	″	″	″	″	″	″	″	″	″	″	″
0	1.3	1.4	1.4	1.5	1.6	1.6	1.6	1.5	1.4	1.4	1.3
50	1.1	1.1	1.2	1.3	1.5	1.5	1.7	1.6	1.5	1.5	1.5
100	0.9	0.9	0.9	1.1	1.3	1.5	1.6	1.7	1.7	1.7	1.7
150	0.7	0.8	0.8	0.9	1.2	1.4	1.6	1.9	1.8	1.8	1.9
200	0.7	0.7	0.6	0.8	1.1	1.2	1.6	1.8	1.8	1.8	1.9
250	0.6	0.6	0.7	0.7	1.0	1.1	1.5	1.7	1.9	1.9	2.0
300	0.7	0.7	0.7	0.7	0.9	1.0	1.4	1.6	1.8	1.9	1.9
350	0.8	0.7	0.7	0.8	0.9	1.0	1.4	1.6	1.6	1.7	1.8
400	1.0	0.9	0.8	0.8	0.9	1.0	1.2	1.4	1.5	1.6	1.6
450	1.1	1.1	1.0	0.9	0.9	0.9	1.0	1.2	1.4	1.4	1.5
500	1.3	1.2	1.2	1.1	0.9	0.9	0.9	1.1	1.2	1.2	1.3
550	1.5	1.4	1.4	1.2	1.0	0.9	0.9	0.9	1.0	1.1	1.1
600	1.6	1.6	1.5	1.4	1.2	1.0	0.8	0.8	0.8	0.9	1.0
650	1.8	1.7	1.6	1.6	1.3	1.1	0.9	0.8	0.7	0.7	0.8
700	1.9	1.8	1.8	1.6	1.4	1.2	0.9	0.7	0.7	0.7	0.7
750	2.0	1.9	1.9	1.7	1.5	1.3	1.0	0.7	0.7	0.6	0.6
800	1.9	1.8	1.8	1.8	1.6	1.4	1.1	0.8	0.6	0.7	0.7
850	1.9	1.8	1.8	1.8	1.6	1.5	1.2	0.9	0.8	0.8	0.7
900	1.7	1.7	1.7	1.7	1.6	1.5	1.3	1.1	0.9	0.9	0.9
950	1.5	1.5	1.5	1.6	1.7	1.6	1.5	1.3	1.2	1.1	1.1
0	1.3	1.4	1.4	1.5	1.6	1.6	1.6	1.5	1.4	1.4	1.3

Constant 1″.2

Perturbations produced by Venus.

Arguments II and III.

III.

II.	0	10	20	30	40	50	60	70	80	90	100	110	120
	"	"	"	"	"	"	"	"	"	"	"	"	"
0	21.6	20.8	19.8	19.0	17.9	16.8	15.9	14.7	14.0	13.2	12.8	12.5	12.2
20	23.1	22.7	21.6	21.0	20.1	19.3	18.4	17.4	16.4	15.5	14.5	13.8	13.4
40	23.5	23.2	22.9	22.7	22.0	21.1	20.4	19.5	18.7	17.9	16.9	16.1	15.3
60	22.2	22.5	23.1	22.7	22.8	22.5	21.9	21.3	20.5	19.9	19.1	18.2	17.4
80	20.0	20.7	21.4	21.7	22.1	22.3	22.2	22.2	21.7	21.3	20.7	19.9	19.3
100	17.6	18.6	19.2	19.9	20.5	21.0	21.6	21.7	21.6	21.6	21.5	21.1	20.5
120	15.3	16.0	16.9	17.7	18.4	19.2	19.8	20.2	20.7	20.8	21.1	21.1	20.8
140	13.6	14.2	14.8	15.5	16.2	17.0	17.6	18.3	19.0	19.4	20.0	20.0	20.4
160	12.7	13.2	13.6	14.1	14.6	15.0	15.7	16.4	17.0	17.3	18.1	18.7	19.2
180	12.7	12.9	13.1	13.5	13.9	14.0	14.5	14.8	15.0	15.8	16.4	16.8	17.2
200	13.2	13.2	13.2	13.4	13.7	13.8	14.1	14.2	14.5	14.5	14.8	15.2	16.0
220	13.5	13.6	13.9	14.1	14.1	14.1	14.2	14.3	14.5	14.6	14.6	14.7	14.8
240	13.6	13.8	14.1	14.4	14.6	14.8	14.8	14.9	15.1	15.1	15.1	14.9	14.8
260	12.8	13.3	13.8	14.2	14.6	15.0	15.3	15.6	15.5	15.5	15.6	15.6	15.6
280	11.5	12.3	13.0	13.4	14.0	14.6	15.1	15.4	16.0	16.2	16.2	16.3	16.2
300	10.1	10.9	11.3	12.1	12.9	13.7	14.2	14.9	15.4	16.0	16.4	16.5	16.7
320	8.2	8.8	9.6	10.6	11.3	12.0	12.9	13.7	14.3	15.0	15.8	16.3	16.8
340	6.9	7.5	8.1	8.4	9.4	10.1	11.1	11.9	12.7	13.6	14.4	15.2	16.0
360	6.5	6.5	6.8	7.4	8.0	8.4	9.1	9.9	10.8	11.5	12.6	13.4	14.4
380	6.8	6.5	6.3	6.4	6.7	7.0	7.6	8.2	8.9	9.6	10.6	11.4	12.4
400	7.5	7.1	6.7	6.4	6.2	6.4	6.5	6.9	7.5	7.9	8.7	9.4	10.3
420	9.1	8.4	7.6	7.1	6.7	6.5	6.3	6.2	6.7	6.8	7.2	7.8	8.4
440	10.6	9.8	9.0	8.6	7.9	7.2	6.7	6.4	6.4	6.4	6.6	6.8	7.1
460	12.1	11.5	10.5	9.6	9.0	8.5	8.0	7.3	6.8	6.6	6.5	6.4	6.5
480	13.6	12.8	11.9	11.0	10.4	9.6	8.8	8.2	7.7	7.2	6.8	6.4	6.5
500	15.1	14.4	13.4	12.4	11.6	10.8	10.1	9.3	8.6	8.1	7.5	7.1	6.8
520	16.5	15.6	14.8	13.9	13.1	12.3	11.3	10.5	9.7	9.1	8.6	7.9	7.4
540	18.1	17.5	16.4	15.5	14.5	13.7	12.8	11.8	11.1	10.4	9.7	8.9	8.2
560	20.4	19.3	18.2	17.6	16.5	15.4	14.4	13.4	12.7	11.6	10.8	10.2	9.2
580	22.8	21.7	20.7	19.7	18.4	17.6	16.6	15.5	14.3	13.4	12.5	11.6	10.6
600	25.2	24.1	23.1	22.2	21.2	19.9	18.6	17.8	16.6	15.6	14.5	13.4	12.6
620	27.3	26.5	25.6	24.7	23.5	22.5	21.6	20.4	19.0	18.1	16.8	15.7	14.7
640	29.0	28.5	27.7	26.9	26.2	25.1	24.1	22.9	21.8	20.8	19.6	18.4	17.2
660	29.8	29.6	29.2	28.5	28.1	27.4	26.5	25.6	24.5	23.4	22.5	21.2	19.8
680	29.7	29.6	29.5	29.5	29.1	28.8	28.2	27.6	27.0	26.0	25.0	23.8	22.8
700	28.8	29.2	29.3	29.5	29.5	29.5	29.2	28.8	28.4	27.8	27.2	26.4	25.2
720	26.9	27.6	28.3	29.0	29.2	29.4	29.4	29.3	29.1	28.9	28.4	27.9	27.3
740	24.7	25.7	26.6	27.3	27.9	28.5	29.1	29.0	29.2	29.3	29.1	28.8	28.4
760	22.2	23.5	24.3	25.3	26.2	27.0	27.6	28.3	28.6	28.7	28.9	29.1	29.0
780	19.6	21.0	22.0	23.2	24.2	25.1	25.9	26.7	27.3	27.8	28.4	28.5	28.7
800	17.2	18.5	19.3	20.9	21.8	22.9	23.9	25.0	25.8	26 4	26.9	27.6	28.1
820	15.2	15.9	17.0	18.4	18.9	20.7	21.7	22.8	23.8	24.8	25.6	26.2	26.6
840	13.2	14.0	15 0	16.0	17.0	18.2	18.8	20.3	21.7	22.7	23.6	24.5	25.3
860	11.5	12.2	13.0	13.9	14.9	15.9	17.1	18.0	18.9	20.3	21.4	22.6	23.5
880	11.0	11.2	11.5	12.2	13.0	13.7	14.8	15.7	16.8	18.1	19.1	20.2	21.1
900	11.2	10.2	10.9	11.5	12.5	12.1	12.8	13.7	14.5	15.5	16.6	17.9	18.5
920	12.1	11.6	11.5	11.1	11.2	11.3	11.7	12.1	12.7	13.4	14 4	15.2	16.4
940	14.0	13.3	12.6	12.3	11.6	11.5	11.3	11.4	11.6	12.0	12.8	13.3	14.2
960	16.7	15.6	14.6	13.7	13.1	12.5	11.9	11.7	11.6	11.4	11.7	12.1	12.6
980	19.5	18.3	17.3	16.4	15.2	14.2	13.4	12.7	12.2	12.0	11.9	11.8	11.8
1000	21,6	20.8	19.8	19.0	17.9	16.8	15.9	14.7	14.0	13.2	12.8	12.5	12.2
	0	10	20	30	40	50	60	70	80	90	00	110	120

Perturbations produced by Venus.

Arguments II and III.

III.

II.	120	130	140	150	160	170	180	190	200	210	220	230	240
	″	″	″	″	″	″	″	″	″	″	″	″	″
0	12.2	12.2	12.3	12.4	12.8	13.3	13.9	14.7	15.6	16.5	17.7	18.8	20.1
20	13.4	12.9	12.6	12.3	12.2	12.4	12.9	13.3	14.0	14.6	15.5	16.4	17.3
40	15.3	14.4	14.0	13.5	13.0	12.9	12.6	12.6	13.1	13.5	14.0	14.4	15.4
60	17.4	16.7	16.0	15.2	14.5	14.0	13.6	13.3	13.2	13.2	13.4	13.5	14.1
80	19.3	18.7	17.7	17.1	16.4	15.9	15.4	14.6	14.3	13.9	13.8	13.7	13.6
100	20.5	20.2	19.5	18.9	18.2	17.5	17.1	16.3	15.9	15.4	14.8	14.6	14.3
120	20.8	20.7	20.4	20.0	19.7	19.2	18.5	18.0	17.3	16.9	16.5	16.2	15.6
140	20.4	20.4	20.2	20.0	20.1	19.7	19.5	19.3	18.8	18.2	17.7	17.4	17.0
160	19.2	19.1	19.4	19.7	19.5	19.6	19.3	19.6	19.2	19.0	18.7	18.4	18.1
180	17.2	17.7	18.5	18.5	18.5	18.8	18.4	18.8	19.0	19.0	18.9	18.6	18.5
200	16.0	16.2	16.6	16.8	17.5	17.6	17.7	17.9	18.1	18.2	18.3	18.3	18.3
220	14.8	15.0	15.3	15.7	16.1	16.2	16.6	16.8	17.1	17.5	17.1	17.4	17.5
240	14.8	14.7	14.8	15.0	15.1	15.4	15.7	15.8	16.0	16.1	16.1	16.2	16.1
260	15.6	15.7	15.3	14.8	15.0	15.0	15.1	15.0	15.1	15.2	15.2	15.1	15.3
280	16.2	16.2	16.2	15.9	15.8	15.8	15.5	15.4	15.1	14.9	14.8	14.7	15.0
300	16.7	17.0	17.1	16.9	16.9	16.6	16.5	16.3	15.9	15.7	15.2	14.9	14.8
320	16.8	17.3	17.5	17.6	17.7	17.6	17.5	17.2	17.0	16.8	16.5	16.1	15.6
340	16.0	16.4	17.2	17.8	17.9	18.1	18.3	18.2	18.2	17.9	17.5	17.3	16.8
360	14.4	15.2	16.0	16.7	17.4	18.1	18.4	18.6	18.8	18.8	18.8	18.7	18.4
380	12.4	13.4	14.3	15.3	16.1	16.9	17.5	18.1	18.6	19.1	19.3	19.5	19.5
400	10.3	11.2	12.3	13.2	14.2	15.1	16.0	16.8	17.8	18.4	18.8	19.3	19.8
420	8.4	9.2	10.0	11.0	12.2	13.0	14.1	15.0	15.9	16.9	17.7	18.5	19.0
440	7.1	7.6	8.4	9.0	9.9	10.9	11.8	12.9	13.8	14.9	16.0	16.7	17.8
460	6.5	6.8	7.2	7.4	8.1	9.0	9.7	10.6	11.7	12.6	13.8	14.6	15.9
480	6.5	6.5	6.4	6.6	7.0	7.5	8.2	8.8	9.6	10.4	11.5	12.5	13.5
500	6.8	6.7	6.5	6.3	6.5	6.6	7.0	7.4	8.2	8.6	9.4	10.4	11.3
520	7.4	7.0	6.8	6.5	6.3	6.1	6.3	6.6	7.0	7.5	8.0	8.8	9.3
540	8.2	7.6	7.2	6.8	6.5	6.3	6.2	6.0	6.2	6.5	6.9	7.4	7.9
560	9.2	8.6	7.9	7.5	6.8	6.6	6.3	6.1	6.0	6.1	6.2	6.5	6.9
580	10.6	9.8	9.1	8.4	7.7	7.3	6.6	6.3	6.1	5.9	5.7	5.9	6.0
600	12.6	11.4	10.5	9.5	8.7	8.1	7.4	7.0	6.4	6.1	5.8	5.5	5.6
620	14.7	13.5	12.4	11.4	10.4	9.5	8.7	7.9	7.3	6.7	6.2	5.6	5.2
640	17.2	16.2	14.9	13.7	12.5	11.4	10.4	9.5	8.7	7.8	7.0	6.5	5.9
660	19.8	19.0	17.6	16.5	15.1	13.9	12.8	11.5	10.5	9.6	8.6	7.7	6.9
680	22.8	21.7	20.4	19.3	18.1	16.8	15.7	14.2	13.0	11.9	10.7	9.6	8.6
700	25.2	24.3	23.3	22.1	20.7	19.7	18.5	17.3	16.0	14.3	13.4	12.1	11.0
720	27.3	26.4	25.7	24.5	23.7	22.5	21.1	20.2	18.8	17.7	16.4	15.3	13.9
740	28.4	27.7	27.4	26.6	25.9	24.9	24.0	22.8	21.5	20.6	19.2	18.1	16.8
760	29.0	28.7	28.3	27.8	27.3	26.8	25.9	25.2	24.3	23.0	21.7	20.7	19.7
780	28.7	28.7	28.8	28.7	28.3	28.0	27.2	26.1	26.1	25.2	24.3	23.3	22.2
800	28.1	28.3	28.4	28.5	28.5	28.4	28.2	27.3	27.3	26.7	25.9	25.1	24.4
820	26.6	27.3	27.8	28.1	28.3	28.1	28.1	28.0	27.9	27.7	27.2	26.5	25.9
840	25.3	26.2	26.7	27.2	27.5	27.9	28.1	28.1	27.9	27.9	27.6	27.3	27.2
860	23.5	24.5	25.1	25.9	26.6	27.1	27.4	27.7	27.9	28.0	27.9	27.7	27.5
880	21.1	22.4	23.3	24.2	25.1	25.8	26.5	27.0	27.3	27.5	27.8	28.0	27.7
900	18.5	20.1	21.3	22.1	23.1	24.7	25.0	25.7	26.3	26.9	27.3	27.5	27.6
920	16.4	17.7	18.4	20.0	21.0	22.2	23.0	23.9	24.9	25.7	26.2	26.9	27.3
940	14.2	14.9	16.1	17.5	18.2	19.6	20.8	21.9	23.0	23.9	24.7	25.7	26.1
960	12.6	13.3	14.1	14.4	15.9	17.2	17.9	19.5	20.5	21.7	22.7	23.9	24.7
980	11.8	12.1	12.7	13.3	14.1	14.8	15.6	16.8	17.6	19.3	20.2	21.4	22.6
1000	12.2	12.2	12.3	12.4	12.8	13.3	13.9	14.7	15.6	16.5	17.6	18.8	20.1
	20	130	140	150	160	170	180	190	200	210	220	230	240

Perturbations produced by Venus.

Arguments II. and III.

III.

II.	240	250	260	270	280	290	300	310	320	330	340	350	360
	″	″	″	″	″	″	″	″	″	″	″	″	″
0	20.1	21.1	22.2	23.4	24.3	25.2	25.8	26.6	27.2	27.6	27.7	27.6	27.6
20	17.3	18.6	19.7	20.9	21.9	23.0	24.2	24.9	25.8	26.6	27.0	27.4	27.7
40	15.4	16.5	17.3	18.3	19.4	20.5	21.6	22.7	23.7	24.9	25.5	26.3	26.9
60	14.1	14.6	15.2	16.3	17.2	18.1	18.9	20.3	21.2	22.3	23.4	24.5	25.3
80	13.6	14.0	14.5	14.9	15.5	16.3	17.3	18.2	19.0	20.0	21.1	22.0	23.1
100	14.3	14.3	14.3	14.4	14.6	15.0	15.5	16.2	16.9	17.7	18.9	19.8	20.8
120	15.6	15.2	14.8	14.8	15.0	14.9	15.0	15.2	15.9	16.3	17.0	17.7	18.5
140	17.0	16.6	16.4	15.8	15.5	15.4	15.6	15.6	15.5	15.6	16.1	16.7	17.1
160	18.1	17.7	17.5	17.3	16.9	16.6	16.3	15.9	16.1	16.3	16.3	16.2	16.5
180	18.5	18.5	18.3	18.1	17.9	17.6	17.5	17.3	17.0	16.9	16.7	16.8	16.9
200	18.3	18.4	18.2	18.2	18.2	18.2	18.1	18.1	17.8	17.7	17.6	17.5	17.7
220	17.5	17.6	17.8	17.8	18.0	18.0	18.2	18.1	18.1	18.3	18.4	18.3	18.3
240	16.4	16.5	16.7	16.9	17.1	17.3	17.3	17.7	17.5	18.0	18.3	18.4	18.6
260	15.3	15.5	15.5	15.6	15.8	16.1	16.4	16.6	16.8	16.9	17.4	17.7	18.2
280	15.0	14.9	14.9	14.9	14.9	14.7	15.0	15.3	15.5	15.9	16.1	16.4	16.8
300	14.8	14.6	14.6	14.2	14.0	14.0	13.9	13.9	14.2	14.5	14.8	15.0	15.5
320	15.6	15.3	14.7	14.5	14.4	13.1	13.6	13.4	13.3	13.1	13.4	13.6	13.8
340	16.8	16.6	16.0	15.5	15.2	14.5	14.3	13.7	13.1	13.0	12.7	12.6	12.6
360	18.4	17.9	17.5	17.0	16.5	15.9	15.4	14.9	14.3	13.7	13.0	12.6	12.3
380	19.5	19.2	18.9	18.5	17.9	17.7	16.9	16.4	15.8	15.0	14.5	13.6	13.1
400	19.8	19.8	20.1	19.7	19.4	19.1	18.6	18.1	17.5	17.0	16.1	15.2	14.8
420	19.0	19.5	20.0	20.3	20.3	20.3	20.1	19.4	19.0	18.9	18.1	17.3	16.5
440	17.8	18.7	19.2	19.7	20.1	20.4	20.7	20.7	20.5	20.2	19.8	19.5	18.6
460	15.9	16.8	17.6	18.6	19.2	19.9	20.3	20.6	21.0	20.9	20.9	20.8	20.3
480	13.5	14.6	15.5	16.6	17.7	18.5	19.3	19.9	20.5	20.8	21.1	21.2	21.2
500	11.3	12.4	13.4	14.4	15.5	15.5	17.7	18.6	19.1	19.9	20.7	21.0	21.4
520	9.3	10.2	11.2	12.2	13.3	14.2	15.4	16.4	17.6	18.4	19.2	19.8	20.6
540	7.9	8.6	9.4	10.1	11.1	12.1	13.1	14.2	15.3	16.3	17.4	18.3	19.2
560	6.9	7.2	7.8	8.4	9.2	10.1	11.0	11.9	13.1	14.1	15.2	16.2	17.2
580	6.0	6.3	6.6	7.0	7.6	8.4	9.1	9.9	10.9	11.9	12.9	14.1	15.0
600	5.6	5.6	5.8	6.1	6.5	6.8	7.4	8.1	8.8	9.9	10.7	11.8	12.8
620	5.2	5.4	5.3	5.3	5.5	5.9	6.3	6.6	7.2	8.0	8.7	9.5	10.6
640	5.9	5.6	5.2	4.9	5.0	5.0	5.2	5.5	5.8	6.4	7.0	7.6	8.5
660	6.9	6.3	5.7	5.4	5.0	4.8	4.5	4.7	4.9	5.1	5.5	6.0	6.8
680	8.6	7.6	6.9	6.2	5.6	5.1	4.8	4.6	4.2	4.2	4.5	4.6	5.1
700	11.0	10.0	8.7	7.8	6.8	6.3	5.6	5.0	4.6	4.2	4.2	4.0	4.2
720	13.9	12.5	11.2	10.3	9.1	7.9	7.1	6.2	5.6	4.8	4.5	4.2	3.8
740	16.8	15.5	14.4	13.0	11.7	10.5	9.4	8.4	7.2	6.5	5.6	5.0	4.3
760	19.7	18.5	17.2	15.9	14.7	13.5	12.2	10.8	9.8	8.9	7.6	6.7	5.9
780	22.2	21.2	20.1	19.0	17.6	16.3	15.1	14.0	12.6	11.6	10.2	9.2	8.1
800	24.4	23.4	22.2	21.3	20.3	19.2	18.0	16.7	15.4	14.3	13.2	11.9	10.8
820	25.9	25.1	24.4	23.3	22.3	21.6	20.4	19.4	18.2	17.2	15.9	14.6	13.6
840	27.2	26.6	25.8	25.0	24.3	23.5	22.4	21.6	20.5	19.4	18.4	17.3	16.4
860	27.5	27.1	26.8	26.4	25.5	24.8	24.3	23.3	22.2	21.5	20.5	19.6	18.4
880	27.7	27.5	27.2	27.0	26.5	26.0	25.5	24.7	24.1	23.2	22.0	21.4	20.4
900	27.6	27.8	27.9	27.6	27.1	26.7	26.5	25.7	25.3	24.6	23.9	23.0	22.0
920	27.3	27.5	27.5	27.6	27.7	27.5	27.2	26.7	26.3	25.7	25.1	24.3	23.6
940	26.1	26.7	27.2	27.4	27.7	27.7	27.6	27.5	27.1	26.6	26.2	25.6	25.5
960	24.7	25.4	26.2	26.6	27.2	27.5	27.7	27.7	27.6	27.4	27.1	27.0	26.2
980	22.6	23.7	24.6	25.3	25.9	26.8	27.2	27.5	27.7	27.8	27.6	27.5	27.1
1000	20.1	21.1	22.2	23.4	24.3	25.2	25.8	26.6	27.2	27.6	27.7	27.6	27.6
	240	250	260	270	280	290	300	310	320	330	340	350	360

Perturbations produced by Venus.

Arguments II. and III.

III.

II.	360	370	380	390	400	410	420	430	440	450	460	470	480
	″	″	″	″	″	″	″	″	″	″	″	″	″
0	27.6	27.7	27.3	26.7	26.2	25.5	24.7	23.8	23.1	22.3	21.3	20.2	19.3
20	27.7	27.8	27.8	27.6	27.4	26.8	26.2	25.6	24.8	24.0	23.1	22.0	20.9
40	26.9	27.3	27.6	27.9	27.9	27.7	27.5	27.1	26.3	25.6	24.9	24.0	23.2
60	25.3	26.0	26.8	27.1	27.5	27.9	27.8	27.7	27.3	27.1	26.7	25.9	25.0
80	23.1	24.0	25.1	25.9	26.5	27.3	27.5	27.9	28.2	28.0	27.6	27.5	27.2
100	20.8	21.8	22.6	23.6	24.6	25.5	26.2	26.7	27.2	27.5	27.6	27.8	27.4
120	18.5	19.6	20.6	21.5	22.4	23.2	24.1	25.1	25.8	26.4	26.9	27.3	27.5
140	17.1	17.9	18.6	19.3	20.3	21.3	23.0	22.9	23.7	24.7	25.5	26.0	26.7
160	16.5	17.1	17.4	18.1	18.8	19.3	20.1	21.0	21.9	22.6	23.5	24.2	25.1
180	16.9	17.0	17.1	17.4	18.0	18.4	18.9	19.4	20.1	20.7	21.2	22.2	23.0
200	17.7	17.5	17.7	17.7	17.6	18.1	18.3	18.7	19.2	19.7	20.1	20.8	21.5
220	18.3	18.2	18.3	18.3	18.3	18.3	18.6	18.7	18.9	19.3	19.5	20.0	20.4
240	18.6	18.8	18.9	18.9	18.9	19.0	19.2	19.1	19.2	19.5	19.6	19.7	19.9
260	18.2	18.5	18.7	18.8	19.0	19.3	19.5	19.6	19.9	19.9	20.0	20.1	20.2
280	16.8	17.4	17.9	18.3	18.7	19.1	19.3	19.8	20.0	20.2	20.4	20.6	20.8
300	15.5	15.8	16.2	16.6	17.6	18.1	18.5	19.2	19.4	19.9	20.6	20.8	20.9
320	13.8	14.2	14.6	15.1	15.6	16.2	16.8	17.7	18.3	18.9	19.5	20.1	20.8
340	12.6	12.9	13.0	13.3	13.7	14.4	14.9	15.5	16.2	17.1	18.0	18.6	19.4
360	12.3	12.1	11.9	12.0	12.3	12.5	13.0	13.4	14.2	14.9	15.7	16.5	17.3
380	13.1	12.5	11.9	11.6	11.5	11.4	11.6	11.7	12.3	12.7	13.3	14.0	15.0
400	14.8	13.9	13.1	12.5	11.7	11.2	11.1	10.9	11.0	11.1	11.4	12.0	12.6
420	16.5	15.7	15.1	14.3	13.4	12.5	11.7	11.1	10.8	10.8	10.5	10.6	10.7
440	18.6	17.9	17.1	16.1	15.6	14.4	13.5	12.8	11.9	11.1	10.6	10.3	10.3
460	20.3	19.8	19.3	18.5	17.6	16.8	15.9	14.7	13.7	12.9	12.0	11.1	10.9
480	21.2	21.1	20.8	20.3	19.7	19.1	18.3	17.4	16.4	15.0	14.1	13.2	12.2
500	21.4	21.4	21.4	21.3	21.1	20.8	20.0	19.5	18.8	17.8	17.0	15.7	14.4
520	20.6	21.2	21.7	21.7	21.5	21.5	21.4	21.1	20.5	19.8	19.1	18.2	17.6
540	19.2	20.0	20.7	21.1	21.8	22.0	21.8	21.7	21.5	21.2	20.9	20.3	19.6
560	17.2	18.4	19.0	20.0	20.8	21.1	22.7	21.9	22.2	22.1	21.9	21.7	21.1
580	15.0	16.0	17.3	18.2	19.1	19.9	20.8	21.1	21.7	22.0	22.2	22.3	22.1
600	12.8	13.9	15.1	15.9	17.2	18.0	19.0	19.9	20.6	21.3	21.8	22.0	22.4
620	10.6	11.5	12.7	13.7	14.9	16.0	17.1	18.3	19.1	19.9	20.8	21.3	22.0
640	8.5	9.5	10.4	11.3	12.3	13.7	14.9	16.0	17.1	18.1	19.0	19.9	20.7
660	6.8	7.4	8.2	9.1	10.1	11.1	12.2	13.6	14.6	15.8	17.1	18.1	19.0
680	5.1	5.7	6.4	7.1	7.9	8.7	9.7	11.0	12.1	13.1	14.1	15.7	16.8
700	4.2	4.4	4.7	5.1	5.8	6.7	7.4	8.4	9.4	10.6	11.5	13.0	14.1
720	3.8	3.8	3.8	4.0	4.4	4.8	5.4	5.9	6.9	8.0	9.1	10.1	11.5
740	4.3	3.9	3.8	3.7	3.6	3.8	3.9	4.4	4.9	5.7	6.4	7.4	8.9
760	5.9	5.1	4.4	4.0	3.6	3.4	3.4	3.5	3.9	4.3	4.7	5.2	5.9
780	8.1	7.1	6.1	5.3	4.6	4.1	3.7	3.3	3.3	3.1	3.4	3.6	4.1
800	10.8	9.7	8.5	7.5	6.5	5.6	4.9	4.2	3.8	3.4	3.2	3.1	3.1
820	13.6	12.5	11.2	10.1	9.0	8.0	6.9	6.1	5.3	4.7	3.9	3.7	3.1
840	16.4	15.1	13.7	12.9	11.7	10.6	9.5	8.6	7.5	6.6	5.7	4.9	4.4
860	18.4	17.5	16.6	15.4	14.3	13.1	12.1	11.1	10.0	9.1	7.9	7.0	6.3
880	20.4	19.6	18.7	17.5	16.6	15.6	14.5	13.6	12.5	11.5	10.4	9.5	8.6
900	22.0	21.1	20.2	19.4	18.7	17.7	16.5	15.7	14.7	13.8	12.5	11.9	10.9
920	23.6	22.7	21.7	21.1	20.1	19.4	18.4	17.5	16.7	15.6	14.8	13.9	13.1
940	25.5	24.1	23.4	22.4	21.4	20.6	19.9	19.0	18.2	17.3	16.6	15.7	14.8
960	26.2	25.6	24.7	24.1	23.3	22.3	21.3	20.6	19.0	18.9	17.9	17.1	16.3
980	27.1	26.7	26.3	25.5	24.9	23.8	23.4	22.2	21.0	20.4	19.4	18.6	17.7
1000	27.6	27.7	27.3	26.7	26.2	25.5	24.7	23.8	23.1	22.3	21.3	20.2	19.3
	360	370	380	390	400	410	420	430	440	450	460	470	480

TABLE XXX.

Perturbations produced by Venus.

Arguments II and III.

III.

II.	480	490	500	510	520	530	540	550	560	570	580	590	600
	″	″	″	″	″	″	″	″	″	″	″	″	″
0	19.3	18.3	17.4	16.6	15.7	15.0	14.2	13.6	13.1	12.3	11.7	11.3	10.8
20	20.9	20.2	19.1	18.2	17.1	16.2	15.5	14.7	14.1	13.3	12.7	12.2	11.5
40	23.2	22.0	20.8	20.1	18.9	17.9	17.1	15.9	15.1	14.4	13.7	13.0	12.3
60	25.0	24.0	23.2	22.0	20.7	19.9	18.9	17.7	16.8	15.8	14.9	14.0	13.3
80	27.2	26.4	25.6	24.1	23.2	22.1	20.8	20.0	18.7	17.9	16.6	15.6	14.8
100	27.4	27.2	26.8	26.3	25.4	24.5	23.5	22.2	20.9	20.0	18.6	17.6	16.6
120	27.5	27.5	27.6	27.1	26.8	26.3	25.4	24.6	23.7	22.4	21.0	20.1	18.8
140	26.7	27.0	27.2	27.4	27.3	27.4	26.9	26.2	25.4	24.6	23.9	22.6	21.1
160	25.1	25.6	26.1	26.7	26.9	27.3	27.1	27.0	26.9	26.4	25.5	24.7	23.9
180	23.0	23.8	24.5	25.0	25.7	26.3	26.7	26.8	27.0	26.8	26.6	26.2	25.6
200	21.5	22.2	22.8	23.5	24.1	24.7	25.5	25.8	26.3	26.6	26.6	26.6	26.4
220	20.4	21.0	21.5	22.0	22.6	23.2	23.8	24.5	25.0	25.4	25.8	26.0	26.2
240	19.9	20.4	20.8	21.2	21.6	21.8	22.2	22.6	23.1	23.3	23.9	24.2	24.6
260	20.2	20.3	20.6	21.2	21.4	21.7	21.9	22.2	22.3	22.7	23.1	23.3	23.6
280	20.8	20.8	21.0	21.1	21.3	21.4	21.5	21.8	22.0	22.2	22.7	23.0	23.3
300	20.9	21.0	21.5	21.7	21.7	22.0	22.0	22.1	22.1	22.2	22.4	22.6	22.8
320	20.8	21.2	21.5	21.6	22.0	22.3	22.5	22.5	22.6	22.7	22.8	22.8	22.9
340	19.4	20.2	20.8	21.5	21.9	22.1	22.6	23.0	23.2	23.4	23.3	23.4	23.5
360	17.3	18.4	19.5	20.0	20.6	21.5	22.2	22.7	23.0	23.7	23.7	24.0	24.2
380	15.0	15.9	16.9	17.8	18.6	19.6	20.6	21.5	22.3	22.9	23.5	23.9	24.5
400	12.6	13.2	14.2	15.4	16.2	17.3	18.1	19.2	20.3	21.4	22.4	23.0	23.7
420	10.7	11.2	12.0	12.5	13.5	14.5	15.6	16.7	17.7	18.7	20.1	21.0	22.0
440	10.3	10.2	10.3	10.5	11.3	12.0	12.9	13.6	14.7	16.0	17.0	18.3	19.5
460	10.9	10.1	9.9	9.9	9.9	10.1	10.7	11.3	12.2	13.0	14.0	15.1	16.5
480	12.2	11.4	10.7	10.1	9.7	9.5	9.7	9.9	10.2	10.7	11.7	12.5	13.4
500	14.4	13.6	12.5	11.6	10.9	10.2	9.8	9.4	9.3	9.6	9.8	10.2	11.1
520	17.6	16.2	15.1	13.9	12.9	11.9	10.9	10.3	9.8	9.5	9.2	9.2	9.6
540	19.6	18.6	18.0	16.7	15.4	14.5	12.2	12.3	11.3	10.5	10.1	9.5	9.3
560	21.1	20.4	19.8	19.0	18.2	17.2	16.0	14.8	13.7	12.7	11.7	10.9	10.2
580	22.1	21.8	21.5	20.9	20.3	19.3	18.6	17.3	16.5	15.4	14.0	12.9	12.2
600	22.4	22.4	22.2	22.2	21.5	21.2	20.6	19.5	19.1	17.7	16.8	15.8	14.4
620	22.0	22.3	22.4	22.4	22.3	22.3	21.9	21.5	20.9	20.0	19.3	18.0	16.9
640	20.7	21.7	22.0	22.3	22.6	22.5	22.6	22.4	22.0	21.6	21.1	20.3	19.6
660	19.0	20.0	20.8	21.3	22.1	22.3	22.6	22.8	22.7	22.6	22.2	21.8	21.3
680	16.8	18.0	19.0	19.9	20.8	21.5	22.1	22.6	22.7	23.0	23.0	22.8	22.4
700	14.1	15.2	16.8	17.9	18.8	20.0	22.1	21.5	22.2	22.6	22.9	23.0	23.2
720	11.5	12.7	13.9	15.0	16.4	17.9	18.6	19.7	20.8	21.6	22.3	22.7	23.0
740	8.9	9.8	10.9	12.2	13.6	14.8	16.2	17.5	18.7	19.5	20.6	21.6	22.3
760	5.9	6.8	8.0	9.3	10.3	11.8	13.2	14.5	15.9	17.4	18.2	19.5	20.5
780	4.1	4.9	5.6	6.4	7.5	8.6	9.9	11.1	12.6	14.0	15.6	16.8	18.1
800	3.1	3.3	4.4	4.8	5.5	6.1	6.9	7.9	9.4	10.7	12.1	13.4	14.9
820	3.1	3.1	3.2	3.1	3.6	3.9	4.8	5.7	6.5	7.5	8.7	10.0	11.5
840	4.4	3.7	3.5	3.2	3.2	3.1	3.4	3.7	4.1	5.0	6.2	7.0	8.2
860	6.3	5.5	4.6	4.1	3.6	3.4	3.3	3.2	3.4	3.4	4.0	4.5	5.6
880	8.6	7.6	6.7	5.9	5.2	4.5	4.1	3.8	3.5	3.4	3.4	3.6	3.9
900	10.9	10.0	9.1	8.3	7.2	6.5	5.8	5.1	4.4	4.2	3.8	3.6	3.6
920	13.1	12.1	11.2	10.3	9.6	8.7	7.7	6.9	6.3	5.8	5.1	4.6	4.2
940	14.8	14.1	13.1	12.4	11.5	10.8	9.8	9.1	8.3	7.6	6.8	6.5	5.9
960	16.3	15.4	14.6	14.0	13.2	12.6	11.7	11.0	10.1	9.6	8.8	8.1	7.5
980	17.7	16.8	16.2	15.2	14.5	13.9	13.1	12.5	11.8	11.2	10.5	9.7	9.3
1000	19.3	18.3	17.4	16.6	15.7	15.0	14.2	13.6	13.1	12.3	11.7	11.3	10.8
	480	490	500	510	520	530	540	550	560	570	580	590	600

Perturbations produced by Venus.

Arguments II. and III.

III.

II.	600	610	620	630	640	650	660	670	680	690	700	710	720
	"	"	"	"	"	"	"	"	"	"	"	"	"
0	10.8	10.2	9.5	9.1	8.4	7.9	7.4	7.0	6.6	6.3	5.9	5.5	5.4
20	11.5	11.3	10.7	10.4	9.8	9.4	8.9	8.5	7.9	7.7	7.3	6.7	6.6
40	12.3	12.0	11.5	11.0	10.7	10.3	10.0	9.6	9.3	8.9	8.5	8.1	7.8
60	13.3	12.7	12.1	11.6	11.2	10.9	10.5	10.2	10.0	9.8	9.5	9.2	8.9
80	14.8	13.6	12.9	12.4	11.8	11.3	10.9	10.7	10.3	9.9	9.8	9.8	9.6
100	16.6	15.4	14.4	13.4	12.6	12.1	11.5	11.0	10.6	10.2	10.0	9.9	9.6
120	18.8	17.7	16.4	15.3	14.3	13.2	12.4	11.6	11.2	10.6	10.1	10.1	9.6
140	21.1	20.1	18.9	17.7	16.5	15.2	14.2	13.0	12.3	11.6	11.1	10.3	9.9
160	23.9	22.9	21.5	20.4	19.2	17.9	16.6	15.3	14.1	13.1	12.0	11.2	10.5
180	25.6	24.8	23.9	22.9	21.6	20.6	19.1	18.0	16.7	15.5	14.3	12.9	12.0
200	26.4	26.0	25.6	24.9	24.0	22.9	21.7	20.8	19.3	18.1	16.9	15.5	14.4
220	26.2	26.3	26.1	25.8	25.3	24.9	24.1	23.1	21.2	20.9	19.7	18.3	17.1
240	24.6	25.1	25.1	25.3	25.2	25.1	24.7	24.3	24.0	23.0	21.9	21.3	20.2
260	23.6	23.9	24.2	24.5	24.7	24.8	24.9	24.6	24.3	23.8	23.4	22.9	21.6
280	23.3	23.6	23.9	24.2	24.7	24.8	25.0	24.9	24.9	24.8	24.4	24.0	23.5
300	22.8	23.0	23.3	23.4	23.8	24.0	24.1	24.5	24.5	24.6	24.5	24.4	24.0
320	22.9	23.0	23.1	23.2	23.4	23.3	23.6	23.8	24.0	23.9	24.2	24.2	24.2
340	23.5	23.5	23.5	23.4	23.5	23.6	23.6	23.5	23.5	23.6	23.9	23.8	23.8
360	24.2	24.2	24.3	24.2	24.2	24.0	23.7	23.9	24.0	23.7	23.7	23.6	23.6
380	24.5	24.6	24.8	25.1	24.8	24.9	25.0	24.9	24.6	24.5	24.5	24.3	24.0
400	23.7	24.3	24.7	25.0	25.4	25.7	25.7	25.5	25.5	25.4	25.2	24.8	24.6
420	22.0	23.0	23.7	24.6	25.0	25.7	26.1	26.2	26.3	26.5	26.2	26.0	25.9
440	19.5	20.8	21.7	22.7	23.7	24.6	25.4	26.0	26.5	26.7	26.9	27.0	26.9
460	16.5	17.8	19.0	20.1	21.4	22.3	23.5	24.8	25.4	26.1	26.7	27.1	27.3
480	13.4	14.5	15.6	17.0	18.5	19.7	20.9	22.1	23.2	24.4	25.4	26.2	26.8
500	11.1	12.0	13.0	13.8	14.9	16.3	17.9	19.1	20.5	21.6	22.9	24.2	25.1
520	9.6	9.8	10.5	11.5	12.4	13.4	14.4	15.5	17.1	18.4	19.9	21.2	22.3
540	9.3	9.0	9.2	9.6	10.3	11.0	11.9	12.8	13.9	15.1	16.5	17.9	19.4
560	10.2	9.7	9.3	9.1	9.1	9.4	10.0	10.6	11.5	12.4	13.3	14.5	16.0
580	12.2	11.3	10.4	9.9	9.4	9.0	9.2	9.3	9.7	10.4	11.0	12.0	12.7
600	14.4	13.3	12.5	11.6	10.8	10.1	9.6	9.4	9.1	9.3	9.9	10.0	10.8
620	16.9	16.1	14.9	13.7	12.7	12.0	11.1	10.4	9.8	9.5	9.5	9.3	9.7
640	19.6	18.4	17.4	16.3	15.2	14.2	13.1	12.1	11.3	10.6	10.1	9.6	9.5
660	21.3	20.6	19.9	18.7	17.8	16.7	15.6	14.4	13.4	12.4	11.7	11.0	10.2
680	22.4	22.0	21.5	20.8	20.2	19.0	18.1	17.0	15.8	14.7	13.7	12.8	12.0
700	23.2	23.2	22.6	22.2	21.7	21.0	20.5	19.3	18.3	17.3	16.0	15.0	14.1
720	23.0	23.3	23.2	23.4	23.1	22.4	21.9	21.3	20.8	19.5	18.5	17.6	16.4
740	22.3	22.8	23.2	23.4	23.6	23.6	23.3	22.8	22.2	21.6	21.1	19.9	18.8
760	20.5	21.4	22.5	22.8	23.3	23.7	23.6	23.8	23.5	23.3	22.7	21.8	21.3
780	18.1	19.2	20.4	21.3	22.3	23.0	23.3	23.7	23.8	24.0	23.8	23.5	23.0
800	14.9	16.4	17.7	19.1	20.1	21.2	21.1	22.9	23.4	23.8	24.1	24.2	23.9
820	11.5	12.9	14.3	15.8	17.8	18.7	20.0	20.9	22.0	22.7	23.5	23.9	24.0
840	8.2	9.5	10.8	12.2	13.8	15.2	16.6	18.1	19.5	20.6	21.7	22.6	23.3
860	5.6	6.8	7.7	8.8	10.2	11.5	13.2	14.7	16.0	17.4	19.0	20.2	21.3
880	3.9	4.4	5.2	6.1	7.2	8.2	9.7	10.9	12.5	14.1	15.4	16.8	18.2
900	3.6	3.6	3.9	4.2	5.0	5.7	6.6	7.8	9.1	10.3	11.8	13.4	14.8
920	4.2	3.8	3.9	3.9	4.0	4.3	4.7	5.4	6.4	7.3	8.6	9.8	11.2
940	5.9	5.1	4.6	4.4	4.2	4.3	4.3	4.3	4.9	5.3	6.3	7.0	8.0
960	7.5	6.9	6.3	5.8	5.3	4.7	4.7	4.6	4.6	4.6	4.9	5.4	6.0
980	9.3	8.7	7.9	7.4	6.8	6.4	6.0	5.6	5.2	5.0	4.9	5.1	5.1
1000	10.8	10.2	9.5	9.1	8.4	7.9	7.4	7.0	6.6	6.3	5.9	5.5	5.4
	600	610	620	630	640	650	660	670	680	690	700	710	720

TABLE XXX.

Perturbations produced by Venus.

Arguments II. and III.

III.

II.	720	730	740	750	760	770	780	790	800	810	820	830	840
	″	″	″	″	″	″	″	″	″	″	″	″	″
0	5.4	5.5	5.8	6.0	6.3	6.8	7.6	8.4	9.3	10.4	11.7	12.9	14.3
20	6.6	6.3	6.0	6.1	6.1	6.2	6.5	6.9	7.7	8.3	9.4	10.2	11.2
40	7.8	7.4	7.1	7.0	6.7	6.6	6.8	6.8	6.9	7.2	7.7	8.5	9.3
60	8.9	8.8	8.3	8.1	7.8	7.6	7.4	7.4	7.3	7.4	7.4	7.7	8.3
80	9.6	9.5	9.1	9.1	9.0	8.8	8.4	8.2	8.1	8.1	8.0	8.1	8.2
100	9.6	9.5	9.6	9.5	9.5	9.3	9.3	9.2	9.2	9.0	8.7	8.7	8.7
120	9.6	9.6	9.5	9.3	9.4	9.6	9.6	9.5	9.5	9.6	9.6	9.6	9.6
140	9.9	9.5	9.6	9.4	9.3	9.3	9.0	9.3	9.5	9.8	9.7	9.8	10.0
160	10.5	9.9	9.5	9.1	8.9	9.0	8.9	9.0	9.0	9.0	9.5	9.6	9.9
180	12.0	11.0	10.1	9.7	9.1	8.8	8.7	8.3	8.5	8.7	8.8	9.0	9.1
200	14.4	13.3	12.0	11.0	10.1	9.4	8.9	8.5	8.2	8.0	8.0	8.3	8.5
220	17.1	15.7	14.6	13.2	12.0	10.9	10.2	9.2	8.7	8.3	7.9	7.7	7.7
240	20.2	19.1	17.8	16.5	14.5	13.4	12.2	11.1	10.0	9.4	8.4	8.0	7.7
260	21.6	21.1	20.1	19.2	17.3	15.9	14.6	13.4	12.4	11.3	10.1	9.1	8.6
280	23.5	22.7	21.6	21.0	19.8	18.8	17.3	16.1	15.0	13.5	12.5	11.5	10.2
300	24.0	23.4	23.2	22.4	21.4	20.5	19.8	18.7	17.5	16.1	15.0	13.7	12.4
320	24.2	23.9	23.5	23.1	22.7	22.2	21.2	20.6	19.6	18.6	17.5	16.3	15.1
340	23.8	23.9	23.7	23.5	23.2	22.8	22.3	21.4	20.9	20.5	19.2	18.6	17.4
360	23.6	23.6	23.6	23.3	23.3	23.1	22.9	22.4	22.0	21.4	20.4	19.9	18.9
380	24.0	24.0	23.7	23.5	23.3	23.1	23.1	22.7	22.4	22.2	21.6	20.8	20.0
400	24.6	24.4	24.4	24.0	23.8	23.4	23.2	23.0	22.8	22.4	22.1	21.6	21.3
420	25.9	25.6	25.2	24.8	24.7	24.3	23.9	23.6	23.3	22.9	22.7	22.3	21.7
440	26.9	26.6	26.4	26.2	25.9	25.5	25.2	24.9	24.5	23.8	23.4	23.0	22.8
460	27.3	27.6	27.6	27.4	27.0	26.9	26.5	26.1	25.6	25.0	24.6	24.2	23.7
480	26.8	27.4	27.6	28.0	28.1	28.2	27.7	27.4	27.3	26.6	26.2	25.7	25.1
500	25.1	26.1	26.8	27.5	28.1	28.2	28.6	28.5	28.4	28.3	27.6	27.2	26.7
520	22.3	23.9	24.8	25.9	26.8	27.5	28.1	28.5	28.7	29.0	28.8	28.6	28.4
540	19.4	20.7	22.1	23.4	24.6	25.6	26.5	27.4	28.0	28.7	28.9	29.1	29.2
560	16.0	17.3	18.6	19.9	21.4	22.9	24.1	25.5	26.4	27.3	28.2	28.6	29.2
580	12.7	14.1	15.5	16.8	18.0	19.3	20.9	22.2	23.5	24.9	26.1	27.0	27.8
600	10.8	11.6	12.7	13.6	14.9	16.2	17.5	18.7	20.2	21.8	23.0	24.4	25.5
620	9.7	10.0	10.5	10.7	12.2	13.2	14.4	15.6	17.0	18.3	19.6	21.2	22.6
640	9.5	9.4	9.6	10.1	10.4	11.1	12.0	13.0	14.0	15.2	16.5	17.9	19.2
660	10.2	10.0	9.7	9.5	9.5	9.9	10.4	11.0	11.7	12.7	13.8	14.9	16.2
680	12.0	11.2	10.5	10.0	9.7	9.5	9.6	10.0	10.4	11.0	11.6	12.5	13.8
700	14.1	13.1	12.3	11.3	10.7	10.1	9.7	9.7	9.9	9.9	10.4	10.9	11.5
720	16.4	15.3	14.4	13.3	12.2	11.6	10.9	10.2	10.1	9.9	10.0	10.1	10.4
740	18.8	17.7	16.7	15.6	14.4	13.5	12.4	11.5	11.1	10.7	10.1	10.0	10.3
760	21.3	20.1	19.2	18.1	16.6	15.6	14.7	13.6	12.8	11.9	11.3	10.7	10.3
780	23.0	22.3	21.5	20.5	19.4	18.4	17.2	15.8	14.9	14.0	13.0	12.2	11.3
800	23.9	23.9	23.4	22.6	21.9	20.7	19.8	18.8	17.5	16.2	15.1	14.2	13.4
820	24.0	24.5	24.2	23.9	23.3	22.6	22.3	21.3	20.3	19.4	18.3	17.3	16.2
840	23.3	24.0	24.3	24.5	24.4	24.3	23.8	23.4	22.7	21.7	20.8	19.6	18.3
860	21.3	22.3	23.3	23.9	24.2	24.7	24.5	24.5	24.3	23.6	23.1	21.9	21.0
880	18.2	19.7	20.9	22.0	22.8	23.8	24.1	24.6	24.8	24.7	24.5	24.0	23.5
900	14.8	16.1	17.6	19.0	20.6	21.5	22.5	23.2	24.1	24.5	24.2	24.8	24.5
920	11.2	12.6	14.0	15.5	17.0	18.4	19.9	21.0	22.0	22.9	23.5	24.5	24.5
940	8.0	9.3	10.7	12.0	13.3	14.8	16.4	17.6	19.1	20.4	21.4	22.4	23.2
960	6.0	6.9	7.8	8.6	10.2	11.5	12.7	14.1	15.6	16.9	18.5	19.5	20.7
980	5.1	5.5	6.0	6.7	7.7	8.5	9.7	10.9	12.2	13.6	14.8	16.1	17.6
1000	5.4	5.5	5.8	5.8	6.3	6.8	7.6	8.4	9.3	10.5	11.7	12.9	14.3
	720	730	740	750	760	770	780	790	800	810	820	830	840

Perturbations produced by Venus.

Arguments II. and III.

III.

II.	840	850	860	870	880	890	900	910	920	930	940	950	960
	′	″	″	″	″	″	″	″	″	″	″	″	″
0	14.3	15.5	16.9	18.2	19.2	20.2	21.4	22.5	23.0	23.5	24.0	24.2	24.2
20	11.2	12.4	13.6	14.9	16.2	17.3	18.6	19.6	20.5	21.5	22.4	23.1	23.6
40	9.3	10.2	10.9	11.8	13.3	14.2	15.5	16.6	17.8	18.8	19.7	20.7	21.6
60	8.3	8.7	9.5	10.1	10.8	11.6	12.7	13.8	14.9	15.9	17.0	18.1	19.1
80	8.2	8.3	8.6	8.9	9.6	10.3	10.7	11.6	12.5	13.3	14.5	15.2	16.2
100	8.7	8.7	8.9	9.0	9.1	9.4	9.9	10.4	11.0	11.7	12.4	12.9	14.0
120	9.6	9.5	9.3	9.6	9.6	9.7	9.9	9.8	10.4	10.9	11.3	11.8	12.3
140	10.0	10.2	10.1	10.2	10.1	10.3	10.4	10.5	10.5	10.6	10.9	11.4	11.5
160	9.9	10.0	10.2	10.4	10.6	11.0	11.0	10.9	11.0	11.3	11.3	11.3	11.6
180	9.1	9.6	9.9	10.1	10.4	10.7	11.0	11.3	11.5	11.7	11.7	11.9	12.2
200	8.5	8.8	9.1	9.5	9.7	10.0	10.5	11.0	11.2	11.6	12.0	12.2	12.4
220	7.7	7.7	8.1	8.4	8.8	9.2	9.7	10.1	10.6	11.0	11.4	11.8	12.3
240	7.7	7.3	7.4	7.4	7.7	8.0	8.4	9.0	9.6	10.0	10.5	11.0	11.5
260	8.6	7.9	7.4	7.2	7.1	7.1	7.3	7.6	8.1	8.5	9.3	10.0	10.4
280	10.2	9.2	8.3	7.9	7.4	7.1	7.0	6.9	7.0	7.3	7.7	8.5	8.8
300	12.4	11.4	10.4	9.3	8.5	7.8	7.4	6.9	6.7	6.8	6.8	7.0	7.5
320	15.1	13.9	12.5	11.4	10.5	9.7	8.6	7.8	7.4	7.0	6.6	6.5	6.7
340	17.4	16.4	15.2	13.9	12.7	11.6	10.6	9.7	8.7	8.0	7.3	6.8	6.6
360	18.9	18.1	17.4	16.3	15.1	13.8	12.8	11.7	10.6	9.8	8.8	8.0	7.4
380	20.0	19.6	18.8	17.7	16.9	13.0	15.1	13.9	12.7	11.8	10.8	9.8	8.9
400	21.3	20.6	19.6	19.4	18.4	17.6	16.5	15.7	14.8	13.7	12.8	11.8	10.9
420	21.7	21.1	20.8	20.3	19.3	18.9	18.2	17.2	16.3	15.3	14.5	13.7	12.6
440	22.8	22.1	21.6	20.8	20.6	19.7	19.0	18.6	17.7	16.6	15.9	15.1	14.2
460	23.7	23.3	22.7	22.0	21.6	20.9	20.2	19.5	18.5	18.1	17.3	16.7	15.7
480	25.1	24.4	23.9	23.3	22.8	22.0	21.4	20.9	20.2	19.3	18.3	17.7	16.9
500	26.7	26.3	25.7	24.9	24.3	23.6	23.0	22.3	21.4	20.7	20.3	19.1	18.1
520	28.4	27.8	27.3	26.8	26.3	25.6	24.7	23.9	23.3	22.6	21.8	20.8	20.1
540	29.2	29.2	28.9	28.5	27.8	27.4	26.8	26.1	25.3	24.4	23.7	23.0	22.0
560	29.2	29.3	29.5	29.6	29.3	29.1	28.8	28.0	27.4	26.9	26.1	25.1	24.3
580	27.8	28.6	29.0	29.4	29.6	29.8	29.8	29.3	28.0	28.7	27.9	27.3	26.6
600	25.5	26.7	27.6	28.4	28.9	29.2	29.6	29.9	29.9	29.8	29.3	29.0	28.5
620	22.6	23.8	25.0	26.2	27.1	27.9	28.8	29.3	29.6	29.8	30.1	29.8	29.6
640	19.2	20.6	21.6	23.3	24.6	25.2	26.6	27.8	28.3	28.9	29.4	29.7	29.9
660	16.2	17.5	18.8	20.2	21.1	22.9	24.0	25.1	26.2	27.1	28.2	28.8	29.2
680	13.8	14.7	15.8	16.9	18.4	19.9	20.6	22.3	23.6	24.9	25.8	26.7	27.5
700	11.5	12.3	13.4	14.6	15.6	16.7	18.0	19.5	20.7	22.0	23.1	24.2	25.1
720	10.4	11.0	11.4	12.3	13.3	14.3	15.6	16.4	17.7	19.3	19.9	21.6	22.6
740	10.3	10.4	10.5	11.0	11.4	12.2	13.3	14.2	15.3	16.5	17.4	18.8	19.5
760	10.3	10.0	10.2	10.3	10.7	11.0	11.5	12.2	13.1	14.2	15.1	16.0	17.3
780	11.3	10.8	10.6	10.2	10.2	10.5	10.7	11.1	11.5	12.3	13.2	14.0	15.0
800	13.4	12.5	11.7	11.0	10.6	10.3	10.3	10.4	10.7	11.0	11.6	11.3	12.2
820	16.2	15.2	14.4	13.5	13.5	11.9	11.4	11.0	10.9	10.8	10.8	11.2	11.4
840	18.3	17.1	16.2	14.9	14.1	13.0	12.4	11.7	11.2	10.7	10.6	11.1	11.2
860	21.0	20.2	18.7	17.7	16.6	15.4	14.3	13.3	12.5	11.9	11.4	11.0	10.9
880	23.5	22.4	21.3	20.4	19.3	18.0	17.0	15.9	14.8	13.7	12.8	12.0	12.6
900	24.5	24.2	23.8	22.7	21.9	19.9	19.7	18.6	17.2	16.4	15.3	14.1	13.3
920	24.5	24.8	24.7	24.3	24.1	23.2	22.3	21.3	20.0	19.3	18.0	16.7	15.7
940	23.2	24.0	24.5	24.6	24.5	24.5	24.2	23.5	22.7	21.8	20.6	19.5	18.4
960	20.7	21.9	22.8	23.6	24.0	24.5	24.5	24.2	24.3	23.7	22.9	22.1	21.0
980	17.6	18.7	20.1	21.2	22.2	23.1	23.6	24.0	24.3	24.3	24.3	23.7	23.0
1000	14.3	15.5	16.9	18.2	19.2	20.2	21.4	22.5	23.0	23.5	24.0	24.2	24.2
	840	850	860	870	880	890	900	910	920	930	940	950	960

Perturbations by Venus. Arguments II and III.

Perturbations by Mars. Arguments II and IV.

II.	III. 960	970	980	990	1000	IV. 0	10	20	30	40	50	60	70
	"	"	"	"	"	"	"	"	"	"	"	"	"
0	24.2	23.7	23.1	22.5	21.6	9.5	10.2	10.8	11.2	11.5	11.7	11.8	11.5
20	23.6	23.7	24.0	23.4	23.1	8.3	9.1	9.8	10.5	10.9	11.2	11.5	11.6
40	21.6	22.4	22.9	23.5	23.5	7.1	7.9	8.8	9.4	10.0	10.6	10.8	11.2
60	19.1	20.1	20.7	21.5	22.2	5.8	6.7	7.6	8.4	9.1	9.8	10.3	10.5
80	16.2	17.3	18.4	19.7	20.0	4.3	5.3	6.4	7.2	8.0	8.9	9.3	9.9
100	14.0	14.8	15.6	16.5	17.6	3.3	4.2	5.0	5.9	6.8	7.6	8.4	9.1
120	12.3	12.9	13.7	14.3	15.3	2.4	3.1	3.9	4.8	5.6	6.4	7.3	8.0
140	11.5	12.0	12.6	12.8	13.6	2.1	2.4	2.9	3.8	4.6	5.5	6.3	7.0
160	11.6	11.8	12.1	12.3	12.7	2.0	2.2	2.4	2.7	3.5	4.4	5.1	5.9
180	12.2	12.2	12.3	12.5	12.7	1.9	2.0	2.3	2.6	2.9	3.4	3.9	4.9
200	12.4	12.7	12.8	13.1	13.2	2.3	2.2	2.2	2.4	2.7	3.0	3.4	3.8
220	12.3	12.7	13.0	13.3	13.5	3.0	2.6	2.5	2.4	2.5	2.7	3.1	3.5
240	11.5	12.1	12.4	13.1	13.6	3.7	3.3	3.0	2.9	2.7	2.8	2.9	3.2
260	10.4	11.0	11.5	12.2	12.8	4.8	4.1	3.7	3.5	3.1	3.1	3.0	3.1
280	8.8	9.6	10.4	10.7	11.5	5.5	5.1	4.6	4.1	3.8	3.5	3.5	3.4
300	7.5	7.9	8.6	9.0	10.1	6.2	5.8	5.6	5.0	4.8	4.2	3.9	3.8
320	6.7	6.8	7.3	7.8	8.3	6.9	6.6	6.1	5.9	5.4	5.1	4.7	4.3
340	6.6	6.4	6.6	6.7	6.2	7.2	7.1	6.9	6.5	6.2	5.8	5.5	5.1
360	7.4	6.9	6.5	6.5	6.5	7.5	7.4	7.1	7.0	6.8	6.4	6.2	5.8
380	8.9	8.2	7.5	6.9	6.8	7.5	7.6	7.3	7.3	7.2	7.1	6.7	6.5
400	10.9	10.0	9.0	8.3	7.5	7.3	7.3	7.5	7.4	7.4	7.4	7.1	7.0
420	12.6	11.6	10.7	9.9	9.1	6.9	7.0	7.3	7.4	7.4	7.4	7.3	7.5
440	14.2	13.3	12.5	11.6	10.6	6.5	6.8	6.8	7.1	7.2	7.3	7.3	7.4
460	15.7	14.8	13.9	13.0	12.1	6.2	6.2	6.5	6.7	6.8	7.1	7.1	7.3
480	16.9	16.3	15.5	14.5	13.6	5.8	5.9	6.0	6.2	6.4	6.5	7.0	6.9
500	18.1	17.6	16.6	15.8	15.1	5.3	5.4	5.7	5.8	6.0	6.0	6.3	6.6
520	20.1	19.2	18.1	17.4	16.5	5.1	5.1	5.1	5.3	5.4	5.6	5.8	6.0
540	22.0	21.0	20.2	19.2	18.1	4.7	4.8	4.8	4.8	5.0	5.1	5.4	5.5
560	24.3	23.5	22.6	21.5	20.6	4.4	4.5	4.6	4.6	4.7	4.8	4.8	5.0
580	26.6	25.7	24.9	23.8	23.0	4.2	4.3	4.4	4.3	4.5	4.4	4.4	4.5
600	28.5	27.8	27.0	26.3	25.4	4.0	4.2	4.3	4.2	4.2	4.2	4.2	4.3
620	29.6	29.2	28.8	28.2	27.4	4.2	4.0	4.1	4.0	4.0	4.0	4.0	3.9
640	29.9	30.0	29.9	29.5	29.5	4.3	4.2	4.1	4.0	4.1	4.0	3.9	3.9
660	29.2	29.5	29.7	29.8	29.9	4.6	4.4	4.3	4.1	4.1	4.1	4.0	3.8
680	27.5	28.6	28.9	29.2	29.7	4.8	4.6	4.5	4.3	4.2	4.1	4.0	3.9
700	25.1	26.4	27.3	27.8	28.7	5.3	5.0	4.8	4.5	4.6	4.0	4.1	4.1
720	22.6	23.9	25.0	26.1	26.8	5.8	5.5	5.1	5.0	4.7	4.5	4.1	4.1
740	19.5	21.3	22.5	23.6	24.6	6.5	6.1	5.7	5.4	5.2	4.9	4.6	4.3
760	17.3	18.6	19.4	21.0	22.1	7.4	6.7	6.4	6.0	5.6	5.3	5.1	5.0
780	15.0	15.8	17.1	18.5	19.3	8.2	7.6	6.9	6.5	6.4	5.8	5.6	5.3
800	12.2	14.1	14.8	15.9	17.0	9.2	8.5	8.0	7.3	6.8	6.5	6.1	5.8
820	11.4	12.0	12.5	13.4	15.4	10.1	9.6	8.8	8.2	7.6	7.1	6.7	6.5
840	11.2	11.3	11.7	12.2	13.2	10.9	10.4	9.8	9.1	8.4	7.9	7.5	6.9
860	10.9	10.8	10.9	11.2	11.5	11.7	11.0	10.4	10.0	9.4	8.7	8.2	7.7
880	12.6	11.3	11.1	10.8	11.0	12.3	11.9	11.3	10.6	10.2	9.7	8.9	8.4
900	13.3	12.3	12.9	11.3	11.2	12.4	12.2	11.8	11.6	10.8	10.3	9.7	9.3
920	15.7	14.6	13.7	12.8	12.1	12.3	12.3	12.2	11.9	11.6	11.0	10.5	9.9
940	18.4	17.3	16.2	14.5	14.0	12.1	12.1	12.2	12.2	11.8	11.4	11.0	10.6
960	21.0	20.0	18.9	17.9	16.7	11.4	11.9	11.9	12.0	12.0	11.7	11.4	11.0
980	23.0	22.4	21.4	20.3	19.5	10.6	11.1	11.6	11.8	11.9	11.9	11.7	11.4
1000	24.2	23.7	23.1	22.5	21.6	9.5	10.2	10.8	11.2	11.5	11.7	11.8	11.5
	960	970	980	990	1000	0	10	20	30	40	50	60	70

Perturbations produced by Mars

Arguments II and IV.

IV.

II.	70	80	90	100	110	120	130	140	150	160	170	180	190	200
	″	″	″	″	″	″	″	″	″	″	″	″	″	″
0	11.5	11.2	11.0	10.6	10.1	9.9	9.5	9.0	8.6	8.2	8.1	7.8	7.6	7.4
20	11.6	11.4	11.0	10.9	10.6	10.2	9.7	9.1	9.1	8.8	8.4	8.1	7.9	7.8
40	11.2	11.3	11.2	11.0	10.8	10.5	10.3	9.8	9.4	9.3	9.1	8.7	8.4	8.2
60	10.5	10.9	11.1	10.9	11.0	10.9	10.4	10.0	9.7	9.5	9.2	8.8	8.7	8.4
80	9.9	10.0	10.5	10.9	10.8	10.7	10.4	10.3	10.0	9.7	9.3	9.0	8.8	8.6
100	9.1	9.5	9.8	10.1	10.6	10.5	10.4	10.3	10.1	9.9	9.6	9.3	9.0	8.8
120	8.0	8.8	9.3	9.5	9.9	10.2	10.2	10.1	10.0	9.8	9.6	9.4	9.1	8.9
140	7.0	7.9	8.4	9.0	9.3	9.6	9.9	9.9	9.9	9.7	9.7	9.4	9.3	8.9
160	5.9	6.5	7.2	8.0	8.5	8.9	9.2	9.6	9.5	9.6	9.5	9.5	9.3	9.1
180	4.9	5.6	6.4	6.9	7.7	8.3	8.6	8.9	9.4	9.3	9.3	9.3	9.2	9.1
200	3.8	4.6	5.3	6.0	6.7	7.4	7.9	8.3	8.0	8.9	9.1	9.0	9.0	8.9
220	3.5	3.9	4.4	5.1	5.8	6.4	7.1	7.6	7.9	8.4	8.6	8.8	8.8	8.7
240	3.2	3.6	4.0	4.4	5.0	5.5	6.2	6.8	7.4	7.6	8.1	8.4	8.4	8.5
260	3.1	3.2	3.8	4.1	4.5	4.9	5.4	5.9	6.6	7.1	7.5	7.7	8.0	8.2
280	3.4	3.4	3.5	3.8	4.2	4.5	4.9	5.5	5.6	6.2	6.8	7.1	7.5	7.8
300	3.8	3.7	3.7	3.7	3.9	4.4	4.7	4.9	5.4	5.7	6.0	6.6	6.9	7.3
320	4.3	4.2	4.1	4.0	4.1	4.2	4.4	4.7	5.0	5.4	5.8	6.0	6.4	6.6
340	5.1	4.9	4.6	4.4	4.4	4.3	4.5	4.5	5.0	5.2	5.5	5.8	6.0	6.3
360	5.8	5.6	5.3	5.0	4.8	4.8	4.7	4.8	4.9	5.1	5.4	5.5	5.9	6.1
380	6.5	6.4	5.9	5.7	5.5	5.4	5.1	5.1	5.1	5.1	5.4	5.5	5.7	5.8
400	7.0	6.7	6.7	6.3	6.1	5.9	5.7	5.6	5.5	5.5	5.5	5.6	5.7	5.9
420	7.4	7.2	6.9	7.1	6.7	6.4	6.3	6.1	6.0	5.9	5.9	5.8	5.8	6.1
440	7.5	7.4	7.4	7.0	7.1	7.4	6.8	6.7	6.5	6.3	6.3	6.4	6.2	6.3
460	7.3	7.4	7.4	7.5	7.4	7.3	7.3	7.2	7.1	7.1	6.7	6.7	6.7	6.7
480	6.9	7.1	7.3	7.4	7.5	7.3	7.6	7.5	7.4	7.5	7.4	7.2	7.1	7.1
500	6.6	6.8	6.9	7.2	7.3	7.5	7.5	7.6	7.8	7.7	7.8	7.7	7.6	7.4
520	6.0	6.3	6.5	6.7	7.1	7.2	7.5	7.5	7.7	7.8	7.9	7.6	7.9	7.9
540	5.5	5.7	6.0	6.3	6.6	6.9	7.1	7.3	7.4	7.7	7.9	8.0	8.2	8.3
560	5.0	5.2	5.4	5.8	5.9	6.2	6.6	6.9	7.1	7.4	7.7	7.8	8.1	8.2
580	4.5	4.7	4.9	5.0	5.3	5.7	6.0	6.6	6.8	7.1	7.2	7.5	7.9	8.2
600	4.3	4.3	4.4	4.6	4.6	5.0	5.3	5.6	5.9	6.5	6.9	7.0	7.4	7.7
620	3.9	4.0	4.0	4.1	4.3	4.4	4.6	4.9	5.3	5.4	6.1	6.6	6.9	7.4
640	3.9	3.8	3.8	3.8	3.9	3.9	4.1	4.3	4.5	5.0	5.2	5.8	6.3	6.7
660	3.8	3.7	3.7	3.6	3.6	3.7	3.8	3.9	4.1	4.2	4.5	5.0	5.3	6.0
680	3.9	3.8	3.6	3.4	3.5	3.4	3.5	3.5	3.6	3.7	3.8	4.2	4.6	4.9
700	4.1	3.9	3.8	3.6	3.5	3.3	3.3	3.2	3.2	3.2	3.5	3.6	3.8	4.2
720	4.1	4.1	4.0	3.8	3.6	3.5	3.3	3.2	3.3	3.2	3.0	3.2	3.4	3.6
740	4.3	4.3	4.2	4.0	3.8	3.7	3.5	3.2	3.0	3.0	2.9	2.8	2.9	3.1
760	5.0	4.7	4.4	4.3	4.1	3.8	3.7	3.4	3.1	3.0	2.9	2.7	2.7	2.8
780	5.3	5.1	4.7	4.6	4.4	4.4	4.0	3.8	3.4	3.2	2.9	2.8	2.7	2.5
800	5.8	5.5	5.4	4.8	4.7	4.7	4.5	4.2	3.9	3.5	3.3	2.9	2.8	2.7
820	6.5	6.1	5.8	5.6	5.0	5.0	4.9	4.6	4.3	4.1	3.6	3.3	3.0	2.9
840	6.9	6.7	6.3	6.1	5.8	5.3	5.2	4.9	4.9	4.5	4.2	3.9	3.5	3.1
860	7.7	7.4	6.9	6.6	6.2	6.2	5.5	5.4	5.2	5.0	4.8	4.4	4.1	3.6
880	8.4	7.9	7.6	7.1	6.9	6.4	6.4	5.8	5.7	5.4	5.2	5.0	4.6	4.3
900	9.3	8.7	8.3	7.7	7.4	7.1	6.7	6.6	6.1	6.0	5.6	5.4	5.2	4.9
920	9.9	9.3	8.8	8.4	7.9	7.7	7.3	6.9	6.6	6.3	6.2	6.1	5.6	5.4
940	10.6	10.1	9.5	8.9	8.7	8.2	7.8	7.6	7.2	7.1	6.5	6.5	6.3	5.9
960	11.0	10.7	10.3	9.7	9.1	8.7	8.4	8.0	7.8	7.4	7.2	6.9	6.7	6.5
980	11.4	11.0	10.6	10.2	9.8	9.2	8.9	8.4	8.1	8.0	7.6	7.3	7.2	6.9
1000	11.5	11.2	11.0	10.6	10.0	9.9	9.5	9.0	8.6	8.2	8.1	7.4	7.6	7.4
	70	80	90	100	110	120	130	140	150	160	170	180	190	200

TABLE XXXI.

Perturbations produced by Mars.

Arguments II. and IV.

IV.

II.	200	210	220	230	240	250	260	270	280	290	300	310	320
	″	″	″	″	″	″	″	″	″	″	″	″	″
0	7.4	7.2	7.0	6.6	6.4	6.2	5.7	5.3	4.9	4.7	4.1	3.8	3.4
20	7.8	7.2	7.3	7.2	7.0	6.6	6.3	6.0	5.7	5.3	5.0	4.4	3.9
40	8.2	8.1	7.6	7.5	7.3	7.2	6.8	6.6	6.2	5.9	5.6	5.2	4.7
60	8.4	8.0	7.9	7.8	7.6	7.5	7.3	7.1	6.8	6.4	6.1	5.8	5.4
80	8.6	8.5	8.2	8.0	7.6	7.7	7.6	7.4	7.1	7.0	6.7	6.3	6.0
100	8.8	8.5	8.6	8.4	8.2	7.6	7.7	7.8	7.6	7.3	7.2	6.9	6.6
120	8.9	8.7	8.4	8.4	8.3	8.3	8.0	7.9	7.7	7.6	7.5	7.3	7.0
140	8.9	8.7	8.4	8.3	8.2	8.1	8.3	8.0	7.9	7.8	7.7	7.5	7.4
160	9.1	8.9	8.7	8.4	8.3	8.3	8.2	8.1	8.0	7.9	7.9	7.7	7.6
180	9.1	8.8	8.7	8.5	8.4	8.2	8.0	8.0	8.1	7.9	7.8	8.0	7.8
200	8.9	8.8	8.6	8.4	8.4	8.3	8.1	8.0	7.9	7.8	7.8	7.9	7.9
220	8.7	8.7	8.6	8.4	8.2	8.1	8.0	7.9	7.8	7.7	7.7	7.6	7.7
240	8.5	8.4	8.5	8.3	8.1	8.0	7.8	7.8	7.8	7.8	7.8	7.8	7.6
260	8.2	8.2	8.1	8.1	8.1	7.8	7.8	7.7	7.6	7.6	7.6	7.5	7.4
280	7.8	7.8	8.0	7.8	7.9	7.9	7.7	7.5	7.5	7.3	7.3	7.4	7.3
300	7.3	7.6	7.5	7.6	7.7	7.6	7.6	7.6	7.4	7.3	7.1	7.0	7.1
320	6.6	7.1	7.3	7.4	7.4	7.3	7.4	7.4	7.3	7.1	7.0	7.0	6.8
340	6.3	6.4	6.7	7.2	7.1	7.2	7.2	7.1	7.1	7.0	6.9	6.8	6.8
360	6.1	6.2	6.4	6.5	6.9	6.9	7.0	7.0	6.9	6.8	6.7	6.6	6.5
380	5.8	6.1	6.3	6.4	6.6	6.7	6.6	6.6	6.7	6.8	6.7	6.6	6.5
400	5.9	6.0	6.2	6.3	6.4	6.5	6.6	6.6	6.5	6.6	6.6	6.5	6.4
420	6.1	6.3	6.2	6.4	6.3	6.4	6.5	6.6	6.5	6.5	6.5	6.5	6.4
440	6.3	6.4	6.4	6.6	6.5	6.6	6.5	6.5	6.5	6.5	6.3	6.3	6.2
460	6.7	6.5	6.5	6.6	6.7	6.9	6.7	6.6	6.6	6.6	6.5	6.3	6.2
480	7.1	7.1	7.0	6.9	6.9	6.9	7.0	7.0	6.8	6.7	6.6	6.5	6.3
500	7.4	7.5	7.4	7.4	7.3	7.2	7.3	7.2	7.1	6.9	6.8	6.8	6.6
520	7.9	7.8	7.8	7.8	7.8	7.6	7.6	7.5	7.5	7.4	7.1	7.0	6.9
540	8.3	8.3	8.3	8.2	8.2	8.1	8.0	7.9	7.9	7.8	7.6	7.5	7.2
560	8.2	8.6	8.4	8.6	8.7	8.5	8.5	8.4	8.2	8.3	8.2	8.0	7.6
580	8.2	8.3	8.6	8.8	8.8	9.0	8.9	8.9	8.7	8.7	8.6	8.4	8.4
600	7.7	8.1	8.5	8.6	8.9	9.1	9.1	9.2	9.2	9.1	9.0	8.8	8.7
620	7.4	7.6	8.0	8.5	8.7	9.0	9.2	9.5	9.5	9.5	9.4	9.3	9.2
640	6.7	7.2	7.5	7.9	8.3	8.7	9.0	9.3	9.5	9.8	9.8	9.7	9.7
660	6.0	6.3	7.0	7.3	7.7	8.2	8.7	9.0	9.4	9.7	9.8	10.1	10.0
680	4.9	5.6	6.0	6.6	7.1	7.7	8.1	8.5	9.0	9.3	9.8	10.0	10.2
700	4.2	4.5	5.2	5.8	6.4	6.8	7.4	8.0	8.5	8.9	9.2	9.8	10.1
720	3.6	3.9	4.3	4.7	5.3	5.9	6.6	7.0	7.8	8.3	8.8	9.1	9.7
740	3.1	3.3	3.6	3.9	4.4	4.8	5.6	6.2	6.9	7.5	8.0	8.7	9.2
760	2.8	2.8	3.0	3.3	3.6	4.0	4.4	5.1	5.8	6.5	7.2	7.8	8.4
780	2.5	2.6	2.5	2.7	3.1	3.3	3.7	4.1	4.8	5.4	6.1	6.9	7.6
800	2.7	2.5	2.5	2.5	2.5	2.7	3.0	3.4	3.8	4.4	5.0	5.6	6.6
820	2.9	2.6	2.4	2.3	2.2	2.3	2.6	2.8	3.1	3.4	4.1	4.7	5.4
840	3.1	2.8	2.6	2.4	2.3	2.2	2.3	2.4	2.6	2.8	3.2	3.8	4.3
860	3.6	3.3	3.0	2.7	2.4	2.3	2.1	2.2	2.3	2.5	2.7	3.0	3.4
880	4.3	3.8	3.6	3.2	2.8	2.5	2.3	2.1	2.0	2.2	2.3	2.5	2.6
900	4.9	4.6	4.2	3.6	3.4	2.9	2.6	2.3	2.2	2.2	2.1	2.2	2.4
920	5.4	5.1	4.6	4.5	3.9	3.5	3.2	2.9	2.6	2.2	2.0	2.1	2.2
940	5.9	5.7	5.3	4.9	4.7	4.3	3.8	3.4	3.0	2.7	2.4	2.1	2.0
960	6.5	6.2	5.9	5.5	5.1	4.9	4.5	4.0	3.4	3.1	2.8	2.4	2.3
980	6.9	6.8	6.4	6.1	5.8	5.4	5.1	4.8	4.3	3.9	3.5	3.0	2.7
1000	7.4	7.2	7.0	6.6	6.4	6.2	5.7	5.3	4.9	4.7	4.1	3.8	3.4
	200	210	220	230	240	250	260	270	280	290	300	310	320

Perturbations produced by Mars.

Arguments II. and IV.

IV.

II.	320	330	340	350	360	370	380	390	400	410	420	430	440
	″	″	″	″	″	″	″	″	″	″	″	″	″
0	3.4	2.8	2.6	2.4	2.2	2.3	2.3	2.5	2.7	2.9	3.4	4.0	4.5
20	3.9	3.5	3.1	2.7	2.6	2.4	2.4	2.3	2.5	2.7	3.0	3.3	3.8
40	4.7	4.2	3.9	3.5	3.0	2.8	2.7	2.6	2.5	2.6	2.8	2.9	3.2
60	5.4	5.0	4.6	4.2	3.8	3.4	3.1	2.8	2.8	2.7	2.7	2.7	3.0
80	6.0	5.7	5.4	4.8	4.4	4.0	3.6	3.4	3.1	2.9	2.9	2.9	2.9
100	6.6	6.3	5.9	5.6	5.2	4.8	4.3	4.0	3.7	3.5	3.2	3.0	3.0
120	7.0	6.9	6.4	6.1	5.8	5.3	5.2	4.6	4.3	4.0	3.8	3.6	3.4
140	7.4	7.2	6.9	6.6	6.5	6.1	5.6	5.4	5.0	4.6	4.3	4.0	3.9
160	7.6	7.5	7.3	7.0	6.8	6.6	6.2	5.9	5.5	5.3	4.9	4.6	4.4
180	7.8	7.7	7.5	7.4	7.3	6.9	6.7	6.5	6.2	5.8	5.6	5.3	5.0
200	7.9	7.8	7.7	7.6	7.5	7.3	7.1	6.9	6.6	6.4	6.1	5.6	5.5
220	7.7	7.7	7.7	7.8	7.7	7.5	7.3	7.2	7.0	6.7	6.5	6.2	5.9
240	7.6	7.6	7.6	7.6	7.7	7.6	7.5	7.3	7.2	7.1	6.9	6.6	6.4
260	7.4	7.3	7.5	7.5	7.5	7.6	7.6	7.5	7.5	7.3	7.1	7.0	6.7
280	7.3	7.4	7.3	7.3	7.4	7.4	7.3	7.4	7.3	7.5	7.2	7.1	6.9
300	7.1	7.1	7.1	7.0	7.2	7.3	7.3	7.3	7.2	7.2	7.3	7.2	7.1
320	6.8	6.8	6.9	6.9	6.8	7.0	7.1	7.1	7.1	7.1	7.1	7.0	7.2
340	6.8	6.7	6.6	6.6	6.6	6.8	6.9	6.9	7.0	7.0	6.9	6.9	6.9
360	6.5	6.5	6.4	6.3	6.4	6.5	6.6	6.7	6.8	6.8	6.8	6.8	6.9
380	6.5	6.3	6.3	6.2	6.2	6.2	6.3	6.3	6.4	6.5	6.6	6.7	6.7
400	6.4	6.2	6.2	6.0	6.1	6.0	6.0	6.0	6.0	6.1	6.2	6.3	6.4
420	6.4	6.2	6.1	6.0	5.9	5.8	5.9	5.9	5.9	5.9	5.9	6.0	6.0
440	6.2	6.1	6.0	5.8	5.8	5.7	5.6	5.6	5.6	5.7	5.7	5.8	5.9
460	6.2	6.0	5.9	5.8	5.7	5.5	5.5	5.4	5.5	5.4	5.5	5.3	5.4
480	6.3	6.2	6.0	5.7	5.6	5.5	5.4	5.3	5.2	5.2	5.2	5.3	5.3
500	6.6	6.4	6.2	6.0	5.7	5.4	5.3	5.2	5.1	5.1	5.1	5.0	5.0
520	6.9	6.7	6.4	6.1	6.1	5.7	5.5	5.1	5.1	5.0	4.9	5.0	4.9
540	7.2	7.1	6.7	6.5	6.2	6.1	5.8	5.5	5.2	5.0	4.9	4.8	4.8
560	7.6	7.4	7.3	7.0	6.6	6.3	6.0	5.8	5.4	5.3	5.0	4.7	4.7
580	8.4	8.0	7.8	7.5	7.0	6.8	6.3	6.2	5.9	5.5	5.3	5.0	4.9
600	8.7	8.6	8.3	8.0	7.8	7.4	7.0	6.6	6.3	6.0	5.6	5.3	5.1
620	9.2	9.1	8.9	8.6	8.4	8.1	7.6	7.2	6.8	6.5	6.1	5.7	5.3
640	9.7	9.6	9.4	9.3	9.0	8.7	8.2	7.8	7.4	7.0	6.6	6.3	5.8
660	10.0	10.0	9.9	9.8	9.6	9.3	8.9	8.5	8.2	7.7	7.2	6.8	6.4
680	10.2	10.4	10.3	10.2	10.1	9.9	9.6	9.3	9.0	8.5	8.1	7.5	7.1
700	10.1	10.3	10.5	10.6	10.4	10.3	10.1	9.8	9.6	9.3	8.9	8.3	7.8
720	9.7	10.1	10.3	10.6	10.7	10.6	10.5	10.5	10.2	10.0	9.6	9.2	8.6
740	9.2	9.6	10.0	10.3	10.6	10.7	10.8	10.9	10.6	10.5	10.2	9.9	9.4
760	8.4	9.0	9.5	9.8	10.2	10.6	10.9	11.0	11.0	11.0	10.7	10.5	10.3
780	7.6	8.2	8.9	9.4	9.9	10.3	10.6	10.9	11.1	11.2	11.0	10.8	10.7
800	6.6	7.3	7.9	8.5	9.2	9.8	10.1	10.6	10.8	11.1	11.3	11.1	11.0
820	5.4	6.0	7.0	7.6	8.2	8.9	9.6	10.0	10.5	10.8	11.0	11.3	11.3
840	4.3	5.0	5.6	6.5	7.2	7.9	9.8	9.2	9.9	10.3	10.7	10.9	11.2
860	3.4	4.0	4.6	5.3	6.1	6.9	7.5	8.4	9.1	9.6	10.1	10.7	10.9
880	2.6	3.1	3.7	4.3	5.0	5.7	6.6	7.1	8.1	8.7	9.4	9.8	10.4
900	2.4	2.7	3.0	3.4	4.0	4.6	5.4	6.1	6.9	7.6	8.4	9.1	9.7
920	2.2	2.3	2.3	2.8	3.3	3.7	4.3	5.0	5.8	6.5	7.2	8.0	8.7
940	2.0	2.1	2.3	2.3	2.7	2.9	3.4	4.1	4.7	5.5	6.1	7.0	7.7
960	2.3	2.2	2.2	2.3	2.3	2.5	2.8	3.2	3.9	4.5	5.1	5.7	6.5
980	2.7	2.4	2.2	2.3	2.3	2.4	2.5	2.8	3.0	3.6	4.1	4.7	5.5
1000	3.4	2.8	2.6	2.4	2.2	2.3	2.3	2.5	2.7	2.9	3.4	4.0	4.5
	320	330	340	350	360	370	380	390	400	410	420	430	440

Perturbations produced by Mars.

Arguments II and IV.

IV.

II.	440	450	460	470	480	490	500	510	520	530	540	550	560
	"	"	"	"	"	"	"	"	"	"	"	"	"
0	4.5	5.2	5.9	6.6	7.3	8.0	8.5	9.0	9.5	10.0	10.4	10.7	10.9
20	3.8	4.3	4.9	5.6	6.2	6.9	7.6	8.2	8.8	9.3	9.7	10.0	11.4
40	3.2	3.7	4.2	4.8	5.4	5.9	6.6	7.3	7.9	8.4	8.9	9.4	9.8
60	3.0	3.2	3.6	4.0	4.5	5.1	5.7	6.3	6.9	7.5	8.0	8.6	9.1
80	2.9	3.1	3.3	3.5	3.9	4.4	4.9	5.4	5.9	6.5	7.1	7.7	8.2
100	3.0	3.1	3.2	3.5	3.6	3.8	4.2	4.8	5.3	5.9	6.4	6.9	7.4
120	3.4	3.3	3.3	3.4	3.5	3.6	3.9	4.2	4.7	5.1	5.6	6.0	6.6
140	3.9	3.8	3.6	3.6	3.6	3.7	4.0	4.0	4.2	4.6	5.0	5.4	5.9
160	4.4	4.2	3.9	4.1	3.8	3.7	4.0	4.1	4.2	4.5	4.6	4.9	5.3
180	5.0	4.8	4.4	4.2	4.2	4.2	4.0	4.1	4.3	4.4	4.4	4.7	5.0
200	5.5	5.2	5.1	4.8	4.6	4.5	4.5	4.4	4.5	4.5	4.7	4.6	4.8
220	5.9	5.7	5.5	5.3	5.1	4.9	4.9	4.8	4.7	4.8	4.8	4.9	5.0
240	6.4	6.2	5.9	5.8	5.6	5.4	5.3	5.2	5.1	5.1	5.1	5.2	5.2
260	6.7	6.6	6.4	6.1	6.0	5.9	5.8	5.7	5.6	5.5	5.4	5.4	5.4
280	6.9	6.8	6.7	6.5	6.3	6.2	6.1	6.0	5.9	5.9	5.9	5.8	5.8
300	7.1	7.0	6.8	6.8	6.6	6.5	6.4	6.3	6.2	6.2	6.2	6.2	6.2
320	7.2	7.1	6.9	6.8	6.8	6.7	6.6	6.5	6.5	6.5	6.5	6.6	6.6
340	6.9	6.9	7.0	6.9	6.9	6.8	6.7	6.8	6.7	6.6	6.7	6.8	6.9
360	6.9	6.8	6.8	6.8	6.8	6.7	6.7	6.6	6.6	6.8	6.8	6.8	6.9
380	6.7	6.5	6.5	6.6	6.7	6.6	6.6	6.7	6.7	6.7	6.8	6.9	6.9
400	6.4	6.4	6.3	6.3	6.4	6.5	6.5	6.5	6.6	6.7	6.7	6.8	6.8
420	6.0	6.2	6.3	6.3	6.2	6.2	6.3	6.3	6.3	6.3	6.5	6.6	6.7
440	5.9	5.9	6.0	6.0	6.0	6.0	6.0	6.1	6.0	6.1	6.2	6.2	6.4
460	5.4	5.5	5.7	5.8	5.8	5.8	5.8	5.8	5.8	5.8	5.9	6.0	6.1
480	5.3	5.3	5.5	5.5	5.5	5.6	5.5	5.6	5.4	5.6	5.7	5.5	5.8
500	5.0	5.0	5.1	5.2	5.3	5.3	5.3	5.2	5.2	5.2	5.3	5.4	5.4
520	4.9	4.9	4.9	4.8	5.0	5.1	5.1	5.1	5.1	5.1	5.0	5.0	5.1
540	4.8	4.8	4.7	4.8	4.8	4.9	4.9	5.0	4.9	4.8	4.8	4.9	4.8
560	4.7	4.6	4.6	4.7	4.7	4.6	4.7	4.7	4.7	4.7	4.6	4.6	4.6
580	4.9	4.6	4.5	4.5	4.6	4.5	4.4	4.4	4.5	4.5	4.5	4.4	4.4
600	5.1	4.9	4.6	4.5	4.4	4.4	4.4	4.3	4.3	4.3	4.3	4.3	4.3
620	5.3	5.1	4.9	4.7	4.6	4.4	4.3	4.1	4.2	4.2	4.2	4.2	4.1
640	5.8	5.4	5.2	5.0	4.7	4.6	4.4	4.1	4.1	4.1	4.2	4.2	4.0
660	6.4	6.0	5.7	5.4	5.0	4.8	4.7	4.5	4.3	4.2	4.2	4.1	4.0
680	7.1	6.6	6.2	5.7	5.4	5.1	4.9	4.7	4.5	4.4	4.3	4.0	3.9
700	7.8	7.2	6.8	6.4	6.0	5.6	5.3	5.0	4.7	4.6	4.6	4.3	4.1
720	8.6	8.0	7.6	7.1	6.6	6.2	5.7	5.5	5.2	4.9	4.6	4.6	4.3
740	9.4	9.0	8.4	8.0	7.4	6.9	6.3	6.0	5.6	5.3	5.0	4.7	4.5
760	10.3	9.7	9.3	8.6	8.1	7.6	7.2	6.5	6.2	5.8	5.5	5.2	4.9
780	10.7	10.5	9.9	9.6	9.0	8.5	7.8	7.4	7.0	6.4	6.1	5.7	5.5
800	11.0	11.0	10.6	10.2	9.9	9.3	8.8	8.1	7.7	7.3	6.7	6.3	5.8
820	11.3	11.1	10.9	10.6	10.3	10.0	9.6	9.1	8.5	7.9	7.4	7.0	6.6
840	11.2	11.3	11.2	11.1	11.0	10.7	10.2	9.9	9.4	8.8	8.2	7.7	7.3
860	10.9	11.1	11.4	11.3	11.3	11.2	10.7	10.4	9.9	9.6	9.2	8.5	7.9
880	10.4	10.8	11.0	11.3	11.2	11.2	11.2	10.9	10.5	10.3	9.8	9.3	8.7
900	9.7	10.1	10.6	11.0	11.2	11.2	11.2	11.0	10.9	10.7	10.2	10.0	9.4
920	8.7	9.3	9.9	10.3	10.8	11.0	11.1	11.2	11.2	11.0	10.7	10.4	10.1
940	7.7	8.2	8.8	9.5	10.1	10.4	10.9	11.0	11.2	11.2	11.0	10.7	10.5
960	6.5	7.3	8.1	8.6	9.3	9.8	10.2	10.6	10.8	11.1	11.2	10.9	10.8
980	5.5	6.2	7.0	7.7	8.3	8.9	9.5	10.0	10.4	10.6	10.8	11.0	10.9
1000	4.5	5.2	5.9	6.6	7.3	8.0	8.5	9.0	9.5	10.0	10.4	10.7	10.9
	440	450	460	470	480	490	500	510	520	530	540	550	560

Perturbations produced by Mars.

Arguments II and IV.

IV.

II.	560	570	580	590	600	610	620	630	640	650	660	670	680
	"	"	"	"	"	"	"	"	"	"	"	"	"
0	10 9	10.8	10.6	10.4	10.3	10.0	9.7	9.2	8.9	8.5	8.1	7.9	7.7
20	11.4	10.6	10.7	10.6	10.4	10.2	9.9	9.7	9.3	9.0	8.8	8.5	8.1
40	9.8	10.1	10.4	10.4	10.5	10.3	10.2	9.9	9.6	9.4	9.1	8.9	8.5
60	9.1	9.4	9.8	10.2	10.2	10.3	10.2	10.1	9.9	9.6	9 3	9.0	8.8
80	8.2	8.7	9.0	9.3	9.6	9.8	10.0	9.9	9.8	9.7	9.5	9.3	9 1
100	7.4	7.9	8.4	8.7	9.0	9.4	9.6	9.7	9.8	9.7	9.7	9.5	9.2
120	6.6	6.9	7.6	8.1	8.3	8.6	9.0	9.2	9.4	9.5	9.5	9.4	9.3
140	5.9	6.3	6.8	7.2	7.7	8.0	8.3	8.7	8.9	9.1	9.2	9.3	9.3
160	5.3	5.8	6.0	6.5	6.9	7.4	7.7	8.0	8.4	8.5	8.8	8.9	9.0
180	5.0	5.2	5.6	6.0	6.3	6.7	7.1	7.2	7.7	8.1	8.1	8.4	8.6
200	4.8	5.0	5.3	5.4	5.8	6.1	6.5	6.7	7.1	7.3	7.7	7.8	8.0
220	5.0	5.0	5.1	5.3	5.5	5.7	6.0	6.3	6.6	6.8	7.0	7.3	7.5
240	5.2	5.2	5.3	5.3	5.4	5.5	5.7	5.9	6.1	6.4	6.6	6.8	7.1
260	5.4	5.5	5.5	5.5	5.5	5.5	5.5	5.7	5.8	6.0	6.3	6.4	6.5
280	5.8	5.8	5.8	5.9	5.8	5.8	5.8	5.9	5.9	5.9	6.0	6.1	6.2
300	6.2	6.1	6.2	6.1	6.1	6.1	6.2	6.1	6.0	5.9	5.9	6.0	6.1
320	6.6	6.5	6.6	6.6	6.5	6.5	6.6	6.5	6.5	6.3	6.1	6.0	6.0
340	6.9	6.9	6.9	7.0	7.0	6.9	6.8	6.9	6.9	6.8	6.6	6.5	6.3
360	6.9	7.0	7.2	7.3	7.3	7.3	7.4	7.3	7.3	7.1	7.1	7.0	6.7
380	6.9	7.0	7.2	7.4	7.5	7.6	7.7	7.7	7.7	7.6	7.5	7.4	7.2
400	6.8	7.0	7.1	7.3	7.6	7.9	8.0	8.0	8.1	8.1	8.1	7.9	7.8
420	6.7	6.9	7.0	7.2	7.6	7.8	8.0	8.2	8.3	8.4	8.4	8.5	8.4
440	6.4	6.6	6.9	7.0	7.3	7.5	7.9	8.2	8.4	8.6	8.8	8.8	8.9
460	6.1	6.2	6.5	6.9	7.1	7.2	7.6	8.0	8.4	8.7	9.0	9.1	9.2
480	5.8	5.9	6.0	6.2	6.7	7.1	7.2	7.6	7.9	8.5	8.9	9.2	9.3
500	5.4	5.5	5.6	5.9	6.1	6.4	6.9	7.2	7.7	7.9	8.4	9.0	9.4
520	5.1	5.2	5.2	5.3	5.6	5.9	6.3	6.7	7.0	7.6	8.0	8.4	9.0
540	4.8	4.8	4.8	5.0	5.1	5.4	5.6	6.0	6.4	6.7	7.5	8.1	8.5
560	4.6	4.5	4.5	4.5	4.7	4.8	5.0	5.3	5.8	6.2	6.6	7.1	7.8
580	4.4	4.3	4.3	4.3	4.3	4.3	4.5	4.7	5.2	5.5	5.9	6.4	6.9
600	4.3	4.3	4.2	4.1	4.0	4.0	4.1	4.2	4.5	4.8	5.1	5.7	6.2
620	4.1	4.0	4.0	3.9	3.9	3.8	3.8	3.8	3.8	4.0	4.4	4.9	5.4
640	4.0	3.9	4.0	3.8	3.8	3.8	3.7	3.5	3.5	3.6	3.8	4.0	4.5
660	4.0	4.0	3.9	3.8	3.7	3.5	3.5	3.4	3.3	3.3	3.4	3.5	3.7
680	3.9	4.0	3.9	3.8	3.6	3.5	3.4	3.3	3.2	3.1	3.1	3.1	3.1
700	4.1	3.9	3.9	3.9	3.7	3.5	3.4	3.3	3.2	3.0	3.0	3.0	2.9
720	4.3	4.1	4.0	3.9	3.8	3.8	3.5	3.4	3.1	2.9	2.9	2.7	2.7
740	4.5	4.2	4.2	4.2	4.0	3.7	3.6	3.4	3.3	3.0	2.8	2.6	2.5
760	4.9	4.7	4.5	4.3	4.2	4.1	3.8	3.6	3.3	3.1	2.9	2.8	2.5
780	5.5	5.1	4.9	4.5	4.4	4.3	4.1	3.9	3.8	3.4	3.2	3.0	2.7
800	5.8	5.6	5.2	5.0	4.6	4.5	4.4	4.3	4.1	3.8	3.5	3.1	2.8
820	6.6	6.1	5.8	5.5	5.3	5.0	4.8	4.6	4.4	4.2	4.0	3.6	3.3
840	7.3	6.8	6.5	6.1	5.7	5.5	5.2	5.0	4.7	4.6	4.3	4.1	3.8
860	7.9	7.5	7.0	6.7	6.4	5.9	5.8	5.4	5.1	5.0	4.8	4.6	4.4
880	8.7	8.2	7.8	7.3	6.9	6.6	6.3	6.0	5.7	5.4	5.2	5.0	4.7
900	9.4	9.0	8.5	8.0	7.6	7.2	6.8	6.6	6.3	5.9	5.6	5.4	5.2
920	10.1	9.8	9.2	8.7	8.3	7.8	7.4	7.0	6.7	6.4	6.0	5.8	5.7
940	10.5	10.2	9.8	9.4	8.8	8.5	8.0	7.6	7.3	6.9	6.6	6.2	6 1
960	10.8	10.5	10.2	10.0	9.5	9.1	8.6	8.2	7.8	7.5	7.1	6.8	6 6
980	10.9	10.7	10.3	10.2	9.9	9.6	9.2	9.0	8.5	8.0	7.7	7.4	7 2
1000	10.9	10.8	10.6	10.4	10.3	10.0	9.7	9.2	8.9	8.5	8.1	7.9	7.7
	560	570	580	590	600	610	620	630	640	650	660	670	680

Perturbations produced by Mars.

Arguments II. and IV.

IV.

II.	680	690	700	710	720	730	740	750	760	770	780	790	800
	″	″	″	″	″	″	″	″	″	″	″	″	″
0	7.7	7.4	6.9	6.8	6.7	6.4	6.1	5.8	5.5	5.2	4.8	4.4	3.7
20	8.1	7.8	7.4	7.0	7.1	6.9	6.7	6.4	6.1	5.8	5.5	5.1	4.7
40	8.5	8.3	7.8	7.5	7.2	7.1	7.0	6.9	6.6	6.4	6.1	5.8	5.3
60	8.8	8.6	8.3	8.1	7.8	7.6	7.5	7.4	7.1	6.9	6.7	6.3	6.0
80	9.1	8.9	8.7	8.4	8.1	8.0	7.8	7.6	7.4	7.3	7.1	6.9	6.5
100	9.2	8.9	8.8	8.7	8.6	8.3	8.0	7.7	7.6	7.6	7.6	7.3	7.0
120	9.3	9.2	9.0	8.7	8.6	8.4	8.2	8.1	7.9	7.8	7.7	7.6	7.5
140	9.3	9.2	9.0	9.0	8.7	8.5	8.4	8.3	8.0	7.8	7.7	7.7	7.7
160	9.0	9.0	8.9	8.8	8.7	8.6	8.5	8.4	8.2	8.0	7.9	7.8	7.8
180	8.6	8.6	8.7	8.7	8.7	8.6	8.5	8.3	8.3	8.0	8.2	7.8	7.9
200	8.0	8.2	8.3	8.3	8.5	8.4	8.4	8.4	8.2	8.1	8.1	8.1	7.9
220	7.5	7.7	7.9	8.1	8.2	8.2	8.1	8.2	8.2	8.0	8.1	8.0	8.0
240	7.1	7.2	7.4	7.5	7.6	7.7	7.8	7.8	7.9	8.0	8.0	7.8	7.8
260	6.5	6.7	6.9	7.1	7.2	7.3	7.4	7.5	7.6	7.6	7.7	7.7	7.8
280	6.2	6.3	6.5	6.7	6.7	6.9	7.1	7.2	7.3	7.3	7.3	7.3	7.4
300	6.1	6.0	6.2	6.4	6.4	6.5	6.6	6.7	6.9	6.9	6.9	7.1	7.1
320	6.0	6.0	6.0	6.0	6.2	6.1	6.2	6.3	6.5	6.5	6.6	6.6	6.8
340	6.3	6.2	6.0	6.0	6.0	6.0	6.1	6.1	6.2	6.2	6.3	6.3	6.4
360	6.7	6.6	6.4	6.1	6.0	5.9	6.0	5.9	5.9	5.9	6.0	6.1	6.2
380	7.2	7.1	6.8	6.6	6.4	6.2	6.1	5.9	5.8	5.7	5.6	5.8	5.9
400	7.8	7.7	7.4	7.1	6.8	6.6	6.4	6.1	6.0	5.8	5.6	5.5	5.6
420	8.4	8.2	8.0	7.8	7.5	7.2	6.8	6.5	6.2	6.0	5.7	5.5	5.4
440	8.9	8.8	8.7	8.4	8.2	7.8	7.5	7.1	6.6	6.2	6.0	5.7	5.6
460	9.2	9.2	9.2	9.0	8.8	8.5	8.2	7.9	7.5	6.9	6.5	6.3	6.0
480	9.3	9.5	9.6	9.6	9.4	9.2	9.1	8.6	8.3	7.8	7.2	6.9	6.5
500	9.4	9.6	9.8	10.0	9.9	9.8	9.6	9.4	9.1	8.7	8.2	7.6	7.2
520	9.0	9.5	9.8	10.1	10.2	10.3	10.3	10.0	9.8	9.5	9.1	8.5	8.0
540	8.5	9.1	9.5	10.0	10.3	10.5	10.6	10.6	10.4	10.1	9.8	9.5	9.0
560	7.8	8.5	9.0	9.5	9.9	10.4	10.8	10.8	10.9	10.8	10.6	10.2	9.9
580	6.9	7.6	8.3	9.0	9.7	10.0	10.4	10.7	11.1	11.2	11.0	11.0	10.6
600	6.2	6.8	7.4	8.0	8.9	9.6	10.1	10.4	10.9	11.3	11.4	11.3	11.2
620	5.4	5.9	6.5	7.1	7.8	8.6	9.4	10.3	10.6	11.0	11.5	11.7	11.7
640	4.5	5.0	5.5	6.2	6.8	7.6	8.4	9.2	10.0	10.7	11.1	11.6	11.8
660	3.7	4.1	4.7	5.2	5.9	6.5	7.3	8.3	9.1	9.8	10.5	11.2	11.5
680	3.1	3.4	3.8	4.3	4.8	5.5	6.2	7.0	7.8	8.7	9.6	10.2	11.0
700	2.9	2.8	3.0	3.4	3.9	4.5	5.2	6.0	6.7	7.5	8.5	9.4	10.1
720	2.7	2.6	2.5	2.7	3.1	3.5	4.0	4.8	5.6	6.4	7.3	8.2	9.1
740	2.5	2.4	2.4	2.4	2.5	2.7	3.1	3.6	4.5	5.2	6.1	6.9	7.8
760	2.5	2.3	2.2	2.1	2.1	2.3	2.4	2.8	3.2	4.1	4.7	5.7	6.6
780	2.7	2.5	2.3	2.1	2.0	1.9	2.1	2.2	2.5	2.9	3.6	4.4	5.2
800	2.8	2.7	2.4	2.2	2.0	1.8	1.8	1.8	2.0	2.3	2.5	3.2	4.0
820	3.3	3.0	2.7	2.3	2.1	1.9	1.8	1.5	1.7	1.7	2.0	2.2	2.9
840	3.8	3.5	3.0	2.6	2.3	2.1	1.9	1.6	1.5	1.5	1.6	1.7	2.2
860	4.4	4.0	3.5	3.2	2.8	2.3	1.9	1.7	1.4	1.3	1.2	1.4	1.6
880	4.7	4.4	4.1	3.7	3.3	3.0	2.5	2.1	1.7	1.4	1.3	1.2	1.2
900	5.2	5.0	4.6	4.3	4.0	3.6	3.2	2.7	2.2	1.6	1.3	1.2	1.1
920	5.7	5.3	5.1	5.0	4.6	4.2	3.8	3.4	2.9	2.3	1.9	1.3	1.1
940	6.1	5.9	5.6	5.4	5.2	4.8	4.5	3.9	3.5	3.1	2.6	2.1	1.5
960	6.6	6.4	6.2	5.9	5.6	5.4	5.1	4.7	4.3	3.7	3.2	2.8	2.3
980	7.2	6.9	6.6	6.4	6.2	5.9	5.6	5.3	5.0	4.6	4.0	3.5	3.0
1000	7.7	7.4	6.9	6.8	6.7	6.4	6.1	5.8	5.5	5.2	4.8	4.4	3.7
	680	690	700	710	720	730	740	750	760	770	780	790	800

TABLE XXXI.

Perturbations produced by Mars

Arguments II. and IV.

IV.

II.	800	810	820	830	840	850	860	870	880	890	900	910	920
	″	″	″	″	″	″	″	″	″	″	″	″	″
0	3.7	3.2	2.6	2.1	1.7	1.3	0.9	0.7	0.7	1.0	1.2	1.6	2.2
20	4.7	4.2	3.6	3.1	2.4	1.9	1.5	1.2	0.8	0.6	0.9	1.2	1.5
40	5.3	4.9	4.5	3.8	3.3	2.7	2.0	1.7	1.4	1.0	0.8	0.9	1.0
60	6.0	5.7	5.2	4.7	4.1	3.6	3.1	2.6	2.0	1.5	1.2	0.9	1.0
80	6.5	6.3	6.0	5.5	5.0	4.6	4.0	3.4	2.7	2.2	1.8	1.5	1.3
100	7.0	6.7	6.5	6.3	5.9	5.3	4.9	4.4	3.7	3.1	2.5	2.1	1.7
120	7.5	7.3	7.0	6.8	6.5	6.2	5.7	5.1	4.7	4.1	3.5	2.9	2.4
140	7.7	7.7	7.5	7.3	7.0	6.7	6.4	6.0	5.6	5.1	4.5	3.8	3.3
160	7.8	7.9	7.7	7.6	7.4	7.2	7.0	6.8	6.3	5.8	5.4	4.8	4.2
180	7.9	7.8	7.9	7.9	7.7	7.6	7.5	7.1	7.0	6.6	6.1	5.7	5.2
200	7.9	7.9	7.8	7.9	7.8	7.7	7.6	7.5	7.5	7.1	6.8	6.3	6.1
220	8.0	7.9	7.8	7.8	7.8	7.8	7.8	7.8	7.6	7.5	7.4	7.1	6.7
240	7.8	7.7	7.7	7.7	7.7	7.7	7.8	7.8	7.7	7.6	7.6	7.5	7.2
260	7.8	7.7	7.7	7.6	7.7	7.7	7.7	7.7	7.7	7.7	7.8	7.8	7.6
280	7.4	7.4	7.5	7.5	7.5	7.5	7.5	7.5	7.5	7.6	7.6	7.8	7.7
300	7.1	7.2	7.3	7.3	7.3	7.3	7.3	7.4	7.5	7.4	7.5	7.5	7.7
320	6.8	6.9	6.8	7.0	7.1	7.1	7.1	7.1	7.3	7.3	7.3	7.4	7.4
340	6.4	6.5	6.6	6.6	6.7	6.7	6.8	6.9	7.0	7.1	7.2	7.2	7.2
360	6.2	6.2	6.2	6.3	6.4	6.4	6.5	6.6	6.7	6.7	6.9	6.9	7.1
380	5.9	5.8	5.8	5.9	6.0	6.1	6.2	6.3	6.4	6.4	6.4	6.6	6.8
400	5.6	5.6	5.6	5.7	5.7	5.7	5.8	5.9	5.9	6.0	6.1	6.2	6.4
420	5.4	5.4	5.5	5.5	5.5	5.5	5.5	5.5	5.6	5.6	5.6	5.7	5.8
440	5.6	5.3	5.3	5.3	5.3	5.2	5.2	5.2	5.2	5.1	5.0	5.3	5.5
460	6.0	5.6	5.4	5.3	5.2	5.2	5.1	5.0	5.1	5.2	5.2	5.2	5.3
480	6.5	6.0	5.7	5.4	5.2	5.2	5.1	4.9	4.9	4.9	4.9	5.0	5.0
500	7.2	6.8	6.3	5.9	5.6	5.3	5.0	4.8	4.9	4.8	4.8	4.8	4.9
520	8.0	7.4	7.0	6.5	6.1	5.5	5.4	5.1	4.9	4.7	4.7	4.7	4.8
540	9.0	8.4	7.8	7.3	6.7	6.3	5.8	5.4	5.2	4.9	4.7	4.7	4.7
560	9.9	9.5	8.8	8.2	7.7	7.1	6.5	6.0	5.7	5.3	5.0	4.8	4.6
580	10.6	10.2	9.8	9.3	8.8	8.1	7.2	6.8	6.4	6.0	5.6	5.1	4.9
600	11.2	11.0	10.7	10.3	9.6	9.1	8.5	7.7	7.1	6.4	6.1	5.6	5.3
620	11.7	11.5	11.4	11.0	10.6	9.9	9.5	8.9	8.1	7.4	6.8	6.3	5.9
640	11.8	11.9	11.8	11.7	11.3	11.0	10.4	9.8	9.3	8.5	7.8	7.1	6.6
660	11.5	11.8	12.0	12.1	11.9	11.6	11.2	10.8	10.2	9.6	8.9	8.2	7.5
680	11.0	11.6	12.1	12.2	12.1	12.2	12.1	11.5	11.1	10.6	10.1	9.2	8.5
700	10.1	10.9	11.6	12.1	12.4	12.3	12.3	12.3	11.9	11.4	10.8	10.4	9.7
720	9.1	10.0	10.6	11.4	11.9	12.4	12.6	12.5	12.4	12.0	11.6	11.2	10.8
740	7.8	8.8	9.7	10.5	11.3	11.8	12.3	12.8	12.6	12.6	12.3	11.9	11.5
760	6.6	7.6	8.5	9.4	10.3	11.0	11.7	12.1	12.6	12.8	12.7	12.5	12.1
780	5.2	6.3	7.1	8.1	9.2	10.1	10.7	11.6	12.0	12.4	12.8	12.9	12.8
800	4.0	4.8	5.7	6.7	7.7	8.7	9.7	10.5	11.3	11.9	12.3	12.5	12.9
820	2.9	3.6	4.4	5.4	6.4	7.3	8.4	9.5	10.3	11.0	11.7	12.1	12.5
840	2.2	2.7	3.3	4.0	4.9	6.0	7.0	8.0	9.1	10.0	10.8	11.4	12.0
860	1.6	1.6	2.2	2.9	3.6	4.6	5.6	6.6	7.6	8.6	9.6	10.5	11.2
880	1.2	1.3	1.5	1.9	2.6	3.3	4.1	5.2	6.1	7.1	8.2	9.2	10.1
900	1.1	1.1	1.2	1.3	1.7	2.2	2.9	3.8	4.8	5.7	6.8	7.9	8.8
920	1.1	1.0	1.0	1.1	1.1	1.4	1.9	2.6	3.4	4.4	5.3	6.3	7.4
940	1.5	1.1	0.8	0.9	1.0	1.1	1.3	1.6	2.3	3.1	3.9	5.0	5.9
960	2.3	1.7	1.3	0.9	0.7	0.8	0.9	1.2	1.4	2.0	2.8	3.5	4.6
980	3.0	2.5	1.9	1.4	1.2	1.0	0.8	0.9	1.2	1.4	1.7	2.4	3.3
1000	3.7	3.2	2.6	2.1	1.7	1.3	0.9	0.7	0.7	1.0	1.2	1.6	2.2
	800	810	820	830	840	850	860	870	880	890	900	910	920

TABLE XXXI. | TABLE XXXII.

Perturbations by Mars. Arguments II. and IV. — *Pert's. by Jupiter* Arg's. II. and V.

II.	IV. 920	930	940	950	960	970	980	990	1000	V. 0	10	20	30
	″	″	″	″	″	″	″	″	″	″	″	″	″
0	2.2	3.0	3.8	4.8	5.8	6.9	7.8	8.4	9.5	15.3	15.1	15.0	15.0
20	1.5	2.1	2.6	3.4	4.4	5.5	6.5	7.6	8.7	14.9	14.9	14.7	14.8
40	1.0	1.4	1.8	2.5	3.2	4.0	5.2	6.0	7.1	14.7	14.6	14.6	14.5
60	1.0	1.1	1.3	1.8	2.3	3.0	3.7	4.8	5.8	14.4	14.4	14.4	14.4
80	1.3	1.1	1.2	1.4	1.6	2.2	2.7	3.6	4.5	13.4	13.9	14.0	14.2
100	1.7	1.3	1.2	1.2	1.3	1.6	2.0	2.6	3.3	13.2	13.4	13.6	13.7
120	2.4	2.0	1.5	1.4	1.4	1.4	1.7	1.9	2.4	12.3	12.7	13.0	13.3
140	3.3	2.8	2.3	2.0	1.7	1.5	1.5	1.8	2.1	11.3	11.8	12.1	12.5
160	4.2	3.6	3.1	2.6	2.1	2.0	1.7	1.7	1.9	10.2	10.7	11.2	11.7
180	5.2	4.6	4.0	3.5	3.1	2.5	2.0	2.0	1.9	9.1	9.6	10.1	10.7
200	6.1	5.5	5.0	4.4	3.9	3.5	2.8	2.7	2.9	7.8	8.3	8.9	9.5
220	6.7	6.3	5.8	5.4	4.9	4.4	3.9	3.2	3.0	6.8	7.2	7.7	8.3
240	7.2	6.9	6.6	6.1	5.6	5.3	4.8	4.2	3.7	5.7	6.2	6.6	7.2
260	7.6	7.5	7.1	6.8	6.5	6.0	5.6	5.2	4.8	4.8	5.2	5.6	6.1
280	7.7	7.7	7.5	7.3	7.1	6.7	6.3	5.9	5.5	3.9	4.1	4.7	5.2
300	7.7	7.7	7.7	7.7	7.4	7.2	7.0	6.6	6.1	3.4	3.5	3.9	4.3
320	7.4	7.4	7.6	7.7	7.6	7.6	7.3	7.1	6.9	3.2	3.1	3.4	3.6
340	7.2	7.2	7.3	7.5	7.7	7.6	7.6	7.6	7.7	3.2	3.0	3.0	3.1
360	7.1	7.1	7.1	7.2	7.2	7.6	7.6	7.6	7.5	3.5	3.2	2.9	2.9
380	6.8	6.9	7.0	7.0	7.0	7.1	7.3	7.5	7.5	4.5	4.0	3.4	3.1
400	6.4	6.6	6.6	6.7	6.7	6.9	7.0	7.1	7.3	5.0	4.3	3.8	3.5
420	5.8	5.9	6.2	6.3	6.6	6.5	6.7	6.7	6.9	6.1	5.2	4.6	4.1
440	5.5	5.6	5.7	5.8	6.0	6.1	6.3	6.5	6.5	7.5	6.6	5.8	4.9
460	5.3	5.3	5.4	5.7	5.7	5.7	5.9	6.1	6.2	9.0	7.9	7.0	6.3
480	5.0	5.0	5.0	5.1	5.3	5.4	5.5	5.6	5.8	10.5	9.5	8.5	7.6
500	4.9	4.9	5.0	5.0	5.0	5.1	5.2	5.3	5.3	12.3	11.3	10.0	9.1
520	4.8	4.8	4.8	4.8	4.8	4.7	4.9	5.0	5.1	14.0	12.7	11.7	10.7
540	4.7	4.7	4.6	4.6	4.6	4.5	4.6	4.6	4.7	15.6	14.5	13.3	12.3
560	4.6	4.5	4.5	4.4	4.5	4.5	4.5	4.5	4.4	17.1	16.1	15.1	14.0
580	4.9	4.7	4.6	4.5	4.4	4.4	4.4	4.4	4.2	18.6	17.4	16.5	15.7
600	5.3	4.9	4.8	4.7	4.5	4.4	4.4	4.3	4.1	19.8	19.0	17.9	17.0
620	5.9	5.5	5.1	4.8	4.6	4.5	4.4	4.3	4.2	20.8	20.1	19.2	18.4
640	6.6	6.1	5.6	5.4	5.0	4.7	4.6	4.5	4.3	21.6	20.9	20.2	19.5
660	7.5	6.8	6.3	5.9	5.5	5.3	4.9	4.8	4.6	22.1	21.6	21.0	20.4
680	8.5	7.8	7.3	6.5	6.1	5.6	5.4	5.1	4.8	22.3	22.0	21.6	21.2
700	9.7	8.9	8.1	7.6	7.0	6.3	5.9	5.6	5.3	22.2	22.0	21.7	21.5
720	10.8	10.0	9.3	8.5	7.9	7.2	6.6	6.1	5.8	22.0	21.9	21.7	21.6
740	11.5	11.0	10.2	9.7	8.9	8.2	7.6	6.9	6.5	21.6	21.6	21.5	21.5
760	12.1	11.8	11.3	10.5	10.0	9.3	8.5	7.9	7.3	21.2	21.1	21.1	21.2
780	12.8	12.3	11.9	11.4	10.9	10.2	9.6	9.0	8.2	20.4	20.5	20.6	20.7
800	12.9	12.9	12.5	12.1	11.7	11.2	10.5	9.8	9.2	19.6	19.8	19.9	20.1
820	12.5	12.7	12.8	12.7	12.2	11.9	11.2	10.7	10.1	18.8	19.0	19.2	19.4
840	12.0	12.4	12.6	12.8	12.6	12.4	12.2	11.5	10.9	18.1	18.2	18.4	18.6
860	11.2	11.8	12.3	12.5	12.7	12.5	12.5	12.3	11.7	17.4	17.5	17.6	17.9
880	10.1	11.0	11.5	12.1	12.3	12.6	12.6	12.4	12.3	16.9	16.9	16.9	17.1
900	8.8	9.8	10.6	11.3	11.8	12.2	12.4	12.5	12.4	16.3	16.4	16.4	16.5
920	7.4	8.4	9.3	10.2	11.0	11.5	12.1	12.2	12.3	16.0	15.9	15.9	16.0
940	5.9	7.1	8.1	8.9	9.9	10.7	11.2	11.7	12.1	15.8	15.7	15.7	15.6
960	4.6	5.6	6.7	7.7	8.7	9.4	10.2	10.9	11.4	15.5	15.4	15.3	15.4
980	3.3	4.2	5.2	6.2	7.3	8.2	8.9	9.9	10.6	15.3	15.2	15.2	15.1
1000	2.2	3.0	3.8	4.8	5.8	6.9	7.8	8.7	9.5	15.3	15.1	15.0	15.0
	920	930	940	950	960	970	980	990	1000	0	10	20	30

Perturbations produced by Jupiter.

Arguments II. and V.

V.

II.	30	40	50	60	70	80	90	100	110	120	130	140	150
	″	″	″	″	″	″	″	″	″	″	″	″	″
0	15.0	14.8	14.7	14.7	14.6	14.5	14.5	14.4	14.5	14.5	14.6	14.7	14.8
20	14.8	14.7	14.6	14.4	14.4	14.2	14.2	14.1	14.1	14.1	14.1	14.1	14.2
40	14.5	14.4	14.4	14.3	14.2	14.1	13.9	13.8	13.8	13.8	13.8	13.8	13.7
60	14.4	14.3	14.3	14.2	14.1	13.9	13.8	13.6	13.5	13.5	13.5	13.4	13.3
80	14.2	14.2	14.1	14.5	14.0	13.8	13.7	13.5	13.4	13.2	13.1	13.0	13.1
100	13.7	13.7	13.9	13.9	13.8	13.7	13.6	13.5	13.4	13.2	13.0	12.8	12.7
120	13.3	13.4	13.4	13.5	13.6	13.5	13.5	13.3	13.3	13.2	13.0	12.8	12.6
140	12.5	12.8	13.0	13.1	13.2	13.2	13.3	13.2	13.1	13.0	12.9	12.8	12.6
160	11.7	12.0	12.4	12.6	12.7	12.8	12.9	12.9	13.0	12.9	12.8	12.7	12.5
180	10.7	11.1	11.6	11.9	12.2	12.3	12.5	12.5	12.6	12.7	12.8	12.6	12.5
200	9.5	10.0	10.6	11.0	11.5	11.7	11.9	12.2	12.2	12.3	12.4	12.3	12.3
220	8.3	8.8	9.5	9.9	10.4	10.8	11.3	11.5	11.8	11.9	12.0	12.0	12.0
240	7.2	7.7	8.2	8.9	9.4	9.8	10.3	10.6	11.0	11.3	11.5	11.7	11.8
260	6.1	6.5	7.1	7.6	8.3	8.8	9.3	9.7	10.1	10.5	10.9	11.0	11.2
280	5.2	5.5	6.0	6.5	7.1	7.6	8.2	8.7	9.2	9.6	10.0	10.4	10.6
300	4.3	4.7	5.1	5.5	6.1	6.6	7.1	7.6	8.1	8.7	9.1	9.4	9.9
320	3.6	3.9	4.3	4.6	5.1	5.4	6.0	6.6	7.2	7.7	8.1	8.5	8.9
340	3.1	3.3	3.5	3.8	4.1	4.5	5.0	5.4	6.1	6.6	7.2	7.6	8.0
360	2.9	3.0	3.1	3.3	3.6	3.8	4.1	4.5	5.0	5.5	6.1	6.6	7.1
380	3.1	2.8	2.8	2.7	2.8	2.9	3.0	3.2	3.5	4.1	4.6	5.0	5.6
400	3.5	3.1	2.9	2.9	2.8	2.8	3.0	3.1	3.4	3.8	4.2	4.7	5.2
420	4.1	3.6	3.3	3.1	2.8	2.7	2.8	2.9	3.1	3.2	3.5	3.8	4.3
440	4.9	4.4	3.9	3.4	3.1	2.7	2.8	2.7	2.8	3.1	3.1	3.2	3.5
460	6.3	5.4	4.8	4.3	3.7	3.2	2.9	2.8	2.8	2.7	2.7	2.8	3.2
480	7.6	6.7	5.9	5.2	4.6	4.1	3.6	3.1	3.0	2.8	2.8	2.6	2.7
500	9.1	8.1	7.2	6.4	5.7	5.0	4.4	3.9	3.4	3.2	3.1	2.9	2.7
520	10.7	9.5	8.7	7.7	6.9	6.1	5.5	4.8	4.2	3.8	3.5	3.2	3.1
540	12.3	11.1	10.2	9.1	8.4	7.4	6.6	5.9	5.3	4.7	4.1	3.8	3.5
560	14.0	13.0	11.9	10.8	9.9	8.7	7.9	7.1	6.4	5.8	5.2	4.5	4.1
580	15.7	14.5	13.6	12.5	11.4	10.4	9.3	8.3	7.7	6.9	6.2	5.5	5.0
600	17.0	16.0	15.0	14.0	13.1	12.0	11.0	10.1	9.2	8.2	7.5	6.7	6.0
620	18.4	17.4	16.5	15.5	14.7	13.6	12.6	11.6	10.7	9.8	9.0	8.0	7.3
640	19.5	18.5	17.9	17.0	16.0	15.1	14.2	13.1	12.2	11.3	10.8	9.4	8.7
660	20.4	19.7	18.9	18.1	17.4	16.3	15.6	14.6	13.7	12.8	11.9	11.0	10.1
680	21.2	20.5	19.9	19.1	18.5	17.6	16.8	16.0	15.1	14.2	13.5	12.5	11.6
700	21.5	21.0	20.6	20.0	19.3	18.7	18.0	17.1	16.5	15.6	14.7	13.8	13.0
720	21.6	21.2	21.0	20.5	20.0	19.3	18.9	18.3	17.5	16.8	16.1	15.1	14.3
740	21.5	21.2	21.1	20.8	20.5	20.0	19.4	18.9	18.4	17.7	17.2	16.8	15.7
760	21.2	21.0	21.0	20.8	20.7	20.3	20.0	19.4	19.0	18.6	17.9	17.4	16.7
780	20.7	20.7	20.7	20.6	20.6	20.3	20.2	19.8	19.4	19.1	18.7	18.1	17.6
800	20.1	20.2	20.3	20.3	20.4	20.3	20.1	19.9	19.7	19.3	19.1	18.7	18.2
820	19.4	19.5	19.7	19.8	19.9	19.9	19.9	19.8	19.8	19.6	19.2	18.9	18.7
840	18.6	18.8	18.9	19.0	19.2	19.3	19.4	19.4	19.4	19.4	19.4	19.0	18.9
860	17.9	18.0	18.3	18.4	18.6	18.7	18.8	18.9	19.0	19.1	19.1	19.0	18.8
880	17.1	17.2	17.5	17.6	17.9	18.0	18.2	18.3	18.5	18.6	18.6	18.6	18.7
900	16.5	16.6	16.8	16.9	17.1	17.1	17.4	17.5	17.7	17.9	18.1	18.2	18.2
920	16.0	16.0	16.1	16.2	16.4	16.5	16.7	16.8	17.0	17.2	17.4	17.5	17.7
940	15.6	15.5	15.6	15.6	15.7	15.8	16.0	16.1	16.3	16.5	16.8	16.8	17.1
960	15.4	15.3	15.3	15.2	15.2	15.2	15.3	15.4	15.6	15.7	15.9	16.0	16.3
980	15.1	15.0	15.0	14.9	14.9	14.8	14.9	14.9	14.9	15.0	15.2	15.3	15.5
1000	15.0	14.8	14.7	14.7	14.6	14.5	14.5	14.4	14.5	14.5	14.6	14.7	14.8
	30	40	50	60	70	80	90	100	110	120	130	140	150

TABLE XXXII.

Perturbations produced by Jupiter.

Arguments II. and V.

V.

II.	150	160	170	180	190	200	210	220	230	240	250	260	270
	"	"	"	"	"	"	"	"	"	"	"	"	"
0	14.8	15.0	15.3	15.5	15.8	15.9	16.2	16.3	16.7	17.0	17.1	17.3	17.5
20	14.2	14.3	14.6	14.8	14.9	15.2	15.5	15.7	15.9	16.2	16.6	16.8	17.1
40	13.7	13.7	13.9	14.1	14.3	14.5	14.8	15.0	15.3	15.5	15.8	16.2	16.4
60	13.3	13.2	13.4	13.5	13.6	13.8	14.1	14.3	14.6	14.8	15.1	15.5	15.8
80	13.1	13.0	13.0	13.0	13.1	13.1	13.3	13.5	13.8	14.1	14.4	14.5	15.1
100	12.7	12.7	12.7	12.6	12.7	12.6	12.8	12.9	13.1	13.4	13.7	14.0	14.2
120	12.6	12.5	12.5	12.4	12.3	12.2	12.3	12.3	12.6	12.8	13.0	13.3	13.6
140	12.6	12.4	12.4	12.3	12.1	12.0	12.0	12.0	12.1	12.1	12.3	12.5	12.8
160	12.5	12.3	12.2	12.1	12.1	11.9	11.8	11.8	11.8	11.8	11.9	12.0	12.2
180	12.5	12.3	12.2	12.1	11.9	11.8	11.7	11.5	11.5	11.5	11.6	11.7	11.8
200	12.3	12.2	12.2	12.0	11.9	11.7	11.7	11.5	11.4	11.3	11.2	11.3	11.5
220	12.0	12.0	12.1	12.0	11.8	11.6	11.6	11.5	11.4	11.3	11.2	11.1	11.1
240	11.8	11.8	11.9	11.9	11.8	11.6	11.5	11.4	11.3	11.2	11.1	11.1	11.0
260	11.2	11.5	11.6	11.6	11.6	11.5	11.3	11.3	11.3	11.2	11.1	11.0	10.9
280	10.6	10.8	11.1	11.2	11.2	11.2	11.3	11.3	11.2	11.2	11.1	11.0	10.9
300	9.9	10.1	10.5	10.8	10.9	11.0	11.1	11.0	11.0	11.0	11.0	11.1	10.9
320	8.9	9.4	9.7	10.1	10.4	10.5	10.7	10.8	10.8	10.8	10.8	10.8	10.9
340	8.0	8.5	9.1	9.3	9.6	9.9	10.2	10.3	10.5	10.6	10.6	10.7	10.7
360	7.1	7.5	8.0	8.4	8.9	9.2	9.5	9.8	10.1	10.3	10.4	10.5	10.5
380	5.6	6.2	6.8	7.3	7.8	8.3	8.9	9.3	9.7	10.0	10.0	10.1	10.2
400	5.2	5.6	6.2	6.6	7.0	7.5	7.9	8.4	8.8	9.1	9.4	9.7	9.9
420	4.3	4.8	5.3	5.8	6.2	6.6	7.1	7.4	7.9	8.4	8.7	9.1	9.4
440	3.5	3.9	4.4	4.9	5.4	5.7	6.2	6.7	7.1	7.6	7.9	8.4	8.7
460	3.2	3.3	3.8	4.1	4.5	4.9	5.4	5.7	6.3	6.7	7.2	7.7	8.0
480	2.7	2.9	3.2	3.6	3.9	4.3	4.7	5.0	5.4	5.9	6.3	6.8	7.3
500	2.7	2.7	2.9	3.1	3.4	3.6	4.0	4.4	4.8	5.2	5.7	5.9	6.4
520	3.1	2.8	2.9	3.0	3.1	3.2	3.5	3.8	4.2	4.7	4.9	5.4	5.7
540	3.5	3.2	3.1	3.0	3.0	3.0	3.3	3.5	3.7	4.1	4.3	4.7	5.1
560	4.1	3.8	3.6	3.3	3.2	3.2	3.2	3.3	3.5	3.7	4.0	4.3	4.5
580	5.0	4.6	4.2	4.0	3.6	3.5	3.3	3.2	3.4	3.5	3.7	4.0	4.2
600	6.0	5.4	5.1	4.6	4.3	3.9	3.7	3.5	3.5	3.6	3.7	3.8	4.0
620	7.3	6.6	6.0	5.6	5.1	4.6	4.2	4.0	3.9	3.8	3.9	3.9	4.0
640	8.7	7.8	7.3	6.6	6.1	5.5	5.2	4.7	4.4	4.2	4.0	4.0	4.1
660	10.1	9.3	8.6	7.7	7.2	6.5	6.2	5.9	5.3	4.9	4.6	4.5	4.4
680	11.6	10.8	10.0	9.3	8.5	7.5	7.3	6.7	6.3	5.8	5.5	5.2	4.9
700	13.0	12.1	11.5	10.7	9.9	9.0	8.5	7.8	7.4	6.9	6.3	6.0	5.8
720	14.3	13.5	12.8	12.1	11.3	10.6	9.8	9.1	8.7	8.0	7.6	7.0	6.6
740	15.7	14.9	14.2	13.4	12.7	12.0	11.2	10.5	9.7	9.3	8.9	8.2	7.7
760	16.7	15.9	15.5	14.7	13.9	13.3	12.6	11.8	11.2	10.5	10.0	9.5	9.0
780	17.6	17.0	16.4	15.7	15.1	14.6	13.8	13.2	12.6	11.9	11.2	10.8	10.2
800	18.2	17.8	17.3	16.8	16.2	16.0	15.0	14.3	13.7	13.1	12.6	12.0	11.5
820	18.7	18.3	18.0	17.6	17.0	16.6	16.0	15.3	14.9	14.3	13.7	13.1	12.6
840	18.9	18.7	18.4	18.2	17.7	17.2	16.8	16.3	15.8	15.3	14.9	14.4	13.8
860	18.8	18.7	18.6	18.4	18.3	17.9	17.4	17.1	16.7	16.3	15.9	15.4	15.0
880	18.7	18.5	18.6	18.5	18.3	18.2	18.0	17.7	17.4	17.1	16.6	16.3	15.9
900	18.2	18.2	18.3	18.3	18.3	18.1	18.1	18.0	17.8	17.6	17.3	17.0	16.7
920	17.7	17.9	18.0	18.0	18.1	18.1	18.0	18.0	18.0	17.8	17.7	17.6	17.3
940	17.1	17.1	17.4	17.6	17.6	17.7	17.8	17.8	17.9	18.0	17.8	17.8	17.7
960	16.3	16.5	16.8	16.9	17.1	17.2	17.4	17.5	17.6	17.8	17.9	18.0	17.9
980	15.5	15.7	16.1	16.3	16.5	16.7	16.8	17.0	17.2	17.3	17.6	17.7	17.9
1000	14.8	15.0	15.3	15.5	15.8	15.9	16.2	16.3	16.7	17.0	17.1	17.3	17.5
	150	160	170	180	190	200	210	220	230	240	250	260	270

Perturbations produced by Jupiter.

Arguments II. and V

V

II.	270	280	290	300	310	320	330	340	350	360	370	380	390
	″	″	″	″	″	″	″	″	″	″	″	″	″
0	17.5	17.5	17.7	17.8	17.9	17.9	18.0	18.0	17.9	17.7	17.6	17.5	17.5
20	17.1	17.3	17.5	17.6	17.8	17.8	18.0	18.1	18.1	18.1	18.0	18.0	18.0
40	16.4	16.8	16.9	17.2	17.6	17.7	17.9	18.1	18.3	18.3	18.4	18.4	18.6
60	15.8	16.0	16.4	16.7	16.9	17.3	17.6	17.9	18.2	18.3	18.5	18.5	18.7
80	15.1	15.4	15.7	16.1	16.4	16.7	17.0	17.5	17.8	18.0	18.3	18.5	18.8
100	14.2	14.6	15.1	15.0	15.8	16.1	16.5	17.0	17.2	17.5	17.9	18.3	18.7
120	13.6	13.7	14.2	14.5	15.0	15.4	15.8	16.2	16.7	17.1	17.3	17.9	18.3
140	12.8	13.1	13.3	13.7	14.2	14.4	15.1	15.5	15.9	16.3	16.8	17.3	17.7
160	12.2	12.4	12.6	12.9	13.4	13.8	14.1	14.6	15.2	15.5	16.0	16.5	17.1
180	11.8	11.9	12.1	12.3	12.5	12.8	13.3	13.7	14.4	14.7	15.2	15.7	16.3
200	11.5	11.5	11.6	11.7	12.0	12.1	12.5	13.0	13.4	13.8	14.3	14.7	15.5
220	11.1	11.1	11.2	11.3	11.6	11.7	11.9	12.3	12.7	13.0	13.5	14.0	14.5
240	11.0	10.9	10.9	11.0	11.2	11.3	11.5	11.8	12.1	12.3	12.8	13.2	13.8
260	10.9	10.8	10.8	10.8	10.9	10.9	11.1	11.3	11.4	11.6	12.0	12.3	13.0
280	10.9	10.8	10.7	10.6	10.7	10.6	10.8	11.0	11.2	11.3	11.5	11.8	12.2
300	10.9	10.8	10.7	10.6	10.6	10.5	10.6	10.7	10.8	10.9	11.1	11.4	11.8
320	10.9	10.7	10.7	10.6	10.6	10.5	10.5	10.6	10.7	10.6	10.7	11.0	11.2
340	10.7	10.7	10.6	10.5	10.5	10.4	10.5	10.5	10.6	10.5	10.6	10.7	10.8
360	10.5	10.5	10.5	10.5	10.5	10.4	10.4	10.4	10.4	10.3	10.5	10.6	10.8
380	10.2	10.3	10.3	10.3	10.4	10.3	10.4	10.4	10.4	10.3	10.3	10.4	10.6
400	9.9	10.0	10.0	10.2	10.3	10.2	10.2	10.3	10.4	10.3	10.3	10.3	10.5
420	9.4	9.6	9.8	9.9	10.1	10.2	10.1	10.2	10.2	10.2	10.3	10.3	10.4
440	8.7	9.0	9.2	9.4	9.7	9.8	10.0	10.1	10.2	10.1	10.1	10.2	10.4
460	8.0	8.4	8.6	8.8	9.1	9.3	9.6	9.9	10.1	10.0	10.0	10.2	10.3
480	7.3	7.6	7.9	8.4	8.7	8.9	9.1	9.4	9.6	9.7	9.8	10.0	10.1
500	6.4	6.9	7.2	7.6	8.0	8.3	8.6	8.9	9.2	9.4	9.5	9.7	9.9
520	5.7	6.1	6.6	6.9	7.3	7.6	7.9	8.3	8.6	8.9	9.1	9.4	9.7
540	5.1	5.4	5.8	6.2	6.7	7.0	7.4	7.7	8.0	8.3	8.6	8.9	9.2
560	4.5	4.9	5.1	5.5	6.0	6.3	6.7	7.2	7.5	7.7	8.0	8.3	8.7
580	4.2	4.4	4.8	5.0	5.3	5.7	6.1	6.6	6.9	7.1	7.4	7.7	8.1
600	4.0	4.2	4.3	4.7	4.9	5.2	5.6	6.0	6.3	6.5	6.8	7.2	7.6
620	4.0	4.0	4.1	4.3	4.7	4.8	5.1	5.5	5.8	6.1	6.4	6.7	7.0
640	4.1	4.1	4.2	4.2	4.4	4.6	4.8	5.1	5.4	5.6	5.9	6.3	6.6
660	4.4	4.3	4.3	4.3	4.5	4.5	4.7	4.9	5.1	5.3	5.5	5.8	6.2
680	4.9	4.9	4.7	4.6	4.7	4.5	4.6	4.8	5.0	5.1	5.3	5.5	5.8
700	5.8	5.4	5.2	5.1	5.0	4.9	4.9	4.9	5.1	5.2	5.3	5.4	5.6
720	6.6	6.2	5.9	5.7	5.6	5.5	5.4	5.3	5.3	5.3	5.3	5.4	5.5
740	7.7	7.2	6.8	6.5	6.4	6.1	6.0	5.9	5.8	5.7	5.6	5.5	5.7
760	9.0	8.2	7.9	7.5	7.2	6.9	6.7	6.5	6.3	6.1	5.9	5.9	6.0
780	10.2	9.7	9.1	8.4	8.2	7.7	7.6	7.4	7.2	6.9	6.6	6.5	6.5
800	11.5	11.0	10.4	9.8	9.4	8.7	8.5	8.3	8.0	7.7	7.6	7.3	7.1
820	12.6	12.1	11.7	11.2	10.6	10.1	9.7	9.2	9.1	8.6	8.3	8.1	7.9
840	13.8	13.2	12.8	12.3	11.9	11.3	10.9	10.5	10.2	9.6	9.4	9.1	8.9
860	15.0	14.4	13.8	13.5	13.1	12.6	12.1	11.7	11.2	10.7	10.4	10.1	10.0
880	15.9	15.4	15.0	14.4	14.2	13.7	13.4	12.9	12.5	12.0	11.5	11.3	11.1
900	16.7	16.4	15.9	15.5	15.2	14.8	14.4	14.1	13.7	13.2	12.8	12.4	12.2
920	17.3	17.1	16.8	16.5	16.2	15.7	15.5	15.2	14.8	14.3	14.0	13.6	13.3
940	17.7	17.5	17.3	17.1	16.9	16.6	16.3	16.1	16.0	15.5	15.0	14.7	14.5
960	17.9	17.8	17.6	17.5	17.4	17.2	17.0	16.9	16.8	16.4	16.2	15.8	15.6
980	17.9	17.8	17.8	17.8	17.8	17.8	17.6	17.5	17.3	17.2	17.0	16.8	16.6
1000	17.5	17.7	17.7	17.8	17.9	17.9	18.0	18.0	17.9	17.7	17.6	17.5	17.5
	270	280	290	300	310	320	330	340	350	360	370	380	390

TABLE XXXII.

Perturbations produced by Jupiter.

Arguments II. and V.

V.

II.	390	400	410	420	430	440	450	460	470	480	490	500	510
	″	″	″	″	″	″	″	″	″	″	″	″	″
0	17.5	17.1	17.0	16.7	16.5	16.3	16.1	15.8	15.6	15.1	14.6	14.3	13.9
20	18.0	18.1	17.7	17.5	17.5	17.2	17.1	16.8	16.7	16.3	16.0	15.6	15.3
40	18.6	18.6	18.5	18.4	18.3	18.1	18.0	17.8	17.6	17.3	17.2	16.8	16.5
60	18.7	18.9	18.9	18.9	18.9	18.7	18.8	18.6	18.7	18.4	18.1	17.9	17.7
80	18.8	18.9	19.2	19.3	19.4	19.3	19.3	19.3	19.3	19.2	19.2	18.9	18.8
100	18.7	18.9	19.1	19.4	19.7	19.8	19.8	19.8	19.8	19.8	19.9	19.7	19.7
120	18.3	18.6	18.9	19.2	19.5	19.8	20.0	20.1	20.3	20.3	20 4	20.4	20.4
140	17.7	18.2	18.6	18.9	19.2	19.6	20.0	20.3	20.5	20.6	20.7	20.8	21.0
160	17.1	17.6	17.9	18.5	19.0	19.3	19.8	20.2	20.5	20.6	20.9	21.1	21.2
180	16.3	16.8	17.3	17.9	18.3	18.8	19.3	19.8	20.3	20.6	20.9	21.1	21.4
200	15.5	16.0	16.5	17.1	17.7	18.2	18.6	19.1	19.8	20.2	20.7	21.0	21.4
220	14.5	15.0	15.6	16.1	16.9	17.4	18.0	18.6	19.0	19.7	20.3	20.7	21.1
240	13.8	14.2	14.7	15.2	15.9	16.5	17.1	17.7	18.4	18.9	19.5	20.1	20.7
260	13.0	13.4	13.9	14.4	15.0	15.5	16.3	16.9	17.5	18.0	18.6	19.3	20.0
280	12.2	12.7	13.0	13.5	14.2	14.7	15.3	15.9	16.7	17.2	17.8	18.4	19.1
300	11.8	11.9	12.4	12 8	13.3	13.8	14.4	14.9	15.7	16.3	17.0	17.6	18.2
320	11.2	11.5	11.8	12.2	12.7	13.0	13.6	14.1	14.7	15.3	16.0	16.6	17.4
340	10.8	11.2	11.4	11.6	12.1	12.4	12.9	13.4	13.9	14.4	15.1	15.7	16.4
360	10.8	10.8	11.0	11.2	11.6	11.9	12.3	12.6	13.2	13.6	14.2	14.8	15.5
380	10.6	10.6	10.7	10.9	11.2	11.4	11.9	12.2	12.6	12.9	13.5	13.9	14.5
400	10.5	10.5	10.6	10.6	10.9	11.1	11.4	11.8	12.2	12.5	12.9	13.3	13.8
420	10.4	10.4	10.5	10.6	10.7	10.9	11.2	11.3	11.7	11.9	12.4	12.8	13.3
440	10.4	10.4	10.4	10.5	10.7	10.8	10.9	11.1	11.3	11.6	11.9	12.2	12.7
460	10.3	10.4	10.4	10.4	10.6	10.6	10.7	10.9	11.2	11.3	11.7	11.9	12.2
480	10.1	10.2	10.3	10.4	10.6	10.6	10.7	10.8	11.0	11.2	11.4	11.7	12.0
500	9.9	10.0	10.1	10.2	10.4	10.5	10.7	10.8	10.9	11.0	11.2	11.3	11.7
520	9.7	9.8	9.8	10.0	10.2	10.3	10.5	10.6	10.9	10.8	11.1	11.3	11.5
540	9.2	9.4	9.6	9.8	10.0	10.2	10.3	10.4	10.6	10.7	10.9	11.1	11.4
560	8.7	8.9	9.1	9.3	9.7	9.8	10.1	10.3	10.5	10.6	10.7	10.8	11.2
580	8.1	8.5	8.7	8.7	9.2	9.4	9.7	9.9	10.2	10.4	10.6	10.7	10.9
600	7.6	7.9	8.2	8.5	8.8	9.0	9.3	9.5	9.8	10.0	10.3	10.5	10.7
620	7.0	7.3	7.6	7.9	8.2	8.5	8.8	9.0	9.4	9.6	10.0	10.1	10.4
640	6.6	6.8	7.1	7.4	7.7	7.9	8.2	8.6	8.9	9.1	9.4	9.7	10.1
660	6.2	6.4	6.6	6.9	7.3	7.6	7.9	8.1	8.3	8.6	8.9	9.2	9.5
680	5.8	6.1	6.2	6.5	6.8	7.0	7.4	7.6	7.9	8.1	8.4	8.7	9.0
700	5.6	5.8	6.0	6.2	6.4	6.6	6.9	7.1	7.4	7.6	7.9	8.2	8.5
720	5.5	5.6	5.7	5.9	6.2	6.3	6.5	6.8	7.1	7.2	7.5	7.7	8.0
740	5.7	5.7	5.7	5.8	6.0	6.1	6.2	6.4	6.7	6.9	7.1	7.2	7.5
760	6.0	6.0	6.0	6.0	6.0	6 1	6.2	6.3	6.4	6.5	6.7	6.8	7.1
780	6.5	6.3	6.2	6.2	6.3	6.3	6.3	6.3	6.4	6.4	6.5	6.7	6.8
800	7.1	7.0	6.7	6.6	6.7	6.5	6.5	6.4	6.5	6.5	6.5	6.6	6.7
820	7.9	7.6	7.5	7.3	7.2	7.0	7.0	6.8	6.8	6.7	6.6	6.6	6.7
840	8.9	8.6	8.3	8.1	7.8	7.7	7.6	7.4	7.3	7.1	7.0	6.8	6.8
860	10.0	9.7	9.3	9.0	8.7	8.4	8.2	8.1	7.9	7.7	7.6	7.3	7.2
880	11.1	10.5	10.4	10.0	9.7	9.5	9.2	8.9	8.7	8.4	8.2	7.9	7.7
900	12.2	11.8	11.5	11.0	10.8	10.5	10.3	9.9	9.7	9.4	9.0	8.8	8.5
920	13.3	13.0	12.6	12.3	12.1	11.5	11.3	11.0	10.6	10.2	10.1	9.7	9.4
940	14.5	14.1	13.8	13.5	13.2	12.8	12.5	11.9	11.8	11.3	11.0	10.7	10.4
960	15.6	15.3	14.9	14.6	14.4	14.0	13.7	13.3	13.0	12.5	12.1	11.8	11.5
980	16.6	16.3	16.0	15.7	15.6	15.2	14.9	14.6	14.2	13.8	13.6	12.9	12.7
1000	17.5	17.1	17.0	16.7	16.5	16.3	16.1	15.8	15.6	15.1	14.6	14.3	13.9
	390	400	410	420	430	440	450	460	470	480	490	500	510

Perturbations produced by Jupiter.

Arguments II. and V.

V.

II.	510	520	530	540	550	560	570	580	590	600	610	620	630
	"	"	"	"	"	"	"	"	"	"	"	"	"
0	13.9	13.4	13.1	12.7	12.1	11.8	11.3	10.8	10.2	9.9	9.4	8.9	8.4
20	15.3	14.9	14.4	13.9	13.5	13.1	12.5	12.1	11.5	11.0	10.4	10.0	9.4
40	16.5	16.3	15.7	15.4	15.0	14.3	13.8	13.4	12.8	12.3	11.7	11.1	10.5
60	17.7	17.3	17.0	16.6	16.1	15.8	15.3	14.7	14.3	13.7	13.0	12.4	11.8
80	18.8	18.5	18.1	17.9	17.4	17.1	16.6	16.2	15.7	15.1	14.5	13.9	13.2
100	19.7	19.5	19.2	19.0	18.8	18.4	17.9	17.6	17.0	16.5	16.0	15.2	14.7
120	20.4	20.3	20.2	20.0	19.7	19.5	19.1	18.8	18.4	18.0	17.3	16.8	16.2
140	21.0	21.1	21.0	20.8	20.7	20.4	20.2	19.9	19.6	19.3	18.8	18.3	17.7
160	21.2	21.5	21.5	21.6	21.5	21.3	21.2	21.0	20.6	20.4	20.1	19.6	19.1
180	21.4	21.6	21.8	22.0	22.0	22.1	21.9	21.8	21.6	21.4	21.1	20.7	20.3
200	21.4	21.7	21.9	22.1	22.3	22.5	22.5	22.5	22.4	22.3	22.1	21.8	21.5
220	21.1	21.5	21.8	22.2	22.5	22.8	23.1	23.1	22.9	22.8	22.9	22.6	22.5
240	20.7	21.1	21.5	21.9	22.3	22.7	23.0	23.3	23.4	23.5	23.4	23.3	23.2
260	20.0	20.6	21.0	21.6	22.0	22.4	22.8	23.2	23.5	23.8	23.8	23.8	23.9
280	19.1	19.9	20.4	20.9	21.5	22.0	22.4	23.0	23.3	23.7	24.0	24.1	24.1
300	18.2	19.0	19.6	20.3	20.7	21.3	21.8	22.3	23.0	23.4	23.8	24.1	24.3
320	17.4	18.9	18.7	19.4	20.0	20.6	21.1	21.8	22.3	22.9	23.3	23.7	24.2
340	16.4	17.0	17.6	18.5	19.2	19.9	20.4	21.1	21.6	22.2	22.8	23.3	23.7
360	15.5	16.2	16.7	17.4	18.2	18.9	19.5	20.1	20.8	21.5	22.0	22.6	23.2
380	14.5	15.2	15.9	16.6	17.1	17.9	18.6	19.3	19.8	20.5	21.1	21.8	22.5
400	13.8	14.4	14.9	15.6	16.2	16.8	17.6	18.4	19.1	19.7	20.3	20.9	21.5
420	13.3	13.7	14.2	14.8	15.3	16.0	16.5	17.4	18.0	18.7	19.4	20.0	20.6
440	12.7	13.1	13.6	14.1	14.6	15.2	15.7	16.4	17.1	17.8	18.4	18.9	19.6
460	12.2	12.7	13.0	13.5	13.9	14.4	15.0	15.6	16.1	16.9	17.5	18.2	18.7
480	12.0	12.2	12.5	13.0	13.4	13.9	14.3	14.8	15.3	15.9	16.6	17.3	17.9
500	11.7	12.0	12.2	12.6	12.9	13.3	13.8	14.3	14.7	15.2	15.7	16.4	16.9
520	11.5	11.9	12.0	12.3	12.6	13.0	13.2	13.8	14.2	14.7	15.1	15.5	16.2
540	11.4	11.6	11.9	12.2	12.4	12.7	12.9	13.3	13.7	14.2	14.6	15.0	15.4
560	11.2	11.4	11.5	11.9	12.1	12.4	12.7	13.1	13.4	13.8	14.1	14.5	14.9
580	10.9	11.2	11.4	11.6	11.9	12.2	12.4	12.8	13.1	13.5	13.8	14.2	14.5
600	10.7	10.8	11.1	11.5	11.7	12.0	12.2	12.5	12.8	13.1	13.4	13.8	14.2
620	10.4	10.7	10.7	11.1	11.4	11.6	12.0	12.3	12.5	12.9	13.1	13.4	13.8
640	10.1	10.4	10.6	10.7	11.0	11.3	11.6	12.0	12.3	12.6	12.9	13.2	13.5
660	9.5	9.9	10.2	10.5	10.6	11.0	11.3	11.6	11.9	12.3	12.6	12.9	13.2
680	9.0	9.3	9.6	10.0	10.3	10.5	10.8	11.3	11.5	11.9	12.2	12.4	12.8
700	8.5	8.9	9.1	9.5	9.8	10.1	10.3	10.7	11.1	11.4	11.8	12.1	12.4
720	8.0	8.3	8.5	9.0	9.2	9.6	9.9	10.2	10.5	10.9	11.3	11.7	12.0
740	7.5	7.8	8.0	8.3	8.6	9.0	9.3	9.7	9.9	10.4	10.8	11.1	11.5
760	7.1	7.3	7.5	7.9	8.1	8.4	8.6	9.1	9.4	9.7	10.1	10.5	10.9
780	6.8	7.0	7.1	7.3	7.6	7.9	8.1	8.5	8.8	9.2	9.4	9.8	10.2
800	6.7	6.8	6.8	7.0	7.1	7.3	7.5	7.8	8.2	8.5	8.8	9.1	9.5
820	6.7	6.8	6.6	6.8	6.9	7.0	7.1	7.4	7.6	7.9	8.1	8.4	8.7
840	6.8	6.8	6.8	6.8	6.8	6.9	6.9	7.1	7.2	7.4	7.6	7.9	8.1
860	7.2	7.1	7.1	7.0	6.9	6.9	6.8	6.8	6.9	7.1	7.2	7.3	7.6
880	7.7	7.5	7.4	7.3	7.1	7.0	6.8	6.8	6.7	6.8	6.8	7.0	7.2
900	8.5	8.2	7.9	7.7	7.5	7.3	7.2	7.1	6.9	6.9	6.8	6.8	6.8
920	9.4	9.2	8.7	8.4	8.1	7.9	7.6	7.4	7.1	7.0	6.9	6.8	6.7
940	10.4	10.0	9.7	9.4	8.9	8.6	8.3	8.1	7.7	7.4	7.1	6.9	6.7
960	11.5	11.2	10.7	10.4	9.8	9.5	9.1	8.8	8.5	8.1	7.7	7.4	7.1
980	12.7	12.3	11.8	11.5	11.1	10.6	10.0	9.7	9.2	8.9	8.5	8.1	7.7
1000	13.9	13.4	13.1	12.7	12.1	11.8	11.3	10.8	10.2	9.9	9.4	8.9	8.4
	510	520	530	540	550	560	570	580	590	600	610	620	630

Perturbations produced by Jupiter.

Arguments II. and V.

V.

II.	630	640	650	660	670	680	690	700	710	720	730	740	750
	″	″	″	″	″	″	″	″	″	″	″	″	″
0	8.4	8.0	7.7	7.3	6.9	6.7	6.5	6.5	6.3	6.2	6.2	6.4	6.5
20	9.4	9.0	8.4	8.0	7.5	7.1	6.9	6.7	6.4	6.3	6.0	6.1	6.1
40	10.5	10.1	9.4	8.9	8.3	7.8	7.4	7.0	6.6	6.4	6.2	5.9	5.8
60	11.8	11.3	10.6	10.1	9.3	8.7	8.2	7.7	7.2	6.8	6.4	6.2	5.8
80	13.2	12.7	12.0	11.3	10.5	9.9	9.2	8.7	8.1	7.6	7.1	6.6	6.2
100	14.7	14.1	13.4	12.8	12.0	11.3	10.6	9.9	9.1	8.5	7.9	7.3	6.8
120	16.2	15.4	14.9	14.2	13.4	12.7	12.0	11.3	10.4	9.8	8.9	8.2	7.6
140	17.7	17.2	16.4	15.6	14.9	14.2	13.4	12.7	11.9	11.1	10.2	9.6	8.8
160	19.1	18.6	17.9	17.3	16.6	15.7	15.0	14.2	13.3	12.6	11.7	10.9	10.0
180	20.3	19.9	19.4	18.8	18.0	17.3	16.7	15.8	15.0	14.1	13.2	12.4	11.5
200	21.5	21.2	20.8	20.2	19.3	18.9	18.1	17.5	16.6	15.7	14.9	14.0	13.1
220	22.5	22.3	21.9	21.5	21.0	20.3	19.7	19.0	18.2	17.5	16.6	15.5	14.7
240	23.2	23.0	22.9	22.5	22.0	21.6	21.1	20.5	19.8	19.1	18.2	17.3	16.4
260	23.9	23.8	23.7	23.5	23.1	22.7	22.3	21.8	21.2	20.6	19.8	19.1	18.1
280	24.1	24.3	24.2	24.2	24.0	23.7	23.5	23.1	22.4	21.8	21.2	20.5	19.8
300	24.3	24.5	24.6	24.6	24.5	24.4	24.2	23.9	23.6	23.1	22.5	21.9	21.2
320	24.2	24.5	24.7	24.9	24.8	24.8	24.8	24.7	24.4	24.1	23.7	23.1	22.5
340	23.7	24.2	24.5	24.7	25.0	25.2	25.1	25.0	25.0	24.9	24.6	24.1	23.7
360	23.2	23.7	24.2	24.5	24.7	25.0	25.1	25.3	25.4	25.3	25.1	24.9	24.5
380	22.5	23.1	23.6	24.1	24.4	24.7	25.1	25.2	25.4	25.5	25.4	25.3	25.2
400	21.5	22.3	22.8	23.4	23.9	24.3	24.7	25.1	25.2	25.4	25.6	25.6	25.5
420	20.6	21.3	22.0	22.6	23.1	23.6	24.1	24.5	25.0	25.2	25.4	25.6	25.7
440	19.6	20.3	21.0	21.8	22.3	22.9	23.4	23.9	24.3	24.8	25.0	25.2	25.6
460	18.7	19.4	20.1	20.7	21.3	21.9	22.6	23.3	23.6	24.1	24.6	24.8	25.1
480	17.9	18.5	19.1	19.7	20.3	21.0	21.6	22.2	22.8	23.3	23.8	24.3	24.6
500	16.9	17.6	18.2	18.8	19.3	19.9	20.7	21.4	21.9	22.5	22.9	23.4	23.9
520	16.2	16.8	17.3	17.9	18.4	19.0	19.7	20.4	21.0	21.6	21.1	22.6	23.0
540	15.4	16.1	16.6	17.2	17.5	18.1	18.7	19.3	19.9	20.5	21.2	22.7	22.2
560	14.9	15.4	16.0	16.5	16.9	17.3	17.9	18.4	18.9	19.6	20.1	20.7	21.3
580	14.5	15.0	15.3	15.9	16.3	16.7	17.1	17.6	18.1	18.7	19.3	19.8	20.3
600	14.2	14.6	14.9	15.3	15.8	16.3	16.6	17.0	17.4	17.9	18.3	18.9	19.4
620	13.8	14.2	14.6	14.9	15.1	15.7	16.2	16.6	16.9	17.3	17.6	18.0	18.5
640	13.5	14.0	14.2	14.6	14.8	15.1	15.6	16.1	16.5	16.8	17.1	17.5	17.9
660	13.2	13.5	13.9	14.3	14.6	14.9	15.2	15.6	15.9	16.4	16.6	17.0	17.3
680	12.8	13.2	13.5	13.9	14.2	14.5	14.9	15.2	15.6	16.0	16.2	16.5	16.8
700	12.4	12.9	13.3	13.5	13.8	14.2	14.5	14.9	15.1	15.6	15.9	16.2	16.4
720	12.0	12.4	12.8	13.2	13 5	13.8	14.2	14.5	14.8	15.1	15.5	15.8	16.1
740	11.5	11.9	12.2	12.6	12.9	13.3	13.8	14.2	14.5	14.8	15.1	15.4	15.7
760	10.9	11.4	11.8	12.2	12.4	12.8	13.2	13.7	14.1	14.5	14.7	15.0	15.4
780	10.2	10.6	11.2	11.6	11.9	12.4	12.8	13.2	13.5	13.9	14.3	14.6	14.9
800	9.5	10.0	10.3	10.9	11.3	11.6	12.1	12.6	12.9	13.4	13.8	14.2	14.5
820	8.7	9.3	9.7	10.0	10.5	10.9	11.4	11.9	12.3	12.8	13.2	13.6	14.0
840	8.1	8.4	8.8	9.3	9.6	10.1	10.6	11.1	11.6	12.1	12.5	13.0	13.4
860	7.6	7.9	8.1	8.5	8.8	9.2	9.7	10.2	10.7	11.2	11.7	12.1	12.6
880	7.2	7.4	7.6	7.8	8.1	8.5	8.8	9.4	9.8	10.2	10.7	11.2	11.8
900	6.8	7.0	7.1	7.3	7.4	7.8	8.2	8.5	8.9	9.4	9.8	10.3	10.8
920	6.7	6.8	6.8	6.9	7.0	7.0	7.4	7.8	8.1	8.6	8.9	9.4	9.9
940	6.7	6.7	6.7	6.8	6.7	6.8	6.8	7.1	7.4	7.7	8.1	8.4	8.9
960	7.1	7.0	6.8	6.7	6.5	6.5	6.6	6.7	6.8	7.1	7.3	7.7	8.0
980	7.7	7.4	7.1	6.9	6.6	6.5	6.4	6.4	6.3	6.5	6.8	6.9	7.3
1000	8.4	8.0	7.7	7.3	6.9	6.7	6.5	6.5	6.3	6.2	6.2	6.4	6.5
	630	640	650	660	670	680	690	700	710	720	730	740	750

Perturbations produced by Jupiter.

Arguments II. and V.

V.

II.	750	760	770	780	790	800	810	820	830	840	850	860	870
	″	″	″	″	″	″	″	″	″	″	″	″	″
0	6.5	6.8	7.2	7.5	8.0	8.4	8.8	9.5	10.1	10.5	11.0	11.6	12.4
20	6.1	6.2	6.5	6.7	7.0	7.4	7.9	8.4	9.0	9.5	10.0	10.6	11.1
40	5.8	5.9	5.9	6.2	6.4	6.6	6.9	7.4	7.8	8.2	8.8	9.5	10.0
60	5.8	5.7	5.7	5.7	5.9	6.1	6.2	6.5	6.9	7.2	7.7	8.3	8.8
80	6.2	5.8	5.7	5.6	5.4	5.6	5.7	5.9	6.1	6.3	6.7	7.3	7.8
100	6.8	6.3	5.9	5.6	5.5	5.3	5.3	5.4	5.4	5.6	5.9	6.3	6.8
120	7.6	7.4	6.5	6.0	5.7	5.5	5.1	5.2	5.1	5.1	5.2	5.5	5.8
140	8.8	8.1	7.4	6.8	6.2	5.8	5.4	5.2	5.0	4.9	4.8	5.0	5.1
160	10.0	9.3	8.5	7.8	7.2	6.5	5.9	5.5	5.1	5.9	4.7	4.7	4.7
180	11.5	10.6	9.7	9.0	8.2	7.5	6.9	6.3	5.8	5.2	4.8	4.7	4.5
200	13.1	12.2	11.2	10.4	9.5	8.8	7.9	7.1	6.5	5.9	5.3	5.0	4.7
220	14.7	13.8	12.9	12.0	11.1	10.2	9.3	8.4	7.5	6.7	6.1	5.5	5.2
240	16.4	15.3	14.5	13.6	12.6	11.7	10.7	9.8	8.8	7.9	7.0	6.5	5.9
260	18.1	17.2	16.3	15.3	14.3	13.3	12.2	11.4	10.4	9.4	8.3	7.7	6.9
280	19.8	18.9	17.9	17.0	16.1	15.0	14.0	13.0	11.9	10.9	9.9	8.9	8.0
300	21.2	20.4	19.6	18.7	17.7	16.8	15.8	14.7	13.7	12.6	11.5	10.5	9.4
320	22.5	21.9	21.2	20.4	19.4	18.5	17.4	16.5	15.5	14.2	13.2	12.3	11.2
340	23.7	23.0	22.4	21.8	21.1	20.2	19.2	18.3	17.1	16.1	15.0	13.9	12.9
360	24.5	24.0	23.6	23.0	22.4	21.6	20.8	19.9	18.9	17.9	16.8	15.9	14.7
380	25.2	24.9	24.5	24.0	23.5	22.8	22.1	21.4	20.5	19.5	18.5	17.6	16.5
400	25.5	25.4	25.1	24.8	24.5	23.9	23.4	22.7	21.9	21.0	20.1	19.2	18.2
420	25.7	25.6	25.5	25.3	25.0	24.5	24.2	23.7	23.2	22.3	21.5	20.7	19.8
440	25.6	25.6	25.7	25.7	25.5	25.3	24.9	24.6	24.1	23.4	22.7	22.0	21.2
460	25.1	25.3	25.5	25.6	25.8	25.7	25.4	25.2	24.8	24.3	23.7	23.1	22.5
480	24.6	24.9	25.2	25.4	25.6	25.6	25.5	25.4	25.2	24.9	24.5	24.1	23.5
500	23.9	24.2	24.7	25.0	25.3	25.4	25.5	25.5	25.4	25.2	24.9	24.7	24.3
520	23.0	23.6	23.9	24.3	24.7	24.9	25.2	25.4	25.4	25.3	25.2	25.1	24.8
540	22.2	22.6	23.2	23.6	24.0	24.4	24.6	24.9	25.1	25.0	25.1	25.1	25.0
560	21.3	21.7	22.2	22.8	23.2	23.7	24.0	24.3	24.6	24.7	24.8	24.9	24.9
580	20.3	20.8	21.3	21.8	22.3	22.7	23.2	23.7	23.9	24.1	24.4	24.6	24.7
600	19.4	19.9	20.4	20.8	21.4	21.9	22.2	22.7	23.1	23.4	23.7	24.1	24.3
620	18.5	19.0	19.5	20.1	20.5	20.9	21.4	21.8	22.2	22.6	22.9	23.3	23.6
640	17.9	18.3	18.7	19.2	19.7	20.1	20.5	22.0	21.3	21.7	22.1	22.5	22.8
660	17.3	17.6	18.1	18.5	18.9	19.4	19.6	20.1	20.5	20.7	21.2	21.7	22.0
680	16.8	17.1	17.4	17.8	18.2	18.6	18.9	19.4	19.7	20.1	20.4	20.7	21.2
700	16.4	16.7	16.9	17.3	17.7	18.0	18.3	18.7	18.9	19.2	19.6	20.0	20.3
720	16.1	16.3	16.5	16.9	17.2	17.6	17.8	18.0	18.3	18.5	18.7	19.2	19.5
740	15.7	16.0	16.2	16.5	16.7	17.0	17.3	17.6	17.8	17.9	18.1	18.5	18.8
760	15.4	15.7	16.0	16.1	16.4	16.6	16.7	17.2	17.4	17.4	17.8	18.0	18.2
780	14.9	15.3	15.6	15.9	16.1	16.3	16.5	16.7	16.9	17.1	17.3	17.6	17.7
800	14.5	14.7	15.2	15.5	15.8	15.9	16.2	16.5	16.6	16.8	16.9	17.1	17.3
820	14.0	14.4	14.7	15.1	15.4	15.7	15.8	16.1	16.3	16.4	16.6	16.9	17.0
840	13.4	13.7	14.1	14.5	15.1	15.4	15.4	15.8	15.9	16.1	16.2	16.6	16.7
860	12.6	13.1	13.5	13.9	14.3	14.8	15.2	15.5	15.6	15.8	16.0	16.3	16.4
880	11.8	12.3	12.8	13.3	13.7	14.1	14.5	15.0	15.3	15.4	15.6	15.9	16.1
900	10.8	11.3	11.9	12.4	13.0	13.4	13.7	14.2	14.7	15.0	15.2	15.5	15.7
920	9.9	10.3	10.8	11.4	12.0	12.5	12.9	13.4	14.0	14.3	14.7	15.0	15.3
940	8.9	9.4	9.9	10.4	11.0	11.6	12.1	12.5	13.0	13.6	13.9	14.4	14.7
960	8.0	8.3	8.8	9.4	10.0	10.6	11.1	11.7	12.2	12.5	13.1	13.7	14.1
980	7.3	7.6	7.9	8.4	8.9	9.5	9.9	10.5	11.1	11.6	12.1	12.8	13.3
1000	6.5	6.8	7.2	7.5	8.0	8.4	8.8	9.5	10.0	10.5	11.0	11.6	12.4
	750	760	770	780	790	800	810	820	830	840	850	860	870

Perturbations produced by Jupiter.

Arguments II. and V.

V.

II.	870	880	890	900	910	920	930	940	950	960	970	980	990	1000
	"	"	"	"	"	"	"	"	"	"	"	"	"	"
0	12.4	12.9	13.2	13.6	13.9	14.2	14.4	14.8	15.0	15.1	15.1	15.2	15.2	15.3
20	11.1	11.7	12.2	12.7	13.2	13.6	13.8	14.1	14.4	14.7	14.8	15.0	14.9	14.9
40	10.0	10.5	11.1	11.7	12.3	12.6	13.0	13.4	13.7	14.1	14.3	14.6	14.7	14.7
60	8.8	9.4	9.9	10.6	11.2	11.8	12.1	12.6	12.9	13.3	13.6	13.9	14.2	14.4
80	7.8	8.3	8.7	9.3	10.0	10.5	11.1	11.6	12.1	12.5	12.8	13.2	13.5	13.8
100	6.8	7.2	7.6	8.1	8.6	9.4	9.9	10·5	10.9	11.4	12.0	12.4	12.8	13.2
120	5.8	6.1	6.6	7.1	7.6	8.1	8.7	9.4	9.9	10.4	10.8	11.4	11.8	12.3
140	5.1	5.3	5.6	6.0	6.5	7.0	7.5	8.2	8.7	9.3	9.7	10.3	10.8	11.3
160	4.7	4.8	4.8	5.2	5.6	5.9	6.3	6.8	7.4	8.0	8.6	9.2	9.7	10.2
180	4.5	4.5	4.4	4.5	4.8	5.1	5.4	5.8	6.2	6.9	7.4	8.0	8.4	9.1
200	4.7	4.5	4.2	4.2	4.2	4.4	4.6	5.0	5.3	5.7	6.3	6.9	7.4	7.8
220	5.2	4.7	4.3	4.2	4.1	4.1	4.0	4.3	4.5	4.8	5.1	5.7	6.2	6.8
240	5.9	5.3	4.7	4.3	4.1	4.0	3.8	3.9	4.0	4.2	4.3	4.7	5.2	5.7
260	6.9	6.1	5.4	4.9	4.4	4.1	3.8	3.7	3.6	3.7	3.8	4.1	4.3	4.9
280	8.0	7.2	6.3	5.7	5.2	4.6	4.1	3.8	3.5	3.5	3.5	3.6	3.7	3.9
300	9.4	8.5	7.5	6.8	6.1	5.4	4.7	4.3	3.9	3.6	3.3	3.3	3.3	3.4
320	11.2	10.1	9.1	8.1	7.3	6.5	5.7	5.0	4.4	4.0	3.6	3.4	3.2	3.2
340	12.9	11.8	10.7	9.6	8.7	7.7	6.8	6.0	5.2	4.6	4.1	3.7	3.4	3.2
360	14.7	13.4	12.3	11.1	10.1	9.2	8.3	7.4	6.4	5.7	4.9	4.3	3.8	3.5
380	16.5	15.4	14.2	13.0	11.8	10.8	9.7	8.7	7.8	6.9	6.1	5.4	4.6	4.1
400	18.2	17.2	16.0	14.9	13.8	12.4	11.4	10.4	9.3	8.3	7.3	6.4	5.6	5.0
420	19.8	18.8	17.7	16.7	15.5	14.4	13.1	11.9	10.9	9.8	8.8	8.0	6.9	6.1
440	21.2	20.3	19.3	18.3	17.3	16.2	14.9	13.8	12.7	11.5	10.5	9.5	8.4	7.5
460	22.5	21.6	20.6	19.7	18.9	17.9	16.7	15.6	14.3	13.3	12.2	10.9	10.0	9.0
480	23.5	22.7	22.0	21.1	20.2	19.3	18.2	17.3	16.2	15.0	13.8	12·8	11.6	10.5
500	24.3	23.8	23.0	22.3	21.6	20.7	19.7	18.8	17.8	16.7	15.4	14.5	13.4	12.3
520	24.8	24.3	23.7	23.2	22.7	21.9	21.1	20.2	19.2	18.3	17.2	16.1	15.0	14.0
540	25.0	24.8	24.3	23.9	23.4	22.8	22.1	21.3	20.6	19.7	18.7	17.6	16.6	15.6
560	24.9	24.8	24.7	24·4	24.0	23.6	22.9	22.4	21.6	20.8	20.0	19.1	18.2	17.1
580	24.7	24.7	24.6	24.5	24.3	23.9	23.5	23.1	22.5	21.9	21.1	20.3	19.5	18.6
600	24.3	24.3	24.3	24.3	24.3	24.1	23.8	23.5	23.0	22.5	22.0	21.4	20.6	19.8
620	23.6	23.7	23.9	24.0	24.1	24.1	23.9	23.7	23.4	23.1	22.6	22.1	21.4	20.8
640	22.8	23.1	23.2	23.4	23.6	23.7	23.8	23.7	23.5	23.2	22.9	22.6	22.1	21.6
660	22.0	22.3	22.5	22.8	23.0	23.2	23.2	23.3	23.2	23.1	23.0	22.8	22.5	22.1
680	21.2	21.5	21.7	22.0	22.3	22.5	22.6	22.8	22.9	22.9	22.8	22.7	22.7	22.3
700	20.3	20.7	20.9	21.2	21.5	21.7	21.9	22.2	22.3	22.5	22.5	22.5	22.4	22.2
720	19.5	19.8	20.1	20.4	20.8	21.1	21.2	21.4	21.6	21.8	21.9	22.0	22.0	22.0
740	18.8	19.0	19.2	19.6	19.9	20.2	20.5	20.7	20.9	21.1	21.2	21.5	21.5	21.6
760	18.2	18.5	18.4	18.8	19.1	19.4	19.6	19.9	20.1	20.3	20.5	20.8	21.0	21.2
780	17.7	17.8	18.0	18.1	18.4	18.7	18.8	19.1	19.3	19.5	19.7	20.0	20.2	20.4
800	17.3	17.4	17.4	17.7	17.9	18.0	18.1	18.4	18.6	18.9	18.9	19.1	19.4	19.6
820	17.0	17.2	17.2	17.2	17.4	17.4	17.6	17.8	17.8	18.1	18.3	18.5	18.6	18.8
840	16.7	16.8	16.8	16.9	17.2	17.2	17.1	17.1	17.3	17.4	17.5	17.8	17.9	18 1
860	16.4	16.5	16.5	16.6	16.6	16.7	16.8	16.9	16.9	17.0	17.0	17.1	17.2	17.4
880	16.1	16.3	16.3	16.5	16.5	16.5	16.6	16.6	16.6	16.6	16.6	16.7	16.7	16.9
900	15.7	15.9	16.1	16.2	16.3	16.4	16.3	16.3	16.2	16.2	16.2	16.3	16.3	16.3
920	15.3	15.5	15.6	15.9	16.0	16.1	16.1	16.1	16.0	16.1	16.1	16.1	16.0	16.0
940	14.7	15.9	15.2	15.4	15.7	15.8	15.8	16.0	15.9	15.9	15.9	15.8	15.7	15.8
960	14.1	14.3	14.5	14.8	15.2	15.5	15.5	15.7	15.7	15.7	15.6	15.6	15.5	15.5
980	13.3	12.7	13.9	14.2	14.5	14.8	15.1	15.3	15.4	15.5	15.4	15.4	15.4	15.3
1000	12.4	12.9	13.2	13.6	13.9	14.2	14.4	14.8	15.0	15.1	15.1	15.2	15.2	15 3
	870	880	890	900	910	920	930	940	950	960	970	980	990	1000

TABLE XXXIII.

Perturbations produced by Saturn.

Arguments II and VII.

VII.

II	0	100	200	300	400	500	600	700	800	900	1000
	″	″	″	″	″	″	″	″	″	″	″
0	1.2	1.5	1.4	1.0	0.7	0.6	0.5	0.5	0.4	0.8	1.2
100	0.9	1.2	1.3	1.1	0.9	0.8	0.7	0.7	0.6	0.7	0.9
200	0.7	0.9	1.0	1.1	1.0	0.9	0.8	0.8	0.9	0.8	0.7
300	0.9	0.8	0.7	0.8	0.9	1.0	1.0	1.0	1.0	1.0	0.9
400	1.0	0.9	0.6	0.4	0.6	0.9	1.0	1.1	1.1	1.1	1.0
500	1.1	1.0	0.8	0.4	0.2	0.5	1.0	1.3	1.3	1.2	1.1
600	1.2	1.1	0.9	0.6	0.2	0.2	0.5	1.1	1.5	1.5	1.2
700	1.4	1.1	1.0	0.8	0.4	0.1	0.3	0.8	1.4	1.7	1.4
800	1.6	1.3	1.0	0.8	0.6	0.4	0.1	0.3	1.0	1.6	1.6
900	1.5	1.4	1.1	0.9	0.7	0.6	0.3	0.2	0.6	1.2	1.5
1000	1.2	1.5	1.4	1.0	0.7	0.6	0.5	0.5	0.4	0.8	1.2

Constant, 1.″0

TABLE XXXIV.

Variable Part of Sun's Aberration.

Argument, Sun's Mean Anomaly.

	0s	Is	IIs	IIIs	IVs	Vs	
°	″	″	″	″	″	″	°
0	0.0	0.0	0.1	0.3	0.5	0.6	30
3	0.0	0.0	0.2	0.3	0.5	0.6	27
6	0.0	0.0	0.2	0.3	0.5	0.6	24
9	0.0	0.0	0.2	0.3	0.5	0.6	21
12	0.0	0.1	0.2	0.4	0.5	0.6	18
15	0.0	0.1	0.2	0.4	0.5	0.6	15
18	0.0	0.1	0.2	0.4	0.5	0.6	12
21	0.0	0.1	0.3	0.4	0.6	0.6	9
24	0.0	0.1	0.3	0.4	0.6	0.6	6
27	0.0	0.1	0.3	0.4	0.6	0.6	3
30	0.0	0.1	0.3	0.5	0.6	0.6	0
	XIs	Xs	IXs	VIIIs	VIIs	VIs	

Constant, 0.″3

TABLE XXXV.

Moon's Epochs.

Years.	1	2	3	4	5	6	7	8	9	10	11	12	13
1830	00174	4541	4461	4638	9885	0635	5979	9921	7623	219	226	458	468
1831	00103	1749	4127	9381	2357	6432	7040	2378	6487	825	587	177	940
1832 B	00032	8957	3793	4125	4829	2229	8100	4835	5351	432	948	897	413
1833	00235	6816	4499	9156	7636	8399	9219	7683	4239	108	340	687	920
1834	00164	4024	4164	3900	0107	4196	0279	0140	3103	715	701	406	393
1835	00093	1232	3830	8644	2579	9993	1340	2598	1967	321	061	125	866
1836 B	00022	8441	3496	3388	5051	5791	2400	5055	0831	928	422	845	339
1837	00224	6299	4202	8419	7858	1960	3518	7903	9719	605	814	635	846
1838	00153	3508	3868	3163	0329	7757	4579	0360	8583	211	175	354	319
1839	00082	0716	3534	7907	2801	3555	5639	2818	7447	818	536	074	792
1840 B	00011	7925	3199	2651	5273	9352	6700	5275	6310	424	896	793	265
1841	00213	5783	3906	7682	8080	5522	7818	8123	5199	101	288	583	772
1842	00142	2991	3571	2425	0551	1319	8879	0580	4062	707	649	302	245
1843	00071	0200	3237	7169	3023	7116	9939	3038	2926	314	010	022	718
1844 B	00000	7408	2903	1913	5495	2914	1000	5495	1790	920	371	741	191
1845	00203	5266	3609	6944	8302	9083	2118	8343	0678	597	763	531	698
1846	00132	2475	3275	1688	0778	4880	3179	0800	9542	203	123	250	171
1847	00061	9683	2941	6432	3245	0678	4239	3257	8406	810	484	970	644
1848 B	99990	6892	2606	1176	5717	6475	5300	5715	7270	416	845	689	117
1849	00192	4750	3312	6207	8524	2644	6418	8563	6158	093	237	479	624
1850	00121	1958	2978	0951	0995	8442	7479	1020	5022	700	597	199	097
1851	00050	9167	2644	5695	3467	4239	8539	3477	3885	306	958	918	570
1852 B	99979	6375	2310	0439	5939	0036	9600	5935	2749	913	319	637	043
1853	00181	4233	3016	5469	8746	6206	0718	8782	1637	589	711	427	550
1854	00110	1442	2681	0213	1217	2003	1778	1240	0501	196	072	147	023
1855	00039	8650	2347	4957	3689	7801	2839	3697	9365	802	432	866	496
1856 B	99968	5859	2013	9701	6160	3598	3899	6155	8229	409	793	586	969
1857	00171	3717	2719	4732	8968	9767	5018	9002	7117	086	185	375	476
1858	00100	0925	2385	9476	1439	5565	6078	1460	5981	692	546	095	949
1859	00029	8134	2051	4220	3911	1362	7139	3917	4845	299	907	814	422
1860 B	99958	5342	1716	8964	6383	7159	8199	6374	3709	905	267	534	895
1861	00160	3200	2423	3995	9190	3329	9317	9222	2597	581	659	323	402
1862	00089	0409	2088	8739	1661	9126	0378	1679	1461	188	020	043	875
1863	00018	7617	1754	3483	4133	4923	1438	4137	0324	795	381	762	348
1864 B	99947	4826	1420	8227	6605	0721	2499	6594	9188	401	742	482	821
1865	00149	2684	2126	3257	9412	6890	3617	9442	8076	078	134	272	328
1866	00078	9893	1792	8001	1883	2687	4678	1899	6940	685	494	991	801
1867	00007	7101	1457	2745	4355	8485	5738	4357	5804	291	855	711	274
1868 B	99936	4309	1123	7489	6827	4282	6799	6814	4668	898	216	431	747
1869	00138	2168	1829	2520	9634	0452	7917	9662	3556	574	608	220	254
1870	00067	9376	1495	7264	2105	6249	8978	2119	2420	181	968	940	727
1871	99996	6585	1161	2008	4577	2046	0038	4576	1283	787	329	659	200
1872 B	99925	3793	0827	6752	7049	7843	1099	7034	0147	394	690	378	673
1873	00127	1651	1533	1782	9856	4013	2217	9881	9035	070	082	168	189
1874	00056	9860	1198	6526	2327	9810	3277	2339	7899	677	443	888	653
1875	99985	6068	0864	1270	4799	5608	4338	4796	6763	283	803	607	126
1876 B	99914	3277	0530	6014	7280	1405	5398	7254	5627	890	164	327	599
1877	00117	1135	1236	1045	0078	7574	6517	0101	4515	567	556	116	106
1878	00046	8343	0902	5789	2549	3372	7577	2559	3379	173	917	836	579
1879	99975	5552	0568	0533	5021	9169	8638	5016	2243	780	278	555	052
1880 B	99904	2760	0233	5277	7493	4966	9698	7473	1107	386	638	275	525
1881	10106	0618	0940	0308	0300	1136	0816	0321	9995	062	030	064	032
1882	00035	7827	0605	5052	2771	6933	1877	2798	8859	669	391	784	505
1883	99964	5035	0271	9796	5243	2730	2937	5236	7722	276	752	503	978
1884 B	99893	2244	9937	4540	7715	8528	3998	7693	6586	882	113	223	451
1885	00095	0042	0643	9570	0522	4697	5116	0541	5474	559	505	013	957

TABLE XXXV.

Moon's Epochs.

Years.	14	15	16	17	18	19	20	21	22	23	24	25	26	27	28	29	30	31
1830	921	392	230	588	462	523	536	52	60	44	94	51	47	98	99	99	89	52
1831	115	532	589	940	937	296	703	30	70	41	65	53	94	48	24	24	51	44
1832 B.	309	673	949	293	412	070	870	07	81	38	36	55	42	97	48	49	14	35
1833	602	844	345	688	913	845	037	85	92	45	07	61	92	53	77	77	77	27
1834	796	984	704	040	388	619	203	62	03	42	77	63	40	03	01	01	39	18
1835	989	124	063	393	863	392	370	39	13	38	48	65	87	51	26	26	02	10
1836 B.	183	265	423	745	338	166	537	17	24	35	19	67	34	01	50	51	64	01
1837	476	436	819	140	840	942	704	94	35	42	90	73	85	58	79	79	27	93
1838	670	576	178	492	315	715	870	72	46	38	60	75	32	07	04	04	89	84
1839	864	716	537	845	790	489	037	49	56	35	31	77	80	56	28	28	52	76
1840 B.	058	857	897	197	265	262	204	26	67	32	02	79	27	06	53	53	14	67
1841	351	028	293	592	766	038	371	04	78	39	73	85	77	62	81	81	77	59
1842	544	168	652	944	241	811	537	81	89	35	43	87	25	12	06	06	40	51
1843	738	308	012	297	716	585	704	58	99	32	14	89	72	61	30	31	02	42
1844 B.	932	449	371	649	191	358	871	36	10	29	85	91	19	10	55	55	65	34
1845	225	620	767	044	692	134	038	13	21	36	56	97	70	67	84	83	27	26
1846	419	760	126	396	167	907	204	91	32	32	26	99	17	16	08	08	90	17
1847	613	901	486	749	643	681	371	68	42	29	97	01	65	65	33	33	52	09
1848 B.	806	041	845	101	118	454	538	45	53	26	68	03	12	15	57	58	15	00
1849	099	212	241	496	619	230	705	23	64	33	39	09	63	71	86	86	77	92
1850	293	352	600	848	094	003	871	00	75	29	09	10	10	20	10	10	40	83
1851	487	493	960	201	569	777	038	78	85	26	80	12	57	70	35	35	02	75
1852 B.	681	633	319	553	044	550	205	55	96	23	51	14	04	19	59	60	65	66
1853	974	804	715	948	545	326	372	33	07	30	22	20	55	76	88	88	28	58
1854	168	944	074	300	020	099	539	10	18	26	93	22	03	25	12	12	90	50
1855	361	085	434	653	495	873	705	87	28	23	63	24	50	74	37	37	53	41
1856 B.	555	225	793	005	970	646	872	65	39	20	34	26	97	23	61	62	15	33
1857	848	396	189	400	471	422	039	42	50	27	05	32	48	30	90	90	78	24
1858	042	537	548	752	947	195	206	20	61	24	76	34	95	29	15	15	40	16
1859	236	677	908	105	422	969	372	97	71	20	46	36	42	79	39	40	03	07
1860 B.	430	817	267	457	897	742	539	74	82	17	17	38	89	28	64	64	65	99
1861	723	988	663	852	398	518	706	52	93	24	88	44	41	84	92	92	28	91
1862	916	129	022	204	873	291	873	29	04	20	60	46	88	34	17	17	91	82
1863	110	269	382	557	348	065	039	06	14	17	29	48	35	82	41	42	53	74
1864 B.	304	409	741	909	823	838	206	84	25	14	00	50	82	32	66	66	16	65
1865	597	580	137	304	324	614	373	61	36	21	71	56	33	89	95	94	78	57
1866	791	721	496	657	799	387	540	39	47	17	42	58	80	38	19	19	41	49
1867	985	861	856	009	274	161	707	16	57	14	12	60	28	87	44	44	03	40
1868 B.	178	001	215	362	749	934	873	93	68	11	83	62	75	37	68	69	66	32
1869	471	172	611	756	251	710	040	71	79	18	54	68	26	93	97	97	28	23
1870	665	313	970	109	726	483	207	48	90	15	26	69	73	43	21	21	91	15
1871	859	454	330	462	201	257	374	26	00	12	97	71	20	93	46	46	53	07
1872 B.	053	594	689	814	676	030	541	03	11	09	68	73	67	42	70	71	16	98
1873	346	765	085	209	177	806	708	81	22	16	39	79	18	99	99	99	79	90
1874	540	905	484	561	652	579	875	58	33	12	10	81	66	48	23	23	41	82
1875	733	046	804	914	127	353	41	35	43	09	80	83	13	97	48	48	04	73
1876 B.	927	186	163	266	602	126	208	13	54	06	51	85	60	46	72	73	66	65
1877	220	357	559	661	103	902	375	90	65	13	22	91	11	03	01	01	29	56
1878	414	498	918	013	579	675	542	68	76	10	93	93	58	52	26	25	91	48
1879	608	638	278	366	54	449	708	45	86	06	63	95	05	02	50	51	54	39
1880 B.	802	778	637	718	529	222	875	22	97	03	34	97	52	51	75	75	16	31
1881	095	949	033	113	30	998	042	00	08	10	05	03	04	07	03	03	79	23
1882	288	90	392	465	505	771	209	77	19	06	77	05	51	57	28	28	42	14
1883	482	230	752	818	980	545	375	54	29	03	46	07	98	05	52	53	04	06
1884 B.	676	370	111	170	455	318	542	32	40	10	17	09	45	55	77	77	67	97
1885	969	541	507	565	956	94	709	09	51	07	88	15	96	12	06	05	29	89

TABLE XXXV.

Moon's Epochs.

Years.	Evection.	Anomaly.	Variation.	Longitude.
	s ° ′ ″	s ° ′ ″	s ° ′ ″	s ° ′ ″
1830	5 17 4 12	11 24 31 4.5	2 13 2 39	11 22 55 37.7
1831	11 7 35 41	2 23 14 24.6	6 22 40 4	4 2 18 42.8
1832 B	4 28 7 11	5 21 57 44.4	11 2 17 28	8 11 41 48.0
1833	10 29 57 40	9 3 44 58.5	3 24 6 21	1 4 15 28.4
1834	4 20 29 11	0 2 28 18.5	8 3 43 45	5 13 38 33.6
1835	10 11 0 40	3 1 11 38.6	0 13 21 10	9 23 1 38.8
1836 B	4 1 32 9	5 29 54 58.7	4 22 58 34	2 2 24 44.0
1837	10 3 22 39	9 11 42 12.8	9 14 47 27	6 24 58 24.5
1838	3 23 54 9	0 10 25 32.9	1 24 24 51	11 4 21 29.8
1839	9 14 25 38	3 9 8 53.1	6 4 2 16	3 13 44 35.0
1840 B	3 4 57 8	6 7 52 13.2	10 13 39 42	7 23 7 40.4
1841	9 6 47 37	9 19 39 27.5	3 5 28 33	0 15 41 20.9
1842	2 27 19 7	0 18 22 47.6	7 15 5 58	4 25 4 26.2
1843	8 17 50 37	3 17 6 7.9	11 24 43 23	9 4 27 31.6
1844 B	2 8 22 7	6 15 49 28.1	4 4 20 48	1 13 50 37.0
1845	8 10 12 36	9 27 36 42.5	8 26 9 40	6 6 24 17.5
1846	2 0 44 6	0 26 20 2.8	1 5 47 5	10 15 47 23.0
1847	7 21 15 35	3 25 3 23.2	5 15 24 30	2 25 10 28.3
1848 B	1 11 47 5	6 23 46 43.5	9 25 1 55	7 4 33 33.7
1849	7 13 37 35	10 5 33 57.9	2 16 50 47	11 27 7 14.5
1850	1 4 9 4	1 4 17 18.3	6 26 28 12	4 6 30 19.9
1851	6 24 40 35	4 3 0 38.6	11 6 5 37	8 15 53 25.4
1852 B	0 15 12 5	7 1 43 59.2	3 15 43 3	0 25 16 31.0
1853	6 17 2 34	10 13 31 13.7	8 7 31 54	5 17 50 11.6
1854	0 7 34 4	1 12 14 34.1	0 17 9 20	9 27 13 17.2
1855	5 28 5 33	4 10 57 54.7	4 26 46 44	2 6 36 22.7
1856 B	11 18 37 3	7 9 41 15.2	9 6 24 10	6 15 59 28.2
1857	5 20 27 33	10 21 28 29.8	1 28 13 2	11 8 33 9.1
1858	11 10 59 2	1 20 11 50.3	6 7 50 27	3 17 56 14.6
1859	5 1 30 33	4 18 55 10.9	10 17 27 53	7 27 19 20.1
1860 B	10 22 2 3	7 17 38 31.4	2 27 5 18	0 6 42 25.8
1861	4 23 52 32	10 29 25 46.1	7 18 54 10	4 29 16 6.6
1862	10 14 24 2	1 28 9 6.6	11 28 31 35	9 8 39 12.2
1863	4 4 55 32	4 26 52 27.8	4 8 9 1	1 18 2 17.9
1864 B	9 25 27 2	7 25 35 48.0	8 17 46 25	5 27 25 23.5
1865	3 27 17 31	11 7 23 2.7	1 9 35 18	10 19 59 4.3
1866	9 17 49 2	2 6 6 23.3	5 19 12 43	2 29 22 10.1
1867	3 8 20 31	5 4 49 44.0	9 28 50 9	7 8 45 15.7
1868 B	8 28 52 2	8 3 33 4.7	2 8 27 34	11 18 8 21.4
1869	3 0 42 33	11 15 20 19.6	7 0 16 26	4 10 42 2.3
1870	8 21 14 2	2 14 3 40.3	11 9 53 51	8 20 5 8.0
1871	2 11 45 33	5 12 47 0.6	3 19 31 16	0 29 28 13.5
1872 B	8 2 17 3	8 11 30 21.2	7 29 8 42	5 8 57 19.1
1873	2 4 7 32	11 23 17 35.7	0 20 57 33	10 1 24 59.7
1874	7 24 39 2	3 22 0 56.1	4 20 34 59	2 10 48 5.3
1875	1 15 10 31	5 20 44 16.7	9 10 12 23	6 20 11 10.8
1876 B	7 5 42 1	8 19 27 37.2	1 19 49 49	10 29 34 16.3
1877	1 7 32 31	0 1 14 51.8	6 11 38 41	3 22 7 57.2
1878	6 28 4 0	2 29 58 12.3	10 21 16 6	8 1 31 2.7
1879	0 18 35 31	5 28 41 32.9	3 0 53 32	0 10 54 8.2
1880 B	6 9 7 1	8 27 24 53.4	7 10 30 57	16 20 17 13.9
1881	0 10 57 30	0 9 12 8.1	0 2 19 49	9 12 50 54.7
1882	6 1 29 0	3 7 55 28.6	4 11 57 14	1 22 14 0.3
1883	11 22 0 30	6 6 38 49.3	8 21 34 40	6 1 37 6.8
1884 B	5 12 32 0	9 5 22 10.0	1 1 12 4	10 11 0 11.6
1885	11 14 22 29	0 17 9 24.7	5 23 0 57	3 8 33 52.4

TABLE XXXV.

Moon's Epochs.

Years.	Supp. of Node.	II	V	VI	VII	VIII	IX	X	XI	XII
	s ° ′ ″	s ° ′								
1830	6 7 7 11.0	10 24 46	498	502	900	904	427	062	025	433
1831	6 26 26 53.3	2 15 18	912	914	208	210	506	001	211	710
1832 B	7 15 46 35.5	6 5 50	326	327	516	516	586	940	397	986
1833	8 5 9 28.4	10 7 31	774	779	852	856	702	885	624	297
1834	8 24 29 10.7	1 28 3	187	191	159	163	782	825	810	573
1835	9 13 48 53.0	5 18 35	601	603	467	469	861	764	996	850
1836 B	10 3 8 35.2	9 9 8	015	016	775	775	941	703	182	127
1837	10 22 31 28.1	1 10 49	463	468	111	116	057	648	409	437
1838	11 11 51 10.4	5 1 21	876	880	419	423	137	588	595	714
1839	0 1 10 52.6	8 21 53	290	292	726	729	217	527	781	991
1840 B	0 20 30 34.9	0 12 25	704	705	034	035	296	466	967	268
1841	1 9 53 27.7	4 14 6	152	157	370	375	412	411	194	578
1842	1 29 13 10.0	8 4 38	566	569	678	682	492	350	380	855
1843	2 18 32 52.2	11 25 10	980	980	986	988	572	290	566	131
1844 B	3 7 52 34.5	3 15 42	393	394	293	294	651	229	752	408
1845	3 27 15 27.4	7 17 23	840	846	629	634	767	174	979	718
1846	4 16 35 9.6	11 7 55	254	258	937	941	847	113	165	995
1847	5 5 54 51.8	2 28 27	668	670	245	247	927	053	351	272
1848 B	5 25 14 34.1	6 18 59	082	083	553	553	006	992	537	549
1849	6 14 37 27.0	10 20 40	531	535	889	893	122	937	764	859
1850	7 3 57 9.2	2 11 12	944	947	196	200	202	876	950	136
1851	7 23 16 51.5	6 1 44	358	359	504	506	282	816	136	413
1852 B	8 12 36 33.6	9 22 17	772	772	812	812	362	755	322	689
1853	9 1 59 26.5	1 23 58	220	223	148	152	477	700	549	000
1854	9 21 19 8.8	5 14 30	634	636	456	459	557	639	735	276
1855	10 10 38 51.1	9 5 2	047	048	763	765	637	579	921	553
1856 B	10 29 58 33.3	0 25 34	461	461	071	071	717	518	107	830
1857	11 19 21 26.2	4 27 15	909	912	407	411	832	463	334	140
1858	0 8 41 8.4	8 17 47	323	325	715	718	912	402	520	417
1859	0 28 0 50.7	0 8 19	736	737	023	024	992	342	706	694
1860 B	1 17 20 32.9	3 28 51	150	150	330	330	072	281	892	971
1861	2 6 43 25.8	8 0 32	598	601	666	670	187	226	119	281
1862	2 26 3 8.0	11 21 4	012	014	974	977	267	165	305	558
1863	3 15 22 50.1	3 11 36	426	426	282	283	347	105	491	834
1864 B	4 4 42 32.3	7 2 8	839	839	590	589	427	044	677	111
1865	4 24 5 25.2	11 3 49	287	291	926	929	542	989	904	422
1866	5 13 25 7.3	2 24 21	701	703	233	236	622	928	090	698
1867	6 2 44 49.5	6 14 53	115	115	541	542	702	868	276	975
1868 B	6 22 4 31.7	10 5 26	529	528	849	848	782	807	462	252
1869	7 11 27 24.6	2 7 7	977	980	185	188	897	752	689	562
1870	8 0 47 6.7	5 27 39	390	392	493	495	977	691	875	839
1871	8 20 6 49.0	9 18 11	804	804	801	801	057	631	061	116
1872 B	9 9 26 31.1	1 8 44	218	217	109	107	137	570	247	392
1873	9 28 49 24.0	5 10 25	666	668	445	447	252	515	474	703
1874	10 18 9 6.3	9 0 57	080	081	753	754	332	454	660	979
1875	11 7 28 48.6	0 21 29	493	493	060	060	412	394	846	256
1876 B	11 26 48 30.8	4 12 1	907	906	368	366	492	333	032	533
1877	0 16 11 23.7	8 13 42	355	357	704	706	607	278	259	843
1878	1 5 31 5.9	0 4 14	769	770	012	013	687	217	445	120
1879	1 24 50 48.2	3 24 46	182	182	320	319	767	157	631	397
1880 B	2 14 10 30.4	7 15 18	596	595	627	623	847	096	817	674
1881	3 3 33 23.3	11 16 59	044	046	963	965	962	041	044	984
1882	3 22 53 5.5	3 7 31	458	459	271	272	042	980	230	261
1883	4 12 12 47.6	6 28 3	872	871	579	578	122	920	416	537
1884 B	5 1 32 29.8	10 18 35	285	284	887	884	202	859	602	814
1885	5 20 55 23.0	2 20 16	733	736	223	224	317	804	829	125

TABLE XXXVI.

Moon's Motions for Months.

Months.		1	2	3	4	5	6	7	8	9	10	11	12	13
January		00000	0000	0000	0000	0000	0000	0000	0000	0000	000	000	000	000
February		08487	0146	2246	8896	0402	1533	1789	2099	0753	175	965	184	059
March	Com.	16153	8343	1371	6931	9797	1951	3404	3027	1433	139	836	157	016
	Bis.	16427	8993	2411	7218	0132	2323	3462	3418	1457	209	868	228	050
April	Com.	24640	8490	3616	5827	0199	3484	5193	5126	2186	314	801	342	076
	Bis.	24914	9140	4657	6114	0534	3856	5251	5517	2210	384	832	412	110
May	Com.	32853	7986	4822	4436	0265	4646	6924	6835	2914	419	735	456	101
	Bis.	33127	8636	5862	4723	0600	5018	6982	7226	2938	489	766	526	135
June	Com.	41340	8133	7067	3332	0666	6179	8713	8934	3667	593	700	640	160
	Bis.	41614	8783	8107	3619	1002	6551	8771	9325	3691	663	731	710	194
July	Com.	49554	7629	8273	1942	0732	7341	0444	0643	4396	698	634	754	185
	Bis.	49828	8279	9313	2228	1068	7713	0502	1034	4420	768	665	824	219
Aug.	Com.	58041	7776	0518	0838	1134	8874	2233	2742	5148	873	599	938	245
	Bis.	58315	8426	1558	1125	1470	9246	2290	3133	5173	943	630	009	279
Sept.	Com.	66528	7922	2764	9734	1536	0408	4021	4842	5901	048	563	123	304
	Bis.	66802	8572	3804	0021	1871	0780	4079	5232	5925	118	595	193	338
Oct.	Com.	74741	7419	3969	8343	1602	1569	5752	6550	6630	152	497	237	329
	Bis.	75015	8069	5009	8630	1938	1941	5810	6941	6654	222	528	307	363
Nov.	Com.	83228	7565	6215	7239	2004	3102	7541	8649	7382	327	462	421	383
	Bis.	83502	8215	7255	7526	2339	3475	7599	9040	7407	397	493	492	423
Dec.	Com.	91442	7062	7420	5848	2070	4264	9272	0358	8111	432	396	535	414
	Bis.	91716	7712	8460	6135	2405	4636	9330	0749	8135	502	427	606	448

TABLE XXXVI.

Moon's Motions for Months.

Months.		Evection.				Anomaly.				Variation.				Longitude.			
		s	°	′	″	s	°	′	″	s	°	′	″	s	°	′	″
January		0	0	0	0	0	0	0	0.0	0	0	0	0	0	0	0	0.0
February		11	20	48	42	1	15	0	53.1	0	17	54	48	1	18	28	5.8
March	Com.	10	7	40	26	1	20	50	4.2	11	29	15	15	1	27	24	26.6
	Bis.	10	18	59	26	2	3	53	58.2	0	11	26	42	2	10	35	1.6
April	Com.	9	28	29	8	3	5	50	57.3	0	17	10	3	3	15	52	32.5
	Bis.	10	9	48	8	3	18	54	51.2	0	29	21	29	3	29	3	7.5
May	Com.	9	7	58	51	4	7	47	56.4	0	22	53	24	4	21	10	3.3
	Bis.	9	19	17	50	4	20	51	50.3	1	5	4	50	5	4	20	38.3
June	Com.	8	28	47	33	5	22	48	49.4	1	10	48	11	6	9	38	9.1
	Bis.	9	10	6	33	6	5	52	43.4	1	22	59	38	6	22	48	44.1
July	Com.	8	8	17	16	6	24	45	48.5	1	16	31	32	7	14	55	39.9
	Bis.	8	19	36	15	7	7	49	42.5	1	28	42	59	7	28	6	15.0
Aug.	Com.	7	29	5	59	8	9	46	41.6	2	4	26	20	9	3	23	45.8
	Bis.	8	10	24	58	8	22	50	35.5	2	16	37	47	9	16	34	20.8
Sept.	Com.	7	19	54	41	9	24	47	34.6	2	22	21	7	10	21	51	51.6
	Bis.	8	1	13	40	10	7	51	28.6	3	4	32	34	11	5	2	26.7
Oct.	Com.	6	29	24	24	10	26	44	33.7	2	28	4	28	11	27	9	22.4
	Bis.	7	10	43	23	11	9	48	27.7	3	10	15	55	0	10	19	57.5
Nov.	Com.	6	20	13	6	0	11	45	26.8	3	15	59	16	1	15	37	28.3
	Bis.	7	1	32	5	0	24	49	20.7	3	28	10	43	1	28	48	3.3
Dec.	Com.	5	29	42	49	1	13	42	25.9	3	21	42	37	2	20	54	59.1
	Bis.	6	11	1	48	1	26	46	19.8	4	3	54	4	3	4	5	34.1

Moon's Motions for Months.

Months.		14	15	16	17	18	19	20	21	22	23	24	25	26	27	28	29	30	31
January		000	000	000	000	000	000	000	00	00	00	00	00	00	00	00	00	00	00
February		074	946	135	304	805	066	014	24	26	14	82	28	14	17	29	96	05	07
March	Com.	851	801	159	482	532	125	027	45	50	98	57	43	18	12	46	82	10	15
	Bis.	950	831	196	524	558	127	027	46	51	08	59	47	21	19	51	85	10	15
April	Com.	925	747	294	786	336	191	041	68	77	12	39	70	32	29	76	77	15	23
	Bis.	024	778	331	828	362	193	042	69	77	22	42	74	36	36	80	80	16	23
May	Com.	899	663	392	047	115	254	055	91	02	15	19	94	43	38	01	70	21	30
	Bis.	999	693	429	089	141	256	055	92	03	26	22	98	47	45	05	73	21	30
June	Com.	973	609	527	351	920	320	069	15	28	29	01	21	57	55	31	65	26	38
	Bis.	073	639	563	393	946	322	069	15	29	40	04	25	61	62	35	68	26	38
July	Com.	948	525	625	613	699	384	083	37	54	33	81	45	68	64	56	58	31	45
	Bis.	047	555	661	655	725	386	083	38	55	43	84	49	72	71	60	61	31	46
Aug.	Com.	022	471	759	917	503	449	097	61	80	47	64	72	82	81	85	53	36	53
	Bis.	121	501	796	959	529	451	097	62	81	57	66	77	86	88	90	56	36	53
Sept.	Com.	096	417	894	221	308	515	111	85	07	61	46	00	97	97	15	49	42	61
	Bis.	195	447	931	263	334	517	111	85	08	71	49	04	01	04	19	52	42	61
Oct.	Com.	071	333	992	483	087	578	125	07	32	65	26	23	08	07	40	41	47	68
	Bis.	170	363	029	525	113	581	126	08	33	75	28	28	11	14	44	44	47	69
Nov.	Com.	145	279	127	787	892	644	139	31	59	79	08	51	22	23	70	37	52	76
	Bis.	244	309	163	829	918	646	140	32	60	89	11	55	26	30	74	40	52	76
Dec.	Com.	120	194	225	049	670	708	153	54	85	83	88	74	33	33	95	29	57	84
	Bis.	219	225	261	091	696	710	153	54	86	93	90	79	37	40	99	32	57	84

TABLE XXXVI.

Moon's Motions for Months.

Months.		Supp. of Node.				II			V	VI	VII	VIII	IX	X	XI	XII
		s	°	′	″	s	°	′								
January		0	0	0	0.0	0	0	0	000	000	000	000	000	000	000	000
February		0	1	38	29.7	11	15	43	054	224	875	045	111	165	290	043
March	Com.	0	3	7	27.5	9	27	59	007	330	666	989	114	313	455	984
	Bis.	0	3	10	38.2	10	9	8	041	369	694	023	150	319	496	018
April	Com.	0	4	45	57.3	9	13	42	061	554	542	034	225	478	745	027
	Bis.	0	4	49	7.9	9	24	51	095	593	570	068	261	484	787	061
May	Com.	0	6	21	16.4	8	18	15	081	738	389	046	300	638	993	036
	Bis.	0	6	24	27.0	8	29	25	115	778	417	080	336	643	034	070
June	Com.	0	7	59	46.1	8	3	58	136	962	264	091	411	802	282	079
	Bis.	0	8	2	56.7	8	15	8	170	002	293	124	447	808	324	113
July	Com.	0	9	35	5.2	7	8	32	156	147	112	103	486	962	531	088
	Bis.	0	9	38	15.9	7	19	41	190	186	140	136	522	967	572	122
Aug.	Com.	0	11	13	35.0	6	24	15	210	371	987	147	597	126	820	131
	Bis.	0	11	16	45.6	7	5	24	244	411	015	182	633	132	862	164
Sept	Com.	0	12	52	4.7	6	9	58	265	595	862	193	708	291	110	173
	Bis.	0	12	55	15.4	6	21	7	299	635	891	227	744	296	152	207
Oct.	Com.	0	14	27	23.8	5	14	32	285	780	710	204	783	451	358	182
	Bis.	0	14	30	34.4	5	25	41	319	819	738	238	819	456	400	216
Nov.	Com.	0	16	5	53.5	5	0	15	339	004	585	250	894	615	648	225
	Bis.	0	16	9	4.2	5	11	24	373	043	613	283	930	621	690	259
Dec.	Com.	0	17	41	12.6	4	4	49	359	188	432	261	969	775	896	234
	Bis.	0	17	44	23.3	4	15	58	393	228	461	295	005	780	938	268

Moon's Motions for Days.

D.	1	2	3	4	5	6	7	8	9	10	11	12	13
1	00000	0000	0000	0000	0000	0000	0000	0000	0000	000	000	000	000
2	00274	0650	1040	0287	0336	0372	0058	0390	0024	070	031	070	034
3	00548	1300	2080	0574	0671	0744	0115	0781	0049	140	062	141	068
4	00821	1950	3121	0861	1007	1116	0173	1171	0073	210	093	211	103
5	01095	2600	4161	1148	1342	1488	0231	1561	0097	281	125	282	137
6	01369	3249	5201	1435	1678	1860	0289	1952	0121	351	156	352	171
7	01643	3899	6241	1722	2013	2232	0346	2342	0146	421	187	423	205
8	01916	4549	7281	2009	2349	2604	0404	2732	0170	491	218	493	239
9	02190	5199	8321	2296	2684	2976	0462	3122	0194	561	249	564	273
10	02464	5849	9362	2583	3020	3348	0519	3513	0219	631	280	634	308
11	02738	6499	0402	2870	3355	3720	0577	3903	0243	702	311	705	342
12	03012	7149	1442	3157	3691	4093	0635	4293	0267	772	342	775	376
13	03285	7799	2482	3444	4026	4465	0692	4684	0291	842	374	845	410
14	03559	8449	3522	3731	4362	4837	0750	5074	0316	912	405	916	444
15	03833	9098	4563	4018	4698	5209	0808	5464	0340	982	436	986	478
16	04107	9748	5603	4305	5033	5581	0866	5854	0364	052	467	057	513
17	04380	0398	6643	4592	5369	5953	0923	6245	0389	122	498	127	547
18	04654	1048	7683	4878	5704	6325	0981	6635	0413	193	529	198	581
19	04928	1698	8723	5165	6040	6697	1039	7025	0437	263	560	268	315
20	05202	2348	9763	5452	6375	7069	1096	7416	0461	333	591	339	649
21	05476	2998	0804	5739	6711	7441	1154	7806	0486	403	623	409	683
22	05749	3648	1844	6026	7046	7813	1212	8196	0510	473	654	480	718
23	06023	4298	2884	6313	7382	8185	1269	8586	0534	543	685	550	752
24	06297	4947	3924	6600	7717	8557	1327	8977	0559	614	716	621	786
25	06571	5597	4964	6887	8053	8929	1385	9367	0583	684	747	691	820
26	06844	6247	6005	7174	8389	9301	1443	9757	0607	754	778	762	854
27	07118	6897	7045	7461	8724	9673	1500	0148	0631	824	809	832	888
28	07392	7547	8085	7748	9060	0045	1558	0538	0656	894	840	903	923
29	07666	8197	9125	8035	9395	0417	1616	0928	0680	964	872	973	957
30	07940	8847	0165	8322	9731	0789	1673	1[illegible]19	0704	034	903	043	991
31	08213	9497	1205	8609	0066	1161	1731	1709	0729	105	934	114	025

Moon's Motion for Days.

D.	14	15	16	17	18	19	20	21	22	23	24	25	26	27	28	29	30	31
1	000	000	000	000	000	000	000	00	00	00	00	00	00	00	00	00	00	00
2	099	031	037	042	026	002	000	01	01	10	03	04	04	07	04	03	00	00
3	198	061	073	084	052	004	001	02	02	20	05	08	07	14	08	06	00	00
4	297	092	110	126	078	006	001	02	03	30	08	12	11	21	13	09	01	01
5	397	122	146	168	104	008	002	03	03	41	11	16	15	28	17	12	01	01
6	496	153	183	210	130	011	002	04	04	51	13	21	18	35	21	15	01	01
7	595	183	220	252	156	013	003	05	05	61	16	25	22	42	25	18	01	01
8	694	214	256	294	182	015	003	05	06	71	19	29	26	49	29	22	01	02
9	793	244	293	336	208	017	004	06	07	81	21	33	30	56	33	25	01	02
10	892	275	329	379	234	019	004	07	08	91	24	37	33	63	38	28	02	02
11	992	305	366	421	260	021	005	08	09	01	27	41	37	70	42	31	02	02
12	091	336	403	463	286	023	005	08	09	11	29	45	41	77	46	34	02	03
13	190	366	439	505	312	025	005	09	10	22	32	49	44	84	50	37	02	03
14	289	397	476	547	337	028	006	10	11	32	34	53	48	91	54	40	02	03
15	388	427	512	589	363	030	006	11	12	42	37	58	52	98	58	43	02	03
16	487	458	549	631	389	032	007	11	13	52	40	62	55	05	63	46	03	04
17	587	488	586	673	415	034	007	12	14	62	42	66	59	12	67	49	03	04
18	686	519	622	715	441	036	008	13	14	72	45	70	63	19	71	52	03	04
19	785	549	659	757	467	038	008	14	15	82	48	74	66	26	75	55	03	04
20	884	580	695	799	493	040	009	14	16	92	50	78	70	33	79	59	03	05
21	983	611	732	841	519	042	009	15	17	03	53	82	74	40	84	62	03	05
22	082	641	769	883	545	044	010	16	18	13	56	86	77	47	88	65	04	05
23	182	672	805	925	571	047	010	17	19	23	58	90	81	54	92	68	04	05
24	281	702	842	967	597	049	011	17	20	33	61	95	85	61	96	71	04	06
25	380	733	878	009	623	051	011	18	20	43	64	99	89	68	00	74	04	06
26	479	763	915	052	649	053	011	19	21	53	66	03	92	75	04	77	04	06
27	578	794	952	094	675	055	012	20	22	63	69	07	96	82	09	80	04	06
28	677	824	988	136	701	057	012	20	23	73	72	11	00	89	13	83	05	06
29	777	855	025	178	727	059	013	21	24	84	74	15	03	96	17	86	05	07
30	876	885	061	220	753	061	013	22	25	94	77	19	07	03	21	89	05	07
31	975	916	098	262	779	064	014	23	26	04	80	23	11	10	25	92	05	07

Moon's Motions for Days.

D.	Evection.	Anomaly.	Variation.	M. Longitude.
	s ° ′	s ° ′ ″	s ° ′ ″	s ° ′ ″
1	0 0 0 0	0 0 0 0 0	0 0 0 0	0 0 0 0 0
2	0 11 18 59	0 13 3 54.0	0 12 11 27	0 13 10 35.0
3	0 22 37 59	0 26 7 47.9	0 24 22 53	0 26 21 10.1
4	1 3 56 58	1 9 11 41.9	1 6 34 20	1 9 31 45.1
5	1 15 15 58	1 22 15 35.9	1 18 45 47	1 22 42 20.1
6	1 26 34 57	2 5 19 29.8	2 0 57 13	2 5 52 55.1
7	2 7 53 57	2 18 23 23.8	2 13 8 40	2 19 3 30.2
8	2 19 12 56	3 1 27 17.8	2 25 20 7	3 2 14 5.2
9	3 0 31 55	3 14 31 11.7	3 7 31 34	3 15 24 40.2
10	3 11 50 55	3 27 35 5.7	3 19 43 0	3 28 35 15.2
11	3 23 9 54	4 10 38 59.7	4 1 54 27	4 11 45 50.3
12	4 4 28 54	4 23 42 53.7	4 14 5 54	4 24 56 25.3
13	4 15 47 53	5 6 46 47.6	4 26 17 20	5 8 7 0.3
14	4 27 6 53	5 19 50 41.6	5 8 28 47	5 21 17 35.4
15	5 8 25 52	6 2 54 35.6	5 20 40 14	6 4 28 10.4
16	5 19 44 51	6 15 58 29.5	6 2 51 40	6 17 38 45.4
17	6 1 3 51	6 29 2 23.5	6 15 3 7	7 0 49 20.4
18	6 12 22 50	7 12 6 17.5	6 27 14 34	7 13 59 55.5
19	6 23 41 50	7 25 10 11.4	7 9 26 1	7 27 10 30.5
20	7 5 0 49	8 8 14 5.4	7 21 37 27	8 10 21 5.5
21	7 16 19 49	8 21 17 59.4	8 3 48 54	8 23 31 40.5
22	7 27 38 48	9 4 21 53.4	8 16 0 21	9 6 42 15.6
23	8 8 57 47	9 17 25 47.3	8 28 11 47	9 19 52 50.6
24	8 20 16 47	10 0 29 41.3	9 10 23 14	10 3 3 25.6
25	9 1 35 46	10 13 33 35.3	9 22 34 41	10 16 14 0.7
26	9 12 54 46	10 26 37 29.2	10 4 46 7	10 29 24 35.7
27	9 24 13 45	11 9 41 23.2	10 16 57 34	11 12 35 10.7
28	10 5 32 45	11 22 45 17.2	10 29 9 1	11 25 45 45.7
29	10 16 51 44	0 5 49 11.1	11 11 20 28	0 8 56 20.8
30	10 28 10 43	0 18 53 5.1	11 23 31 54	0 22 6 55.8
31	11 9 29 43	1 1 56 59.1	0 5 43 21	1 5 17 30.8

Moon's Motions for Days.

D.	Supp. of Node.	II	V	VI	VII	VIII	IX	X	XI	XII
	s ° ′ ″	s ° ′								
1	0 0 0 0.0	0 0 0	000	000	000	000	000	000	000	000
2	0 0 3 10.6	0 11 9	034	039	028	034	036	005	042	034
3	0 0 6 21.3	0 22 18	068	079	056	067	072	011	083	067
4	0 0 9 31.9	1 3 27	102	118	085	101	108	016	125	101
5	0 0 12 42.5	1 14 37	136	158	113	135	143	021	166	135
6	0 0 15 53.2	1 25 46	170	197	141	169	179	027	208	168
7	0 0 19 3.8	2 6 55	204	237	169	202	215	032	250	202
8	0 0 22 14.5	2 18 4	238	276	198	236	251	037	291	235
9	0 0 25 25.1	2 29 13	272	316	226	270	287	043	333	269
10	0 0 28 35.7	3 10 22	306	355	254	303	323	048	374	303
11	0 0 31 46.4	3 21 31	340	395	282	337	358	053	416	336
12	0 0 34 57.0	4 2 40	374	434	311	371	394	058	458	370
13	0 0 38 7.6	4 13 50	408	474	339	405	430	064	499	404
14	0 0 41 18.3	4 24 59	442	513	367	438	466	069	541	437
15	0 0 44 28.9	5 6 8	476	553	395	472	502	074	583	471
16	0 0 47 39.5	5 17 17	510	592	424	506	538	080	624	505
17	0 0 50 50.2	5 28 26	544	632	452	539	573	085	666	538
18	0 0 54 0.8	6 9 35	578	671	480	573	609	090	707	572
19	0 0 57 11.5	6 20 44	612	711	508	607	645	096	749	605
20	0 1 0 22.1	7 1 53	646	750	537	641	681	101	791	639
21	0 1 3 32.7	7 13 3	680	790	565	674	717	106	832	673
22	0 1 6 43.4	7 24 12	714	829	593	708	753	112	874	706
23	0 1 9 54.0	8 5 21	748	869	621	742	788	117	915	740
24	0 1 13 4.6	8 16 30	782	908	650	775	824	122	957	774
25	0 1 16 15.3	8 27 39	816	948	678	809	860	128	999	807
26	0 1 19 25.9	9 8 48	850	987	706	843	896	133	040	841
27	0 1 22 36.5	9 19 57	884	027	734	877	932	138	082	875
28	0 1 25 47.2	10 1 6	918	066	762	910	968	143	123	908
29	0 1 28 57.8	10 12 16	952	106	791	944	003	149	165	942
30	0 1 32 8.5	10 23 25	986	145	819	978	039	154	207	975
31	0 1 35 19.1	11 4 34	020	185	847	011	075	159	248	009

Moon's Motions for Hours

H.	1	2	3	4	5	6	7	8	9	10	11	12	13
1	11	27	43	12	14	16	2	16	1	3	1	3	1
2	23	54	87	24	28	31	5	33	2	6	3	6	3
3	34	81	130	36	42	47	7	49	3	9	4	9	4
4	46	108	173	48	56	62	10	65	4	12	5	12	6
5	57	135	217	60	70	78	12	81	5	15	6	15	7
6	68	162	260	72	84	93	14	98	6	18	8	18	9
7	80	190	303	84	98	109	17	114	7	20	9	20	10
8	91	217	347	96	112	124	19	130	8	23	10	23	11
9	103	244	390	108	126	140	22	146	9	26	12	26	13
10	114	271	433	120	140	155	24	163	10	29	13	29	14
11	125	298	477	131	154	171	26	179	11	32	14	32	16
12	137	325	520	143	168	186	29	195	12	35	16	35	17
13	148	352	563	155	182	202	31	211	13	38	17	38	18
14	160	379	607	167	196	217	34	228	14	41	18	41	20
15	171	406	650	179	210	233	36	244	15	44	19	44	21
16	182	433	693	191	224	248	38	260	16	47	21	47	23
17	194	460	737	203	238	264	41	276	17	50	22	50	24
18	205	487	780	215	252	279	43	293	18	53	23	53	25
19	217	515	823	227	266	295	46	309	19	56	25	56	27
20	228	542	867	239	280	310	48	325	20	58	26	58	28
21	239	569	910	251	294	326	50	341	21	61	27	61	30
22	251	596	953	263	308	341	53	358	22	64	28	64	31
23	262	623	997	275	322	357	55	374	23	67	30	67	33
24	274	650	1040	287	336	372	58	390	24	70	31	70	34

Hours.	Evection.	Anomaly.	Variation.	Longitude.
	° ′ ″	° ′ ″	° ′ ″	° ′ ″
1	0 28 17	0 32 39.7	0 30 29	0 32 56.5
2	0 56 35	1 5 19.5	1 0 57	1 5 52.9
3	1 24 52	1 37 59.2	1 31 26	1 38 49.4
4	1 53 10	2 10 39.0	2 1 54	2 11 45.8
5	2 21 27	2 43 18.7	2 32 23	2 44 42.3
6	2 49 45	3 15 58.5	3 2 52	3 17 38.8
7	3 18 2	3 48 38.2	3 33 20	3 50 35.2
8	3 46 20	4 21 18.0	4 3 49	4 23 31.7
9	4 14 37	4 53 57.7	4 34 17	4 56 28.1
10	4 42 55	5 26 37.5	5 4 46	5 29 24.6
11	5 11 12	5 59 17.2	5 35 15	6 2 21.0
12	5 39 30	6 31 57.0	6 5 43	6 35 17.5
13	6 7 47	7 4 36.7	6 36 12	7 8 14.0
14	6 36 5	7 37 16.5	7 6 40	7 41 10.4
15	7 4 22	8 9 56.2	7 37 9	8 14 6.9
16	7 32 40	8 42 36.0	8 7 38	8 47 3.4
17	8 0 57	9 15 15.7	8 38 6	9 19 59.8
18	8 29 15	9 47 55.5	9 8 35	9 52 56.3
19	8 57 32	10 20 35.2	9 39 3	10 25 52.7
20	9 25 50	10 53 15.0	10 9 32	10 58 49.2
21	9 54 7	11 25 54.7	10 40 1	11 31 45.6
22	10 22 24	11 58 34.5	11 10 29	12 4 42.1
23	10 50 42	12 31 14.2	11 40 58	12 37 38.6
24	11 18 59	13 3 54.0	12 11 27	13 10 35.0

Moon's Motions for Hours.

H.	14	15	16	17	18	19	20	21	22	23	24	25	26	27	28	29
1	4	1	2	2	1	0	0	0	0	0	0	0	0	0	0	0
2	8	3	3	4	2	0	0	0	0	1	0	0	0	1	0	0
3	12	4	5	5	3	0	0	0	0	1	0	1	0	1	1	0
4	16	5	6	7	4	0	0	0	0	2	0	1	1	1	1	1
5	21	6	8	9	5	0	0	0	0	2	1	1	1	1	1	1
6	25	8	9	11	6	0	0	0	0	3	1	1	1	2	1	1
7	29	9	11	12	8	1	0	0	0	3	1	1	1	2	1	1
8	33	10	12	14	9	1	0	0	0	3	1	1	1	2	1	1
9	37	11	14	16	10	1	0	0	0	4	1	2	1	3	1	1
10	41	13	15	18	11	1	0	0	0	4	1	2	2	3	2	1
11	45	14	17	19	12	1	0	0	0	5	1	2	2	3	2	1
12	49	15	18	21	13	1	0	0	0	5	1	2	2	3	2	2
13	54	16	20	23	14	1	0	0	0	5	1	2	2	4	2	2
14	58	18	21	25	15	1	0	0	0	6	2	2	2	4	2	2
15	62	19	23	26	16	1	0	0	0	6	2	3	2	4	3	2
16	66	20	25	28	17	1	0	1	1	7	2	3	2	5	3	2
17	70	21	26	30	18	1	0	1	1	7	2	3	3	5	3	2
18	74	23	28	32	19	2	0	1	1	8	2	3	3	5	3	2
19	78	24	29	33	21	2	0	1	1	8	2	3	3	6	3	3
20	83	25	31	35	22	2	0	1	1	8	2	3	3	6	3	3
21	87	26	32	37	23	2	0	1	1	9	2	4	3	6	4	3
22	91	28	34	39	24	2	0	1	1	9	2	4	3	6	4	3
23	95	29	35	40	25	2	0	1	1	10	3	4	4	7	4	3
24	99	31	37	42	26	2	0	1	1	10	3	4	4	7	4	3

H.	Sup. of Nod.	II	V	VI	VII	VIII	IX	X	XI	XII
	′ ″	° ′								
1	0 7.9	0 28	1	2	1	1	1	0	2	1
2	0 15.9	0 56	3	3	2	3	3	0	3	3
3	0 23.8	1 24	4	5	4	4	4	1	5	4
4	0 31.8	1 52	6	7	5	6	6	1	7	6
5	0 39.7	2 19	7	8	6	7	7	1	9	7
6	0 47.7	2 47	9	10	7	9	9	1	10	9
7	0 55.6	3 15	10	12	8	10	10	2	12	10
8	1 3.6	3 43	11	13	9	11	12	2	14	11
9	1 11.5	4 11	13	15	11	13	13	2	15	13
10	1 19.4	4 39	14	16	12	14	15	2	17	14
11	1 27.4	5 7	16	18	13	15	16	2	19	15
12	1 35.3	5 35	17	20	14	17	18	3	21	17
13	1 43.3	6 2	18	21	15	18	19	3	23	18
14	1 51.2	6 30	20	23	16	19	21	3	24	19
15	1 59.2	6 58	21	25	18	21	22	3	26	21
16	2 7.1	7 26	23	26	19	22	24	4	28	22
17	2 15.0	7 54	24	28	20	24	25	4	29	24
18	2 23.0	8 22	26	29	21	25	27	4	31	25
19	2 30.9	8 50	27	31	22	27	28	4	33	27
20	2 38.9	9 18	28	32	24	28	30	4	35	28
21	2 46.8	9 45	30	34	25	29	31	5	37	29
22	2 54.8	10 13	31	36	26	31	33	5	38	31
23	3 2.7	10 41	33	38	27	32	34	5	40	32
24	3 10.6	11 9	34	39	28	34	36	5	42	34

Moon's Motions for Minutes.

Min.	1	2	3	4	5	6	7	8	9	10	11	12	13	14	15	16	17	18
31	6	14	22	6	7	8	1	8	0	1	1	1	1	2	1	1	1	1
32	6	14	23	6	7	8	1	9	1	2	1	2	1	2	1	1	1	1
33	6	15	24	7	8	9	1	9	1	2	1	2	1	2	1	1	1	1
34	6	15	25	7	8	9	1	9	1	2	1	2	1	2	1	1	1	1
35	7	16	25	7	8	9	1	10	1	2	1	2	1	2	1	1	1	1
36	7	16	26	7	8	9	1	10	1	2	1	2	1	3	1	1	1	1
37	7	17	27	7	9	10	1	10	1	2	1	2	1	3	1	1	1	1
38	7	17	27	8	9	10	2	10	1	2	1	2	1	3	1	1	1	1
39	7	18	28	8	9	10	2	11	1	2	1	2	1	3	1	1	1	1
40	8	18	29	8	9	10	2	11	1	2	1	2	1	3	1	1	1	1
41	8	19	30	8	10	11	2	11	1	2	1	2	1	3	1	1	1	1
42	8	19	30	8	10	11	2	11	1	2	1	2	1	3	1	1	1	1
43	8	19	31	9	10	11	2	12	1	2	1	2	1	3	1	1	1	1
44	8	20	32	9	10	11	2	12	1	2	1	2	1	3	1	1	1	1
45	9	20	32	9	10	12	2	12	1	2	1	2	1	3	1	1	1	1
46	9	21	33	9	11	12	2	12	1	2	1	2	1	3	1	1	1	1
47	9	21	34	9	11	12	2	13	1	2	1	2	1	3	1	1	1	1
48	9	22	35	10	11	12	2	13	1	2	1	2	1	3	1	1	1	1
49	9	22	35	10	11	13	2	13	1	2	1	2	1	3	1	1	1	1
50	9	23	36	10	11	13	2	13	1	2	1	2	1	3	1	1	1	1
51	10	23	37	10	12	13	2	14	1	2	1	2	1	4	1	1	1	1
52	10	24	38	10	12	13	2	14	1	3	1	3	1	4	1	1	1	1
53	10	24	38	11	12	14	2	14	1	3	1	3	1	4	1	1	1	1
54	10	24	39	11	12	14	2	14	1	3	1	3	1	4	1	1	2	1
55	10	25	40	11	13	14	2	15	1	3	1	3	1	4	1	1	2	1
56	11	25	40	11	13	14	2	15	1	3	1	3	1	4	1	1	2	1
57	11	26	41	11	13	15	2	15	1	3	1	3	1	4	1	1	2	1
58	11	26	42	12	13	15	2	16	1	3	1	3	1	4	1	2	2	1
59	11	27	43	12	14	15	2	16	1	3	1	3	1	4	1	2	2	1
60	11	27	43	12	14	15	2	16	1	3	1	3	1	4	1	2	2	1

Moon's Motions for Minutes.

Min.	Evec.	Anom.	Varia.	Long.	Sup. Nod.	II	V	VI	VII	VIII	IX	XI	XII
	′ ″	′ ″	′ ″	′ ″	″								
1	0 28	0 32.7	0 30	0 32.9	0.1	0	0	0	0	0	0	0	0
2	0 57	1 5.3	1 1	1 5.9	0.3	1	0	0	0	0	0	0	0
3	1 25	1 38.0	1 31	1 38.8	0.4	1	0	0	0	0	0	0	0
4	1 53	2 10.6	2 2	2 11.8	0.5	2	0	0	0	0	0	0	0
5	2 21	2 43.3	2 32	2 44.7	0.7	2	0	0	0	0	0	0	0
6	2 50	3 16.0	3 3	3 17.6	0.8	3	0	0	0	0	0	0	0
7	3 18	3 48.6	3 33	3 50.6	0.9	3	0	0	0	0	0	0	0
8	3 46	4 21.3	4 4	4 23.5	1.1	4	0	0	0	0	0	0	0
9	4 15	4 54.0	4 34	4 56.5	1.2	4	0	0	0	0	0	0	0
10	4 43	5 26.6	5 5	5 29.4	1.3	5	0	0	0	0	0	0	0
11	5 11	5 59.3	5 35	6 2.4	1.5	5	0	0	0	0	0	0	0
12	5 40	6 31.9	6 6	6 35.3	1.6	6	0	0	0	0	0	0	0
13	6 8	7 4.6	6 36	7 8.2	1.7	6	0	0	0	0	0	0	0
14	6 36	7 37.3	7 7	7 41.2	1.9	7	0	0	0	0	0	0	0
15	7 4	8 9.9	7 37	8 14.1	2.0	7	0	0	0	0	0	0	0
16	7 33	8 42.6	8 8	8 47.1	2.1	7	0	0	0	0	0	0	0
17	8 1	9 15.3	8 38	9 20.0	2.3	8	0	0	0	0	0	0	0
18	8 29	9 47.9	9 9	9 52.9	2.4	8	0	0	0	0	0	1	0
19	8 58	10 20.6	9 39	10 25.9	2.5	9	0	0	0	0	0	1	0
20	9 26	10 53.2	10 10	10 58.8	2.6	9	0	1	0	0	0	1	0
21	9 54	11 25.9	10 40	11 31.8	2.8	10	0	1	0	0	0	1	0
22	10 22	11 58.6	11 11	12 4.7	2.9	10	1	1	0	0	1	1	0
23	10 51	12 31.2	11 41	12 37.6	3.0	11	1	1	0	0	1	1	0
24	11 19	13 3.9	12 12	13 10.6	3.2	11	1	1	0	1	1	1	1
25	11 47	13 36.6	12 42	13 43.5	3.3	12	1	1	0	1	1	1	1
26	12 16	14 9.2	13 13	14 16.5	3.4	12	1	1	1	1	1	1	1
27	12 44	14 41.9	13 43	14 49.4	3.6	13	1	1	1	1	1	1	1
28	13 12	15 14.6	14 13	15 22.3	3.7	13	1	1	1	1	1	1	1
29	13 40	15 47.2	14 44	15 55.3	3.8	13	1	1	1	1	1	1	1
30	14 9	16 19.9	15 14	16 28.2	4.0	14	1	1	1	1	1	1	1

Moon's Motions for Minutes.

Min.	1	2	3	4	5	6	7	8	9	10	11	12	13	14	15	16	17	18
1	0	0	1	0	0	0	0	0	0	0	0	0	0	0	0	0	0	0
2	0	1	1	0	0	1	0	1	0	0	0	0	0	0	0	0	0	0
3	1	1	2	1	1	1	0	1	0	0	0	0	0	0	0	0	0	0
4	1	2	3	1	1	1	0	1	0	0	0	0	0	0	0	0	0	0
5	1	2	4	1	1	1	0	1	0	0	0	0	0	0	0	0	0	0
6	1	3	4	1	1	2	0	2	0	0	0	0	0	0	0	0	0	0
7	1	3	5	1	2	2	0	2	0	0	0	0	0	0	0	0	0	0
8	2	4	6	2	2	2	0	2	0	0	0	0	0	1	0	0	0	0
9	2	4	6	2	2	2	0	2	0	0	0	0	0	1	0	0	0	0
10	2	5	7	2	2	3	0	3	0	0	0	0	0	1	0	0	0	0
11	2	5	8	2	3	3	0	3	0	1	0	1	0	1	0	0	0	0
12	2	5	9	2	3	3	0	3	0	1	0	1	0	1	0	0	0	0
13	2	6	9	3	3	3	1	4	0	1	0	1	0	1	0	0	0	0
14	3	6	10	3	3	4	1	4	0	1	0	1	0	1	0	0	0	0
15	3	7	11	3	3	4	1	4	0	1	0	1	0	1	0	0	0	0
16	3	7	12	3	4	4	1	4	0	1	0	1	0	1	0	0	0	0
17	3	8	12	3	4	4	1	5	0	1	0	1	0	1	0	0	0	0
18	3	8	13	4	4	5	1	5	0	1	0	1	0	1	0	0	1	0
19	4	9	14	4	4	5	1	5	0	1	0	1	0	1	0	0	1	0
20	4	9	14	4	5	5	1	5	0	1	0	1	0	1	0	1	1	0
21	4	10	15	4	5	5	1	6	0	1	0	1	0	1	0	1	1	0
22	4	10	16	4	5	6	1	6	0	1	0	1	1	2	0	1	1	0
23	4	10	17	5	5	6	1	6	0	1	0	1	1	2	0	1	1	0
24	5	11	17	5	6	6	1	7	0	1	1	1	1	2	1	1	1	0
25	5	11	18	5	6	6	1	7	0	1	1	1	1	2	1	1	1	0
26	5	12	19	5	6	7	1	7	0	1	1	1	1	2	1	1	1	0
27	5	12	19	5	6	7	1	7	0	1	1	1	1	2	1	1	1	0
28	5	13	20	6	7	7	1	8	0	1	1	1	1	2	1	1	1	0
29	6	13	21	6	7	7	1	8	0	1	1	1	1	2	1	1	1	0
30	6	14	22	6	7	8	1	8	0	1	1	1	1	2	1	1	1	0

Moon's Motions for Minutes.

Min.	Evec.	Anom.	Varia.	Long.	Sup. Nod.	II	V	VI	VII	VIII	IX	XI	XII
	′ ″	′ ″	′ ″	′ ″	″								
31	14 37	16 52.5	15 45	17 1.2	4.1	14	1	1	1	1	1	1	1
32	15 5	17 25.2	16 15	17 34.1	4.2	15	1	1	1	1	1	1	1
33	15 34	17 57.9	16 46	18 7.1	4.4	15	1	1	1	1	1	1	1
34	16 2	18 30.5	17 16	18 40.0	4.5	16	1	1	1	1	1	1	1
35	16 30	19 3.2	17 47	19 12.9	4.7	16	1	1	1	1	1	1	1
36	16 58	19 35.8	18 17	19 45.9	4.8	17	1	1	1	1	1	1	1
37	17 27	20 8.5	18 48	20 18.8	4.9	17	1	1	1	1	1	1	1
38	17 55	20 41.2	19 18	20 51.8	5.0	18	1	1	1	1	1	1	1
39	18 23	21 13.8	19 49	21 24.7	5.2	18	1	1	1	1	1	1	1
40	18 52	21 46.5	20 19	21 57.6	5.3	19	1	1	1	1	1	1	1
41	19 20	22 19.2	20 50	22 30.6	5.4	19	1	1	1	1	1	1	1
42	19 48	22 51.8	21 20	23 3.5	5.6	20	1	1	1	1	1	1	1
43	20 16	23 24.5	21 51	23 36.5	5.7	20	1	1	1	1	1	1	1
44	20 45	23 57.1	22 21	24 9.4	5.8	21	1	1	1	1	1	1	1
45	21 13	24 29.8	22 52	24 42.3	6.0	21	1	1	1	1	1	1	1
46	21 41	25 2.5	23 22	25 15.3	6.1	21	1	1	1	1	1	1	1
47	22 10	25 35.1	23 53	25 48.2	6.2	22	1	1	1	1	1	1	1
48	22 38	26 7.8	24 23	26 21.2	6.4	22	1	1	1	1	1	1	1
49	23 6	26 40.5	24 54	26 54.1	6.5	23	1	1	1	1	1	1	1
50	23 34	27 13.1	25 24	27 27.0	6.6	23	1	1	1	1	1	1	1
51	24 3	27 45.8	25 55	28 0.0	6.8	24	1	1	1	1	1	1	1
52	24 31	28 18.5	26 25	28 32.9	6.9	24	1	1	1	1	1	1	1
53	24 59	28 51.1	26 56	29 5.9	7.0	25	1	1	1	1	1	1	1
54	25 28	29 23.8	27 26	29 38.8	7.1	25	1	1	1	1	1	2	1
55	25 56	29 56.4	27 56	30 11.8	7.3	26	1	1	1	1	1	2	1
56	26 24	30 29.1	28 27	30 44.7	7.4	26	1	1	1	1	1	2	1
57	26 52	31 1.8	28 57	31 17.6	7.5	27	1	2	1	1	1	2	1
58	27 21	31 34.4	29 28	31 50.6	7.7	27	1	2	1	1	1	2	1
59	27 49	32 7.1	29 58	32 23.5	7.8	28	1	2	1	1	1	2	1
60	28 17	32 39.8	30 29	32 56.5	7.9	28	1	2	1	1	1	2	1

Moon's Motions for Seconds.

Sec.	Evec.	Anom.	Var.	Long.	Sec.	Evec.	Anom.	Var.	Long.
	″	″	″	″		″	″	″	″
1	0	0.5	1	0.5	31	15	16.9	16	17.0
2	1	1.1	1	1.1	32	15	17.4	16	17.6
3	1	1.6	2	1.6	33	16	18.0	17	18.1
4	2	2.2	2	2.2	34	16	18.5	17	18.7
5	2	2.7	3	2.7	35	17	19.1	18	19.2
6	3	3.3	3	3.3	36	17	19.6	18	19.8
7	3	3.8	4	3.8	37	18	20.1	19	20.3
8	4	4.3	4	4.4	38	18	20.7	19	20.9
9	4	4.9	5	4.9	39	18	21.2	20	21.4
10	5	5.4	5	5.5	40	19	21.8	20	22.0
11	5	6.0	6	6.0	41	19	22.3	21	22.5
12	6	6.5	6	6.6	42	20	22.9	21	23.1
13	6	7.1	7	7.1	43	20	23.4	22	23.6
14	7	7.6	7	7.7	44	21	24.0	22	24.2
15	7	8.2	8	8.2	45	21	24.5	23	24.7
16	8	8.7	8	8.8	46	22	25.0	23	25.3
17	8	9.2	9	9.3	47	22	25.6	24	25.8
18	9	9.8	9	9.9	48	23	26.1	24	26.4
19	9	10.3	10	10.4	49	23	26.7	25	26.9
20	9	10.9	10	11.0	50	24	27.2	25	27.4
21	10	11.4	11	11.5	51	24	27.8	26	28.0
22	10	12.0	11	12.1	52	25	28.3	26	28.5
23	11	12.5	12	12.6	53	25	28.9	27	29.1
24	11	13.1	12	13.2	54	26	29.4	27	29.6
25	12	13.6	13	13.7	55	26	29.9	28	30.2
26	12	14.1	13	14.3	56	26	30.5	28	30.7
27	13	14.7	14	14.8	57	27	31.0	29	31.3
28	13	15.2	14	15.4	58	27	31.6	29	31.8
29	14	15.8	15	15.9	59	28	32.1	30	32.4
30	14	16.3	15	16.5	60	28	32.7	30	32.9

First Equation of Moon's Longitude.—Argument 1.

Arg.	1 ′	″	Diff. for 10 ″	Arg.	1 ′	″	Diff. for 10 ″	Arg.	1 ′	″	Diff. for 10 ″	Arg.	1 ′	″	Diff. for 10 ″
0	12	40.0	4.24	2500	1	40.7	0.16	5000	12	40.0	4.06	7500	23	39.3	0.02
50	12	18.8	4.22	2550	1	41.5	0.28	5050	13	0.3	4.04	7550	23	39.4	0.10
100	11	57.7	4.22	2600	1	42.9	0.42	5100	13	20.5	4.04	7600	23	38.9	0.24
150	11	36.6	4.20	2650	1	45.0	0.54	5150	13	40.7	4.04	7650	23	37.7	0.38
200	11	15.6	4.18	2700	1	47.7	0.66	5200	14	0.9	4.00	7700	23	35.8	0.50
250	10	54.7	4.16	2750	1	51.0	0.80	5250	14	20.9	4.00	7750	23	33.3	0.62
300	10	33.9	4.14	2800	1	55.0	0.92	5300	14	40.9	3.98	7800	23	30.2	0.76
350	10	13.2	4.12	2850	1	59.6	1.04	5350	15	0.8	3.94	7850	23	26.4	0.88
400	9	52.6	4.06	2900	2	4.8	1.18	5400	15	20.5	3.92	7900	23	22.0	1.02
450	9	32.3	4.04	2950	2	10.7	1.28	5450	15	40.1	3.90	7950	23	16.9	1.14
500	9	12.1	4.00	3000	2	17.1	1.42	5500	15	59.6	3.84	8000	23	11.2	1.26
550	8	52.1	3.94	3050	2	24.2	1.54	5550	16	18.8	3.80	8050	23	4.9	1.40
600	8	32.4	3.88	3100	2	31.9	1.64	5600	16	37.8	3.78	8100	22	57.9	1.52
650	8	13.0	3.84	3150	2	40.1	1.76	5650	16	56.7	3.72	8150	22	50.3	1.66
700	7	53.8	3.78	3200	2	48.9	1.88	5700	17	15.3	3.66	8200	22	42.0	1.76
750	7	34.9	3.70	3250	2	58.3	1.98	5750	17	33.6	3.60	8250	22	33.2	1.90
800	7	16.4	3.64	3300	3	8.2	2.10	5800	17	51.6	3.56	8300	22	23.7	2.00
850	6	58.2	3.58	3350	3	18.7	2.20	5850	18	9.4	3.50	8350	22	13.7	2.12
900	6	40.3	3.50	3400	3	29.7	2.32	5900	18	26.9	3.42	8400	22	3.1	2.24
950	6	22.8	3.42	3450	3	41.3	2.42	5950	18	44.0	3.36	8450	21	51.9	2.36
1000	6	5.7	3.34	3500	3	53.4	2.50	6000	19	0.8	3.28	8500	21	40.1	2.46
1050	5	49.0	3.24	3550	4	5.9	2.62	6050	19	17.2	3.22	8550	21	27.8	2.56
1100	5	32.8	3.16	3600	4	19.0	2.70	6100	19	33.3	3.14	8600	21	15.0	2.68
1150	5	17.0	3.08	3650	4	32.5	2.80	6150	19	49.0	3.04	8650	21	1.6	2.78
1200	5	1.6	2.98	3700	4	46.5	2.88	6200	20	4.2	2.98	8700	20	47.7	2.88
1250	4	46.7	2.88	3750	5	0.9	2.98	6250	20	19.1	2.88	8750	20	33.3	2.98
1300	4	32.3	2.78	3800	5	15.8	3.04	6300	20	33.5	2.80	8800	20	18.4	3.08
1350	4	18.4	2.68	3850	5	31.0	3.14	6350	20	47.5	2.70	8850	20	3.0	3.16
1400	4	5.0	2.56	3900	5	46.7	3.22	6400	21	1.0	2.62	8900	19	47.2	3.24
1450	3	52.2	2.46	3950	6	2.8	3.28	6450	21	14.1	2.50	8950	19	31.0	3.34
1500	3	39.9	2.36	4000	6	19.2	3.36	6500	21	26.6	2.42	9000	19	14.3	3.42
1550	3	28.1	2.24	4050	6	36.0	3.42	6550	21	38.7	2.32	9050	18	57.2	3.50
1600	3	16.9	2.12	4100	6	53.1	3.50	6600	21	50.3	2.20	9100	18	39.7	3.58
1650	3	6.3	2.00	4150	7	10.6	3.56	6650	22	1.3	2.10	9150	18	21.8	3.64
1700	2	56.3	1.90	4200	7	28.4	3.60	6700	22	11.8	1.98	9200	18	3.6	3.70
1750	2	46.8	1.76	4250	7	46.4	3.66	6750	22	21.7	1.88	9250	17	45.1	3.78
1800	2	38.0	1.66	4300	8	4.7	3.72	6800	22	31.1	1.76	9300	17	26.2	3.84
1850	2	29.7	1.52	4350	8	23.3	3.78	6850	22	39.9	1.64	9350	17	7.0	3.88
1900	2	22.1	1.40	4400	8	42.2	3.80	6900	22	48.1	1.54	9400	16	47.6	3.94
1950	2	15.1	1.26	4450	9	1.2	3.84	6950	22	55.8	1.42	9450	16	27.9	4.00
2000	2	8.8	1.14	4500	9	20.4	3.90	7000	23	2.9	1.28	9500	16	7.9	4.04
2050	2	3.1	1.02	4550	9	39.9	3.92	7050	23	9.3	1.18	9550	15	47.7	4.06
2100	1	58.0	0.88	4600	9	59.5	3.94	7100	23	15.2	1.04	9600	15	27.4	4.12
2150	1	53.6	0.76	4650	10	19.2	3.98	7150	23	20.4	0.92	9650	15	6.8	4.14
2200	1	49.8	0.62	4700	10	39.1	4.00	7200	23	25.0	0.80	9700	14	46.1	4.16
2250	1	46.7	0.50	4750	10	59.1	4.00	7250	23	29.0	0.66	9750	14	25.3	4.18
2300	1	44.2	0.38	4800	11	19.1	4.04	7300	23	32.3	0.54	9800	14	4.4	4.20
2350	1	42.3	0.24	4850	11	39.3	4.04	7350	23	35.0	0.42	9850	13	43.4	4.22
2400	1	41.1	0.10	4900	11	59.5	4.04	7400	23	37.1	0.28	9900	13	22.3	4.22
2450	1	40.6	0.02	4950	12	19.7	4.06	7450	23	38.5	0.16	9950	13	1.2	4.24
2500	1	40.7		5000	12	40.0		7500	23	39.3		10000	12	40.0	

Equations 2 *to* 7 *of Moon's Longitude.* Arguments 2 to 7

Arg.	2	diff	3	diff	4	diff	5	diff	6	diff	7	diff	Arg.
	′ ″	″	′ ″	″	′ ″	″	′ ″	″	′ ″	″	′ ″	″	
2500	4 57.3	0.3	0 2.3	0.1	6 30.3	0.4	3 39.4	0.2	0 6.2	0.2	0 0.8	0.1	2500
2600	4 57.0	0.9	0 2.4	0.4	6 29.9	1.1	3 39.2	0.7	0 6.4	0.5	0 0.9	0.4	2400
2700	4 56.1	1.4	0 2.8	0.5	6 28.8	1.9	3 38.5	1.0	0 6.9	0.8	0 1.3	0.5	2300
2800	4 54.7	2.0	0 3.3	0.8	6 26.9	2.6	3 37.5	1.5	0 7.7	1.1	0 1.8	0.9	2200
2900	4 52.7	2.6	0 4.1	1.0	6 24.3	3.3	3 36.0	1.9	0 8.8	1.5	0 2.7	1.0	2100
3000	4 50.1	3.1	0 5.1	1.3	6 21.0	4.1	3 34.1	2.4	0 10.3	1.8	0 3.7	1.3	2000
3100	4 47.0	3.7	0 6.4	1.4	6 16.9	4.7	3 31.7	2.7	0 12.1	2.1	0 5.0	1.4	1900
3200	4 43.3	4.2	0 7.8	1.6	6 12.2	5.4	3 29.0	3.1	0 14.2	2.4	0 6.4	1.7	1800
3300	4 39.1	4.7	0 9.4	1.9	6 6.8	6.1	3 25.9	3.5	0 16.6	2.6	0 8.1	1.9	1700
3400	4 34.4	5.2	0 11.3	2.0	6 0.7	6.7	3 22.4	3.9	0 19.2	3.0	0 10.0	2.1	1600
3500	4 29 2	5.7	0 13.3	2.2	5 54.0	7.4	3 18.5	4.2	0 22.2	3.2	0 12.1	2.3	1500
3600	4 23 5	6.1	0 15.5	2.4	5 46.6	7.9	3 14.3	4.6	0 25.4	3.5	0 14.4	2.4	1400
3700	4 17.4	6.6	0 17.9	2.6	5 38.7	8.4	3 9.7	4.8	0 28.9	3.8	0 16.8	2.7	1300
3800	4 10.8	6.9	0 20.5	2.7	5 30.3	9.0	3 4.9	5.2	0 32.7	3.9	0 19.5	2.8	1200
3900	4 3.9	7.3	0 23.2	2.9	5 21.3	9.4	2 59.7	5.4	0 36.6	4.1	0 22.3	2.9	1100
4000	3 56.6	7.7	0 26.1	3.0	5 11.9	9.9	2 54 3	5.7	0 40.7	4.4	0 25.2	3.1	1000
4100	3 48.9	7.9	0 29.1	3.1	5 2.0	10.3	2 48.6	5.9	0 45.1	4.5	0 28.3	3.2	900
4200	3 41.0	8.3	0 32.2	3.2	4 51.7	10.7	2 42.7	6.1	0 49.6	4.7	0 31.5	3.3	800
4300	3 32.7	8.5	0 35.4	3.4	4 41.0	10.9	2 36.6	6.3	0 54.3	4.9	0 34.8	3.4	700
4400	3 24.2	8.7	0 38.8	3.4	4 30.1	11.3	2 30.3	6.5	0 59.2	4.9	0 38.2	3.5	600
4500	3 15.5	8.9	0 42.2	3.5	4 18.8	11.5	2 23.8	6.6	1 4.1	5.1	0 41.7	3.6	500
4600	3 6.6	9.0	0 45.7	3.5	4 7.3	11.6	2 17.2	6.7	1 9.2	5.1	0 45.3	3.6	400
4700	2 57.6	9.1	0 49.2	3.6	3 55.7	11.8	2 10.5	6.8	1 14 3	5.2	0 48.9	3.7	300
4800	2 48.5	9.3	0 52.8	3.6	3 43.9	12.0	2 3.7	6.8	1 19.5	5.2	0 52.6	3.7	200
4900	2 39.2	9.2	0 56.4	3.6	3 31.9	11.9	1 56.9	6.9	1 24.7	5.3	0 56.3	3.7	100
5000	2 30.0	9.2	1 0.0	3.6	3 20.0	11.9	1 50.0	6.9	1 30.0	5.3	1 0.0	3.7	0
5100	2 20.8	9.3	1 3.6	3.6	3 8.1	12.0	1 43.1	6.8	1 35.3	5.2	1 3.7	3.7	9900
5200	2 11.5	9.1	1 7.2	3.6	2 56.1	11.8	1 36.3	6.8	1 40.5	5.2	1 7.4	3.7	9800
5300	2 2.4	9.0	1 10.8	3.5	2 44.3	11.6	1 29.5	6.7	1 45.7	5.1	1 11.1	3.6	9700
5400	1 53.4	8.9	1 14.3	3.5	2 32.7	11.5	1 22.8	6.6	1 50.8	5.1	1 14.7	3.6	9600
5500	1 44.5	8.7	1 17.8	3.4	2 21.2	11.3	1 16.2	6.5	1 55.9	4.9	1 18 3	3.5	9500
5600	1 35.8	8.5	1 21.2	3.4	2 9.9	10.9	1 9.7	6.3	2 0 8	4.9	1 21.8	3.4	9400
5700	1 27.3	8.3	1 24.6	3.2	1 59.0	10.7	1 3.4	6.1	2 5.7	4.7	1 25.2	3.3	9300
5800	1 19.0	7.9	1 27.8	3.1	1 48.3	10.3	0 57.3	5.9	2 10.4	4.5	1 28.5	3.2	9200
5900	1 11.1	7.7	1 30.9	3.0	1 38.0	9.9	0 51.4	5.7	2 14.9	4.4	1 31.7	3.1	9100
6000	1 3.4	7.3	1 33.9	2.9	1 28.1	9.4	0 45.7	5.4	2 19 3	4.1	1 34.8	2.9	9000
6100	0 56.1	6.9	1 36.8	2.7	1 18.7	9.0	0 40 3	5.2	2 23.4	3.9	1 37.7	2.8	8900
6200	0 49.2	6.6	1 39.5	2.6	1 9.7	8.4	0 35.1	4.8	2 27.3	3.8	1 40.5	2.7	8800
6300	0 42.6	6.1	1 42.1	2.4	1 1.3	7.9	0 30.3	4.6	2 31.1	3.5	1 43.2	2.4	8700
6400	0 36.5	5.7	1 44.5	2.2	0 53.4	7.4	0 25.7	4.2	2 34.6	3.2	1 45.6	2.3	8600
6500	0 30.8	5.2	1 46.7	2.0	0 46.0	6.7	0 21.5	3.9	2 37.8	3.0	1 47.9	2.1	8500
6600	0 25.6	4 7	1 48.7	1.9	0 39.3	6.1	0 17.6	3.5	2 40.8	2.6	1 50.0	1.9	8400
6700	0 20.9	4.2	1 50.6	1.6	0 33.2	5.4	0 14.1	3.1	2 43.4	2.4	1 51.9	1.7	8300
6800	0 16.7	3.7	1 52.2	1.4	0 27.8	4.7	0 11.0	2.7	2 45.8	2.1	1 53.6	1.4	8200
6900	0 13.0	3.1	1 53.6	1.3	0 23.1	4.1	0 8.3	2.4	2 47.9	1.8	1 55.0	1.3	8100
7000	0 9.9	2.6	1 54.9	1.0	0 19.0	3.3	0 5.9	1.9	2 49.7	1.5	1 56.3	1.0	8000
7100	0 7.3	2.0	1 55.9	0.8	0 15.7	2.6	0 4.0	1.5	2 51.2	1.1	1 57.3	0.9	7900
7200	0 5.3	1.4	1 56.7	0.5	0 13.1	1.9	0 2.5	1.0	2 52.3	0.8	1 58.2	0.5	7800
7300	0 3.9	0.9	1 57.2	0.4	0 11.2	1.1	0 1.5	0.7	2 53.1	0.5	1 58.7	0.4	7700
7400	0 3.0	0.3	1 57.6	0.1	0 10.1	0.4	0 0.8	0.2	2 53.6	0.2	1 59.1	0.1	7600
7500	0 2.7		1 57.7		0 9.7		0 0 6		2 53.8		1 59.2		7500

TABLE XLIII.

Equations 8 and 9.

Arg.	8	9	Arg.	8	9
	′ ″	′ ″		′ ″	′ ″
0	1 20.0	1 20.0	5000	1 20.0	1 20.0
100	1 15.5	1 28.7	5100	1 24.4	1 25.8
200	1 11.1	1 37.3	5200	1 28.8	1 31.4
300	1 6.7	1 45.7	5300	1 33.1	1 36.9
400	1 2.3	1 53.7	5400	1 37.4	1 42.0
500	0 58.0	2 1.3	5500	1 41.6	1 46.8
600	0 53.8	2 8.3	5600	1 45.8	1 51.0
700	0 49.7	2 14.7	5700	1 49.8	1 54.6
800	0 45.7	2 20.2	5800	1 53.8	1 57.6
900	0 41.9	2 25.0	5900	1 57.6	1 59.8
1000	0 38.2	2 28.9	6000	2 1.2	2 1.3
1100	0 34.7	2 31.9	6100	2 4.7	2 1.9
1200	0 31.4	2 33.9	6200	2 8.0	2 1.7
1300	0 28.2	2 34.9	6300	2 11.2	2 0.7
1400	0 25.3	2 35.0	6400	2 14.1	1 58.8
1500	0 22.6	2 34.1	6500	2 16.8	1 56.1
1600	0 20.1	2 32.2	6600	2 19.3	1 52.5
1700	0 17.9	2 29.5	6700	2 21.6	1 48.3
1800	0 15.9	2 25.9	6800	2 23.7	1 43.4
1900	0 14.2	2 21.5	6900	2 25.4	1 37.8
2000	0 12.7	2 16.4	7000	2 27.0	1 31.7
2100	0 11.5	2 10.7	7100	2 28.2	1 25.1
2200	0 10.5	2 4.4	7200	2 29.2	1 18.2
2300	0 9.9	1 57.7	7300	2 30.0	1 11.1
2400	0 9.5	1 50.7	7400	2 30.4	1 3.8
2500	0 9.4	1 43.5	7500	2 30.6	0 56.5
2600	0 9.6	1 36.2	7600	2 30.5	0 49.3
2700	0 10.1	1 28.9	7700	2 30.1	0 42.3
2800	0 10.8	1 21.8	7800	2 29.5	0 35.6
2900	0 11.8	1 14.9	7900	2 28.5	0 29.3
3000	0 13.0	1 8.3	8000	2 27.3	0 23.6
3100	0 14.6	1 2.2	8100	2 25.8	0 18.5
3200	0 16.3	0 56.6	8200	2 24.1	0 14.1
3300	0 18.4	0 51.7	8300	2 22.1	0 10.5
3400	0 20.7	0 47.5	8400	2 19.9	0 7.8
3500	0 23.2	0 43.9	8500	2 17.4	0 5.9
3600	0 25.9	0 41.2	8600	2 14.7	0 5.0
3700	0 28.8	0 39.3	8700	2 11.8	0 5.1
3800	0 32.0	0 38.3	8800	2 8.6	0 6.1
3900	0 35.3	0 38.1	8900	2 5.3	0 8.1
4000	0 38.8	0 38.7	9000	2 1.8	0 11.1
4100	0 42.4	0 40.2	9100	1 58.1	0 15.0
4200	0 46.2	0 42.4	9200	1 54.3	0 19.8
4300	0 50.2	0 45.4	9300	1 50.3	0 25.3
4400	0 54.2	0 49.0	9400	1 46.2	0 31.7
4500	0 58.4	0 53.2	9500	1 42.0	0 38.7
4600	1 2.6	0 58.0	9600	1 37.7	0 46.3
4700	1 6.9	1 3.1	9700	1 33.3	0 54.3
4800	1 11.2	1 8.6	9800	1 28.9	1 2.7
4900	1 15.6	1 14.2	9900	1 24.5	1 11.3
5000	1 20.0	1 20.0	10000	1 20.0	1 20.0

TABLE XLIV.

Equations 10 and 11.

Arg.	10	11	Arg.	10	11
	″	″		″	″
0	10.0	10.0	500	10.0	10.0
10	9.3	11.1	510	9.6	10.8
20	8.6	12.1	520	9.2	11.5
30	8.0	13.1	530	8.9	12.3
40	7.4	14.1	540	8.5	12.9
50	6.8	15.0	550	8.2	13.6
60	6.2	15.8	560	7.9	14.2
70	5.7	16.6	570	7.7	14.6
80	5.3	17.3	580	7.5	15.0
90	4.9	17.9	590	7.4	15.4
100	4.6	18.3	600	7.3	15.6
110	4.3	18.6	610	7.2	15.7
120	4.1	18.9	620	7.3	15.7
130	4.0	19.0	630	7.4	15.6
140	4.0	18.9	640	7.5	15.4
150	4.0	18.8	650	7.8	15.1
160	4.2	18.6	660	8.1	14.7
170	4.4	18.2	670	8.4	14.2
180	4.6	17.7	680	8.7	13.5
190	4.9	17.1	690	9.2	12.8
200	5.3	16.5	700	9.7	12.1
210	5.7	15.7	710	10.2	11.3
220	6.2	14.9	720	10.7	10.4
230	6.7	14.1	730	11.2	9.5
240	7.2	13.2	740	11.7	8.6
250	7.7	12.3	750	12.3	7.7
260	8.3	11.4	760	12.8	6.8
270	8.8	10.5	770	13.3	5.9
280	9.3	9.6	780	13.8	5.1
290	9.8	8.7	790	14.3	4.3
300	10.3	7.9	800	14.7	3.5
310	10.8	7.2	810	15.1	2.9
320	11.3	6.5	820	15.4	2.3
330	11.6	5.8	830	15.6	1.8
340	11.9	5.3	840	15.8	1.4
350	12.2	4.9	850	16.0	1.2
360	12.5	4.6	860	16.0	1.1
370	12.6	4.4	870	16.0	1.0
380	12.7	4.3	880	15.9	1.1
390	12.8	4.3	890	15.7	1.4
400	12.7	4.4	900	15.4	1.7
410	12.6	4.6	910	15.1	2.1
420	12.5	5.0	920	14.7	2.7
430	12.3	5.4	930	14.3	3.4
440	12.1	5.8	940	13.8	4.2
450	11.8	6.4	950	13.2	5.0
460	11.5	7.1	960	12.6	5.9
470	11.1	7.7	970	12.0	6.9
480	10.8	8.5	980	11.4	7.9
490	10.4	9.2	990	10.7	8.9
500	10.0	10.0	1000	10.0	10.0

TABLE XLV.

Equations 12 to 19.

Arg.	12	13	14	15	16	17	18	19	Arg.
	″	″	″	″	″	″	″		
250	2.3	1.6	7.8	0.0	33.7	3.4	16.7	0.4	250
260	2.3	1.6	7.8	0.0	33.7	3.4	16.7	0.4	240
270	2.4	1.7	7.9	0.1	33.6	3.5	16.6	0.4	230
280	2.6	1.9	8.0	0.2	33.5	3.5	16.6	0.5	220
290	2.9	2.2	8.2	0.3	33.2	3.6	16.5	0.5	210
300	3.2	2.5	8.4	0.5	33.0	3.7	16.4	0.6	200
310	3.5	2.9	8.7	0.7	32.7	3.9	16.2	0.7	190
320	4.0	3.4	9.0	1.0	32.4	4.0	16.1	0.8	180
330	4.5	3.9	9.3	1.2	32.0	4.2	15.9	1.0	170
340	5.1	4.4	9.7	1.6	31.6	4.4	15.7	1.1	160
350	5.7	5.1	10.1	1.9	31.1	4.7	15.4	1.3	150
360	6.4	5.8	10.6	2.3	30.6	4.9	15.2	1.5	140
370	7.1	6.6	11.1	2.7	30.1	5.2	14.9	1.7	130
380	7.9	7.4	11.7	3.2	29.4	5.5	14.6	1.9	120
390	8.7	8.3	12.2	3.6	28.7	5.8	14.3	2.1	110
400	9.6	9.2	12.8	4.1	28.0	6.1	13.9	2.3	100
410	10.5	10.1	13.5	4.6	27.3	6.5	13.6	2.5	90
420	11.5	11.1	14.1	5.2	26.6	6.8	13.2	2.8	80
430	12.5	12.2	14.8	5.7	25.8	7.2	12.9	3.1	70
440	13.5	13.2	15.5	6.3	25.0	7.6	12.5	3.3	60
450	14.5	14.3	16.2	6.9	24.2	8.0	12.1	3.6	50
460	15.6	15.4	17.0	7.5	23.4	8.4	11.7	3.9	40
470	16.7	16.5	17.7	8.1	22.6	8.8	11.3	4.1	30
480	17.8	17.7	18.5	8.7	21.7	9.2	10.8	4.4	20
490	18.9	18.8	19.2	9.4	20.9	9.6	10.4	4.7	10
500	20.0	20.0	20.0	10.0	20.0	10.0	10.0	5.0	0
510	21.1	21.2	20.8	10.6	19.1	10.4	9.6	5.3	990
520	22.2	22.3	21.5	11.3	18.3	10.8	9.2	5.6	980
530	23.3	23.5	22.3	11.9	17.4	11.2	8.7	5.9	970
540	24.4	24.6	23.0	12.5	16.6	11.6	8.3	6.1	960
550	25.5	25.7	23.8	13.1	15.8	12.0	7.9	6.4	950
560	26.5	26.8	24.5	13.7	15.0	12.4	7.5	6.7	940
570	27.5	27.8	25.2	14.3	14.2	12.8	7.1	6.9	930
580	28.5	28.9	25.9	14.8	13.4	13.2	6.8	7.2	920
590	29.5	29.9	26.5	15.4	12.7	13.5	6.4	7.5	910
600	30.4	30.8	27.2	15.9	12.0	13.9	6.1	7.7	900
610	31.3	31.7	27.8	16.4	11.3	14.2	5.7	7.9	890
620	32.1	32.6	28.3	16.8	10.6	14.5	5.4	8.1	880
630	32.9	33.4	28.9	17.3	9.9	14.8	5.1	8.3	870
640	33.6	34.2	29.4	17.7	9.4	15.1	4.8	8.5	860
650	34.3	34.9	29.9	18.1	8.9	15.3	4.6	8.7	850
660	34.9	35.6	30.3	18.4	8.4	15.6	4.3	8.9	840
670	35.5	36.1	30.7	18.8	8.0	15.8	4.1	9.0	830
680	36.0	36.6	31.0	19.0	7.6	16.0	3.9	9.2	820
690	36.5	37.1	31.3	19.3	7.3	16.1	3.8	9.3	810
700	36.8	37.5	31.6	19.5	7.0	16.3	3.6	9.4	800
710	37.1	37.8	31.8	19.7	6.8	16.4	3.5	9.5	790
720	37.4	38.1	32.0	19.8	6.5	16.5	3.4	9.5	780
730	37.6	38.3	32.1	19.9	6.4	16.5	3.4	9.6	770
740	37.7	38.4	32.2	20.0	6.3	16.6	3.3	9.6	760
750	37.7	38.4	32.2	20.0	6.3	16.6	3.3	9.6	750

TABLE XLVI.

Equation 20.

Arg.	20	Arg.
	″	
0	10.0	500
10	10.9	510
20	11.8	520
30	12.7	530
40	13.5	540
50	14.3	550
60	15.0	560
70	15.7	570
80	16.2	580
90	16.7	590
100	17.0	600
110	17.2	610
120	17.4	620
130	17.4	630
140	17.2	640
150	17.0	650
160	16.7	660
170	16.2	670
180	15.7	680
190	15.0	690
200	14.3	700
210	13.5	710
220	12.7	720
230	11.8	730
240	10.9	740
250	10.0	750
260	9.1	760
270	8.2	770
280	7.3	780
290	6.5	790
300	5.7	800
310	5.0	810
320	4.3	820
330	3.8	830
340	3.3	840
350	3.0	850
360	2.8	860
370	2.6	870
380	2.6	880
390	2.8	890
400	3.0	900
410	3.3	910
420	3.8	920
430	4.3	930
440	5.0	940
450	5.7	950
460	6.5	960
470	7.3	970
480	8.2	980
490	9.1	990
500	10.0	1000

TABLE XLVII.

Equations 21 *to* 29.

Arg	21	22	23	24	25	26	27	28	29	Arg
	″	″	″	″	″	″	″	″	″	
25	7.8	3.2	7.1	6.1	5.9	4.1	5.8	4.3	5.7	25
27	7.8	3.2	7.1	6.1	5.9	4.1	5.8	4.3	5.7	23
29	7.7	3.3	7.0	6.1	5.9	4.1	5.8	4.3	5.7	21
31	7.6	3.3	7.0	6.0	5.8	4.2	5.7	4.3	5.7	19
33	7.5	3.4	6.8	6.0	5.8	4.2	5.7	4.4	5.6	17
35	7.3	3.5	6.7	5.9	5.7	4.3	5.6	4.4	5.6	15
37	7.0	3.7	6.5	5.8	5.7	4.3	5.6	4.5	5.5	13
39	6.8	3.9	6.3	5.7	5.6	4.4	5.5	4.6	5.4	11
41	6.5	4.0	6.1	5.6	5.5	4.5	5.4	4.6	5.4	09
43	6.2	4.2	5.9	5.5	5.4	4.6	5.3	4.7	5.3	07
45	5.9	4.4	5.6	5.3	5.3	4.7	5.2	4.8	5.2	05
47	5.5	4.7	5.4	5.2	5.2	4.8	5.1	4.9	5.1	03
49	5.2	4.9	5.1	5.1	5.1	4.9	5.0	5.0	5.0	01
51	4.8	5.1	4.9	4.9	4.9	5.1	5.0	5.0	5.0	99
53	4.5	5.3	4.6	4.8	4.8	5.2	4.9	5.1	4.9	97
55	4.1	5.6	4.4	4.7	4.7	5.3	4.8	5.2	4.8	95
57	3.8	5.8	4.1	4.5	4.6	5.4	4.7	5.3	4.7	93
59	3.5	6.0	3.9	4.4	4.5	5.5	4.6	5.4	4.6	91
61	3.2	6.1	3.7	4.3	4.4	5.6	4.5	5.4	4.6	89
63	3.0	6.3	3.5	4.2	4.3	5.7	4.4	5.5	4.5	87
65	2.7	6.5	3.3	4.1	4.3	5.7	4.4	5.6	4.4	85
67	2.5	6.6	3.2	4.0	4.2	5.8	4.3	5.6	4.4	83
69	2.4	6.7	3.0	4.0	4.2	5.8	4.3	5.7	4.3	81
71	2.3	6.7	3.0	3.9	4.1	5.9	4.2	5.7	4.3	79
73	2.2	6.8	2.9	3.9	4.1	5.9	4.2	5.7	4.3	77
75	2.2	6.8	2.9	3.9	4.1	5.9	4.2	5.7	4.3	75

TABLE XLVIII.

Equations 30 *and* 31

Arg.	30	31
	″	″
0	5.0	5.0
2	5.0	5.0
4	4.9	5.1
6	4.9	5.1
8	4.8	5.2
10	4.8	5.2
12	4.7	5.3
14	4.6	5.4
16	4.5	5.5
18	4.4	5.5
20	4.2	5.6
22	4.1	5.7
24	4.0	5.8
26	3.9	5.8
28	3.8	5.9
30	3.7	5.9
32	3.7	5.9
34	3.7	5.9
36	3.7	5.9
38	3.8	5.8
40	3.9	5.7
42	4.1	5.6
44	4.3	5.5
46	4.5	5.3
48	4.8	5.2
50	5.0	5.0
52	5.2	4.8
54	5.5	4.7
56	5.7	4.5
58	5.9	4.4
60	6.1	4.3
62	6.2	4.2
64	6.3	4.1
66	6.3	4.1
68	6.3	4.1
70	6.3	4.1
72	6.2	4.1
74	6.2	4.2
76	6.0	4.2
78	5.9	4.3
80	5.8	4.4
82	5.7	4.5
84	5.5	4.6
86	5.4	4.6
88	5.3	4.7
90	5.2	4.8
92	5.1	4 8
94	5.1	4 9
96	5.0	4.9
98	5.0	5.0
100	5.0	5.0

TABLE XLIX.

Equation 32. Argument, Supp. of Node.

	IIIs	IVs	Vs	VIs	VIIs	VIIIs	
°	″	″	″	″	″	″	°
0	3.1	4.0	6.5	10.0	13.5	16.0	30
2	3.1	4.2	6.8	10.2	13.7	16.1	28
4	3.1	4.3	7.0	10.5	13.8	16.2	26
6	3.1	4.4	7.2	10.7	14.0	16.3	24
8	3.2	4.6	7.4	11.0	14.2	16.4	22
10	3.2	4.7	7.6	11.2	14.4	16.5	20
12	3.3	4.9	7.9	11.4	14.6	16.6	18
14	3.3	5.0	8.1	11.7	14.8	16.6	16
16	3.4	5.2	8.3	11.9	15.0	16.7	14
18	3.4	5.4	8.6	12.1	15.1	16.7	12
20	3.5	5.6	8.8	12.4	15.3	16.8	10
22	3.6	5.8	9.0	12.6	15.4	16.8	8
24	3.7	6.0	9.3	12.8	15.6	16.9	6
26	3.8	6.2	9.5	13.0	15.7	16.9	4
28	3.9	6.3	9.8	13.2	15.8	16.9	2
30	4.0	6.5	10.0	13.5	16.0	16.9	0
	IIs	Is	Os	XIs	Xs	IXs	

Constant 55″

Evection.

Argument. Evection, corrected.

Deg.	0s		Is		IIs		IIIs		IVs		Vs	
	1°	Diff.	2°	Diff.	2°	Diff.	2°	Diff.	2°	Diff.	2°	Diff.
	′ ″	″	′ ″	″	′ ″	″	′ ″	″	′ ″	″	′ ″	″
0	30 00.0	85.5	10 43.5	73.2	40 9.7	40.9	50 25.5	2.0	39 8.3	43.4	9 42.0	72.7
1	31 25.5	85.4	11 56.7	72.3	40 50.6	39.5	50 23.5	3.4	38 24.9	44.5	8 29.3	73.3
2	32 50.9	85.4	13 9.0	71.6	41 30.1	38.2	50 20.1	4.9	37 40.4	45.8	7 16.0	74.0
3	34 16.3	85.3	14 20.6	70.7	42 8.3	36.8	50 15.2	6.4	36 54.6	47.0	6 2.0	74.6
4	35 41.6	85.1	15 31.3	69.8	42 45.1	35.5	50 8.8	7.8	36 7.6	48.1	4 47.4	75.2
5	37 6.7	85.1	16 41.1	69.0	43 20.6	34.1	50 1.0	9.3	35 19.5	49.3	3 32.2	75.9
6	38 31.8	84.9	17 50.1	68.1	43 54.7	32.7	49 51.7	10.7	34 30.2	50.5	2 16.3	76.4
7	39 56.7	84.7	18 58.2	67.1	44 27.4	31.4	49 41.0	12.2	33 39.7	51.6	0 59.9	76.9
8	41 21.4	84.4	20 5.3	66.2	44 58.8	29.9	49 28.8	13.7	32 48.1	52.7	59 43.0	77.4
9	42 45.8	84.3	21 11.5	65.2	45 28.7	28.6	49 15.1	14.9	31 55.4	53.8	58 25.6	78.0
10	44 10.1	83.9	22 16.7	64.3	45 57.3	27.2	49 0.2	16.7	31 1.6	54.9	57 7.6	78.4
11	45 34.0	83.7	23 21.0	63.2	46 24.5	25.7	48 43.5	17.9	30 6.7	56.0	55 49.2	78.9
12	46 57.7	83.4	24 24.2	62.2	46 50.2	24.3	48 25.6	19.3	29 10.7	57.0	54 30.3	79.3
13	48 21.1	83.0	25 26.4	61.2	47 14.5	22.9	48 6.3	20.8	28 13.7	58.0	53 11.0	79.7
14	49 44.1	82.6	26 27.6	60.0	47 37.4	21.4	47 45.5	22.2	27 15.7	59.1	51 51.3	80.1
15	51 6.7	82.2	27 27.6	59.0	47 58.8	20.0	47 23.3	23.5	26 16.6	60.0	50 31.2	80.5
16	52 28.9	81.8	28 26.6	58.0	48 18.8	18.6	46 59.8	25.0	25 16.6	61.0	49 10.7	80.8
17	53 50.7	81.3	29 24.6	56.8	48 37.4	17.1	46 34.8	26.3	24 15.6	62.0	47 49.9	81.1
18	55 12.0	80.9	30 21.4	55.6	48 54.5	15.6	46 8.5	27.8	23 13.6	62.9	46 28.8	81.3
19	56 32.9	80.3	31 17.0	54.5	49 10.1	14.3	45 40.7	29.1	22 10.7	63.9	45 7.5	81.7
20	57 53.2	79.8	32 11.5	53.3	49 24.4	12.7	45 11.6	30.4	21 6.8	64.7	43 45.8	81.9
21	59 13.0	79.3	33 4.8	52.2	49 37.1	11.2	44 41.2	31.7	20 2.1	65.7	42 23.9	82.1
22	0 32.3	78.7	33 57.0	50.9	49 48.3	9.8	44 9.5	33.1	18 56.4	66.5	41 1.8	82.3
23	1 51.0	78.1	34 47.9	49.8	49 58.1	8.3	43 36.4	34.5	17 49.9	67.3	39 39.5	82.5
24	3 9.1	77.4	35 37.7	48.5	50 6.4	6.9	43 1.9	35.7	16 42.6	68.2	38 17.0	82.6
25	4 26.5	76.8	36 26.2	47.2	50 13.3	5.4	42 26.2	37.0	15 34.4	68.9	36 54.4	82.7
26	5 43.3	76.1	37 13.4	46.0	50 18.7	3.9	41 49.2	38.4	14 25.5	69.8	35 31.7	82.9
27	6 59.4	75.5	37 59.4	44.8	50 22.6	2.4	41 10.8	39.6	13 15.7	70.5	34 8.8	82.9
28	8 14.9	74.7	38 44.2	43.4	50 25.0	1.0	40 31.2	40.8	12 5.2	71.2	32 45.9	82.9
29	9 29.6	73.9	39 27.6	42.1	50 26.0	0.5	39 50.4	42.1	10 54.0	72.0	31 23.0	83.0
30	10 43.5		40 9.7		50 25.5		39 8.3		9 42.0		30 0.0	
	2°		2°		2°		2°		2°		1°	

Evection.

Argument. Evection, corrected.

Deg	VI^s		VII^s		VIII^s		IX^s		X^s		XI^s	
	1°	Diff.	0°	Diff.	0°	Diff.	0°	Diff.	0°	Diff.	0°	Diff.
	′ ″	″	′ ″	″	′ ″	″	′ ″	″	′ ″	″	′ ″	″
0	30 0.0	83.0	50 18.0	72.0	20 51.7	42.1	9 34.5	0.5	19 50.3	42.1	49 16.5	73.9
1	28 37.0	82.9	49 6.0	71.2	20 9.6	40.8	9 34.0	1.0	20 32.4	43.4	50 30.4	74.7
2	27 14.1	82.9	47 54.8	70.5	19 28.8	39.6	9 35.0	2.4	21 15.8	44.8	51 45.1	75.5
3	25 51.2	82.9	46 44.3	69.8	18 49.2	38.4	9 37.4	3.9	22 0.6	46.0	53 0.6	76.1
4	24 28.3	82.7	45 34.5	68.9	18 10.8	37.0	9 41.3	5.4	22 46.6	47.2	54 16.7	76.8
5	23 5.6		44 25.6		17 33.8		9 46.7		23 33.8		55 33.5	
		82.6		68 2		35.7		6.9		48.5		77.4
6	21 43.0	82.5	43 17.4	67.3	16 58.1	34.5	9 53.6	8.3	24 22.3	49.8	56 50.9	78.1
7	20 20.5	82.3	42 10.1	66.5	16 23.6	33.1	10 1.9	9.8	25 12.1	50.9	58 9.0	78.7
8	18 58.2	82.1	41 3.6	65.7	15 50.5	31.7	10 11.7	11.2	26 3.0	52.2	59 27.7	79.3
9	17 36.1	81.9	39 57.9	64.7	15 18.8	30.4	10 22.9	12.7	26 55.2	53.3	0 47.0	79.8
10	16 14.2		38 53.2		14 48.4		10 35.6		27 48.5		2 6.8	
		81.7		63.9		29.1		14.3		54.5		80.3
11	14 52.5	81.3	37 49.3	62.9	14 19.3	27.8	10 49.9	15.6	28 43.0	55.6	3 27.1	80.9
12	13 31.2	81.1	36 46.4	62.0	13 51.5	26.3	11 5.5	17.1	29 38.6	56.8	4 48.0	81.3
13	12 10.1	80.8	35 44.4	61.0	13 25.2	25.0	11 22.6	18.6	30 35.4	58.0	6 9.3	81.8
14	10 49.3	80.5	34 43.4	60.0	13 0.2	23.5	11 41.2	20.0	31 33 4	59.0	7 31.1	82.2
15	9 28.8		33 43.4		12 36.7		12 1.2		32 32.4		8 53.3	
		80.1		59.1		22.2		21.4		60.0		82.6
16	8 8.7	79.7	32 44.3	58.0	12 14.5	20.8	12 22.6	22.9	33 32.4	61.2	10 15.9	83.0
17	6 49.0	79.3	31 46.3	57.0	11 53.7	19.3	12 45.5	24.3	34 33.6	62.2	11 38.9	83.4
18	5 29.7	78.9	30 49.3	56.0	11 34.4	17.9	13 9.8	25.7	35 35.8	63.2	13 2.3	83.7
19	4 10.8	78.4	29 53.3	54.9	11 16.5	16.7	13 35.5	27.2	36 39.0	64.3	14 26.0	83.9
20	2 52.4		28 58.4		10 59.8		14 2.7		37 43.3		15 49.9	
		78.0		53.8		14.9		28.6		65.2		84.3
21	1 34.4	77.4	28 4.6	52.7	10 44.9	13.7	14 31.3	29.9	38 48.5	66.2	17 14.2	84.4
22	0 17.0	76.9	27 11.9	51.6	10 31.2	12.2	15 1.2	31.4	39 54.7	67.1	18 38.6	84.7
23	59 0.1	76.4	26 20.3	50.5	10 19.0	10.7	15 32.6	32.7	41 1.8	68.1	20 3.3	84.9
24	57 43.7	75.9	25 29.8	49.3	10 8.3	9.3	16 5.3	34.1	42 9.9	69.0	21 28.2	85.1
25	56 27.8		24 40.5		9 59.0		16 39.4		43 18.9		22 53.3	
		75.2		48.1		7.8		35.5		69.8		85.1
26	55 12.6	74.6	23 52.4	47.0	9 51.2	6.4	17 14.9	36.8	44 28.7	70.7	24 18.4	85.3
27	53 58.0	74.0	23 5.4	45.8	9 44.8	4.9	17 51.7	38.2	45 39.4	71.6	25 43.7	85.4
28	52 44.0	73.3	22 19.6	44.5	9 39.9	3.4	18 29.9	39.5	46 51.0	72.3	27 9.1	85.4
29	51 30.7	72.7	21 35.1	43.4	9 36.5	2.0	19 9.4	40.9	48 3.3	73.2	28 34.5	85.5
30	50 18.0		20 51.7		9 34.5		19 50.3		49 16.5		30 0.0	
	0°		0°		0°		0°		0°		1°	

Equation of Moon's Centre.

Argument. Anomaly corrected.

	0s		Is		IIs		IIIs		IVs		Vs	
	7°	Diff for10	10°	Diff for10	12°	Diff for10	13°	Diff for10	12°	Diff for10	9°	Diff for10
° ′	′ ″	″	′ ″	″	′ ″	″	′ ″	″	′ ″	″	′ ″	″
0 0	0 0.0	70.9	20 57.9	59.2	38 43.6	30.1	17 35.2	4.8	16 20.8	35.2	58 28.9	55.0
30	3 32.6	70.9	23 55.6	58.9	40 14.0	29.6	17 20.9	5.4	14 35.3	35.6	55 43.8	55.3
1 0	7 5.2	70.9	26 52.2	58.5	41 42.7	29.0	17 4.8	5.9	12 48.5	36.0	52 58.0	55.5
30	10 37.8	70.8	29 47.7	58.1	43 9.6	28.4	16 47.1	6.5	11 0.4	36.4	50 11.6	55.7
2 0	14 10.3	70 8	32 42.0	57.7	44 34.9	27.8	16 27.6	7.0	9 11.1	36.9	47 24.5	55.9
30	17 42.7	70.8	35 35.2	57.3	45 58.4	27.3	16 6.5	7.6	7 20.5	37.3	44 36.8	56.1
3 0	21 15.0	70.8	38 27.1	57.0	47 20.2	26.7	15 43.7	8.2	5 28.7	37.7	41 48.5	56.3
30	24 47.3	70.7	41 18.0	56.5	48 40.3	26.1	15 19.2	8.7	3 35.6	38.1	38 59.5	56.5
4 0	28 19.4	70.6	44 7.6	56.1	49 58.7	25.5	14 53.1	9.3	1 41.3	38.5	36 10.0	56.7
30	31 51.2	70.6	46 56.0	55.7	51 15.3	25.0	14 25.2	9.8	59 45.8	38.9	33 19.8	56.9
5 0	35 23.0	70.5	49 43.2	55.3	52 30.2	24.4	13 55.8	10.4	57 49.1	39.3	30 29.1	57.1
30	38 54.5	70.4	52 29.1	54.9	53 43.3	23.8	13 24.7	10.9	55 51.1	39.7	27 37.8	57.3
6 0	42 25.8	70.4	55 13.8	54.5	54 54.7	23.2	12 51.9	11.5	53 52.0	40.1	24 45.9	57.5
30	45 56.9	70.3	57 57.2	54.0	56 4.4	22.6	12 17.4	12.0	51 51.7	40.5	21 53.5	57.6
7 0	49 27.7	70.2	0 39.3	53.6	57 12.3	22.1	11 41.4	12.6	49 50.3	40.9	19 0.6	57.8
30	52 58.2	70.1	3 20.1	53.2	58 18.5	21.5	11 3.7	13.1	47 47.6	41.3	16 7.1	58.0
8 0	56 28.5	70.0	5 59.7	52.7	59 22.9	20.9	10 24.3	13.6	45 43.8	41.7	13 13.1	58.2
30	59 58.4	69.9	8 37.9	52.3	0 25.6	20.3	9 43.4	14.2	43 38.9	42.0	10 18.6	58.3
9 0	3 28.0	69.7	11 14.8	51.8	1 26.5	19.7	9 0.8	14.7	41 32.8	42.4	7 23.6	58.5
30	6 57.2	69.6	13 50.3	51.4	2 25.7	19.1	8 16.6	15.3	39 25.6	42.8	4 28.1	58.6
10 0	10 26.0	69.5	16 24.5	50.9	3 23.0	18.6	7 30.8	15.8	37 17.3	43.1	1 32.2	58.8
30	13 54.5	69.3	18 57.3	50.5	4 18.7	17.9	6 43.4	16.3	35 7.9	43.5	58 35.8	59.0
11 0	17 22.5	69.2	21 28.8	50.0	5 12.5	17.4	5 54.4	16.8	32 57.4	43.9	55 38.9	59.1
30	20 50.1	69.1	23 58.8	49.6	6 4.6	16.8	5 3.9	17.4	30 45.8	44.2	52 41.7	59.3
12 0	24 17.3	68.9	26 27.5	49.1	6 54.9	16.2	4 11.7	17.9	28 33.1	44.6	49 43.9	59.4
30	27 44.0	68.7	28 54.7	48.6	7 43.5	15.6	3 18.0	18.4	26 19.4	44.9	46 45.8	59.5
13 0	31 10.2	68.5	31 20.5	48.1	8 30.3	15.0	2 22.7	19.0	24 4.6	45.3	43 47.3	59.6
30	34 35.8	68.4	33 44.9	47.7	9 15.4	14.4	1 25.8	19.5	21 48.8	45.6	40 48.4	59.8
14 0	38 1.0	68.2	36 7.9	47.2	9 58.6	13.8	0 27.4	20.0	19 31.9	45.9	37 49.1	59.9
30	41 25.6	68.0	38 29.4	46.6	10 40.1	13.3	59 27.4	20.5	17 14.1	46.3	34 49.5	60.0
15 0	44 49.6		40 49.3		11 19.9		58 25.9		14 55.2		31 49 4	
	8°		11°		13°		12°		11°		8°	

TABLE LI.

Equation of Moon's Centre.

Argument. Anomaly corrected.

	VIˢ		VIIˢ		VIIIˢ		IXˢ		Xˢ		XIˢ	
	7°	Diff for 10	4°	Diff for 10	1°	Diff for 10	0°	Diff for 10	1°	Diff for 10	3°	Diff for 10
° ′	′ ″	″	′ ″	″	′ ″	″	′ ″	″	′ ″	″	′ ″	″
0 0	0 0.0	61.8	1 31.1	54.8	43 39.2	34.7	42 24.8	4.2	21 16.4	30.7	39 2.1	59.6
30	56 54.6	61.8	58 46.7	54 6	41 55.0	34.3	42 12.1	3.6	22 48.5	31.2	42 0.8	60.0
1 0	53 49.2	61.8	56 3.0	54 3	40 12.0	33.8	42 1.2	3.1	24 22.2	31.8	4[illegible] [illegible].7	60.3
30	50 43.9	61.8	53 20.0	54.1	38 30.5	33.4	41 52.0	2.5	25 57.7	32.4	4[illegible] 1.7	60.7
2 0	47 38.6	61.7	50 37.7	53.8	36 50.3	33.0	41 44.4	1.9	27 34.8	33.0	51 3.7	61.0
30	44 33.4	61.8	47 56.2	53.6	35 11.3	32.5	41 38.7	1.4	29 13.7	33.5	54 6.7	61.3
3 0	41 28.1	61.7	45 15.4	53.4	33 33.7	32.1	41 34.6	0.8	30 54.2	34.0	57 0.7	61.7
30	38 23.0	61.7	42 35.3	53.1	31 57.5	31.6	41 32.2	0.2	32 36.3	34.6	0 15.8	62.0
4 0	35 18.0	61.7	39 56.0	52.9	30 22.6	31.2	41 31.6	0.4	34 20.2	35.1	3 21.8	62.3
30	32 13.0	61.6	37 17.4	52.6	28 49.0	30.7	41 32.7	1.0	36 5.6	35.7	6 28.8	62.7
5 0	29 8.1	61.6	34 39.6	52.3	27 16.8	30.2	41 35.6	1.5	37 52.8	36.2	9 36.8	63.0
30	26 3.4	61.5	32 2.7	52.1	25 46.1	29.8	41 40.1	2.1	39 41.5	36.8	12 45.7	63.3
6 0	22 58.8	61.5	29 26.5	51.8	24 16.7	29.3	41 46.4	2.7	41 32.0	37.3	15 55.5	63.6
30	19 54.3	61.4	26 51.1	51.5	22 48.7	28.9	41 54.5	3.3	43 24.0	37.9	19 6.2	63.9
7 0	16 50.0	61.4	24 16.6	51.2	21 22.1	28.4	42 4.3	3.9	45 17.7	38.4	22 17.8	64.2
30	13 45.8	61.3	21 42.9	51.0	19 56.9	27.9	42 15.9	4.4	47 12.9	39.0	25 30.3	64.5
8 0	10 41.9	61.3	19 10.0	50.7	18 33.1	27.4	42 29.2	5.0	49 9.8	39.5	28 43.7	64.7
30	7 38.0	61.2	16 33.0	50.4	17 10.8	27.0	42 44.2	5.6	51 8.3	40.0	31 57.8	65.0
9 0	4 34.4	61.1	14 6.9	50.1	15 49.8	26.5	43 1.1	6.2	53 8.4	40.6	35 12.9	65.3
30	1 31.0	61.1	11 36.6	49.8	14 30.4	26.0	43 19.6	6.8	55 10.1	41.1	38 28.7	65.5
10 0	58 27.8	61.0	9 7.3	49.5	13 12.5	25.5	43 39.9	7.4	57 13.3	41.6	41 45.2	65.8
30	55 24.9	60.9	6 38.9	49.2	11 55.9	25.0	44 2.0	8.0	59 18.2	42.1	45 2.6	66.0
11 0	52 22.2	60.8	4 11.3	48.9	10 40.9	24.5	44 25.9	8.5	1 24.5	42.6	48 20.7	66.3
30	49 19.7	60.7	1 44.7	48.6	9 27.3	24.0	44 51.5	9.1	3 32.4	43.2	51 39.6	66.5
12 0	46 17.5	60.6	59 18.9	48.2	8 15.2	23.5	45 18.8	9.7	5 41.9	43.7	54 59.1	66.7
30	43 15.6	60.5	56 54.2	47.9	7 4.6	23.1	45 48.0	10.3	7 52.9	44.2	58 19.3	67.0
13 0	40 14.0	60.5	54 30.4	47.6	5 55.4	22.5	46 18.9	10.9	10 5.5	44.7	1 40.3	67.2
30	37 12.6	60.3	52 7.5	47.3	4 47.8	22.0	46 51.5	11.5	12 19.5	45.2	5 1.9	67.4
14 0	34 11.6	60.2	49 45.6	47.0	3 41.7	21.5	47 26.0	12.1	14 35.1	45.7	8 24.1	67.6
30	31 10.9	60.1	47 24.7	46.6	2 37.1	21.0	48 2.2	12.6	16 52.1	46.2	11 46.9	67.8
15 0	28 10.6		45 4.8		1 34.1		48 40.1		19 10.7		15 10.4	
	5°		2°		1°		0°		2°		5°	

TABLE LI.

Equation of Moon's Centre

Argument. Anomaly corrected.

	0s		Is		IIs		IIIs		IVs		Vs	
	8°	Diff for 10	11°	Diff for 10	13°	Diff for 10	12°	Diff for 10	11°	Diff for 10	8°	Diff for 10
° ′	′ ″	″	′ ″	″	′ ″	″	′ ″	″	′ ″	″	′ ″	″
15 0	44 49.6	67.8	40 49.3	46.2	11 19.9	12.6	58 25.9	21.0	14 55.2	46.6	31 49.4	60.1
30	48 13.1	67.6	43 7.9	45.7	11 57.8	12.1	57 22.9	21.5	12 35.3	47.0	28 49.1	60.2
16 0	51 35.9	67.4	45 24.9	45.2	12 34.0	11.5	56 18.3	22.0	10 14.4	47.3	25 48.4	60.3
30	54 58.1	67.2	47 40.5	44.7	13 8.5	10.9	55 12.2	22.5	7 52.5	47.7	22 47.4	60.5
17 0	58 19 7	67.0	49 54.5	44.2	13 41.1	10.3	54 4.6	23.1	5 29.6	47.9	19 46.0	60.5
30	1 40.7	66.7	52 7.1	43.7	14 12.0	9.7	52 55.4	23.5	3 5.8	48.3	16 44.4	60.6
18 0	5 0.9	66.5	54 18.1	43.2	14 41.2	9.1	51 44.8	24.0	0 41.1	48.6	13 42.5	60.7
30	8 20.4	66.3	56 27.6	42.6	15 8.5	8.5	50 32.7	24.5	58 15.3	48.9	10 40.3	60.8
19 0	11 39.3	66.0	58 35.5	42.1	15 34.1	8.0	49 19.1	25.0	55 48.7	49.2	7 37.8	60.9
30	14 57.4	65.8	0 41.8	41.6	15 58.0	7.4	48 4.1	25.5	53 21.1	49.5	4 35.1	61.0
20 0	18 14.8	65.5	2 46.7	41.1	16 20.1	6.8	46 47.5	26.0	50 52.7	49.8	1 32.2	61.1
30	21 31.3	65.3	4 49.9	40.6	16 40.4	6.2	45 29.6	26.5	48 23.4	50.1	58 29.0	61.1
21 0	24 47.1	65.0	6 51.6	40.0	16 58.9	5.6	44 10.2	27.0	45 53.1	50.4	55 25.6	61.2
30	28 2.2	64.7	8 51.7	39.5	17 15.8	5.0	42 49.2	27.4	43 22.0	50.7	52 22.0	61.3
22 0	31 16.3	64.5	10 50.2	39.0	17 30.8	4 4	41 26.9	27.9	40 50.0	51.0	49 18.1	61.3
30	34 29.7	64.2	12 47.1	38.4	17 44.1	3.9	40 3.1	28.4	38 17.1	51.2	46 14.2	61.4
23 0	37 42.2	63.9	14 42.3	37.9	17 55.7	3.3	38 37.9	28.9	35 43.4	51.5	43 10.0	61.4
30	40 53.8	63.6	16 36.0	37.3	18 5.5	2.7	37 11.3	29.3	33 8.9	51.8	40 5.7	61.5
24 0	44 4.5	63.2	18 28.0	36.8	18 13.6	2.1	35 43.3	29.8	30 33.5	52.1	37 1.2	61.5
30	47 14.3	63.0	20 18.5	36.2	18 19.9	1.5	34 13.9	30.2	27 57.3	52.3	33 56.6	61.6
25 0	50 23.2	62.7	22 7.2	35.7	18 24.4	1.0	32 43.2	30.7	25 20.4	52.6	30 51.9	61.6
30	53 31.2	62.3	23 54.4	35.1	18 27.3	0.4	31 11.0	31.2	22 42.6	52.9	27 47.0	61.7
26 0	56 38.2	62.0	25 39.8	34.6	18 28.4	0.2	29 37.4	31.6	20 4.0	53.1	24 42.0	61.7
30	59 44.2	61.7	27 23.7	34.1	18 27.8	0.8	28 2.5	32.1	17 24.7	3.3	21 37.0	61.7
27 0	2 49.3	61.3	29 5.8	33.5	18 25.4	1.4	26 26.3	32.5	14 44.7	53.6	18 31.8	61.7
30	5 53.3	61.0	30 46.3	33.0	18 21.3	1.9	24 48.7	33.0	12 3.8	53.8	15 26.6	61.7
28 0	8 56.3	60.7	32 25.2	32.4	18 15.6	2.5	23 9.7	33.4	9 22.3	54.1	12 21.4	61.8
30	11 58.3	60.3	34 2.3	31.8	18 8.0	3.1	21 29.5	33.8	6 40.0	54.3	9 16.1	61.8
29 0	14 59.3	60.0	35 37.8	31.2	17 58.8	3.6	19 48.0	34.3	3 57.0	54.6	6 10.8	61.8
30	17 59.2	59.6	37 11.5	30.7	17 47.9	4.2	18 5.0	34.7	1 13.3	54.8	3 5.4	61.8
30 0	20 57.9		38 43.6		17 35.2		16 20.8		58 28.9		0 0.0	
	10°		12°		13°		12°		9°		7°	

Equation of Moon's Centre.

Argument. Anomaly corrected.

	VIˢ		VIIˢ		VIIIˢ		IXˢ		Xˢ		XIˢ	
	5°	Diff for 10	2°	Diff for 10	1°	Diff for 10	0°	Diff for 10	2°	Diff for 10	5°	Diff for 10
° ′	′ ″	″	′ ″	″	′ ″	″	′ ″	″	′ ″	″	′ ″	″
15 0	28 10.6	60.0	45 4.8	46.3	1 34.1	20.5	48 40.1	13.3	19 10.7	46.6	15 10.4	68.0
30	25 10.5	59.9	42 45.9	45.9	0 32.6	20.0	49 19.9	13.8	21 30.6	47.2	8 34.4	68.2
16 0	22 10.9	59.8	40 28.1	45.6	59 32.6	19.5	50 1.4	14.4	23 52.1	47.7	21 59.0	68.4
30	19 11.6	59.6	38 11.2	45.3	58 34.2	19.0	50 44.6	15.0	26 15.1	48.1	25 24.2	68.5
17 0	16 12.7	59.5	35 55.4	44.9	57 37.3	18.4	51 29.7	15.6	28 39.5	48.6	28 49.8	68.7
30	13 14.2	59.4	33 40.6	44.6	56 42.0	17.9	52 16.5	16.2	31 5.3	49.1	32 16.0	68.9
18 0	10 16.1	59.3	31 26.9	44.2	55 48.3	17.4	53 5.1	16.8	33 32.5	49.6	35 42.7	69.1
30	7 18.3	59.1	29 14.2	43.9	54 56.1	16.8	53 55.4	17.4	36 1.2	50.0	39 9.9	69.2
19 0	4 21.1	59.0	27 2.6	43.5	54 5.6	16.3	54 47.5	17.9	38 31.2	50.5	42 37.5	69.3
30	1 24.2	58.8	24 52.1	43.1	53 16.6	15.8	55 41.3	18.6	41 2.7	50.9	46 5.5	69.5
20 0	58 27.8	58.6	22 42.7	42.8	52 29.2	15.3	56 37.0	19.1	43 35.5	51.4	49 34.0	69.6
30	55 31.9	58.5	20 34.4	42.4	51 43.4	14.7	57 34.3	19.7	46 9.7	51.8	53 2.8	69.7
21 0	52 36.4	58.3	18 27.2	42.0	50 59.2	14.2	58 33.5	20.3	48 45.2	52.3	56 31.9	69.9
30	49 41.4	58.2	16 21.1	41.6	50 16.6	13.6	59 34.4	20.9	51 22.1	52.7	0 1.6	70.0
22 0	46 46.9	58.0	14 16.2	41.3	49 35.7	13.1	0 37.1	21.5	54 0.3	53.2	3 31.5	70.1
30	43 52.9	57.8	12 12.4	40.9	48 56.3	12.6	1 41.5	22.1	56 39.9	53.6	7 1.8	70.2
23 0	40 59.4	57.6	10 9.7	40.5	48 18.6	12.0	2 47.7	22.6	59 20.7	54.0	10 32.3	70.3
30	38 6.5	57.5	8 8.3	40.1	47 42.6	11.5	3 55.6	23.2	2 2.8	54.5	14 3.1	70.4
24 0	35 14.1	57.3	6 8.0	39.7	47 8.1	10.9	5 5.3	23.8	4 46.2	54.9	17 34.2	70.4
30	32 22.2	57.1	4 8.9	39.3	46 35.3	10.4	6 16.7	24.4	7 30.9	55.3	21 5.5	70.5
25 0	29 30.9	56.9	2 10.9	38.9	46 4.2	9.8	7 29.8	25.0	10 16.8	55.7	24 37.0	70.6
30	26 40.2	56.7	0 14.2	38.5	45 34.8	9.3	8 44.7	25.5	13 4.0	56.1	28 8.8	70.6
26 0	23 50.0	56.5	58 18.7	38.1	45 6.9	8.7	10 1.3	26.1	15 52.4	56.5	31 40.7	70.7
30	21 0.5	56.3	56 24.4	37.7	44 40.8	8.2	11 19.7	26.7	18 42.0	57.0	35 12.8	70.7
27 0	18 11.5	56.1	54 31.3	37.3	44 16.3	7.6	12 39.8	27.3	21 32.9	57.3	38 45.1	70.8
30	15 23.2	55.9	52 39.5	36.9	43 53.5	7.0	14 1.6	27.8	24 24.8	57.7	42 17.3	70.8
28 0	12 35.5	55.7	50 48.9	36.4	43 32.4	6.5	15 25.1	28.4	27 18.0	58.1	45 49.7	70.8
30	9 48.4	55.5	48 59.6	36.0	43 12.9	5.9	16 50.4	29.0	30 12.3	58.5	49 22.2	70.9
29 0	7 2.0	55.3	47 11.5	35.6	42 55.2	5.4	18 17.3	29.6	33 7.8	58.9	52 54.8	70.9
30	4 16.2	55.0	45 24.7	35.2	42 39.1	4.8	19 46.0	30.1	36 4.4	59.2	56 27.4	70.9
30 0	1 31.1		43 39.2		42 24.8		21 16.4		39 2.1		0 0.0	
	4°		1°		0°		1°		3°		7°	

TABLE LII.

Variation.

Argument. Variation, corrected.

	0s		Is		IIs		IIIs		IVs		Vs	
Degr.	0°	Diff.	1°	Diff.	1°	Diff.	0°	Diff.	0°	Diff.	0°	Diff.
	′ ″	″	′ ″	″	′ ″	″	′ ″	″	′ ″	″	′ ″	″
0	38 0.0	73.3	8 1.5	34.0	6 57.9	39.9	35 54.4	74.0	5 29.5	35.3	6 1.6	40.0
1	39 13.3	73.3	8 35.5	31.7	6 18.0	42.1	34 40.4	73.8	4 54.2	32.9	6 41.6	42.3
2	40 26.5	73.0	9 7.2	29.3	5 35.9	44.2	33 26.6	73.6	4 21.3	30.7	7 23.9	44.5
3	41 39.5	72.7	9 36.5	26.9	4 51.7	46.2	32 13.0	73.4	3 50.6	28.3	8 8.4	46.6
4	42 52.2	72.3	10 3.4	24.5	4 5.5	48.2	30 59.6	72.9	3 22.3	25.8	8 55.0	48.7
5	44 4.5	71.9	10 27.9	22.0	3 17.3	50.1	29 46.7	72.4	2 56.5	23.4	9 43.7	50.8
6	45 16.4	71.3	10 49.9	19.5	2 27.2	51.9	28 34.3	71.9	2 33.1	21.0	10 34.5	52.8
7	46 27.7	70.7	11 9.4	17.0	1 35.3	53.7	27 22.4	71.2	2 12.1	18.4	11 27.3	54.7
8	47 38.4	69.9	11 26.4	14.5	0 41.6	55.5	26 11.2	70.5	1 53.7	15.9	12 22.0	56.6
9	48 48.3	69.1	11 40.9	12.0	59 46.1	57.1	25 0.7	69.6	1 37.8	13.3	13 18.6	58.3
10	49 57.4	68.2	11 52.9	9.3	58 49.0	58.8	23 51.1	68.8	1 24.5	10.8	14 16.9	60.1
11	51 5.6	67.2	12 2.2	6.8	57 50.2	60.2	22 42.3	67.8	1 13.7	8.2	15 17.0	61.7
12	52 12.8	66.1	12 9.0	4.2	56 50.0	61.7	21 34.5	66.6	1 5.5	5.5	16 18.7	63.3
13	53 18.9	64.9	12 13.2	1.6	55 48.3	63.1	20 27.9	65.6	1 0.0	3.0	17 22.0	64.9
14	54 23.8	63.7	12 14.8	0.9	54 45.2	64.3	19 22.3	64.3	0 57.0	0.3	18 26.9	66.2
15	55 27.5	62.3	12 13.9	3.6	53 40.9	65.6	18 18.0	63.0	0 56.7	2.3	19 33.1	67.6
16	56 29.8	60.9	12 10.3	6.1	52 35.3	66.8	17 15.0	61.6	0 59.0	4.9	20 40.7	68.9
17	57 30.7	59.4	12 4.2	8.7	51 28.5	67.8	16 13.4	60.2	1 3.9	7.6	21 49.6	70.0
18	58 30.1	57.9	11 55.5	11.3	50 20.7	68.8	15 13.2	58.6	1 11.5	10.1	22 59.6	71.2
19	59 28.0	56.2	11 44.2	13.7	49 11.9	69.7	14 14.6	57.1	1 21.6	12.8	24 10.8	72.1
20	0 24.2	54.5	11 30.5	16.4	48 2.2	70.5	13 17.5	55.3	1 34.4	15.4	25 22.9	73.0
21	1 18.7	52.7	11 14.1	18.8	46 51.7	71.2	12 22.2	53.7	1 49.8	18.0	26 35.9	73.9
22	2 11.4	50.9	10 55.3	21.3	45 40.5	71.9	11 28.5	51.8	2 7.8	20.5	27 49.8	74.7
23	3 2.3	48.9	10 34.0	23.8	44 28.6	72.5	10 36.7	49.9	2 28.3	23.1	29 4.5	75.2
24	3 51.2	47.0	10 10.2	26.2	43 16.1	72.9	9 46.8	48.0	2 51.4	25.5	30 19.7	75.9
25	4 38.2	44.9	9 44.0	28.6	42 3.2	73.3	8 58.8	46.1	3 16.9	28.1	31 35.6	76.3
26	5 23.1	42.9	9 15.4	30.9	40 49.9	73.7	8 12.7	44.0	3 45.0	30.6	32 51.9	76.7
27	6 6.0	40.7	8 44.5	33.3	39 36.2	73.8	7 28.7	41.9	4 15.6	32.9	34 8.6	77.0
28	6 46.7	38.5	8 11.2	35.5	38 22.4	74.0	6 46.8	39 7	4 48.5	35.4	35 25.6	77.1
29	7 25.2	36.3	7 35.7	37.8	37 8.4	74.0	6 7.1	37.6	5 23.9	37.7	36 42.7	77.3
30	8 1.5		6 57.9		35 54.4		5 29.5		6 1.6		38 0.0	
	1°		1°		0°		0°		0°		0°	

Variation.

Argument. Variation corrected.

Deg	VI^s 0°	Diff.	VII^s 1°	Diff.	VIII^s 1°	Diff.	IX^s 0°	Diff.	X^s 0°	Diff.	XI^s 0°	Diff.
	′ ″	″	′ ″	″	′ ″	″	′ ″	″	′ ″	″	′ ″	″
0	38 0.0	77.3	9 58.4	37.7	10 30.5	37.6	40 5.6	74.0	9 2.1	37.8	7 58.5	36.3
1	39 17.3	77.1	10 36.1	35.4	9 52.9	39.7	38 51.6	74.0	8 24.3	35.5	8 34.8	38.5
2	40 34.4	77.0	11 11.5	32.9	9 13.2	41.9	37 37.6	73.8	7 48.8	33.3	9 13.3	40.7
3	41 51.4	76.7	11 44.4	30.6	8 31.3	44.0	36 23.8	73.7	7 15.5	30.9	9 54.0	42.9
4	43 8.1	76.3	12 15.0	28.1	7 47.3	46.1	35 10.1	73.3	6 44.6	28.6	10 36.9	44.9
5	44 24.4	75.9	12 43.1	25.5	7 1.2	48.0	33 56.8	72.9	6 16.0	26.2	11 21.8	47.0
6	45 40.3	75.2	13 8.6	23.1	6 13.2	49.9	32 43.9	72.5	5 49.8	23.8	12 8.8	48.9
7	46 55.5	74.7	13 31.7	20.5	5 23.3	51.8	31 31.4	71.9	5 26.0	21.3	12 57.7	50.9
8	48 10.2	73.9	13 52.2	18.0	4 31.5	53.7	30 19.5	71.2	5 4.7	18.8	13 48.6	52.7
9	49 24.1	73.0	14 10.2	15.4	3 37.8	55.3	29 8.3	70.5	4 45.9	16.4	14 41.3	54.5
10	50 37.1	72.1	14 25.6	12.8	2 42.5	57.1	27 57.8	69.7	4 29.5	13.7	15 35.8	56.2
11	51 49.2	71.2	14 38.4	10.1	1 45.4	58.6	26 48.1	68.8	4 15.8	11.3	16 32.0	57.9
12	53 0.4	70.0	14 48.5	7.6	0 46.8	60.2	25 39.3	67.8	4 4.5	8.7	17 29.9	59.4
13	54 10.4	68.9	14 56.1	4.9	59 46.6	61.6	24 31.5	66.8	3 55.8	6.1	18 29.3	60.9
14	55 19.3	67.6	15 1.0	2.3	58 45.0	63.0	23 24.7	65.6	3 49.7	3.6	19 30.2	62.3
15	56 26.9	66.2	15 3.3	0.3	57 42.0	64.3	22 19.1	64.3	3 46.1	0.9	20 32.5	63.7
16	57 33.1	64.9	15 3.0	3.0	56 37.7	65.6	21 14.8	63.1	3 45.2	1.6	21 36.2	64.9
17	58 38.0	63.3	15 0.0	5.5	55 32.1	66.6	20 11.7	61.7	3 46.8	4.2	22 41.1	66.1
18	59 41.3	61.7	14 54.5	8.2	54 25.5	67.8	19 10.0	60.2	3 51.0	6.8	23 47.2	67.2
19	0 43.0	60.1	14 46.3	10.8	53 17.7	68.8	18 9.8	58.8	3 57.8	9.3	24 54.4	68.2
20	1 43.1	58.3	14 35.5	13.3	52 8.9	69.6	17 11.0	57.1	4 7.1	12.0	26 2.6	69.1
21	2 41.4	56.6	14 22.2	15.9	50 59.3	70.5	16 13.9	55.5	4 19 1	14.5	27 11.7	69.9
22	3 38.0	54.7	14 6.3	18.4	49 48.8	71.2	15 18.4	53.7	4 33.6	17.0	28 21.6	70.7
23	4 32.7	52.8	13 47.9	21.0	48 37.6	71.9	14 24.7	51.9	4 50.6	19.5	29 32.3	71.3
24	5 25.5	50.8	13 26.9	23.4	47 25.7	72.4	13 32.8	50.1	5 10.1	22.0	30 43.6	71.9
25	6 16.3	48.7	13 3.5	25.8	46 13.3	72.9	12 42.7	48.2	5 32.1	24.5	31 55.5	72.3
26	7 5.0	46.6	12 37.7	28.3	45 0.4	73.4	11 54.5	46.2	5 56.6	26.9	33 7.8	72.7
27	7 51.6	44.5	12 9.4	30.7	43 47.0	73.6	11 8.3	44.2	6 23.5	29.3	34 20.5	73.0
28	8 36.1	42.3	11 38.7	32.9	42 33.4	73.8	10 24.1	42.1	6 52.8	31.7	35 33.5	73.2
29	9 18.4	40.0	11 5.8	35.3	41 19.6	74.0	9 42.0	39.9	7 24.5	34.0	36 46.7	73.3
30	9 58.4		10 30.5		40 5.6		9 2.1		7 58.5		38 0.0	
	1°		1°		0°		0°		0°		0°	

TABLE LIII. *Reduction.*

Argument. Supplement of Node+Moon's Orbit Longitude.

	0s VIs	Diff.	Is VIIs	Diff.	IIs VIIIs	Diff.	IIIs IXs	Diff.	IVs Xs	Diff.	Vs XIs	Diff.
°	′ ″	″	′ ″	″	′ ″	″	′ ″	″	′ ″	″	′ ″	″
0	7 0.0	14.4	1 3.0	7.0	1 3.0	7.4	7 0.0	14.4	12 57.0	7.0	12 57.0	7.4
1	6 45.6	14.4	0 56.0	6.5	1 10.4	7.9	7 14.4	14.4	13 4.0	6.5	12 49.6	7.9
2	6 31.2	14.3	0 49.5	6.1	1 18.3	8.2	7 28.8	14.3	13 10.5	6.1	12 41.7	8.2
3	6 16.9	14.3	0 43.4	5.6	1 26.5	8.7	7 43.1	14.3	13 16.6	5.6	12 33.5	8.7
4	6 2.6	14.2	0 37.8	5.1	1 35.2	9.0	7 57.4	14.2	13 22.2	5.1	12 24.8	9.0
5	5 48.4	14.1	0 32.7	4.5	1 44.2	9.5	8 11.6	14.1	13 27.3	4.5	12 15.8	9.5
6	5 34.3	14.0	0 28.2	4.3	1 53.7	9.8	8 25.7	14.0	13 31.8	4.3	12 6.3	9.8
7	5 20.3	13.9	0 23.9	3.9	2 3.5	10.2	8 39.7	13.9	13 36.1	3.9	11 56.5	10.2
8	5 6.4	13.8	0 20.0	3.2	2 13.7	10.5	8 53.6	13.8	13 40.0	3.2	11 46.3	10.5
9	4 52.6	13.6	0 16.8	2.7	2 24.2	10.8	9 7.4	13.6	13 43.2	2.7	11 35.8	10.8
10	4 39.0	13.4	0 14.1	2.3	2 35.0	11.2	9 21.0	13.4	13 45.9	2.3	11 25.0	11.2
11	4 25.6	13.3	0 11.8	1.7	2 46.2	11.5	9 34.4	13.3	13 48.2	1.7	11 13.8	11.5
12	4 12.3	13.0	0 10.1	1.3	2 57.7	11.8	9 47.7	13.0	13 49.9	1.3	11 2.3	11.8
13	3 59.3	12.8	0 8.8	0.7	3 9.5	12.1	10 0.7	12.8	13 51.2	0.7	10 50.5	12.1
14	3 46.5	12.6	0 8.1	0.3	3 21.6	12.3	10 13.5	12.6	13 51.9	0.3	10 38.4	12.3
15	3 33.9	12.3	0 7.8	0.3	3 33.9	12.6	10 26.1	12.3	13 52.2	0.3	10 26.1	12.6
16	3 21.6	12.1	0 8.1	0.7	3 46.5	12.8	10 38.4	12.1	13 51.9	0.7	10 13.5	12.8
17	3 9.5	11.8	0 8.8	1.3	3 59.3	13.0	10 50.5	11.8	13 51.2	1.3	10 0.7	13.0
18	2 57.7	11.5	0 10.1	1.7	4 12.3	13.3	11 2.3	11.5	13 49.9	1.7	9 47.7	13.3
19	2 46.2	11.2	0 11.8	2.3	4 25.6	13.4	11 13.8	11.2	13 48.2	2.3	9 34.4	13.4
20	2 35.0	10.8	0 14.1	2.7	4 39.0	13.6	11 25.0	10.8	13 45.9	2.7	9 21.0	13.6
21	2 24.2	10.5	0 16.8	3.2	4 52.6	13.8	11 35.8	10.5	13 43.2	3.2	9 7.4	13.8
22	2 13.7	10.2	0 20.0	3.9	5 6.4	13.9	11 46.3	10.2	13 40.0	3.9	8 53.6	13.9
23	2 3.5	9.8	0 23.9	4.3	5 20.3	14.0	11 56.5	9.8	13 36.1	4.3	8 39.7	14.0
24	1 53.7	9.5	0 28.2	4.5	5 34.3	14.1	12 6.3	9.5	13 31.8	4.5	8 25.7	14.1
25	1 44.2	9.0	0 32.7	5.1	5 48.4	14.2	12 15.8	9.0	13 27.3	5.1	8 11.6	14.2
26	1 35.2	8.7	0 37.8	5.6	6 2.6	14.3	12 24.8	8.7	13 22.2	5.6	7 57.4	14.3
27	1 26.5	8.2	0 43.4	6.1	6 16.9	14.3	12 33.5	8.2	13 16.6	6.1	7 43.1	14.3
28	1 18.3	7.9	0 49.5	6.5	6 31.2	14.4	12 41.7	7.9	13 10.5	6.5	7 28.8	14.4
29	1 10.4	7.4	0 56.0	7.0	6 45.6	14.4	12 49.6	7.4	13 4.0	7.0	7 14.4	14.4
30	1 3.0		1 3.0		7 0.0		12 57.0		12 57.0		7 0.0	

TABLE LIV. *Lunar Nutation in Longitude.*

Argument. Supplement of the Node.

	0s +	Is +	IIs +	IIIs +	IVs +	Vs +	
°	″	″	″	″	″	″	°
0	0.0	8.5	14.8	17.3	15.2	8.8	30
2	0.6	9.0	15.1	17.2	14.9	8.1	28
4	1.2	9.4	15.4	17.2	14.5	7.7	26
6	1.7	10.0	15.6	17.2	14.2	7.2	24
8	2.3	10.4	15.9	17.2	13.8	6.5	22
10	2.9	10.9	16.4	17.1	13.5	6.1	20
12	3.5	11.4	16.3	17.0	13.0	5.4	18
14	4.1	11.8	16.5	16.9	12.6	4.8	16
16	4.6	12.2	16.7	16.7	12.2	4.3	14
18	5.2	12.6	16.8	16.5	11.8	3.7	12
20	5.8	13.1	16.9	16.4	11.3	3.0	10
22	6.2	13.4	17.1	16.2	10.9	2.4	8
24	6.9	13.8	17.1	15.9	10.4	1.8	6
26	7.4	14.1	17.2	15.7	9.8	1.3	4
28	7.8	14.5	17.2	15.4	9.4	0.6	2
30	8.5	14.8	17.3	15.2	8.8	0.0	0
	— XIs	— Xs	— IXs	— VIIIs	— VIIs	— VIs	

Moon's Distance from the North Pole of the Ecliptic

Argument. Supplement of Node+Moon's Orbit Longitude.

	III^s	IV^s		V^s		VI^s		VII^s		VIII^s	
	84°	85°	Diff. for 10	87°	Diff. for 10	89°	Diff. for 10	92°	Diff. for 10	94°	
s ′	′ ″	′ ″	″	′ ″	″	′ ″	″	′ ″	″	′ ″	° ′
0 0	39 16.0	20 42.7	27.2	13 46.6	46.8	48 0.0	53.8	22 13.4	46.6	15 17.3	30 0
30	39 16.7	22 4.2	27.6	16 6.9	47.0	50 41.4	53.8	24 33.1	46.4	16 37.7	30
1 0	39 18.8	23 27.0	28.0	18 27.8	47.2	53 22.9	53.8	26 52.2	46.0	17 56.8	29 0
30	39 22.4	24 51.0	28.4	20 49.5	47.4	56 4.3	53.8	29 10.2	45.8	19 14.6	30
2 0	39 27.3	26 16.2	28.8	23 11.8	47.7	58 45.7	53.8	31 27.5	45.6	20 31.3	28 0
30	39 33.7	27 42.6	29.2	25 34.8	47.9	1 27.0	53.8	33 44.2	45.3	21 46.7	30
3 0	39 41.5	29 10.1	29.6	27 58.5	48.1	4 8.3	53.7	36 0.2	45.0	23 0.8	27 0
30	39 50.6	30 38.9	30.0	30 22.8	48.3	6 49.5	53.7	38 15.3	44.8	24 13.7	30
4 0	40 1.2	32 8.8	30.4	32 47.7	48.5	9 30.6	53.7	40 29.7	44.5	25 25.3	26 0
30	40 13.2	33 39.9	30.8	35 13.2	48.7	12 11.6	53.6	42 43.3	44.3	26 35.7	30
5 0	40 26.7	35 12.2	31.1	37 39.3	48.9	14 52.5	53.6	44 56.2	44.0	27 44.8	25 0
30	40 41.5	36 45.6	31.5	40 6.1	49.1	17 33.3	53.6	47 8.1	43.8	28 52.6	30
6 0	40 57.7	38 20.1	31.9	42 33.4	49.3	20 14.0	53.5	49 19.4	43.4	29 59.0	24 0
30	41 15.4	39 55.8	32.3	45 1.2	49.5	22 54.4	53.5	51 29.7	43.2	31 4.3	30
7 0	41 34.4	41 32.7	32.6	47 29.6	49.7	25 34.8	53.4	53 39.3	42.9	32 8.2	23 0
30	41 54.8	43 10.6	33.0	49 58.6	49.8	28 14.9	53.3	55 48.0	42.6	33 10.9	30
8 0	42 16.7	44 49.7	33.4	52 28.1	50.0	30 54.9	53.3	57 55.8	42.3	34 12.2	22 0
30	42 39.9	46 29.9	33.8	54 58.2	50.2	33 34.7	53.2	0 2.8	42.0	35 12.2	30
9 0	43 4.6	48 11.2	34.1	57 28.7	50.4	36 14.3	53.1	2 8.9	41.7	36 10.9	21 0
30	43 30.6	49 53.5	34.5	59 59.8	50.5	38 53.7	53.0	4 14.1	41.5	37 8.3	30
10 0	43 58.1	51 37.0	34.9	2 31.3	50.7	41 32.8	53.0	6 18.4	41.1	38 4.4	20 0
30	44 26.9	53 21.6	35.2	5 3.3	50.8	44 11.7	52.9	8 21.8	40.8	38 59.1	30
11 0	44 57.1	55 7.1	35.7	7 35.8	51.0	46 50.4	52.8	10 24.3	40.5	39 52.5	19 0
30	45 28.8	56 53.8	35.9	10 8.8	51.1	49 28.7	52.7	12 25.9	40.2	40 44.6	30
12 0	46 1.8	58 41.6	36.2	12 42.1	51.3	52 6.8	52.6	14 26.6	39.9	41 35.3	18 0
30	46 36.1	0 30.3	36.6	15 16.0	51.4	54 44.6	52.5	16 26.3	39.6	42 24.7	30
13 0	47 11.9	2 20.1	37.0	17 50.2	51.6	57 22.1	52.4	18 25.0	39.3	43 12.7	17 0
30	47 49.0	4 11.0	37.3	20 24.9	51.7	59 59.3	52.3	20 22.8	38.0	43 59.4	30
14 0	48 27.5	6 2.9	37.6	22 59.9	51.8	2 36.2	52.2	22 19.7	38.6	44 44.7	16 0
30	49 7.4	7 55.7	38.0	25 35.3	51.9	5 12.7	52.1	24 15.5	38.3	45 28.7	30
15 0	49 48.7	9 49.6		28 11.1		7 48.9		26 10.4		46 11.3	15 0
	84°	86°		88°		91°		93°		94°	
	II^s	I^s		O^s		XI^s		X^s		IX^s	

Moon's Distance from the North Pole of the Ecliptic.

Argument. Supplement of Node+Moon's Orbit Longitude.

	IIIs	IVs		Vs		VIs		VIIs		VIIIs	
	84°	86°	Diff. for 10	88°	Diff. for 10	91°	Diff. for 10	93°	Diff. for 10	94°	
° ′	′ ″	′ ″	″	′ ″	″	′ ″	″	′ ″	″	′ ″	° ′
15 0	49 48.7	9 49.6	38.3	28 11.1	52.1	7 48.9	51.9	26 10.4	38.0	46 11.3	15 0
30	50 31.3	11 44.5	38.6	30 47.3	52.2	10 24.7	51.8	28 4.3	37.6	46 52.6	30
16 0	51 15.3	13 40.3	39.0	33 23.8	52.3	13 0.1	51.7	29 57.1	37.3	47 32.5	14 0
30	52 0.6	15 37.2	39.3	36 0.7	52.4	15 35.1	51.6	31 49.0	37.0	48 11.0	30
17 0	52 47.3	17 35.0	39.6	38 37.9	52.5	18 9.8	51.4	33 39.9	36.6	48 48.1	13 0
30	53 35.3	19 33.7	39.9	41 15.4	52.6	20 44.0	51.3	35 29.7	36.2	49 23.9	30
18 0	54 24.7	21 33.4	40.2	43 53.2	52.7	23 17.9	51.1	37 18.4	35.9	49 58.2	12 0
30	55 15.4	23 34.1	40.5	46 31.3	52.8	25 51.2	51.0	39 6.2	35.6	50 31.2	30
19 0	56 7.5	25 35.7	40.8	49 9.6	52.9	28 24.2	50.8	40 52.9	35.2	51 2.9	11 0
30	57 0.9	27 38.2	41.1	51 48.3	53.0	30 56.7	50.7	42 38.4	34.9	51 33.1	30
20 0	57 55.6	29 41.6	41.4	54 27.2	53.0	33 28.7	50.5	44 23.0	34.5	52 1.9	10 0
30	58 51.7	31 45.9	41.7	57 6.3	53.1	36 0.2	50.4	46 6.5	34.1	52 29.4	30
21 0	59 49.1	33 51.1	42.0	59 45.7	53.2	38 31.3	50.2	47 48.8	33.8	52 55.4	9 0
30	0 47.8	35 57.2	42.3	2 25.3	53.3	41 1.8	50.0	49 30.1	33.4	53 20.1	30
22 0	1 47.8	38 4.2	42.6	5 5.1	53.3	43 31.9	49.8	51 10.3	33.0	53 43.3	8 0
30	2 49.1	40 12.0	42.9	7 45.1	53.4	46 1.4	49.7	52 49.4	32.6	54 5.2	30
23 0	3 51.8	42 20.7	43.2	10 25.2	53.5	48 30.4	49.5	54 27.3	32.3	54 25.6	7 0
30	4 55.7	44 30.3	43.4	13 5.6	53.5	50 58.8	49.3	56 4.2	31.9	54 44.6	30
24 0	6 1.0	46 40.6	43.6	15 46.0	53.6	53 26.6	49.1	57 39.9	31.5	55 2.3	6 0
30	7 7.4	48 51.9	44.0	18 26.7	53.6	55 53.9	48.9	59 14.4	31.1	55 18.5	30
25 0	8 15.2	51 3.8	44.3	21 7.5	53.6	58 20.7	48.7	0 47.8	30.8	55 33.3	5 0
30	9 24.3	53 16.7	44.5	23 48.4	53.7	0 46.8	48.5	2 20.1	30.4	55 46.8	30
26 0	10 34.7	55 30.3	44.8	26 29.4	53.7	3 12.3	48.3	3 51.2	30.0	55 58.8	4 0
30	11 46.3	57 44.7	45.0	29 10.5	53.7	5 37.2	48.2	5 21.1	29.6	56 9.4	30
27 0	12 59.2	59 59.8	45.3	31 51.7	53.7	8 1.5	47.9	6 49.9	29.2	56 18.5	3 0
30	14 13.3	2 15.8	45.6	34 33.0	53.8	10 25.2	47.7	8 17.4	28.8	56 26.3	30
28 0	15 28.7	4 32.5	45.8	37 14.3	53.8	12 48.2	47.4	9 43.8	28.4	56 32.7	2 0
30	16 45.4	6 49.8	46.0	39 55.7	53.8	15 10.5	47.2	11 9.0	28.0	56 37.6	30
29 0	18 3.2	9 7.8	46.4	42 37.1	53.8	17 32.2	47.0	12 33.0	27.6	56 41.2	1 0
30	19 22.3	11 26.9	46.6	45 18.6	53.8	19 53.1	46.7	13 55.9	27.2	56 43.3	30
30 0	20 42.7	13 46.6		48 0.0		22 13.4		15 17.3		56 44.0	0 0
	85°	87°		89°		92°		94°		94°	
	IIs	Is		0s		XIs		Xs		IXs	

Equation II *of the Moon's Polar Distance.*

Argument II, corrected.

	IIIs	diff.	IVs	diff.	Vs	diff.	VIs	diff.	VIIs	diff.	VIIIs	diff.	
°	″	″	′ ″	″	′ ″	″	′ ″	″	′ ″	″	′ ″	″	°
0	0 13.8	0.1	1 24.4	4.6	4 36.9	8.0	9 0.0	9.2	13 23.1	7.9	16 35.6	4.6	30
1	0 13.9	0.2	1 29.0	4.8	4 44.9	8.1	9 9.2	9.2	13 31.0	7.8	16 40.2	4.4	29
2	0 14.1	0.4	1 33.8	4.9	4 53.0	8.1	9 18.4	9.1	13 38.8	7.8	16 44.6	4.3	28
3	0 14.5	0.6	1 38.7	5.1	5 1.1	8.2	9 27.5	9.2	13 46.6	7.6	16 48.9	4.1	27
4	0 15.1	0.7	1 43.8	5.2	5 9.3	8.3	9 36.7	9.2	13 54.2	7.6	16 53.0	3.9	26
5	0 15.8	0.9	1 49.0	5.3	5 17.6	8.4	9 45.9	9.1	14 1.8	7.5	16 56.9	3.8	25
6	0 16.7	1.0	1 54.3	5.5	5 26.0	8.4	9 55.0	9.1	14 9.3	7.4	17 0.7	3.7	24
7	0 17.7	1.2	1 59.8	5.6	5 34.4	8.5	10 4.1	9.1	14 16.7	7.3	17 4.4	3.5	23
8	0 18.9	1.4	2 5.4	5.7	5 42.9	8.5	10 13.2	9.1	14 24.0	7.2	17 7.9	3.4	22
9	0 20.3	1.5	2 11.1	5.8	5 51.4	8.6	10 22.3	9.1	14 31.2	7.0	17 11.3	3.2	21
10	0 21.8	1.7	2 16.9	6.0	6 0.0	8.7	10 31.4	9.0	14 38.2	7.0	17 14.5	3.0	20
11	0 23.5	1.8	2 22.9	6.1	6 8.7	8.7	10 40.4	9.0	14 45.2	6.9	17 17.5	2.9	19
12	0 25.3	2.0	2 29.0	6.2	6 17.4	8.8	10 49.4	9.0	14 52.1	6.8	17 20.4	2.8	18
13	0 27.3	2.1	2 35.2	6.3	6 26.2	8.8	10 58.4	8.9	14 58.9	6.6	17 23.2	2.6	17
14	0 29.4	2.3	2 41.5	6.4	6 35.0	8.8	11 7.3	8.9	15 5.5	6.6	17 25.8	2.5	16
15	0 31.7	2.5	2 47.9	6.6	6 43.8	8.9	11 16.2	8.8	15 12.1	6.4	17 28.3	2.3	15
16	0 34.2	2.6	2 54.5	6.6	6 52.7	8.9	11 25.0	8.8	15 18.5	6.3	17 30.6	2.1	14
17	0 36.8	2.8	3 1.1	6.8	7 1.6	9.0	11 33.8	8.8	15 24.8	6.2	17 32.7	2.0	13
18	0 39.6	2.9	3 7.9	6.9	7 10.6	9.0	11 42.6	8.7	15 31.0	6.1	17 34.7	1.8	12
19	0 42.5	3.0	3 14.8	7.0	7 19.6	9.0	11 51.3	8.7	15 37.1	6.0	17 36.5	1.7	11
20	0 45.5	3.2	3 21.8	7.0	7 28.6	9.1	12 0.0	8.6	15 43.1	5.8	17 38.2	1.5	10
21	0 48.7	3.4	3 28.8	7.2	7 37.7	9.1	12 8.6	8.5	15 48.9	5.7	17 39.7	1.4	9
22	0 52.1	3.5	3 36.0	7.3	7 46.8	9.1	12 17.1	8.5	15 54.6	5.6	17 41.1	1.2	8
23	0 55.6	3.7	3 43.3	7.4	7 55.9	9.1	12 25.6	8.4	16 0.2	5.5	17 42.3	1.0	7
24	0 59.3	3.8	3 50.7	7.5	8 5.0	9.1	12 34.0	8.4	16 5.7	5.3	17 43.3	0.9	6
25	1 3.1	3.9	3 58.2	7.6	8 14.1	9.2	12 42.4	8.3	16 11.0	5.2	17 44.2	0.7	5
26	1 7.0	4.1	4 5.8	7.6	8 23.3	9.2	12 50.7	8.2	16 16.2	5.1	17 44.9	0.6	4
27	1 11.1	4.3	4 13.4	7.8	8 32.5	9.1	12 58.9	8.1	16 21.3	4.9	17 45.5	0.4	3
28	1 15.4	4.4	4 21.2	7.8	8 41.6	9.2	13 7.0	8.1	16 26.2	4.8	17 45.9	0.2	2
29	1 19.8	4.6	4 29.0	7.9	8 50.8	9.2	13 15.1	8.0	16 31.0	4.6	17 46.1	0.1	1
30	1 24.4		4 36.9		9 0.0		13 23.1		16 35.6		17 46.2		0
	IIs		Is		Os		XIs		Xs		IXs		

TABLE LVII.

Equation III *of Moon's Polar Distance.*

Argument. Moon's True Longitude.

	IIIs	IVs	Vs	VIs	VIIs	VIIIs	
°	″	″	″	″	″	″	°
0	16.0	14.9	12.0	8.0	4.0	1.1	30
6	16.0	14.5	11.3	7.2	3.3	0.7	24
12	15.8	13.9	10.5	6.3	2.6	0.4	18
18	15.6	13.4	9.7	5.5	2.1	0.2	12
24	15.3	12.7	8.8	4.7	1.5	0.0	6
30	14.9	12.0	8.0	4.0	1.1	0.0	0
	IIs	Is	Os	XIs	Xs	IXs	

TABLE LVIII.

To convert Degrees and Minutes into Decimal Parts.

Deg. &Min.	Dec parts.
° ′	
1 5	003
1 26	4
1 48	5
2 10	6
2 31	7
2 53	8
3 14	9
3 36	10
3 58	11
4 19	12
4 41	13
5 2	14
5 24	15
5 46	16
6 7	17
6 29	18
6 50	19
7 12	20
7 34	21
7 55	22
8 17	23
8 38	24
9 0	25
9 22	26
9 43	27
10 5	28
10 26	29
10 48	30
11 10	31
11 31	32
11 53	33
12 14	34
12 36	35
12 58	36
13 19	37
13 41	38
14 2	39
14 24	40
14 46	41
15 7	42
15 29	43
15 50	44
16 12	45
16 34	46
16 55	47
17 17	48
17 38	49
18 0	50
18 22	51
18 43	52
19 5	53

TABLE LIX.

Equations of Moon's Polar Distance.

Arguments, Arg. 20 of Long.; V to IX corrected; X not corrected; and XI and XII corrected.

Arg.	20	V	VI	VII	VIII	IX	X	XI	Arg.	Arg	XII	Arg.
	″	″	″	″	″	″	″	″			″	
250	0.3	55.9	6.1	2.6	25.1	3.0	0.7	0.9	250	0	4.0	500
260	0.3	55.8	6.2	2.7	25.1	3.1	0.7	0.9	240	10	3.7	510
270	0.4	55.7	6.3	2.8	25.0	3.2	0.8	1.0	230	20	3.4	520
280	0.6	55.4	6.5	3.0	24.9	3.5	1.0	1.0	220	30	3.1	530
290	0.8	55.1	6.9	3.3	24.8	3.8	1.2	1.1	210	40	2.8	540
300	1.0	54.6	7.3	3.7	24.7	4.3	1.5	1.2	200	50	2.5	550
310	1.3	54.1	7.8	4.2	24.4	4.9	1.8	1.3	190	60	2.3	560
320	1.7	53.4	8.4	4.7	24.1	5.6	2.2	1.4	180	70	2.1	570
330	2.1	52.7	9.1	5.4	23.8	6.4	2.7	1.5	170	80	1.9	580
340	2.6	51.9	9.8	6.1	23.5	7.2	3.2	1.7	160	90	1.7	590
350	3.1	51.0	10.7	6.9	23.2	8.2	3.8	1.9	150	100	1.6	600
360	3.7	50.0	11.6	7.7	22.8	9.2	4.4	2.1	140	110	1.5	610
370	4.3	48.9	12.6	8.7	22.4	10.3	5.1	2.3	130	120	1.5	620
380	4.9	47.7	13.6	9.7	21.9	11.5	5.8	2.5	120	130	1.5	630
390	5.6	46.5	14.8	10.7	21.4	12.8	6.6	2.8	110	140	1.5	640
400	6.4	45.2	16.0	11.8	20.9	14.1	7.4	3.0	100	150	1.6	650
410	7.1	43.9	17.2	13.0	20.4	15.5	8.3	3.3	90	160	1.7	660
420	7.9	42.5	18.5	14.2	19.9	17.0	9.1	3.5	80	170	1.9	670
430	8.8	41.0	19.8	15.5	19.3	18.5	10.1	3.8	70	180	2.1	680
440	9.6	39.5	21.2	16.8	18.7	20.1	11.0	4.1	60	190	2.3	690
450	10.5	38.0	22.6	18.1	18.1	21.7	12.0	4.4	50	200	2.5	700
460	11.3	36.4	24.1	19.4	17.5	23.3	12.9	4.7	40	210	2.8	710
470	12.2	34.9	25.5	20.8	16.9	24.9	13.9	5.0	30	220	3.1	720
480	13.2	33.2	27.0	22.2	16.3	26.6	15.0	5.4	20	230	3.4	730
490	14.1	31.6	28.5	23.6	15.6	28.3	16.0	5.7	10	240	3.7	740
500	15.0	30.0	30.0	25.0	15.0	30.0	17.0	6.0	0	250	4.0	750
510	15.9	28.4	31.5	26.4	14.4	31.7	18.0	6.3	990	260	4.3	760
520	16.8	26.8	33.0	27.8	13.7	33.4	19.0	6.6	980	270	4.6	770
530	17.8	25.1	34.5	29.2	13.1	35.1	20.1	7.0	970	280	4.9	780
540	18.7	23.6	35.9	30.6	12.5	36.7	21.1	7.3	960	290	5.2	790
550	19.5	22.0	37.4	31.9	11.9	38.3	22.0	7.6	950	300	5.5	800
560	20.4	20.5	38.8	33.2	11.3	39.9	23.0	7.9	940	310	5.7	810
570	21.2	19.0	40.2	34.5	10.7	41.5	23.9	8.2	930	320	5.9	820
580	22.1	17.5	41.5	35.8	10.1	43.0	24.9	8.5	920	330	6.1	830
590	22.9	16.1	42.8	37.0	9.6	44.5	25.7	8.7	910	340	6.3	840
600	23.6	14.8	44.0	38.2	9.1	45.9	26.6	9.0	900	350	6.4	850
610	24.4	13.5	45.2	39.3	8.6	47.2	27.4	9.2	890	360	6.5	860
620	25.1	12.3	46.4	40.3	8.1	48.5	28.2	9.5	880	370	6.5	870
630	25.7	11.1	47.4	41.3	7.6	49.7	28.9	9.7	870	380	6.5	880
640	26.3	10.0	48.4	42.3	7.2	50.8	29.6	9.9	860	390	6.5	890
650	26.9	9.0	49.3	43.1	6.8	51.8	30.2	10.1	850	400	6.4	900
660	27.4	8.1	50.2	43.9	6.5	52.8	30.8	10.3	840	410	6.3	910
670	27.9	7.3	50.9	44.6	6.2	53.6	31.3	10.5	830	420	6.1	920
680	28.3	6.6	51.6	45.3	5.9	54.4	31.8	10.6	820	430	5.9	930
690	28.7	5.9	52.2	45.8	5.6	55.1	32.2	10.7	810	440	5.7	940
700	29.0	5.4	52.7	46.3	5.3	55.7	32.5	10.8	800	450	5.5	950
710	29.2	4.9	53.1	46.7	5.2	56.2	32.8	10.9	790	460	5.2	960
720	29.4	4.6	53.5	47.0	5.1	56.5	33.0	11.0	780	470	4.9	970
730	29.6	4.3	53.7	47.2	5.0	56.8	33.2	11.0	770	480	4.6	980
740	29.7	4.2	53.8	47.3	4.9	56.9	33.3	11.1	760	490	4.3	990
750	29.7	4.1	53.9	47.4	4.9	57.0	33.3	11.1	750	500	4.0	1000

Constant 10″

TABLE LX.

Small Equations of Moon's Parallax.

Args., 1, 2, 4, 5, 6, 8, 9, 12, 13, of Long.

A.	1	2	4	5	6	8	9	12	13	A.
		"	"	"	"	"	"	"	"	
0	0.0	1.6	0.6	1.6	1.9	0.0	3.6	1.4	2.0	100
3	0.0	1.6	0.6	1.6	1.9	0.0	3.5	1.4	2.0	97
6	0.0	1.5	0.6	1.5	1.8	0.0	3.1	1.4	1.9	94
9	0.1	1.5	0.6	1.5	1.8	0.1	2.6	1.3	1.8	91
12	0.1	1.4	0.5	1.4	1.7	0.2	1.9	1.2	1.7	88
15	0.1	1.3	0.5	1.3	1.6	0.2	1.3	1.1	1.6	85
18	0.2	1.1	0.4	1.1	1.4	0.3	0.7	1.0	1.4	82
21	0.3	1.0	0.4	1.0	1.3	0.5	0.2	0.9	1.2	79
24	0.4	0.9	0.3	0.9	1.2	0.6	0.0	0.7	1.0	76
27	0.5	0.7	0.3	0.7	1.0	0.7	0.1	0.6	0.9	73
30	0.5	0.6	0.2	0.6	0.9	0.8	0.4	0.5	0.7	70
33	0.6	0.4	0.2	0.4	0.7	0.9	0.8	0.4	0.5	67
36	0.7	0.3	0.1	0.3	0.6	1.0	1.5	0.3	0.4	64
39	0.7	0.2	0.1	0.2	0.5	1.1	2.1	0.2	0.2	61
42	0.8	0.1	0.0	0.1	0.4	1.1	2.8	0.1	0.1	58
45	0.8	0.0	0.0	0.0	0.3	1.2	3.2	0.0	0.0	55
48	0.8	0.0	0.0	0.0	0.3	1.2	3.5	0.0	0.0	52
50	0.8	0.0	0.0	0.0	0.3	1.2	3.6	0.0	0.0	50

Constant 7"

The first two figures only of the Arguments are taken.

TABLE LXI.

Moon's Equatorial Parallax.

Argument. Arg. of Evection.

	0^s	I^s	II^s	III^s	IV^s	V^s	
°	' "	' "	' "	"	"	"	°
0	1 20.8	1 15.6	1 1.5	42.6	24.1	10.8	30
1	1 20.8	1 15.2	1 0.9	41.9	23.6	10.5	29
2	1 20.8	1 14.9	1 0.3	41.3	23.0	10.2	28
3	1 20.7	1 14.5	59.7	40.6	22.5	9.9	27
4	1 20.7	1 14.2	59.2	40.0	21.9	9.6	26
5	1 20.6	1 13.8	58.6	39.4	21.4	9.4	25
6	1 20.6	1 13.4	57.9	38.7	20.9	9.1	24
7	1 20.5	1 13.0	57.3	38.1	20.4	8.8	23
8	1 20.4	1 12.6	56.7	37.4	19.9	8.6	22
9	1 20.3	1 12.2	56.1	36.8	19.4	8.4	21
10	1 20.2	1 11.7	55.5	36.1	18.9	8.2	20
11	1 20.1	1 11.3	54.9	35.5	18.4	8.0	19
12	1 19.9	1 10.8	54.2	34.9	17.9	7.8	18
13	1 19.8	1 10.4	53.6	34.2	17.5	7.6	17
14	1 19.6	1 9.9	53.0	33.6	17.0	7.4	16
15	1 19.5	1 9.4	52.3	33.0	16.6	7.2	15
16	1 19.3	1 9.0	51.7	32.4	16.1	7.1	14
17	1 19.1	1 8.5	51.1	31.7	15.7	6.9	13
18	1 18.9	1 8.0	50.4	31.1	15.2	6.8	12
19	1 18.7	1 7.5	49.8	30.5	14.8	6.7	11
20	1 18.4	1 7.0	49.1	29.9	14.4	6.5	10
21	1 18.2	1 6.5	48.5	29.3	14.0	6.4	9
22	1 18.0	1 5.9	47.8	28.7	13.6	6.3	8
23	1 17.7	1 5.4	47.2	28.1	13.2	6.3	7
24	1 17.4	1 4.8	46.5	27.5	12.9	6.2	6
25	1 17.1	1 4.3	45.9	26.9	12.5	6.1	5
26	1 16.9	1 3.8	45.2	26.3	12.1	6.1	4
27	1 16.6	1 3.2	44.6	25.8	11.8	6.1	3
28	1 16.2	1 2.6	43.9	25.2	11.5	6.0	2
29	1 15.9	1 2.1	43.3	24.7	11.1	6.0	1
30	1 15.6	1 1.5	42.6	24.1	10.8	6.0	0
	XI^s	X^s	IX^s	VIII^s	VII^s	VI^s	

Moon's Equatorial Parallax.

Argument. Anomaly.

	0^s	diff	I^s	diff	II^s	diff	III^s	diff	IV^s	diff	V^s	diff	
°	′ ″	″	′ ″	″	′ ″	″	′ ″	″	′ ″	″	′ ″	″	°
0	58 57.7	0.0	58 27.0	2.0	57 7.9	3.1	55 29.8	3.2	54 1.9	2.5	53 3.2	1.4	30
1	58 57.7	0.1	58 25.0	2.0	57 4.8	3.2	55 26.6	3.2	53 59.4	2.5	53 1.8	1.3	29
2	58 57.6	0.2	58 23.0	2.1	57 1.6	3.2	55 23.4	3.2	53 56.9	2.4	53 0.5	1.2	28
3	58 57.4	0.3	58 20.9	2.2	56 58.4	3.2	55 20.2	3.2	53 54.5	2.4	52 59.3	1.2	27
4	58 57.1	0.3	58 18.7	2.2	56 55.2	3.2	55 17.0	3.2	53 52.1	2.4	52 58.1	1.1	26
5	58 56.8	0.4	58 16.5	2.2	56 52.0	3.2	55 13.8	3.2	53 49.7	2.3	52 57.0	1.2	25
6	58 56.4	0.4	58 14.3	2.3	56 48.8	3.3	55 10.6	3.1	53 47.4	2.3	52 55.8	1.0	24
7	58 56.0	0.6	58 12.0	2.4	56 45.5	3.2	55 7.5	3.1	53 45.1	2.2	52 54.8	1.0	23
8	58 55.4	0.6	58 9.6	2.4	56 42.3	3.3	55 4.4	3.1	53 42.9	2.3	52 53.8	1.0	22
9	58 54.8	0.6	58 7.2	2.4	56 39.0	3.3	55 1.3	3.1	53 40.6	2.1	52 52.8	0.9	21
10	58 54.2	0.8	58 4.8	2.5	56 35.7	3.3	54 58.2	3.1	53 38.5	2.2	52 51.9	0.9	20
11	58 53.4	0.8	58 2.3	2.5	56 32.4	3.3	54 55.1	3.0	53 36.3	2.1	52 51.0	0.9	19
12	58 52.6	0.8	57 59.8	2.6	56 29.1	3.3	54 52.1	3.0	53 34.2	2.1	52 50.1	0.8	18
13	58 51.8	1.0	57 57.2	2.6	56 25.8	3.3	54 49.1	3.0	53 32.1	2.0	52 49.3	0.7	17
14	58 50.8	1.0	57 54.6	2.7	56 22.5	3.3	54 46.1	3.0	53 30.1	2.0	52 48.6	0.7	16
15	58 49.8	1.1	57 51.9	2.7	56 19.2	3.3	54 43.1	2.9	53 28.1	1.9	52 47.9	0.7	15
16	58 48.7	1.1	57 49.2	2.8	56 15.9	3.3	54 40.2	2.9	53 26.2	1.9	52 47.2	0.6	14
17	58 47.6	1.2	57 46.4	2.7	56 12.6	3.3	54 37.3	2.9	53 24.3	1.9	52 46.6	0.6	13
18	58 46.4	1.3	57 43.7	2.9	56 9.3	3.3	54 34.4	2.9	53 22.4	1.8	52 46.0	0.5	12
19	58 45.1	1.3	57 40.8	2.8	56 6.0	3.3	54 31.5	2.8	53 20.6	1.8	52 45.5	0.5	11
20	58 43.8	1.4	57 38.0	2.9	56 2.7	3.4	54 28.7	2.8	53 18.8	1.8	52 45.0	0.4	10
21	58 42.4	1.5	57 35 1	2.9	55 59.3	3.3	54 25.9	2.8	53 17.0	1.7	52 44.6	0.4	9
22	58 40.9	1.5	57 32.2	2.9	55 56.0	3.3	54 23.1	2.8	53 15.3	1.6	52 44.2	0.4	8
23	58 39.4	1.6	57 29.3	3.0	55 52.7	3.3	54 20.3	2.7	53 13.7	1.7	52 43.8	0.3	7
24	58 37.8	1.6	57 26.3	3.0	55 49.4	3.3	54 17.6	2.7	53 12.0	1.6	52 43.5	0.2	6
25	58 36.2	1.8	57 23.3	3.0	55 46.1	3.3	54 14.9	2.7	53 10.4	1.5	52 43.3	0.2	5
26	58 34.4	1.7	57 20.2	3.0	55 42.8	3.2	54 12.2	2.6	53 8.9	1.5	52 43.1	0.2	4
27	58 32.7	1.8	57 17.2	3.1	55 39.6	3.2	54 9.6	2.6	53 7.4	1.5	52 42.9	0.1	3
28	58 30.9	1.9	57 14.1	3.1	55 36.4	3.3	54 7.0	2.6	53 5.9	1.4	52 42.8	0.1	2
29	58 29.0	2.0	57 11.0	3.1	55 33.1	3.3	54 4.4	2.5	53 4.5	1.3	52 42.7	0.0	1
30	58 27.0		57 7.9		55 29.8		54 1.9		53 3.2		52 42.7		0
	XI^s		X^s		IX^s		VIII^s		VII^s		VI^s		

Moon's Equatorial Parallax.

Argument. Argument of the Variation.

	0ˢ	Iˢ	IIˢ	IIIˢ	IVˢ	Vˢ	
°	″	″	″	″	″	″	°
0	55.6	42.3	16.0	3.7	17.6	44.0	30
1	55.6	41.5	15.3	3.8	18.5	44.8	29
2	55.5	40.7	14.5	3.8	19.3	45.6	28
3	55.5	39.8	13.8	3.9	20.1	46.3	27
4	55.3	39.0	13.1	4.1	21.0	47.0	26
5	55.2	38.1	12.4	4.3	21.9	47.7	25
6	55.0	37.2	11.7	4.5	22.7	48.4	24
7	54.8	36.3	11.1	4.7	23.6	49.1	23
8	54.6	35.5	10.4	5.0	24.5	49.7	22
9	54.3	34.6	9.8	5.3	25.4	50.3	21
10	54.0	33.7	9.2	5.6	26.3	50.9	20
11	53.7	32.7	8.7	6.0	27.2	51.5	19
12	53.3	31.8	8.2	6.3	28.2	52.1	18
13	52.9	30.9	7.7	6.8	29.1	52.6	17
14	52.5	30.0	7.2	7.2	30.0	53.1	16
15	52.0	29.1	6.7	7.7	30.9	53.5	15
16	51.5	28.2	6.3	8.2	31.8	54.0	14
17	51.0	27.2	5.9	8.7	32.8	54.4	13
18	50.5	26.3	5.6	9.3	33.7	54.8	12
19	49.9	25.4	5.3	9.8	34.6	55.1	11
20	49.4	24.5	5.0	10.5	35.5	55.4	10
21	48.8	23.6	4.7	11.1	36.4	55.7	9
22	48.1	22.7	4.5	11.7	37.3	56.0	8
23	47.4	21.9	4.3	12.4	38.2	56.2	7
24	46.8	21.0	4.1	13.1	39.0	56.4	6
25	46.1	20.1	3.9	13.8	39.9	56.6	5
26	45.4	19.3	3.8	14.5	40.8	56.8	4
27	44.6	18.5	3.7	15.3	41.6	56.9	3
28	43.9	17.6	3.7	16.1	42.4	56.9	2
29	43.1	16.8	3.7	16.8	43.2	57.0	1
30	42.3	16.0	3.7	17.6	44.0	57.0	0
	XIˢ	Xˢ	IXˢ	VIIIˢ	VIIˢ	VIˢ	

TABLE LXIV.

Reduction of the Parallax, and also of the Latitude.

Argument. Latitude.

Lat.	Red. of par	Red. of Lat.
°	″	′ ″
0	0.0	0 0.0
3	0.0	1 11.8
6	0.1	2 22.7
9	0.3	3 32.1
12	0.5	4 39.3
15	0.7	5 43.4
18	1.0	6 43.7
21	1.4	7 39.7
24	1.8	8 30.7
27	2.3	9 16.1
30	2.7	9 55.4
33	3.3	10 28.3
36	3.8	10 54.3
39	4.4	11 13.2
42	4.9	11 24.7
45	5.5	11 28.7
48	6.1	11 25.2
51	6.7	11 14.1
54	7.2	10 55.7
57	7.8	10 30.0
60	8.3	9 57.4
63	8.8	9 18.3
66	9.2	8 32.9
69	9.7	7 42.0
72	10.0	6 45.9
75	10.3	5 45.4
78	10.6	4 41.0
81	10.8	3 33.5
84	11.0	2 23.7
87	11.1	1 12.3
90	11.1	0 0.0

Subsidiary Table.

Lat.	+3′	—3′
°	″	″
0	+0.0	—0.0
6	0.0	0.0
12	0.0	0.0
15	0.0	0.0
18	0.1	0.1
24	0.1	0.1
30	0.1	0.1
36	0.2	0.2
42	0.2	0.2
48	0.3	0.3
54	0.3	0.3
60	0.4	0.4
72	0.5	0.5
78	0.6	0.6
84	0.6	0.6
90	+0.6	—0.6

TABLE LXV.

Moon's Semi-diameter.

Argument. Equatorial Parallax.

Eq.Par	Semidia.	Eq.Par	Semidia.	Eq.Par	Semidia.	sec	Pro. Par.
′ ″	′ ″	′ ″	′ ″	′ ″	′ ″		″
53 0	14 26.5	56 0	15 15.6	59 0	16 4.6	1	0.3
53 10	14 29.3	56 10	15 18.3	59 10	16 7.4	2	0.5
53 20	14 32.0	56 20	15 21.0	59 20	16 10.1	3	0.8
53 30	14 34.7	56 30	15 23.8	59 30	16 12.8	4	1.1
53 40	14 37.4	56 40	15 26.5	59 40	16 15.6	5	1.4
53 50	14 40.2	56 50	15 29.2	59 50	16 18.3	6	1.6
54 0	14 42.9	57 0	15 31.9	60 0	16 21.0	7	1.9
54 10	14 45.6	57 10	15 34.7	60 10	16 23.7	8	2.2
54 20	14 48.3	57 20	15 37.4	60 20	16 26.4	9	2.4
54 30	14 51.1	57 30	15 40.1	60 30	16 29.2	10	2.7
54 40	14 53.8	57 40	15 42.8	60 40	16 31.9		
54 50	14 56.5	57 50	15 45.6	60 50	16 34.6		
55 0	14 59.2	58 0	15 48.3	61 0	16 37.3		
55 10	15 2.0	58 10	15 51.0	61 10	16 40.1		
55 20	15 4.7	58 20	15 53.7	61 20	16 42.8		
55 30	15 7.4	58 30	15 56.5	61 30	16 45.5		
55 40	15 10.1	58 40	15 59.2	61 40	16 48.2		
55 50	15 12.9	58 50	16 1.9	61 50	16 51.0		
56 0	15 15.6	59 0	16 4.6	62 0	16 53.7		

TABLE LXVI.

Augmentation of Moon's Semi-diameter.

Alt.	Horizon. Semi-diameter. 14′30″	15′	16′	17′	Alt.	Horizon. Semi-diameter. 14′ 30″	15′	16′	17
°	″	″	″	″	°	″	″	″	″
2	0.6	0.6	0.7	0.8	42	9.2	9.8	11.2	12.6
4	1.0	1.1	1.3	1.5	45	9.7	10.4	11.8	13.3
6	1.5	1.6	1.9	2.1	48	10.2	10.9	12.4	14.0
8	2.0	2.1	2.4	2.7	51	10.6	11.4	13.0	14.7
10	2.4	2.6	3.0	3.4	54	11.1	11.8	13.5	15.2
12	2.9	3.1	3.6	4.0	57	11.5	12.3	14.0	15.8
14	3.4	3.6	4.1	4.7	60	11 8	12.7	14.4	16.3
16	3.8	4.1	4.7	5.3	63	12.2	13.0	14.9	16.8
18	4.3	4.6	5.2	5.9	66	12.5	13.4	15.2	17.2
21	4.9	5.3	6.0	6.8	69	12.8	13.7	15.6	17.6
24	5.6	6.0	6.8	7.7	72	13.0	13.9	15.9	17.9
27	6.2	6.7	7.6	8.6	75	13.2	14.1	16.1	18.2
30	6.9	7.3	8.4	9.5	78	13.4	14.3	16.3	18.4
33	7.5	8.0	9.1	10.3	81	13.5	14.4	16.5	18.6
36	8.1	8.6	9.8	11.1	84	13.6	14.5	16.6	18.7
39	8.6	9.2	10.5	11.9	90	13.7	14.6	16.7	18.8

Moon's Horary Motion in Longitude.

Arguments. 1 to 18 of Longitude.

Arg.	2	3	4	5	6	1	7	8	9	Arg.
	″	″	″	″	″	″	″	″	″	
0	5.0	0.0	2.9	1.9	0.0	0.00	0.00	0.00	0.16	100
2	5.0	0.0	2.8	1.9	0.0	0.00	0.00	0.00	0.15	98
4	4.9	0.0	2.8	1.9	0.0	0.01	0.00	0.02	0.15	96
6	4.8	0.1	2.8	1.9	0.1	0.03	0.01	0.05	0.14	94
8	4.7	0.2	2.7	1.8	0.1	0.06	0.01	0.09	0.12	92
10	4.5	0.3	2.6	1.7	0.2	0.09	0.02	0.14	0.10	90
12	4.3	0.4	2.5	1.7	0.2	0.13	0.02	0.19	0.09	88
14	4.1	0.6	2.3	1.6	0.3	0.18	0.03	0.26	0.07	86
16	3.8	0.7	2.2	1.5	0.4	0.23	0.04	0.33	0.05	84
18	3.6	0.9	2.0	1.4	0.5	0.28	0.05	0.41	0.03	82
20	3.3	1.1	1.9	1.3	0.6	0.34	0.06	0.50	0.02	80
22	3.0	1.3	1.7	1.1	0.7	0.40	0.07	0.58	0.01	78
24	2.7	1.5	1.5	1.0	0.8	0.46	0.08	0.67	0.00	76
26	2.3	1.7	1.3	0.9	0.9	0.52	0.10	0.77	0.00	74
28	2.0	1.9	1.2	0.8	1.0	0.58	0.11	0.86	0.00	72
30	1.7	2.1	1.0	0.7	1.1	0.63	0.12	0.94	0.01	70
32	1.4	2.2	0.8	0.5	1.2	0.69	0.13	1.03	0.01	68
34	1.2	2.4	0.7	0.4	1.3	0.74	0.14	1.11	0.03	66
36	0.9	2.6	0.5	0.3	1.3	0.78	0.15	1.18	0.05	64
38	0.7	2.7	0.4	0.3	1.4	0.82	0.16	1.25	0.06	62
40	0.5	2.8	0.3	0.2	1.5	0.86	0.16	1.30	0.08	60
42	0.3	2.9	0.2	0.1	1.5	0.89	0.17	1.35	0.10	58
44	0.2	3.0	0.1	0.1	1.6	0.91	0.17	1.39	0.11	56
46	0.1	3.1	0.0	0.0	1.6	0.93	0.18	1.42	0.12	54
48	0.0	3.1	0.0	0.0	1.6	0.94	0.18	1.44	0.13	52
50	0.0	3.1	0.0	0.0	1.6	0.94	0.18	1.44	0.13	50

Arg.	10	11	12	13	14	15	16	17	18	Arg.
	″	″	″	″	″	″	″	″	″	
0	0.00	0.26	0.00	0.00	0.00	0.00	0.26	0.00	0.21	100
2	0.00	0.25	0.00	0.00	0.00	0.00	0.26	0.00	0.20	98
4	0.02	0.24	0.01	0.00	0.01	0.00	0.26	0.00	0.20	96
6	0.04	0.22	0.03	0.01	0.02	0.01	0.25	0.00	0.20	94
8	0.08	0.20	0.04	0.02	0.04	0.01	0.25	0.01	0.20	92
10	0.12	0.17	0.07	0.03	0.06	0.02	0.24	0.01	0.20	90
12	0.16	0.14	0.09	0.04	0.09	0.02	0.22	0.02	0.19	88
14	0.20	0.11	0.12	0.06	0.12	0.03	0.21	0.02	0.19	86
16	0.24	0.08	0.16	0.07	0.15	0.04	0.20	0.03	0.18	84
18	0.28	0.05	0.19	0.09	0.19	0.05	0.19	0.04	0.18	82
20	0.31	0.03	0.23	0.11	0.22	0.06	0.17	0.05	0.17	80
22	0.34	0.01	0.27	0.13	0.26	0.07	0.15	0.06	0.17	78
24	0.35	0.00	0.31	0.15	0.30	0.08	0.14	0.07	0.16	76
26	0.36	0.00	0.35	0.17	0.34	0.08	0.12	0.07	0.16	74
28	0.35	0.01	0.39	0.19	0.38	0.09	0.11	0.08	0.15	72
30	0.34	0.02	0.43	0.21	0.42	0.10	0.09	0.09	0.15	70
32	0.32	0.04	0.47	0.23	0.45	0.11	0.07	0.10	0.14	68
34	0.29	0.06	0.50	0.25	0.49	0.12	0.06	0.11	0.14	66
36	0.26	0.09	0.54	0.26	0.52	0.13	0.05	0.12	0.13	64
38	0.22	0.11	0.57	0.28	0.55	0.14	0.04	0.12	0.13	62
40	0.18	0.14	0.59	0.29	0.58	0.14	0.02	0.13	0.12	60
42	0.15	0.16	0.62	0.30	0.60	0.15	0.01	0.13	0.12	58
44	0.12	0.19	0.63	0.31	0.62	0.15	0.01	0.14	0.12	56
46	0.10	0.21	0.65	0.32	0.63	0.16	0.00	0.14	0.12	54
48	0.09	0.22	0.66	0.32	0.64	0.16	0.00	0.14	0.12	52
50	0.08	0.22	0.66	0.32	0.64	0.16	0.00	0.14	0.11	50

TABLE LXVIII.

Moon's Horary Motion in Longitude.

Argument. Argument of the Evection.

	0s	Is	IIs	IIIs	IVs	Vs	
°	"	"	"	"	"	"	°
0	80.3	74.7	59.6	39.4	19.8	5.9	30
1	80.3	74.3	58.9	38.7	19.3	5.6	29
2	80.3	73.9	58.3	38.0	18.7	5.3	28
3	80.2	73.5	57.7	37.3	18.1	5.0	27
4	80.2	73.1	57.1	36.6	17.6	4.7	26
5	80.1	72.7	56.4	36.0	17.0	4.4	25
6	80.1	72.3	55.8	35.3	16.5	4.1	24
7	80.0	71.9	55.1	34.6	15.9	3.8	23
8	79.9	71.4	54.5	33.9	15.4	3.6	22
9	79.8	71.0	53.8	33.2	14.9	3.4	21
10	79.7	70.5	53.1	32.5	14.4	3.1	20
11	79.5	70.1	52.5	31.9	13.9	2.9	19
12	79.4	69.6	51.8	31.2	13.4	2.7	18
13	79.2	69.1	51.1	30.5	12.9	2.5	17
14	79.1	68.6	50.5	29.9	12.4	2.3	16
15	78.9	68.1	49.8	29.2	11.9	2.1	15
16	78.7	67.6	49.1	28.6	11.4	2.0	14
17	78.5	67.0	48.4	27.9	11.0	1.8	13
18	78.2	66.5	47.7	27.2	10.5	1.7	12
19	78.0	66.0	47.0	26.6	10.1	1.6	11
20	77.8	65.4	46.4	26.0	9.7	1.4	10
21	77.5	64.9	45.7	25.3	9.3	1.3	9
22	77.2	64.3	45.0	24.7	8.8	1.2	8
23	77.0	63.7	44.3	24.1	8.4	1.2	7
24	76.7	63.2	43.6	23.5	8.0	1.1	6
25	76.4	62.6	42.9	22.8	7.7	1.0	5
26	76.1	62.0	42.2	22.2	7.3	1.0	4
27	75.7	61.4	41.5	21.6	6.9	0.9	3
28	75.4	60.8	40.8	21.0	6.6	0.9	2
29	75.0	60.2	40.1	20.4	6.2	0.9	1
30	74.7	59.6	39.4	19.8	5.9	0.9	0
	XIs	Xs	IXs	VIIIs	VIIs	VIs	

TABLE LXIX.

Moon's Horary Motion in Longitude.

Arguments. Sum of Equations, 2, 3, &c., and Evection corrected

		0"	10"	20"		
s	°				s	°
O	0	0.0	0.2	0.5	XII	0
I	0	0.0	0.2	0.4	XI	0
II	0	0.1	0.2	0.3	X	0
III	0	0.2	0.2	0.2	IX	0
IV	0	0.3	0.2	0.1	VIII	0
V	0	0.4	0.2	0.0	VII	0
VI	0	0.5	0.2	0.0	VI	0
		0"	10"	20"		

Moon's Horary Motion in Longitude.

Arguments. Sum of preceding equations, and Anomaly corrected

s	°	0″	10″	20″	30″	40″	50″	60″	70″	80″	90″	100″	s	°
		″	″	″	″	″	″	″	″	″	″	″		
O	0	4.1	5.3	6.5	7.6	8.8	10.0	11.2	12.4	13.5	14.7	15.9	XII	0
	5	4.1	5.3	6.5	7.7	8.8	10.0	11.2	12.3	13.5	14.7	15.9		25
	10	4.2	5.4	6.5	7.7	8.8	10.0	11.2	12.3	13.5	14.6	15.8		20
	15	4.3	5.5	6.6	7.7	8.9	10.0	11.1	12.3	13.4	14.5	15.7		15
	20	4.5	5.6	6.7	7.8	8.9	10.0	11.1	12.2	13.3	14.4	15.5		10
	25	4.8	5.8	6.9	7.9	9.0	10.0	11.0	12.1	13.1	14.2	15.2		5
I	0	5.1	6.0	7.0	8.0	9.0	10.0	11.0	12.0	13.0	14.0	14.9	XI	0
	5	5.4	6.3	7.2	8.2	9.1	10.0	10.9	11.8	12.8	13.7	14.6		25
	10	5.7	6.6	7.4	8.3	9.2	10.0	10.8	11.7	12.6	13.4	14.3		20
	15	6.1	6.9	7.7	8.5	9.2	10.0	10.8	11.5	12.3	13.1	13.9		15
	20	6.6	7.2	7.9	8.6	9.3	10.0	10.7	11.4	12.1	12.8	13.4		10
	25	7.0	7.6	8.2	8.8	9.4	10.0	10.6	11.2	11.8	12.4	13.0		5
II	0	7.5	8.0	8.5	9.0	9.5	10.0	10.5	11.0	11.5	12.0	12.5	X	0
	5	7.9	8.4	8.8	9.2	9.6	10.0	10.4	10.8	11.2	11.6	12.1		25
	10	8.4	8.7	9.1	9.4	9.7	10.0	10.3	10.6	10.9	11.3	11.6		20
	15	8.9	9.1	9.4	9.6	9.8	10.0	10.2	10.4	10.6	10.9	11.1		15
	20	9.4	9.5	9.7	9.8	9.9	10.0	10.1	10.2	10.3	10.5	10.6		10
	25	9.9	9.9	9.9	10.0	10.0	10.0	10.0	10.0	10.1	10.1	10.1		5
III	0	10.4	10.3	10.2	10.1	10.1	10.0	9.9	9.9	9.8	9.7	9.6	IX	0
	5	10.8	10.7	10.5	10.3	10.2	10.0	9.8	9.7	9.5	9.3	9.2		25
	10	11.3	11.0	10.8	10.5	10.3	10.0	9.7	9.5	9.2	9.0	8.7		20
	15	11.7	11.4	11.0	10.7	10.3	10.0	9.7	9.3	9.0	8.6	8.3		15
	20	12.1	11.7	11.3	10.9	10.4	10.0	9.6	9.1	8.7	8.3	7.9		10
	25	12.5	12.0	11.5	11.0	10.5	10.0	9.5	9.0	8.5	8.0	7.5		5
IV	0	12.9	12.3	11.7	11.2	10.6	10.0	9.4	8.8	8.3	7.7	7.1	VIII	0
	5	13.3	12.6	11.9	11.3	10.6	10.0	9.4	8.7	8.1	7.4	6.7		25
	10	13.6	12.9	12.1	11.4	10.7	10.0	9.3	8.6	7.9	7.1	6.4		20
	15	13.9	13.1	12.3	11.5	10.8	10.0	9.2	8.5	7.7	6.9	6.1		15
	20	14.1	13.3	12.5	11.6	10.8	10.0	9.2	8.4	7.5	6.7	5.9		10
	25	14.4	13.5	12.6	11.7	10.9	10.0	9.1	8.3	7.4	6.5	5.6		5
V	0	14.6	13.7	12.7	11.8	10.9	10.0	9.1	8.2	7.3	6.3	5.4	VII	0
	5	14.7	13.8	12.8	11.9	10.9	10.0	9.1	8.1	7.2	6.2	5.3		25
	10	14.9	13.9	12.9	12.0	11.0	10.0	9.0	8.0	7.1	6.1	5.1		20
	15	15.0	14.0	13.0	12.0	11.0	10.0	9.0	8.0	7.0	6.0	5.0		15
	20	15.1	14.1	13.0	12.0	11.0	10.0	9.0	8.0	7.0	5.9	4.9		10
	25	15.1	14.1	13.1	12.0	11.0	10.0	9.0	8.0	6.9	5.9	4.9		5
VI	0	15.1	14.1	13.1	12.1	11.0	10.0	9.0	8.0	6.9	5.9	4.9	VI	0
		0″	10″	20″	30″	40″	50″	60″	70″	80″	90″	100″		

Moon's Horary Motion in Longitude.

Argument. Anomaly corrected.

	0s	diff.	Is	diff.	IIs	diff.	IIIs	diff.	IVs	diff.	Vs	diff.	
°	″	″	″	″	″	″	″	″	″	″	″	″	°
0	441.5	0.0	404.1	2.5	309.3	3.7	195.3	3.7	95.8	2.8	30.6	1.4	30
1	441.5	0.1	401.6	2.4	305.6	3.7	191.6	3.7	93.0	2.8	29.2	1.4	29
2	441.3	0.2	399.2	2.6	301.9	3.8	187.9	3.6	90.2	2.6	27.8	1.4	28
3	441.1	0.3	396.6	2.6	298.1	3.7	184.3	3.7	87.6	2.7	26.4	1.3	27
4	440.8	0.4	394.0	2.7	294.4	3.8	180.6	3.6	84.9	2.6	25.1	1.3	26
5	440.4	0.5	391.3	2.7	290.6	3.8	177.0	3.6	82.3	2.6	23.8	1.2	25
6	439.9	0.5	388.6	2.8	286.8	3.8	173.4	3.6	79.7	2.6	22.6	1.2	24
7	439.4	0.7	385.8	2.8	283.0	3.8	169.8	3.5	77.1	2.5	21.4	1.1	23
8	438.7	0.7	383.0	2.9	279.2	3.8	166.3	3.5	74.6	2.5	20.3	1.1	22
9	438.0	0.8	380.1	3.0	275.4	3.9	162.8	3.5	72.1	2.4	19.2	1.0	21
10	437.2	0.9	377.1	3.0	271.5	3.8	159.3	3.5	69.7	2.4	18.2	1.0	20
11	436.3	1.0	374.1	3.0	267.7	3.9	155.8	3.4	67.3	2.3	17.2	0.9	19
12	435.3	1.1	371.1	3.1	263.8	3.8	152.4	3.5	65.0	2.3	16.3	0.9	18
13	434.2	1.1	368.0	3.2	260.0	3.8	148.9	3.4	62.7	2.3	15.4	0.8	17
14	433.1	1.3	364.8	3.2	256.2	3.9	145.5	3.3	60.4	2.2	14.6	0.8	16
15	431.8	1.3	361.6	3.2	252.3	3.8	142.2	3.3	58.2	2.1	13.8	0.7	15
16	430.5	1.4	358.4	3.3	248.5	3.9	138.9	3.3	56.1	2.2	13.1	0.7	14
17	429.1	1.5	355.1	3.3	244.6	3.8	135.6	3.3	53.9	2.0	12.4	0.6	13
18	427.6	1.5	351.8	3.4	240.8	3.9	132.3	3.2	51.9	2.1	11.8	0.6	12
19	426.1	1.6	348.4	3.4	236.9	3.8	129.1	3.2	49.8	1.9	11.2	0.5	11
20	424.5	1.7	345.0	3.4	233.1	3.8	125.9	3.2	47.9	2.0	10.7	0.5	10
21	422.7	1.7	341.6	3.5	229.3	3.9	122.7	3.1	45.9	1.9	10.2	0.4	9
22	421.0	1.9	338.1	3.5	225.4	3.8	119.6	3.1	44.0	1.8	9.8	0.4	8
23	419.1	1.9	334.6	3.5	221.6	3.8	116.5	3.1	42.2	1.8	9.4	0.3	7
24	417.2	2.0	331.1	3.6	217.8	3.8	113.4	3.0	40.4	1.7	9.1	0.3	6
25	415.2	2.1	327.5	3.5	214.0	3.7	110.4	3.0	38.7	1.7	8.8	0.2	5
26	413.1	2.2	324.0	3.7	210.3	3.8	107.4	2.9	37.0	1.7	8.6	0.2	4
27	410.9	2.2	320.3	3.6	206.5	3.7	104.5	2.9	35.3	1.6	8.4	0.1	3
28	408.7	2.3	316.7	3.7	202.8	3.8	101.6	2.9	33.7	1.6	8.3	0.1	2
29	406.4	2.3	313.0	3.7	199.0	3.7	98.7	2.9	32.1	1.5	8.2	0.0	1
30	404.1		309.3		195.3		95.8		30.6		8.2		0
	XIs		Xs		IXs		VIIIs		VIIs		VIs		

Moon's Horary Motion in Longitude.

Arguments. Sum of preceding Equations, and Arg. of Variation.

	" 0	" 50	" 100	" 150	" 200	" 250	" 300	" 350	" 400	" 450	" 500	" 550	" 600	
s °	"	"	"	"	"	"	"	"	"	"	"	"	"	s °
O 0	4.5	5.5	6.5	7.6	8.6	9.6	10.6	11.6	12.6	13.7	14.7	15.7	16.7	XII 0
5	4.6	5.6	6.6	7.6	8.6	9.6	10.6	11.6	12.6	13.6	14.6	15.6	16.6	25
10	4.8	5.8	6.8	7.7	8.7	9.6	10.6	11.5	12.5	13.4	14.4	15.3	16.3	20
15	5.3	6.1	7.0	7.9	8.8	9.7	10.5	11.4	12.3	13.1	14.0	14.9	15.8	15
20	5.8	6.6	7.4	8.2	8.9	9.7	10.5	11.2	12.0	12.8	13.5	14.3	15.1	10
25	6.6	7.2	7.8	8.5	9.1	9.7	10.4	11.0	11.7	12.3	12.9	13.6	14.2	5
I 0	7.4	7.8	8.3	8.8	9.3	9.8	10.3	10.8	11.3	11.8	12.3	12.7	13.2	XI 0
5	8.3	8.6	8.9	9.2	9.5	9.9	10.2	10.5	10.8	11.2	11.5	11.8	12.1	25
10	9.2	9.3	9.5	9.6	9.8	9.9	10.1	10.2	10.4	10.5	10.7	10.8	11.0	20
15	10.2	10.1	10.1	10.1	10.0	10.0	10.0	10.0	9.9	9.9	9.9	9.8	9.8	15
20	11.1	10.9	10.7	10.5	10.3	10.1	9.9	9.7	9.5	9.2	9.0	8.8	8.6	10
25	12.1	11.7	11.3	10.9	10.5	10.2	9.8	9.4	9.0	8.6	8.3	7.9	7.5	5
II 0	12.9	12.4	11.8	11.3	10.8	10.2	9.7	9.1	8.6	8.1	7.5	7.0	6.4	X 0
5	13.7	13.0	12.3	11.6	11.0	10.3	9.6	8.9	8.2	7.5	6.9	6.2	5.5	25
10	14.3	13.5	12.7	11.9	11.1	10.3	9.5	8.7	7.9	7.1	6.3	5.5	4.7	20
15	14.9	14.0	13.1	12.2	11.3	10.4	9.5	8.6	7.7	6.8	5.8	4.9	4.0	15
20	15.3	14.3	13.3	12.3	11.4	10.4	9.4	8.4	7.5	6.5	5.5	4.5	3.6	10
25	15.5	14.5	13.5	12.4	11.4	10.4	9.4	8.4	7.4	6.3	5.3	4.3	3.3	5
III 0	15.6	14.5	13.5	12.5	11.4	10.4	9.4	8.4	7.3	6.3	5.3	4.2	3.2	IX 0
5	15.4	14.4	13.4	12.4	11.4	10.4	9.4	8.4	7.4	6.4	5.4	4.4	3.3	25
10	15.2	14.2	13.3	12.3	11.3	10.4	9.4	8.5	7.5	6.5	5.6	4.6	3.6	20
15	14.8	13.9	13.0	12.1	11.2	10.4	9.5	8.6	7.7	6.8	5.9	5.1	4.2	15
20	14.2	13.4	12.6	11.9	11.1	10.3	9.5	8.8	8.0	7.2	6.4	5.6	4.9	10
25	13.5	12.9	12.2	11.6	10.9	10.3	9.6	9.0	8.4	7.6	7.0	6.3	5.7	5
IV 0	12.7	12.2	11.7	11.2	10.7	10.2	9.7	9.2	8.7	8.2	7.7	7.2	6.7	VIII 0
5	11.9	11.5	11.2	10.8	10.5	10.1	9.8	9.5	9.1	8.8	8.4	8.1	7.7	25
10	10.9	10.7	10.6	10.4	10.2	10.1	9.9	9.7	9.6	9.4	9.2	9.1	8.9	20
15	9.9	9.9	10.0	10.0	10.0	10.0	10.0	10.0	10.0	10.0	10.1	10.1	10.1	15
20	8.9	9.1	9.3	9.5	9.7	9.9	10.1	10.3	10.5	10.7	10.9	11.1	11.3	10
25	8.0	8.4	8.7	9.1	9.5	9.9	10.2	10.6	11.0	11.3	11.7	12.1	12.5	5
V 0	7.1	7.6	8.2	8.7	9.2	9.8	10.3	10.9	11.4	11.9	12.5	13.0	13.6	VII 0
5	6.3	7.0	7.6	8.3	9.0	9.7	10.4	11.1	11.8	12.5	13.2	13.9	14.6	25
10	5.6	6.4	7.2	8.0	8.8	9.7	10.5	11.3	12.1	13.0	13.8	14.6	15.4	20
15	5.0	5.9	6.8	7.8	8.7	9.6	10.6	11.5	12.4	13.3	14.3	15.2	16.1	15
20	4.6	5.6	6.6	7.6	8.6	9.6	10.6	11.6	12.6	13.6	14.6	15.7	16.7	10
25	4.3	5.4	6.4	7.5	8.5	9.6	10.6	11.7	12.7	13.8	14.9	15.9	17.0	5
VI 0	4.2	5.3	6.4	7.4	8.5	9.6	10.6	11.7	12.8	13.9	14.9	16.0	17.1	VI 0
	" 0	" 50	" 100	" 150	" 200	" 250	" 300	" 350	" 400	" 450	" 500	" 550	" 600	

TABLE LXXIII.

Moon's Horary Motion in Longitude.

Argument. Argument of the Variation.

	0s	Is	IIs	IIIs	IVs	Vs	
°	"	"	"	"	"	"	°
0	77.2	57.8	20.3	2.4	21.5	59.7	30
1	77.2	56.7	19.2	2.5	22.7	60.9	29
2	77.1	55.5	18.1	2.6	23.8	62.0	28
3	77.0	54.3	17.0	2.7	25.0	63.1	27
4	76.8	53.1	16.0	2.9	26.2	64.2	26
5	76.6	51.8	15.0	3.1	27.5	65.3	25
6	76.4	50.5	14.1	3.3	28.7	66.3	24
7	76.1	49.3	13.2	3.7	30.0	67.3	23
8	75.7	48.0	12.3	4.0	31.3	68.3	22
9	75.3	46.7	11.4	4.4	32.6	69.2	21
10	74.9	45.4	10.6	4.9	33.9	70.1	20
11	74.4	44.1	9.8	5.3	35.2	70.9	19
12	73.9	42.8	9.0	5.9	36.5	71.7	18
13	73.3	41.5	8.3	6.4	37.8	72.5	17
14	72.7	40.2	7.6	7.0	39.2	73.3	16
15	72.0	38.9	7.0	7.7	40.5	74.0	15
16	71.3	37.5	6.4	8.3	41.8	74.7	14
17	70.6	36.2	5.8	9.1	43.2	75.3	13
18	69.8	34.9	5.3	9.8	44.5	75.8	12
19	69.0	33.6	4.8	10.6	45.8	76.4	11
20	68.1	32.3	4.4	11.5	47.2	76.9	10
21	67.2	31.1	4.0	12.3	48.5	77.3	9
22	66.3	29.8	3.7	13.2	49.8	77.7	8
23	65.3	28.6	3.3	14.2	51.1	78.1	7
24	64.4	27.3	3.1	15.1	52.4	78.4	6
25	63.4	26.1	2.9	16.1	53.6	78.6	5
26	62.3	24.9	2.7	17.1	54.9	78.9	4
27	61.2	23.7	2.5	18.2	56.1	79.0	3
28	60.1	22.5	2.5	19.3	57.3	79.2	2
29	59.0	21.4	2.4	20.4	58.5	79.2	1
30	57.8	20.3	2.4	21.5	59.7	79.2	0
	XIs	Xs	IXs	VIIIs	VIIs	VIs	

Moon's Horary Motion in Longitude.

Arguments. Arg. of Reduction and Sum of preceding Equations

	″ 0	″ 50	″ 100	″ 150	″ 200	″ 250	″ 300	″ 350	″ 400	″ 450	″ 500	″ 550	″ 600	″ 650	
s °	″	″	″	″	″	″	″	″	″	″	″	″	″	″	s °
O 0	3.3	3.1	2.9	2.7	2.5	2.3	2.1	1.9	1.7	1.5	1.3	1.1	0.9	0.7	XII 0
5	3.3	3.1	2.9	2.7	2.5	2.3	2.1	1.9	1.7	1.5	1.3	1.1	0.9	0.7	25
10	3.2	3.0	2.8	2.6	2.4	2.3	2.1	1.9	1.7	1.5	1.3	1.1	1.0	0.8	20
15	3.1	2.9	2.8	2.6	2.4	2.2	2.1	1.9	1.7	1.5	1.4	1.2	1.0	0.9	15
20	3.0	2.8	2.7	2.5	2.4	2.2	2.1	1.9	1.8	1.6	1.5	1.3	1.1	1.0	10
25	2.8	2.7	2.6	2.4	2.3	2.2	2.1	1.9	1.8	1.7	1.5	1.4	1.3	1.2	5
I 0	2.6	2.5	2.4	2.3	2.2	2.1	2.0	1.9	1.8	1.7	1.6	1.5	1.4	1.3	XI 0
5	2.4	2.4	2.3	2.2	2.2	2.1	2.0	2.0	1.9	1.8	1.8	1.7	1.6	1.6	25
10	2.2	2.2	2.2	2.1	2.1	2.0	2.0	2.0	1.9	1.9	1.9	1.8	1.8	1.8	20
15	2.0	2.0	2.0	2.0	2.0	2.0	2.0	2.0	2.0	2.0	2.0	2.0	2.0	2.0	15
20	1.8	1.8	1.8	1.9	1.9	1.9	2.0	2.0	2.1	2.1	2.1	2.2	2.2	2.2	10
25	1.6	1.6	1.7	1.8	1.8	1.9	2.0	2.0	2.1	2.2	2.2	2.3	2.4	2.4	5
II 0	1.4	1.5	1.6	1.7	1.8	1.9	2.0	2.1	2.2	2.3	2.4	2.5	2.6	2.7	X 0
5	1.2	1.3	1.4	1.6	1.7	1.8	1.9	2.1	2.2	2.3	2.5	2.6	2.7	2.8	25
10	1.0	1.2	1.3	1.5	1.6	1.8	1.9	2.1	2.2	2.4	2.5	2.7	2.9	3.0	20
15	0.9	1.1	1.2	1.4	1.6	1.8	1.9	2.1	2.3	2.5	2.6	2.8	3.0	3.1	15
20	0.8	1.0	1.2	1.4	1.6	1.7	1.9	2.1	2.3	2.5	2.7	2.9	3.0	3.2	10
25	0.7	0.9	1.1	1.3	1.5	1.7	1.9	2.1	2.3	2.5	2.7	2.9	3.1	3.3	5
III 0	0.7	0.9	1.1	1.3	1.5	1.7	1.9	2.1	2.3	2.5	2.7	2.9	3.1	3.3	IX 0
5	0.7	0.9	1.1	1.3	1.5	1.7	1.9	2.1	2.3	2.5	2.7	2.9	3.1	3.3	25
10	0.8	1.0	1.2	1.4	1.6	1.7	1.9	2.1	2.3	2.5	2.7	2.9	3.0	3.2	20
15	0.9	1.1	1.2	1.4	1.6	1.8	1.9	2.1	2.3	2.5	2.6	2.8	3.0	3.1	15
20	1.0	1.2	1.3	1.5	1.6	1.8	1.9	2.1	2.2	2.4	2.5	2.7	2.9	3.0	10
25	1.2	1.3	1.4	1.6	1.7	1.8	1.9	2.1	2.2	2.3	2.5	2.6	2.7	2.8	5
IV 0	1.4	1.5	1.6	1.7	1.8	1.9	2.0	2.1	2.2	2.3	2.4	2.5	2.6	2.7	VIII 0
5	1.6	1.6	1.7	1.8	1.8	1.9	2.0	2.0	2.1	2.2	2.2	2.3	2.4	2.4	25
10	1.8	1.8	1.8	1.9	1.9	1.9	2.0	2.0	2.1	2.1	2.1	2.2	2.2	2.2	20
15	2.0	2.0	2.0	2.0	2.0	2.0	2.0	2.0	2.0	2.0	2.0	2.0	2.0	2.0	15
20	2.2	2.2	2.2	2.1	2.1	2.0	2.0	2.0	1.9	1.9	1.9	1.8	1.8	1.8	10
25	2.4	2.4	2.3	2.2	2.2	2.1	2.0	2.0	1.9	1.8	1.8	1.7	1.6	1.6	5
V 0	2.6	2.5	2.4	2.3	2.2	2.1	2.0	1.9	1.8	1.7	1.6	1.5	1.4	1.3	VII 0
5	2.8	2.7	2.6	2.4	2.3	2.2	2.1	1.9	1.8	1.7	1.5	1.4	1.3	1.2	25
10	3.0	2.8	2.7	2.5	2.4	2.2	2.1	1.9	1.8	1.6	1.5	1.3	1.1	1.0	20
15	3.1	2.9	2.8	2.6	2.4	2.2	2.1	1.9	1.7	1.5	1.4	1.2	1.0	0.9	15
20	3.2	3.0	2.8	2.6	2.4	2.3	2.1	1.9	1.7	1.5	1.3	1.1	1.0	0.8	10
25	3.3	3.1	2.9	2.7	2.5	2.3	2.1	1.9	1.7	1.5	1.3	1.1	0.9	0.7	5
VI 0	3.3	3.1	2.9	2.7	2.5	2.3	2.1	1.9	1.7	1.5	1.3	1.1	0.9	0.7	VI 0
	″ 0	″ 50	″ 100	″ 150	″ 200	″ 250	″ 300	″ 350	″ 400	″ 450	″ 500	″ 550	″ 600	″ 650	

TABLE LXXV.

Moon's Horary Motion in Long.

Arg. Arg. of Reduction.

	0s VIs	Is VIIs	IIs VIIIs	
°	"	"	"	°
0	2.1	6.0	14.0	30
1	2.1	6.3	14.2	29
2	2.1	6.5	14.4	28
3	2.1	6.8	14.7	27
4	2.2	7.0	14.9	26
5	2.2	7.3	15.1	25
6	2.2	7.5	15.3	24
7	2.3	7.8	15.5	23
8	2.4	8.1	15.7	22
9	2.5	8.4	15.9	21
10	2.5	8.6	16.1	20
11	2.6	8.9	16.2	19
12	2.7	9.2	16.4	18
13	2.9	9.4	16.6	17
14	3.0	9.7	16.7	16
15	3.1	10.0	16.9	15
16	3.3	10.3	17.0	14
17	3.4	10.6	17.1	13
18	3.6	10.8	17.3	12
19	3.8	11.1	17.4	11
20	3.9	11.4	17.5	10
21	4.1	11.6	17.5	9
22	4.3	11.9	17.6	8
23	4.5	12.2	17.7	7
24	4.7	12.5	17.8	6
25	4.9	12.7	17.8	5
26	5.1	13.0	17.8	4
27	5.3	13.2	17.9	3
28	5.6	13.5	17.9	2
29	5.8	13.7	17.9	1
30	6.0	14.0	17.9	0
	XIs Vs	Xs IVs	IXs IIIs	

Constant to be added 27′24″.0.

TABLE LXXVI.

Moon's Horary Motion in Long.

(Equation of the second order.)

Arguments. Arg's. of Table LXX

Arg.		" 0	" 50	" 100
s	°	"	"	"
O	0	0.05	0.05	0.05
I	0	0.08	0.05	0.02
II	0	0.10	0.05	0.00
III	0	0.10	0.05	0.00
IV	0	0.09	0.05	0.01
V	0	0.07	0.05	0.03
VI	0	0.05	0.05	0.05
VII	0	0.03	0.05	0.07
VIII	0	0.01	0.05	0.09
IX	0	0.00	0.05	0.10
X	0	0.00	0.05	0.10
XI	0	0.02	0.05	0.08
XII	0	0.05	0.05	0.05
		" 0	" 50	" 100

TABLE LXXVII.

Moon's Horary Motion in Longitude.

(Equations of the second order.)

Arguments. Arguments of Tables LXXII and LXXIV.

			Variation.							Reduction.	
			" 0	" 100	" 200	" 300	" 400	" 500	" 600	" 0	" 600
s	s	°	"	"	"	"	"	"	"	"	"
O.	VI.	0	0.14	0.14	0.14	0.14	0.14	0.14	0.14	0.03	0.03
I.	VII.	0	0.22	0.19	0.16	0.13	0.10	0.06	0.02	0.01	0.05
I.	VII.	15	0.23	0.20	0.17	0.13	0.10	0.05	0.01	0.01	0.06
II.	VIII.	0	0.22	0.19	0.16	0.13	0.10	0.07	0.03	0.01	0.05
III.	IX.	0	0.14	0.14	0.14	0.14	0.14	0.14	0.14	0.03	0.03
IV.	X.	0	0.06	0.09	0.12	0.15	0.18	0.21	0.26	0.05	0.01
IV.	X.	15	0.05	0.08	0.11	0.15	0.18	0.23	0.28	0.05	0.00
V.	XI.	0	0.06	0.09	0.12	0.15	0.18	0.22	0.26	0.05	0.01
VI.	XII.	0	0.14	0.14	0.14	0.14	0.14	0.14	0.14	0.03	0.03

Moon's Horary Motion in Longitude

(Equations of the second order.)

Arguments. Args. of Evection, Anomaly, Variation, Reduction.

		Evec.	Anom.	Var.	Red.	Evec.	Anom.	Var.	Red.		
s	°	"	"	"	"	"	"	"	"	s	°
0	0	0.16	1.05	0.34	0.08	0.16	1.05	0.34	0.08	XII	0
	5	0.15	0.93	0.28	0.09	0.18	1.17	0.40	0.06		25
	10	0.13	0.81	0.22	0.10	0.19	1.28	0.46	0.05		20
	15	0.12	0.70	0.17	0.11	0.21	1.40	0.51	0.04		15
	20	0.10	0.59	0.12	0.12	0.22	1.50	0.56	0.03		10
	25	0.09	0.49	0.08	0.13	0.24	1.60	0.60	0.02		5
I	0	0.08	0.40	0.05	0.14	0.25	1.70	0.63	0.01	XI	0
	5	0.07	0.31	0.02	0.15	0.26	1.78	0.66	0.01		25
	10	0.05	0.24	0.01	0.15	0.27	1.86	0.67	0.00		20
	15	0.04	0.17	0.01	0.15	0.28	1.92	0.67	0.00		15
	20	0.03	0.12	0.01	0.15	0.29	1.98	0.67	0.00		10
	25	0.03	0.07	0.03	0.15	0.30	2.02	0.65	0.01		5
II	0	0.02	0.04	0.06	0.14	0.31	2.05	0.62	0.01	X	0
	5	0.01	0.02	0.09	0.13	0.32	2.08	0.59	0.02		25
	10	0.01	0.00	0.13	0.12	0.32	2.09	0.54	0.03		20
	15	0.00	0.00	0.18	0.11	0.32	2.10	0.50	0.04		15
	20	0.00	0.00	0.24	0.10	0.33	2.09	0.44	0.05		10
	25	0.00	0.02	0.29	0.09	0.33	2.08	0.39	0.06		5
III	0	0.00	0.04	0.35	0.08	0.33	2.06	0.33	0.08	IX	0
	5	0.00	0.07	0.40	0.06	0.33	2.03	0.27	0.09		25
	10	0.01	0.10	0.46	0.05	0.32	2.00	0.22	0.10		20
	15	0.01	0.14	0.51	0.04	0.32	1.96	0.17	0.11		15
	20	0·01	0.18	0.56	0.03	0.31	1.91	0.12	0.12		10
	25	0.02	0.23	0.60	0.02	0.31	1.87	0.08	0.13		5
IV	0	0.03	0.28	0.63	0.01	0.30	1.82	0.05	0.14	VIII	0
	5	0.03	0.34	0.66	0.01	0.29	1.76	0.02	0.15		25
	10	0.04	0.39	0.67	0.00	0.28	1.70	0.01	0.15		20
	15	0.05	0.45	0.68	0.00	0.27	1.64	0.00	0.15		15
	20	0.06	0.52	0.67	0.00	0.26	1.58	0.00	0.15		10
	25	0.08	0.58	0.66	0.01	0.25	1.52	0.02	0.15		5
V	0	0.09	0.64	0.64	0.01	0.24	1.45	0.04	0.14	VII	0
	5	0.10	0.71	0.60	0.02	0.23	1.39	0.08	0.13		25
	10	0.11	0.78	0.56	0.03	0.22	1.32	0.12	0.12		20
	15	0.12	0.84	0.51	0.04	0.20	1.25	0.16	0.11		15
	20	0.14	0.91	0.46	0.05	0.19	1.18	0.22	0.10		10
	25	0.15	0.98	0.40	0.06	0.18	1.12	0.28	0.09		5
VI	0	0.16	1.05	0.34	0.08	0.16	1.05	0.34	0.08	VI	0

TABLE LXXIX.

Moon's Horary Motion in Latitude.

Argument. Arg. I of Latitude.

	0s	Is	IIs	IIIs	IVs	Vs	
°	"	"	"	"	"	"	°
0	378.0	354.3	289.2	200.0	110.8	45.7	30
1	378.0	352.7	286.5	196.9	108.1	44.2	29
2	377.9	351.1	283.8	193.8	105.4	42.7	28
3	377 8	349.4	281.0	190.7	102.8	41.3	27
4	377.6	347.7	278.3	187.5	100.2	39.9	26
5	377.3	346.0	275.5	184.4	97.7	38.6	25
6	377.0	344.2	272.6	181.3	95.1	37.3	24
7	376.7	342.3	269.8	178.2	92.6	36.1	23
8	376.3	340.5	266.9	175.1	90.2	34.9	22
9	375.8	338.5	264.0	172.1	87.7	33.8	21
10	375.3	336.6	261.1	169.0	85.3	32.7	20
11	374.7	334.5	258.1	165.9	83.0	31.6	19
12	374.1	332.5	255.2	162.9	80.7	30.7	18
13	373.5	330.4	252.2	159.8	78.1	29.7	17
14	372.7	328.3	249.2	156.8	76.1	28.9	16
15	372.0	326.1	246.2	153.8	73.9	28.0	15
16	371.1	323.9	243.2	150.8	71.7	27.3	14
17	370.3	321.9	240.2	147.8	69.6	26.5	13
18	369.3	319.3	237.1	144.8	67.5	25.9	12
19	368.4	317.0	234.1	141.9	65.5	25.3	11
20	367.3	314.7	231.0	138.9	63.4	24.7	10
21	366.2	312.3	227.9	136.0	61.5	24.2	9
22	365.1	309.8	224.9	133.1	59.5	23.7	8
23	363.9	307.4	221.8	130.2	57.7	23.3	7
24	362.7	304.9	218.7	127.4	55.8	23.0	6
25	361.4	302.3	215.6	124.5	54.0	22.7	5
26	360.1	299.8	212.5	121.7	52.3	22.4	4
27	358.7	297.2	209.3	119.0	50.6	22.2	3
28	357.3	294.6	206.2	116.2	48.9	22.1	2
29	355.8	291.9	203.1	113.5	47.3	22.0	1
30	354.3	289.2	200.0	110.8	45.7	22.0	0
	XIs	Xs	IXs	VIIIs	VIIs	VIs	

TABLE LXXX.

Moon's Horary Motion in Latitude.

Arguments. Args. V, VI, VII, VIII, IX, X, XI, and XII, of Latitude.

Arg.	V	VI	VII	VIII	IX	X	XI	XII	Arg.
	"	"	"	"	"	"	"	"	
0	0.00	0.50	0.34	0.00	0.50	0.04	0.12	0.08	1000
50	0.01	0.49	0.33	0.00	0.49	0.04	0.12	0.07	950
100	0.04	0.45	0.30	0.02	0.45	0.04	0.11	0.05	900
150	0.09	0.40	0.27	0.04	0.40	0.03	0.10	0.03	850
200	0.16	0.33	0.22	0.06	0.33	0.03	0.08	0.01	800
250	0.23	0.25	0.17	0.09	0.25	0.02	0.06	0.00	750
300	0.30	0 17	0.12	0.12	0.17	0.01	0.04	0.01	700
350	0.37	0.10	0.07	0.14	0.10	0.01	0.02	0.03	650
400	0.42	0.05	0.04	0.16	0.05	0.00	0.01	0.05	600
450	0.45	0.01	0.01	0.18	0.01	0.00	0.00	0.07	550
500	0.46	0.00	0.00	0.18	0.00	0.00	0.00	0.08	500

Arguments. Preceding equation, and Sum of equations of Horary Motion in Longitude, except the last two.

Pr. eq.	0″	50″	100″	150″	200″	250″	300″	350″	400″	450″	500″	550″	600″	650″	
	1″.6	1″.4	1″.1	0″.9	0″.6	0″.4	0″.1	0″.2	0″.4	0″.7	0″.9	1″.2	1″.4	1″.7	Diff.
″	″	″	″	″	″	″	″	″	″	″	″	″	″	″	″
20	59.0	54.5	50.0	45.4	40.9	36.4	31.8	27.3	22.8	18.2	13.7	9.1	4.6	0.1	4.5
30	57.4	53.1	48.9	44.6	40.3	36.0	31.7	27.4	23.2	18.9	14.6	10.3	6.0	1.7	4.3
40	55.8	51.8	47.7	43.7	39.7	35.6	31.6	27.6	23.6	19.5	15.5	11.5	7.4	3.4	4.0
50	54.2	50.4	46.6	42.9	39.1	35.3	31.5	27.7	24.0	20.2	16.4	12.6	8.8	5.1	3.8
60	52.6	49.1	45.5	42.0	38.5	34.9	31.4	27.9	24.4	20.8	17.3	13.8	10.2	6.7	3.5
70	51.0	47.7	44.4	41.1	37.9	34.6	31.3	28.0	24.8	21.5	18.2	14.9	11.7	8.4	3.3
80	49.3	46.3	43.3	40.3	37.3	34.2	31.2	28.2	25.2	22.1	19.1	16.1	13.1	10.0	3.0
90	47.7	45.0	42.2	39.4	36.7	33.9	31.1	28.3	25.6	22.8	20.0	17.3	14.5	11.7	2.8
100	46.1	43.6	41.1	38.6	36.0	33.5	31.0	28.5	26.0	23.4	20.9	18.4	15.9	13.4	2.5
110	44.5	42.2	40.0	37.7	35.4	33.2	30.9	28.6	26.4	24.1	21.8	19.6	17.3	15.0	2.3
120	42.9	40.9	38.9	36.9	34.8	32.8	30.8	28.8	26.8	24.8	22.7	20.7	18.7	16.7	2.0
130	41.3	39.5	37.8	36.0	34.2	32.5	30.7	28.9	27.2	25.4	23.7	21.9	20.1	18.4	1.8
140	39.7	38.2	36.7	35.1	33.6	32.1	30.6	29.1	27.6	26.1	24.6	23.0	21.5	20.0	1.5
150	38.1	36.8	35.5	34.3	33.0	31.8	30.5	29.2	28.0	26.7	25.5	24.2	23.0	21.7	1.3
160	36.5	35.4	34.4	33.4	32.4	31.4	30.4	29.4	28.4	27.4	26.4	25.4	24.4	23.3	1.0
170	34.8	34.1	33.3	32.6	31.8	31.1	30.3	29.5	28.8	28.0	27.3	26.5	25.8	25.0	0.8
180	33.2	32.7	32.2	31.7	31.2	30.7	30.2	29.7	29.2	28.7	28.2	27.7	27.2	26.7	0.5
190	31.6	31.4	31.1	30.9	30.6	30.4	30.1	29.8	29.6	29.3	29.1	28.8	28.6	28.3	0.3
200	30.0	30.0	30.0	30.0	30.0	30.0	30.0	30.0	30.0	30.0	30.0	30.0	30.0	30.0	0.0
210	28.4	28.6	28.9	29.1	29.4	29.6	29.9	30.2	30.4	30.7	30.9	31.2	31.4	31.7	0.3
220	26.8	27.3	27.8	28.3	28.8	29.3	29.8	30.3	30.8	31.3	31.8	32.3	32.8	33.3	0.5
230	25.2	25.9	26.7	27.4	28.2	28.9	29.7	30.5	31.2	32.0	32.7	33.5	34.2	35.0	0.8
240	23.5	24.6	25.6	26.6	27.6	28.6	29.6	30.6	31.6	32.6	33.6	34.6	35.6	36.7	1.0
250	21.9	23.2	24.5	25.7	27.0	28.2	29.5	30.8	32.0	33.3	34.5	35.8	37.1	38.3	1.3
260	20.3	21.8	23.3	24.9	26.4	27.9	29.4	30.9	32.4	33.9	35.4	37.0	38.5	40.0	1.5
270	18.7	20.5	22.2	24.0	25.8	27.5	29.3	31.1	32.8	34.6	36.3	38.1	39.9	41.6	1.8
280	17.1	19.1	21.1	23.1	25.2	27.2	29.2	31.2	33.2	35.2	37.3	39.3	41.3	43.3	2.0
290	15.5	17.8	20.0	22.3	24.6	26.8	29.1	31.4	33.6	35.9	38.2	40.4	42.7	45.0	2.3
300	13.9	16.4	18.9	21.4	24.0	26.5	29.0	31.5	34.0	36.6	39.1	41.6	44.1	46.6	2.5
310	12.3	15.0	17.8	20.6	23.3	26.1	28.9	31.7	34.4	37.2	40.0	42.7	45.5	48.3	2.8
320	10.7	13.7	16.7	19.7	22.7	25.8	28.8	31.8	34.8	37.9	40.9	43.9	46.9	50.0	3.0
330	9.0	12.3	15.6	18.9	22.1	25.4	28.7	32.0	35.2	38.5	41.8	45.1	48.3	51.6	3.3
340	7.4	10.9	14.5	18.0	21.5	25.1	28.6	32.1	35.6	39.2	42.7	46.2	49.8	53.3	3.5
350	5.8	9.6	13.4	17.1	20.9	24.7	28.5	32.3	36.0	39.8	43.6	47.4	51.2	54.9	3.8
360	4.2	8.2	12.3	16.3	20.3	24.4	28.4	32.4	36.4	40.5	44.5	48.5	52.6	56.6	4.0
370	2.6	6.9	11.1	15.4	19.7	24.0	28.3	32.6	36.8	41.1	45.4	49.7	54.0	58.3	4.3
380	1.0	5.5	10.0	14.6	19.1	23.6	28.2	32.7	37.2	41.8	46.3	50.9	55.4	59.9	4.5
	0″	50″	100″	150″	200″	250″	300″	350″	400″	450″	500″	550″	600″	650″	

TABLE LXXXII. *Moon's Horary Motion in Latitude.*

Argument. Arg II. of Latitude.

	0s	Is	IIs	IIIs	IVs	Vs	
°	″	″	″	″	″	″	°
0	9.3	8.7	7.1	5.0	2.9	1.3	30
3	9.3	8.6	6.9	4.8	2.7	1.2	27
6	9.2	8.5	6.7	4.6	2.5	1.1	24
9	9.2	8.3	6.5	4.3	2.3	1.0	21
12	9.2	8.2	6.3	4.1	2.1	0.9	18
15	9.1	8.0	6.1	3.9	2.0	0.9	15
18	9.1	7.9	5.9	3.7	1.8	0.8	12
21	9.0	7.7	5.7	3.5	1.7	0.8	9
24	8.9	7.5	5.4	3.3	1.5	0.8	6
27	8.8	7.3	5.2	3.1	1.4	0.7	3
30	8.7	7.1	5.0	2.9	1.3	0.7	0
	XIs	Xs	IXs	VIIIs	VIIs	VIs	

TABLE LXXXIII.

Moon's Horary Motion in Latitude.

Arguments. Preceding equation, and Sum of equations of Horary Motion in Longitude, except the last two.

Prec. equ.	″ 0	″ 100	″ 200	″ 300	″ 400	″ 500	″ 600	″ 700
″	″	″	″	″	″	″	″	″
0	2.1	1.8	1.5	1.2	0.9	0.6	0.3	0.0
1	1.9	1.6	1.4	1.1	0.9	0.7	0.4	0.2
2	1.7	1.5	1.3	1.1	1.0	0.8	0.6	0.3
3	1.5	1.4	1.2	1.1	1.0	0.9	0.8	0.6
4	1.3	1.2	1.2	1.1	1.1	1.0	0.9	0.9
5	1.1	1.1	1.1	1.1	1.1	1.1	1.1	1.1
6	0.9	1.0	1.0	1.1	1.1	1.2	1.3	1.3
7	0.7	0.8	1.0	1.1	1.2	1.3	1.4	1.6
8	0.5	0.7	0.9	1.1	1.2	1.4	1.6	1.9
9	0.3	0.6	0.8	1.1	1.3	1.5	1.8	2.0
10	0.1	0.4	0.7	1.0	1.3	1.6	1.9	2.2
	″ 0	″ 100	″ 200	″ 300	″ 400	″ 500	″ 600	″ 700

Constant to be subtracted 237″.2.

TABLE LXXXIV.

Moon's Hor. Motion in Lat

(Equa. of second order.)

Argument. Arg. I of Lat.

s °	I ″	I ″	s °
O 0	0.90	0.90	XII 0
5	0.83	0.97	25
10	0.75	1.05	20
15	0.68	1.12	15
20	0.61	1.19	10
25	0.54	1.26	5
I 0	0.47	1.33	XI 0
5	0.41	1.39	25
10	0.35	1.45	20
15	0.29	1.51	15
20	0.24	1.56	10
25	0.20	1.60	5
II 0	0.16	1.64	X 0
5	0.12	1.68	25
10	0.09	1.71	20
15	0.07	1.73	15
20	0.05	1.75	10
25	0.04	1.76	5
III 0	0.04	1.76	IX 0
5	0.04	1.76	25
10	0.05	1.75	20
15	0.07	1.73	15
20	0.09	1.71	10
25	0.12	1.68	5
IV 0	0.16	1.64	VIII 0
5	0.20	1.60	25
10	0.24	1.56	20
15	0.29	1.51	15
20	0.35	1.45	10
25	0.41	1.39	5
V 0	0.47	1.33	VII 0
5	0.54	1.26	25
10	0.61	1.19	20
15	0.68	1.12	15
20	0.75	1.05	10
25	0.83	0.97	5
VI 0	0.90	0.90	VI 0

TABLE LXXXV.

Moon's Horary Motion in Latitude.

(Equations of second order.)

Arguments. Preceding equation, and Sum of equations of Horary Motion in Longitude, except the last two.

Prec. equ.	″ 0	″ 100	″ 200	″ 300	″ 400	″ 500	″ 600	″ 700
″	″	″	″	″	″	″	″	″
0.00	0.65	0.57	0.48	0.39	0.31	0.21	0.12	0.00
0.10	0.62	0.55	0.47	0.39	0.31	0.23	0.15	0.04
0.20	0.69	0.53	0.46	0.39	0.32	0.25	0.18	0.09
0.30	0.66	0.51	0.45	0.39	0.33	0.27	0.21	0.13
0.40	0.63	0.48	0.44	0.39	0.34	0.29	0.24	0.17
0.50	0.50	0.46	0.43	0.38	0.35	0.30	0.27	0.21
0.60	0.47	0.44	0.42	0.38	0.36	0.32	0.29	0.25
0.70	0.44	0.42	0.40	0.38	0.36	0.34	0.32	0.30
0.80	0.41	0.40	0.39	0.38	0.37	0.36	0.35	0.34
0.90	0.38	0.38	0.38	0.38	0.38	0.38	0.38	0.38
1.00	0.35	0.36	0.37	0.38	0.39	0.40	0.41	0.42
1.10	0.32	0.34	0.36	0.38	0.40	0.42	0.44	0.46
1.20	0.29	0.32	0.34	0.38	0.40	0.44	0.47	0.51
1.30	0.26	0.30	0.33	0.38	0.41	0.46	0.49	0.55
1.40	0.23	0.28	0.32	0.37	0.42	0.47	0.52	0.59
1.50	0.20	0.25	0.31	0.37	0.43	0.49	0.55	0.63
1.60	0.17	0.23	0.30	0.37	0.44	0.51	0.58	0.67
1.70	0.14	0.21	0.29	0.37	0.45	0.53	0.61	0.72
1.80	0.11	0.19	0.28	0.37	0.45	0.55	0.64	0.76
	″ 0	″ 100	″ 200	″ 300	″ 400	″ 500	″ 600	″ 700

Mean New Moons and Arguments, in January.

Years.	Mean New Moon in January.	I.	II.	III.	IV.	N.
	d. h. m.					
1836 B	17 10 32	0469	1246	17	08	669
1837	5 19 20	0171	9852	00	97	692
1838	24 16 53	0681	9175	99	85	799
1839	14 1 42	0383	7780	82	74	822
1840 B	3 10 30	0085	6386	65	63	844
1841	21 8 3	0595	5709	63	51	951
1842	10 16 51	0297	4314	46	40	974
1843	29 14 24	0807	3637	44	28	081
1844 B	18 23 13	0509	2243	28	17	104
1845	7 8 1	0211	0848	11	06	126
1846	26 5 34	0721	0171	09	94	234
1847	15 14 22	0423	8777	92	84	256
1848 B	4 23 11	0125	7382	75	73	278
1849	22 20 43	0635	6705	73	61	386
1850	12 5 32	0337	5311	56	50	408
1851	1 14 21	0038	3916	40	39	431
1852 B	20 11 53	0549	3239	38	27	538
1853	8 20 42	0251	1845	21	16	560
1854	27 18 14	0761	1168	19	04	668
1855	17 3 3	0463	9773	02	93	690
1856 B	6 11 51	0164	8379	85	82	713
1857	24 9 24	0675	7702	84	70	820
1858	13 18 13	0376	6307	67	59	843
1859	3 3 1	0078	4913	50	48	865
1860 B	22 0 34	0588	4236	48	36	972
1861	10 9 23	290	2842	31	25	994
1862	29 6 55	800	2164	29	13	102
1863	18 15 44	502	770	12	2	124
1864 B	8 0 32	204	9376	95	91	146
1865	25 22 5	714	8699	94	79	254
1866	15 6 54	416	7304	77	68	276
1867	4 15 42	117	5910	60	57	299
1868 B	23 13 15	628	5234	58	46	406
1869	11 22 3	329	3838	41	35	428
1870	30 19 36	840	3161	40	23	536
1871	20 4 25	541	1767	23	12	558
1872 B	9 13 13	243	372	6	1	581
1873	27 10 46	753	9695	4	89	688
1874	16 19 34	455	8301	87	78	710
1875	6 4 23	157	6906	70	67	733
1876 B	25 1 55	667	6229	69	55	840
1877	13 10 44	369	4835	52	44	862
1878	2 19 33	71	3441	35	33	885
1879	21 17 5	581	2763	33	21	993
1880 B	11 1 54	283	1369	16	10	15
1881	28 23 27	793	692	14	99	123
1882	18 8 15	495	9297	98	88	145
1883	7 17 4	197	7903	81	77	167
1884 B	26 14 36	707	7226	79	65	275
1885	14 23 25	409	5832	62	54	297

Mean Lunations and Changes of the Arguments.

Num	Lunations.	I.	II.	III.	IV.	N.
	d. h m					
½	14 18 22	404	5359	58	50	43
1	29 12 44	808	717	15	99	85
2	59 1 28	1617	1434	31	98	170
3	88 14 12	2425	2151	46	97	256
4	118 2 56	3234	2869	61	96	341
5	147 15 40	4042	3586	76	95	426
6	177 4 24	4851	4303	92	95	511
7	206 17 8	5659	5020	7	94	596
8	236 5 52	6468	5737	22	93	682
9	265 18 36	7276	6454	37	92	767
10	295 7 20	8085	7171	53	91	852
11	324 20 5	8893	7889	68	90	937
12	354 8 49	9702	8606	83	89	22
13	383 21 33	510	9323	98	88	108

TABLE LXXXVIII.

Number of Days from the commencement of the year to the first of each month.

Months.	Com.	Bis.
	Days.	Days.
January . . .	0	0
February . . .	31	31
March	59	60
April	90	91
May	120	121
June	151	152
July	181	182
August	212	213
September . .	243	244
October . . .	273	274
November . .	304	305
December . .	334	335

Equations for New and Full Moon.

Arg.	I	II	Arg.	I	II	Arg	III	IV	Arg
	h m	h m		h m	h m		m	m	
0	4 20	10 10	5000	4 20	10 10	25	3	31	25
100	4 36	9 36	5100	4 5	10 50	26	3	31	24
200	4 52	9 2	5200	3 49	11 30	27	3	30	23
300	5 8	8 28	5300	3 34	12 9	28	3	30	22
400	5 24	7 55	5400	3 19	12 48	29	3	30	21
500	5 40	7 [illegible]	5500	3 4	13 26	30	3	30	20
600	5 55	6 49	5600	2 49	14 3	31	3	30	19
700	6 10	6 17	5700	2 35	14 39	32	4	30	18
800	6 24	5 46	5800	2 21	15 13	33	4	29	17
900	6 38	5 15	5900	2 8	15 46	34	4	29	16
1000	6 51	4 46	6000	1 55	16 18	35	4	29	15
1100	7 4	4 17	6100	1 42	16 48	36	5	28	14
1200	7 15	3 50	6200	1 31	17 16	37	5	28	13
1300	7 27	3 24	6300	1 19	17 42	38	5	27	12
1400	7 37	2 59	6400	1 9	18 6	39	5	27	11
1500	7 47	2 35	6500	0 59	18 28	40	6	26	10
1600	7 55	2 14	6600	0 50	18 48	41	6	26	9
1700	8 3	1 53	6700	0 42	19 6	42	7	25	8
1800	8 10	1 35	6800	0 34	19 21	43	7	25	7
1900	8 16	1 18	6900	0 28	19 33	44	7	24	6
2000	8 21	1 3	7000	0 22	19 44	45	8	23	5
2100	8 25	0 51	7100	0 17	19 52	46	8	23	4
2200	8 29	0 40	7200	0 14	19 57	47	9	22	3
2300	8 31	0 32	7300	0 11	20 0	48	9	21	2
2400	8 32	0 25	7400	0 9	20 1	49	10	21	1
2500	8 32	0 21	7500	0 8	19 59	50	10	20	0
2600	8 31	0 19	7600	0 8	19 55	51	10	19	99
2700	8 29	0 20	7700	0 9	19 48	52	11	19	98
2800	8 26	0 23	7800	0 11	19 40	53	11	18	97
2900	8 23	0 28	7900	0 15	19 29	54	12	17	96
3000	8 18	0 36	8000	0 19	19 17	55	12	17	95
3100	8 12	0 47	8100	0 24	19 2	56	13	16	94
3200	8 6	0 59	8200	0 30	18 45	57	13	15	93
3300	7 58	1 14	8300	0 37	18 27	58	13	15	92
3400	7 50	1 32	8400	0 45	18 6	59	14	14	91
3500	7 41	1 52	8500	0 53	17 45	60	14	14	90
3600	7 31	2 14	8600	1 3	17 21	61	15	13	89
3700	7 21	2 38	8700	1 13	16 56	62	15	13	88
3800	7 9	3 4	8800	1 25	16 30	63	15	12	87
3900	6 58	3 32	8900	1 36	16 3	64	15	12	86
4000	6 45	4 2	9000	1 49	15 34	65	16	11	85
4100	6 32	4 34	9100	2 2	15 5	66	16	11	84
4200	6 19	5 7	9200	2 16	14 34	67	16	11	83
4300	6 5	5 41	9300	2 30	14 3	68	16	10	82
4400	5 51	6 17	9400	2 45	13 31	69	17	10	81
4500	5 36	6 54	9500	3 0	12 58	70	17	10	80
4600	5 21	7 32	9600	3 16	12 25	71	17	10	79
4700	5 6	8 11	9700	3 32	11 52	72	17	10	78
4800	4 51	8 50	9800	3 48	11 18	73	17	10	77
4900	4 35	9 30	9900	4 4	10 44	74	17	9	76
5000	4 20	10 10	10000	4 20	10 10	75	17	9	75

Mean Right Ascensions and Declinations of 50 *principal Fixed Stars, for the beginning of* 1840.

	Stars' Name.	Mag	Right Ascen.	AnnualVar.	Declination.	Ann. Var.
			h m s	s	° ′ ″	″
1	*Algenib*	2.3	0 5 0.31	+ 3.0775	14 17 38.82 N	+ 20.051
2	β Andromedae	2	1 0 46.7	3.309	34 46 17.2 N	19.35
3	*Polaris*	2.3	1 2 10.38	16.1962	88 27 21.96 N	19.339
4	*Achernar*	1	1 31 44.88	2.2351	58 3 5.13 S	— 18.473
5	α Arietis	3	1 58 9.94	3.3457	22 42 11.81 N	+ 17.455
6	α Ceti	2.3	2 53 55.34	+ 3.1257	3 27 30.09 N	+ 14.561
7	α Persei	2.3	3 12 55.97	4.2280	49 17 8.74 N	13.371
8	*Aldebaran*	1	4 26 44.77	3.4264	16 10 56.82 N	7.949
9	*Capella*	1	5 4 52.67	4.4066	45 49 42.81 N	4.793
10	*Rigel*	1	5 6 51.09	2.8783	8 23 29.29 S	— 4.620
11	β Tauri	2	5 16 10.96	+ 3.7820	28 27 58.20 N	+ 3.825
12	γ Orionis	2	5 16 33.1	3.210	6 11 55.3 N	+ 3.82
13	α Columbae	2	5 33 51.52	2.1688	34 9 47.41 S	— 2.291
14	α Orionis	1	5 46 30.71	3.2430	7 22 17.14 N	+ 1.191
15	*Canopus*	1	6 20 24.18	1.3278	52 36 38.42 S	1.778
16	*Sirius*	1	6 38 5.76	+ 2.6458	16 30 4.79 S	+ 4.449
17	*Castor*	3	7 24 23.06	3.8572	32 13 58.89 N	— 7.206
18	*Procyon*	1.2	7 30 55.53	3.1448	5 37 48.92 N	8.720
19	*Pollux*	2	7 35 31.07	3.6840	28 24 25.57 N	8.107
20	α Hydrae	2	9 19 43.57	2.9500	7 58 4.83 S	+ 15.341
21	*Regulus*	1	9 59 50.93	+ 3.2220	12 44 49.70 N	— 17.356
22	α Ursae Majoris	1.2	10 53 47.98	3.8077	62 36 48.93 N	19.221
23	β Leonis	2.3	11 40 53.69	3.0660	15 28 1.16 N	19.985
24	β Virginis	3.4	11 42 21.4	3.124	2 40 2 6 N	19.98
25	γ Ursae Majoris	2	11 45 22.93	3.1914	54 35 4.67 N	20.014
26	α 2 Crucis	2	12 17 43.7	+ 3.258	62 12 47. 9 S	+ 19.99
27	*Spica*	1	13 16 46.36	3.1502	10 19 24.39 S	18.945
28	θ Centauri	2	13 57 18.0	3.491	35 34 41.9 S	17.499
29	α Draconis	3.4	14 0 2.8	1.625	65 8 32.1 N	— 17.37
30	*Arcturus*	1	14 8 21.96	2.7335	20 1 7.67 N	18.956
31	α 2 Centauri	1	14 28 47.84	+ 4.0086	60 10 6.24 S	+ 15.152
32	α 2 Librae	3	14 42 2.44	3.3088	15 22 18.25 S	15.256
33	β Ursae Minoris	3	14 51 14.66	— 0.2787	74 48 34.18 N	— 14.712
34	γ 2 Ursae Minoris	3.4	15 21 1.3	— 0.179	72 24 14.1 N	12.81
35	α Coronae Borealis	2	15 27 54.87	+ 2.5277	27 15 27.71 N	12.361
36	α Serpentis	2.3	15 36 23.43	+ 2.9386	6 56 2.80 N	— 11.770
37	β Scorpii	2	15 56 8.68	3.4729	19 21 38.82 S	+ 10.330
38	*Antares*	1	16 19 36.49	3.6625	26 4 13.13 S	8.519
39	α Herculis	3.4	17 7 21.30	2.7317	14 34 41.43 N	— 4.576
40	α Ophiuchi	2	17 27 30.56	2.7724	12 40 58.65 N	2.844
41	δ Ursae Minoris	3	18 23 56.48	— 19.[illegible]072	86 35 28.89 N	+ 2.161
42	*Vega*	1	18 31 31.19	+ 2.0116	38 38 16.85 N	2.742
43	*Altair*	1	19 42 58.61	2.9255	8 27 0.21 N	8.701
44	α 2 Capricorni	3	20 9 10.34	3.3323	13 2 5.57 S	— 10.705
45	α Cygni	1	20 35 58.80	2.0416	44 42 41.38 N	+ 12.614
46	α Aquarii	3	21 57 33.93	+ 3.0835	1 5 38.00 S	— 17.256
47	*Fomalhaut*	1	22 48 47.67	3.3114	30 28 4.91 S	19.092
48	β Pegasi	2	22 56 1.1	2.878	27 13 1.7 N	+ 19.255
49	*Markab*	2	22 56 47.75	2.9771	14 20 46.92 N	19.295
50	α Andromedae	1	24 0 7.72	3.0704	28 12 27.06 N	20.056

Constants for the Aberration and Nutation in Right Ascension and Declination of the Stars in the preceding Catalogue

	Aberration.				Nutation.			
	φ	M	θ	N	φ′	M′	θ′	N′
	s ° ′		s ° ′		s ° ′		s ° ′	
1	8 28 47	0.1087	7 27 12	0.9657	6 8 24	0.0300	5 28 30	0.8381
2	8 13 39	0.1830	6 19 12	1.0740	6 19 53	0.0838	5 10 8	0.8496
3	8 13 51	1.6526	5 16 57	1.3052	8 16 7	1.3427	5 10 22	0.8493
4	8 5 20	0.3801	10 26 46	1.2798	4 10 12	0.0775	5 0 31	0.8629
5	7 28 26	0.1397	7 0 2	0.8972	6 11 1	0.0695	4 22 53	0.8765
6	7 14 11	0.1149	8 23 8	0.8678	6 1 26	0.0322	4 8 16	0.9078
7	7 9 30	0.3020	5 3 5	1.0630	6 18 13	0.1849	4 3 47	0.9179
8	6 21 43	0.1447	7 23 12	0.5760	6 3 27	0.0726	3 17 54	0.9502
9	6 12 51	0.2875	3 25 37	0.9112	6 5 46	0.1830	3 10 29	0.9605
10	6 12 20	0.1355	9 3 42	1.0300	5 28 47	$\bar{1}$.9966	3 10 4	0.9608
11	6 10 13	0.1873	4 19 21	0.3917	6 2 52	0.1008	3 8 19	0.9626
12	6 10 6	0.1340	8 26 4	0.7851	6 0 40	0.0441	3 8 14	0.9626
13	6 6 5	0.2145	9 4 24	1.2348	5 26 18	$\bar{1}$.8750	3 4 57	0.9648
14	6 3 13	0.1361	8 28 23	0.7521	6 0 15	0.0481	3 2 37	0.9657
15	5 25 22	0.3491	8 25 53	1.2960	6 8 46	$\bar{1}$.6679	2 26 15	0.9657
16	5 21 21	0.1501	8 25 51	1.1152	6 1 51	1.9658	2 22 58	0.9636
17	5 10 40	0.2010	1 2 17	0.6620	5 24 2	0.1257	2 14 6	0.9535
18	5 9 6	0.1297	9 6 54	0.8071	5 28 47	0.0414	2 12 47	0.9513
19	5 8 2	0.1829	0 14 32	0.6052	5 24 2	0.1114	2 11 53	0.9499
20	4 12 39	0.1158	8 17 31	0.9967	6 3 41	0.0081	1 18 37	0.9007
21	4 2 22	0.1162	10 3 47	0.8457	5 23 47	0.0480	1 7 59	0.8782
22	3 18 7	0.4366	0 3 28	1.2394	4 18 58	0.2407	0 21 57	0.8520
23	3 5 21	0.1117	10 6 20	0.9621	5 20 56	0.0344	0 6 35	0.8393
24	3 4 57	0.0958	9 6 51	0.9075	5 28 25	0.0253	0 6 5	0.8390
25	3 4 8	0.3229	11 17 28	1.2298	4 21 46	0.1465	0 5 5	0.8388
26	2 25 19	0.4261	6 8 5	1.2585	7 16 2	0.2089	11 24 14	0.8390
27	2 9 22	0.1066	8 3 31	0.8862	6 5 51	0.0154	11 5 6	0.8559
28	1 28 40	0.1942	6 7 12	1.0176	6 17 31	0.1062	10 23 8	0.8760
29	1 27 53	0.4824	10 23 28	1.2995	3 25 50	0.1090	10 22 16	0.8777
30	1 25 46	0.1336	9 28 18	1.0974	5 18 49	1.9937	10 20 1	0.8822
31	1 20 32	0.4123	5 7 54	1.1820	6 29 6	0.2460	10 14 36	0.8937
32	1 17 26	0.1273	7 18 24	0.6923	6 6 29	0.0593	10 11 28	0.9006
33	1 14 42	0.6961	10 15 5	1.3087	2 26 45	0.2235	10 8 47	0.9066
34	1 7 20	0.6386	10 7 33	1.3087	2 27 7	0.0960	10 1 45	0.9225
35	1 5 45	0.1704	9 22 28	1.1785	5 17 18	$\bar{1}$.9510	10 0 18	0.9257
36	1 3 43	0.1237	9 8 22	0.9994	5 27 30	0.0058	9 28 26	0.9298
37	0 28 58	0.1485	7 4 4	0.6237	6 5 20	0.0795	9 24 12	0.9386
38	0 23 24	0.1728	5 27 59	0.5816	6 5 49	0.1029	9 19 21	0.9478
39	0 12 13	0.1451	9 5 25	1.0962	5 27 45	$\bar{1}$.9742	9 9 58	0.9610
40	0 7 34	0.1427	9 3 4	1.0786	5 28 48	1.9803	9 6 9	0.9642
41	11 23 47	1.3571	8 22 49	1.2821	11 19 31	0.8257	8 24 57	0.9650
42	11 22 50	0.2393	8 24 29	1.2545	6 5 31	$\bar{1}$.8436	8 24 10	0.9644
43	11 6 15	0.1309	8 22 59	1.0237	6 2 16	$\bar{1}$.9988	8 10 21	0.9472
44	11 0 2	0.1341	9 29 33	0.6961	5 26 12	0.0609	8 4 55	0.9368
45	10 23 29	0.2668	8 0 39	1.2634	6 28 32	$\bar{1}$.9042	7 29 0	0.9242
46	10 2 57	0.1057	9 2 31	0.8988	5 29 26	0.0264	7 8 37	0.8794
47	9 19 26	0.1638	11 7 34	1.0271	5 13 8	0.0765	6 23 30	0.8540
48	9 17 29	0.1491	7 17 0	1.1171	6 17 2	0.0162	6 21 13	0.8511
49	9 17 17	0.1120	8 2 5	1.0138	6 8 23	0.0157	6 20 58	0.8508
50	9 0 6	0.1495	7 6 42	1.0785	6 17 20	0.0444	6 0 8	0.8380

Mean Longitudes and Latitudes of some of the principal Fixed Stars for the beginning of 1840, *with their Annual Variations*

Stars' Name.	Mag	Longitude.	Annual Var.	Latitude.	Annual Var.
		s ° ′ ″	″	° ′ ″	″
α Arietis	3	1 5 25 27.6	50.277	9 57 40.9 N	+ 0.161
Aldebaran	1	2 7 33 5.9	50.210	5 28 38.0 S	— 0.335
Capella	1	2 19 37 17.8	50.302	22 51 44.4 N	— 0.052
Polaris	2.3	2 26 19 20.1	47.959	66 4 59.5 N	+ 0.552
Sirius	1	3 11 52 32.9	49.488	39 34 4.3 S	+ 0.319
Canopus	1	3 12 44 59.6	49.366	75 50 57.6 S	+ 0.459
Pollux	2	3 21 0 22.0	49.502	6 40 20.2 N	+ 0.255
Regulus	1	4 27 36 13.2	49.946	0 27 38.3 N	+ 0.220
Spica	1	6 21 36 29.2	50.085	2 2 29.7 S	+ 0.171
Arcturus	1	6 22 0 4.7	50.711	30 51 17.5 N	+ 0.214
Antares	1	8 7 31 45.2	50.120	4 32 51.6 S	+ 0.424
Altair	1.2	9 29 31 5.9	50.795	29 18 37.3 N	+ 0.080
Fomalhaut	1	11 1 36 22.0	50.595	21 6 49.7 S	+ 0.213
Achernar	1	11 13 2 5.3	50.346	17 6 17.3 S	— 0.083
α Pegasi	2	11 21 15 24.7	50.112	19 24 40.9 N	+ 0.098

TABLE added to TABLE XC.

Mean Right Ascensions and Declinations of Polaris and δ Ursae Minoris for 1830, 1840, 1850, and 1860.

Stars.	Years	Right Asc.	Ann. Var.	Declination.	Ann. Var.
		° ′ ″	″	° ′ ″	″
Polaris	1830	0 59 30.76	+ 15.478	88 24 8.82	+ 19.371
	1840	1 2 10.32	16.470	88 27 22.43	19.309
	1850	1 5 0.29	17.567	88 30 35.40	19.240
	1860	1 8 1.79	18.784	88 33 47.64	19.163
δ Ursae Minoris	1830	18 27 5.13	— 19.167	86 35 5.70	+ 2.363
	1840	18 23 53.03	19.241	86 35 27.93	2.085
	1850	18 20 40.21	19.305	86 35 47.36	1.805
	1860	18 17 26.77	19.360	86 36 3.97	1.523

Mean Places of 50 *Principal Fixed Stars.*

For January 0d., 1870.

Star's Name.	Mag.	Right Ascen.	Annual Var.	Declination.	Annual Var.
		h. m. s.	s.		
α Andromedæ	2	0 1 40.238	+ 3.0864	N. 28° 22′ 21″.62	+19.″899
γ Pegasi (*Algenib*)..	3.2	0 6 32.548	3.0811	N. 14 27 38 .40	20 .027
α Urs. Min. (*Polaris*)	2	1 11 16.990	20.1966	N. 88 36 58 .74	19 .091
α Eridani (*Achernar*)	1	1 32 52.026	2.2349	S. 57 53 51 .51	18 .419
α Arietis	2	1 59 50.914	3.3665	N. 22 50 46 .85	17 .224
α Ceti	2.3	2 55 29.039	+ 3.1273	N. 3 34 40 .00	+14 .344
α Persei...........	2	3 15 3.176	4.2481	N. 49 23 44 .88	13 .169
α Tauri (*Aldebaran*).	1	4 28 27.782	3.4353	N. 16 14 44 .14	7 .622
α Aurigæ (*Capella*).	1	5 7 5.338	4.4217	N. 45 51 44 .60	4 .154
β Orionis (*Rigel*)....	1	5 8 17.411	2.8799	S. 8 21 15 .13	4 .464
β Tauri............	2	5 18 4.487	+ 3.7873	N. 28 29 40 .42	+ 3 .446
δ Orionis..........	2	5 25 21.975	3.0641	S. 0 23 52 .63	2 .980
α Columbæ........	2	5 34 56.721	2.1778	S. 34 8 40 .13	2 .187
α Orionis..........	Var.	5 48 8.026	3.2462	N. 7 22 48 .64	+ 1 .035
α Argûs (*Canopus*)..	1	6 21 6.083	1.3303	S. 52 37 32 .06	— 1 .842
α Canis Maj. (*Sirius*)	1	6 39 25.283	+ 2.6452	S. 16 32 24 .53	— 6 .657
α^2 Geminor (*Castor*).	2.1	7 26 18.152	3.8417	N. 32 10 14 .97	7 .651
α CanisMin.(*Procyon*)	1	7 32 29.688	3.1446	N. 5 33 22 .44	8 .908
β Geminor (*Pollux*).	1.2	7 37 21.465	3.6812	N. 28 20 15 .56	8 .324
α Hydræ..........	2	9 21 11.889	2.9485	S. 8 5 47 .54	15 .397
α Leonis (*Regulus*)..	1.2	10 1 26.776	+ 3.2023	N. 12 36 5 .31	—17 .423
α Ursæ Majoris.....	2	10 55 41.171	3.7653	N. 62 27 7 .09	19 .360
β Leonis..........	2	11 42 25.582	3.0648	N. 15 17 55 .39	20 .099
γ Ursæ Majoris.....	2.3	11 46 58.914	3.1887	N. 54 25 2 .81	20 .027
η Virginis.........	3.4	12 13 15.244	3.0650	N. 0 3 21 .65	20 .054
α^1 Crucis..........	1	12 19 22.701	+ 3.2650	S. 62 22 38 .17	—19 .932
α Virginis (*Spica*) ..	1	13 18 20.737	3.1506	S. 10 28 55 .45	18 .932
ζ Virginis	3.4	13 28 4.232	3.0523	N. 0 4 11 .64	18 .527
α Bootis (*Arcturus*).	1	14 9 43.897	2.7338	N. 19 51 37 .41	18 .903
α^2 Centauri	1	14 30 48.325	4.0367	S. 60 17 39 .13	15 .022
ε Bootis	2.3	14 39 18.503	+ 2.6194	N. 27 37 24 .86	—15 .395
α^2 Libræ...........	2.3	14 43 41.335	+ 3.3058	S. 15 29 59 .61	15 .211
β Ursæ Minoris.....	2	14 51 6.857	— 0.2489	N. 74 41 11 .24	14 .757
β Libræ	2	15 10 0.758	+ 3.2188	S. 8 54 5 .16	13 .562
α Coronæ Borealis..	2	15 29 10.990	2.5377	N. 27 9 13 .63	12 .335
α Serpentis........	2.3	15 37 51.859	+ 2.9492	N. 6 50 11 .26	—11 .598
β^1 Scorpii	2	15 57 52.802	3.4776	S. 19 26 50 .27	10 .206
α Scorpii (*Antares*)..	1.2	16 21 26.333	3.6668	S. 26 8 27 .62	8 .387
α Herculis	Var.	17 8 43.136	2.7322	N. 14 32 26 .06	4 .404
α Ophiuchi	2	17 28 53.949	+ 2.7808	N. 12 39 24 .42	— 2 .921
δ Ursæ Minoris.....	4.5	18 14 16.673	—19.3995	N. 86 36 21 .06	+ 1 .260
α Lyræ (*Vega*).....	1	18 32 32.155	+ 2.0304	N. 38 39 51 .25	3 .126
α Aquilæ (*Altair*)...	1.2	19 44 26.344	2.9272	N. 8 31 37 .06	9 .210
α^2 Capricorni	3.4	20 10 50.295	3.3324	S. 12 56 45 .08	10 .840
α Cygni...........	2.1	20 36 59.960	2.0430	N. 44 49 0 .98	13 .690
61^1 Cygni....... .	5.6	21 1 3.949	+ 2.6737	N. 38 6 41 .36	+17 .495
α Aquarii..........	3	21 59 6.275	3.0823	S. 0 57 1 .81	17 .317
α Pis.Aus(*Fomalhaut*)	1.2	22 50 27.642	3.3288	S. 30 18 38 .37	18 .966
α Pegasi (*Markab*) .	2	22 58 17.136	2.9831	N. 14 30 23 .04	19 .312
γ Cephei..........	3.4	23 34 1.860	2.4018	N. 76 54 24 .86	20 .077

Second Differences.

Hours & Minutes.		1′	2′	3′	4′	5′	6′	7′	8′	9′	10′	11′
h *m*	*h* *m*	″	″	″	″	″	″	″	″	″	″	″
0 0	12 0	0.0	0.0	0.0	0.0	0.0	0.0	0.0	0.0	0.0	0.0	0.0
0 10	11 50	0.4	0.8	1.2	1.6	2.0	2.4	2.9	3.3	3.7	4.1	4.5
0 20	11 40	0.8	1.6	2.4	3.2	4.1	4.9	5.7	6.5	7.3	8 1	8.9
0 30	11 30	1.2	2.4	3.6	4.8	6.0	7.2	8.4	9.6	10.8	12.0	13.2
0 40	11 20	1.6	3.1	4.7	6.3	7.9	9.4	11.0	12.6	14.2	15.7	17.3
0 50	11 10	1.9	3.9	5.8	7.8	9.7	11.6	13.6	15.5	17.4	19.4	21.4
1 0	11 0	2.3	4.6	6.9	9.2	11.5	13.8	16.0	18.3	20.6	22.9	25.2
1 10	10 50	2.6	5.3	7.9	10.5	13.2	15.8	18.4	21.1	23.7	26.3	29.0
1 20	10 40	3.0	5.9	8.9	11.9	14.8	17.8	20.7	23.7	26.7	29.6	32.6
1 30	10 30	3.3	6.6	9.8	13.1	16.4	19.7	23.0	26.3	29.5	32.8	36.1
1 40	10 20	3.6	7.2	10.8	14.4	17.9	21.5	25.1	28.7	32.3	35.9	39.5
1 50	10 10	3.9	7.8	11.6	15.5	19.4	23.3	27.2	31.0	34.9	38.8	42.7
2 0	10 0	4.2	8.3	12.5	16.7	20.8	25.0	29.2	33.3	37.5	41.7	45.8
2 10	9 50	4.4	8.9	13.3	17.8	22.2	26.6	31.1	35.5	40.0	44.4	48.8
2 20	9 40	4.7	9.4	14.1	18.8	23.5	28.2	32.9	37.6	42.3	47.0	51.7
2 30	9 30	4.9	9.9	14.8	19.8	24.7	29.7	34.6	39.6	44.5	49.5	54.4
2 40	9 20	5.2	10.4	15.6	20.7	25.9	31.1	36.3	41.5	46.7	51.9	57.0
2 50	9 10	5.4	10.8	16.2	21.6	27.1	32.5	37.9	43.3	48.7	54.1	59.5
3 0	9 0	5.6	11.3	16.9	22.5	28.1	33.8	39.4	45.0	50.6	56.3	61.9
3 10	8 50	5.8	11.7	17.5	23.3	29.1	35.0	40.8	46.6	52.4	58.3	64.1
3 20	8 40	6.0	12.0	18.1	24.1	30.1	36.1	42.1	48.1	54.2	60.2	66.2
3 30	8 30	6.2	12.4	18.6	24.8	31.0	37.2	43.4	49.6	55.8	62.0	68.2
3 40	8 20	6.4	12.7	19.1	25.5	31.8	38.2	44.6	50.9	57.3	63.7	70.0
3 50	8 10	6.5	13.0	19.6	26.1	32.6	39.1	45.7	52.2	58.7	65.2	71.7
4 0	8 0	6.7	13.3	20.0	26.7	33.3	40.0	46.7	53.3	60.0	66.7	73.3
4 10	7 50	6.8	13.6	20.4	27.2	34.0	40.8	47.6	54.4	61.2	68.0	74.8
4 20	7 40	6.9	13.8	20.8	27.7	34.6	41.5	48.4	55.4	62.3	69.2	76.1
4 30	7 30	7.0	14.1	21.1	28.1	35.2	42.2	49.2	56.2	63.3	70.3	77.3
4 40	7 20	7.1	14.3	21.4	28.5	35.6	42.8	49.9	57.0	64.2	71.3	78.4
4 50	7 10	7.2	14.4	21.6	28.9	36.1	43.3	50.5	57.7	64.9	72.2	79.4
5 0	7 0	7.3	14.6	21.9	29.2	36.5	43.8	51.0	58.3	65.6	72.9	80.2
5 10	6 50	7.4	14.7	22.1	29.4	36.8	44.1	51.5	58.8	66.2	73.6	80.9
5 20	6 40	7.4	14.8	22.2	29.6	37.0	44.4	51.9	59.3	66.7	74.1	81.5
5 30	6 30	7.4	14.9	22.3	29.8	37.2	44.7	52.1	59.6	67.0	74.5	81.9
5 40	6 20	7.5	15.0	22.4	29.9	37.4	44.9	52.3	59.8	67.3	74.8	82.2
5 50	6 10	7.5	15.0	22.5	30.0	37.5	45.0	52.5	60.0	67.4	74.9	82.4
6 0	6 0	7.5	15.0	22.5	30.0	37.5	45.0	52.5	60.0	67.5	75.0	82.5

Second Differences.

Hours & Min.				10″	20″	30″	40″	50″	1″	2″	3″	4″	5″	6″	7″	8″	9″
h	*m*	*h*	*m*	″	″	″	″	″	″	″	″	″	″	″	″	″	″
0	0	12	0	0.0	0.0	0.0	0.0	0.0	0.0	0.0	0.0	0.0	0.0	0.0	0.0	0.0	0.0
0	10	11	50	0.1	0.1	0.2	0.3	0.3	0.0	0.0	0.0	0.0	0.0	0.0	0.0	0.1	0.1
0	20	11	40	0.1	0.3	0.4	0.5	0.7	0.0	0.0	0.0	0.1	0.1	0.1	0.1	0.1	0.1
0	30	11	30	0.2	0.4	0.6	0.8	1.0	0.0	0.0	0.1	0.1	0.1	0.1	0.1	0.2	0.2
0	40	11	20	0.3	0.5	0.8	1.0	1.3	0.0	0.1	0.1	0.1	0.1	0.2	0.2	0.2	0.2
0	50	11	10	0.3	0.6	1.0	1.3	1.6	0.0	0.1	0.1	0.1	0.2	0.2	0.2	0.3	0.3
1	0	11	0	0.4	0.8	1.1	1.5	1.9	0.0	0.1	0.1	0.2	0.2	0.2	0.3	0.3	0.3
1	10	10	50	0.4	0.9	1.3	1.8	2.2	0.0	0.1	0.1	0.2	0.2	0.3	0.3	0.4	0.4
1	20	10	40	0.5	1.0	1.5	2.0	2.5	0.0	0.1	0.1	0.2	0.2	0.3	0.3	0.4	0.4
1	30	10	30	0.5	1.1	1.6	2.2	2.7	0.1	0.1	0.2	0.2	0.3	0.3	0.4	0.4	0.5
1	40	10	20	0.6	1.2	1.8	2.4	3.0	0.1	0.1	0.2	0.2	0.3	0.4	0.4	0.5	0.5
1	50	10	10	0.6	1.3	1.9	2.6	3.2	0.1	0.1	0.2	0.3	0.3	0.4	0.5	0.5	0.6
2	0	10	0	0.7	1.4	2.1	2.8	3.5	0.1	0.1	0.2	0.3	0.3	0.4	0.5	0.6	0.6
2	10	9	50	0.7	1.5	2.2	3.0	3.7	0.1	0.1	0.2	0.3	0.4	0.4	0.5	0.6	0.7
2	20	9	40	0.8	1.6	2.3	3.1	3.9	0.1	0.2	0.2	0.3	0.4	0.5	0.5	0.6	0.7
2	30	9	30	0.8	1.6	2.5	3.3	4.1	0.1	0.2	0.2	0.3	0.4	0.5	0.6	0.7	0.7
2	40	9	20	0.9	1.7	2.6	3.5	4.3	0.1	0.2	0.3	0.3	0.4	0.5	0.6	0.7	0.8
2	50	9	10	0.9	1.8	2.7	3.6	4.5	0.1	0.2	0.3	0.4	0.5	0.5	0.6	0.7	0.8
3	0	9	0	0.9	1.9	2.8	3.8	4.7	0.1	0.2	0.3	0.4	0.5	0.6	0.7	0.7	0.8
3	10	8	50	1.0	1.9	2.9	3.9	4.9	0.1	0.2	0.3	0.4	0.5	0.6	0.7	0.8	0.9
3	20	8	40	1.0	2.0	3.0	4.0	5.0	0.1	0.2	0.3	0.4	0.5	0.6	0.7	0.8	0.9
3	30	8	30	1.0	2.1	3.1	4.1	5.2	0.1	0.2	0.3	0.4	0.5	0.6	0.7	0.8	0.9
3	40	8	20	1.1	2.1	3.2	4.2	5.3	0.1	0.2	0.3	0.4	0.5	0.6	0.7	0.8	1.0
3	50	8	10	1.1	2.2	3.3	4.3	5.4	0.1	0.2	0.3	0.4	0.5	0.7	0.8	0.9	1.0
4	0	8	0	1.1	2.2	3.3	4.4	5.6	0.1	0.2	0.3	0.4	0.6	0.7	0.8	0.9	1.0
4	10	7	50	1.1	2.3	3.4	4.5	5.7	0.1	0.2	0.3	0.5	0.6	0.7	0.8	0.9	1.0
4	20	7	40	1.2	2.3	3.5	4.6	5.8	0.1	0.2	0.3	0.5	0.6	0.7	0.8	0.9	1.0
4	30	7	30	1.2	2.3	3.5	4.7	5.9	0.1	0.2	0.4	0.5	0.6	0.7	0.8	0.9	1.1
4	40	7	20	1.2	2.4	3.6	4.8	5.9	0.1	0.2	0.4	0.5	0.6	0.7	0.8	1.0	1.1
4	50	7	10	1.2	2.4	3.6	4.8	6.0	0.1	0.2	0.4	0.5	0.6	0.7	0.8	1.0	1.1
5	0	7	0	1.2	2.4	3.6	4.9	6.1	0.1	0.2	0.4	0.5	0.6	0.7	0.9	1.0	1.1
5	10	6	50	1.2	2.5	3.7	4.9	6.1	0.1	0.2	0.4	0.5	0.6	0.7	0.9	1.0	1.1
5	20	6	40	1.2	2.5	3.7	4.9	6.1	0.1	0.2	0.4	0.5	0.6	0.7	0.9	1.0	1.1
5	30	6	30	1.2	2.5	3.7	5.0	6.2	0.1	0.2	0.4	0.5	0.6	0.7	0.9	1.0	1.1
5	40	6	20	1.2	2.5	3.7	5.0	6.2	0.1	0.2	0.4	0.5	0.6	0.7	0.9	1.0	1.1
5	50	6	10	1.2	2.5	3.7	5.0	6.2	0.1	0.2	0.4	0.5	0.6	0.7	0.9	1.0	1.1
6	0	6	0	1.3	2.6	3.8	5.0	6.3	0.1	0.2	0.4	0.5	0.6	0.7	0.9	1.0	1.1

Third Differences.

Time after noon or midnight.	10″	20″	30″	40″	50″	1′	2′	3′	4′	5′	Time after noon or midnight.
+	″	″		″	″	″	″	″	″	″	—
0h. 0m.	0.0	0.0	0.0	0.0	0.0	0.0	0.0	0.0	0.0	0.0	12h. 0m.
0 30	0.0	0.1	0.1	0.1	0.2	0.2	0.4	0.5	0.7	0.9	11 30
1 0	0.1	0.1	0.2	0.2	0.3	0.3	0.6	1.0	1.3	1.5	11 0
1 30	0.1	0.1	0.2	0.3	0.3	0.4	0.8	1.2	1.6	2.1	10 30
2 0	0.1	0.2	0.2	0.3	0.4	0.5	0.9	1.4	1.9	2.3	10 0
2 30	0.1	0.2	0.2	0.3	0.4	0.5	1.0	1.4	1.9	2.4	9 30
3 0	0.1	0.2	0.2	0.3	0.4	0.5	0.9	1.4	1.9	2.3	9 0
3 30	0.1	0.1	0.2	0.3	0.4	0.4	0.9	1.3	1.7	2.2	8 30
4 0	0.1	0.1	0.2	0.2	0.3	0.4	0.7	1.1	1.5	1.9	8 0
4 30	0.0	0.1	0.1	0.2	0.2	0.3	0.6	0.9	1.2	1.5	7 30
5 0	0.0	0.1	0.1	0.1	0.2	0.2	0.4	0.6	0.8	1.0	7 0
5 30	0.0	0.0	0.1	0.1	0.1	0.1	0.2	0.3	0.4	0.5	6 30
6 0	0.0	0.0	0.0	0.0	0.0	0.0	0.0	0.0	0.0	0.0	6 0
+											—

TABLE XCV.

Fourth Differences.

Time after noon or midnight.	10″	20″	30″	40″	50″	1′	2′	3′	Time after noon or midnight.
h. m.	″	″	″	″	″	″	″	″	h. m.
0 0	0.0	0.0	0.0	0.0	0.0	0.0	0.0	0.0	12 0
0 30	0.0	0.1	0.1	0.1	0.2	0.2	0.4	0.6	11 30
1 0	0.1	0.1	0.2	0.3	0.3	0.4	0.8	1.2	11 0
1 30	0.1	0.2	0.3	0.4	0.5	0.6	1.2	1.7	10 30
2 0	0.1	0.2	0.4	0.5	0.6	0.7	1.5	2.2	10 0
2 30	0.1	0.3	0.4	0.6	0.7	0.9	1.8	2.7	9 30
3 0	0.2	0.3	0.5	0.7	0.9	1.0	2.1	3.1	9 0
3 30	0.2	0.4	0.6	0.8	0.9	1.1	2.3	3.4	8 30
4 0	0.2	0.4	0.6	0.8	1.0	1.2	2.5	3.7	8 0
4 30	0.2	0.4	0.7	0.9	1.1	1.3	2.6	3.9	7 30
5 0	0.2	0.5	0.7	0.9	1.1	1.4	2.7	4.1	7 0
5 30	0.2	0.5	0.7	0.9	1.2	1.4	2.8	4.2	6 30
6 0	0.2	0.5	0.7	0.9	1.2	1.4	2.8	4.2	6 0

′	0	1	2	3	4	5	6	7	8	9
″	0	60	120	180	240	300	360	420	480	540
0		1.7782	1.4771	1.3010	1.1761	1.0792	1.0000	9331	8751	8239
1	3.5563	1.7710	1.4735	1.2986	1.1743	1.0777	9988	9320	8742	8231
2	3.2553	1.7639	1.4699	1.2962	1.1725	1.0763	[illegible]	9310	8733	8223
3	3.0792	1.7570	1.4664	1.2939	1.1707	1.0749	9964	9300	8724	8215
4	2.9542	1.7501	1.4629	1.2915	1.1689	1.0734	9952	9289	8715	8207
5	2.8573	1.7434	1.4594	1.2891	1.1671	1.0720	9940	9279	8706	8199
6	2.7782	1.7368	1.4559	1.2868	1.1654	1.0706	9928	9269	8697	8191
7	2.7112	1.7302	1.4525	1.2845	1.1636	1.0692	9916	9259	8688	8183
8	2.6532	1.7238	1.4491	1.2821	1.1619	1.0678	9905	9249	8679	8175
9	2.6021	1.7175	1.4457	1.2798	1.1601	1.0663	9893	9238	8670	8167
10	2.5563	1.7112	1.4424	1.2775	1.1584	1.0649	9881	9228	8661	8159
11	2.5149	1.7050	1.4390	1.2753	1.1566	1.0635	9869	9218	8652	8152
12	2.4771	1.6990	1.4357	1.2730	1.1549	1.0621	9858	9208	8643	8144
13	2.4424	1.6930	1.4325	1.2707	1.1532	1.0608	9846	9198	8635	8136
14	2.4102	1.6871	1.4292	1.2685	1.1515	1.0594	9834	9188	8626	8128
15	2.3802	1.6812	1.4260	1.2663	1.1498	1.0580	9823	9178	8617	8120
16	2.3522	1.6755	1.4228	1.2640	1.1481	1.0566	9811	9168	8608	8112
17	2.3259	1.6698	1.4196	1.2618	1.1464	1.0552	9800	9158	8599	8104
18	2.3010	1.6642	1.4165	1.2596	1.1447	1.0539	9788	9148	8591	8097
19	2.2775	1.6587	1.4133	1.2574	1.1430	1.0525	9777	9138	8582	8089
20	2.2553	1.6532	1.4102	1.2553	1.1413	1.0512	9765	9128	8573	8081
21	2.2341	1.6478	1.4071	1.2531	1.1397	1.0498	9754	9119	8565	8073
22	2.2139	1.6425	1.4040	1.2510	1.1380	1.0484	9742	9109	8556	8066
23	2.1946	1.6372	1.4010	1.2488	1.1363	1.0471	9731	9099	8547	8058
24	2.1761	1.6320	1.3979	1.2467	1.1347	1.0458	9720	9089	8539	8050
25	2.1584	1.6269	1.3949	1.2445	1.1331	1.0444	9708	9079	8530	8043
26	2.1413	1.6218	1.3919	1.2424	1.1314	1.0431	9697	9070	8522	8035
27	2.1249	1.6168	1.3890	1.2403	1.1298	1.0418	9686	9060	8513	8027
28	2.1091	1.6118	1.3860	1.2382	1.1282	1.0404	9675	9050	8504	8020
29	2.0939	1.6069	1.3831	1.2362	1.1266	1.0391	9664	9041	8496	8012
30	2.0792	1.6021	1.3802	1.2341	1.1249	1.0378	9652	9031	8487	8004
31	2.0649	1.5973	1.3773	1.2320	1.1233	1.0365	9641	9021	8479	7997
32	2.0512	1.5925	1.3745	1.2300	1.1217	1.0352	9630	9012	8470	7989
33	2.0378	1.5878	1.3716	1.2279	1.1201	1.0339	9619	9002	8462	7981
34	2.0248	1.5832	1.3688	1.2259	[illegible]	1.0326	9608	8992	8453	7974
35	2.0122	1.5786	1.3660	1.2239	1.1170	1.0313	9597	8983	8445	7966
36	2.0000	1.5740	1.3632	1.2218	1.1154	1.0300	9586	8973	8437	7959
37	1.9881	1.5695	1.3604	1.2198	1.1138	1.0287	9575	8964	8428	7951
38	1.9765	1.5651	1.3576	1.2178	1.1123	1.0274	9564	8954	8420	7944
39	1.9652	1.5607	1.3549	1.2159	1.1107	1.0261	9553	8945	8411	7936
40	1.9542	1.5563	1.3522	1.2139	1.1091	1.0248	9542	8935	8403	7929
41	1.9435	1.5520	1.3495	1.2119	1.1076	1.0235	9532	8926	8395	7921
42	1.9331	1.5477	1.3468	1.2099	1.1061	1.0223	9521	8917	8386	7914
43	1.9228	1.5435	1.3441	1.2080	1.1045	1.0210	9510	8907	8378	7906
44	1.9128	1.5393	1.3415	1.2061	1.1030	1.0197	9499	8898	8370	7899
45	1.9031	1.5351	1.3388	1.2041	1.1015	1.0185	9488	8888	8361	7891
46	1.8935	1.5310	1.3362	1.2022	1.0999	1.0172	9478	8879	8353	7884
47	1.8842	1.5269	1.3336	1.2003	1.0984	1.0160	9467	8870	8345	7877
48	1.8751	1.5229	1.3310	1.1984	1.0969	1.0147	9456	8861	8337	7869
49	1.8661	1.5189	1.3284	1.1965	1.0954	1.0135	9446	8851	8328	7862
50	1.8573	1.5149	1.3259	1.1946	1.0939	1.0122	9435	8842	8320	7855
51	1.8487	1.5110	1.3233	1.1927	1.0924	1.0110	9425	8833	8312	7847
52	1.8403	1.5071	1.3208	1.1908	1.0909	1.0098	9414	8824	8304	7840
53	1.8320	1.5032	1.3183	1.1889	1.0894	1.0085	9404	8814	8296	7832
54	1.8239	1.4994	1.3158	1.1871	1.0880	1.0073	9393	8805	8288	7825
55	1.8159	1.4956	1.3133	1.1852	1.0865	1.0061	9383	8796	8279	7818
56	1.8081	1.4918	1.3108	1.1834	1.0850	1.0049	9372	8787	8271	7811
57	1.8004	1.4881	1.3083	1.1816	1.0835	1.0036	9362	8778	8263	7803
58	1.7929	1.4844	1.3059	1.1797	1.0821	1.0024	9351	8769	8255	7796
59	1.7855	1.4808	1.3034	1.1779	1.0806	1.0012	9341	8760	8247	7789
60	1.7782	1.4771	1.3010	1.1761	1.0792	1.0000	9331	8751	8239	7782

′	10	11	12	13	14	15	16	17	18	19	20	21
″	600	660	720	780	840	900	960	1020	1080	1140	1200	1260
0	7782	7368	6990	6642	6320	6021	5740	5477	5229	4994	4771	4559
1	7774	7361	6984	6637	6315	6016	5736	5473	5225	4990	4768	4556
2	7767	7354	6978	6631	6310	6011	5731	5469	5221	4986	4764	4552
3	7760	7348	6972	6625	6305	6006	5727	5464	5217	4983	4760	4549
4	7753	7341	6966	6620	6300	6001	5722	5460	5213	4979	4757	4546
5	7745	7335	6960	6614	6294	5997	5718	5456	5209	4975	4753	4542
6	7738	7328	6954	6609	6289	5992	5713	5452	5205	4971	4750	4539
7	7731	7322	6948	6603	6284	5987	5709	5447	5201	4967	4746	4535
8	7724	7315	6942	6598	6279	5982	5704	5443	5197	4964	4742	4532
9	7717	7309	6936	6592	6274	5977	5700	5439	5193	4960	4739	4528
10	7710	7302	6930	6587	6269	5973	5695	5435	5189	4956	4735	4525
11	7703	7296	6924	6581	6264	5968	5691	5430	5185	4952	4732	4522
12	7696	7289	6918	6576	6259	5963	5686	5426	5181	4949	4728	4518
13	7688	7283	6912	6570	6254	5958	5682	5422	5177	4945	4724	4515
14	7681	7276	6906	6565	6248	5954	5677	5418	5173	4941	4721	4511
15	7674	7270	6900	6559	6243	5949	5673	5414	5169	4937	4717	4508
16	7667	7264	6894	6554	6238	5944	5669	5409	5165	4933	4714	4505
17	7660	7257	6888	6548	6233	5939	5664	5405	5161	4930	4710	4501
18	7653	7251	6882	6543	6228	5935	5660	5401	5157	4926	4707	4498
19	7646	7244	6877	6538	6223	5930	5655	5397	5153	4922	4703	4494
20	7639	7238	6871	6532	6218	5925	5651	5393	5149	4918	4699	4491
21	7632	7232	6865	6527	6213	5920	5646	5389	5145	4915	4696	4488
22	7625	7225	6859	6521	6208	5916	5642	5384	5141	4911	4692	4484
23	7618	7219	6853	6516	6203	5911	5637	5380	5137	4907	4689	4481
24	7611	7212	6847	6510	6198	5906	5633	5376	5133	4903	4685	4477
25	7604	7206	6841	6505	6193	5902	5629	5372	5129	4900	4682	4474
26	7597	7200	6836	6500	6188	5897	5624	5368	5125	4896	4678	4471
27	7590	7193	6830	6494	6183	5892	5620	5364	5122	4892	4675	4467
28	7583	7187	6824	6489	6178	5888	5615	5359	5118	4889	4671	4464
29	7577	7181	6818	6484	6173	5883	5611	5355	5114	4885	4668	4460
30	7570	7175	6812	6478	6168	5878	5607	5351	5110	4881	4664	4457
31	7563	7168	6807	6473	6163	5874	5602	5347	5106	4877	4660	4454
32	7556	7162	6801	6467	6158	5869	5598	5343	5102	4874	4657	4450
33	7549	7156	6795	6462	6153	5864	5594	5339	5098	4870	4653	4447
34	7542	7149	6789	6457	6148	5860	5589	5335	5094	4866	4650	4444
35	7535	7143	6784	6451	6143	5855	5585	5331	5090	4863	4646	4440
36	7528	7137	6778	6446	6138	5850	5580	5326	5086	4859	4643	4437
37	7522	7131	6772	6441	6133	5846	5576	5322	5082	4855	4639	4434
38	7515	7124	6766	6435	6128	5841	5572	5318	5079	4852	4636	4430
39	7508	7118	6761	6430	6123	5836	5567	5314	5075	4848	4632	4427
40	7501	7112	6755	6425	6118	5832	5563	5310	5071	4844	4629	4424
41	7494	7106	6749	6420	6113	5827	5559	5306	5067	4841	4625	4420
42	7488	7100	6743	6414	6108	5823	5554	5302	5063	4837	4622	4417
43	7481	7093	6738	6409	6103	5818	5550	5298	5059	4833	4618	4414
44	7474	7087	6732	6404	6099	5813	5546	5294	5055	4830	4615	4410
45	7467	7081	6726	6398	6094	5809	5541	5290	5051	4826	4611	4407
46	7461	7075	6721	6393	6089	5804	5537	5285	5048	4822	4608	4404
47	7454	7069	6715	6388	6084	5800	5533	5281	5044	4819	4604	4400
48	7447	7063	6709	6383	6079	5795	5528	5277	5040	4815	4601	4397
49	7441	7057	6704	6377	6074	5790	5524	5273	5036	4811	4597	4394
50	7434	7050	6698	6372	6069	5786	5520	5269	5032	4808	4594	4390
51	7427	7044	6692	6367	6064	5781	5516	5265	5028	4804	4590	4387
52	7421	7038	6687	6362	6059	5777	5511	5261	5025	4800	4587	4384
53	7414	7032	6681	6357	6055	5772	5507	5257	5021	4797	4584	4380
54	7407	7026	6676	6351	6050	5768	5503	5253	5017	4793	4580	4377
55	7401	7020	6670	6346	6045	5763	5498	5249	5013	4789	4577	4374
56	7394	7014	6664	6341	6040	5758	5494	5245	5009	4786	4573	4370
57	7387	7008	6659	6336	6035	5754	5490	5241	5005	4782	4570	4367
58	7381	7002	6653	6331	6030	5749	5486	5237	5002	4778	4566	4364
59	7374	6996	6648	6325	6025	5745	5481	5233	4998	4775	4563	4361
60	7368	6990	6642	6320	6021	5740	5477	5229	4994	4771	4559	4357

′	22	23	24	25	26	27	28	29	30	31	32	33
″	1320	1380	1440	1500	1560	1620	1680	1740	1800	1860	1920	1980
0	4357	4164	3979	3802	3632	3468	3310	3158	3010	2868	2730	2596
1	4354	4161	3976	3799	3629	3465	3307	3155	3008	2866	2728	2594
2	4351	4158	3973	3796	3626	3463	3305	3153	3005	2863	2725	2592
3	4347	4155	3970	3793	3623	3460	3302	3150	3003	2861	2723	2590
4	4344	4152	3967	3791	3621	3457	3300	3148	3001	2859	2721	2588
5	4341	4149	3964	3788	3618	3454	3297	3145	2998	2856	2719	2585
6	4338	4145	3961	3785	3615	3452	3294	3143	2996	2854	2716	2583
7	4334	4142	3958	3782	3612	3449	3292	3140	2993	2852	2714	2581
8	4331	4139	3955	3779	3610	3446	3289	3138	2991	2849	2712	2579
9	4328	4136	3952	3776	3607	3444	3287	3135	2989	2847	2710	2577
10	4325	4133	3949	3773	3604	3441	3284	3133	2986	2845	2707	2574
11	4321	4130	3946	3770	3601	3438	3282	3130	2984	2842	2705	2572
12	4318	4127	3943	3768	3598	3436	3279	3128	2981	2840	2703	2570
13	4315	4124	3940	3765	3596	3433	3276	3125	2979	2838	2701	2568
14	4311	4120	3937	3762	3593	3431	3274	3123	2977	2835	2698	2566
15	4308	4117	3934	3759	3590	3428	3271	3120	2974	2833	2696	2564
16	4305	4114	3931	3756	3587	3425	3269	3118	2972	2831	2694	2561
17	4302	4111	3928	3753	3585	3423	3266	3115	2969	2828	2692	2559
18	4298	4108	3925	3750	3582	3420	3264	3113	2967	2826	2689	2557
19	4295	4105	3922	3747	3579	3417	3261	3110	2965	2824	2687	2555
20	4292	4102	3919	3745	3576	3415	3259	3108	2962	2821	2685	2553
21	4289	4099	3917	3742	3574	3412	3256	3105	2960	2819	2683	2551
22	4285	4096	3914	3739	3571	3409	3253	3103	2958	2817	2681	2548
23	4282	4092	3911	3736	3568	3407	3251	3101	2955	2815	2678	2546
24	4279	4089	3908	3733	3565	3404	3248	3098	2953	2812	2676	2544
25	4276	4086	3905	3730	3563	3401	3246	3096	2950	2810	2674	2542
26	4273	4083	3902	3727	3560	3399	3243	3093	2948	2808	2672	2540
27	4269	4080	3899	3725	3557	3396	3241	3091	2946	2805	2669	2538
28	4266	4077	3896	3722	3555	3393	3238	3088	2943	2803	2667	2535
29	4263	4074	3893	3719	3552	3391	3236	3086	2941	2801	2665	2533
30	4260	4071	3890	3716	3549	3388	3233	3083	2939	2798	2663	2531
31	4256	4068	3887	3713	3546	3386	3231	3081	2936	2796	2660	2529
32	4253	4065	3884	3710	3544	3383	3228	3078	2934	2794	2658	2527
33	4250	4062	3881	3708	3541	3380	3225	3076	2931	2792	2656	2525
34	4247	4059	3878	3705	3538	3378	3223	3073	2929	2789	2654	2522
35	4244	4055	3875	3702	3535	3375	3220	3071	2927	2787	2652	2520
36	4240	4052	3872	3699	3533	3372	3218	3069	2924	2785	2649	2518
37	4237	4049	3869	3696	3530	3370	3215	3066	2922	2782	2647	2516
38	4234	4046	3866	3693	3527	3367	3213	3064	2920	2780	2645	2514
39	4231	4043	3863	3691	3525	3365	3210	3061	2917	2778	2643	2512
40	4228	4040	3860	3688	3522	3362	3208	3059	2915	2775	2640	2510
41	4224	4037	3857	3685	3519	3359	3205	3056	2912	2773	2638	2507
42	4221	4034	3855	3682	3516	3357	3203	3054	2910	2771	2636	2505
43	4218	4031	3852	3679	3514	3354	3200	3052	2908	2769	2634	2503
44	4215	4028	3849	3677	3511	3351	3198	3049	2905	2766	2632	2501
45	4212	4025	3846	3674	3508	3349	3195	3047	2903	2764	2629	2499
46	4209	4022	3843	3671	3506	3346	3193	3044	2901	2762	2627	2497
47	4205	4019	3840	3668	3503	3344	3190	3042	2898	2760	2625	2494
48	4202	4016	3837	3665	3500	3341	3188	3039	2896	2757	2623	2492
49	4199	4013	3834	3663	3497	3338	3185	3037	2894	2755	2621	2490
50	4196	4010	3831	3660	3495	3336	3183	3034	2891	2753	2618	2488
51	4193	4007	3828	3657	3492	3333	3180	3032	2889	2750	2616	2486
52	4189	4004	3825	3654	3489	3331	3178	3030	2887	2748	2614	2484
53	4186	4001	3822	3651	3487	3328	3175	3027	2884	2746	2612	2482
54	4183	3998	3820	3649	3484	3325	3173	3025	2882	2744	2610	2480
55	4180	3995	3817	3646	3481	3323	3170	3022	2880	2741	2607	2477
56	4177	3991	3814	3643	3479	3320	3168	3020	2877	2739	2605	2475
57	4174	3988	3811	3640	3476	3318	3165	3018	2875	2737	2603	2473
58	4171	3985	3808	3637	3473	3315	3163	3015	2873	2735	2601	2471
59	4167	3982	3805	3635	3471	3313	3160	3013	2870	2732	2599	2469
60	4164	3979	3802	3632	3468	3310	3158	3010	2868	2730	2596	2467

′	34	35	36	37	38	39	40	41	42	43	44	45
″	2040	2100	2160	2220	2280	2340	2400	2460	2520	2580	2640	2700
0	2467	2341	2218	2099	1984	1871	1761	1654	1549	1447	1347	1249
1	2465	2339	2216	2098	1982	1869	1759	1652	1547	1445	1345	1248
2	2462	2337	2214	2096	1980	1867	1757	1650	1546	1443	1344	1246
3	2460	2335	2212	2094	1978	1865	1755	1648	1544	1442	1342	1245
4	2458	2333	2210	2092	1976	1863	1754	1647	1542	1440	1340	1243
5	2456	2331	2208	2090	1974	1862	1752	1645	1540	1438	1339	1241
6	2454	2328	2206	2088	1972	1860	1750	1643	1539	1437	1337	1240
7	2452	2326	2204	2086	1970	1858	1748	1641	1537	1435	1335	1238
8	2450	2324	2202	2084	1968	1856	1746	1640	1535	1433	1334	1237
9	2448	2322	2200	2082	1967	1854	1745	1638	1534	1432	1332	1235
10	2445	2320	2198	2080	1965	1852	1743	1636	1532	1430	1331	1233
11	2443	2318	2196	2078	1963	1850	1741	1634	1530	1428	1329	1232
12	2441	2316	2194	2076	1961	1849	1739	1633	1528	1427	1327	1230
13	2439	2314	2192	2074	1959	1847	1737	1631	1527	1425	1326	1229
14	2437	2312	2190	2072	1957	1845	1736	1629	1525	1423	1324	1227
15	2435	2310	2188	2070	1955	1843	1734	1627	1523	1422	1322	1225
16	2433	2308	2186	2068	1953	1841	1732	1626	1522	1420	1321	1224
17	2431	2306	2184	2066	1951	1839	1730	1624	1520	1418	1319	1222
18	2429	2304	2182	2064	1950	1838	1728	1622	1518	1417	1317	1221
19	2426	2302	2180	2062	1948	1836	1727	1620	1516	1415	1316	1219
20	2424	2300	2178	2061	1946	1834	1725	1619	1515	1413	1314	1217
21	2422	2298	2176	2059	1944	1832	1723	1617	1513	1412	1313	1216
22	2420	2296	2174	2057	1942	1830	1721	1615	1511	1410	1311	1214
23	2418	2294	2172	2055	1940	1828	1719	1613	1510	1408	1309	1213
24	2416	2291	2170	2053	1938	1827	1718	1612	1508	1407	1308	1211
25	2414	2289	2169	2051	1936	1825	1716	1610	1506	1405	1306	1209
26	2412	2287	2167	2049	1934	1823	1714	1608	1504	1403	1304	1208
27	2410	2285	2165	2047	1933	1821	1712	1606	1503	1402	1303	1206
28	2408	2283	2163	2045	1931	1819	1711	1605	1501	1400	1301	1205
29	2405	2281	2161	2043	1929	1817	1709	1603	1499	1398	1300	1203
30	2403	2279	2159	2041	1927	1816	1707	1601	1498	1397	1298	1201
31	2401	2277	2157	2039	1925	1814	1705	1599	1496	1395	1296	1200
32	2399	2275	2155	2037	1923	1812	1703	1598	1494	1393	1295	1198
33	2397	2273	2153	2035	1921	1810	1702	1596	1493	1392	1293	1197
34	2395	2271	2151	2033	1919	1808	1700	1594	1491	1390	1291	1195
35	2393	2269	2149	2032	1918	1806	1698	1592	1489	1388	1290	1193
36	2391	2267	2147	2030	1916	1805	1696	1591	1487	1387	1288	1192
37	2389	2265	2145	2028	1914	1803	1694	1589	1486	1385	1287	1190
38	2387	2263	2143	2026	1912	1801	1693	1587	1484	1383	1285	1189
39	2384	2261	2141	2024	1910	1799	1691	1585	1482	1382	1283	1187
40	2382	2259	2139	2022	1908	1797	1689	1584	1481	1380	1282	1186
41	2380	2257	2137	2020	1906	1795	1687	1582	1479	1378	1280	1184
42	2378	2255	2135	2018	1904	1794	1686	1580	1477	1377	1278	1182
43	2376	2253	2133	2016	1903	1792	1684	1578	1476	1375	1277	1181
44	2374	2251	2131	2014	1901	1790	1682	1577	1474	1373	1275	1179
45	2372	2249	2129	2012	1899	1788	1680	1575	1472	1372	1274	1178
46	2370	2247	2127	2010	1897	1786	1678	1573	1470	1370	1272	1176
47	2368	2245	2125	2009	1895	1785	1677	1571	1469	1368	1270	1174
48	2366	2243	2123	2007	1893	1783	1675	1570	1467	1367	1269	1173
49	2364	2241	2121	2005	1891	1781	1673	1568	1465	1365	1267	1171
50	2362	2239	2119	2003	1889	1779	1671	1566	1464	1363	1266	1170
51	2359	2237	2117	2001	1888	1777	1670	1565	1462	1362	1264	1168
52	2357	2235	2115	1999	1886	1775	1668	1563	1460	1360	1262	1167
53	2355	2233	2113	1997	1884	1774	1666	1561	1459	1359	1261	1165
54	2353	2231	2111	1995	1882	1772	1664	1559	1457	1357	1259	1163
55	2351	2229	2109	1993	1880	1770	1663	1558	1455	1355	1257	1162
56	2349	2227	2107	1991	1878	1768	1661	1556	1454	1354	1256	1160
57	2347	2225	2105	1989	1876	1766	1659	1554	1452	1352	1254	1159
58	2345	2223	2103	1987	1875	1765	1657	1552	1450	1350	1253	1157
59	2343	2220	2101	1986	1873	1763	1655	1551	1449	1349	1251	1156
60	2341	2218	2099	1984	1871	1761	1654	1549	1447	1347	1249	1154

′	46	47	48	49	50	51	52	53	54	55	56	57	58	59
″	2760	2820	2880	2940	3000	3060	3120	3180	3240	3300	3360	3420	3480	3540
0	1154	1061	0969	0880	0792	0706	0621	0539	0458	0378	0300	0223	0147	0073
1	1152	1059	0968	0878	0790	0704	0620	0537	0456	0377	0298	0221	0146	0072
2	1151	1057	0966	0877	0789	0703	0619	0536	0455	0375	0297	0220	0145	0071
3	1149	1056	0965	0875	0787	0702	0617	0535	0454	0374	0296	0219	0143	0069
4	1148	1054	0963	0874	0786	0700	0616	0533	0452	0373	0294	0218	0142	0068
5	1146	1053	0962	0872	0785	0699	0615	0532	0451	0371	0293	0216	0141	0067
6	1145	1051	0960	0871	0783	0697	0613	0531	0450	0370	0292	0215	0140	0066
7	1143	1050	0959	0869	0782	0696	0612	0529	0448	0369	0291	0214	0139	0064
8	1141	1048	0957	0868	0780	0694	0610	0528	0447	0367	0289	0213	0137	0063
9	1140	1047	0956	0866	0779	0693	0609	0526	0446	0366	0288	0211	0136	0062
10	1138	1045	0954	0865	0777	0692	0608	0525	0444	0365	0287	0210	0135	0061
11	1137	1044	0953	0863	0776	0690	0606	0524	0443	0363	0285	0209	0134	0060
12	1135	1042	0951	0862	0774	0689	0605	0522	0442	0362	0284	0208	0132	0058
13	1134	1041	0950	0860	0773	0687	0603	0521	0440	0361	0283	0206	0131	0057
14	1132	1039	0948	0859	0772	0686	0602	0520	0439	0359	0282	0205	0130	0056
15	1130	1037	0947	0857	0770	0685	0601	0518	0438	0358	0280	0204	0129	0055
16	1129	1036	0945	0856	0769	0683	0599	0517	0436	0357	0279	0202	0127	0053
17	1127	1034	0944	0855	0767	0682	0598	0516	0435	0356	0278	0201	0126	0052
18	1126	1033	0942	0853	0766	0680	0596	0514	0434	0354	0276	0200	0125	0051
19	1124	1031	0941	0852	0764	0679	0595	0513	0432	0353	0275	0199	0124	0050
20	1123	1030	0939	0850	0763	0678	0594	0512	0431	0352	0274	0197	0122	0049
21	1121	1028	0938	0849	0762	0676	0592	0510	0430	0350	0273	0196	0121	0047
22	1119	1027	0936	0847	0760	0675	0591	0509	0428	0349	0271	0195	0120	0046
23	1118	1025	0935	0846	0759	0673	0590	0507	0427	0348	0270	0194	0119	0045
24	1116	1024	0933	0844	0757	0672	0588	0506	0426	0346	0269	0192	0117	0044
25	1115	1022	0932	0843	0756	0670	0587	0505	0424	0345	0267	0191	0116	0042
26	1113	1021	0930	0841	0754	0669	0585	0503	0423	0344	0266	0190	0115	0041
27	1112	1019	0929	0840	0753	0668	0584	0502	0422	0342	0265	0189	0114	0040
28	1110	1018	0927	0838	0751	0666	0583	0501	0420	0341	0264	0187	0112	0039
29	1109	1016	0926	0837	0750	0665	0581	0499	0419	0340	0262	0186	0111	0038
30	1107	1015	0924	0835	0749	0663	0580	0498	0418	0339	0261	0185	0110	0036
31	1105	1013	0923	0834	0747	0662	0579	0497	0416	0337	0260	0184	0109	0035
32	1104	1012	0921	0833	0746	0661	0577	0495	0415	0336	0258	0182	0107	0034
33	1102	1010	0920	0831	0744	0659	0576	0494	0414	0335	0257	0181	0106	0033
34	1101	1008	0918	0830	0743	0658	0574	0493	0412	0333	0256	0180	0105	0031
35	1099	1007	0917	0828	0741	0656	0573	0491	0411	0332	0255	0179	0104	0030
36	1098	1005	0915	0827	0740	0655	0572	0490	0410	0331	0253	0177	0103	0029
37	1096	1004	0914	0825	0739	0654	0570	0489	0408	0329	0252	0176	0101	0028
38	1095	1002	0912	0824	0737	0652	0569	0487	0407	0328	0251	0175	0100	0027
39	1093	1001	0911	0822	0736	0651	0568	0486	0406	0327	0250	0174	0099	0025
40	1091	0999	0909	0821	0734	0649	0566	0484	0404	0326	0248	0172	0098	0024
41	1090	0998	0908	0819	0733	0648	0565	0483	0403	0324	0247	0171	0096	0023
42	1088	0996	0906	0818	0731	0647	0563	0482	0402	0323	0246	0170	0095	0022
43	1087	0995	0905	0816	0730	0645	0562	0480	0400	0322	0244	0169	0094	0021
44	1085	0993	0903	0815	0729	0644	0561	0479	0399	0320	0243	0167	0093	0019
45	1084	0992	0902	0814	0727	0642	0559	0478	0398	0319	0242	0166	0091	0018
46	1082	0990	0900	0812	0726	0641	0558	0476	0396	0318	0241	0165	0090	0017
47	1081	0989	0899	0811	0724	0640	0557	0475	0395	0316	0239	0163	0089	0016
48	1079	0987	0897	0809	0723	0638	0555	0474	0394	0315	0238	0162	0088	0015
49	1078	0986	0896	0808	0721	0637	0554	0472	0392	0314	0237	0161	0087	0013
50	1076	0984	0894	0806	0720	0635	0552	0471	0391	0313	0235	0160	0085	0012
51	1074	0983	0893	0805	0719	0634	0551	0470	0390	0311	0234	0158	0084	0011
52	1073	0981	0891	0803	0717	0633	0550	0468	0388	0310	0233	0157	0083	0010
53	1071	0980	0890	0802	0716	0631	0548	0467	0387	0309	0232	0156	0082	0008
54	1070	0978	0888	0801	0714	0630	0547	0466	0386	0307	0230	0155	0080	0007
55	1068	0977	0887	0799	0713	0628	0546	0464	0384	0306	0229	0153	0079	0006
56	1067	0975	0885	0798	0711	0627	0544	0463	0383	0305	0228	0152	0078	0005
57	1065	0974	0884	0796	0710	0626	0543	0462	0382	0304	0227	0151	0077	0004
58	1064	0972	0883	0795	0709	0624	0541	0460	0381	0302	0225	0150	0075	0002
59	1062	0971	0881	0793	0707	0623	0540	0459	0379	0301	0224	0148	0074	0001
60	1061	0969	0880	0792	0706	0621	0539	0458	0378	0300	0223	0147	0073	0000

PLATE I.--EQUATORIAL TELESCOPE OF THE OBSERVATORY OF HARVARD COLLEGE.

PLATE II.—TOTAL ECLIPSE OF THE SUN, OF JULY 18, 1860, AS OBSERVED BY DR. FEILITZSCH, AT CASTELLON DE LA PLANA.

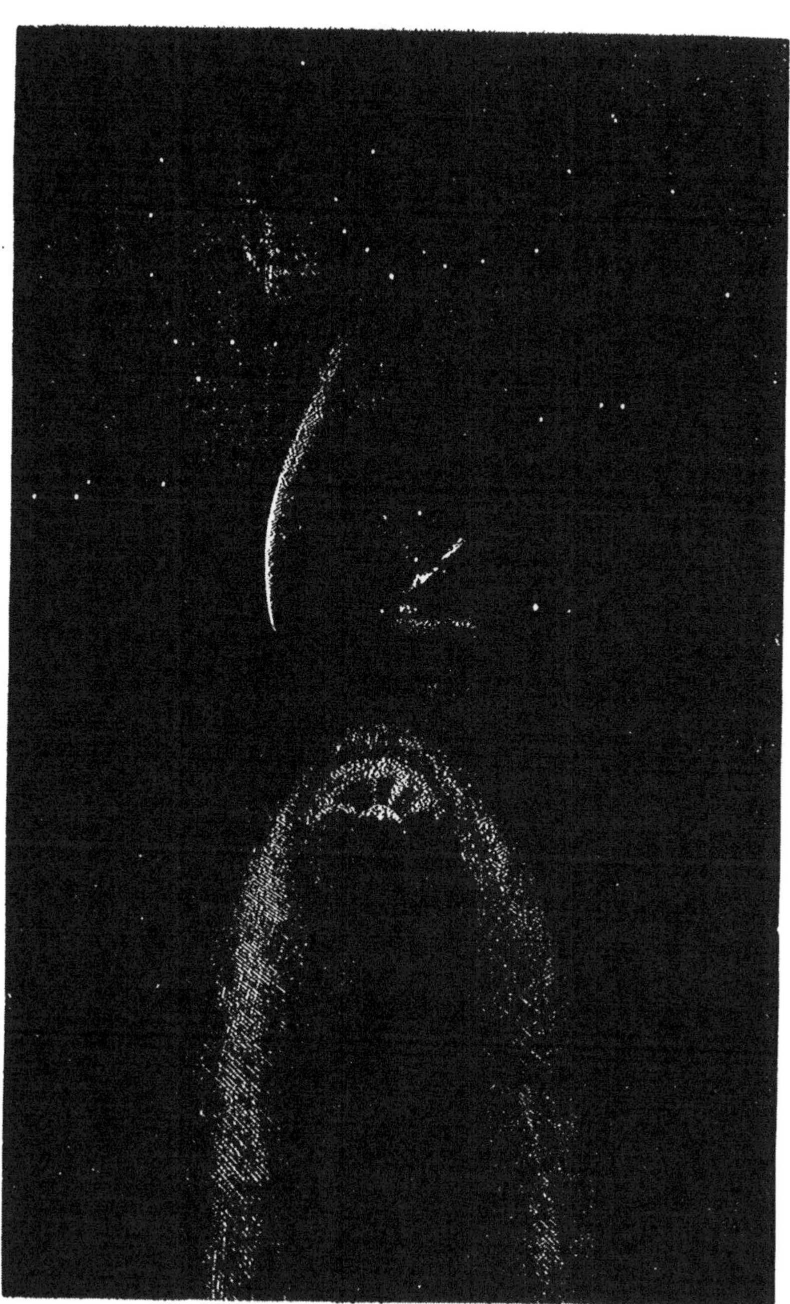

Plate III.—Donati's Comet.

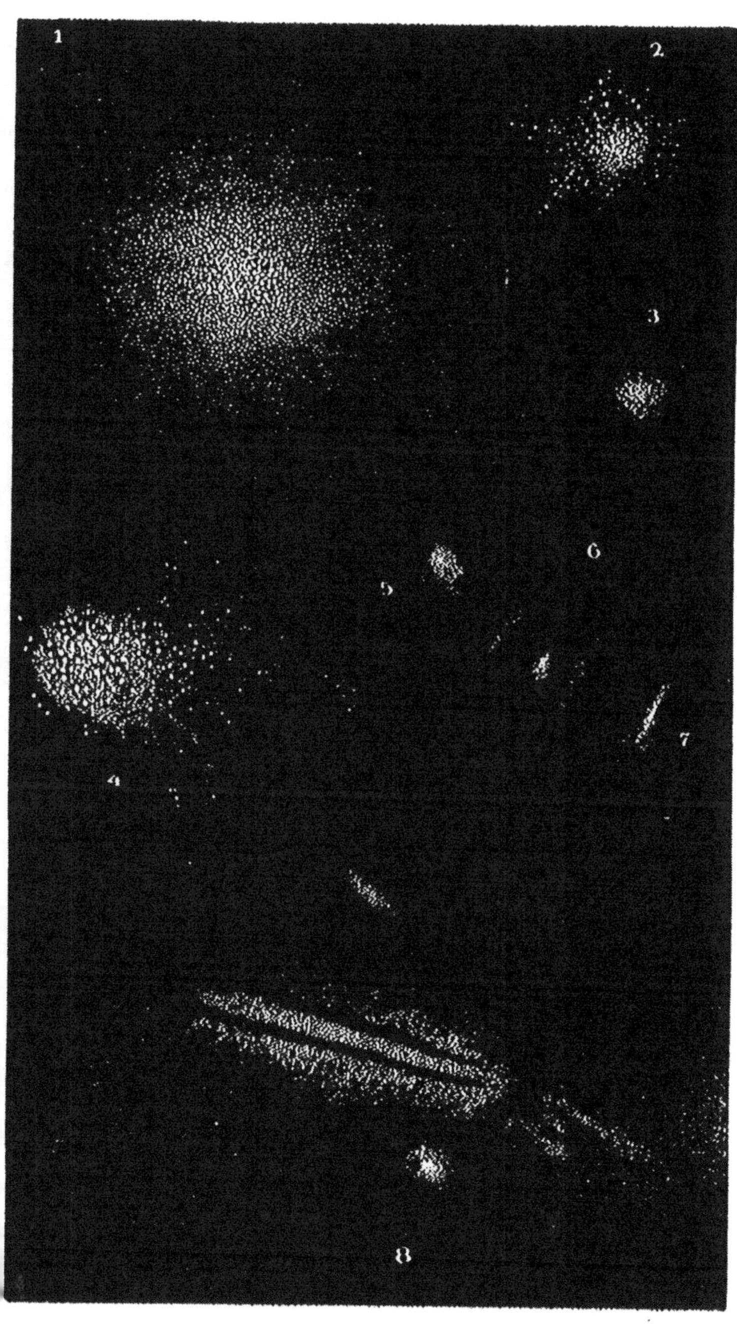

Plate IV.—Clusters and Nebulæ.

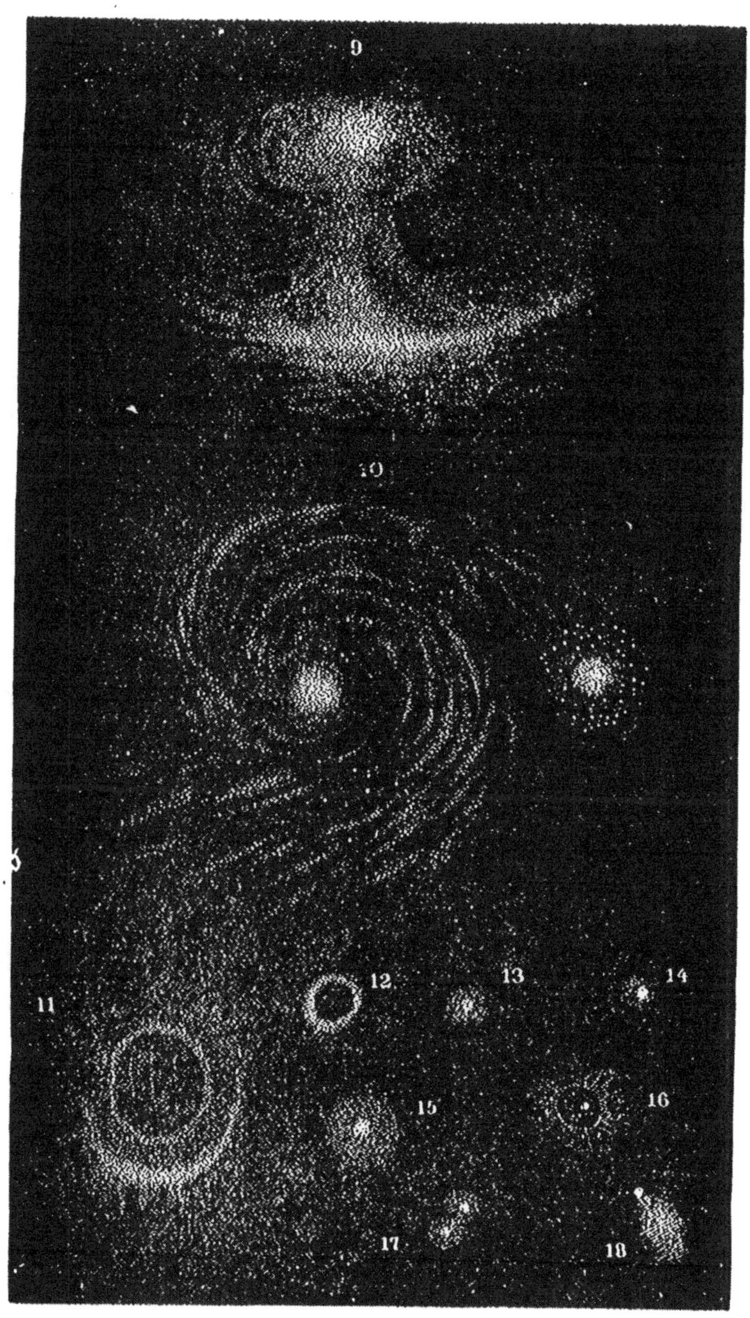

PLATE V.—NEBULÆ.

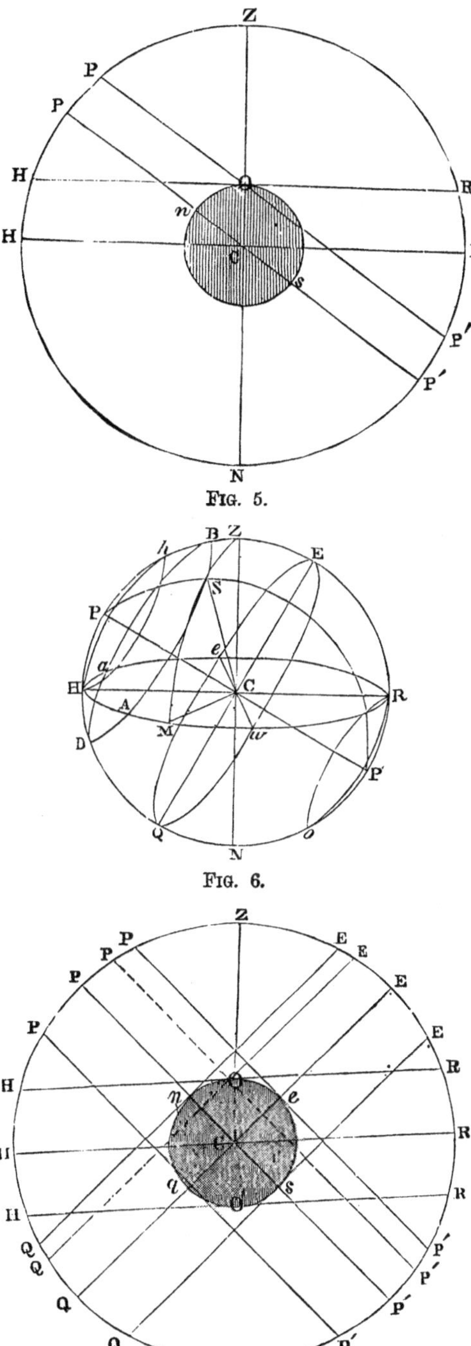

FIG. 5.

FIG. 6.

FIG. 7.

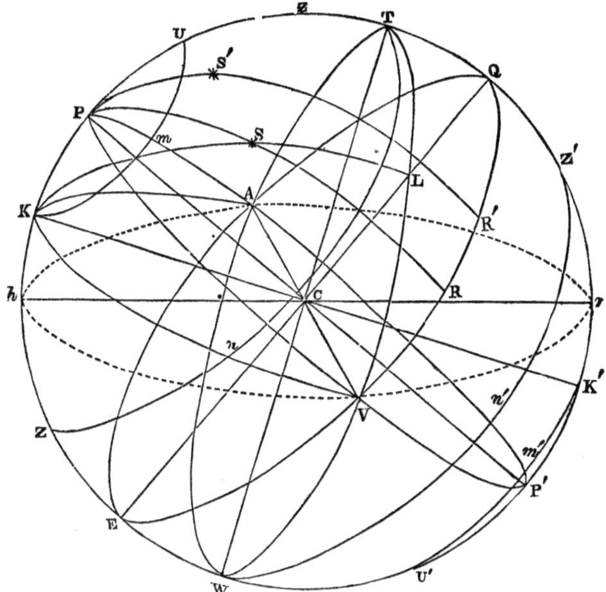

FIG. 8.

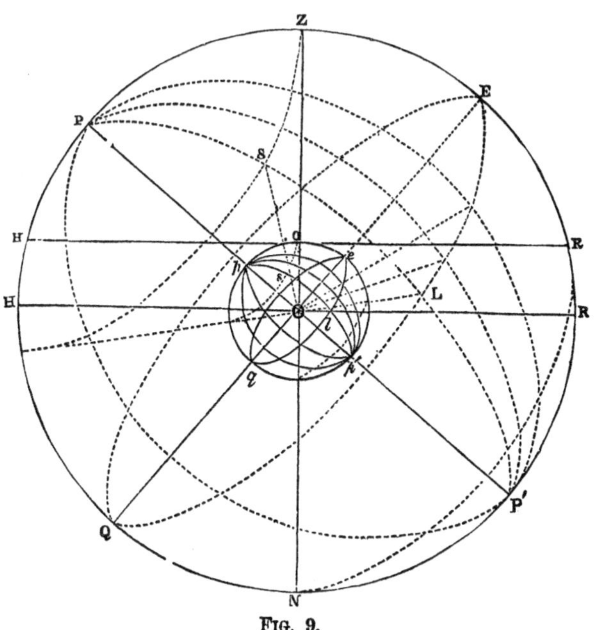

FIG. 9.

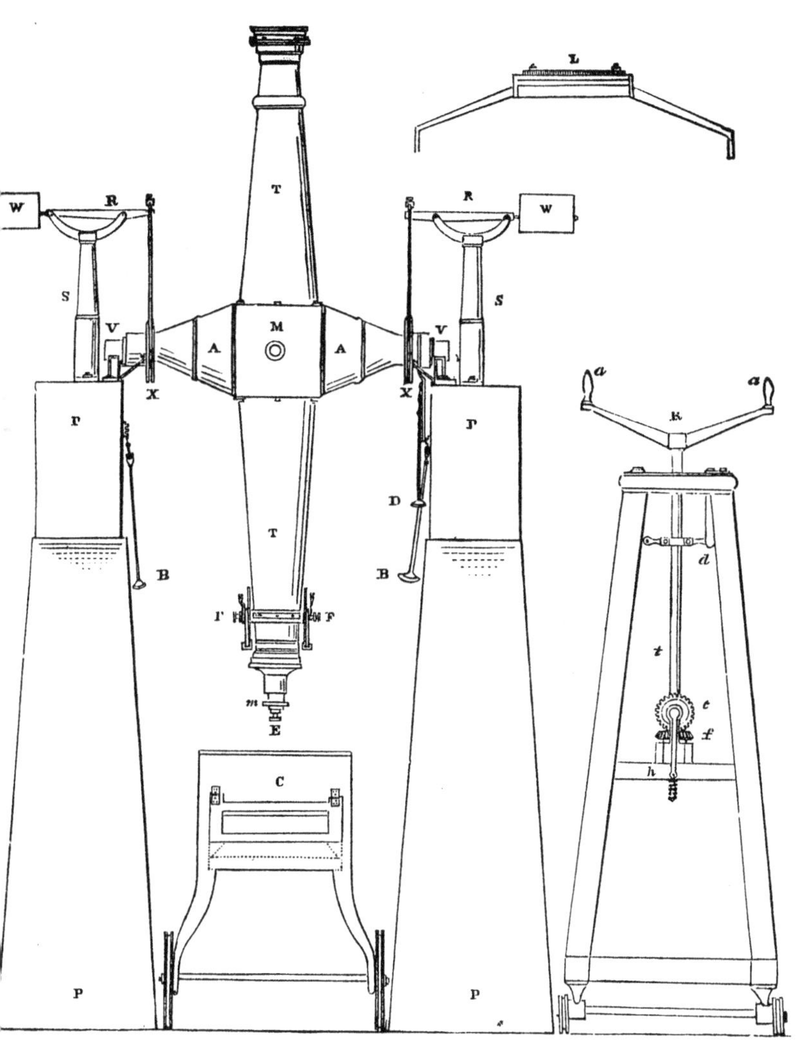

FIG. 19.

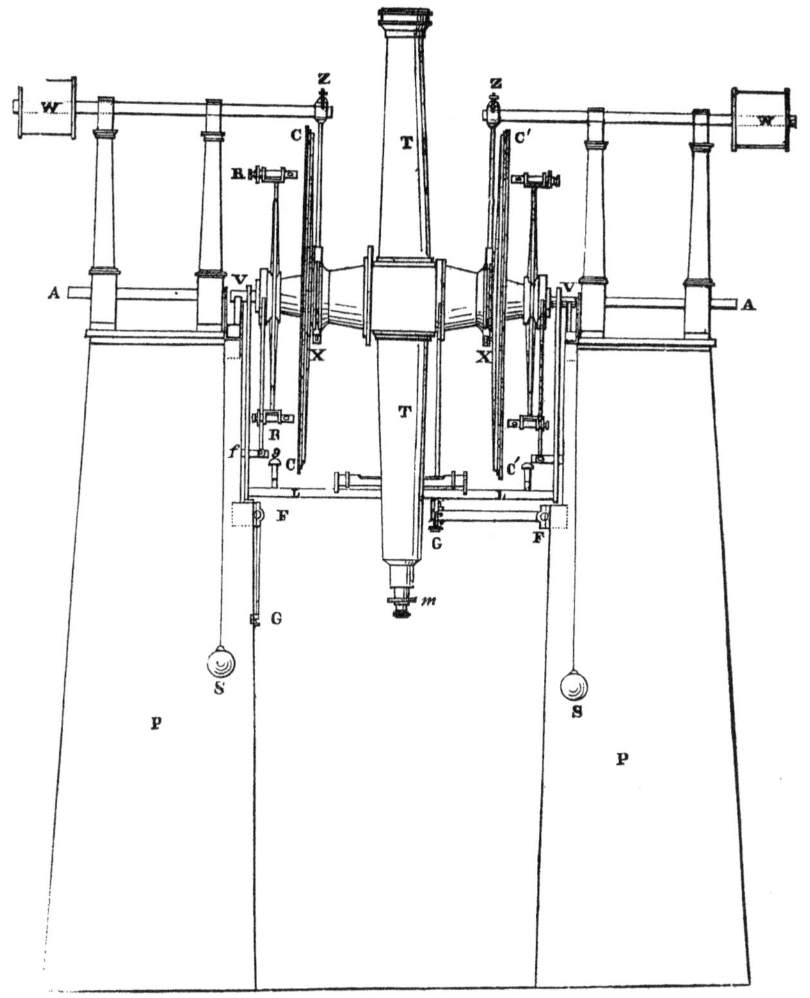

Fig. 23.

I
S
C
S
S
F
K
T
N
E
H
M
140
130
120
110
100
90
80
70
D
50
40
30
20
10
0
G
B

FIG. 28.

P
E''
N
C
n
B
L
A
V
X
D
n'
C'
E'
V
V
E

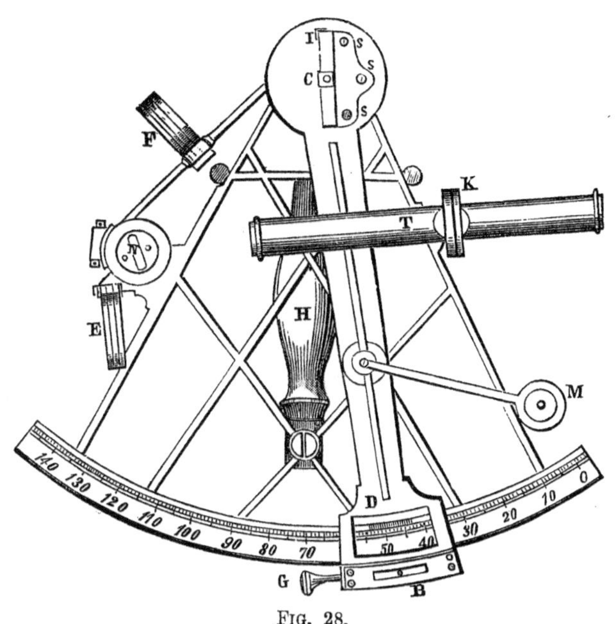

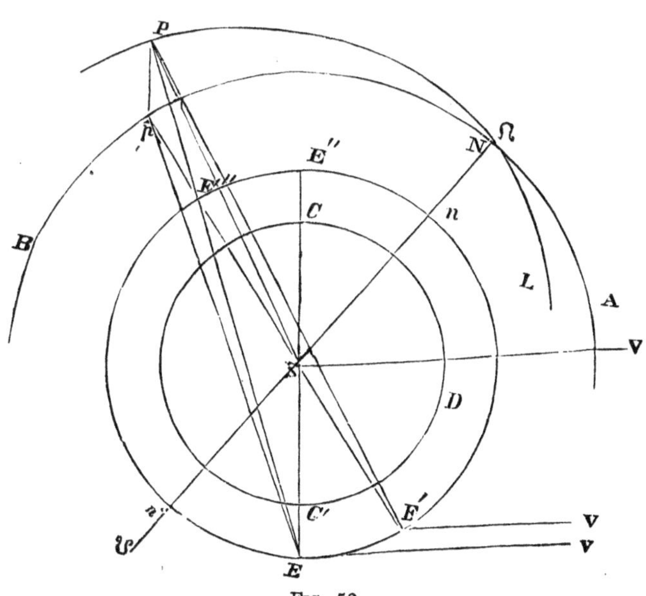

FIG. 52.

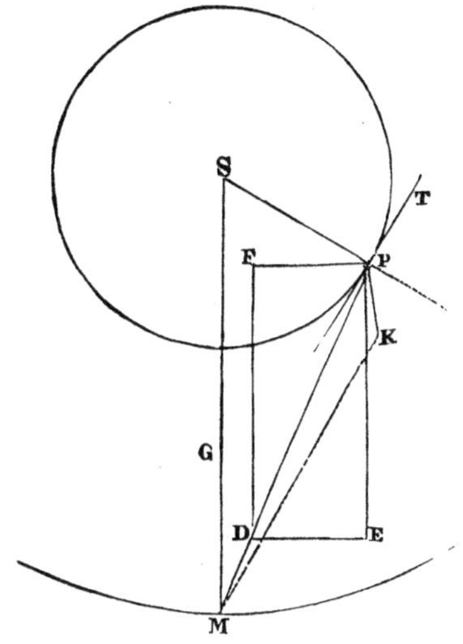

Fig. 61.

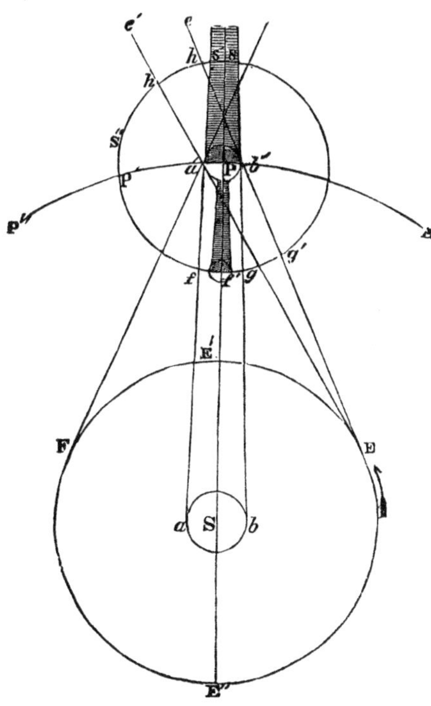

Fig. 65.

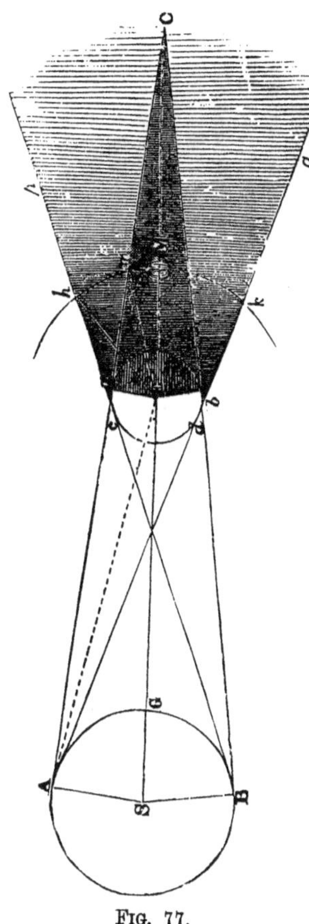

FIG. 77.

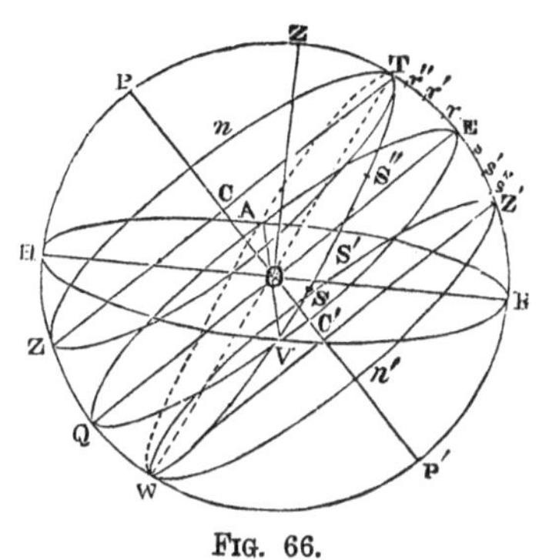

FIG. 66.

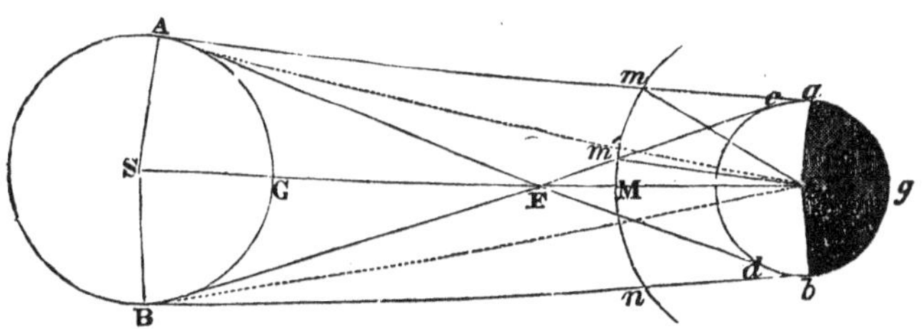

FIG. 82.

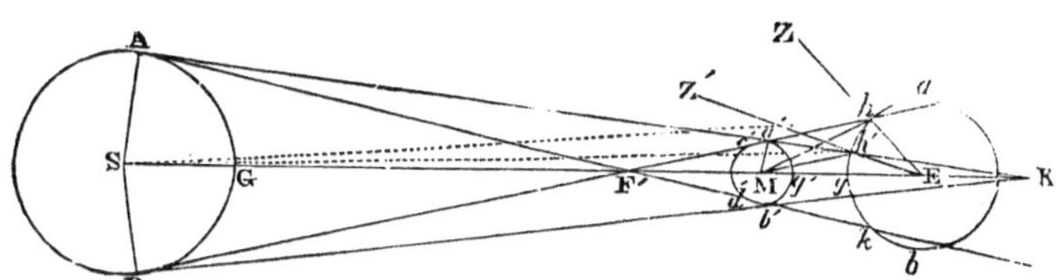

FIG. 84.

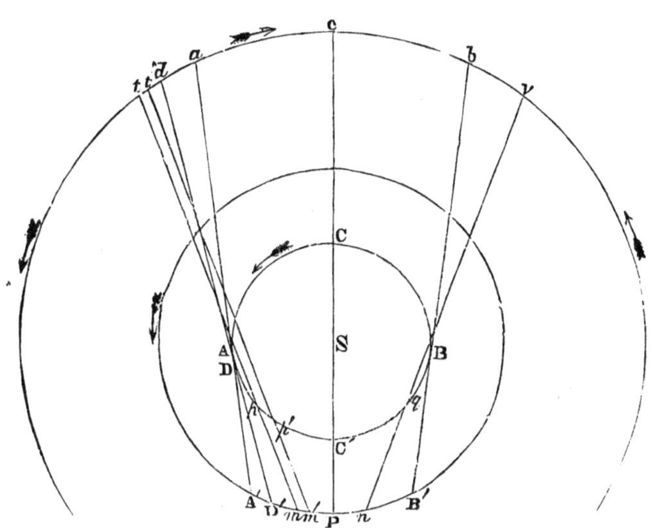

FIG. 86.

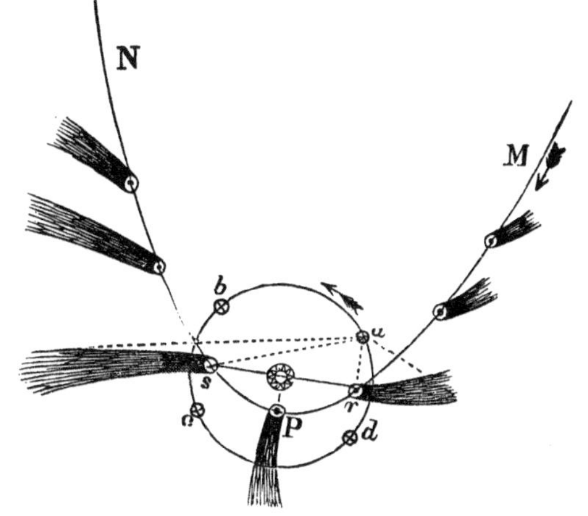

FIG. 98.

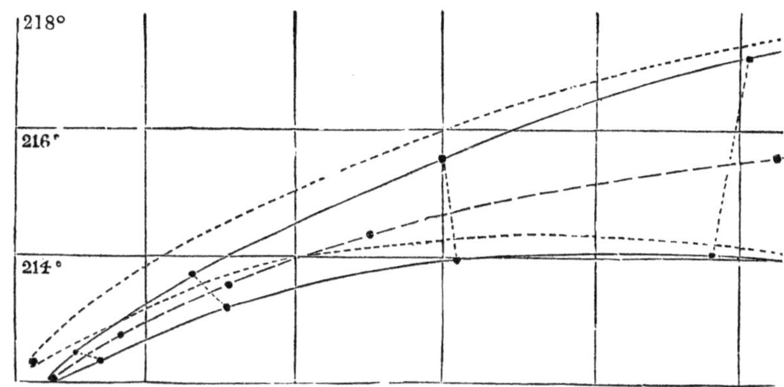

Fig. 101.

Fig. 110.

Fig. 112.

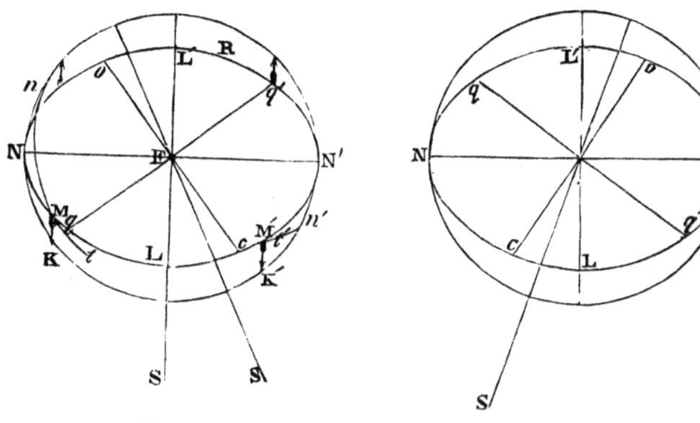

FIG. 116.

FIG. 117.

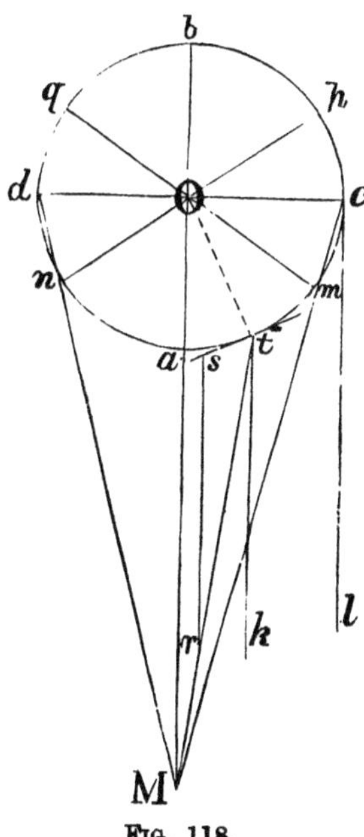

FIG. 118.

New York, August, 1873.

JOHN WILEY & SON'S

LIST OF PUBLICATIONS,

15 ASTOR PLACE,

Under the Mercantile Library and Trade Salerooms.

AGRICULTURE.

DOWNING. **FRUITS AND FRUIT-TREES OF AMERICA**; or the Culture, Propagation, and Management in the Garden and Orchard, of Fruit-trees generally, with descriptions of all the finest varieties of Fruit, Native and Foreign, cultivated in this country. By A. J. Downing. Second revision and correction, with large additions. By Chas. Downing. 1 vol. 8vo, over 1100 pages, with several hundred outline engravings. Price, with Supplement for 1872..........................$5 00

"As a work of reference it has no equal in this country, and deserves a place in the Library of every Pomologist in America."—*Marshall P. Wilder.*

" **ENCYCLOPEDIA OF FRUITS**; or, **Fruits and Fruit-Trees of America.** Part 1.—APPLES. With an Appendix containing many new varieties, and brought down to 1872. By Chas. Downing. With numerous outline engravings. 8vo, full cloth..$2 50

" **ENCYCLOPEDIA OF FRUITS**; or, **Fruits and Fruit-Trees of America.** Part 2.—CHERRIES, GRAPES, PEACHES, PEARS, &c. With an Appendix containing many new varieties, and brought down to 1872. By Chas. Downing. With numerous outline engravings. 8vo, full cloth.........$2 50

" **FRUITS AND FRUIT-TREES OF AMERICA.** By A. J. Downing. First revised edition. By Chas. Downing. 12mo, cloth...$2 00

" **SELECTED FRUITS.** From Downing's Fruits and Fruit-Trees of America. With some new varieties, including their Culture, Propagation, and Management in the Garden and Orchard, with a Guide to the selection of Fruits, with reference to the Time of Ripening. By Chas. Downing. Illustrated with upwards of four hundred outlines of Apples, Cherries, Grapes, Plums, Pears, &c. 1 vol., 12mo....$2 50

" **LOUDON'S GARDENING FOR LADIES, AND COMPANION TO THE FLOWER-GARDEN.** Second American from third London edition. Edited by A. J. Downing. 1 vol., 12mo..........................$2 00

DOWNING & LINDLEY. **THE THEORY OF HORTICULTURE.** By J. Lindley. With additions by A. J. Downing. 12mo, cloth.......$2 00

DOWNING. **COTTAGE RESIDENCES.** A Series of Designs for Rural Cottages and Cottage Villas, with Garden Grounds. By A. J. Downing. Containing a revised List of Trees, Shrubs, and Plants, and the most recent and best selected Fruit, with some account of the newer style of Gardens. By Henry Winthrop Sargent and Charles Downing. With many new designs in Rural Architecture. By George E Harney, Architect. 1 vol. 4to................................$6 00

DOWNING & WIGHTWICK. **HINTS TO PERSONS ABOUT BUILDING IN THE COUNTRY.** By A. J. Downing. And **HINTS TO YOUNG ARCHITECTS,** calculated to facilitate their practical operations. By George Wightwick, Architect. Wood engravings. 8vo, cloth........................$2 00

KEMP. **LANDSCAPE GARDENING; or, How to Lay Out a Garden.** Intended as a general guide in choosing, forming, or improving an estate (from a quarter of an acre to a hundred acres in extent), with reference to both design and execution. With numerous fine wood engravings. By Edward Kemp. 1 vol. 12mo, cloth........................$2 50

LIEBIG **CHEMISTRY IN ITS APPLICATION TO AGRICULTURE, &c.** By Justus Von Liebig. 12mo, cloth....$1 00

" **LETTERS ON MODERN AGRICULTURE.** By Baron Von Liebig. Edited by John Blyth, M.D. With addenda by a practical Agriculturist, embracing valuable suggestions, adapted to the wants of American Farmers. 1 vol. 12mo, cloth..$1 00

" **PRINCIPLES OF AGRICULTURAL CHEMISTRY,** with special reference to the late researches made in England. By Justus Von Liebig. 1 vol. 12mo ....75 cents.

PARSONS. **HISTORY AND CULTURE OF THE ROSE.** By S. B. Parsons. 1 vol. 12mo $1 25

ARCHITECTURE.

DOWNING. **COTTAGE RESIDENCES;** or, a Series of Designs for Rural Cottages and Cottage Villas and their Gardens and Grounds, adapted to North America. By A. J. Downing. Containing a revised List of Trees, Shrubs, Plants, and the most recent and best selected Fruits. With some account of the newer style of Gardens, by Henry Wentworth Sargent and Charles Downing. With many new designs in Rural Architecture by George E. Harney, Architect........................$6 00

DOWNING & WIGHTWICK. **HINTS TO PERSONS ABOUT BUILDING IN THE COUNTRY.** By A. J. Downing. And **HINTS TO YOUNG ARCHITECTS,** calculated to facilitate their practical operations. By George Wightwick, Architect. With many wood-cuts. 8vo, cloth..................$2 00

HATFIELD. **THE AMERICAN HOUSE CARPENTER.** A Treatise upon Architecture, Cornices, and Mouldings, Framing, Doors, Windows, and Stairs; together with the most important principles of Practical Geometry. New, thoroughly revised, and improved edition, with about 150 additional pages, and numerous additional plates. By R. G. Hatfield. 1 vol. 8vo..$3 50

NOTICES OF THE WORK.

"The clearest and most thoroughly practical work on the subject."

"This work is a most excellent one, very comprehensive, and lucidly arranged."

"This work commends itself by its practical excellence."

"It is a valuable addition to the library of the architect, and almost indispensable to every scientific master-mechanic."—*R. R. Journal.*

HOLLY **CARPENTERS' AND JOINERS' HAND-BOOK,** containing a Treatise on Framing, Roofs, etc., and useful Rules and Tables. By H. W. Holly. 1 vol. 18mo, cloth........$0 75

" **THE ART OF SAW-FILING SCIENTIFICALLY TREATED AND EXPLAINED.** With Directions for putting in order all kinds of Saws. By H. W. Holly. 18mo, cloth..$0 75

RUSKIN **SEVEN LAMPS OF ARCHITECTURE.** 1 vol. 12mo, cloth, plates......................................$1 75

RUSKIN. **LECTURES ON ARCHITECTURE AND PAINTING.** 1 vol. 12mo, cloth, plates..........................$1 50

" **LECTURE BEFORE SOCIETY OF ARCHITECTS.** 0 15

WOOD. **A TREATISE ON THE RESISTANCE OF MATERIALS,** and an Appendix on the Preservation of Timber. By De Volson Wood, Prof. of Engineering, University of Michigan. 1 vol. 8vo, cloth........................$2 50

This work is used as a Text-Book in Iowa University, Iowa Agricultural College, Illinois Industrial University, Sheffield Scientific School, New Haven, Cooper Institute, New York, Polytechnic College, Brooklyn, University of Michigan, and other institutions.

" **A TREATISE ON BRIDGES.** Designed as a Text-book and for Practical Use. By De Volson Wood. 1 vol. 8vo, numerous illustrations, ...$3 00

ASSAYING—ASTRONOMY.

BODEMANN. **A TREATISE ON THE ASSAYING OF LEAD, SILVER, COPPER, GOLD, AND MERCURY.** By Bodemann and Kerl. Translated by W. A. Goodyear. 1 vol. 12mo, cloth $2 50

MITCHELL. **A MANUAL OF PRACTICAL ASSAYING.** By John Mitchell. Third edition, edited by William Crookes. 1 vol. thick 8vo, cloth.................................$10 00

NORTON. **A TREATISE ON ASTRONOMY, SPHERICAL AND PHYSICAL,** with Astronomical Problems and Solar, Lunar, and other Astronomical Tables for the use of Colleges and Scientific Schools. By William A. Norton. Fourth edition, revised, remodelled, and enlarged. Numerous plates. 8vo, cloth $3 50

BIBLES, &c.

BAGSTER. **THE COMMENTARY WHOLLY BIBLICAL.** Contents: —The Commentary: an Exposition of the Old and New Testaments in the very words of Scripture. 2264 pp. II. An outline of the Geography and History of the Nations mentioned in Scripture. III. Tables of Measures, Weights, and Coins. IV. An Itinerary of the Children of Israel from Egypt to the Promised Land. V. A Chronological comparative Table of the Kings and Prophets of Israel and Judah. VI. A Chart of the World's History from Adam to the Third Century, A. D. VII. A complete Series of Illustrative Maps. IX. A Chronological Arrangement of the Old and New Testaments. X. An Index to Doctrines and Subjects, with numerous Selected Passages, quoted in full. XI. An Index to the Names of Persons mentioned in Scripture. XII. An Index to the Names of Places found in Scripture. XIII. The Names, Titles, and Characters of Jesus Christ our Lord, as revealed in the Scriptures, methodically arranged.

2 volumes 4to, cloth.............................$19 50
2 volumes 4to, half morocco, gilt edges............. 26 00
2 volumes 4to, morocco, gilt edges.................. 35 00
3 volumes 4to, cloth................................ 20 00
3 volumes 4to, half morocco, gilt edges............. 33 00
3 volumes 4to, morocco, gilt edges.................. 40 00

BLANK-PAGED BIBLE. **THE HOLY SCRIPTURES OF THE OLD AND NEW TESTAMENTS;** with copious references to parallel and illustrative passages, and the alternate pages ruled for MS. notes.

This edition of the Scriptures contains the Authorized Version, illustrated by the references of "Bagster's Polyglot Bible," and enriched with accurate maps, useful tables, and an Index of Subjects.

1 vol. 8vo, half morocco..........................$9 00
1 vol. 8vo, morocco extra..........................10 50
1 vol. 8vo, full morocco...........................12 00

THE TREASURY BIBLE. Containing the authorized English version of the Holy Scriptures, interleaved with a Treasury of more than 500,000 Parallel Passages from Canne, Brown, Blayney, Scott, and others. With numerous illustrative notes.

1 vol., half bound....................$7 50
1 vol., morocco....................10 00

COMMON PRAYER, 48mo Size.

(Done in London expressly for us.)

COMMON PRAYER.
No. 1. Gilt and red edges, imitation morocco..........$0 62½
No. 2. Gilt and red edges, rims.................... 87½
No. 3. Gilt and red edges, best morocco and calf....... 1 25
No. 4. Gilt and red edges, best morocco and calf, rims.. 1 50

BOOK-KEEPING.

JONES. **BOOKKEEPING AND ACCOUNTANTSHIP.** Elementary and Practical. In two parts, with a Key for Teachers. By Thomas Jones, Accountant and Teacher. 1 volume 8vo, cloth....................$2 50

" **BOOKKEEPING AND ACCOUNTANTSHIP.** School Edition. By Thomas Jones. 1 vol. 8vo, half roan.......$1 50

" **BOOKKEEPING AND ACCOUNTANTSHIP.** Set of Blanks. In 6 parts. By Thomas Jones..............$1 50

" **BOOKKEEPING AND ACCOUNTANTSHIP.** Double Entry; Results obtained from Single Entry; Equation of Payments, etc. By Thomas Jones. 1 vol. thin 8vo...$0 75

CHEMISTRY.

CRAFTS. **A SHORT COURSE IN QUALITATIVE ANALYSIS;** with the new notation. By Prof. J. M. Crafts. Second edition. 1 vol. 12mo, cloth....................$1 50

JOHNSON'S FRESENIUS. **A MANUAL OF QUALITATIVE CHEMICAL ANALYSIS.** By C. R. Fresenius. Edited by S. W. Johnson, Professor in Sheffield Scientific School, Yale College. With Chemical Notation and Nomenclature, old and new. 1 vol. 8vo, cloth....................$4 50

" **A SYSTEM OF INSTRUCTION IN QUANTITATIVE CHEMICAL ANALYSIS.** By C. R. Fresenius. From latest editions, edited, with additions, by Prof. S. W. Johnson. With Chemical Notation and Nomenclature, old and new....................$6 00

KIRKWOOD **COLLECTION OF REPORTS (CONDENSED) AND OPINIONS OF CHEMISTS IN REGARD TO THE USE OF LEAD PIPE FOR SERVICE PIPE,** in the Distribution of Water for the Supply of Cities. By Jas. P. Kirkwood. 8vo, cloth....................$1 50

MILLER. **ELEMENTS OF CHEMISTRY, THEORETICAL AND PRACTICAL.** By Wm. Allen Miller. 3 vols. 8vo..$18 00

" Part I.—**CHEMICAL PHYSICS.** 1 vol. 8vo..........$4 00

" Part II.—**INORGANIC CHEMISTRY.** 1 vol. 8vo..... 6 00

" Part III.—**ORGANIC CHEMISTRY.** 1 vol. 8vo...... 10 00

"Dr. Miller's Chemistry is a work of which the author has every reason to feel proud. It is now by far the largest and most accurately written Treatise on Chemistry in the English language," etc.—*Dublin Med. Journal.*

" **MAGNETISM AND ELECTRICITY.** By Wm. Allen Miller. 1 vol. 8vo....................$2 50

MUSPRATT. **CHEMISTRY—THEORETICAL, PRACTICAL, AND ANALYTICAL**—as applied and relating to the Arts and Manufactures. By Dr. Sheridan Muspratt. 2 vols. 8vo, cloth, $18.00; half russia........................$24 00

NOAD. **A MANUAL OF QUALITATIVE AND QUANTITATIVE CHEMICAL ANALYSIS.** For the use of Students. By H. M. Noad, author of "Manual of Electricity." 1 vol. 12mo. (London.) Complete........................$6 00

" **QUANTITATIVE ANALYSIS.** 1 vol. cloth.......... 4 00

PERKINS. **AN ELEMENTARY MANUAL OF QUALITATIVE CHEMICAL ANALYSIS.** By Maurice Perkins. 12mo, cloth..$1 00

DRAWING AND PAINTING.

BOUVIER AND OTHERS. **HANDBOOK ON OIL PAINTING.** Handbook of Young Artists and Amateurs in Oil Painting; being chiefly a condensed compilation from the celebrated Manual of Bouvier, with additional matter selected from the labors of Merriwell, De Montalbert, and other distinguished Continental writers on the art. In 7 parts. Adapted for a Text-Book in Academies of both sexes, as well as for self-instruction. Appended, a new Explanatory and Critical Vocabulary. By an American Artist. 12mo, cloth..................$2 00

COE. **PROGRESSIVE DRAWING BOOK.** By Benj. H. Coe. One vol., cloth..................................$3 50

" **DRAWING FOR LITTLE FOLKS**; or, First Lessons for the Nursery. 30 drawings. Neat cover............$0 20

" **FIRST STUDIES IN DRAWING.** Containing Elementary Exercises, Drawings from Objects, Animals, and Rustic Figures. Complete in *three numbers* of 18 studies each, in neat covers. Each.............................$0.20

" **COTTAGES.** An Introduction to Landscape Drawing. *Containing 72 Studies.* Complete in four numbers of 18 studies each, in neat covers. Each........................$0.20

" **EASY LESSONS IN LANDSCAPE.** Complete in four numbers of 10 Studies each. In neat 8vo cover. Each, $0 20

" **HEADS, ANIMALS, AND FIGURES.** Adapted to Pencil Drawing. Complete in three numbers of 10 Studies each. In neat 8vo covers. Each............................$0 20

" **COPY BOOK, WITH INSTRUCTIONS**.............$0 37½

RUSKIN. **THE ELEMENTS OF DRAWING.** In Three Letters to Beginners. By John Ruskin. 1 vol. 12mo...........$1 00

" **THE ELEMENTS OF PERSPECTIVE.** Arranged for the use of Schools. By John Ruskin...................$1 00

SMITH. **A MANUAL OF TOPOGRAPHICAL DRAWING.** By Prof. R. S. Smith. Second edition. 1 vol. 8vo, cloth, plates...$2 00

" **MANUAL OF LINEAR PERSPECTIVE.** Form, Shade, Shadow, and Reflection. By Prof. R. S. Smith. 1 vol. 8vo, plates, cloth......................................$2 00

WARREN. **CONSTRUCTIVE GEOMETRY AND INDUSTRIAL DRAWING.** By S. Edward Warren, Professor in the Massachusetts Institute of Technology, Boston:—

I. ELEMENTARY WORKS.

1. ELEMENTARY FREE-HAND GEOMETRICAL DRAWING. A series of progressive exercises on regular lines and forms, including systematic instruction in lettering; a training of the eye and hand for all who are learning to draw. 12mo, cloth, many cuts...................................... 75 cts.
Vols. 1 and 3, bound in 1 vol............................$1 75

ELEMENTARY WORKS.—Continued.

WARREN

2. PLANE PROBLEMS IN ELEMENTARY GEOMETRY. With numerous wood-cuts. 12mo, cloth..................$1 25

3. DRAFTING INSTRUMENTS AND OPERATIONS. Containing full information about all the instruments and materials used by the draftsmen, with full directions for their use. With plates and wood-cuts. One vol. 12mo, cloth, $1 25

4. ELEMENTARY PROJECTION DRAWING. Revised and enlarged edition. In five divisions. This and the last volume are favorite text-books, especially valuable to all Mechanical Artisans, and are particularly recommended for the use of all higher public and private schools. New revised and enlarged edition, with numerous wood-cuts and plates. (1872.) 12mo, cloth.......................................$1 50

5. ELEMENTARY LINEAR PERSPECTIVE OF FORMS AND SHADOWS. Part I.—Primitive Methods, with an Introduction. Part II.—Derivative Methods, with Notes on Aerial Perspective, and many Practical Examples. Numerous wood-cuts. 1 vol. 12mo, cloth..........................$1 00

II. HIGHER WORKS.

These are designed principally for Schools of Engineering and Architecture, and for the members generally of those professions; and the first three are also designed for use in those colleges which provide courses of study adapted to the preliminary general training of candidates for the scientific professions, as well as for those technical schools which undertake that training themselves.

1. GENERAL PROBLEMS OF ORTHOGRAPHIC PROJECTIONS. The foundation course for the subsequent theoretical and practical works. A new edition of this work will soon appear.

2. GENERAL PROBLEMS OF SHADES AND SHADOWS. A wider range of problems than can elsewhere be found in English, and the principles of shading. 1 vol. 8vo, with numerous plates. Cloth.......................... $3 50

3. HIGHER LINEAR PERSPECTIVE. Distinguished by its concise summary of various methods of perspective construction; a full set of standard problems, and a careful discussion of special higher ones. With numerous large plates. 8vo, cloth....................................... $4 00

4. ELEMENTS OF MACHINE CONSTRUCTION AND DRAWING; or, Machine Drawings. With some elements of descriptive and rational cinematics. A Text-Book for Schools of Civil and Mechanical Engineering, and for the use of Mechanical Establishments, Artisans, and Inventors. Containing the principles of gearings, screw propellers, valve motions, and governors, and many standard and novel examples, mostly from present American practice. By S. Edward Warren. 2 vols. 8vo. 1 vol. text and cuts, and 1 vol. large plates...... $7 50

A FEW FROM MANY TESTIMONIALS.

"It seems to me that your Works only need a thorough examination to be introduced and permanently used in all the Scientific and Engineering Schools." —Prof. J. G. FOX, *Collegiate and Engineering Institute, New York City.*

"I have used several of your Elementary Works, and believe them to be better adapted to the purposes of instruction than any others with which I am acquainted."—H. F. WALLING, *Prof. of Civil and Topographical Engineering, Lafayette College, Easton, Pa.*

"Your Works appear to me to fill a very important gap in the literature of the subjects treated. Any effort to draw Artisans, etc., away from the 'rule of thumb,' and give them an insight into principles, is in the right direction, and meets my heartiest approval. This is the distinguishing feature of your Elementary Works."—Prof. H. L. EUSTIS, *Lawrence Scientific School, Cambridge, Mass.*

"The author has happily divided the subjects into two great portions: the former embracing those processes and problems proper to be taught to all students in Institutions of Elementary Instruction; the latter, those suited to advanced students preparing for technical purposes. The Elementary Books ought to be used in all High Schools and Academies; the Higher ones in Schools of Technology."—WM. W. FOLWELL, *President of University of Minnesota.*

DYEING, &c.

MACFARLANE. **A PRACTICAL TREATISE ON DYEING AND CALICO-PRINTING.** Including the latest Inventions and Improvements. With an Appendix, comprising definitions of chemical terms, with tables of Weights, Measures, &c. By an experienced Dyer. With a supplement, containing the most recent discoveries in color chemistry. By Robert Macfarlane. 1 vol. 8vo.......................................$5 00

REIMANN. **A TREATISE ON THE MANUFACTURE OF ANILINE AND ANILINE COLORS.** By M. Reimann. To which is added the Report on the Coloring Matters derived from Coal Tar, as shown at the French Exhibition, 1867. By Dr. Hofmann. Edited by Wm. Crookes. 1 vol. 8vo, cloth, $2 50

"Dr. Reimann's portion of the Treatise, written in concise language, is profoundly practical, giving the minutest details of the processes for obtaining all the more important colors, with woodcuts of apparatus. Taken in conjunction with Hofmann's Report, we have now a complete history of Coal Tar Dyes, both theoretical and practical."—*Chemist and Druggist.*

ENGINEERING.

AUSTIN. **A PRACTICAL TREATISE ON THE PREPARATION, COMBINATION, AND APPLICATION OF CALCAREOUS AND HYDRAULIC LIMES AND CEMENTS.** To which is added many useful recipes for various scientific, mercantile, and domestic purposes. By James G. Austin. 1 vol. 12mo.......................................$2 00

COLBURN **LOCOMOTIVE ENGINEERING AND THE MECHANISM OF RAILWAYS.** A Treatise on the Principles and Construction of the Locomotive Engine, Railway Carriages, and Railway Plant, with examples. Illustrated by Sixty-four large engravings and two hundred and forty woodcuts. By Zerah Colburn. Complete, 20 parts, $15.00; or 2 vols. cloth.......................................$16 00
Or, half morocco, gilt top.......................................$20 00

KNIGHT. **THE MECHANICIAN AND CONSTRUCTOR FOR ENGINEERS.** Comprising Forging, Planing, Lining, Slotting, Shaping, Turning, Screw-cutting, &c. Illustrated with ninety-six plates. By Cameron Knight. 1 vol. 4to, half morocco.......................................$15 00

MAHAN. **AN ELEMENTARY COURSE OF CIVIL ENGINEERING,** for the use of the Cadets of the U. S. Military Academy. By D. H. Mahan. 1 vol. 8vo, with numerous woodcuts. New edition. Edited by Prof. De Volson Wood. Full cloth.......................................$5 00

" **DESCRIPTIVE GEOMETRY,** as applied to the Drawing of Fortifications and Stone-Cutting. For the use of the Cadets of the U. S. Military Academy. By Prof. D. H. Mahan. 1 vol. 8vo. Plates.......................................$1 50

" **INDUSTRIAL DRAWING.** Comprising the Description and Uses of Drawing Instruments, the Construction of Plane Figures, the Projections and Sections of Geometrical Solids, Architectural Elements, Mechanism, and Topographical Drawing. With remarks on the method of Teaching the subject. For the use of Academies and Common Schools. By Prof. D. H. Mahan. 1 vol. 8vo. Twenty steel plates. Full cloth.......................................$3 00

" **A TREATISE ON FIELD FORTIFICATIONS.** Containing instructions on the Methods of Laying Out, Constructing, Defending, and Attacking Entrenchments. With the General Outlines, also, of the Arrangement, the Attack, and Defence of Permanent Fortifications. By Prof. D. H. Mahan. New edition, revised and enlarged. 1 vol. 8vo, full cloth, with plates.......................................$3 50

" **ELEMENTS OF PERMANENT FORTIFICATIONS.** By Prof. D. H. Mahan. 1 vol. 8vo, with numerous large plates. Cloth.......................................$6 50

MAHAN. **ADVANCED GUARD, OUT-POST,** and Detachment Service of Troops, with the Essential Principles of Strategy and Grand Tactics. For the use of Officers of the Militia and Volunteers. By Prof. D. H. Mahan. New edition, with large additions and 12 plates. 1 vol. 18mo, cloth......$1 50

MAHAN & MOSELY. **MECHANICAL PRINCIPLES OF ENGINEERING AND ARCHITECTURE.** By Henry Mosely, M.A., F.R.S. From last London edition, with considerable additions, by Prof. D. H. Mahan, LL.D., of the U. S. Military Academy. 1 vol. 8vo, 700 pages. With numerous cuts. Cloth...$5 00

MAHAN & BRESSE. **HYDRAULIC MOTORS.** Translated from the French Cours de Mecanique, appliquée par M. Bresse. By Lieut. F. A. Mahan, and revised by Prof. D. H. Mahan. 1 vol. 8vo, plates..$2 50

WOOD. **A TREATISE ON THE RESISTANCE OF MATERIALS,** and an Appendix on the Preservation of Timber. By De Volson Wood, Professor of Engineering, University of Michigan. 1 vol. 8vo, cloth..........................$2 50

A TREATISE ON BRIDGES. Designed as a Text-book and for Practical Use. By De Volson Wood. 1 vol. 8vo, numerous illustrations, cloth $3 00

GREEK.

BAGSTER. **GREEK TESTAMENT, ETC.** The Critical Greek and English New Testament in Parallel Columns, consisting of the Greek Text of Scholz, readings of Griesbach, etc., etc. 1 vol. 18mo, half morocco...........................$3 00

" ——— do. Full morocco, gilt edges................. 4 50

" ——— With Lexicon, by T S. Green. Half-bound........ 4 50

" ——— do. Full morocco, gilt edges.................. 6 00

" ——— do. With Concordance and Lexicon. Half mor., 6 00

" ——— do. Limp morocco.......................... 7 50

" **THE ANALYTICAL GREEK LEXICON TO THE NEW TESTAMENT.** In which, by an alphabetical arrangement, is found every word in the Greek text *in every form in which it appears*—that is to say, every occurrent person, number, tense or mood of verbs, every case and number of nouns, pronouns, &c., is placed in its alphabetical order, fully explained by a careful grammatical analysis and referred to its root, so that no uncertainty as to the grammatical structure of any word can perplex the beginner, but, assured of the precise grammatical force of any word he may desire to interpret, he is able immediately to apply his knowledge of the English meaning of the root with accuracy and satisfaction. 1 vol. small 4to, half bound.................................$6 50

" **GREEK-ENGLISH LEXICON TO TESTAMENT.** By T. S. Green. Half morocco $1 50

HEBREW.

GREEN. **A GRAMMAR OF THE HEBREW LANGUAGE.** With copious Appendixes. By W. H. Green, D.D., Professor in Princeton Theological Seminary. 1 vol. 8vo, cloth....$3 50

" **AN ELEMENTARY HEBREW GRAMMAR.** With Tables, Reading Exercises, and Vocabulary. By Prof. W. H. Green, D.D. 1 vol. 12mo, cloth.......................$1 50

" **HEBREW CHRESTOMATHY**; or, Lessons in Reading and Writing Hebrew. By Prof. W. H. Green, D.D. 1 vol. 8vo, cloth...$2 00

LETTERIS. **A NEW AND BEAUTIFUL EDITION OF THE HEBREW BIBLE.** Revised and carefully examined by Myer Levi Letteris. 1 vol. 8vo, with key, marble edges.....$2 50

"This edition has a large and much more legible type than the known one volume editions, and the print is excellent, while the name of LETTERIS is a sufficient guarantee for correctness." —*Rev. Dr.* J. M. WISE, *Editor of the* ISRAELITE.

BAGSTER'S GESENIUS. **BAGSTER'S COMPLETE EDITION OF GESENIUS' HEBREW AND CHALDEE LEXICON.** In large, clear, and perfect type. Translated and edited with additions and corrections, by S. P. Tregelles, LL.D.

In this edition great care has been taken to guard the student from Neologian tendencies by suitable remarks whenever needed.

"The careful revisal to which the Lexicon has been subjected by a faithful and Orthodox translator exceedingly enhances the practical value of this edition." —*Edinburgh Ecclesiastical Journal.*

Small 4to, half bound............................$7 50

BAGSTER'S **NEW POCKET HEBREW AND ENGLISH LEXICON.** The arrangement of this Manual Lexicon combines two things—the etymological order of roots and the alphabetical order of words. This arrangement tends to lead the learner onward; for, as he becomes more at home with roots and derivatives, he learns to turn at once to the root, without first searching for the particular word in its alphabetic order. 1 vol. 18mo, cloth....................................$2 00

"This is the most beautiful, and at the same time the most correct and perfect Manual Hebrew Lexicon we have ever used."—*Eclectic Review.*

IRON, METALLURGY, &c.

BODEMANN. **A TREATISE ON THE ASSAYING OF LEAD, SILVER, COPPER, GOLD, AND MERCURY.** By Bodemann & Kerl. Translated by W. A. Goodyear. 1 vol. 12mo, $2 50

CROOKES. **A PRACTICAL TREATISE ON METALLURGY.** Adapted from the last German edition of Prof. Kerl's Metallurgy. By William Crookes and Ernst Rohrig. In three vols. thick 8vo. Price....................................$30 00

Separately. Vol. 1. Lead, Silver, Zinc, Cadmium, Tin, Mercury, Bismuth, Antimony, Nickel, Arsenic, Gold, Platinum, and Sulphur....................................$10 00

Vol. 2. Copper and Iron.......................... 10 00

Vol. 3. Steel, Fuel, and Supplement................ 10 00

FAIRBAIRN. **CAST AND WROUGHT IRON FOR BUILDING.** By Wm. Fairbairn. 8vo, cloth..........................$2 00

FRENCH. **HISTORY OF IRON TRADE, FROM 1621 TO 1857.** By B. F. French. 8vo, cloth..........................$2 00

KIRKWOOD **COLLECTION OF REPORTS (CONDENSED) AND OPINIONS OF CHEMISTS IN REGARD TO THE USE OF LEAD PIPE FOR SERVICE PIPE,** in the Distribution of Water for the Supply of Cities. By I. P. Kirkwood, C.E. 8vo, cloth.......................$1 50

LESLEY. **THE IRON MANUFACTURER'S GUIDE TO THE FURNACES, FORGES, AND ROLLING-MILLS OF THE UNITED STATES.** By J. P. Lesley. With maps and plates. 1 vol. 8vo, cloth.......................$8 00

MACHINISTS—MECHANICS.

FITZGERALD. **THE BOSTON MACHINIST.** A complete School for the Apprentice and Advanced Machinist. By W. Fitzgerald. 1 vol. 18mo, cloth..................................$0 75

HOLLY. **SAW FILING.** The Art of Saw Filing Scientifically Treated and Explained. With Directions for putting in order all kinds of Saws, from a Jeweller's Saw to a Steam Saw-mill. Illustrated by forty-four engravings. Third edition. By H. W. Holly. 1 vol. 18mo, cloth..........................$0 75

KNIGHT. **THE MECHANISM AND ENGINEER INSTRUCTOR.** Comprising Forging, Planing, Lining, Slotting, Shaping, Turning, Screw-Cutting, etc., etc. By Cameron Knight. 1 vol. 4to, half morocco...........................$15 00

TURNING, &c. **LATHE, THE, AND ITS USES, ETC.; or, Instruction in the Art of Turning Wood and Metal.** Including a description of the most modern appliances for the ornamentation of plane and curved surfaces, with a description also of an entirely novel form of *Lathe* for Eccentric and Rose Engine Turning, a Lathe and Turning Machine combined, and other valuable matter relating to the Art. 1 vol. 8vo, copiously illustrated. Including Supplement. 8vo, cloth......$7 00

"The most complete work on the subject ever published."—*American Artisan.*

"Here is an invaluable book to the practical workman and amateur."—*London Weekly Times.*

TURNING, &c. **SUPPLEMENT AND INDEX TO LATHE AND ITS USES.** Large type. Paper, 8vo....................$0 90

WILLIS. **PRINCIPLES OF MECHANISM.** Designed for the use of Students in the Universities and for Engineering Students generally. By Robert Willis, M.D., F.R.S., President of the British Association for the Advancement of Science, &c., &c. Second edition, enlarged. 1 vol. 8vo, cloth...........$7 50

*** It ought to be in every large Machine Workshop Office, in every School of Mechanical Engineering at least, and in the hands of every Professor of Mechanics, &c.—Prof. S. EDWARD WARREN.

MANUFACTURES.

BOOTH. **NEW AND COMPLETE CLOCK AND WATCH MAKERS' MANUAL.** Comprising descriptions of the various gearings, escapements, and Compensations now in use in French, Swiss, and English clocks and watches, Patents, Tools, etc., with directions for cleaning and repairing. With numerous engravings. Compiled from the French, with an Appendix containing a History of Clock and Watch Making in America. By Mary L. Booth. With numerous plates. 1 vol. 12mo, cloth..................................$2 00

GELDARD. **HANDBOOK ON COTTON MANUFACTURE; or, A Guide to Machine-Building, Spinning, and Weaving.** With practical examples, all needful calculations, and many useful and important tables. The whole intended to be a complete yet compact authority for the manufacture of cotton. By James Geldard. With steel engravings. 1 vol. 12mo, cloth.......................................$2 50

MEDICAL, &c.

BULL. **HINTS TO MOTHERS FOR THE MANAGEMENT OF HEALTH DURING THE PERIOD OF PREGNANCY, AND IN THE LYING-IN ROOM.** With an exposure of popular errors in connection with those subjects. By Thomas Bull, M.D. 1 vol. 12mo, cloth...........$1 00

FRANCKE. **OUTLINES OF A NEW THEORY OF DISEASE**, applied to Hydropathy, showing that water is the only true remedy. With observations on the errors committed in the practice of Hydropathy, notes on the cure of cholera by cold water, and a critique on Priessnitz's mode of treatment. Intended for popular use. By the late H. Francke. Translated from the German by Robert Blakie, M.D. 1 vol. 12mo, cloth...$1 50

GREEN. **A TREATISE ON DISEASES OF THE AIR PASSAGES.** Comprising an inquiry into the History, Pathology, Causes, and Treatment of those Affections of the Throat called Bronchitis, Chronic Laryngitis, Clergyman's Sore Throat, etc., etc. By Horace Green, M.D. Fourth edition, revised and enlarged. 1 vol. 8vo, cloth..................................$3 00

" **A PRACTICAL TREATISE ON PULMONARY TUBERCULOSIS**, embracing its History, Pathology, and Treatment. By Horace Green, M.D. Colored plates. 1 vol. 8vo, cloth.................. ....$5 00

GREEN. **OBSERVATIONS ON THE PATHOLOGY OF CROUP** With Remarks on its Treatment by Topical Medications. By Horace Green, M.D. 1 vol. 8vo, cloth...............$1 25

" **ON THE SURGICAL TREATMENT OF POLYPI OF THE LARYNX, AND ŒDEMA OF THE GLOTTIS.** By Horace Green, M.D. 1 vol. 8vo..................$1 25

" **FAVORITE PRESCRIPTIONS OF LIVING PRACTITIONERS.** With a Toxicological Table, exhibiting the Symptoms of Poisoning, the Antidotes for each Poison, and the Test proper for their detection. By Horace Green. 1 vol. 8vo, cloth..............................$2 50

TILT. **ON THE PRESERVATION OF THE HEALTH OF WOMEN AT THE CRITICAL PERIODS OF LIFE.** By E. G. Tilt, M.D. 1 vol. 18mo, cloth..............$0 50

VON DUBEN. **GUSTAF VON DUBEN'S TREATISE ON MICROSCOPICAL DIAGNOSIS.** With 71 engravings. Translated, with additions, by Prof. Louis Bauer, M.D. 1 vol. 8vo, cloth...$1 00

MINERALOGY.

BRUSH. **ON BLOW-PIPE ANALYSIS.** By Prof. Geo. J. Brush. (In preparation.)

DANA. **DESCRIPTIVE MINERALOGY.** Comprising the most recent Discoveries. Fifth edition. Almost entirely re-written and greatly enlarged. Containing nearly 900 pages 8vo, and upwards of 600 wood engravings. By Prof. J. Dana. Cloth...$10 00

"We have used a good many works on Mineralogy, but have met with none that begin to compare with this in fulness of plan, detail, and execution."—*American Journal of Mining.*

DANA & BRUSH. **APPENDIX TO DANA'S MINERALOGY,** bringing the work down to 1872. By Prof. G. J. Brush. 8vo.....$0 50

DANA. **DETERMINATIVE MINERALOGY.** 1 vol. (In preparation.)

" **A TEXT-BOOK OF MINERALOGY.** 1 vol. (In preparation.)

MISCELLANEOUS.

BAILEY. **THE NEW TALE OF A TUB.** An adventure in verse. By F. W. N. Bailey. With illustrations. 1 vol. 8vo......$0 75

CARLYLE. **ON HEROES, HERO-WORSHIP, AND THE HEROIC IN HISTORY.** Six Lectures. Reported, with emendations and additions. By Thomas Carlyle. 1 vol. 12mo, cloth...$0 75

CATLIN. **THE BREATH OF LIFE; or, Mal-Respiration and its Effects upon the Enjoyments and Life of Man.** By Geo. Catlin. With numerous wood engravings. 1 vol. 8vo, $0 75

CHEEVER. **CAPITAL PUNISHMENT.** A Defence of. By Rev. George B. Cheever, D.D. Cloth..........................$0 50

" **HILL DIFFICULTY,** and other Miscellanies. By Rev. George B. Cheever, D.D. 1 vol. 12mo, cloth.........$1 00

" **JOURNAL OF THE PILGRIMS AT PLYMOUTH ROCK.** By Geo. B. Cheever, D.D. 1 vol. 12mo, cloth........$1 00

" **WANDERINGS OF A PILGRIM IN THE ALPS.** By George B. Cheever, D.D. 1 vol. 12mo, cloth.........$1 00

" **WANDERINGS OF THE RIVER OF THE WATER OF LIFE.** By Rev. Dr. George B. Cheever. 1 vol. 12mo, cloth...$1 00

CONYBEARE. **ON INFIDELITY.** 12mo, cloth...................... 1 00

CHILD'S BOOK **OF FAVORITE STORIES.** Large colored plates. 4to, cloth...$1 50

EDWARDS. **FREE TOWN LIBRARIES.** The Formation, Management and History in Britain, France, Germany, and America. Together with brief notices of book-collectors, and of the respective places of deposit of their surviving collections. By Edward Edwards. 1 vol. thick 8vo..............$4 00

GREEN. **THE PENTATEUCH VINDICATED FROM THE ASPERSIONS OF BISHOP COLENSO.** By Wm. Henry Green, Prof. Theological Seminary, Princeton, N. J. 1 vol. 12mo, cloth..................................$1 25

GOURAUD. **PHRENO-MNEMOTECHNY; or, The Art of Memory.** The series of Lectures explanatory of the principles of the system. By Francis Fauvel-Gouraud. 1 vol. 8vo, cloth, $2 00

" **PHRENO-MNEMOTECHNIC DICTIONARY.** Being a Philosophical Classification of all the Homophonic Words of the English Language. To be used in the application of the Phreno-Mnemotechnic Principles. By Francis Fauvel-Gouraud. 1 vol. 8vo, cloth..............................$2 00

HEIGHWAY. **LEILA ADA.** 12mo, cloth.......................... 1 00

" **LEILA ADA'S RELATIVES.** 12mo, cloth.......... 1 00

KELLY. **CATALOGUE OF AMERICAN BOOKS.** The American Catalogue of Books, from January, 1861, to January, 1866. Compiled by James Kelly. 1 vol. 8vo, net cash.......$5 00

" **CATALOGUE OF AMERICAN BOOKS.** The American Catalogue of Books from January, 1866, to January, 1871. Compiled by James Kelly. 1 vol. 8vo, net...........$7 50

MAVER'S **COLLECTION OF GENUINE SCOTTISH MELODIES.** For the Piano-Forte or Harmonium, in keys suitable for the voice. Harmonized by C. H. Morine. Edited by Geo. Alexander. 1 vol. 4to, half calf.......................$10 00

NOTLEY. **A COMPARATIVE GRAMMAR OF THE FRENCH, ITALIAN, SPANISH, AND PORTUGUESE LANGUAGES.** By Edwin A. Notley. 1 vol., cloth......$5 00

PARKER. **POLAR MAGNETISM.** First and Second Lectures. By John A. Parker. Each..................................$0 25

" **NON-EXISTENCE OF PROJECTILE FORCES IN NATURE.** By John A. Parker.......................$0 25

STORY OF **A POCKET BIBLE.** Illustrated. 12mo, cloth........$1 00

TUPPER **PROVERBIAL PHILOSOPHY.** 12mo................ 1 00

WALTON & COTTON. **THE COMPLETE ANGLER; or, The Contemplative Man's Recreation, by Isaac Walton, and Instructions how to Angle for a Trout or Grayling in a Clear Stream,** by Charles Cotton, with copious notes, for the most part original. A bibliographical preface, giving an account of fishing and Fishing Books, from the earliest antiquity to the time of Walton, and a notice of Cotton and his writings, by Rev. Dr. Bethune. To which is added an appendix, including the most complete catalogue of books in angling ever printed, &c. Also a general index to the whole work. 1 vol. 12mo, cloth..$3 00

WARREN. **NOTES ON POLYTECHNIC OR SCIENTIFIC SCHOOLS IN THE UNITED STATES.** Their Nature, Position, Aims, and Wants. By S. Edward Warren. Paper....$0 40

WILLIAMS. **THE MIDDLE KINGDOM.** A Survey of the Geography, Government, Education, Social Life, Arts, Religion, &c., of the Chinese Empire and its Inhabitants. With a new map of the Empire. By S. Wells Williams. Fourth edition, in 2 vols.......................................$4 00

RUSKIN'S WORKS.

Uniform in size and style.

RUSKIN. **MODERN PAINTERS.** 5 vols. tinted paper, bevelled boards, plates, in box....$18 00

" **MODERN PAINTERS.** 5 vols. half calf.... 27 00

" " " " without plates.... 12 00

" " " " " half calf, 20 00

Vol. 1.—Part 1. General Principles. Part 2. Truth.
Vol. 2.—Part 3. Of Ideas of Beauty.
Vol. 3.—Part 4. Of Many Things.
Vol. 4.—Part 5. Of Mountain Beauty.
Vol. 5.—Part 6. Leaf Beauty. Part 7. Of Cloud Beauty. Part 8. Ideas of Relation of Invention, Formal. Part 9. Ideas of Relation of Invention, Spiritual.

" **STONES OF VENICE.** 3 vols., on tinted paper, bevelled boards, in box....$7 00

" **STONES OF VENICE.** 3 vols., on tinted paper, half calf....$12 00

" **STONES OF VENICE.** 3 vols., cloth.... 6 00

Vol. 1.—The Foundations.
Vol. 2.—The Sea Stories.
Vol. 3.—The Fall.

" **SEVEN LAMPS OF ARCHITECTURE.** With illustrations, drawn and etched by the authors. 1 vol. 12mo, cloth, $1 75

" **LECTURES ON ARCHITECTURE AND PAINTING.** With illustrations drawn by the author. 1 vol. 12mo, cloth....$1 50

" **THE TWO PATHS.** Being Lectures on Art, and its Application to Decoration and Manufacture. With plates and cuts. 1 vol. 12mo, cloth....$1 25

" **THE ELEMENTS OF DRAWING.** In Three Letters to Beginners. With illustrations drawn by the author. 1 vol. 12mo, cloth....$1 00

" **THE ELEMENTS OF PERSPECTIVE.** Arranged for the use of Schools. 1 vol. 12mo, cloth....$1 00

" **THE POLITICAL ECONOMY OF ART.** 1 vol. 12mo, cloth....$1 00

" **PRE-RAPHAELITISM.**
NOTES ON THE CONSTRUCTION OF SHEEPFOLDS.
KING OF THE GOLDEN RIVER; or, The Black Brothers. A Legend of Stiria.
} 1 vol. 12mo, cloth, $1 00

RUSKIN **SESAME AND LILIES.** Three Lectures on Books, Women, &c. 1. Of Kings' Treasuries. 2. Of Queens' Gardens. 3. Of the Mystery of Life. 1 vol. 12mo, cloth....$1 50

" **AN INQUIRY INTO SOME OF THE CONDITIONS AT PRESENT AFFECTING "THE STUDY OF ARCHITECTURE" IN OUR SCHOOLS.** 1 vol. 12mo, paper....$0 15

" **THE ETHICS OF THE DUST.** Ten Lectures to Little Housewives, on the Elements of Crystallization. 1 vol. 12mo, cloth....$1 25

" **"UNTO THIS LAST."** Four Essays on the First Principles of Political Economy. 1 vol. 12mo, cloth.. ....$1 00

RUSKIN **THE CROWN OF WILD OLIVE.** Three Lectures on Work, Traffic, and War. 1 vol. 12mo. cloth................$1 00

" **TIME AND TIDE BY WEARE AND TYNE.** Twenty-five Letters to a Workingman on the Laws of Work. 1 vol. 12mo, cloth.....................................$1 00

" **THE QUEEN OF THE AIR.** Being a Study of the Greek Myths of Cloud and Storm. 1 vol. 12mo, cloth $1 00

" **LECTURES ON ART.** 1 vol. 12mo, cloth........... 1 00

" **FORS CLAVIGERA.** Letters to the Workmen and Labourers of Great Britain. Part 1. 1 vol. 12mo, cloth, plates, $1 00

" **FORS CLAVIGERA.** Letters to the Workmen and Labourers of Great Britain. Part 2. 1 vol. 12mo, cloth, plates, $1 00

MUNERA PULVERIS. Six Essays on the Elements of Political Economy. 1 vol. 12mo, cloth...............$1 00

" **ARATRA PENTELICI.** Six Lectures on the Elements of Sculpture, given before the University of Oxford. By John Ruskin. 12mo, cloth, $1 50, or with plates. $3 00

" **THE EAGLE'S NEST.** Ten Lectures on the relation of Natural Science to Art. 1 vol. 12mo...............$1 50

BEAUTIFUL PRESENTATION VOLUMES.

Printed on tinted paper, and elegantly bound in crape cloth extra, bevelled boards, gilt head.

RUSKIN. **THE TRUE AND THE BEAUTIFUL IN NATURE, ART, MORALS, AND RELIGION.** Selected from the Works of John Ruskin, A.M. With a notice of the author by Mrs. L. C. Tuthill. Portrait. 1 vol. 12mo, cloth, plain, $2.00; cloth extra, gilt head.......................$2 50

" **ART CULTURE.** Consisting of the Laws of Art selected from the Works of John Ruskin, and compiled by Rev. W. H. Platt, for the use of Schools and Colleges as well as the general public. A beautiful volume, with many illustrations. 1 vol. 12mo, cloth (shortly).

" **PRECIOUS THOUGHTS:** Moral and Religious. Gathered from the Works of John Ruskin, A.M. By Mrs. L. C. Tuthill. 1 vol. 12mo, cloth, plain, $1.50. Extra cloth, gilt head.......................................$2 00

" **SELECTIONS FROM THE WRITINGS OF JOHN RUSKIN.** 1 vol. 12mo, cloth extra, gilt head........$2 50

" **SELECTIONS FROM THE WRITINGS OF JOHN RUSKIN.** 1 vol. 12mo, plain cloth.................$2 00

" **SESAME AND LILIES.** 1 vol. 12mo.................$1 75

" **ETHICS OF THE DUST.** 12mo..................... 1 75

" **CROWN OF WILD OLIVE.** 12mo.................. 1 50

RUSKIN'S BEAUTIES.

" **THE TRUE AND BEAUTIFUL**
PRECIOUS THOUGHTS.
CHOICE SELECTIONS.
} 3 vols., in box, cloth extra, gilt head.........$6 00
do., half calf...10 00

RUSKIN'S POPULAR VOLUMES.

CROWN OF WILD OLIVE. SESAME AND LILIES. QUEEN OF THE AIR. ETHICS OF THE DUST. 4 vols. in box, cloth extra, gilt head................$6 00

RUSKIN'S WORKS.

Revised edition.

RUSKIN Vol. 1.—**SESAME AND LILIES.** Three Lectures. By John Ruskin, LL.D. 1. Of King's Treasuries. 2. Of Queens' Gardens. 3. Of the Mystery of Life. 1 vol. 8vo, cloth, $2.00. Large paper...............................$2 50

RUSKIN'S WORKS.

Revised edition.

" Vol. 2.—**MUNERA PULVERIS.** Six Essays on the Elements of Political Economy. By John Ruskin. 1 volume 8vo, cloth. $2 00
Large paper.................... 2 50

" Vol. 3.—**ARATRA PENTELICI.** Six Lectures on the Elements of Sculpture, given before the University of Oxford. By John Ruskin. 1 vol. 8vo....................$4 00
Large paper.................... 4 50

" **THE POETRY OF ARCHITECTURE:** Villa and Cottage. With numerous plates. By Kata Phusin. 1 vol. 12mo, cloth....................$1 50

Kata Phusin is the supposed Nom de Plume of John Ruskin.

" **FORS CLAVIGERA.** Letters to the Workmen and Laborers of great Britain. Part 3. 1 vol. 12mo, cloth........$1 50

RUSKIN—COMPLETE WORKS.

THE COMPLETE WORKS OF JOHN RUSKIN. 27 vols., extra cloth, in a box..$40 00
Ditto 27 vols., extra cloth. *Plates*... 48 00
Ditto Bound in 17 vols., half calf. do.... 70 00

SHIP-BUILDING, &c.

BOURNE. **A TREATISE ON THE SCREW PROPELLER, SCREW VESSELS, AND SCREW ENGINES,** as adapted for Purposes of Peace and War. Illustrated by numerous wood-cuts and engravings. By John Bourne. New edition. 1867. 1 vol. 4to, cloth, $18.00; half russia................$24 00

WATTS. **RANKINE (W. J. M.) AND OTHERS.** Ship-Building, Theoretical and Practical, consisting of the Hydraulics of Ship-Building, or Buoyancy, Stability, Speed and Design—The Geometry of Ship-Building, or Modelling, Drawing, and Laying Off—Strength of Materials as applied to Ship-Building—Practical Ship-Building—Masts, Sails, and Rigging—Marine Steam Engineering—Ship-Building for Purposes of War. By Isaac Watts, C.B., W. J. M. Rankine, C.B., Frederick K. Barnes, James Robert Napier, etc. Illustrated with numerous fine engravings and woodcuts. Complete in 30 numbers, boards, $35.00; 1 vol. folio, cloth, $37.50; half russia, $40 00

WILSON (T. D.) **SHIP-BUILDING, THEORETICAL AND PRACTICAL.** In Five Divisions.—Division I. Naval Architecture. II. Laying Down and Taking off Ships. III. Ship-Building IV. Masts and Spar Making. V. Vocabulary of Terms used—intended as a Text-Book and for Practical Use in Public and Private Ship-Yards. By Theo. D. Wilson, Assistant Naval Constructor, U. S. Navy; Instructor of Naval Construction, U. S. Naval Academy; Member of the Institution of Naval Architects, England. With numerous plates, lithographic and wood. 1 vol. 8vo. $7 50

SOAP.

MORFIT. **A PRACTICAL TREATISE ON THE MANUFACTURE OF SOAPS.** With numerous wood-cuts and elaborate working drawings. By Campbell Morfit, M.D., F.C.S. 1 vol. 8vo....................$20 00

STEAM ENGINE.

TROWBRIDGE. **TABLES, WITH EXPLANATIONS, OF THE NON-CONDENSING STATIONERY STEAM ENGINE,** and of High-Pressure Steam Boilers. By Prof. W. P. Trowbridge, of Yale College Scientific School. 1 vol. 4to plates....................$2 50

" **TREATISE ON THE GENERATION AND UTILIZATION OF HEAT THROUGH THE MEDIUM OF STEAM AND STEAM BOILERS.** Designed as a Text-Book and for Practical use. By Prof. W. P. Trowbridge Very fully illustrated. 1 vol. 8vo (shortly).

TURNING, &c.

THE LATHE, **AND ITS USES, ETC.** On Instructions in the Art of Turning Wood and Metal. Including a description of the most modern appliances for the ornamentation of plane and curved surfaces. With a description, also, of an entirely novel form of Lathe for Eccentric and Rose Engine Turning, a Lathe and Turning Machine combined, and other valuable matter relating to the Art. 1 vol. 8vo, copiously illustrated, cloth..........$7 00

" **SUPPLEMENT AND INDEX TO SAME.** Paper...$0 90

VENTILATION.

LEEDS (L. W.). **A TREATISE ON VENTILATION.** Comprising Seven Lectures delivered before the Franklin Institute, showing the great want of improved methods of Ventilation in our buildings, giving the chemical and physiological process of respiration, comparing the effects of the various methods of heating and lighting upon the ventilation, &c. Illustrated by many plans of all classes of public and private buildings, showing their present defects, and the best means of improving them. By Lewis W. Leeds. 1 vol. 8vo, with numerous wood-cuts and colored plates. Cloth.........$2 50

"It ought to be in the hands of every family in the country."—*Technologist.*

"Nothing could be clearer than the author's exposition of the principles of the principles and practice of both good and bad ventilation."—*Van Nostrand's Engineering Magazine.*

"The work is every way worthy of the widest circulation."—*Scientific American.*

REID. **VENTILATION IN AMERICAN DWELLINGS.** With a series of diagrams presenting examples in different classes of habitations. By David Boswell Reid, M.D. To which is added an introductory outline of the progress of improvement in ventilation. By Elisha Harris, M.D. 1 vol. 12mo, $1 50

WEIGHTS, MEASURES, AND COINS.

TABLES OF WEIGHTS, MEASURES, COINS, &c., OF THE UNITED STATES AND ENGLAND, with their Equivalents in the French Decimal System. Arranged by T. T. Egleston, Professor of Mineralogy, School of Mines, Columbia College. 1 vol. 18mo.....................$0 75

"It is a most useful work for all chemists and others who have occasion to make the conversions from one system to another."—*American Chemist.*

"Every mechanic should have these tables at hand."—*American Horological Journal.*

www.ingramcontent.com/pod-product-compliance
Lightning Source LLC
LaVergne TN
LVHW021101110826
845150LV00001B/145

9781425565800